International Handbook of Research in History, Philosophy and Science Teaching

Michael R. Matthews
Editor

International Handbook of Research in History, Philosophy and Science Teaching

Volume III

Editor
Michael R. Matthews
School of Education
University of New South Wales
Sydney, NSW, Australia

ISBN 978-94-007-7653-1 ISBN 978-94-007-7654-8 (eBook)
DOI 10.1007/978-94-007-7654-8
Springer Dordrecht Heidelberg New York London

Library of Congress Control Number: 2013958394

Printed on acid-free paper

Springer is part of Springer Science+Business Media (www.springer.com)

Contents

Volume 1

Part IX Theoretical Studies: Features of Science and Education

Part X Theoretical Studies: Teaching, Learning and Understanding Science

Volume 3

Chapter 52
Rejecting Materialism: Responses to Modern Science in the Muslim Middle East

Taner Edis and Saouma BouJaoude

52.1 Islam, Science, and "Materialism"

Contemporary Islamic responses to science are similar to responses from within other world religions – they exhibit a degree of ambivalence. Supernatural agents no longer play any explanatory role in the natural sciences, which challenges religious perceptions of divine design in the universe (Edis 2002). At the same time, modern science is closely coupled with technological prowess, making science appear indispensable to Muslim populations seeking to avoid commercial and military subordination in postcolonial times. This puts pressure on religious thinkers to try to achieve harmony between science and Islam while guarding against the materialist influences they perceive in the conceptual frameworks of Western science.

Since Islam is a diverse religion, and since Muslims constantly dispute what is the best interpretation of their religious tradition, there are always various competing visions about how to attain harmony. Some general descriptions are nonetheless possible.

Most popular versions of Islam harbor some ideas about harmony between science and religion, even though these ideas tend to be very superficial. Muslim populations worldwide sustain some very popular pseudoscientific beliefs (Edis 2007; Guessoum 2008). More complex interpretations of religious doctrines are readily available in more intellectual circles, but the level of engagement with science in

T. Edis (✉)
Department of Physics, Truman State University, Kirksville, MO 63501, USA
e-mail: edis@truman.edu

S. BouJaoude
Center for Teaching and Learning and Science and Math Education Center,
American University of Beirut, P. O. Box 11-0236, Beirut, Lebanon
e-mail: boujaoud@aub.edu.lb

M.R. Matthews (ed.), *International Handbook of Research in History, Philosophy and Science Teaching*, DOI 10.1007/978-94-007-7654-8_52,

more sophisticated theologies varies greatly. Most religious preoccupations, such as those about matters of social morality, have little to do with science, and in turn, most routine science does not touch on religion. But some scientific claims associated with the more ambitious theories of modern physics and biology invite friction with traditional beliefs about the supernatural governance of nature. Evolution in particular is a perennial source of conflict (Edis 2007).

Especially in the popular literature, "materialism" has become a common negative term expressing discomfort with modern ideas that emphasize impersonal, purposeless processes in explanations of phenomena. Historically, this is rooted in responses to mechanically inspired nineteenth-century materialist philosophy (e.g., Büchner 1884), together with later opposition to Marxist notions of materialism in the realm of social thought. Current developments of philosophical materialism, such as physicalism (Melnyk 2003), would naturally also be deemed unacceptable, due to their exclusion of supernatural agency and divine design. In popular writings, however, "materialism" is usually a generic term of abuse with little specificity. Nonetheless, as an umbrella term encompassing rejection of visible supernatural agency, opposition to "materialism" captures a central theme in Muslim concerns about modern science.

None of this is distinctively Muslim. Similar pictures can be drawn of Christian, Jewish, and even Hindu responses to theories such as evolution (Brown 2012). And when going beyond broad generalizations, it is very hard to identify a common, coherent Islamic philosophical framework through which Muslims tend to approach science. There is, perhaps, a distinctly Islamic preoccupation with questions about occasionalism and natural causality in Muslim thinking about science, reflecting the particular philosophical heritage of Islamic thought (Fakhry 2004; Leaman 2002). But such features of an Islamic philosophy of science are of minor importance. Both Islam and modern science are too varied for any common framework to dominate in Muslim thinking about science.

Instead, the distinctive aspects of present Muslim responses to science derive from the *history* of Muslim lands, especially the Middle Eastern heartlands of Islam. For Western Christendom, modern science, however revolutionary, still emerged as an organic development within its intellectual culture. When science caused religious discomfort, it did so as an endogenous heresy. For Muslim societies, modern science has been a Western import, significantly detached from their own intellectual tradition, including the highly developed medieval science of Islamicate societies. And for those Muslim thinkers most concerned about developing a religious response to modernity, modern science has often appeared as something to assimilate, to Islamize, and to guard against when it appears to challenge traditionally religious perceptions of nature (Edis 2007).

52.2 Medieval Islam and Modern Science

Natural science and doctrinally traditional forms of Islam have an uneasy relationship today. Muslim countries devote few resources to natural science, and the Muslim contribution to research in natural sciences such as physics is negligible

(Hoodbhoy 2007). Even in applied science and engineering, where the picture is not as dismal, Muslim countries suffer from relative backwardness. While Muslim populations have usually been enthusiastic about adopting modern technology, sustained creativity regarding new technologies still tends to characterize advanced Western and East Asian economies.

Muslims did not always have to play catch-up. Islamic empires famously were home to the best of medieval science. Though participation in serious intellectual life was inevitably confined to a small elite, Muslim scholars typically enjoyed better intellectual resources compared to their medieval European counterparts. Whether in medicine and public health, astronomy, optics, mathematics, or philosophy in the Greek tradition, Muslims were at the forefront of intellectual life. Historians of science continue to address the puzzle of why Europe became the site of the modern scientific revolution, rather than China or the lands of Islam (Huff 2003).

There are many myths about medieval Islamic science. Western observers used to describe Muslim scholarship as merely transmitting the heritage of Greek and Roman antiquity. This is not correct: Muslims further developed the ideas they inherited. Muslim science is often described as stagnating. Some Muslim reformers today add that this was due to the suppression of rationalist theologies, such as that of the Mutazila, and the influence of religious conservatives, such as al-Ghazali, who argued against philosophy (Hoodbhoy 1991). Yet those "foreign sciences" that were perceived as having practical utility were decoupled from the Greek philosophical tradition and continued to slowly develop even when philosophy faced a hostile climate.

Nevertheless, a large scientific and technological gap did open up between Europe and Muslim empires such as the Ottomans, Iran, and Mughal India. Europe made spectacular advances while Muslim conceptions of science remained closely anchored in the medieval tradition. With hindsight, it is easy to admire the accomplishments of medieval Muslim science. But this science operated within a premodern conceptual framework. Muslim medicine, for example, could boast interesting surgical techniques and impressive provisions for public health and hospital complexes. At the same time, the medical texts Muslims used were also full of bloodletting, humor theory, religious and magical practices, and no end of ideas that have a closer affinity to occult thinking than what became natural science as understood today. Furthermore, the medieval natural philosophy practiced in Islamicate societies, like its Catholic European counterpart, did not develop explanations within the sorts of theoretical framework comparable to, for example, early modern physics. Science remained an activity of collecting mostly isolated facts, and these facts were liberally mixed with what later scientists would come to call superstition. In this regard, modern science is radically different compared to its medieval precursors, in terms of explanatory power as well as its more immediately visible technological applications (Edis 2007). When Europeans, by historical accident, stumbled upon the new ways of thinking about nature that became modern science, Muslims would be left behind (Huff 2003).

Muslim intellectuals first became conscious of lagging in their knowledge of nature as a result of persistent military defeats of Muslim empires. The Ottoman Empire, which was nominally sovereign over today's Turkey and most Arab countries up until the twentieth century, was closest to Europe. And in the seventeenth century dire reversals in the Ottoman-controlled Balkan territories forced Ottoman intellectuals to ask why the conquering armies of the faithful were suddenly so weak. Scholars called for intellectual and institutional renewal among Muslims. These calls gathered strength as it became clear that science-based technology and modern forms of social organization gave the Europeans their advantage. Similar experiences were to follow for Mughal India, Iran, and Indonesia – all but the most isolated Muslim countries. Muslims everywhere would face a similar challenge from the industrialized West, and Ottoman intellectuals would craft responses that would find echoes across the Muslim world.

52.3 Ottomans Respond to Modern Europe

Confronted with western European military and commercial superiority, Ottoman intellectuals became preoccupied with rescuing the Empire. Many insisted that if only Muslims would redouble their commitment to faith and proper observation of divine laws, Islam would be restored to its proper dominance. But most, including most religious scholars, came to recognize that technology was the key to European power and that Muslims would have to acquire and master the new forms of knowledge. How this was to be best accomplished, and especially how the religious nature of Muslim civilization could be protected, became matters of contention (İhsanoğlu 2004). This debate shaped intellectual life in the last century of the Ottoman Empire, spilled over into successor states such as the Republic of Turkey and Egypt, and continues vigorously to the present day. Indeed, many of the themes running through earlier Muslim ideas about renewal remain at the forefront of debates today.

In the seventeenth century, Ottoman intellectuals such as Katib Çelebi argued that Muslims needed to be more open to foreign scientific knowledge (Demir 2001, pp. 48–52; Lewis 1993, Chap. 16). Obviously, Western Europe could no longer be ignored; Muslims needed to learn as much as possible about European technical accomplishments. Ottoman officialdom would adapt, for example, sending representatives to European capitals rather than relying only on European diplomats stationed in Istanbul. By the early nineteenth century, as the Ottoman Empire's European territories continued to shrink, diplomats, travelers, and students became a significant conduit of information about Western ways. Translations of Western technological and scientific works became available, and westernized forms of education attracted the children of elites.

Together with Western Europe, however, Ottoman Muslims also looked toward their own history. Especially before the nineteenth century, almost all intellectuals were grounded in the classical tradition of religious scholarship. And for religious scholars, reforming Muslim attitudes toward knowledge made sense in the context of reviving the specifically Muslim "religious sciences." Even Katib Çelebi had expressed a perception that Muslims had become "imitators," relying too much on precedents set by earlier scholars. Assimilating the foreign sciences could be best accomplished if all the Muslim sciences were to become more vibrant. After all, the story went, in their glory days Muslims had enjoyed the best of the foreign sciences, anchored to a God-centered perception of nature, all mediated by a devout engagement with the sacred sources.

Achieving scientific progress in the context of a revival of religious scholarship remains a prominent theme in doctrinally conservative Muslim discourse today (Edis 2007). But the more general notion of revival resonates among more liberal parties to the debate as well. Liberals call for reviving rationalist currents of theology and accuse conservatives of preventing vital social and technological changes (e.g., Mernissi 1992). Most parties agree that recovering the intellectual practices that led to the golden age of medieval Muslim science is vital. Almost everyone overlooks the enormous differences between medieval and modern science.

Calls for revival aside, the immediate task faced in the crumbling Ottoman Empire was to transfer European technology and acquire modern organizational skills. Religious intellectuals felt a need to reconcile this dependence on Europe with the superiority due to Islam. İbrahim Müteferrika, a Hungarian convert to Islam responsible for the first Muslim-operated Ottoman printing press in 1727, stated that Europeans had advanced in rational science precisely because they needed to compensate for the inferiority of their religion. In social and religious matters, Islam was unquestionably superior (Black 2001, p. 268). Religious scholars typically argued that it was permissible to learn the technical knowledge possessed by infidels in order to use this knowledge against them in a military confrontation. More generally, scholars agreed that borrowing knowledge was necessary to defend Islam. However, this borrowing would have to take place in such a way as to guard against cultural contamination.

These remain major themes today. Conservative Muslims often temper their enthusiasm for technology with a conviction of religious superiority and highlight the need to protect the core values of Islamic civilization while modernizing. Indeed, even very traditional religious communities can sometimes adapt quickly to a highly technological environment, strengthening their commitment to orthodoxy in the process (Blank 2001). But more traditional religious commitments remain in tension with institutional structures that are aimed toward contributions to progress within a global scientific enterprise, rather than mainly transferring existing technology.

Just what this protected core of Islam should be is also, naturally, open to dispute. The modernization of Muslim countries over the last few centuries has brought large-scale social changes, including the erosion of the power and social position of the class of religious scholars. In the nineteenth century, this opened up possibilities for more modernist interpretations of Islam (Alperen 2003). Jamal al-Din al-Afghani in Ottoman lands, Sayyid Ahmad Khan in British India, and later Muhammad Abduh in Egypt advocated an interpretation of Islam that was more open to modern education and argued that Islam was in its essence a scientific religion. They downplayed the more miraculous and magical elements in popular piety, joining traditional Sunni orthodoxy in their suspicion of Sufi enthusiasm.

Muslim modernism, however, also harbors a conservative impulse. For example, while modernists downplayed overtly supernatural aspects of popular piety, they did not challenge traditional views of the sacred sources. Modernists produced an apologetic response to modernity rather than a coherent endorsement of modern values. In the twentieth century, with increased literacy in Muslim populations, religious movements that emphasized individual access to sacred sources, bypassing the mediation of traditional scholarship, became more popular. Often this implied a more literal reading of sacred texts and a fundamentalist flavor of populism. Modernist themes of moving beyond traditional forms of religious scholarship are today endorsed by fundamentalists who feel at home in a globalized economy (Roy 2004), as well as by more liberal-minded religious thinkers.

Most doctrinally conservative Muslim responses to Western encroachment have included an element of cultural defensiveness. Nevertheless, westernizers and secular-oriented thinkers have also taken part in the debate. In the Ottoman Empire, such intellectuals were more often rooted in the military and imperial administration rather than in traditional religious scholarship. The Ottoman military went through a wrenching westernization in the late eighteenth and early nineteenth centuries, and some military intellectuals began to think that a more comprehensive westernization of Ottoman society was unavoidable if the Empire was to avoid annihilation in the hands of European powers. A small minority of thinkers educated in a European style even expressed secular ideas. The more radical westernizers finally got a chance to implement their agenda in the 1920s, with the Turkish Revolution founding a new republic in the Anatolian heartland of the Ottoman Empire.

Secular westernizers proposed to solve the problem of adopting Western science by forcing religion to become a private affair. While personally remaining Muslims, modern Turks would become as Westerners in public life, joining the enterprise of creating modern knowledge on equal terms with Europeans (Kazdağlı 2001). Commitment to such a comprehensive ideal of secularization did not take root beyond a relatively small social elite, but due to its command of state power in Turkey and influence on the elites of many other Muslim countries, a secular outlook was a major party to Muslim debates over science and religion in the twentieth century. Even today, when Islamist politics dominate the public imagination in the Muslim Middle East, a secular westernizing view retains some influence due to its overrepresentation among elites and in those scientific institutions established in Muslim countries.

Today, the Muslim world is in a period of extensive religious experimentation. The desire to achieve a distinctly Muslim way of being modern has produced many competing ideas. So modernism, hopes for revival, westernization, or cultural defense are not the only themes that surface in current controversies over science and Islam. For example, Sufism, the mystical strand of Islam, has met with disapproval from both westernizers and conservative scholars and thus has been greatly disadvantaged during the modernization process. Recently, a kind of neo-Sufism has begun to appeal to modern professionals in an urban environment, influencing debates over miracles or spiritual realities accessed through mysticism. Occult notions, then, can find support from those who want to revive medieval perceptions of nature, those who defend survivals of such beliefs in existing religious orders, but also from neo-Sufi's who bring a westernized, New Age flavor to the arguments (Edis 2007; Raudvere and Stenberg 2009).

Even with this diversity of themes, however, it is possible to make some generalizations about the recent history of responses to Western science in the Ottoman Empire and its successor states. Both among religiously conservative intellectuals and much of the general public, enthusiasm for technology has been combined with an insistence that religious convictions should remain at the center of how Muslims perceive nature. This brings up the potential of friction with modern science, which has tended to disenchant the world. Moreover, this friction usually surfaces in the context of institutional and social conflicts between more religiously oriented and westernized segments of Muslim populations.

52.4 Against Materialism

Centuries of military disadvantage taught the Ottomans and their successors that their problems could not be solved by immediate technology transfer or by better techniques of organization. As long as Muslims were borrowers rather than creators of practical knowledge, they would always lag behind.

Most Muslim countries have been considered "underdeveloped" during the twentieth century. Today, countries such as Turkey have attained a middle-income status in part due to industrial manufacturing enterprises migrating away from Europe, but they remain in a peripheral position in the global economy, especially where the cutting edge of science and technology is concerned. Therefore technologically, Muslim-majority countries typically remain in a position of playing catch-up. This means that as with the past century, their educational and industrial policies continue to emphasize immediate applications and technology transfer over basic science. In Turkey and the Arab Middle East, the institutional infrastructure for basic science is poor, and engineering programs typically attract more talented students than natural sciences such as physics. Moreover, an emphasis on applied science has been easy to reconcile with demands for cultural defense. Both in technologically advanced Western countries and among Muslims, the culture of engineering is notoriously more politically and religiously conservative compared to natural science (Gambetta and Herzog 2007). Engineers and medical doctors who reject evolution are commonplace but a rarity among physicists and biologists. And in Muslim countries, engineers are well known to be heavily represented in the leadership of political Islamist movements. But relative success in adopting foreign technologies cannot solve the deeper problem: Muslim populations still lag in the creation of new knowledge, even when restricted to immediate practical knowledge.

This puts pressure on Muslims to master natural science, not just applied science and engineering. Religious responses to natural science, therefore, become important, particularly in areas where modern science appears to challenge traditional beliefs.

In the nineteenth century, most Muslim thinkers continued to take it for granted that Islam was religiously superior to the superficial Christianity of urban Western Europe. Some travelers and diplomats greatly admired Europe, adopting a "the West is the best" mentality. But many Muslim observers of Europe described lands that harnessed vast power through industrialization but were also moral disaster zones. Europeans cared about material things, were good at achieving worldly power and wealth for their elites, but were spiritually lacking. In the twentieth century, visible secularization in Europe would further bolster perceptions that Christian piety was inadequate compared to Muslim faith (Aydın 2000, pp. 43–50). In any case, the European example only intensified concerns that acquiring new knowledge should be done with care to avoid cultural contamination. Yet it was also clear that creating new knowledge required new intellectual habits and adopting those new ways of perceiving nature that were ushered in by the new natural sciences. Culture could not remain immune to change.

Ottoman perceptions of the natural science they imported were colored by the institutional struggle of European science to become independent of religion. Conflict with church authority was not a concern; even today, religious thinkers proudly declare that Islam has no ecclesiastical structure that can impede scientific inquiry (Aydın 2000, p. 86; Şahin 2001, pp. 177–182). But especially in the nineteenth century, materialist currents in European intellectual life influenced Muslim perceptions of science (Gümüşoğlu 2012). Popular materialists such as Ludwig Büchner not only publicly argued against supernatural realities – unthinkable in a Muslim environment – they made it clear that their philosophy was inspired by scientific developments (Büchner 1884). Furthermore, materialism also had radical political implications, even before Marxism developed a significant following. When Darwinian evolution appeared on the intellectual scene, it immediately became part of the larger Western debate over science and the supernatural, rather than just a theory of interest to biologists.

European materialist literature aimed at a middle-class market also attracted attention from Ottoman elites in intellectual centers such as Istanbul and Egypt. Official censorship would not allow the full extent of materialist skepticism about the supernatural to become public, but large portions of books by materialist and anticlerical figures such as Ludwig Büchner, Ernst Haeckel, and Andrew Dickson White were translated and sold reasonably well. Darwinian evolution briefly became a matter of debate. Ottoman elites were in no position to contribute to scientific research, but many desired to learn about science, and in the late nineteenth century, popular science imported from Europe included a dose of materialism (Özervarlı 2003; Ziadat 1986, p. 23).

The direct influence of those Ottoman intellectuals who took a positive view of materialist ideas never extended beyond a small elite. Nevertheless, materialism took a symbolic role in debates over the role of religion and the power of the class of religious scholars in the changing empire. The handful of Ottoman materialists may have been a negligible extreme, but they were not the only ones who suspected Muslim lands were held back by religious obscurantism. And since materialism was a blatant example of infidelity, for conservatives the blasphemy materialists fell into could also serve as a warning against imitating Western ways blindly. Indeed, since conservative Muslims already held Europe in suspicion as a spiritually inferior civilization given to materialism in a crass sense, it seemed that a depraved denial of all that was godly was a logical endpoint of the Western approach to knowledge. Denouncing a symbolic "materialism" could serve as part of a warning against westernizing so far as to lose a Muslim identity. Even today, some Turkish conservatives remember the nineteenth century as a time when a materialist threat first surfaced and demanded a vigorous response from the faithful (Akyol 2009; Hanioğlu 2008; Gümüşoğlu 2012).

Early twentieth-century reform-minded scholars, such as Said Nursi in Turkey, developed apologetic responses to nineteenth-century materialism. Nursi attacked ideas such as natural causality and evolution, while at the same time endorsing science as a way of understanding God's creation. Today, a vaguely defined materialism remains a symbolic enemy, especially for many conservative Muslims – even though, especially after the waning of Marxism, explicitly materialist points of view do not have a significant presence in public debates. The path to reconciling science and religion still goes through defending against a threat of materialism.

52.5 Making Science Acceptable

Some Muslims have sometimes had moral objections to an application of science (Sardar 1984). This is, however, very different from rejecting a "fact" discovered by scientists. Indeed, conservative Muslims have typically been comfortable with science as a means of making an extensive catalogue of facts, presumably documenting the glories of God's creation.

Modern science potentially causes trouble, not because of any result produced in a laboratory but because it is *not* a catalogue of facts collected like stamps. Mature sciences such as physics and biology gain much of their power from the compelling conceptual frameworks they establish, explaining a very wide range of phenomena. Classical physics, later corrected and expanded into modern physics with relativity and quantum mechanics, are ambitious attempts to capture the fundamentals of how the world works. Darwinian evolution claims to present and explain a pattern of descent common to all of life. In such conceptual frameworks, supernatural agents are conspicuous by their absence. Moreover, important aspects of these frameworks are, at face value, hard to reconcile with traditional religious beliefs about how God must have a role in our understanding of nature. Quantum randomness challenges the notion that the universe is controlled by a divine purpose that is distinguishable from chance. Darwinian evolution does not just contradict Quranic stories about the special creation of Adam and Eve; it undercuts the common religious intuition that creativity must always be due to an overarching intelligence rather than a product of mindless processes (Edis 2002, 2006).

Western Christians have responded to such challenges not just by fundamentalist resistance but also by liberal reinterpretations of theology. But from a conservative

Muslim perspective, liberal theology can easily look like capitulation to materialism – another illustration of how Europeans have been unable to defend the core religious truths revealed in the Abrahamic faiths (Aydın 2000, p. 61). Moreover, science is not just a source of worry for intellectuals debating abstract matters. Most ordinary Muslims, even in technologically relatively modern environments, continue to understand their faith in a context where divine action directly and immediately shapes the world. Victims of an earthquake, for example, will usually interpret the earthquake as the will of God, very often as a divine punishment (Küçükcan and Köse 2000) – a view supported by Quranic stories about God punishing wayward peoples by earthquakes. As modern technology gives people more control over aspects of their lives, and state-mandated education includes a more scientific picture of nature, questions about science and religion take on a public significance beyond their merely intellectual interest.

The present Muslim debate over science and religion is significantly different compared to the debate taking place in advanced Western countries. Scientific institutions are weaker in Muslim lands. They are less able to assert their independence from cultural politics. Moreover, liberal theological options, which help grant science an independent sphere of operation, do not enjoy as strong a social presence. Indeed, observing the secularizing trajectory of Western societies, conservative Muslims are likely to be even warier about repeating the mistakes Christians have made.

Doctrinally conservative Muslims, then, often feel a need for an apologetic response to the ways modern science can engender skepticism about supernatural claims. Their literature on science and religion is broadly similar to Christian equivalents; after all, beyond a number of specific doctrinal matters, all Abrahamic faiths face similar challenges from the naturalistic tendencies within science. Like their Christian counterparts, conservative Muslims typically rhetorically endorse science and defend a perception of modern science in complete harmony with traditional beliefs. Where harmony is strained, however, making science acceptable requires some work. Especially in the popular apologetic literature, three approaches stand out: ignoring the challenge, co-opting science, and outright rejection.

Some questions that have become staples in the Western discussion over science and religion have little connection to the concerns of ordinary Muslims. Popular apologetics therefore tends to ignore them, and without public controversy, religious intellectuals devote little attention to such topics. Modern physics is rarely the subject of sustained reflection, with a few exceptions. Some thinkers try to validate the medieval Muslim philosophical position of occasionalism through quantum mechanics, though the typical result is a distortion of the physics. More often, quantum physics gets invoked in a neo-Sufi, even parapsychological context. This almost never rises above the level of popular pseudoscience (Edis 2007). A recent area of scientific development that has drawn the fire of Western opponents of materialism, such as the intelligent design movement, is cognitive neuroscience (e.g., Beauregard and O'Leary 2007). It is hard to find any distinctively Muslim response to the currently dominant materialist point of view in neuroscience.

Another common strategy, especially in popular apologetics, is to co-opt science, arguing that modern science supports traditional beliefs. For example, there is now an extensive literature that is very popular throughout the Muslim world, purporting to demonstrate how the Quran is full of miraculous anticipations of modern science and technology (e.g., Bucaille 1979; Nurbaki 1998). This parallels the way many conservative Protestants claim fulfilled prophesies as a way of providing concrete support for their belief that the Bible is the word of God. More sophisticated Muslim thinkers often have misgivings about this style of apologetics (Rehman 2003), but its popularity remains very strong.

Co-opting science extends to the way that some Muslims take a conciliatory attitude toward evolution, accepting an explicitly guided, non-Darwinian form of biological change. In this connection, there are occasional arguments that medieval Muslim scientists had already thought of the concept of evolution (Bayrakdar 1987). Much of the literature arguing that modern science supports traditional Islam is opportunistic. For example, while many conservative Muslims rely on various pseudobiological arguments to support traditional gender roles, some selectively quote from the popular sociobiological literature to that end while ignoring the evolutionary context of such arguments (Edis and Bix 2005).

Some scientific and scholarly ideas, however, appear to be both worthy of attention and too tainted with materialism to assimilate easily. One example is modern Quranic criticism, which is practiced almost exclusively outside of Muslim lands, and is nearly universally rejected by devout Muslims when they are aware of it. At least one European scholar has found it necessary to publish under a pseudonym to avoid repercussions (Luxenberg 2000). In natural science, the clearest example is Darwinian evolution.

Darwinian evolution – that is, a completely naturalistic understanding of common descent as driven by purposeless material processes such as variation and selection – has usually faced rejection from Muslim populations. Ideas about evolution first reached Ottoman intellectuals in the context of the European controversy about science versus revelation. And the first Ottoman defenders of Darwin came from among westernizers who were less interested in biology than evolution as an example of materialism triumphing over clerical obscurantism. Traditional religious scholars naturally reacted with condemnation. Interestingly, the prominent modernist Jamal al-Din al-Afghani also rejected evolution as an anti-religious philosophy that was absurd on the face of it. He only made concessions to a possibility of a non-Darwinian, guided version of evolution toward the end of his life (Keddie 1968, pp. 130–174; Bezirgan 1988).

As a result of the rejection of Darwin even by reformist Turks and Arabs keen to avoid the taint of materialism, the nineteenth-century Ottoman debate about evolution was stillborn. Most Middle Eastern Muslims, even many among educated elites, remained creationists by default. Indeed, until recently, Darwinian ideas usually did not have enough of a public presence to inspire an elaborate creationist pseudoscience as a reaction. As a result, conservative Muslim attitudes toward evolution today are often roughly comparable to those of more fundamentalist and Pentecostal Christians worldwide. Evolution is clearly unscriptural and obviously a materialist idea; therefore, it cannot be correct – nothing more need be said.

52.6 The Nur Movement

Muslim responses to materialist currents in science do not trickle down from academic theologians. They do not arise from traditional religious scholarship, which has little scope for analysis aside from condemning ideas that go against orthodox readings of the sacred sources. Muslim responses typically come from devout intellectuals facing immediate practical problems, and even more important, they achieve prominence if popular religious movements adopt them.

In this regard, the example of Said Nursi and the "Nur movement" he inspired is particularly important. Western scholarship on Islam has tended to be driven by political questions such as the potential for democracy in Muslim countries, and therefore figures such as the early modernists and later theorists of political Islam such as Sayyid Qutb and Abul A'la Maududi have drawn much attention. They all have writings that usefully represent aspects of modern Muslim thought on science, reason, and religion (Euben 1999). Nursi's movement, however, has been instrumental in shaping popular Turkish attitudes about science and religion today. It pioneered forms of popular apologetics that are now very common throughout the Muslim world. The Nur movement deserves more attention.

Nursi was trained as a traditional scholar in the Ottoman provinces and was most active in the first half of the twentieth century. Much of his thought repeats modernist themes, including an emphasis on reforming Muslim education, a positive view of technology, and a desire to combat harmful intellectual influences associated with nineteenth-century materialism. He argues that mystical illumination complements scientific investigation, that all useful arts and sciences emanate from the "names of God," and that the Quran anticipates modern technological possibilities (Abu Rabi' 2003).

None of these are remarkable ideas for Nursi's time and circumstances. But unlike many other reformist scholars, Nursi started a successful popular movement. He produced the *Epistles of Light*, revered by the Nur movement almost as a secondary scripture. After his death, Nur adherents continued to organize on the basis of studying the *Epistles*. The movement has been notable for its modernizing emphasis, capitalist mentality, and nontraditional forms of organization and religious authority structure. Social scientists have pointed to the Nur movement as an example of a very modern, pro-technology Islamic movement with a strong popular constituency. They have argued that it has been instrumental both in laying the groundwork for provincial economic development and in spearheading the success of modern Islamic forms of thinking in Turkey (Mardin 1989; Yavuz 2003).

The Nur movement is too large to be unified. Its many splinter groups today include the followers of Fethullah Gülen, one of the internationally best known Muslim leaders of recent times, plus others that enjoy considerable political influence in Turkey (Çalışlar and Çelik 2000). Yet a common denominator of Nur-inspired movements for decades has been a form of popular apologetics that places great emphasis on harmony between Islam and science. Most pseudoscience in Turkey, other than superficial Western-derived and media-driven fads such as astrology and UFOs, has the signature of the Nur movement.

In pre-Internet times, the Nur movement produced popular science magazines that regularly included articles on how Muslims should embrace technology and how the Quran miraculously anticipates modern developments, containing knowledge of astrophysics or embryology. They opposed evolution. The immediate social influence of the movement fluctuated according to changing political circumstances within Turkey, but over the long term, the Nur style of apologetics appealed beyond the study circles focused on the *Epistles of Light*. Today, some of the old popular science and religion magazines survive, but the Nur style of apologetics has acquired a much broader influence, becoming embedded in popular culture and spreading through new forms of media.

In today's Turkey, anyone with views on science and religion – for example, an academic theologian proposing a more liberal view concerning evolution – has to contend with the widespread influence of the Nur movement. The Nur version of harmony between science and Islam is entrenched not just in popular culture and popular pseudoscientific beliefs, but it also enjoys elite influence. Academic theologians and scientists who either have direct connections to the Nur movement or have absorbed its ethos are significant voices in the Turkish debate over science and religion (Şahin 2001; Tatlı 1992). Through such channels, opposition to evolution, for example, has a presence in the intellectual high culture as well as the mass media.

52.7 The Growth of Creationism in Turkey

Said Nursi was disappointed by the Turkish Revolution of the 1920s. The westernizers in control of the new Turkish Republic established official secularism, suppressing political expressions of Islam. Nursi spent time in prison. The intellectual influences upon political leaders such as Mustafa Kemal Atatürk tended to be secular, anticlerical, and even included then-current versions of materialism and positivism (Parla and Davison 2004). Science, not religion, was to light the path to a brighter future. Evolution entered the curriculum. Still, evolution remained a relatively small offense against religion in a state-controlled educational system that aimed to make religious sentiment a private matter rather than a reference for public policy.

Until the 1960s, like other Middle Eastern Muslim countries, Turkey presented a picture of grudging but gradual secularization. The Nur movement and similar religious conservative groups produced some creationist literature, but their market was limited. Anti-evolutionary activity stayed confined to the subculture of a strictly observant, self-consciously orthodox minority. It is harder to know what the bulk of the population thought of evolution in the textbooks, in the days before constant opinion polling. Most likely, evolution inspired passive resistance at most, when it drew attention at all.

In the 1970s, together with the rest of the Muslim world, a newer form of political Islam started to gain strength in Turkey. Evolution became a minor culture war item, a way for Islamists to demonstrate opposition to secularization without naming official secularism as a target. In parliament, an Islamist political party attacked the presence of evolution in education but produced only a minor media stir (Atay 2004, pp. 136–137).

Creationism came into its own in the aftermath of the conservative military dictatorship of 1980–1983. Religious conservatives, many with Nur movement connections, gained control of the Turkish Ministry of Education in the first quasi-civilian government. They were convinced that evolutionary ideas were morally corrosive, and yet they were very aware that science commanded significant cognitive authority. So they needed a way to show that evolution was a scientifically dubious idea, maybe even a fraud. They found the resources they needed in Protestant "scientific creationism." Adding an odd chapter to the history of Muslim intellectual borrowing from the West, religious conservatives invoked Christian creationists in a mirror image of the way secularists tend to rely on Western scientific authorities. While the Turkish creationists downplayed some important features of Protestant creationism such as a young earth and flood geology, they adopted the bulk of the anti-evolutionary debating points developed by their Christian counterparts. Indeed, the Ministry of Education had samples of "scientific creationist" literature officially translated and made available to high schools and teachers (Edis 1994).

Since this creationist breakthrough, Turkish secondary school textbooks have often contained anti-Darwinian or explicitly creationist material (Yalçınoğlu 2009). Islamists and conservatives favor a religious identity politics, so even though opposing evolution is not a leading item on political agendas, it tends to be a background commitment. Since 2002 a moderate Islamist party has held power. The Ministry of Education under its administration is perceived as sympathetic to creationism, due to incidents such as retaining creationist material in the curriculum in the face of academicians petitioning for the removal of unscientific material (Kotan 2006).

Islamist politics in Turkey relies on alliances between working class constituencies such as recent immigrants to large cities and a newly prosperous provincially based business class, often united by antagonism toward longer-established secular elites. Religious populists present political conflicts over religion as a clash between a debased, inauthentic secularism and the traditional piety of the common people. Nevertheless, the primary constituency for Turkish creationism is not traditionalists but modern believers, even though they are theologically conservative. It is precisely because they want to take their place in the modern world, where mastering technology is the key to success, that creationists fashion a pseudoscience that harmonizes science and their religious convictions.

Since 1997, the popular appeal of Turkish creationism has deepened. Indeed, the Turkish style of creationism has spread internationally, throughout other Muslim countries and the Muslim diaspora in Europe and North America. The central figure in this development is Harun Yahya, a pseudonym that serves as a brand name for a ubiquitous, well-funded, and media-intensive form of creationist propaganda (Edis 1999; Edis 2003; Riexinger 2008; Sayın and Kence 1999). There is not much new about the content of the Harun Yahya material: it consists of arguments that have no scientific substance and distortions of science often copied from Christian anti-evolution literature, presented with a conservative Muslim emphasis. The range and production quality of this material, however, is impressive. Creationism in the Turkish Ministry of Education resulted mainly in some translations and a few paragraphs expressing opposition to Darwinian evolution in some otherwise unremarkable textbooks. The Yahya operation is much better suited to a postmodern media environment. Large numbers of glossy books, magazines, videos, websites, public events, and television interviews make Yahya's simple, intuitively appealing creationism available to a large public. None of this material is marked out as being religious literature of interest only to a conservative Muslim subculture; from its presentation style to its use of everyday language, Harun Yahya material is designed to be marketed to ordinary, modern Muslims who need not be attracted to strictly observant varieties of Islam. Furthermore, Yahya material is artificially cheap and is often distributed free of cost. Clearly the Harun Yahya enterprise has considerable financial backing, though the source of these funds is not entirely clear.

Turkish scientists have tried to counter such popular creationism, but in the public arena, the creationists hold the upper hand. One reason for the weakness of the scientific position is that Turkish scientists not been able to present an organized response, in part due to other conflicts with the government due to neoliberal policies that would further weaken the position of basic science. In addition, there is also some opposition to evolution within Turkish academia, especially in newly established provincial universities. Even some biologists can go in search of an "alternative biology" more similar to intelligent design than Darwinian evolution (e.g., Yılmaz and Uzunoğlu 1995). Moreover, scientists, especially in the more prestigious universities, represent a very westernized population. They have been most comfortable phrasing their opposition to creationism in the idiom of defending the secular nature of the Turkish state (Sayın and Kence 1999). Since republican secularism has been discredited in popular politics, this has been a strategic blunder.

Islamist rule may also have affected the structure of support for science in the Turkish state. In March of 2009, *Bilim ve Teknik*, the popular science and technology magazine published by the Scientific and Technological Research Council of Turkey, was supposed to have been released with a cover story about the two hundredth birthday of Darwin. A political appointee in upper management, an engineering Ph.D., intervened to change the cover story and delete the Darwinian material (Abbott 2009). This led to another round of the creation-evolution wars in the popular press. Religious columnists charged "Darwinists" and "materialists" with being the real censors, disallowing alternative scientific views favorable to creation from enjoying a proper hearing. Secularist writers interpreted the event as evidence of growing Islamist entrenchment in scientific institutions.

Building on its success in Turkey, the Harun Yahya brand of creationism has now gone global. Today, Yahya material is available in languages spoken by Muslim populations all over the world. Yahya books are prominently displayed in Islamic bookstores in London, appear in schools in Pakistan, and are promoted by speaking tours in Indonesia. As a publicity stunt, Yahya's publisher mailed copies of a volume of a typically lavishly produced encyclopedia called the *Atlas of Creation* to scientists and educators in Europe and North America, drawing media attention outside of Islamic circles (Yahya 2007). There is now a global variety of popular Islamic creationism that goes beyond long-standing but usually passive Muslim resistance to Darwinian ideas. Many modern Muslims are attracted to claims that evolution is scientifically false and that science, properly done, supports Quranic notions of special creation.

52.8 Accommodating Evolution

Strict creationism, whether based on outright rejection of science that contradicts conservative readings of the sacred sources or expressed in the form of a Harun Yahya-style pseudoscience, is not the only option available to Muslims. A minority consisting of theologically liberal and secular Muslims, for example, tends to accept evolution due to their general trust of modern education and science as a cognitive authority. Such acceptance of evolution need not imply more than a superficial knowledge either of biology or of the religious worries about evolution articulated by more conservative Muslims. Nevertheless, there is a constituency for efforts to interpret Islam in a way that is compatible with at least some minimal form of evolution.

The easiest option for harmonizing evolution and traditional beliefs is to downplay Darwinian explanations of the evolutionary process, affirming common descent while portraying biological evolution as a divinely guided progression toward higher forms of life (Ateş 1991). A particular concern is to interpret the Quran in such a way as to allow for a degree of evolution. Some theologians, for example, read 24:45 in the Quran, speaking of God creating all animals from water, as a statement that life emerged from the oceans, just as the scientific history of life on Earth has it. Verses that describe the special creation of Adam and Eve, however, need more strenuous attempts at reinterpretation. Very few Muslims will countenance the idea that the Quran is anything but the direct and unadulterated word of God. Therefore, while many Muslims think that considerable evolution under divine guidance may be applicable to nonhuman forms of life, devout believers typically consider humanity to be a separate creation.

Evolution can also become more acceptable if similar ideas can be located in a Muslim intellectual tradition. Some Turkish theologians have proposed reviving such ideas under a label of "evolutionary creation theory" (Altaytaş 2001; Bayrakdar 1987). This appears to be partly based on questionable readings of medieval philosophical reflections on the ancient Greek concept of the great chain of being. Still, in a cultural climate that privileges Islamic authenticity, such historical connections, even if forced, may help make guided evolution a more attractive view. In any case, efforts to compromise between Darwinian evolution and creationism are common. If the religious experimentation in the Muslim world today should lead to a liberalizing trend that can appeal to a broader base than a westernized elite, guided evolution will become an even more popular option.

Many influential Muslim intellectuals also avoid naive Quranic literalism and strict creationism while also expressing skepticism about the Darwinian, naturalistic form of evolution that is the established position in natural science. Some internationally known scholars defend such views. For example, Seyyed Hossein Nasr and Osman Bakar are well known for their outline and defense of a specifically Islamic philosophy of science. Both Bakar and Nasr allow for limited biological changes over time but deny that purely natural mechanisms can ever account for the creativity seen in the history of life. Like popular creationists, they describe evolutionary biology as a manifestation of materialist philosophy rather than a real science with a true empirical foundation. They go further, however, contrasting a Darwinian view of life with a God-centered perception of nature that hearkens back to classical Islamic conceptions of knowledge and creation. They aspire to restore the Islamic religious sciences to a position of preeminence and to have revelation provide the framework for all knowledge claims, including investigations of the natural world (Bakar 1987, 1999; Nasr 1989).

Proposals to "Islamize science" or otherwise reconstruct modern knowledge in a more Islamic fashion attract much attention in Muslim academic and intellectual circles (AbuSulayman 1989). Bakar and Nasr's views are similar, and they continue to resonate among Muslim thinkers concerned about science and Islam. In scientifically advanced countries, there is very little opposition to evolution in the academic and mainstream intellectual environments, and even the more sophisticated varieties of anti-Darwinian views tend to be muted. In contrast, the Muslim intellectual environment is much more hospitable to ideas hostile to Darwinian evolution. The view that the functional complexity exhibited by life must be due to intelligent design remains deeply embedded in Muslim intellectual culture (Edis 2004).

Therefore, it is not surprising that intelligent design, the latest version of anti-evolutionary thought developed in the United States, has attracted attention from Muslims inclined to be suspicious of evolution. In Turkey, where the public controversy over evolution has been most intense, the major books defending intelligent design have been translated and have been favorably reviewed in the Islamic press (Akyol 2005). The intelligent design literature sets scripture aside and concentrates on claiming that mindless processes such as variation and selection cannot create the information-rich structures seen in biology (Meyer 2009). This approach validates intuitions about divine design in the world common to all Abrahamic religions, including Islam.

Though Muslims have a wide variety of views on creation and evolution, the views of even politically liberal and modernist Muslims tend to gravitate toward explicit divine design. Opinion polls in Muslim countries show strong public sentiment against evolution (Hameed 2008; Miller et al. 2006). And even the non-negligible minority acceptance of evolution in such polls does not necessarily signify agreement with a naturalistic conception of evolution. Muslims who agree with common descent very often hold non-Darwinian views of evolution; it is, at present, impossible to use survey data to differentiate between acceptance of explicitly guided or progressive views of evolution and evolution as understood in mainstream science.

52.9 Evolution Education in the Middle East

Evolution becomes controversial especially in the context of education. Therefore, educational policy concerning evolution is a good indicator of official standpoints and their conflicts and alignments with popular social and political views concerning evolution. Given the wide variety of political contexts in which Middle Eastern science education policies take shape, a diverse array of outcomes can be expected. Examining evolution education provides an opportunity to see how some more general themes about Muslim responses to modern science, such as unease with apparently materialist aspects of science, play out in varying local circumstances.

Together with Turkey, the secondary biology curricula of Lebanon, Egypt, Iran, Saudi Arabia, and Israel have been relatively well investigated. Curricular studies, when combined with surveys indicating views on creation and evolution, give good snapshots of responses to controversial aspects of science at the level of students and educators. More detailed research on local histories is needed to reveal linkages to the broader ongoing debates among Muslims concerning science and religion.

Evolution education in Turkey is marked by changes imposed by different governments since the military coup of 1980, tending toward a more culturally conservative point of stability with the moderate Islamist government in power since 2002. The prominent presence of creationism for decades has clearly affected students and has led to a low level of acceptance of evolution among surveyed undergraduates (Peker et al. 2010). Research also suggests that state education policies have an important effect on the acceptance of evolution in Turkey. For example, a survey has shown that recently trained younger teachers reject evolution significantly more frequently than their older colleagues (Somel et al. 2007). The teachers' views then naturally affect their students. Turkey is a case study exhibiting a very successful anti-evolutionary movement that has found state support as well as a popular base, affecting both public opinion and educational policies (Yalçınoğlu 2009).

The teaching of evolution is treated differently in Egypt and Lebanon: while the theory of evolution is included as one complete unit in the Egyptian secondary level biology curriculum, it was eliminated from the official Lebanese curriculum because of pressures from religious authorities. However, many Lebanese students are still exposed to at least some ideas and concepts associated with the theory of evolution. Many schools in Lebanon implement international curricula such as the International Baccalaureate, the French Baccalaureate, and a variety of American curricula; many Lebanese schools adopt American or French textbooks. The secondary level Egyptian biology textbook required in all public and local private schools includes a unit entitled "Change in living organisms (evolution)" that is taught in Grade 10. As presented in the required textbook published by the Egyptian Ministry of Education (Duwaider et al. 2005–2006), the learning outcomes of the unit expect students to be able to define evolution and "improvement," differentiate evolution from "improvement," explain and critique Lamarck's theory, explain and critique Darwin's theory, explain the modern evolutionary synthesis, define the concept of hereditary balance, explain variability, define and give examples of natural selection, list different types of evidence in support of the theory of evolution, and understand that science is tentative. A close reading of the unit shows that the authors seem neutral – they present content matter and evidence in support of evolution without taking sides. The authors suggest, in the introduction to the unit on evolution, that there have always been different explanations of variability in living things and the appearance of life. They go on to explain the differences between special creation and evolution and state that the theory of evolution is accepted by most biologists, who use evolution to explain a wide range of biological phenomena.

Other Middle Eastern countries show considerable variation. Burton (2010, 2011) investigated the extent to which evolution is emphasized in the science curricula of Iran, Saudi Arabia, and Israel. Burton (2011) found that in Iran "science is not described as simply an outgrowth of Islam or subject to preconceived doctrines of any religion – rather it is affirmed as a separate valid field of knowledge, and one crucial to individual and social welfare" (p. 27). Consequently, according to Burton, the coverage of evolution does not seem to be a controversial topic, resulting in a thorough coverage of the topic at different grade levels. For example, in grade five, students are introduced to the history of the earth and life. This history is based on the work of geologists and other scientists with an emphasis on evidence in support of evolution. The same thing happens at the grade 8 and grade 12 levels where students are exposed to a thorough treatment of evolution and an assertion that almost all modern biologists have accepted Darwin's theory. The only topic that does not receive complete coverage in the Iranian textbooks is the most religiously controversial topic, human evolution.

In contrast to Iran, the science curriculum of Saudi Arabia asserts that science education is grounded in an Islamic view of the universe, humanity, and life and in a strong belief in the harmony between Islam and science (Burton 2011). As a result of the centrality of Islam in education, the Saudi science curriculum and textbooks provide ample evidence from the Quran in support of creation and emphasize the necessity of rejecting the theory of evolution because of its blasphemous and fraudulent nature (Burton 2010).

According to Burton (2010), the situation regarding the teaching of evolution in Israel seems to be in-between Iran and Saudi Arabia. While the publicly supported Israeli religious and secular schools share the same biology curriculum, religious schools are allowed to use educational materials produced by religious authorities that include references to creation, the Creator, and the special status of humans who are created in the image of God. The emphasis on creation in publicly supported religious schools appears to persist even after they study biology at the secondary school level.

Several studies have been conducted in Lebanon and Egypt to investigate students', teachers', and university faculty members' positions regarding evolution. In Lebanon, the positions of Muslims (Sunni, Shiite, and Druze) and Christians were investigated, while in Egypt Muslims (only Sunnis) and Christians were involved in the studies.

Dagher and BouJaoude (1997, 2005) explored how a number of biology majors attending a university in Beirut, Lebanon, accommodated the concept of evolution with their existing religious beliefs. Sixty-two university students enrolled in a required senior biology seminar responded to open-ended questions that addressed (a) their understanding of evolutionary theory, (b) their perception of conflict between this evolutionary science and religion, and (c) whether the concept of

Table 52.1 Students' personal positions toward the theory of evolution represented in relation to their religious affiliation

Position	Christian		Muslim		Total/position	
For evolution	82 %	(n=14)	35 %	(n=16)	48 %	(n=30)
Against evolution	0 %	(n=0)	47 %	(n=21)	34 %	(n=21)
Compromise	6 %	(n=1)	18 %	(n=8)	15 %	(n=9)
Neutral	12 %	(n=2)	0 %	(n=0)	3 %	(n=2)
Total/religion	100 %	(n=17)	100 %	(n=45)	100 %	(n=62)

evolution clashed with their beliefs. Based on their responses, 15 students were selected for an in-depth exploration of their written responses. Students' answers clustered under one of four main positions: for evolution, against evolution, compromise, and neutral. As indicated in Table 52.1 above, more Christian than Muslim students were supportive of evolutionary ideas.

BouJaoude and Kamel (2009) found that Muslim and Christian secondary school students in Egypt and Lebanon had inadequate understandings of the nature of theories and of the scientific bases of evolution. Moreover, they found that there were significant differences between Lebanese Christian and Muslim students regarding their perceptions of the relationship between science and religion, with Muslims being in general more influenced by their religious beliefs than Christians. Also, while more Muslim than Christian Lebanese students rejected evolutionary science, these differences were not as pronounced in Egypt where Muslim and Christian students differed on only a few items on a survey that evaluated their conceptions of evolution.

BouJaoude and colleagues (2011b) investigated distinctions among the diversity of religious traditions represented by Lebanese and Egyptian Muslim high school students regarding their understanding and acceptance of biological evolution and how they relate the science to their religious beliefs. The researchers explored secondary students' conceptions of evolution among members of three Muslim sects – Sunni, Shiite, and Druze – in two cultural contexts, one in which the overwhelming majority of the population is Muslim (Egypt) and another in which there is a sizable Christian community (Lebanon). Data were collected via surveys that examined students' scientific and religious understandings of evolution among 162 Egyptian students (all Sunni Muslims; 63 % females and 37 % males) and 629 Lebanese students (38.5 % Sunni, 38 % Shiite, and 23.5 % Druze; 49 % females and 51 % males). Additional data were collected via semi-structured interviews with 30 Lebanese students to allow triangulation of data for accuracy and authenticity. Results indicate that many Egyptian and Lebanese Muslim students have misconceptions about evolution and the nature of science, which often lead to rejection of evolution. Also, Lebanese Sunni and Shiite students and Egyptian Sunni students tend to exhibit high levels of religiosity, and these students report that their religious beliefs influence their positions regarding evolution. Finally, Sunni and Shiite Lebanese students have religious beliefs, conceptions of evolution, and positions regarding evolution similar to those of Sunni Egyptian students.

BouJaoude and colleagues (2011a) investigated biology professors' and teachers' positions regarding biological evolution and evolution education in Lebanon.

Table 52.2 University professors' positions regarding evolution categorized by religious affiliation

	Muslim					
	Shiite	Sunni	Druze	Christian	Agnostic	Total
Accept	1			2	1	4
Selectively accept		2	1			3

Table 52.3 Teachers' positions regarding evolution categorized by religious affiliation

	Muslim				
	Shiite	Sunni	Druze	Christian	Total
Accept			5	4	9
Reject	3	1		1	5
Selectively accept	2	1			3
Neutral (noncommitted or confused)		2		1	3

Participants were 20 (13 private and 7 public) secondary school biology teachers (16 females) and seven university biology professors (two females and five males) teaching at a private, American-style university. Data came from 25 to 30 min, semi-structured interviews with the teachers and the professors. As shown in the tables above, university faculty members (Table 52.2) were divided between those who accept (4 out of 7) or who selectively accept (3 out of 7) the theory of evolution with more Muslims being selective in their acceptance. As for teachers (Table 52.3), the positions ranged from acceptance to total rejection with more Muslim teachers rejecting the theory than Christian teachers.

There is considerable variation in science education policies among Muslim countries and populations in the Middle East, reflecting a wide variety of local histories and religious influences on the politics of education. Nonetheless, there are some important commonalities. Those aspects of modern science that have a materialist association, especially biological evolution, meet with opposition at many levels. Therefore, even when evolution is taught in conservative Muslim environments, this usually takes place in a context where a creationist alternative is a prominent implicit presence. Iran is a partial exception, perhaps due to local Shia dominance allowing a degree of intellectual independence from Sunni resistance to evolution and the explicit Iranian state support for developing advanced biotechnological capabilities.

52.10 Separating Science and Religion?

The recent Turkish and Arab experiences with friction between science and religion partly derive from a common history as successor states to the Ottoman Empire, having undergone broadly similar historical experiences with Western colonial powers and facing a similar need to import science and technology from non-Muslim sources. In formulating religious intellectual responses to a perceived materialism in parts of modern science, and in developing highly successful varieties of opposition to evolution, Turkey has perhaps been at the cutting edge, due to proximity to Europe and the history of the Turkish experiment with secularism. But much of what can be said about Turkey applies to the Arab Middle East and beyond. For example, while Islamic creationism has acquired a Turkish flavor of late, this has not reduced its international appeal. Conservative Muslims worldwide, in South Asia as well as the Middle East, denounce materialism, sometimes using the stimulus of Harun Yahya to activate local objections to how modern science has removed divine purpose from its conceptual frameworks (Riexinger 2009).

Indeed, Harun Yahya creationism as an international phenomenon illustrates how, in the age of the Internet, doctrinally conservative Muslims concerned about materialist aspects of modern science rapidly interchange ideas and formulate responses that have global resonance. In some respects, the Harun Yahya corpus shows the marks of its Middle Eastern and specifically Turkish origin, as in its many books and pamphlets devoted to praise of Said Nursi and themes popularized by the Nur movement. And yet, especially the creationist Harun Yahya material, though it originates in Turkish, is immediately translated into English and languages of the Muslim diaspora as well as Muslim-majority countries. It is made available globally through well-designed websites and advertised and popularized throughout the Muslim world.

More serious Muslim thinkers about science and religion also have an international audience. Seyyed Hossein Nasr is not an obscure academic from Iran who now teaches in the United States – his views on achieving an Islamic philosophy of science are known and debated by interested Muslim intellectuals worldwide. Notions such as "Islamizing science" or using the anti-Enlightenment aspects of postmodern philosophy to defend Islamic traditions against science-based critiques (Aydın 2008) are, again, put to use and discussed globally. There are, naturally, local differences of emphasis, and local education policies are affected in various manners depending on diverse political circumstances. But the *intellectual* options available to Muslim thinkers today are everywhere alike. With the rise of a globalized Islam constructing a universal religious identity transcending local variations (Roy 2004), the discussions young Muslims enter into online concerning science and religion also sound similar themes. Indeed, populations that operate in a globalized economy, and most keenly feel the effects of technology on their lives, are the strongest constituency for today's efforts to harmonize modern science and traditional religion. They have the most at stake.

So a broad-brush description of conservative Muslim views of science and religion today should be appropriate. Muslim populations typically are positive toward technology. Though religiously informed criticisms of the uses of technology are not unknown (Sardar 1984), doctrinally conservative Muslim intellectuals – who usually take modern social conditions for granted – almost always support science rhetorically. But many also harbor distrust toward the present conceptual frameworks of science, which appear as "only theories" compared to the compelling facticity of the products of applied science. Therefore, conservative Muslims often find it easy to reject aspects of modern science that appear tainted with materialism or that otherwise challenge traditional beliefs.

Broad-brush descriptions must always overlook local variety and important details. But more important, even if it is true that modern science and popular forms of Islam have significant points of friction, the consequences for science and education are not immediately clear. After all, the United States also exhibits a strong degree of religious conservatism in its population, including a high and steady level of support for creationism. Scientists have plenty of occasions to complain about how Americans are scientifically illiterate, how they see science as a way to collect "facts" like stamps, and how they are enthusiastic about technology but often take an anti-intellectual attitude toward the conceptual frameworks of science. Some scientists even directly blame American religiosity for all this (Coyne 2012). And yet the United States supports world-class scientific and educational enterprises.

But the United States is also different from Muslim lands. In the United States, populist opposition to knowledge-based elites, such as that expressed by Christian creationism, is isolated from the intellectual high culture and has little effect on scientific institutions. The strength of liberal religious options also helps to protect science from religious populism, supporting a conventional wisdom according to which science and religion have separate spheres that do not interfere with one another.

In that case, perhaps a Muslim version of a separate spheres or "nonoverlapping magisteria" doctrine (Gould 1997) could help satisfy both desires for religious authenticity and scientific interests in describing how the natural world works. After all, there is no shortage of Muslim scientists who insist that their scientific commitments, including evolutionary biology, are perfectly compatible with their faith (Guessoum 2011). When Turkish academics publicly defend evolution, they usually express views similar to positions taken by American organizations such as the National Center for Science Education, endorsing liberal theological stances and advocating separate spheres (Aydın 2007). A few internationally known religious intellectuals, such as Abdolkarim Soroush, argue on a religious basis that science should be independent of religion (Soroush 2000).

The separation option is still possible. As the world of Islam continues to change rapidly, a strict separate spheres view may become more prominent in the future. But at present, this option is structurally weak and discredited by its associations with political secularism. In Turkey, defenders of evolution are disorganized and demoralized. And throughout the Muslim world, liberal theologians invite conservative reactions. Soroush, for example, was pressured to leave Iran; he now teaches in the United States.

Moreover, emphasizing the prospects for a separation of science and religion in Islam may lead to a distorted picture of specifically Muslim responses to science. The notion of separate spheres has been a successful device to keep the peace between science and religion in the post-Christian West. It may not be as applicable to a Muslim world where religiously conservative intellectuals are determined that they should not come to live in a post-Muslim environment. An insistence on separate spheres would be an imposition of a Western perspective onto Muslim concerns, when Muslims are trying to achieve a non-Western way of being modern. There is a strong tendency to see the present intellectual distance between science and religion as being a Western solution to a Western problem caused by factors inherent to Christianity (Aydın 2000; Şahin 2001). Very often Muslims gravitate toward the notion that since science and religion must coexist in harmony in an

Islamic context, perhaps as in the supposed golden age of medieval Muslim science (e.g., Al-Hassani 2012), there is no problem here to solve.

So conservative Muslim resistance to materialist conceptual frameworks in science deserves to be taken seriously. It is easy to observe that popular Muslim apologetics typically has very low intellectual quality. Recurrent worries about materialism can also seem odd. After all, Marxism has nearly vanished. There are some readers of Richard Dawkins translations, and among secular philosophers and scientists in Muslim lands, there are no doubt even outright physicalists. But all such materialists have negligible wider social influence. Nonetheless, there is also some real intellectual substance articulated in Muslim concerns about materialism.

The conventional wisdom among Western liberals both accepts and conceals conflicts between robust Abrahamic theisms and the naturalism that has come to characterize modern science. If divine purpose has no explanatory role in physics or biology or neuroscience, it becomes hard to see how to make sense of a claim of divine creation. Western liberals accept this difficulty but usually propose to overcome it by having supernatural belief retreat to a metaphysical realm. Religion handles ultimate meaning and purpose, while science investigates the details of how nature operates. This is an intellectually unstable position, since both religious and scientific thinkers usually have ambitions that cross over into each other's territory (Edis 2006). Moreover, Western liberals have inadvertently ended up giving science primacy over religion, in the sense that when a conflict has appeared, as over creation and evolution, it is always theology that has had to retreat and reinterpret. Many Muslims who are aware of the liberal Christian response to science cannot help but perceive it as a bowdlerization of revealed truths.

Conservative Muslim intellectuals affirm science as a form of worldly learning, but they often insist that science should be anchored in Islam. Inquiry should be free but only as long as it respects the boundaries set by faith. Muhammad Abduh, the Egyptian modernist, praised Islam as a religion of reason, saying Islam "did not impose any conditions on reason other than that of maintaining the faith" (Euben 1999, p. 106). This is a common sentiment, surfacing in legal reasoning as well as intellectual debates (Kamali 1997). To Western liberal eyes, it looks like free inquiry has a very important exception, since core Muslim beliefs are not open to question. But many Muslims would reject that framing of their views. By protecting Islamic beliefs from criticism, they see themselves not as carving out an intellectual exception but as protecting reason itself. Both in intellectual culture and at a popular level, Muslims very often think that some awareness of divine laws is constitutive of rationality itself (Rosen 2002). Critically examining religious beliefs invites doubt. And casting doubt on God or revelation can lead to nothing but irrationality and social disintegration.

So even with the religious experimentation taking place among Muslims today, it remains very uncertain whether Muslim views of science and religion will follow a trajectory toward a separate spheres style of accommodation. It is, for example, possible that Muslim populations will continue to gravitate toward a different equilibrium between supernatural beliefs and modern knowledge, one that emphasizes applied science while downplaying the conceptual frameworks of natural science. Whether this is sustainable will depend very much on future opportunities for technological creativity independent of developments in basic science.

References

Abbott, A. (2009). Turkish scientists claim Darwin censorship. *Nature* http://www.nature.com/news/2009/090310/full/news.2009.150.html, Accessed 20 February 2012.

AbuSulayman, 'A. (Ed.) (1989). *Islamization of Knowledge: General Principles and Work Plan* (2nd edition). International Institute of Islamic Thought, Herndon.

Abu-Rabi', I. M. (Ed.) (2003). *Islam at the Crossroads: On the Life and Thought of Bediuzzaman Said Nursi*. State University of New York Press, Albany.

Akyol, M. (2005). Intelligent Design could be a bridge between civilizations. *National Review Online*. http://old.nationalreview.com/comment/akyol200512020813.asp . Accessed 28 May 2012.

Akyol, M. (2009). Dini Siyasete Alet Etmek. *Star*. http://www.stargazete.com/yazar/mustafa-akyol/dini-siyasete-alet-etmek-haber-167851.htm . Accessed 28 May 2012.

Al-Hassani, S.T.S. (Ed.) (2012). *1001 Inventions. The Enduring Legacy of Muslim Civilization*. 3rd Edition. National Geographic, Washington D.C.

Alperen, A. (2003). *Sosyolojik Açıdan Türkiye'de İslam ve Modernleşme*. Karahan Kitabevi, Adana.

Altaytaş, M. (2001). *Hangi Din?* Eylül Yayınları, İstanbul.

Atay, T. (2004). *Din Hayattan Çıkar: Antropolojik Denemeler*. İletişim Yayınları, İstanbul.

Ateş, S. (1991). *Gerçek Din Bu,* vol. 1. Yeni Ufuklar Neşriyat, İstanbul.

Aydın, H. (2007). Evrim (bilim) – Yaratılış (din) çatışması üzerine kimi düşünceler. *Bilim ve Gelecek* 39, 20–25.

Aydın, H. (2008). *Postmodern Çağda İslam ve Bilim*. Bilim ve Gelecek Kitaplığı, İstanbul.

Aydın, M. S. (2000). *İslâmın Evrenselliği*. Ufuk Kitapları, İstanbul.

Bakar, O. (1987). *Critique of Evolutionary Theory: A Collection of Essays*. The Islamic Academy of Science, Kuala Lumpur.

Bakar, O. (1999). *The History and Philosophy of Islamic Science*. The Islamic Texts Society, Cambridge.

Bayrakdar, M. (1987). *İslam'da Evrimci Yaradılış Teorisi*. İnsan Yayınları, İstanbul.

Beauregard, M. & O'Leary, D. (2007). *The Spiritual Brain: A Neuroscientist's Case for the Existence of the Soul*. New York: HarperOne.

Bezirgan, N. A. (1988). The Islamic World. In T. F. Glick (Ed.), *The Comparative Reception of Darwinism*. The University of Chicago Press, Chicago.

Black, A. (2001). *The History of Islamic Political Thought: From the Prophet to the Present*. Routledge, New York.

Blank, J. (2001). *Mullahs on the Mainframe: Islam and Modernity among the Daudi Bohras*. The University of Chicago Press, Chicago.

BouJaoude, S. & Kamel, R. (2009). Muslim Egyptian and Lebanese students' conceptions of biological evolution. Presented at the symposium on Darwin and Evolution in the Muslim World, a conference at Hampshire College, Amherst, MA October 2–3, 2009.

BouJaoude, S., Asghar, A., Wiles, J. R., Jaber, L., Sarieddine, D. & Alters, B. (2011). Biology professors' and teachers' positions regarding biological evolution and evolution education in a Middle Eastern society. *International Journal of Science Education*, *33*, 979–1000.

BouJaoude, S., Wiles, J. Asghar, A. & Alters, B. (2011). Muslim Egyptian and Lebanese students' conceptions of biological evolution. *Science & Education*, *20*, 895–915.

Brown, C. M. (2012). *Hindu Perspectives on Evolution: Darwin, Dharma, and Design*. Routledge, New York.

Bucaille, M. (1979). *The Bible, The Qur'an and Science*. American Trust Publications, Indianapolis.

Büchner, L. (1884). *Force and Matter, or, Principles of the Natural Order of the Universe. With a System of Morality Based Thereupon*. Translated from the 15th German edition; 4th English edition, Asher and Co, London.

Burton, E. K. (2010). Teaching evolution in Muslim states: Iran and Saudi Arabia compared. *Reports of the National Center for Science Education*, 30, 28–32.

Burton, E. K. (2011). Evolution and creationism in Middle Eastern education: a new perspective. *Evolution*, 65, 301–304.

Çalışlar, O. & Çelik, T. (2000). *Erbakan-Fethullah Gülen Kavgası: Cemaat ve Tarikatların Siyasetteki 40 Yılı*. Sıfır Noktası Yayınları, İstanbul.

Coyne, J. A. (2012). Science, Religion, and Society: The Problem of Evolution in America. *Evolution*, doi: 10.1111/j.1558-5646.2012.01664.x

Dagher, Z. & BouJaoude, S. (1997). Scientific views and religious beliefs of college students: the case of biological evolution. *Journal of Research in Science Teaching, 34*, 429–455.

Dagher, Z. & BouJaoude, S. (2005). Students' perceptions of the nature of evolutionary theory. *Science Education, 89*, 378–391.

Demir, R. (2001). *Osmanlılar'da Bilimsel Düşüncenin Yapısı*. Epos Yayınları, Ankara.

Duwaider, A., Harass, H., & Farag, A. (2005–2006). Life science. Cairo, Egypt: Egyptian Ministry of Education, Textbook Section.

Edis, T. (1994). Islamic Creationism in Turkey. *Creation/Evolution* 34, 1–12.

Edis, T. (1999). Cloning Creationism in Turkey. *Reports of the National Center for Science Education* 19:6, 30–35.

Edis, T. (2002). *The Ghost in The Universe: God in Light of Modern Science*. Prometheus Books, Amherst.

Edis, T. (2003). Harun Yahya and Islamic Creationism. In Chesworth, A. et al. (Eds.), *Darwin Day Collection One*. Tangled Bank, Albuquerque.

Edis, T. (2004). Grand Themes, Narrow Constituency. In Young, M. & Edis, T. (Eds.), *Why Intelligent Design Fails: A Scientific Critique of the New Creationism*. Rutgers University Press, New Brunswick.

Edis, T. (2006). *Science and Nonbelief*. Greenwood Press, Westport.

Edis, T. (2007). *An Illusion of Harmony: Science and Religion in Islam*. Prometheus Books, Amherst.

Edis, T. & Bix, A. S. (2005). Biology and 'Created Nature': Gender and the Body in Popular Islamic Literature from Modern Turkey and the West. *Arab Studies Journal*, 12:2/13:1, 140–58.

Euben, R. L. (1999). *Enemy in the Mirror: Islamic Fundamentalism and the Limits of Modern Rationalism*. Princeton University Press, Princeton.

Fakhry, M. 2004. *A History of Islamic Philosophy, 3rd edition*. Columbia University Press, New York.

Gambetta, D. & Hertog, S. (2007). Engineers of Jihad. *Oxford Department of Sociology Working Papers*. http://www.nuff.ox.ac.uk/users/gambetta/engineers%20of%20jihad.pdf . Accessed 28 May 2012.

Gould, S. J. (1997). Nonoverlapping Magisteria. *Natural History* 106, 16–22, 60–62.

Guessoum, N. (2008). The Qur'an, Science, and the (Related) Contemporary Muslim Discourse. *Zygon* 43:2, 411–431.

Guessoum, N. (2011). *Islam's Quantum Question: Reconciling Muslim Tradition and Modern Science*. I.B. Tauris, London.

Gümüşoğlu, H. (2012). *Modernizm'in İnanç Hayatına Etkileri ve Jön Türklük*. Kayıhan Yayınları, İstanbul.

Hameed, S. (2008). Bracing for Islamic Creationism. *Science* 322, 1637–38.

Hanioğlu, M. Ş. (2008). Erken Cumhuriyet ideolojisi ve Vülgermateryalizm. *Zaman*. http://www.zaman.com.tr/haber.do?haberno=763214 . Accessed 28 May 2012.

Hoodbhoy, P. (1991). *Islam and Science: Religious Orthodoxy and the Battle for Rationality*. Zed Books, London.

Hoodbhoy, P. (2007). Science and the Islamic world—The quest for rapprochement. *Physics Today* 60:8, 49.

Huff, T. E. (2003). *The Rise of Early Modern Science: Islam, China, and the West*. 2nd ed. Cambridge University Press, New York.

İhsanoğlu, E. (2004). *Science, Technology, and Learning in the Ottoman Empire: Western Influence, Local Institutions, and the Transfer of Knowledge*. Ashgate/Variorum, Burlington, VT.

Kamali, M. H. (1997). *Freedom of Expression in Islam*. Revised edition. Islamic Texts Society, Cambridge.

Kazdağlı, G. (2001). *Atatürk ve Bilim*. TÜBİTAK Yayınları, Ankara.

Keddie, N. (1968). *An Islamic Response to Imperialism: Political and Religious Writings of Sayyid Jamal al-Din 'al-Afghani'* . University of California Press, Berkeley.

Kotan, B. (2006). Bakan Çelik: Yaratılış aynen kalacak. *Radikal*, http://www.radikal.com.tr/haber.php?haberno=180408 . Accessed 26 May 2012.

Küçükcan, T. & Köse, A. (2000). *Doğal Afetler ve Din: Marmara Depremi Üzerine Psiko-Sosyolojik Bir İnceleme.* Türkiye Diyanet Vakfı, İstanbul.
Leaman, O. (2002). *An Introduction to Classical Islamic Philosophy, 2nd edition.* Cambridge University Press, Cambridge.
Lewis, B. (1993). *Islam in History: Ideas, People, and Events in the Middle East.* Open Court, La Salle.
Luxenberg, C. (2000). *Die Syro-aramäische Lesart des Koran: ein Beitrag zur Entschlüsselung der Koransprache.* Das Arabische Buch, Berlin.
Mardin, Ş. (1989). *Religion and Social Change in Modern Turkey: The Case of Bediüzzaman Said Nursi.* State University of New York Press, Albany.
Melnyk, A. (2003). *A Physicalist Manifesto: Thoroughly Modern Materialism.* Cambridge University Press, New York.
Mernissi, F. (1992). *Islam and Democracy: Fear of the Modern World.* Addison-Wesley, Reading.
Meyer, S.C. (2009). *Signature in the Cell: DNA and the Evidence for Intelligent Design.* HarperOne, New York.
Miller, J.D., Scott, E.C. & Okamoto, S. (2006). Public Acceptance of Evolution. *Science* 313, 765–766.
Nasr, S. H. (1989). *Knowledge and the Sacred.* State University of New York Press, Albany.
Nurbaki, H. (1998). *Kur'an-ı Kerim'den Ayetler ve İlmi Gerçekler.* 7th edition, Türkiye Diyanet Vakfı, Ankara.
Özervarlı, M. S. (2003). Said Nursi's Project of Revitalizing Contemporary Islamic Thought. In Abu-Rabi' 2003.
Parla, T. & Davison, A. (2004). *Corporatist Ideology in Kemalist Turkey.* Syracuse University Press, New York.
Peker, D., Cömert, G. G. & Kence, A. (2010). Three Decades of Anti-evolution Campaign and its Results: Turkish Undergraduates' Acceptance and Understanding of the Biological Evolution Theory. *Science & Education* 19:6–8, 739-755.
Raudvere, C. & Stenberg, L. (Eds.) (2009). *Sufism Today: Heritage and Tradition in the Global Community.* I.B. Tauris, London.
Rehman, J. (2003). Searching for Scientific Facts in the Qur'an: Islamization of Knowledge or a New Form of Scientism? *Islam & Science* 1:2, 247–254.
Riexinger, M. (2008). Propagating Islamic creationism on the Internet. *Masaryk University Journal of Law and Technology* 2:2, 99–112.
Riexinger, M. (2009). Reactions of South Asian Muslims to the Theory of Evolution. *Welt des Islams* 49:2, 212–247.
Rosen, L. (2002). *The Culture of Islam: Changing Aspects of Contemporary Muslim Life.* The University of Chicago Press, Chicago.
Roy, O. (2004). *Globalized Islam: The Search for a New Ummah.* Columbia University Press, New York.
Sardar, Z. (Ed.) (1984). *The Touch of Midas: Science, Values and Environment in Islam and the West.* Manchester University Press, Manchester.
Sayın, Ü. & Kence, A. (1999). Islamic scientific creationism. *Reports of the National Center for Science Education* 19:6, 18–20, 25–29.
Somel, M., Somel, R. N. Ö. & Kence, A. (2007). Turks fighting back against anti-evolution forces. *Nature* 445:147 doi:10.1038/445147c.
Soroush, A. (2000). *Reason, Freedom, and Democracy in Islam: Essential Writings of Abdolkarim Soroush.* Sadri, M. & Sadri, A. (Eds.), Oxford University Press, New York.
Şahin, A. (2001). *İslam ve Sosyoloji Açısından İlim ve Din Bütünlüğü.* Bilge Yayıncılık, İstanbul.
Tatlı, A. (1992). *Evrim ve Yaratılış.* Damla Matbaacılık, Konya.
Yahya, H. (2007). *Atlas of Creation*, vol. 1. Global Publishing, İstanbul.
Yalçınoğlu, P. (2009). Impacts of Anti-Evolutionist Movements on Educational Policies and Practices in USA and Turkey. *İlköğretim Online*, 8(1), 254–267.
Yavuz, M. H. (2003). *Islamic Political Identity in Turkey.* Oxford University Press, Oxford.
Yılmaz, İ. & Uzunoğlu, S. (1995). *Alternatif Biyolojiye Doğru*, TÖV, İzmir.
Ziadat, A. A. (1986). *Western Science in the Arab World: The Impact of Darwinism.* McMillan, London.

Taner Edis is professor of physics at Truman State University, Kirksville, MO, USA. He was born and educated in Turkey, moving to the United States for graduate studies in 1987. While his background is in theoretical physics, he has also pursued interests in the history and philosophy of science, particularly concerning the conflicts between modern science and supernatural belief systems. His books include *The Ghost in the Universe: God in Light of Modern Science*, *Why Intelligent Design Fails: A Scientific Critique of the New Creationism* coedited with Matt Young, and *Science and Nonbelief.* Most recently, he has concentrated on examining the interactions of science, religion, and pseudoscience in an Islamic context. His book *An Illusion of Harmony: Science and Religion in Islam* and numerous articles, especially on Islamic creationism, contribute to understanding the role of science in modernizing Muslim cultures.

Saouma BouJaoude is professor of science education at the American University of Beirut (AUB). He graduated from the University of Cincinnati, Cincinnati, Ohio, USA in 1988 with a doctorate in Curriculum and Instruction with emphasis on science education. From 1988 to 1993 he was assistant professor of science education at Syracuse University, Syracuse, New York, USA. In 1993 he joined the American University of Beirut where he served as director of the Science and Math Education Center (SMEC) (1994–2003), as Chair of the Department of Education (2003–2009), and since 2009 as director of SMEC and the Center of Teaching and Learning. He has published in international journals such as the *Journal of Research in Science Teaching*, *Science Education*, *International Journal of Science Education*, *Journal of Science Teacher Education*, *The Science Teacher*, and *School Science Review*, among others. In addition, he has written chapters in edited books in English and Arabic and has been an active presenter at local, regional, and international education and science education conferences. He is member of several international science education research associations and has served as the International Coordinator and a member of the Executive Board of National Association for Research in Science Teaching. Presently he serves on the editorial boards a number of science education journals and is a reviewer for others. In 2009 he was appointed as a member of the Executive Committee of Supreme Education Council of the State Qatar.

Chapter 53
Indian Experiences with Science: Considerations for History, Philosophy, and Science Education

Sundar Sarukkai

The topic of science and science education in the Indian context encompasses many themes. This chapter will mainly focus on how the disciplines of the history and philosophy of science (HPS) can contribute to the debates in science education (SE) with specific reference to the Indian context. The extensive literature on the role of HPS in science education is comparatively recent. As Turner and Sullenger (1999, p. 10) point out, it is only in the nineties that there was a sudden proliferation of publications in this field. However, very few, if any, deal with the specific relationship between HPS and SE in the Indian context.

Why should this extension to the Indian context be of any interest to the community of science educators? Here are two answers: the first is that in the case of SE in India, it is primarily HPS which can help question the enduring claims that science is a Western enterprise and one that is unique to the West. By doing so, we bring in other cultural practices within the boundaries of science thereby extending the ownership of the essential characteristics of science to other non-Western cultures. Secondly, the Indian experience with science exhibits many stark differences from the European experience, and thus the lessons from a HPS that is responsive to these experiences will potentially offer new contributions to global SE. This chapter will illustrate how both these modes of intervention are possible. Invoking HPS is a powerful way to "localize" science and give a sense of ownership to scientists and science teachers. But in doing this, the HPS community itself is challenged for HPS traditionally has ignored the historical and philosophical contributions from the non-West.

The fundamental questions which frame the discussion here are the following: Does history and philosophy of science matter to science education? Do insights from Indian experiences with science matter to this debate?

There is a larger context in which these questions should be placed and that is the aims of science education. Does science education have the same aims as

S. Sarukkai (✉)
Manipal Centre for Philosophy and Humanities, Manipal University, Manipal, India
e-mail: sundar.sarukkai@manipal.edu

M.R. Matthews (ed.), *International Handbook of Research in History, Philosophy and Science Teaching*, DOI 10.1007/978-94-007-7654-8_53,

education? This cannot be – or at least is not seen to be – given that the larger aims of education include moral education and cultivation of human sensibilities. Science education seems to have become completely about the subject matter of science alone. It has been reduced to the teaching of the content of the various disciplines of science. Even history and philosophy of science are banished from science teaching. In this context, the real question that should be asked is this: Should science education be a handmaiden to the aims of science or should it have an autonomous agenda of its own, one that is more in tune with the aims of education in general?

This chapter makes the following argument. Even if restricted to the subject matter of science, it is possible to make the following observations. The decision of what subject matter to teach, the way to teach it, the background information on that content, and the many decisions of inclusion and exclusion of the content, and evaluation – all of these are based on certain presuppositions. There are hidden assumptions about the nature of science as well as the "invisible hand" of specific histories and philosophies of science in every science lesson.

To claim that HPS is not relevant in the content of science teaching is to ignore these hidden presuppositions. Thus, a call to include HPS in SE is only a call to make these hidden assumptions visible. Once this is done, the immediate relevance of the Indian (and other cultural) experiences to science teaching becomes obvious.

53.1 The Role of History and Philosophy of Science in Science Education

There have been many writers who have suggested the use of history and philosophy of science in science education.[1] This topic is central to science education for it confronts the basic questions of education: What does one teach in science and how does one teach science? It has also led to the question of whether nature of science (NOS) is relevant for the teaching of science, a theme which will be pursued later in this chapter. However, the impact of HPS on science education has been quite limited. In the Indian case, it is even more so. The reasons underlying this indifference illustrate the unique challenges of science and science education in India. Some of these reasons have to do with the very understanding of what constitutes science, the task of education in general and that of science education in particular, the complex histories of the idea of science, and its relationship to tradition and State. This and the following sections will address some of these issues with the fundamental objective being the following: Can the Indian experience with science from ancient times to the present add new insights and new themes to the debate on the role of HPS in science education?

Matthews (1989) argues for the importance of HPS in science teaching (ST). However, he also notes an enduring problem in establishing the relevance of HPS

[1] See Duschl (1985, 1994), Hodson (1988, 1991), and Matthews (1989, 2004).

for ST, namely, the "failure of science teaching to disturb ingrained beliefs" (ibid., p. 5). As he points out, it seems to be the case that students study science for their exams but use different belief systems for "everyday life." This is a "problem" that is very much a part of Indian school system. Moreover, the tendency to bifurcate the domain in which one practices science and the other in which practices contradictory to science are prevalent is not limited to students alone. In India, this is an integral part of scientific institutions also.

As far as the impact of philosophy of science on SE is concerned, it is arguably Kuhn who has had the greatest influence. Matthews (2004) discusses how Kuhn was ignored early on after the publication of his book, *Structure of Scientific Revolutions*, but in the post-seventies, the science educationists caught on to Kuhnian ideas. However, given the skewed understanding of the nature and practice of science, these ideas were most successfully used in the program of constructivism. The SE community seems to exhibit this constant tension of drawing on HPS but at the same time not being sufficiently trained to draw on it properly (ibid., p. 112). So it seems as if the community of science teachers either ignores HPS or jumps on to certain trends such as constructivism and relativism. One might think that perhaps history of science is better accepted as compared to philosophy of science. But, as Rodríguez and Niaz (2004) point out, science textbooks continue to ignore lessons from history of science. By analyzing the description of atomic structure in general physics textbooks published in the USA, they show how science textbooks continue to ignore the rich narratives of experiments available in the history of science on this topic.

Two of the more successful themes drawn from HPS relate to the NOS debate and a richer description of rationality, and their relationship to both science and education.[2] Although certain notions of rationality have come under attack, nevertheless there has been a sustained belief in the rationality of science as well as its importance in the idea of education. Siegel (1989) points out that critical thinking is the essence of education and through the idea of rationality he brings HPS and science education closer to each other. He considers critical thinking as the "educational correlate" of rationality (ibid., p. 21) and as an "educational ideal" (p. 27). Furthermore, he identifies "reason" as the common element of rationality: "a critical thinker is one who appreciates and accepts the importance and convicting force of reasons" (ibid., p. 21). The invocation of reason and rationality as important themes in the relationship between HPS and SE will be useful when we consider Indian philosophical traditions later on in the chapter.

The second theme – the debates on NOS – has been largely catalyzed by work in HPS, but there remains an open question as to whether NOS is relevant at all to science education. Turner and Sullenger (1999) point out how the divergent descriptions of NOS within science studies has had a negative impact in that it has led teachers to draw on this theme in a fragmented manner. There are many reasons why HPS does not make an appreciable dent among science teachers: for example, teachers resist them because they take away time from core teaching of science subjects (p. 13), science teachers themselves have little knowledge of or training in

[2] See Carey et al. (1989), Jenkins (1996), and Machamer (1988).

HPS, and because "teachers share many scientists' disdain for what they regard as soft add-ons" (p. 13). Science and technology studies (STS) are also not accepted by the community of science teachers because of the belief that discussions on science (through NOS) often end up as discussions on "law, economics, religion and power politics" (p. 16). This has led to teachers in many countries rejecting the role of HPS in ST. As these authors note, "Educational theory requires consideration of at least three factors: curricular content, instructional methods and approaches, and learning theory" (p. 17). They believe that HPS and STS have "emphasized" the first two but have ignored the third. In India too, there is continued indifference to HPS among science educationists. Although the National Curriculum Framework (2005) engages with the theme of nature of science and science education, there is little progress on the ground.

What is most surprising is that in the extensive global literature in this area, there is little that draws on other scientific traditions in order to pose the question about the nature of science. Did the Arabs, Indians, and Chinese do science the same way as the Europeans did? Did the Europeans have a unified idea of science at any time? Do the practices and notions of science in Asian and Arabic civilizations matter to the debate on NOS and science education? Does the HPS formulation of NOS continue to propagate the belief that "modern" science arose only in Europe since it draws only on a "Western" history and philosophy of science?

These questions become more urgent given some worthwhile historical challenges to the traditional view of science. The argument for the multicultural origins of science, a view that is still at the periphery of the mainstream history of science, necessitates changes that should be introduced in the nature of science debate. As far as philosophy of science is concerned, philosophical insights drawn from a consideration of Indian science and technology can illustrate how differing views of science and scientific methodology are articulated in other cultures.

Will science education get enriched by drawing on these diverse HPS traditions? Given the unique characteristics of the Indian society as well as its experiences with the notion of science, it should not surprise us to discover that a dialogue between HPS (one that is sensitive to these experiences) and SE leads to a new set of issues, some of which will be discussed below. If there is one particular issue that is at the heart of this dialogue, it is one on the definition of science. In a fundamental sense, unless we come to an agreement as to what constitutes science, it is not really possible to broaden the discussion on SE. It is precisely this question of demarcating science that is at the core of debates in non-Western cultures. Did these cultures "possess" science and technology before "modern science?" If so, how should these traditions be factored in SE? The sections that follow offer challenges to traditional views on science that locate it as part of a European tradition, following which their relevance for the NOS debates in SE is discussed. These arguments are placed within the larger context of defining science and the challenges of taking ownership of science, both of which have implications for science curriculum and pedagogy. Finally, the last two sections give an indication of some specific ways of "using" insights from Indian philosophical practices for SE.

53.2 The "Invisible Hand" of the Histories and Philosophies of Science in Science Texts

The most important rationale for taking HPS seriously in SE is that there is already a hidden domain of HPS in SE. There are many levels of the unarticulated presence of HPS in science texts. First of all, the content in a science text is already a product of a judgment of what constitutes science. For example, a textbook in physics presents certain subject matter as if it belongs unquestionably to the discipline of physics. On what basis is this judgment made? What content in physics, chemistry, and biology, for example, is chosen and on what basis? Why are certain subjects chosen as illustrations of science? A quick answer to these questions is that the scientific community decides what is science, what comes under physics, and so on. But this is to make science education secondary to the aims and ideology of the scientific community. If this is the case, then teachers function as mere conduits in a larger project of international science.

Consider a text in physics which contains Newton's laws of motion. Typically, there will be no historical or philosophical considerations which lead to the statement of these laws. Science texts, even today, present these laws as if they are ahistorical and appeared to Newton suddenly. There is no mention of a long history of motion, including interesting theories from Indian philosophies. Concepts that are used in the text including seminal ones such as mass, force, and energy, usually appear without any historical description. The fact that these concepts appear in other non-Western cultures is forgotten. In a science text, the choice of the content, the way the content is presented and ordered, and the choice of concepts and that of their definitions are all based on particular views of science, which themselves reflect particular histories and philosophies of science. For example, one such implicit assumption is that non-Western science is not really science.

Consider the following presuppositions that are at the core of science textbooks. Firstly, there is a standard history of science that is implicitly accepted. This history begins normally with Copernicus and goes on to Galileo and then Newton, with some mention of Aristotle and some "Greeks" thrown in. This history influences – greatly – the choice of the content of the text. And even within this story, there are many acts of omissions based on what story of science is chosen to be presented in the text. Similarly, the break between science and religion is an important part of this narrative. This is linked to the standard narrative that relates science to Western Enlightenment. The idea of the conflict with religion is often a selective rendering of a much more complex history of modern science's relationship with religion, as are the specific history of concepts that are presented in these texts.

There are larger themes which are also presupposed in these choices. For example, the philosophical assumptions of Platonism are an integral part of mathematics teaching. This worldview intrudes into the construction of the text in as many ways, including the structure of the problems, modes of evaluation, and the way mathematics is introduced. Mathematics is developed in a completely different manner in Indian and Chinese mathematics, but this material will not be found in textbooks. For example, the ways by which irrational numbers and negative numbers are conceived in Indian and Chinese mathematics illustrates alternate approaches to Platonism.[3]

Another invisible theme is the belief that European rationality characterizes science. There are many ways by which this idea is encoded into textbooks. One obvious manifestation of it lies in the almost complete absence of the mention of any non-western historical figure in the list of scientists and mathematicians mentioned in these texts. Along with this absence, there is an explicit emphasis on the figures of Western Enlightenment.

Yet another major theme is the belief that scientific content is universal and that science texts need not take into account the locale specificity of the students or the teacher. This also allows the propagation of these Eurocentric myths without second thought; this has led to generations of students in non-Western societies to believe that their cultures have had no contribution to the science of the modern world.

One may rationalize these choices by pointing to constraints of space in a textbook, access to material, and so on. However, it is the case that these constraints also influence what is taught *as* science. These constraints are also product in a textbook of specific choices which often reflect hidden presuppositions about science. All these choices reflect specific beliefs based on particular histories and philosophies of science and thus can be challenged on many counts as listed below.

First is the choice related to the material that is seen to constitute science. Should one choose to mention the rich domain of Indian metallurgy and technology in ancient times as well as later Chinese technologies? The extensive material on Indian mathematics, including Kerala mathematics, the complex and rich classification of botanical information in various indigenous traditions, and the vibrant practices of chemistry which rival similar practices in Europe of that time are all examples of science that are potential subject matter for science texts.[4] Anybody who is exposed to the multicultural histories of science will be able to appreciate the claim that the origins of science in Europe are indebted to the transmission of knowledge from different traditions including the Chinese, Arabic, Indian, and the Greek. To not integrate these histories into the content of science is to take a particular philosophical position about science and its subject matter.

Next, consider the history of fundamental concepts that are taught in texts. There is almost no recognition or acknowledgement that some of these concepts like mass, inertia, motion, energy, force, and causality have different formulations in non-Western cultures. Textbooks still talk of ideas like mass and force as if they are apples that dropped on Newton and other scientists from the tree of knowledge.

Moreover, the prevailing background of all discourse on science is based on some naive ideas of European rationality. Even today, scientists (and many science educationists) continue to believe that there is something special to this rationality and that alternative rationalities are not possible. Under tremendous pressure from extreme relativism, they might now acknowledge that perhaps other cultures

[3] See Dani (2010) and Mumford (2010).

[4] Details are given in the next sections that follow.

possessed some rationality but none of this has been factored into the basic texts in science. But, if European rationality is privileged, then the full story must also be noted. There is another use of rationality that should be taken into account in these claims: the notion of rationality was central to the European intellectual project as a way to distinguish European cultures from other cultures. Their claim to rationality was fundamental to the colonial project for through this category they could articulate their claim to being superior to other colonized cultures. In a powerful metaphor, they constructed themselves – because of this rationality – as adults and the colonized as children who needed the guidance of adults. So when the idea of rationality is repeatedly invoked, these two problematical characteristics are forgotten: one is the hegemonic work which rationality did and does, and the other is the existence of other equally important structures of rationality in other cultures. The examples of Indian logic and its mathematics are enough to indicate these possibilities; they will be discussed in greater detail in this chapter.

So the fundamental question is this: On what "rational" basis do science educationists make a choice when confronted with these new global histories and philosophies of science? If we look at how the Indian State confronted this issue, one can see how the ideology of universal, European science influenced and continues to influence science education in India.

53.3 Science and the State

History of education has as much influence on SE as does history of science. In India, students are taught science in a culture in which science is highly valorized. They tend to believe that they learn science because it is associated with virtues such as truth, knowledge, rationality, and power, virtues which are reinforced in the curricula. These virtues are those that are transmitted to the students by different agencies including the government, public space, peer groups, teachers, and parents. The strengths as well as the weaknesses inherent in SE are catalyzed by these different interest groups. Science educationists may find it useful to understand the context in which students are being taught science as well as the context of the educational methods to which they are subjected. In the Indian case, it is therefore necessary to consider the historical and ideological trajectory that sets into place educational practices in science. The most important influence on the institution of education is the colonial one, later followed by the independent nation.

The State (both pre and postindependence) was the primary patron of science and education. Sangwan (1990) analyzes some of the complexities in understanding the changing views on education in British India. He identifies three phases: in the first phase, the rulers did not change the system in India; in the second phase, they actually supported "oriental literature and science"; and in the third phase, they provided stronger support for science education. It is important to note that there was also a growing support among Indians for the Anglicization of education. As Sangwan notes, a group of influential Indians in Bengal "exhibited its revulsion" when the British supported Indian educational practices (particularly Sanskritic education) at the expense of "modern" education. But the British continued their support of "oriental education" in spite of protest from influential members of Indian society. Interestingly, the third phase, when European education was established, was also based on the belief that "modern" education would train Indians to support the activities and interests of the company rule, for example, to fulfill the demand for engineers and training of people to run the railways. Moreover, following the ideological interests of Macaulay, there was an overemphasis on teaching European literature and science, at the expense of indigenous science and knowledge systems. This was not conducive to science education since, as Sangwan notes, "what made things worse was their attempt to propagate English education rather than science education in India" (ibid., p. 89).[5]

There were also important changes introduced in the colonial era in the field of education. These changes, according to Kumar (2009), continue to influence educational practices even today. One such was the emphasis on textbooks and the introduction of a particular kind of textbook culture in Indian education. Kumar points out how textbook dominates curriculum and traces this attitude to the British colonial administration who saw education as a way to train Indians in "European attitudes" as well as "imparting to them the skills required for working in colonial administration." The insistence on textbook replaced traditional forms of education in which the teacher had freedom to choose what to teach. Kumar points out that in this colonial system, teachers also had to function as administrators. There were also punitive measures on teachers. Similar attitudes towards examinations and evaluative practices which were introduced by the British continue even today. For many educationists, these practices have been the bane of education in India today, particularly in science education.[6]

Postindependence, there was a calculated emphasis on science education. In so doing, the State was following colonial and modernist beliefs that science and rationality were external to Indian culture and that Indian society was characterized by irrationality and superstition.[7] Science education was to be the cure for this cultural "disease." The belief that science would be the vehicle for development meant that education in India would be skewed towards science education. Nothing defines the Indian State's engagement with science as well as the Constitution of India. At the dawn of the Indian independence, the Constitution was framed with the help of some of the most important intellectuals and freedom fighters at that time. Like most other Constitutions, the Indian Constitution is perhaps the most important public document in modern India. But unlike other Constitutions, this one has an intriguing clause listed in the constitutional duties of all citizens. One of

[5] See also Kumar (1980).

[6] See Kumar (1988). See also Dharampal (1983) for a brief discussion on education practices in eighteenth-century India.

[7] See Arnold (2000) for a discussion on the relation between modernity and science in India. Prakash (1996) discusses the support of the Indian elite to science as an exemplar of modernity. See also Habib (2004) for Islamic science in nineteenth-century India.

the constitutional duties of citizens is to cultivate scientific temper. Article 51A(h) of the Constitution of India states the following under Fundamental Duty: "To develop the scientific temper, humanism and the spirit of inquiry and reform."

To understand science education in India it is important to recognize the factors which made the authors of the Constitution to introduce this clause in the Constitution. First of all, this constitutional duty to "develop the scientific temper" reflects a fundamental aim of a new India, which was to reform its traditional societies and practices. Science becomes a way to reform a society which was seemingly suffused with superstition and irrational practices. Science, along with a particular view of the Western civilization, would liberate Indians from their past. Nehru's observation, as the first Prime Minister of free India, that industries would be the temples of the new India has often been quoted to point to a fundamental shift towards science and technology as the harbinger of a new society, one in which science would be used to better the economic state of the country. The government's support for the sciences immediately postindependence led to a large number of scientific institutes under the support of the government, including the atomic energy program under Homi Bhabha.

Since education was also primarily under the government, this view of science, as an agent of social change and development, as well as a reflection of a superior intellectual virtue special to the West, began to get propagated in schools and colleges. Even in the public sphere, science continues to be projected in these terms thereby adding to the social value of doing science. Since science was viewed as the agent to get rid of traditional beliefs, this led to the piquant situation of using science to erase any vestiges of ancient and medieval Indian intellectual traditions in the educational system, particularly in SE. This erasure is reflected in education at all levels in India leading to a chasm between postindependent India and precolonial India which remains to this day unbridged.

Alongside these ideological connotations of science, some influential Indian politicians too believed that national development meant promotion of science. The early models of development were all technology driven: thus the big dam projects, atomic energy program, the agricultural "revolution" through the use of technology, and so on. All these ideas of development clashed with traditional practices, particularly in the fields of agriculture and medicine. However, the State's commitment to a particular mode of development, one in tune with the rest of the developed world, meant a concomitant support to science and science education.

In spite of the early fervor of the leaders of the new nation, the public reception of science has been quite ambiguous. The claims that science education will eradicate social inequalities, remove caste system, and in general make people "rational" have remained just that. In fact, the greatest challenge to scientific temper in India comes from many of the scientists themselves, who while being good scientists are also committed religious believers. Such a dichotomy is well exemplified by the problematic example of offering prayers to machines.[8] Embodying the spirit of

[8] See Sarukkai (2008).

contradiction which seems to characterize an Indian way of thinking, scientists too easily wear contradictory roles of being a scientist as well as a religious person.[9] Many of them consult astrologers and visit religious institutions for family functions without that reducing their competence and standing as a scientist in their disciplines.

In the Indian scenario, the public space in which science gets articulated is a complex one. Interestingly, this public space also has religious elements within it. The relationship between religion and science in India is a complex one and historically challenges some standard history of science accounts of this relationship.[10] In the long history of science in India, it would be very difficult to actually find any cogently argued opposition to religion in doing and practicing science. Postindependent India, this opposition has been articulated by the Left as well as people science movements, but very often, this has been done with a certain amount of modernist understanding of science where science is an exemplar of rationality and religion that of superstition. Concomitant to this is the belief that ordinary citizens of the country were immersed in such superstition and it was the task of science (supported by the State) to "liberate" these citizens. Unfortunately, the scientists themselves were not up to the task demanded by these groups.

The popularity of science and technology education in India has largely been based on the easier availability of jobs from such education. However, the fact that science institutions, and science education in general, in India are excessively dependent on English means that there is a lack of democratic access to science education. This problem is compounded given the lack of good translations of science texts into other languages. These and other related factors contribute to the belief that science is not something "Indian"; they might also explain why the larger culture continues to ignore artificial expectations such as scientific temper, why the society does not buy into the story that science and religion have to be contradictory, and why there is an embarrassing lack of innovation in science and technology in India today.

Paradoxically, this is also a time when the government has increased its support to science education in an unprecedented manner, particularly in higher education. But almost all scientists and educationists point to certain well-established pedagogical practices as the bane of science education in the country. Rote learning continues to remain the dominant mode of learning in schools and many times in higher education as well. Critical thinking, including the capacity to critique, to build, and to innovate, is largely ignored in pedagogy, whether in science or other fields.

Ironically, the impact of science education on the larger public seems to be quite divorced from the cultivation of scientific temper. India, arguably more than in any other period postindependence, is in the throes of religious revivalism. There is a perceptible shift towards many traditional modes of learning, traditional practices of medicine, and also of some traditional social norms. Some feel that this is a reaction to the "imposition" of science and a corresponding rise in social problems related to

---

[9] A.K. Ramanujan's essay (1989) is often quoted in this context.

[10] See Kapila (2010)

modernity and technological development. Earlier views of science as liberating society from traditional practices, which gave way to the belief that science would be the harbinger of national development, has now changed to the belief that science is a pathway to jobs in the global market. Science education reflects this confusion at the highest level: What really is the goal of science as a national enterprise? An answer to this question is not under the purview of this chapter but an awareness of this issue is important for future State interventions in science education. With this larger background of science in India, the remaining sections will deal with the relationship between HPS and SE, and through the contours of this debate develop some India-specific themes that could be of interest to the global SE community.

53.4 Science Education and the Definition of Science

One of the problems in the understanding of science in India lies in an enduring cultural belief that modern science is a Western invention. Moreover, a sustained claim from European scholars that Asian cultures did not possess the requisites for being scientific (or equivalently, did not possess the capacity for being rational or theoretical) reinforced the outsider status of science in these societies. It is indeed surprising that scholars ranging from Locke to the German philosophers, Husserl and Gadamer, repeatedly question the availability of rationality and science to Asiatic cultures. This claim also becomes an integral part of the colonial discourse; thus for a few centuries, both within India and outside, there is a well-established tradition of scholarship that not only questions the existence of science in India but also questions the capacity of Indians to practice science.[11] Thus, it was no wonder that some science texts still continue to propagate these beliefs by emphasizing science as a Western "invention" and reinforcing this by primarily citing non-Indians as examples of scientists. One of the greatest challenges for science education in India is to convert the status of science from an outsider status to an insider one. While science teachers might see this issue as being irrelevant to the teaching of the content of science, in fact it is the exact opposite. If students do not take cultural ownership over science and if they do not think that they have a stake in the production of science, then it affects how they learn science and what they do with it. Learning has to have a sense of cultural confidence associated with it. Many scientists in India tend to believe that the lack of pioneering research (as exemplified by the lack of Nobel Prizes in the sciences or Fields Medal for scientists working in India postindependence), even after decades of massive support by the State to science in India, is an illustration of the lack of confidence and not taking ownership of science.

[11] See Gadamer (2001), Ganeri (2001) for the early European response to Indian logic; Adas (1989) and Alvares (1991) for the Western view on Indian science and technology; and Halbfass (1988) on the encounter of civilizations.

While there will be disagreements on this prognosis, it is nevertheless the case that science continues to inhabit the space between the inside and the outside in Indian society. This has serious ramifications on science teaching. For example, it is well known that scientists as role models influence young students. Science as a collective enterprise has recognized that creating legends out of scientists is a powerful way of attracting students to the field. Through the stories of these scientists, many popular notions of science, scientific method, and the idea of genius get communicated to the students and the larger public. The creation of the image of science in popular media is an important part of its success as a global enterprise. In this context, what happens to students who find few Indian models of great scientists? What is the lesson that is being communicated when all the scientists they encounter – the ones who have created new disciplines, new theories, and discovered new worlds of science – are predominantly non-Indian? This empirical fact coupled with the enduring hint that non-Western societies do not have an intrinsic capacity for being rational and scientific leads to a crisis of confidence in science learning. The often quoted comment from across all layers of society that a scientist is legitimized when she/he is first recognized in the West has become a cultural truism for other fields also.[12]

As much as research, this ambiguity about science has affected science teaching and learning, and this problem is primarily about the definition of science. Although over the past few decades there have been some seminal works on ancient and medieval Indian science (Adas 1989; Alvares 1991; Dharampal 2000), they are rarely incorporated into science textbooks. Part of the reason is the discomfort in viewing ancient or medieval discoveries and inventions as belonging to science. The problem of defining science is part of the larger theme of the nature of science. In India, this problem has two faces: one is the claim that some "disciplines" like astrology are also science, and the other is the scientificity of certain ancient and medieval practices and traditions. On both these counts, there is much denigration, particularly from the scientific community, and this is reflected in the general absence of any reference to Indian traditions of science within mainstream science.

One response has been to create a category called "indigenous science"; another is the category of "ancient science." Indigenous science has largely been reduced to innovations and modifications to "technologies" of the home and the community. It is largely artisanship and lacks some of the essential characteristics that define science – at least the definitions that constitute "mainstream" science. As innovations, they are largely experimental and technical, and do not exhibit any theoretical inclinations. This very category of "indigenous science" however illustrates an interesting way to appropriate the name of science into activities that have largely been excluded from science. This has also led to significant public movements on alternate technologies based on these indigenous and ancient technologies. However, the absorption of these movements into mainstream science education is still largely absent.

[12] A report in 1992 points out how around 80% of scientists in research institutes in India preferred to publish abroad and "hardly published" in India. See http://dspace.rri.res.in/bitstream/2289/3643/15/Chapter%208.pdf

But what really was the status of science and technology in ancient India? Should those practices be called science? Should they be called technology? And does all this matter to the problem of defining science, definitions that are relevant to teaching science? These questions are important for our understanding of science. Most science books begin with narratives of science in which Thales, Democritus, Socrates, Plato, and Aristotle are invoked as the fathers of science. Sometimes, they are even referred to as scientists. The absence of other cultural representatives in this story of science is inimical to the teaching of science, especially in those cultures which are excluded from a stake in the ownership of science and scientific rationality.

This is ironical, especially when we consider that the theoretical foundation of the sciences was integral to the philosophical traditions of the Hindus, Buddhists, and Jains in ancient India. All these traditions were fundamentally empirical in character and focused primarily on the nature of knowledge. Description of logic was an essential part of this analysis, and even logic was not immune to empirical inputs. The rich contribution of the Hindu, Buddhist, and Jaina philosophers illustrates their attempt to give a foundation for what we would call as scientific methodology today.[13] Alongside, there was a flourishing technological world which saw the world's first invention of steel, zinc, and alloys (such as the five-alloy process).[14] Indian mathematics not only gave the foundations of mathematics to the Arabs which then found its way into Europe, but the Kerala mathematicians also described the first modified heliocentric model (like the one proposed by Tycho Brahe) and the first conceptual ideas related to calculus much before they were known in Europe.[15] Another important marker of science was in the Indian medical systems such as Ayurveda, a practice that flourishes even today in India. These were all exemplars of science, but modern science education in India has largely ignored these ideas and practices of earlier science.

The British and European response to the discovery of these traditions was one of skepticism and derision. But confronted with the empirical evidence of these traditions, including a living tradition of the successful medical system called Ayurveda, they took recourse to the argument that such practices in India were primarily artisanal in nature. According to them, what made an approach scientific were the notions of theory and method, which presumably these earlier traditions in Asia did not possess.[16] Thus, these discoveries were seen to be "accidental," empirical, and not a product of a "method." Similar arguments were part of the European response to Indian philosophical systems, particularly the logical tradition that was part of almost all these systems.[17] It was important to deny the capacity for

[13] See Sarukkai (2005) for a discussion of how Indian logic shows conceptual similarities with methodologies that we would call as scientific methodology today.

[14] See Adas (1989) and Alvares (1991) for a detailed description of Indian technology.

[15] The text listing some of these results is available now with translation and with a symbolic rewriting of the text. See Sarma (2008).

[16] See Adas (1989) and Alvares (1991).

[17] See Matilal (1985), Mohanty (1992), and Sarukkai (2005) for a discussion on the nature of Indian logic and rationality.

logic to the Indians since rationality, a trait that defined the modern European mind, was a value that was used to hierarchize civilizations. European scholars took the view that Indian civilization did not develop logic and therefore did not have access to rationality.[18]

In understanding the relation between Indian culture and science education, we need to engage with the practice of science in ancient and medieval India.[19] This is not only to set right the asymmetry in invoking the ancient Greek or modern Europe in the context of science but also to begin to look at new features and characteristics of what could be called science. The only way these ideas and arguments can enter science education is through HPS. Adding the story of ancient and medieval scientific practices as part of a longer and multicultural history of science will have an immediate impact on the way students learn science. This approach should be of interest to science teachers in other countries as well. While science educators have discussed the importance and effectiveness of teaching scientific concepts through a historical trajectory, they have in general ignored the possibility of other histories of these concepts drawn from non-Western traditions. Almost all concepts such as mass, matter, energy, cause, effect, substance, chemical, and material are described in great detail in different cultures and are articulated in different ways.[20] When history of science draws only upon one historical narrative, it reinforces the myth that all seminal ideas in science (and even philosophy) occurred in the "West." But this is a limited reading of history of science itself. Thus, an encounter with the Indian context forces a change not only in science education (in terms of curriculum as well as pedagogy) but also in the mainstream history of science community. Through this intervention and broadening of their view of history of science, historians of science can actually make a stronger claim to the relevance of HPS to SE across the world.

53.5 Does Multicultural Origin of Science Matter?

In contemporary history of science, there is a strong claim to the multicultural origins of science (Bala 2006). There are some variations in this theme, but all of them agree that to claim modern science began in Europe is to ignore the many cultural influences that made this origin possible. The gist of this argument is that ideas from the Greek, Arabic, Chinese, and Indian cultures reached Europe in various ways – both in the form of texts and instruments. Europe afforded a geographical

[18] See Ganeri (2001) for the early European response to Indian logic.

[19] See Arnold (2000) for more on medieval India. He also points to rich development in science in the Mughul period, particularly in medicine.

[20] Indian metaphysics understands some of these concepts in quite different ways compared to the modern West. For example, the mind-body duality following Descartes is not found in these schools, since mind is seen to be a species of "matter." There are many interesting theories of causation in the various schools. The point is that some of this diversity may actually be useful for alternate understanding of scientific methodology and processes.

location where all these confluences could meet and develop, and it is in this milieu that modern science originated. There are deeper undercurrents to this claim – the possibility that knowledge had actually been transmitted to Europe but which had not been acknowledged. Examples of how Copernicus drew on the Maragha school of astronomy, how Kepler depended on Chinese observations, or how the possibility that the first ideas of calculus went from Kerala in India to Europe raise troubling issues about the beginnings of modern science and their true "origin." General principles that define science such as the amalgamation of the theoretical and the experimental, following Bacon and Galileo, are principles that inform earlier intellectual traditions in India and China. As Bala points out, "solar, lunar and planetary models of al-Shatir are mathematically identical to those proposed by Copernicus some 150 years later" (ibid., p. 83). Similarly, optical revolution in Europe was influenced to a significant extent by the work of al-Haytham (Alhazen). Bala further argues that mathematization of nature – a profoundly important moment in the origin of modern science – is indebted to the "meeting in Europe of Arabic philosophy and science with Chinese mechanical discoveries" (ibid., p. 122).

In the context of SE in India, the discussion on the multicultural origins of science, derived from history of science, will have a significant impact on the reception of science by students. There are two dimensions to this reception: one is the shared cultural and national histories of science available through multicultural histories of science, and the other is the availability of shared vocabularies and practices that might make the student more "culturally comfortable" or "culturally contiguous" to science that is taught today. In almost all the textbooks of science, the heliocentric model is associated with Copernicus. Where a little history of this model is available, it often begins with Tycho Brahe and ends with Galileo. The heliocentric model is an extremely important model of science – both for the history and origin of science and also for communicating the power of science because the claim that the Earth revolves around the sun is now such a cultural truism that it is part of commonsense now.

However, this is an incomplete version of history. There is sufficient literature to illustrate how the model of Tycho Brahe was described by the Kerala mathematicians quite some time before Brahe. The modified heliocentric model developed by these astronomers/mathematicians was described in texts that were known to the Jesuits in Kerala in the early sixteenth century.[21] Whether these ideas got transmitted across continents is not really the issue. What is however at stake is the placement of these seminal ideas within the capacity of other cultures. Does this mean that an Indian student will better understand and appreciate heliocentric motion once she is taught about the contribution of the Kerala mathematicians? There are no simple answers to this question, but this question needs to be considered for it is possible that awareness of a history of "their own" scientific theories might allow students

[21] See Bala (2006), Raju (2001), Ramasubramanian and Srinivas (2010), and Sarma (2008). See also Plofker (2009) for a detailed introduction to Indian mathematics.

to have a different stakeholdership towards that subject thereby influencing their learnability of science.

Teaching multicultural theories of the origin of science relates science and scientific ideas to subjects such as history and geography. It grounds science not only in the empirical world but also in a discursive world where it shares some common stories across disciplines. In the Indian context, one might say that the students face two kinds of alienation with respect to science: one is an experiential strangeness and the other is a cultural strangeness. An ironic illustration of this alienation lies in the history of technology. As mentioned earlier, there is a well-established scholarship now that describes the enormous technological advances of ancient Indian societies.[22] The first examples of metallurgy, including the invention of steel and zinc, were to be found in India. These processes actually found their way to England in the colonial era and catalyzed modern production of these metals. However, this rich and long history of metallurgy, chemistry, and other pioneering innovations in India and China are rarely even mentioned in science textbooks. Teaching science by drawing on multicultural origins of science is a good way to reduce the strangeness on both these counts. However, like in the case of HPS in SE, teachers have to first engage with these new histories.

Finally, we should note that introducing multicultural origins of science in SE is but one step in allowing multiple histories of science to enter the classroom and science texts. In the Indian case, there are such multiple histories available such as the subaltern histories and artisan histories. In all these alternate histories, there are some profound insights into the nature of the world and the universe, as well as important elements of the nature of enquiry, observation, and experimentation.

These arguments are but an extension of arguments in education that suggest that learning becomes easier when ideas are expressed closer to the contexts in which the children are immersed. For example, the attempt to teach abstract ideas through examples that are common to the cognitive experiences of the students is but one variation of the above arguments. What is being suggested here is that scientific ideas become more easily accepted and understood if a common history is exhibited. As a motivation, we only need to note the rich historical debates on the origin of science in Europe, a topic that has become so contentious that it inspires Dear (2005) to remark on the identity crisis confronting historians of science because of the difficulty in establishing strict boundaries that demarcate science from nonscience. The recognition of a stable subject matter for history of science has become difficult given that the "very category of "science" has become historicized – and hence very slippery" (ibid., p. 391). This observation, as well as the ones on the multicultural origins of science, should not be read as supporting the claim that "everything is a science." What it merely does is expand the space within HPS to interrogate the diverse forms of science[23] and take this into account in the larger debates in SE.

[22] See Adas (1989), Alvares (1991), and Dharampal (2000).

[23] We should remember that the diverse forms of science are already exhibited in the contemporary typology of science. Today, science includes quite different disciplines ranging from physics, chemistry, and biology to economics, library science, management science, and so on.

53.6 Nature of Science and Science Education: Lessons from Premodern Indian Science

The above discussion is related to a more commonly discussed theme in relation to HPS, namely, the NOS debate in SE. Turner and Sullenger (1999) offer a good overview of the sociology of the NOS debate within science education. But like other reviews in this field, there is no mention of the possibility of articulating new themes in NOS that are catalyzed by non-Western philosophical traditions.

One of the important ways to argue for the relevance of NOS is to show how awareness of the process of science (e.g., as a combination of the experimental, observational, and the conceptual) adds to the students' understanding of the concepts of science. Matthews (2001) describes how the study of pendulum in a class illustrates the importance of NOS to science education. Much of the impact of NOS has been discussed within the cognitive domain. However, a major influence of NOS on science learning also has to do with various psychological aspects of learning. A robust and complex theory of the nature of science demystifies the inherent ideology of a particular image of science. An outline of how the debate on NOS can be influenced through this engagement with Indian experiences follows; how these issues can directly affect SE is, at this moment, only speculative.[24]

As discussed in an earlier section, science and technology were well-established social practices in ancient and medieval India. All philosophical traditions were fundamentally concerned with the various means of acquiring knowledge. Hence, the study of epistemology was the fundamental concern for these philosophers (Matilal 1986). Thus, it should not be surprising that much of their work impinges on philosophy of science. In exploring the nature of knowledge, extensive work was done on the nature of perception, inference, language (testimony), and other related modes of learning. The Indian theories on inference make two important contributions of relevance to this chapter. One is the five-step process of inference in the early Nyāya tradition, which basically explains one's inference to another through a process of reasoning (Matilal 1999). Let us say that one infers fire on seeing smoke. The five-step process describes how we could effectively explain why we infer fire on seeing smoke. It begins by first noting the empirical event of seeing smoke, then invokes a universal reason (where there is smoke, there is fire), then uses a common example known to the speaker and hearer (such as smoke-fire complex in a kitchen), and finally concludes that the inference of fire is indeed correct. This five-step process at one stroke combines the rhetoric of communication, the process of inference, and also the process of cognition involved in the inference. It is also a system which has

[24] One of the sustained efforts in this direction in India is the series of conferences called epiSTEME organized by the Homi Bhabha Centre for Science Education, Mumbai. The publications of each of their conferences contribute significantly to this debate in India, although the difficulty of directly drawing on Indian intellectual traditions for science education remains.

striking parallels with the deductive-nomological model of scientific explanation (Sarukkai 2005).[25]

The second contribution to the study of inference comes from the Buddhists who offer a semiotic model of inference. They analyze the conditions that will enable one to know when a sign stands for a signified, such as smoke as a sign that stands for fire. This relation between sign and signified is also at the heart of instrumental observation in science. The way by which these Indian logicians analyze these inferences offers new ways of understanding the relation between semiotics and logic as well as between semiotics and science (a topic that has rich resonances in the interpretation of experimental results as well as in the use of mathematical writing in the sciences).[26]

There is an important lesson about NOS in these Indian approaches to inference. First of all, there is a completely different approach to the idea of "theory" in Indian thought. European scholars who encountered Indian logic found the combination of the formal and the empirical very troubling (Ganeri 2001). Similar arguments also legitimized the British appropriation of Indian technology by claiming that the Indians knew how to make steel, for example, but did not understand the theory behind it (Adas 1989; Alvares 1991). Do these alternate approaches have any implication for science and for science education today?

Any account of modern science has to engage with the idea of the theoretical. As is well known, what we call science today was referred to as natural philosophy at least till the seventeenth century. Natural philosophy was seen to be a speculative, theoretical act. The distinction of theorica/practica drawn from Greek thought was extremely influential in the way disciplines such as medicine, astronomy, and other "arts" were understood (Dear 2005, p. 393). Till the early seventeenth century, natural philosophy was immune to this distinction as it was seen to be mainly speculative philosophy with no practical applications. It was Bacon who constructed natural philosophy as a discipline which also had practical utility. Moreover, the influential distinction between mathematics and natural philosophy continued till the eighteenth century. These lead to the distinction between the categories of "pure" and "applied" in the nineteenth century which continues till today. This short historical interlude is merely to point out how such a tradition of understanding science does not occur in Indian thought. At the core of different Indian intellectual traditions there is a suspicion about such distinctions. There is no clear distinction between logic and epistemology, between metaphysics and epistemology, between ethics and epistemology, between formal and the empirical, and between mathematics and the sciences.[27] This approach to theory and practice in Indian thought is of great significance to science teaching of these subjects. For example, drawing on the unique nature of Indian mathematics – both in its empirical grounding and in its

---

[25] This model of scientific explanation suggested by Hempel and Oppenheimer argues that the explanandum is a deductive conclusion from a set of premises which consist of lawlike as well as empirical statements. This structure does a similar task as the Nyāya process.

[26] For a detailed analysis of these semiotic elements, see Sarukkai (2005). See also Sarukkai (2011).

[27] See Bhattacharya (1958) and Mohanty (2002).

textual and discursive practices – might actually help address some problems of science and mathematics education.

Ayurveda, the enduring medical tradition in India, is another classic example of the inherent mix of the theoretical and the empirical. In this case, there is also a strong component of the experimental and not merely the observational. Ayurveda is a classic example of a scientific tradition which exemplifies many of the virtues of modern scientific method including observation, theory, experimentation, and intervention. Yet it is rarely taught as such to students and in school science textbooks; it is rarely mentioned even though the so-called modern (allopathic) medicine is given as an exemplar of science. Ayurveda, as a scientific practice, involves some real observational biology. Its classification of plants is exhaustive and scientific in the modern sense of botany.

What then are the implications for science teaching if such contiguous, local traditions are taught as illustrations of the nature of science? Equally importantly, how will science education in the West learn to draw on these Indian scientific traditions in their own teaching of science? What is being suggested here is that HPS' contribution to the nature of science debate has been too Eurocentric. Drawing on another history of science from another culture – one in which the idea of "science" does not exhibit the standard fault lines of theory and practice, as well as the metaphysical baggage of this distinction – might actually contribute to the development of new ideas and practices in SE.

In the Indian context, whether in philosophy, art, or mathematics, the primary emphasis is on making the theoretical always answerable to the empirical. Thus, there is no idea of pure rationalism, pure mathematics, etc., which characterize some dominant traditions of Western thought. Now this has been interpreted by some European scholars, as pointed out in the earlier sections, to suggest that Indian culture has no understanding of theoretical rationality that is often associated with science. There has also been a long colonial project of interpreting Indian culture as being primarily religious and this along with the claim that modern science begins from a conflict with the Church has led to the claim that Indian culture had no links to modern scientific thought. This myth continues in spite of a large amount of literature on these aspects – both of Indian systems and on the nature of science. In particular, history, philosophy, and sociology of science have illustrated so well the complex relationships which science has with logic, religion, method, and so on (Sarukkai 2012). But the complete absence of these disciplines in science teaching means that certain traditional ideologies about science, such as those discussed in the earlier sections, continue to be propagated in science textbooks even today in India.

The implications of drawing on these "indigenous" modes of understanding the nature of science and mathematics are many. First of all, the metaphysical foundations of these disciplines are quite different in the Indian context when compared to the Greek and the modern "West." For example, Platonism seems to be unthematized in Indian thought. Mathematics and the sciences were always empirically grounded. Since their presuppositions are different, they create the possibility of a different method of teaching science. Secondly, when science is taught along this trajectory, students begin to understand science as a social process. Now, science is dominantly understood as a packaged knowledge system which seems to have been given to us miraculously – if not by gods, then by "great scientists." In contrast, teaching science as a social and historical process, with its own special cultural moorings, may actually interest more students in science. Drawing on Indian logic gives us new methods of communicating the practice of reason and of thinking – arguably the two most important characteristics of learning. It also illustrates the importance of rhetorical communication in matters of reason such as using the nature of evidence and reason on which to base conclusions, details of which are given later in this chapter.

There is another important aspect of NOS from these Indian traditions, which can be briefly alluded to here. This has to do with the relationship between language and science. Almost all Indian philosophical traditions were deeply concerned with the nature of language and its relation to the world and human cognition. There is a constant attempt to create languages which capture truth and knowledge – Sanskrit itself is a very good example of a semi-technical language (Staal 1995). Later philosophers of the Nyāya tradition created a modified form of Sanskrit in order to remove the ambiguities present in ordinary Sanskrit (Bhattacharyya 1987). This engagement with language is wonderfully illustrated in the ways by which mathematics gets written in prose and poetic forms. Although a very important topic, the role of language in science[28] has not been sufficiently dealt with in the HPS literature. The richness of language use in science and science learning attests to the possibilities of drawing on these rich Indian philosophies of language.

Consider the teaching of mathematics. There is a long tradition of mathematics in India, one that includes the seminal text by Āryabhaṭa called the Āryabhaṭīya. What happens when a student is also given these historical descriptions of these alternate forms of doing and writing mathematics? Here are some speculative (but hopefully reasonable) outcomes: one, the student recognizes that there are different ways of doing and describing trigonometry – this automatically decreases the anxiety of doing something the right way or following the "right" method; two, she would recognize that such ideas were not really alien to the larger cultural world which she belonged to; and three, she might even recognize (with some help from the teacher) that mathematics has an interesting relationship with language, perhaps a language which she might be "culturally familiar" with. Such approaches might help in getting rid of the perennial and enduring fear of symbolism that seems to strike school children across the world.

53.7 Constructivism and Indian Philosophy

There is another characteristic of Indian philosophical traditions that may have a direct impact on SE. Indian philosophies have a unique way of describing the human interaction with the world. They, without significant exception, describe these

[28] See Sarukkai (2002).

interactions in terms of cognitive episodes.[29] Thus, perception is defined in terms of cognitive states that occur when somebody sees or perceives something. Inference is also defined in terms of cognitive states. So a description of a process is through a series of cognitive states. This mode of cognitive description for processes including that of logic is a good model for teaching subjects such as logic, mathematics, and science. The example discussed below will illustrate how this method of teaching science has metaphysical overlaps with constructivism while at the same time negating the excessive subjectivities which are potentially inherent in constructivism.

Consider a classic example of inference from Indian logic, the inference of fire on a hill from seeing smoke on the hill. Almost all schools of Indian philosophy considered inference as one of the valid means of knowing and so expended a great deal of effort in trying to describe the processes of valid inference. An example of valid inference would be the inference of fire from seeing smoke, but how can we be certain of this inference? Indian logic, pioneers of which include the Nyāya school and the Buddhists, described the inferential mechanism through a series of cognitive processes. For example, one such description followed by Nyāya is as follows: when a person sees smoke, there is a cognitive state corresponding to that perception. This state is followed by another state in which she remembers a universal rule – the principle or the reason for the inference – that where there is smoke there is fire. This cognitive state is followed by the next one which, through this rule, recognizes the smoke as standing for fire. This leads to a cognitive state which is the inferentially knowing state that there is fire in the hill.

Many Western scholars were puzzled at the cognitive descriptions of the logical process of inference and thus concluded that Indian logicians were committing the fallacy of psychologism.[30] This is a common mistake in responses to cognitive descriptions and is catalyzed by the belief that descriptions of the world should refer entirely to terms of the world and not to terms of personal experiences of the world. But what is intriguing is that the logical school, Nyāya, is a realist one. They do not in any way accept that the perception of smoke and the inference of fire are personal, subjective ones. They are committed to a realist metaphysics which acknowledges the "reality" of the smoke and fire. Yet, they are also committed to the reality of cognitive states, and the challenge in this philosophical approach is to retain the descriptive power of cognitive states and yet retain the possibility of a common reality accessible to the many cognitive subjects. It is this effort that really distinguishes and explains the unique nature of Indian philosophical discourses. As Mohanty points out, these philosophers do not end up with psychologism because "a cognition has a logical structure which allows for being exemplified in another numerically distinct episode belonging to another ego and/or another temporal location" (Mohanty 2010, p. 437).

Why would this matter to constructivism? While constructivism seems to be a popular movement in science education, there have also been strong critiques. For example, Matthews (2002) argues that although constructivism has been very

[29] See Matilal (1986) and Mohanty (1992, pp. 100–132 in particular).

[30] See Mohanty (1992) for a discussion on this theme.

popular in SE, there are nevertheless many points of disagreement and worry. He also notes that there are a variety of ways in which constructivism is invoked in SE, namely, constructivism as a theory of learning, teaching, education, cognition, personal knowledge, scientific knowledge, educational ethics and politics, and worldview (ibid., p. 124). In Matthews (2004, p. 107), he lists various claims of constructivism such as the lack of objective knowledge of reality and experience as the primary basis of scientific knowledge.

It seems to be the case that many of the conclusions of constructivism are grounded in the experiential mode and the cognitive primacy of learning, which situates knowledge/truth within the learner. Dominant analytical traditions in Western philosophy do not have the wherewithal to cope with this form of grounding. (There are other traditions which offer different approaches to "extracting" the objective from the subjective – Husserlian phenomenology is one such.) In Indian philosophy, this grounding in the cognitive states of the experience is a given for all processes related to knowledge of something real. But this does not reduce these descriptions to a naïve form of psychologism and individual subjectivities. That is, the mere fact of a cognitive description of the learning process, *in itself*, does not make the whole process subjective. In other words, there is really nothing in the metaphysics of constructivism (when viewed from the Nyāya perspective or the phenomenological perspective) that necessitates the shift to claims that all truth is within the learner and not "outside," that scientific knowledge and truth are all relative to each learner, and so on. The conclusion that constructivism is necessarily subjective is based on a flawed understanding of the nature of subjectivity as well as of the nature of cognitive states. The Indian logicians' analysis of inference shows how it is possible to hold onto a robust realism while at the same time accepting the primacy of individual cognitive processes of perceiving, inferring, and learning. Through this approach, there is a good possibility to marry constructivism with realism; this can be one example of "using" Indian philosophy for a contemporary debate in SE.

53.8 Rhetoric and Methods of Teaching

It would not be an exaggeration to say that rote education is seen as the bane of Indian education. Whether in newspaper reports or the last National Curriculum Framework, there are repeated references to the effect of rote learning and the prevalence of this mode in Indian schools and colleges. This mode of learning is particularly problematical for science education since rote education is often contrasted with critical and creative thinking, and problem-solving, which are the hallmarks of a "good" science education.

It is a mystery as to why the Indian educational system is saddled with such a strong component of rote learning. Like many other ancient cultures, there is a strong sense of orality in Indian culture.[31] Texts are memorized and recited. Even

[31] See Fuller (2001) for an interesting case study on orality.

when Sanskrit is taught today by traditionalists, the most important component of teaching that language lies in pronunciation. For a long time, texts were transmitted entirely through oral means, and this meant that the language as well as strategies of memorizing made it easier for oral transmission. It is interesting to note how even philosophy and mathematics texts were written in poetic form, thus making them amenable to an oral discourse. For some, this cultural practice of orality seems to have transformed into rote learning in contemporary education.

Although one might be tempted to ascribe the prevalence of rote learning to ancient cultural practices (such as language learning and oral transmission of texts – none of which necessarily implies rote learning), it is the educational policies and practices that are obviously a bigger culprit.[32] The most enduring reason for the continuation of this mode of learning is the examination system. Every year, one only has to see newspaper reports during exams and during admissions to read about the menace of "rote learning."[33] Even the government bodies and their representatives keep bemoaning this characteristic of Indian education. One of the most influential documents on education and education policies in India, the National Curriculum Framework 2005, notes the problem of rote learning and the means to curtail it. The National Curriculum Framework for Teacher Education 2009/2010, makes pointed references to the problem of rote education.[34] Many alternate private schools promote their system by claiming that they teach critical and creative thinking as against rote learning, which they claim is endemic to public education.

This entry into the discussion on rote learning is to offer a perspective from Indian philosophical and intellectual traditions on critical thinking. Critical, reflective thinking is often seen as the antidote to rote learning. The inculcation of these virtues is fundamentally the task of philosophy, and it is here that HPS becomes immediately relevant to SE. Philosophers over the ages have viewed philosophy as being concerned fundamentally with the nature and processes of thinking. It is impossible to engage with philosophy and not be involved in modes of reflection and thought. In the Indian context, there is an interesting practice related to critical thinking. All Indian philosophical traditions use "debate" as their fundamental rhetorical strategy. In these traditions, debate is classified into different types, and a student in these schools has to master these debating strategies. The influence of debate is so integral to Indian philosophies that even texts follow the form of a debate. Thus, a standard philosophical text will first describe the opponent's view and then counter it step by step. Almost all seminal philosophical texts follow this method.

Debate is an important rhetorical strategy. It is a great training for critical thinking and a powerful pedagogical tool. A good illustration of this can be found in the practices of Buddhist monastic training even today, where the students are trained not only in the theories of debate but also in the "performance" of debate. One can

---

[32] See Chitnis (1993) for an analysis of the changes needed in the Indian education system.

[33] See, for example, TOI http://articles.timesofindia.indiatimes.com/2011-06-19/education/29676322_1_high-scorers-evaluation-board-examination.

[34] See http://www.ncte-india.org/publicnotice/NCFTE_2010.pdf.

see these forms of "performative" debates between different schools in Buddhism in many monastic schools.

Debate is the founding principle of all Indian philosophical traditions (Matilal 1999). Logic in India arises out of debate. Every philosophical system classifies the various forms of debates; the intellectual battles between different schools were to be won on the court of debate. Interesting historical anecdotes about debates are prevalent in narratives about the different philosophical traditions. Debates were a common occurrence in the courts of kings, and there were well-established mechanisms for judging the winner of these debates. Debate was the forerunner not only of Indian logic but also of rhetoric and communication praxis.

As Matilal points out, the Buddhist canons of debate illustrated fundamental logical principles such as modus ponens and modus tollens. Debates themselves were classified into "good" and "bad" debates or "amicable" and "hostile." Nyāya classifies debates into three types: one between a student and teacher, whereas the other two are primarily "hostile" in that their basic aim is about being victorious in the debate. The good or "honest" debate between a student and teacher is characterized by the following properties:

1. Establishment (of the thesis) and refutation (of the counter-thesis) should be based upon adequate evidence or means for knowledge *(pramāna)* as well as upon (proper) "hypothetical" or "indirect" reasoning *(tarka).*
2. The conclusion should not entail contradiction with any tenet or accepted doctrine *(siddhānta).*
3. Each side should use the well-known five steps of the demonstration of an argument explicitly.
4. They should clearly recognize a thesis to be defended and a counter thesis to be refuted. (Matilal 1999, p. 45).

All educational instruction followed these steps. These not only establish a communicative mode to discuss and analyze propositions, they are also essential to what we call as critical thinking today. Nyāya also has a classification of the other two kinds of "hostile" debates and lists in great detail the elements constituting these kinds of debates. Since victory (by any means) is the goal in these "hostile" debates, a debater can use the following strategies: quibbling, illegitimate rejoinders, and clinchers. Moreover, there is an extensive classification of the types of quibbling (3), illegitimate rejoinders (24), and clinchers (22) (ibid., p. 47). It is this complex system of argumentation that is at the heart of Indian rationality, and interestingly, it has pedagogy at its center. An illustration of these methods for students will enrich not only science education but also critical thinking and communicative processes in all aspects of education.

Although such practices have not become part of education in India, they can nevertheless be effectively used to teach critical thinking. The performative tradition of debates in Buddhist monasteries illustrates one way of incorporating the skill of critical thinking as part of educational practice. In schools today, such methods can easily be adapted. Here is one possible way to do this. The students will debate a particular hypothesis but will have to follow the rules of debate as enumerated in these philosophical traditions. For example, consider a simple thesis: all matter is made up of atoms. How will the students debate this question? What kind of rules of debate will they invoke? The rules of debate drawn from these Indian traditions of debate can be given to the students, and they can be asked to follow these rules in the course of the debate. For example, the Nyāya describe many kinds of checks, since for them, "debate was like a game of chess, in that the opponent and the proponent make their moves and at the end there is a clincher, when one side will be checkmated" (Matilal 1999, p. 81). They list "twenty-two types of defeat-situations," which are arguments which will show the opponent's arguments to be wrong. One such is abandoning the thesis in the course of argument; thus, if a student suddenly invokes the existence of space and time while debating the existence of atoms, she could be countered by saying that she has abandoned the thesis of the original debate. Some of the other clinchers against the opponent include "irrelevant speech," "incomprehensible speech," "adding unnecessary steps," and "repetition." (Examples of using jargon or dropping names of scientists to justify the thesis can fall under one of these categories.) Students trained in such manual of debates will be able to analyze what they hear and evaluate the faults of arguments based on specific categories. One can modify these rules for the present-day context. The bottom line is that such exercises in thinking and debating can enhance the skills of learning and thinking in a profound manner and there is much that the rational traditions of India and China, for example, can contribute to this project. Such an exercise will be of relevance to students across cultures.

53.9 Conclusion

This chapter explored new ways of engaging with HPS in the context of SE with a specific focus on Indian experiences of science and science education. One of the challenges to science education in India lies in the contestation about the very notion of science. Given that the Indian civilization (like other non-Western ones) had a long engagement with themes and practices of what seems like science, it is necessary to understand why science education in India is almost completely silent about engaging with these traditions. Thus, the sections on multicultural origins of science and the nature of science debate discuss the ramifications of understanding these practices as science and also the importance of including them in the curricula. One of the consequences of including them in the curricula is that students begin to develop a sense of ownership and confidence with respect to scientific thinking and practice. This step allows the possibility of broadening our understanding of science without also, at the same time, claiming that everything is science.

The two sections on constructivism and rhetoric are attempts to illustrate how one can draw on alternate philosophies that can contribute to some of the important contemporary debates in science education. By doing so, attention is also drawn to the exclusiveness of the global HPS community, which has steadfastly ignored the historical and philosophical contributions of the non-West. Hopefully, this chapter can partly loosen the intellectual strings of this community and through it that of the community of science educationists.

This chapter does not address issues related to learning scientific content and the philosophical issues associated with it. This is primarily because of the India centricity of this chapter. Much of the debate may seem unnecessarily focused on certain aspects of earlier science or on cultural implications of ignoring non-Western traditions. However, as argued in this chapter, it may not be desirable to so completely demarcate the learning of content from the learning of the concept and methodologies of science.

In drawing on the earlier intellectual traditions, some contemporary contributions have been kept out of the purview of this chapter. In particular, there is no engagement with two influential thinkers/educationists of contemporary India, Gandhi and Tagore. Both of them were deeply interested in education and both of them had deep insights into the nature of education. They are of particular interest because of their attempts to fundamentally integrate educational practices and ethics. Gandhi (1951, 1953) proposed his ideas of education as part of a new system called the *Nai Talim*. This schooling is radically different from mainstream, government-supported education system. It is important to realize that these approaches emphasized fundamental ethical principles as being integral to the education process, content, and pedagogy. These principles include social justice and nonviolence, the defining principles for Gandhi. Gandhi's views on education make ethical action integral to the process of learning; it is not possible to divorce the content of what we learn from the set of acts to which ethics is applicable. That is, the Gandhian challenge is to make the ethical an integral part of the cognitive. But the insights from Gandhi and Tagore are not specific to science education, and hence, this chapter concentrates on alternate engagements between HPS and SE.

Finally, is it important to take into account other HPS traditions in SE? The unequivocal answer to this question has to be in the affirmative. The major reason is that many of these beliefs about science which are at the foundation of curricula and teaching of science are just plainly mistaken. They represent a particular rhetoric of science, and this rhetoric is abundant in science texts. Those who claim that texts are only about content of science and nothing else are only echoing a particular rhetoric of science which is itself based on some specific history and philosophy of science. These particular views have been challenged enough in HPS literature. Ironically, the alternate formulations related to science are not just from non-Western cultures. There is absence of many small traditions within Europe itself, those which have not found their way into these science texts.

This is ironical given that HPS has generated enough material, especially in recent times. This is especially true of history of science which has been able to engage far more deeply with multicultural and global histories of science. They have also been able to disentangle the many small science traditions within Europe. However, the scientific community not only ignores these narratives but many times is also inimical towards them. The question that science educationists should ask is this: Can they afford to similarly ignore these contributions in teaching science? Should they be subservient to the dominant ideologies of the mainstream scientists? Most importantly, can SE be independent of the ideology of science propagated by the scientists, an ideology which is deeply encoded in the content of science in various ways?

The enduring question as to whether students will do better in science if they are taught the history and philosophy of science has to be addressed. Rather than take this position, it is more useful – at this juncture – to emphasize the importance of HPS to science teachers. The question that is relevant now is whether teachers will be able to teach science better if they are exposed to HPS. There are good reasons to believe that the answer to this question is an unqualified yes.

Acknowledgments I am grateful to Anu Joy who has deeply contributed to my understanding of science education. Comments and suggestions from the many anonymous reviewers of this chapter, as well as from Michael Matthews, were very useful and I thank them for taking the effort to read the text so carefully.

References

Adas, M. (1989). *Machines as the measure of men*. Ithaca: Cornell University Press.
Alvares, C. (1991). *Decolonising history*. New York: The Apex Press.
Arnold, D. (2000). *The new Cambridge history of India III.5: Science, technology and medicine in colonial India*. Cambridge: Cambridge University Press.
Bala, A. (2006). *The dialogue of civilizations in the birth of modern science*. New York: Palgrave Macmillan.
Bhattacharya, K. (1958). Classical philosophies of India and the west. *Philosophy East and West* 8, 17–36.
Bhattacharyya, S. (1987). *Doubt, belief and knowledge*. New Delhi: Indian Council of Philosophical Research.
Carey, S., Evans, R., Honda, M., & Unger, C. (1989). An experiment is when you try it and see if it works: A study of grade 7 students' understanding of the construction of scientific knowledge. *International Journal of Science Education* 11, 514–529.
Chitnis, S. (1993). Gearing a colonial system of education to take independent India towards development. *Higher Education* 26(1), 21–41.
Dani, S. G. (2010). Geometry in the Sulvasutras. In C. S. Seshadri (Ed.), *Studies in the history of Indian mathematics*. New Delhi: Hindustan Book Agency.
Dear, P. (2005). What is the history of science the history of? Early modern roots of the ideology of modern science. *Isis* 96(3), 390–406.
Dharampal. (1983). *The beautiful tree*. Goa: Other India Press.
Dharampal. (2000). *Collected writings, 5 Volumes*. Mapusa: Other India Press.
Duschl, R.A. (1985). Science education and philosophy of science: Twenty-five years of mutually exclusive development. *School Science and Mathematics*, 87, 541–555.
Duschl, R. A. (1994). Research on the history and philosophy of science. In D. L. Gabel (Ed.), *Handbook of research on science teaching*. New York: MacMillan.
Fuller, C. (2001). Orality, literacy and memorization: Priestly education in contemporary South India. *Modern Asian Studies* 35 (1), 1–31.
Gadamer, H. (2001). *The beginning of knowledge*. Trans. R. Coltman. New York: Continuum.
Gandhi, M. K. (1951). *Basic education*. Ed. B. Kumarappa. Ahmedabad: Navjivan.
Gandhi, M. K. (1953). *Towards new education*. Ed. B. Kumarappa. Ahmedabad: Navjivan.
Ganeri, J (Ed.). (2001). *Indian logic: A reader*. Surrey: Curzon Press.

Habib, I. (2004). Viability of Islamic science: Some insights from 19th century India. *Economic and Political Weekly* 39(23), 2351–2355.
Halbfass, W. (1988). *India and Europe: An essay in understanding*. Albany: SUNY.
Hodson, D. (1988). Toward a philosophically more valid science curriculum. *Science Education* 72, 19–40.
Hodson, D. (1991). The role of philosophy in science teaching. In M. R. Matthews (Ed.), *History, philosophy, and science teaching: Selected reading*. New York: Teachers College Press.
Jenkins, E. W. (1996). The nature of science as a curriculum component. *Journal of Curriculum Standards* 28, 137–150.
Kapila, S. (2010). The enchantment of science in India. *Isis* 101(1), 120–132.
Kumar, Deepak. (1980). Patterns of colonial science in India. *Indian Journal of History of Science* 15 (1), 105–113
Kumar, K. (1988). Origins of India's textbook culture. *Comparative Education Review* 32 (4), 452–464.
Kumar, K. (2009). *What is worth teaching? (revised edition)*. Hyderabad: Orient Blackswan.
Machamer, P. (1988). Philosophy of science: An overview for education. *Science & Education* 7, 1–11.
Matilal, B. K. (1985). *Logic, language and reality: Indian philosophy and contemporary issues*. Delhi: Motilal Banarsidass.
Matilal, B. K. (1986). *Perception: An essay on classical Indian theories of knowledge*. Delhi: Oxford University Press.
Matilal, B. K. (1999). *The character of logic in India*. Eds. J. Ganeri and H. Tiwari. Delhi: Oxford University Press.
Matthews, M. R. (1989). History, philosophy, and science teaching: A brief review. *Synthese* 80 (1), 1–7.
Matthews, M. R. (2001). How pendulum studies can promote knowledge of the nature of science. *Journal of Science Education and Technology* 10 (4), 359–368.
Matthews, M. R. (2002). Constructivism and science education: A further appraisal. *Journal of Science Education and Technology* 11 (2), 121–134.
Matthews, M. R. (2004). Thomas Kuhn's impact on science education: What lessons can be learned? *Science Education* 88, 90–118.
Mohanty, J. N. (1992). *Reason and tradition in Indian thought: An essay on the nature of Indian philosophical thinking*. Oxford: Clarendon Press.
Mohanty, J. (2002). *Classical Indian philosophy*. Delhi: Oxford University Press.
Mohanty, J. N. (2010). Indian epistemology. In J. Dancy, E. Sosa and M. Steup (Eds.), *A companion to epistemology, Volume 4*. Malden, MA: Wiley Blackwell.
Mumford, D. (2010). What's so baffling about negative numbers? A cross-cultural comparison. In C. S. Seshadri (Ed.), *Studies in the history of Indian mathematics*. New Delhi: Hindustan Book Agency.
National Curriculum Framework. (2005). New Delhi: NCERT.
Plofker, K. (2009). *Mathematics in India*. Princeton: Princeton University Press.
Prakash, G. (1996). Science between the lines. In S. Amin and D. Chakrabarty (Eds.), *Subaltern studies IX*. Delhi: Oxford University Press.
Raju, C. K. (2001). Computers, mathematics education, and the alternative epistemology of the calculus in the Yuktibhāṣā. *Philosophy East and West* 51, 325–362.
Ramanujan, A. K. (1989). Is there an Indian way of thinking? *Contributions to Indian Sociology* 23 (1), 41–58.
Ramasubramanian. K & M. D. Srinivas. (2010). Development of calculus in India. In C. S. Seshadri (Ed.), *Studies in the history of Indian mathematics*. Delhi: Hindusthan Book Agency.
Rodríguez, M & M. Niaz (2004). A reconstruction of structure of the atom and its implications for general physics textbooks: A history and philosophy of science perspective. *Journal of Science Education and Technology*. 13 (3), 409–424.
Sangwan, S. (1990). Science education in India under colonial constraints, 1792–1857. *Oxford Review of Education* 16 (1), 81–95.

Sarma, K. V (Ed.). (2008). *Ganita-Yukti-Bhāṣā, Vol 1: Mathematics*. Explanatory notes by K. Ramasubramanian, M. D. Srinivas and M. S. Sriram. Delhi: Hindustan Book Agency.
Sarukkai, S. (2002). *Translating the world: Science and language*. Lanham: University Press of America.
Sarukkai, S. (2005). *Indian philosophy and philosophy of science*. Delhi: CSC/Motilal Banarsidass.
Sarukkai, S. (2008). Culture of technology and ICTs. In A. Saith, M. Vijayabaskar and V. Gayathri (Eds.), *ICTs and Indian social change*. London and Delhi: Sage.
Sarukkai, S. (2011). Indian logic and philosophy of science: The logic-epistemology link. In J. van Benthem, A. Gupta and R. Parikh (Eds.), *Proof, computation and agency: Logic at the crossroads*. Synthese Library, Vol 352. Dordrecht: Springer.
Sarukkai, S. (2012). *What is science?* Delhi: National Book Trust.
Siegel, H. (1989). The rationality of science, critical thinking, and science education. *Synthese* 80, 9–41.
Staal, F. (1995). The Sanskrit of science. *Journal of Indian Philosophy* 23, 73–127.
Turner, S & Karen S. (1999). Kuhn in the classroom, Lakatos in the lab: Science educators confront the Nature-of-Science debate. *Science, Technology & Human Values* 24 (1), 5–30.

Sundar Sarukkai is trained in physics and philosophy, and has a Ph.D. from Purdue University, USA. His research interests include philosophy of science and mathematics, phenomenology and philosophy of language and art, and drawing on both Indian and Western philosophical traditions. He is the author of the following books: *Translating the World: Science and Language* (University Press of America, 2002), *Philosophy of Symmetry* (Indian Institute of Advanced Studies, 2004), *Indian Philosophy and Philosophy of Science* (CSC/Motilal Banarsidass, 2005), *What is Science?* (National Book Trust, 2012), and *The Cracked Mirror: An Indian Debate on Experience and Theory* (Oxford University Press, 2012, coauthored with Gopal Guru). He is also a coeditor of *Logic, Navya-Nyaya & Applications: Studies in Logic Series 15* (Eds. M. Chakraborty, B. Lowe, M.N. Mitra, S. Sarukkai. London: College Publications, 2008) and *Logic and its Applications: Springer Lecture Notes in Artificial Intelligence, 5378* (Eds. R. Ramanujam and S. Sarukkai. Berlin: Springer-Verlag, 2009). Presently, he is the Director of the Manipal Centre for Philosophy & Humanities, Manipal University, India.

Chapter 54
Historical Interactions Between Judaism and Science and Their Influence on Science Teaching and Learning

Jeff Dodick and Raphael B. Shuchat

54.1 Introduction

To the eye of the layman, Jews and science seem to have a definite association. To support such claims, some point to the large number of Jews who have won Nobel prizes in the sciences and Fields Medals in mathematics (Efron 2007) and to the numerous scientists of Jewish origin teaching at US-based universities (Lipsett and Raab 1995). However, as Efron (2007, p. 2) rightly points out, statistics such as these are "crude" given that most practicing scientists of Jewish origin are not usually guided by the tenets of Judaism, so it is a misconception to argue that Judaism, in of itself, is the reason for these scientists' interest or even association with their respective fields. The question, therefore, is what does Judaism have to say about science?

In this chapter, we will examine the historical and philosophical meeting between Judaism and science and how it in turn has influenced the teaching and learning of science. In so doing, we will be asking the following questions: How has the relationship between science and Judaism developed over history? What are the philosophical approaches that have developed in Judaism for dealing with the challenges that science sometimes poses? What are the subjects of science that most specifically create such challenges for Judaism? And most important for this book chapter: How has this meeting between Judaism and science affected the teaching and learning of science?

The authors contributed equally to this manuscript.

J. Dodick, Ph.D. (✉)
Science Teaching Center, The Hebrew University of Jerusalem, Givat Ram Campus, Jerusalem 91904, Israel
e-mail: jeff.dodick@gmail.com

R.B. Shuchat, Ph.D.
The Center for Basic Jewish Studies, Bar-Ilan University, Ramat Gan 52900, Israel
e-mail: shubr@zahav.net.il

M.R. Matthews (ed.), *International Handbook of Research in History, Philosophy and Science Teaching*, DOI 10.1007/978-94-007-7654-8_54,

In order to answers these questions, we will first provide a brief review of the historical interactions between Judaism and science, the goal being to examine the major trends that have developed through time. Based on these historical trends, we will define the major philosophical approaches (or models) that have developed within Judaism for dealing with the possible challenges posed by the domain of science. By understanding how Jewish thinkers have coped with these perceived challenges, it will be possible to analyze how this relationship has affected the modern science education system (specifically in Israel or in educational systems, outside of Israel, where Jews are the majority). The answers we provide to these questions may serve as a guide towards improving the quality of science education in Jewish school systems around the world.

54.2 Judaism and Science: A Historical Overview

To give a comprehensive analysis of the interactions of Judaism with science over the ages is beyond the scope of such a short chapter. Instead, as this chapter is found within a book on the role of history and philosophy of science in science teaching, we specifically define those interactions between Judaism and science that have relevance for science teaching. In this way, we will be better positioned to understand how Jewish teachers and students of science understand and even cope with potential conflicts between the two perspectives.

In order to understand the Jewish position on science, we will begin by defining what we mean by science for this chapter. In early Western-Greek culture, the philosopher was de facto the scientist: the physicist, the astronomer, and the medical doctor. Therefore, in order to gain a broad understanding of Judaism and science throughout the ages, we have to relate to three categories which characterize the scientific enterprise over the ages and their interface with Judaism: (1) technology, (2) exact sciences (most notably, astronomy and biology[1]), and (3) natural philosophy in antiquity. In the modern era, with the separation of science and philosophy, we have to relate as well to the fields of (4) cosmology and cosmogony (most notably, evolution). The first two categories represent the products

[1] In this historical discussion, we address, among other things, the simple question of the rejection or the acceptance of biology, in general, as a science in antiquity. In the twentieth century, the issue of biomedical ethics has developed tremendously and so have the discussions concerning medical ethics and Jewish law. Much has been written about, including organ transplants and the definition of death, fertility issues, machines for prolonging life and disconnecting terminal patients from them, and cloning. Steinberg (2003) wrote an *Encyclopaedia of Jewish Medical Ethics*; moreover, *Shaare Zedek Hospital* in Israel has a journal dealing with such issues titled, *Asiya*, and much discussion can be found in legal journals as well. However, there are two reasons why this issue is not part of this short chapter: First, the research is all done on the graduate level and by experts and does not find its way to the classroom at the high school or even undergraduate level. Secondly, this issue has nothing to do with acceptance of biology or medicine and how it effects education, but very specific ethical issues within that realm.

of human reason and the third and forth, human speculation and inquiry about life and the (formation of the) universe.

To understand any Biblical Jewish position, we must use the oral tradition, or as it is also known, the rabbinic tradition, to interpret the Masoretic text of the Bible; this rabbinic tradition commences with the Talmudic and Midrashic period (100 BCE–600 CE) and represents the classical period of Jewish literature.[2]

54.2.1 Technology

In the Book of Genesis, Noah, was so named by his father Lemekh to mean: "This one shall comfort us for our work and the toil of our hands because of the ground which the Lord has cursed" (Genesis 5, 29). The curse of the ground is mentioned twice before in Genesis: the first time as part of Adam's punishment for partaking of the Tree of Knowledge (Genesis 3, 19) and the second time when Cain kills Abel (Genesis 4, 11–12). The rabbinic understanding of the curse was that the land would not produce food so easily, so man would have to sweat and toil to produce something which is not just thorns and thistles. However, this curse is not insurmountable; it takes human initiative and cooperation to overcome it. The Midrash (Tanhuma Genesis, 11) writes that until Noah was born, one planted wheat and barley but harvested mostly thorns. After Noah was born, "They harvested that which they had planted; not only that, but until Noah they did the work by hand. [However, after] Noah was born he invented ploughs, scythes, and shovels and all their work tools" (Poupko, 1990).

[2] By necessity, we focus on primary sources of the Jewish literary tradition, such as the Bible (the Masoretic text), the Talmud (Preisler & Havlin, 1998) and the Midrash, and their interpretations. We do so because it is these source texts and their interpretations that have been used authoritatively by Jewish thinkers, and in addition it is these texts which have been used for coping with different scientific positions. In turn, this has affected the modern science education curricula in many Jewish school systems. The Talmud is the authoritative body of Jewish law and lore accumulated over a period of six centuries (c.100 BCE–c.500 CE) in both Israel and Babylonia. The Talmud has two components: the Mishnah (Kehati, 1991), the first written compendium of Judaism's Oral Law redacted by Rabbi Judah the Prince in 200 CE, and the Gemara, an in-depth discussion of the theoretical base of the laws of the Mishnah. In addition, the Gemara includes nonlegal discussions and interpretations of Biblical texts called Aggadah as well as stories with moral implications to human behavior. The Gemara written in Babylonia is the more popular corpus and is also referred to as the Babylonian Talmud. There was a parallel Gemara written in Palestine, and it is referred to as the Palestinian or the Jerusalem Talmud (Rozenboim 2010). If "Jerusalem Talmud" is not mentioned by name in the references in this chapter, then one can assume that the Babylonian Talmud is the version being referenced. Midrashim (pl.) are rabbinic interpretations of the Hebrew Bible consisting of homily and exegesis, on both its legal ramifications and its lore. Much of the Midrashic teachings are attributed to the Tannaim (rabbinical scholars of the period of the Mishnah who lived between 100 BCE and 200 CE). Individual Midrashic commentaries continued to be composed by rabbis after 200 CE until the Middle Ages. The Talmudic and Midrashic texts are seen as the classical period of Judaism in which the oral traditions and interpretations were put to text. This literature is referred to as classical rabbinic (or *Hazal* in Hebrew) literature. All denominations of Judaism are in a dialogue with this classical literature whether they see it as authoritative (as does Orthodox Judaism) or not (as does Reform Judaism).

For the rabbinic mind, human ingenuity and the technology it produced is not only a positive thing, but it is how humankind is expected to overcome "the curse of the ground." Instead of taking a passive position of accepting a Divine punishment, the rabbinic literature saw this "curse of the ground" as something that humankind brought about through misguided human behavior and therefore had become an issue that had to be resolved. This is comparable with the Biblical story of Moses breaking the tablets of the law after which God told him to make new ones. The idea of fixing what you break is how the rabbis of the Talmudic period interpreted this story. The resultant technology was and is the human attempt to rectify the flaws of nature caused by their own wrong actions.

A second way of viewing technology (and science) is from a practical point of view: providing one with the practical means to have an occupation. The Talmud (Makkot, 8b) says that it is incumbent upon a father to teach his son an occupation. Thus, the rabbis learned from the verse "And you shall live by them" (Deuteronomy 30, 19) that one may take time away from Torah study to study an occupation (Jerusalem Talmud, Peah 1, 1).

This second approach is also reflected in the famous debate between Rabbi Ishmael and Rabbi Simeon Bar Yohai, in which Rabbi Ishmael said that one must take off time from Torah study for a livelihood, whereas Simeon Bar Yochai thought that one should strive to devote all one's time to Torah (Talmud Brakhot, 36). The Talmud, however, preferred the view of Rabbi Ishmael finding it more practical and applicable. From the Biblical period until modern times, Jewish religious authority has largely remained positive towards the role of technology in society.[3]

54.2.2 Exact Sciences

The Talmud did not limit science's role to the practical task of insuring one's livelihood. Exact sciences, such as astronomy, were seen as bringing one to recognize the wonders of God's world, as seen by the Talmudic statement:

> Rabbi Joshua Ben Pazi in the name of Bar Kapara said: anyone who can calculate the seasons and the astral [movements of the heavens] and does not, about him the verse says: 'and the acts of God he does not behold and the works of His hands they did not see' (Isaiah 5, 12). Rabbi Samuel Ben Nahmani said: How do I know that it is a mitzvah [a Divine commandment] for one to calculate seasons and astral [movements]? For it says: 'For you shall keep and do [these commandments] for this is your wisdom and knowledge in the eyes of the nations (Deuteronomy 4, 6). Which wisdom is considered by the nations? This is the calculations of seasons and astral [movements] (Talmud Shabbat, 75a).

[3] In recent times, members of the ultra-Orthodox camp have raised concerns over the access that some computer technology gives to the media that is not in accord with their (Jewish) philosophy. As an example, some 40,000 ultra-Orthodox, US-based Jews attended a meeting at Citi Field (in New York, NY) to hear lectures about the dangers of the Internet (Grynbaum 2012). Similarly, in Israel, public calls are sometimes made to ban home computers in ultra-Orthodox communities due to their "spiritual dangers" (Ettinger 2007).

Rabbi Samuel Edeles (known as Maharsha), the sixteenth-century Biblical commentator, argued that this rabbinic statement is not speaking about calculating the Jewish calendar, since this is a calculation by the moon, but rather we are speaking here of the mathematical calculation of the movements of the heavenly bodies. Thus, Rabbi Josef Karo (2009) in his *Code of Jewish Law* (published 1565) allows one to look into an astrolabe on the Sabbath since Rabbi Karo understood that there is a rabbinic ordinance to study the heavens.

Both of these rabbis based their opinions on the rabbis of the Talmud who held scholars of astronomy in great esteem and had no problem admitting a mistake if proven wrong by non-Jewish scientists in this issue (Talmud Pesahim, 94b). In general the feeling was that there was wisdom to be gained from the scholars of the nations in this field, as the rabbinic dictum states: "If they tell you there is wisdom among the nations, believe them" (Midrash Rabbah Eikhah, 2, 17) (Freedman & Simon, 1939).

Aside from the high regard, the rabbis had for astronomy, the relation to the exact sciences was seen in quite a practical sense, similar to the attitudes towards technology. In the area of biology, we have a few sources for the study of zoology or botany in order to better understand the commandments. The Talmud states that Rav, the third-century head of the Talmudic academy in Sura, Babylonia, spent 18 years with shepherds in order to be able to differentiate between temporary and permanent wounds in animals. This he did to identify which animal qualified as a first born (sacrifice) for the Temple (Talmud Sanhedrin, 5b).

The students of Rabbi Ishmael dissected the dead body of a criminal to understand issues of purity and non-purity (Talmud Bekhorot, 45a). Again, the Talmud accepts the opinion of non-Jewish botanists when deciding an issue concerning the agricultural laws; its regard for the science of biology can be seen from the very fact that it allows one to go to a physician claiming that medicine is a legitimate science and furthermore claims that a doctor who does not charge for his skills is probably not worth seeing (Talmud Bava Kama, 85a).

54.2.3 Philosophy

In order to correctly understand the rabbinic attitude towards general knowledge, one (also) needs to understand the rabbinic attitude towards philosophy, especially Greek philosophy, which was the forerunner of scientific knowledge in the West; and in order to understand their attitude towards philosophy, it is important to introduce the subject with a brief discussion of the Talmud's attitude to Greek culture, as Greece was the birthplace of philosophy.

In this discussion we differentiate between two issues: Greek language on the one hand and Greek philosophy on the other. The Mishnah (Megilla 1, 9) states that Rabbi Gamliel permitted the translation of the Torah from Hebrew into Greek. In the Talmud, Bar Kapara added that speaking Greek was appropriate for a Jew since the beauty of Jephet (father of *Yavan* in Genesis 10,2 which is the Hebrew name for Greece) should be in "the tents of Shem" (i.e., the Jewish people) (Jerusalem Talmud Megilla, 1, 9).

Rabbi Simeon Ben Gamliel claimed that only Greek could capture the meaning of the Torah in translation (Jerusalem Talmud Megilla, 1, 9).

Despite the Talmud's positive attitude towards the Greek language (and the exact sciences) it saw the Greek use of verbal intimation negative light but was generally silent about Greek philosophy (See *Hokhma Yevanit* in Zevin (1963)). In the post-Talnudic period Greek philosophy was an issue debated for centuries among Jewish thinkers that set the tone for some of the modern Jewish attitudes towards philosophy and science.

Rabbi Hai Gaon of eighth-century Babylonia saw Greek philosophy as something which could sway one from the path of truth. Saadiah Gaon (882–942 CE), however, embraced the Islamic Philosophy of the Kalam,[4] which was strongly based on the Greek philosophical model, and was well versed in the sciences of his day. Isaac Israeli (855–955 CE) in Kairouan (modern Tunisia) also drew heavily on the philosophy and science of his day.

In Spain, the attitude of most rabbinic figures began with the acceptance of the value of Greek philosophy. Solomon Ibn Gabirol (1020–1057 CE) took a neo-platonic stance in his *Fons Vitae* (the Latin edition of what has been shown to be the original *Mekor Hayim – Source of life*) as well in his classic poem *Keter Malkhut* (*The Crown of the King*). Abraham Ibn Daud (1110–1180 CE) wrote an astronomical work and was the first to create a Jewish philosophical work based on the writings of Aristotle, titled *Emunah Ramah* (*The Sublime Faith*). Bahya Ibn Paquda, in the late eleventh century saw the study of Greek philosophy as an important tool for understanding nature and metaphysics incorporating these ideas in his *Duties of the Heart* (Ibn Paquda 1970). Abraham Bar-Hiyya (1070–1136 CE) embraced Aristotelian thinking openly in his *Higayon Henefesh* (*Meditation of the Soul*) and wrote works on astronomy, mathematics, and geometry (Bar-Hiyya 1968). Similarly, Abraham Ibn Ezra (1089–1164 CE) incorporated Aristotelian ideas and astronomy into his Torah commentary. He wrote a work entitled *Lukhot* (*Tables*) entailing astronomical tables and wrote a work on the astrolabe entitled *Keli Nehoshet* (*The Copper Instrument*) as well as *Yesod Mispar* (*Basic Numbers*) on arithmetic.

Maimonides (1135–1204 CE) was an avid believer in the importance of studying Greek philosophy and science and formulated in his *Commentary on the Mishnah*, the famous statement: "Accept truth from whoever offers it" (Maimon 1961). This echoes the (previously discussed) Talmudic respect for all knowledge, even that which originates outside the Jewish world. In addition to discussing issues of Greek philosophy and cosmology in his philosophic work, *Guide for the Perplexed* (Maimon 1956), Maimonides even incorporates ideas on cosmology into the first volume of his Halakhic[5] work, the *Mishneh Torah* (*Repetition of the Torah*) (Maimon 1987). Maimonides saw human reason and faith as inseparable. After all, if God created

[4] Kalam is an Islamic school of philosophy that seeks theological principles through dialectic; it flourished in what is today modern Iraq, from the eighth to tenth century CE (Wolfson 1976).

[5] Halakha is the collective body of Jewish religious law, including Biblical law and later Talmudic and rabbinic law, as well as customs and traditions. Judaism classically draws no distinction in its laws between religious and ostensibly nonreligious life. Hence, Halakha guides not only Jewish religious practices and beliefs but also numerous aspects of day-to-day life.

humankind with the faculty for reason, then it cannot be that this God-given gift is at odds with revelation. The faculty of human reason is the "image of God" through which He created us (*Guide for the Perplexed*, I, I). Therefore, we need to use this faculty to understand revelation correctly. The need for harmony between reason and revelation, he states clearly: "We always attempt to integrate Torah and reason, and therefore will always explain issues (of faith) from a natural point of view. Only that which is clearly described as a miracle (by the Bible) without any other possible explanation, will we grant it the name of miracle" (as cited by Shilat 1995). Therefore, in issues of science and philosophy, Maimonides goes to great lengths to demonstrate how the scientific thinking of his day is in total harmony with Jewish faith.

Even with the difficult issue of Aristotle's theory of the eternity of the universe, which appears totally opposed to the Biblical notion of creation, Maimonides defends the (Biblical) act of creation by using Aristotelian logic and arguments from nature (*Guide for the Perplexed* 2, 13–32). The place of logic is so important in Maimonides thinking that he argued that logical deductions from revelation are part of the original intention of the revelation (*Guide for the Perplexed* 3, introduction); therefore, revelation and reason can never contradict each other. Maimonides believed in the inseparability of revelation and reason (and its derivative, science). This is best demonstrated by his statement that "if he would have been convinced that science had proven that the earth was created differently than our understanding of the Biblical text, he would have had no problem reinterpreting Genesis 1, 1" (based on *Guide For the Perplexed* 2, 25 as cited by Sacks (2011, pp. 219–220)).

The philosopher Gersonides (1288–1344 CE) accepted the Aristotelian ideas as filtered through Islamic philosophy and was an avid student of the sciences himself writing on arithmetic, geometry, trigonometry, and astronomy (Touati and Goldstein 2007). He is said to have invented a marine navigational tool called Jacob's ladder (Stanford Encyclopedia of Philosophy, http://plato.stanford.edu/entries/gersonides).

Even Judah Halevi (1075–1141 CE) who claimed in his *Kuzari* that philosophy was limited in its ability to prove religious belief was still well-versed in philosophy and the sciences (Halevi 1998). In addition, the Raavad of Posquieres, in the twelfth century, who was a contemporary of, a commentator on, and fierce opponent of Maimonides, is still quite silent concerning Maimonides' embrace of philosophy. In addition, Nahmanides (1194–1270 CE) despite his leaning towards Kabbalah, defended the study of the sciences and Maimonides' *Guide to the Perplexed* in face of French rabbinic opposition (Shavel 1963).

In general, the entire Spanish era (900–1391 CE), prior to the inquisition was an age of acculturation in which rabbis openly embraced Western thought and culture while remaining faithful to their religious beliefs. It was also the most creative period of religious philosophy in which the three monotheistic religions, Judaism, Christianity, and Islam, stood side by side on the Iberian Peninsula. Despite any ongoing political struggle, the thinkers of all three religions openly borrowed ideas from each other in the common battle against problems arising from Aristotelian thinking.

Maimonides borrowed ideas from Al-Farabi and Avicenna. Gersonides borrowed openly from Averroes and Al-Farabi and Thomas Aquinas borrowed openly from Maimonides and Al-Farabi. In fact, the common front and common issues of the three religions were so vast that the Jewish philosopher Ibn Gabirol's book, *Mekor*

Hayim (*Source of Life*), translated into Latin, was mistakenly thought to be the product of an Arab-Christian scholastic philosopher by the name of Avicebron until the Hebrew original was discovered in 1846 by Solomon Munk.

In general, as Shuchat (2008) noted, Jewish philosophy evolved when two events occurred: (1) a meeting between Judaism and Western culture took place and (2) a period in which the Jewish community enjoyed at least minimal civil rights as a minority. This occurred during three time periods: (A) the Hellenistic period from about the second century BCE in Israel and Egypt until the end of the revolt against the Roman empire in 115 CE in Alexandria; (B) the Muslim period, from the eighth century until the expulsion of Jews from Spain in 1492; and (C) the modern period, from the emancipation of the late eighteenth century until today.

In these three periods Jews experienced both Western culture and felt accepted enough to ask themselves how their neighbors saw them and took interest in the surrounding culture and thought. In the interim periods, where the Jews of the Western world did not enjoy these rights, they usually limited their study to Jewish legal writings and Kabbalah.[6]

54.2.3.1 The Opposition to Philosophy

The controversy over Maimonides writings saw the growth of an anti-philosophical movement in Provence and Spain. The Maimonidean controversy began during Maimonides' lifetime but turned into an anti-rationalist debate only in its second stage (1230–1235 CE). Solomon B. Abraham of Montpellier, David B. Saul, and Rabbi Jonah Gerondi led the anti-philosophy movement in 1232 CE. Their argument seems to have been more that the Jewish philosophers were compromising on the observance of the law and allegorizing scripture and Biblical miracles, than an attack on philosophy per se. With the burning of Maimonides books by the church in 1232 CE, the shock brought Rabbi Jonah Gerondi to retract and the controversy ended (Ben Sasson et al. 2007).

The third stage of the controversy (1288–1290 CE) was short lived, but the fourth and final controversy (1300–1306 CE) seems to have erupted again due to renewed allegations that the rationalists gave allegorical interpretations of the Bible, were lax in observance of the law, and denied Biblical miracles.

Rabbi Moses Aba Mari Astruc of Lunel persuaded Rabbi Solomon Ben Adret (also known as Rashba) to join forces. The Rashba was willing only to ban the study of philosophy or the natural sciences before the age of 25 (Ben Adret 2000).

[6] Kabbalah (literally "receiving") is a discipline and school of thought discussing the mystical aspects of Judaism. It is a set of esoteric teachings meant to define the inner meaning of both the Bible and the traditional rabbinic literature (including Midrash and Talmud) as well as to explain the significance of Jewish religious observances in light of the inner soul and upper spiritual worlds. The term Kabbalah, meaning Jewish mysticism, is a term from the twelfth century CE and afterwards. However, Jewish mystical texts date back to at least the second temple period if not earlier. The best-known Kabbalistic work is the book of Zohar or more correctly Zoharic literature, which first appeared in Spain in the late thirteenth century (Dodick et al. 2010).

However, even Rashba, who opposed philosophy, neither had no problem with the study of Greek medicine (Reponsa part 1, letter 415) nor was he actually against studying the exact sciences.

It is possible that the political changes in Spain helped create the anti-rationalist movement. With the re-conquest of Spain by the Christians, Jews were suffering from the crusades and from the impact of martyrdom in their wake. The Maimonidean synthesis with Greek culture seemed less appealing and a move to mysticism was being felt. After the massive conversion of Jews to Christianity during the months of Spanish rioting against the Jewish communities in 1391, many Jewish scholars regarded the adherence to philosophic doctrine as a threat to the Jewish community; this included Hasdai Crescas (1340–1412 CE) who criticized Aristotelian physics in what was to be one of the first serious attacks on the system (Wolfson 1929).

After the Spanish expulsion, the interest in philosophy dwindled in the Jewish world. With the exception of scholars such as Joseph Solomon Delmedigo (1591–1655) in his book *Sefer Elim*; R. Moses Isserles (1520–1572) of Cracow and R. Abraham de Herrera (1570–1635), who combined philosophy and Kabbalah; R. Menasseh Ben Israel (1604–1657); R. Moses Zacuto (1625–1697), a kabbalist who was in contact with Spinoza; and R. Loew of Prague (1520–1609, known as the Maharal), Kabbalah took over from philosophy as the main intellectual interest.

David Nieto (1654–1728 CE), in his *Second Kuzari*, claimed that the rabbis of the Talmud were never against philosophy (Nieto 1993). In the eighteenth century, scholars like Rabbis Elijah Ben Solomon Zalman (1720–1797, better known as the Vilna Gaon), Jacob Emden, and Jonathan Eibeshitz still held the sciences in great respect. The Vilna Gaon was quoted as saying that for every amount that one lacks knowledge of the general sciences, he lacks one-hundred fold in the study of Torah (Baruch ben Jacob (1780) of Shklov, *Introduction to his translation of Euclid* in Hebrew). The Vilna Gaon even wrote his own treatise on algebra titled, *Ayil Meshulash* (*The Three Rams*), but concurrently he was rather cold towards philosophy (Shuchat 1996).

With the onset of the Jewish emancipation (from the later eighteenth to twentieth century) in Europe and Russia came the rise of the Jewish *haskalah* (*or enlightenment*) movement, which saw its goal to reintroduce secular education to the traditional Jewish masses. In Western Europe, the father of the Jewish *haskalah* movement was Moses Mendelssohn (1729–1786), a traditional and observant Jew well versed and acculturated in German intellectual society. Mendelssohn set out to portray Judaism as a religion of reason in his work *Jerusalem*, using religion and philosophical reasoning hand in hand as Maimonides did before him. The first period of the haskalah (in the late eighteenth century and the beginning of the nineteenth century) saw many religious Jews, especially in Eastern Europe, even rabbis, embracing the message of secular studies alongside Torah studies; however, with the secularization of the Russian and Eastern European haskalah movement, rabbinic leaders disassociated themselves with it and even became antagonistic to it.

Since then, ultra-Orthodox[7] Jewish thinkers tended to disassociate themselves from the study of secular knowledge, especially philosophy and the humanities, even if they had no overall opposition to the exact sciences. An example of such was the voluntary closing of the Volozhin Yeshiva (Seminary) in Lithuania after the government forced its students to include secular studies into its syllabus in the second half of the nineteenth century (Stampfer 2005).

In the Hassidic[8] (ultra-Orthodox) camp as well, there was a feeling of suspicion towards philosophy. However, even Rabbi Nahman Ben Simha of Breslov, the famous anti-rationalist Hassidic leader, who shunned philosophy (Ben Simha 1990) claimed that there was some good in all of the sciences (Likutei Mohran 18).

The revival of secular studies within Orthodoxy in Western Europe is attributed to Rabbi Samson Raphael Hirsch (1808–1888) who coined the term *Torah Im Derech Eretz* (or Torah *with secular knowledge*). Later in the twentieth century Rabbi A. I. Kook (first chief rabbi of prestate Israel) believed that all Torah scholars should have a basic knowledge of general culture and science. This has become the position of the modern-Orthodox stream in Judaism. However, within the ultra-Orthodox world, the exact sciences, i.e., physics, chemistry, and biology (excluding evolutionary biology), are tolerated but the humanities (including philosophy) are viewed with profound distrust.[9]

---

[7] In general, Orthodox Judaism is the approach to Judaism that adheres to the rabbinic interpretation and application of the laws and ethics of the Bible as found in the Talmudic literature. In the early nineteenth century, Orthodox Judaism divided into two different camps, the modern Orthodox and ultra-Orthodox which encompass a wide spectrum of beliefs. Nonetheless, Waxman (1998) details three major differences separating modern Orthodoxy and ultra-Orthodoxy. The first involves the ultra-Orthodox stance towards the larger society in general and the larger Jewish community, which is essentially an attitude of isolation, as opposed to the inclusive attitude of the modern Orthodox. The second is in reference to modernity, general scholarship and science, with the ultra-Orthodox being antagonistic and modern Orthodoxy being accommodating, if not always welcoming. Third, there is a basic difference between the two in their attitudes towards Zionism and active involvement in the rebirth and development of Israel, with the ultra-Orthodox being antagonistic and the modern Orthodox welcoming Zionism as a religious value. In this chapter we will use the English term ultra-Orthodox (even with its political connotations) as opposed to the Hebrew term, Haredi, as this is the more common term in English sources.

[8] Hasidism is a branch of ultra-Orthodox Judaism that promotes spirituality and joy through the internalization of Jewish mysticism as the fundamental aspect of the Jewish faith. It was founded in the eighteenth-century Eastern Europe by Rabbi Israel Baal Shem Tov (1698–1760) as a reaction against overly legalistic Judaism identified with Orthodox Jewry in Lithuania (sometimes called Mitnagdim (pl.) or "the opposition"). Today, the ultra-Orthodox community is comprised of both Hassidim and Mitnagdim.

[9] The ultra-Orthodox community saw support for their position in the opposition of the rabbis of the Middle Ages to philosophy. In the argument between the anti-rationalists and the Maimonidean school, for instance, they saw themselves as siding with the anti-rationalists against the Maimonidean embracing of philosophy and secular science.

54.2.4 Cosmology and Cosmogony (In the Modern Period)

54.2.4.1 Cosmology

Despite the seeming open-mindedness towards science in the nineteenth-century Western Europe, there was a greater ambivalence towards science in Eastern Europe indicated by the pain that the parting with the geocentric system of planets had on a few Jewish thinkers. The Vilna Gaon in the late eighteenth century still spoke of a Ptolemaic astronomical system in his commentary to the mystical *Sefer Yetzirah* (*Book of Creation*). Is this due to opposition to the new astronomy or just a lack of awareness? The Gaon studied philosophy and science from Hebrew texts; thus, it is possible that these texts were outdated and therefore could have had antiquated views of science. It is also possible that his ideas were just commentaries on the views in the *Sefer Yetzirah* which was written close to the Ptolemaic period.

Another scholar from Vilna, Rabbi Pinchas Elijah Horowitz (1765–1821 CE), in 1797 CE, published *Sefer Habrit* (*Book of the Covenant*), which acted as a Jewish encyclopedia of science. This volume needs special attention since it was extremely popular in the nineteenth and early twentieth century among Eastern Europe Jewish scholars, as evidenced by its more than 26 editions published between 1897 and 1925 CE in the original Hebrew as well as in Yiddish and Ladino (Robinson 1989).[10] An unusual aspect of this work was its attempt to create a synthesis between science and Kabbalah (Robinson 1989). In the chapters on astronomy Horowitz displays sympathy to the Copernican system but ultimately rejects it in favor of the geocentric position (Rosenbloom 1996).[11]

Rabbi Reuven Landau (as cited by Robinson 1983) of Romania wrote books on trigonometry (*Middah Berurah* or *Clear Measurement*) and astronomy (*Mahalakh ha-Kokavim* or *The Movement of the Planets*). In them, he tried to explain to the reader all the fundamentals of these fields, but he was also careful to integrate an explanation of how the Divine force permeates all of nature (Brown 2008). Despite his knowledge of the new cosmology, he raises objections to Copernicus' proofs and sides with the geocentric universe for spiritual reasons; if the Earth was not the center of creation, possibly humanity was not the center either. Landau, as with Horowitz's *Sefer Habrit* before him, adopts Tycho Brahe's system in which the sun and the moon revolve around the Earth but the other planets revolve around the sun. However, it should be mentioned that in the second edition of this book in 1818 CE, the publisher writes that it is possible for a believing Jew to adopt the Copernican view if he so chooses (Brown 2008).

---

[10] Solomon Schechter, the noted scholar of the Cairo Genizah in the early twentieth century, admits that in his youth in a village in Romania, he heard of America through Sefer Habrit (Robinson 1989).

[11] Nussbaum (2002, 2006) shows that in recent years, there has been a revival of geocentrism among some in the Orthodox community including rabbis and scientists; it is unknown how widespread this phenomenon is.

In Bialystok, Hayyim Selig Slonimski (1810–1904 CE) was a talmudist, a mathematician, and a popularizer of science for traditional Jews. Coming from the same mind-set of the Vilna Gaon and *Sefer Habrit* that secular knowledge is needed for the proper comprehension of Torah, he published his first volume on mathematics titled, *Mosdei Hokhma* (*Foundations of Wisdom*) in 1834 with rabbinic approbations (Robinson 1983). In 1838, he published a book on astronomy entitled *Toldot Ha-shamayim* (*The Heavenly Hosts*). He was one of the first to explain that the six days of creation are really six eons and therefore came closer to the ideas of the geology of the time than those relying on Biblically based calculations for the Earth's origins (Robinson 1983). In sum, Slominski looked for synthesis between science and rabbinic literature.

In Western Europe, however, Jewish thinkers seem to have been quicker to accept this new worldview. Raphael Halevi of Hanover (1685–1779 CE) a mathematician and philosopher, who had studied with Leibnitz, published two books in astronomy in 1756 CE. In his *Tekhunat Ha-Shamayim* (*Astronomy of the Heavens*), he openly embraces the Copernican system. It is of interest that Rabbi Landau read this work and quoted from it, without adopting this position (Brown 2008). Similarly, Joseph Ginsburg in his *Ittim La-Bina* (*Wisdom of our Days*) explained that one could accept the Copernican model and remain a faithful Jew. Dov Ber Rukenstein in his two-part series on astronomy entitled *Mesilot Ha-Meorot* (*Pathways of the Heavenly Bodies*) (as cited in Robinson 1983) talked of Copernicus' model as being accepted by all scientists of his day. Therefore, writing in the late nineteenth century, Rabbi Samson Raphael Hirsch could say:

> What Judaism does consider vitally important is the acceptance of the premise that all the hosts of heaven move only in accordance with the laws of the one, sole God. But whether we view these laws from the Ptolemaic or Copernican vantage point is a matter of total indifference to the purely moral objectives of Judaism. Judaism never made a credo of these or similar notions (Hirsch 1992, p. 263).

54.2.4.2 Cosmogony (Including Evolution)

In classical Jewish philosophy, Aristotelian physics and cosmology were seen as a challenge, rather than as an overt threat; thus, although being diametrically opposed to the book of Genesis, great effort was invested in order to reach a synthesis between the Aristotelian theory of the eternity of the universe and the Biblical creation narrative. In the twelfth century, Maimonides in his *Guide for the Perplexed* took great pains to explain how one can explain creation with the same Aristotelian hypothesis but with some alterations.

In the nineteenth century, rabbinic thinkers dealing with the new theories of cosmogony and particularly evolution acted in a similar same way. Orthodox Rabbi Israel Lipschutz of Danzig was a learned legalist who had a great interest in the science. Writing in the 1800s, before Darwin's *On The Origin of Species* was published, Lipschutz was familiar with the "evolutionary" theories of Lamarck. Rather than seeing the new theories as a threat to Biblical belief, he sees the idea of an ice age and the regeneration of life as a proof for the Jewish belief in the eventual resurrection of the dead. No criticism of the theory can be found in his writings, just a great enthusiasm that science is now proving the age old Kabbalistic theory that there were earlier worlds than ours (Shuchat 2005).

In the post-Darwinian world of the nineteenth century, there was still no major change. Jewish thinkers like Rabbis Elijah Benamozegh of Italy and Samson Raphael Hirsch of Germany, writing at the same time that the Church and the scientists of Europe were battling each other verbally, did not see evolution as a major threat to Jewish belief. For example, Rabbi Hirsch writes:

> Judaism is not frightened even by the hundred of thousands and millions of years which the geological theory of the earth's development bandies about so freely. Judaism would have nothing to fear from that theory even if it were based on something more than mere hypothesis, on the still unproven presumption that the forces we see at work in our world today are the same as those that were in existence, with the same degree of potency, when the world was first created. Our rabbis, the Sages of Judaism, discuss [Bereshit Rabbah 9, 2 and Mishna Hagigah 16a] the possibility that earlier worlds were brought into existence and subsequently destroyed by the Creator before He made our own earth in its present form and order. However, the rabbis never made the acceptance of this and similar possibilities an article of faith binding on all Jews. They were willing to live with any theory that did not reject the basic truth that every beginning is from God (Hirsch 1992, p. 265).

Rabbi Elijah Benamozegh (1862) saw the new scientific discoveries as proving the Midrashic and Kabbalistic notion of earlier worlds which God created before our own:

> In conclusion, this belief in earlier worlds is an ancient one in our nation and it stands as a proof for the divine nature of the Torah, which natural science now confirms. . . . And I finish [this discussion] with the dear words of the scholar in the Kuzari [1, 40] who said: 'If a believer in Torah had to admit to the existence of primordial matter of earlier worlds that predated us, this would not blemish our faith' (Benamozegh 1862).

In addition, Benamozegh saw the new theories of evolution as a proof of human potential and ultimately of the resurrection:

> I believe, as science teaches, that animal forms appeared on the earth and evolved into more perfect beings, either as Cuvier said, by revolutions and cataclysms, or by slow evolutionary processes, like the opinion of the modernist Lyell, or Darwin and others. More and more perfect species have developed, one after the other, over the course of millions of years on the face of the earth. The most perfect form is Man. But will nature stop here? This would indeed be strange. Present humankind, as Renan [French expert of Middle East ancient languages and civilizations] says, will evolve into another, more perfect human being. But Renan and the others stop here. They do not say that the order that reigns in the physical world has to reign in the moral one as well, and that there is no reason to believe that the 'I' that force which created the actual human, does not have to create the future human as well. They do not say that the 14 monads, the atoms, which are minuscule forces, are indestructible (as science teaches) for it is inevitable to believe that they will compose the future Man on a regenerated earth. All this is stated by Judaism, and is called the Resurrection (Benamozegh 1877, pp. 276–277).

Similarly, Rabbi A. I. Kook's writing in the early twentieth century also displayed an optimistic view of evolution claiming that it is closer to the Kabbalistic notion of creation than the philosophical idea of creation ex nihilo.[12] Despite this enthusiasm in his more philosophical writings, in his public letters, Rabbi Kook writes more cautiously. After explaining to a correspondent why the new theories of evolution do not contradict the Torah, he writes:

> We do not have to accept theories as certainties, no matter how widely accepted, for they are like blossoms that fade. Very soon science will be developed further and all of today's new theories will be derided and scorned and the well-respected wisdom of our day will seem small-minded Feldman (1986, p. 6).

Continuing this trend, Rabbi Isaac Halevi Herzog (the first chief rabbi of Israel), writing in the mid-twentieth century, displayed the discomfort that many later rabbinic figures were to have with the theory of evolution. This discomfort was caused not just by the challenge which this theory posed to Biblical exegesis but by the fact that it was considered to be one of the paradigms of modern secular scientific thought, which many of these rabbinic figures felt was in opposition to all organized religion (Robinson 2006).

Some rabbis of the second half of the twentieth century began, like their Christian contemporaries, to see the theory of evolution as a threat. As Orthodox Jews entered the arena of the sciences, many of them entered the battle against evolution, arguing from a scientific standpoint, rather than a Biblical or Talmudic point of view, and looked to those who opposed evolution as their comrades in arms (Cherry 2006). Rabbi Herzog's attempt to look for ways to harmonize the simple meaning of Genesis 1 with evolution without the multitude of rabbinic commentaries reflects this new attitude:

> How can the Torah chronology be scientifically defended, in view of the aeons which science postulates for the existence of man upon this earth? There is, of course, the well known Midrash, 'boneh olamotu-maharivan' [he built his worlds from annihilation] [Midrash Genesis Rabbah 3, 7; Ecclesiastes Rabbah 3:11], but this can only help if we assume that "maharivan" does not mean annihilation, so that we can assume that fossils of man asserted by science to be so many hundreds of thousands of years old are relics of a previous earth. Yet anthropology seems to assert upon internal evidence that the present man is already hundreds of thousands of years old! [...] Of course, strictly literal interpretation of the Pentateuchal text is out of the question. But super literary interpretation should be resorted to only when reason absolutely rules the literary sense being utterly impossible... (as cited in Shuchat 2008–2009, p. 155).

Rabbinic scholars are not detached from the world around them. During periods of social turmoil, when the thinkers of the age begin to doubt the validity of the scientific order of the day, Jewish thinkers do so as well. The events of the Second World War proved both the supreme power of scientific technology as well as the threatening implications of the misuse of that power. The subconscious social impact of the atom bomb attacks on Japan and a war that used modern technology to claim millions of lives cannot be underestimated. Although faith in science

[12] The kabbalists had a different take on creationism seeing it more as an act of emanation rather than creation ex nihilo. They also differed on the question of the time that it took to create the universe (Shuchat 2009).

remained unscathed for the first decade and a half after the war, and the scientific community emerged from the war with enhanced prestige, these events planted the seed for the disillusionment with science, in general, and more specifically evolution that put it on the defensive in the 1960s and 1970s (Ben-David 1991).

The technological boom of the nineteenth and early twentieth centuries led to a belief in the omnipotence of science, and religious fundamentalist voices against the theory of evolution were stifled, out of respect for science; by the 1970s, however, attacks on science gained legitimacy, and the popular reaction to science was now a mixture of enthusiastic support and profound mistrust (Ben-David 1991).

In the Jewish world, a second element contributed to increased disdain for science. After the destruction of European Jewry, including all major institutions of Jewish learning and culture, some of the Orthodox rabbinic leadership did everything possible to hold on to what remained and held suspect any new way of thinking that might pose some type of threat to religious survival. These feelings of suspicion towards all new ways of thinking became more manifest in the seventies, as society as a whole became critical of science. As a result, the late twentieth century saw the Jewish attitude to science take on different voices. The theory of evolution, in particular, which was seen as one of the paradigms of modern, secular (scientific) thinking, became representative of how various elements in Judaism see religion and science. The syntheses of classical Jewish philosophy were therefore at times forgotten.

Approaches of Reform and Conservative Thinkers Towards Evolution in the Twentieth Century

The historical picture for Judaism in the nineteenth and twentieth centuries becomes even more complex with the rise of non-Orthodox movements in Europe, which eventually made their way to North America in the late nineteenth century. The non-Orthodox rabbis of the early twentieth century were very committed to finding a way to synthesize between science and Judaism, specifically the modern theory of evolution. The theological debates, which arose in light of the Scopes trial in 1925 over the legality of teaching evolution in public schools in Tennessee (Numbers 1998), generated a discussion among leading Reform rabbis in the United States of how to treat this sensitive issue.

The view of Reform rabbis of the 1920s was identical to their predecessors, Rabbis Kaufmann Kohler and Emil Hirsch of the late nineteenth century, in their belief that fundamentalists had erred in understanding the first verses of Genesis, literally, and in assuming that evolution denied a creator (Swelitz 2006). They argued that Genesis is not a textbook for science and literal interpretations of it were not acceptable. Reform rabbis went as far as claiming that progressive change and design were an inherent part of evolution and therefore provided a case for God as a creator (Swelitz 2006). In the 1930s, Rabbis Cohon, Brickner, and Felix Levy saw the new physics as supporting the view of intelligent design making the evolution of life possible (Swelitz 2006).

Conservative rabbis in the 1920s and 1930s like Levinthal and Finkelstein took the same position as the Reform on this issue (Swelitz 2006). Rabbi Mordecai Kaplan was somewhat of an anomaly at this point adopting a naturalist approach to God that disregarded the theological arguments leading from evolution to God.

In the postwar era of the 1950s and 1960s, Reform Rabbi Emil Fackenheim, Rabbi Abraham Joshua Heschel of the Jewish Theological Seminary, and theological scholar Will Herberg believed that an excessive reliance on science and reason had distorted the proper understanding of Judaism (Swelitz 2006). No evolutionary argument can explain a personal God. It is necessary to demarcate the boundaries between science and religion, they argued.

By the end of the 1960s, evolution was generally ignored by most among the Conservative and Reform, except in the writings of Reform Rabbi W. Gunther Plaut. This position, separating science from religion, was challenged by Reform Rabbis Levi Olan and Roland Gittelsohn and Conservative Rabbi Robert Gordis, who defended the centrality of reason and science in Jewish theology. Gittelsohn was personally interested in evolutionary biology and advocated what he called "religious naturalism" invoking the new science to aid one in proving the existence of God (Swelitz 2006). Gordis shared Milton Steinberg's belief that religion has to provide a philosophy of life which includes the conclusions of science.

In the 1980s there was a renewed interest in evolution, with the attempt by creationists in the United States to gain equal time in public schools for teaching Biblical creation. However this time, Reform rabbis, like William Leffler and Jack Luxemburg, maintained the need to emphasize the limitations of science in proving or disproving God (Swelitz 2006). This apparent divorce of science from religion in the 1980s was followed by evolution reentering Jewish theology with the renewed interest in Kabbalah. The idea of cosmic evolution was adopted by Reform Rabbi Lawrence Kushner as well as Rabbi Zalman Schachter-Shalomi of the Jewish Revival Movement and Prof. Arthur Green.

Approaches of Ultra and Modern-Orthodox Thinkers Towards Evolution in the Twentieth Century

Turning to the postwar Orthodox world of North America, we see that the situation was different. It was mentioned previously that Rabbi Herzog was hesitant in utilizing the rabbinic notion of earlier worlds and preferred to see if there were scientists who held other views. The second half of the twentieth century saw Orthodox responses to evolution, which were much different than those of the late nineteenth and early twentieth centuries. The ultra-Orthodox saw evolution as representing a secular alternative to the religious weltanschauung and therefore saw it as stepping over its legitimate boundaries. Rabbi Moses Feinstein claimed that

> Textbooks of secular studies that contain matters of heresy with respect to the creation of the world… are forbidden to be taught…If it is not possible to obtain other books, it is necessary to tear out those pages from the textbook (Feinstein 1982).

More modern-Orthodox thinkers looked for inroads to recreate the syntheses of earlier days. Like his predecessor Rabbi Herzog, Rabbi Aaron Lichtenstein wrote in the late twentieth century:

> Confronted by evident contradiction [between Torah and science] one would... initially strive to ascertain whether it is apparent or real... whether indeed the methodology of madda [science] does inevitable lead to a given conclusion, and ... whether... Torah can be interpreted... so as to avert a collision (Robinson 2006, p. 78).

An interesting phenomenon that developed in the second half of the twentieth century, with the entry of Orthodox Jews into Western universities, was the place of the Orthodox Jewish scientist. In 1948, some of these scientists founded a group they called the *Association of Orthodox Jewish Scientists* (AOJS). One of its aims was to resolve "apparent challenges of scientific theory to Orthodox Judaism" (Robinson 2006, p. 79), and evolution, specifically, was an important issue that they needed to deal with.

In the late twentieth century, three Orthodox Jewish physicists can be seen as representing three different approaches to evolution: Prof. Hermann Branover, Prof. Nathan Aviezer, and Prof. Gerald Schroeder. The American-trained Prof. Aviezer (1990) of Bar-Ilan University in Israel, in his work *In the Beginning*, took a nonliteral attitude to the 6 days of creation, seeing them as epochs rather than days of 24 h, but then continues to read into the literal text a novel interpretation in which he claims that the main elements of the Biblical story harmonize with all the main elements of modern scientific cosmogony (Cherry 2006). Aviezer also takes a non-chronological reading of the creation story in which he sees the 6 days of creation as representing two stages: days 1–4 which represent the formation of the structure of the universe and days 5–6 which represent the inhabitants of the universe which begin while the universe is being formed.

The second approach is from another American-trained physicist from Israel, Gerald Schroeder. Schroeder (1998) accepts, as does Aviezer (2002), the evolutionary timetable; however, in a novel literary hermeneutic, he claims that the 6 days of creation were 6 days of 24 h, but claims that according to Einstein's theory of relativity and time dilation, from the perspective of the forward rushing cosmos ("God's perspective"), 6 days is equivalent to 15 billion years looking backwards.

A third perspective is that of Russian-educated Prof. Hermann Branover of Israel. Associating himself with the ultra-Orthodox Hassidic community of Lubavitch, he holds a literalist view of the creation story. He uses alternative scientific views to argue against scientific evolution.

Ultra-orthodox groups such as the Israeli outreach organization "Arachim" feel more comfortable with these more aggressive fundamentalist anti-evolution positions. The open attacks of these fundamentalists against scientific thinking have gone so far as to find among them those who now are even questioning again Copernican heliocentricity in the twenty-first century. Moderate elements in the ultra-Orthodox world, such as the Aish Hatorah outreach organization, see Schroeder's position as saving both creationism and science, whereas modern-Orthodox Jews feel comfortable with Aviezer's ideas or just accept a nonliteral interpretation of the creation story (Sacks 2011).

Evolution aside, mainstream Orthodox Jewish rabbinic thinkers tend to adopt a generally positive attitude to science. This view of the legitimacy of science to overcome the Biblical curse of the ground mentioned earlier or to heal the sick is the age-old Jewish view which sees the idea of scientific progress as a way of mending the world when used for the good.

54.3 Historical Summary

This brief historical overview shows a somewhat complicated relationship between Judaism and science, but certain tendencies can be deduced from it. As we have seen, from earliest times, technology was seen positively as something that can help mankind overcome the difficulties of life. The Talmud praises the study of astronomy and sees biology and medicine as legitimate fields of study. The Jewish rationalists of the middle ages, especially in Spain, were particularly open to general studies and well versed in the sciences, medicine, and philosophy of their day.

The debate over philosophy in the post-Maimonidean era seems to have been more of an attack against lax observance, as well as the non-Orthodox ideas of the rationalists, than a ban on science per se. Philosophy was often seen as the culprit which brought in foreign ideas to Judaism. This same style of controversy can be seen in the middle to late nineteenth century Eastern Europe between the secular exponents of the Haskalah (or enlightenment) and their rabbinic counterparts.

The late nineteenth century saw the rise of Darwinian evolution and its entrance into Jewish thinking. Early thinkers until the First World War had an open and even accepting attitude, but in the post Second World War period, suspicion arose and the fear of foreign elements challenging Jewish faith renewed the debate over the relationship of science and Judaism. Most modern orthodox, as well as well as almost all Conservative and Reform thinkers, showed an attitude of acceptance; in contrast, the postwar ultra-Orthodox camp, suspicious of most modern concepts, showed antagonism to these ideas, even if they did not oppose the study of the sciences for the need of a livelihood or to practice medicine. Jewish educators abroad and in the educational system of the State of Israel struggle to this day to accommodate these different philosophical approaches, as we will see in the next sections of this chapter.

54.4 Philosophical Approaches Towards the Interaction Between Science and Judaism

Our brief historical survey confirms what Efron (2007) previously noted about the attitudes of Jewish thinkers towards science, in that historically it was "never subject to consensus." Certainly, we have seen that there were specific periods and regions where (rabbinical) authorities were worried about how secular science might affect Jewish piety and so strongly opposed contact with secular learning, including science, or specific scientific disciplines. At the same time, Judaism has often looked positively upon science, and its precursor, the study of nature and astronomy in antiquity, not just in its applied form where it benefits man's ability to derive a living or protect one's health but also in order to gain a better understanding of the natural world.

Moreover, Efron (2007) suggests that Judaism has avoided many of the science-religion clashes that have occurred among the Christians. In part, this was due to the fact that Jews never developed institutions with the coercive power to declare an idea or a book to be an anathema.[13] More importantly, in his view, the long exegetical tradition within Judaism of reading and interpreting texts meant that Jews by their nature did not sanctify the ideal of consensus. In fact, Jewish exegetes actively sought to multiply interpretations to arrive at deeper understandings of a text. Indeed, we can see this tradition of multiple interpretations operating in our brief historical review in the previous section of this chapter.

Thus, even if there might have been a mainstream trend during any period of Jewish history concerning how Judaism saw science, or any of its disciplines, from a practical viewpoint, rather than looking for consensus, it is better to discuss a spectrum of philosophical approaches that were developed to classify the (multiple) positions of Judaism towards science. In this section, we will discuss these approaches in order to create a set of definitions that can be applied to our discussion about Judaism and its interaction with science education.

Much of the work dealing with the philosophical interaction between religion and science has focused on Christian perspectives. The four most comprehensive works on this interaction include the books of Barbour (1997), Brooke (1998), Haught (1995), and McGrath (1999).

Among sources dealing with Judaism's interaction with science, there are two comprehensive works: Lamm's (2010) *Torah Umadda* and Rosenberg's (1988) *Science and Religion in the New Jewish Philosophy* (published in Hebrew). Both works are important but emphasize different approaches. Lamm's (2010) work is somewhat broader in that it deals with how Jewish thought has dealt with worldly knowledge, in general, rather than just science, which is Rosenberg's focus. From a practical perspective, Rosenberg's (1988) work has been used in a number of science education studies to classify the positions held by religiously Jewish teachers (Dodick et al. 2010) and students (Allouch 2010) in Israeli schools, and so we will examine his approaches here, as a precursor to our discussion of science education; nonetheless, whenever possible, we will integrate Lamm's (2010) discussion.

In structure, Rosenberg's (1988) approaches are somewhat similar to those mentioned in Barbour (1997) albeit the number of categories he developed was larger. Moreover, both Rosenberg (1988) and Lamm (2010) develop a set of approaches or models based on Jewish thinkers and their interpretation of classical

[13] One of the most famous historical examples of the use of coercive power in the Christian world was the Church's imprisonment of Galileo as a heretic in 1613 for his support of the heliocentric theory. Bronowski (1973, p. 218) argues that "the effect of the trial and the imprisonment was to put a total stop to the scientific tradition in the Mediterranean."

Jewish texts (such as the Talmud and Midrash) which contrasts with Barbour's approach in which he delineates a set of historical-based Christian attitudes towards science. Thus, Lamm (2010) and Rosenberg (1988) provide us with greater insight than Barbour (1997) when we examined science education and its interaction with Judaism. In his book, Rosenberg (1988) talks about four main approaches.

54.4.1 Limiting Approach

This approach opposes any attempt at integrating secular knowledge with Jewish thought. From this point of view, such a mixture creates the chance that heresy may infect the student of Torah; therefore, from a practical perspective, there was no room in the curriculum of a Torah student for such lesser knowledge (Lamm 2010). When faced with a scientific approach to problematic issues such as creation, those adopting this approach reject the scientific approach, as it challenges the primacy of the Bible's literal meaning. An example of this approach can be found in the writings of the late Rabbi Menachem Schneerson the former leader of the Lubavitch Hassidic movement in his commentary concerning geologic time and evolution:

> In view of the unknown conditions which existed in prehistoric times (atmospheric pressures, radioactivity) conditions which could have caused reactions of an entirely different nature and tempo from those known under present-day processes of nature, one cannot exclude the possibility that dinosaurs existed 5,722 years ago, and became fossilized under terrific natural cataclysms in the course of a few years rather than millions of years (Schneerson 1972).

In philosophical terms, Rabbi Schneerson (1972) was rejecting the principle of uniformity which states that the laws of nature remain unvarying throughout time. This approach to secular learning, in general, and science specifically is most common among the ultra-Orthodox. Such explanations also *seem* to match most closely with a Christian fundamentalist view of religion and its relationship to science, most notably those issues connected to creation and evolution.[14]

54.4.2 Explanatory Approach

In this approach, Biblical texts are not understood literally, but rather are explained so that religion and science can be brought closer together. Contradictions are

[14] Regarding evolution, Robinson (2006) argues that care should be taken in blindly comparing ultra-Orthodox attitudes to fundamentalist Christians too closely. The ultra-Orthodox are united in their opposition to Christian creationism as it is based on the King James Bible and not on traditional Jewish texts, which incorporate the cumulative perspectives obtained from (a large number of) traditional Torah commentaries and interpretations. In fact, Robinson (2006) could only find one source written from an ultra-Orthodox perspective whose author identifies as a creationist. Thus, at least in philosophy, if not deed, the ultra-Orthodox do differ from fundamentalist Christians.

viewed as a misunderstanding of the Bible and simply require proper interpretation. For example, with regard to the Earth's age, some Jewish Biblical commentators explain that the days of creation went far beyond a 24-h period of time, or as Rabbi Abbahu states in the Midrash that "God created [many] worlds and destroyed them until he created this one" (Rabba Bereshit, 3, Sect. 7). Thus, according to this interpretation, there were cycles of destruction and creation culminating in this world, such that the age of this world far exceeds the 6-day period of creation.

Among the most important exponents of the explanatory approach was Maimonides. More than that, his attitude to secular studies, in general, was not just that it was permissible but that there was an "obligation to pursue them as an act of mitzvah" (i.e., religious command) (Lamm 2010, p. 67).

54.4.3 Parallel Approach

This approach sees contradictions between science and religion as being derived from not clearly separating between the domains, as the former deals with rational explanations of nature, while the latter focuses on religious belief which illuminates human purpose, meanings, and values. Each domain has value for human experience, but they should not be integrated. Scientist and philosopher Yeshayahu Leibowitz is a noted exponent of this approach:

> There is no mutual dependency between scientific knowledge and decisions about [religious] values. What can the immense achievement of science contribute to these decisions on values? Science cannot contribute anything because concerning the problem addressed by these decisions, such as to be a believer, not only does science have nothing to contribute, but these questions cannot even be posed because these concepts do not appear in the lexicon of science (Leibowitz 1985, p. 35).

Historically, one of the more important exponents of the parallel approach in the Jewish world of education is Rabbi Samson Raphael Hirsch's Torah Im Derech Eretz ("Torah with secular knowledge") (Lamm 2010) whom we discussed previously. This approach also represents, as we have seen, the position of Reform Rabbi Emil Fackenheim, Rabbi Abraham Joshua Heschel of the Jewish Theological Seminary, and the theological scholar Will Herberg. Philosophically, the parallel approach is also equivalent to scientist Stephen J. Gould's (1997, 1998, 1999) principle of "respectful noninterference" between the worlds of science and religion or *NOMA* (Nonoverlapping Magisteria).

54.4.4 Complementary Approach

This approach suggests that science complements religion, creating a synthesis of the sacred and secular. Supporters of this approach see a strong (though not necessarily literal) fit between scientific discoveries and what is described in the Bible (Lamm 2010). This approach is personified by Rabbi A. I. Kook who viewed the theory of evolution as a model for spiritual growth; thus, he did not see it posing a threat to religion:

> The theory of evolution that is presently gaining acceptance in the world has a greater affinity with the secret teachings of the Kabbalah better than all other philosophies. Evolution which proceeds on a course of improvement offers us the basis for optimism in the world. How can we despair when we realize that everything evolves and immediately improves? In probing the inner meaning of evolution toward an improved state, we find here an explanation of the divine concepts with absolute clarity. Evolution sheds a light on all the ways of God (Kook 1938, p. 555).

54.4.5 Conflict Approach

This approach was not found in Rosenberg (1988), but it emerged as a consequence of interviews that were held by Dodick et al. (2010) with religiously observant Jewish teachers (in the Israeli high school system) and scientists (in the Israeli university system); it was therefore added to the taxonomy used by Dodick et al. (2010) to classify the philosophical approaches of religiously oriented, Jewish teachers and scientists. Conflict emphasizes the understanding that there sometimes exists a contradiction between science and religion because of the overlap between the two domains such as occurs with evolution. Such conflict largely arises because of the open, unanswerable questions that occur due to this overlap. Nonetheless, although some are affected by this conflict, they are willing to live with the situation and do not reject science as is the case with the limiting approach.

54.5 Judaism and Its Interaction with Science Education

In discussing Judaism and its interaction with science education, it should be understood that prior to the emancipation period in Europe, Jewish contact with general secular learning and science learning, specifically, was largely limited to those rabbis who approved of and conducted such learning (Efron 2007). Thus, it is impossible to talk of the interaction between Judaism and science education on a large scale before that time period. Even post-emancipation, there is no published research dealing with science education until we enter the twentieth century. Therefore, we will confine our discussion to recent times because all of the education research that has been conducted on this topic has been done in the last 20 years or so.

Unfortunately, there are only a small number of studies dealing with the interaction between Judaism and science education, especially when compared to the larger number of studies from a Christian perspective. This is due to a number of interrelated factors. Most science education studies dealing with the interaction between science and religion emanate from Western countries, where the dominant religious background (measured by population) is Christian. Therefore, by default, such studies are strongly flavored by a Christian perspective because the majority of school-age students come from a Christian background.

Hence, if we are to understand how Jewish attitudes towards science influence science education, we need to discuss the situation where Jews represent the majority and thus influence the school system. If we are talking geographically, we must focus on Israel, the only country with a Jewish majority. If we are talking systemically, we can also include the extensive private Jewish school systems that have developed in Western countries, most notably in the United States, which contains the world's second-largest Jewish population after Israel (DellaPergola 2010).[15]

Historically, for those groups of Jews who were not opposed to the integration of secular knowledge into the Jewish domain, the scientific issues that are most challenging to Judaism emanate from subjects touching upon Biblical creation including cosmology, geologic time, and most notably evolution. Not coincidentally, these issues have also had the greatest impact on the interaction between Judaism and science education, and it will be a discussion of these conflicts that will dominate this chapter.

However, before analyzing this conflict, we must discuss Jewish school systems, both within and without Israel, because their structure affects how controversy is dealt with. Indeed, school systems that serve the Jewish public are guided by a specific religious philosophy and in turn this philosophy guides the school system's interaction with secular subjects such as science, so it is important that we discuss their basic structures.

In Israel, the school system is divided between Hebrew and Arabic speakers. The two largest divisions among the Hebrew-speaking system are the Secular State and National Religious systems, respectively (Dodick et al. 2010). The Secular State system teaches a population of primarily secular and traditional students with many of its teachers coming from secular backgrounds. Its matriculation system is designed so that students have the possibility for continuing to higher academic studies. The only component of "religion" in this system is that the Bible is one of the core subjects (and is taught as part of the cultural background of Israel, and not from a religious perspective).

In contrast, the National Religious school system's philosophy is to integrate secular and (Orthodox-based) religious studies, making it possible for its students to pursue both secular studies at a university and religious studies at a Jewish seminary.[16] Many of the teachers are religious in orientation and have a professional background from a university, college, or seminary, depending on the subjects they teach.

A third system of schools in Israel, termed independent, focuses exclusively on the ultra-Orthodox population. For male students, the focus is on religious studies with little to no secular studies, including science learning; the ultimate goal is to prepare them for higher religious studies in a *Kollel*.[17] There is more flexibility in

[15] Of the approximately 13.5 million Jews in the world in 2010, Israel's Jewish population accounted for 42.5 %, and the United States' Jewish population accounted for 40 % of the total (DellaPergola 2010).

[16] Philosophically, Israeli-based National Religious schools are most similar to modern-Orthodox day schools outside of Israel.

[17] A Kollel is a Yeshiva learning program for married men.

the education of the female population who study secular subjects, including some subjects in science, out of a practical need to secure their families' financial futures. However, studies in the female ultra-Orthodox system do not usually lead to higher academic learning.

Outside Israel, concerns about inculcating youth in the practices and religious literature of Judaism spurred on the development of private "day" schools as well as afternoon schools, among the various denominations of Judaism in many Western countries. Day schools offer a "dual" curriculum, offering a secular program including science and math, as well as course offerings in traditional Jewish subjects such as the Bible, Hebrew, and Jewish history. With the exception of ultra-Orthodox schools, secular subjects share equal time with religious subjects including science (as part of a longer school day). This means that the majority of graduates from (k-12) Jewish-based schools are well displaced to tackle higher education, if they so desire. Afternoon schools offer various Jewish subjects and are attended by students who attend secular schools during the day.

In the case of the ultra-Orthodox within Israel (as well as sometimes outside of it) secular learning is (largely) omitted for its male population because religious subjects take priority; all the more so, scientific issues of creation are not taught because they challenge the belief in God's creation. In simple terms, they have strongly adopted a limiting philosophical approach. In the exact opposite way, scientific issues of creation pose much less difficulty to the liberal branches of Judaism, including the Conservative and Reform movements[18]; thus, they are taught as a usual part of the science curriculum.

However, the situation is different within modern-Orthodox schools. Although most are committed to a *Torah Umadda* (*Torah and Science*) philosophy, which believes in the integration of religion and secular learning, creation issues can test that resolve, creating conflict, so this group will be a prominent feature of our discussion.[19]

We begin this discussion by examining both the philosophical approaches of the schools and the teachers that work within these schools. We start here, because the roots of students' approaches to the conflict between science and religion are

[18]As we have seen, Swelitz (2006) extensively explored the historical responses of Conservative and Reform rabbis towards evolution. Using Rosenberg's (1988) system, their responses can be classified as falling within the parallel, explanatory, and complementary approaches. They do not adopt a limiting approach, in contrast to some among the Orthodox. Looking at the Conservative movement today, although they appear to have no official position, many of their Rabbis have adopted the idea of theistic evolution. Rabbi David Fine, who has authorized official responsa for the Conservative movement's committee on *Jewish Law and Standards*, expressed this idea as the following: "Did God create the world, or not? Is it God's handiwork? Many of the people who accept evolution, even many scientists, believe in what is called 'theistic evolution,' that is, that behind the billions of years of cosmic and biological evolution, there is room for belief in a creator, God, who set everything into motion, and who stands outside the universe as the cause and reason for life" (http://www.jewishvirtuallibrary.org/jsource/Judaism/jewsevolution.html).

[19]It will also be seen that almost all of the empirical studies that examine the relationship between Judaism and science education focus on the modern Orthodox, so this is another reason for this focus.

strongly shaped by how they understand science, its nature, and its relationship to religion and it is the schools and their teachers that most strongly shape this understanding. It might also be added that school choice both reflects and is influenced by informal sources – parents and religious authority.

Outside of Israel, Selya (2006) has completed the only major study concerning modern-Orthodox day high schools and their perspectives towards evolution. Surveying 12 such schools in the United States and Canada, she discovered four approaches for teaching evolution (that largely match the philosophical approaches of Rosenberg 1988). These approaches include curricula where evolution is taught in class without a religious discussion, whatsoever (parallel approach). In other schools, teachers teach evolution with the aid of a religious teacher or rabbi who interprets the creation story either in a nonliteral way (explanatory) or from an intelligent design perspective (limiting/complementary). A third perspective is assigning students' readings on evolution without discussing them in the class (sometimes because it was part of the mandated final year examinations) (parallel approach). Finally, evolution was not taught at all (limiting approach).

There were no instances of substituting a creation-science curriculum to replace the standard biology texts or of school administrators removing the chapter from the science textbooks as reported by Wolowesky (1997) and Landa (1991); the latter, as we have seen, was recommended by Rabbi Feinstein (1982), one of the most important ultra-Orthodox rabbinic decisors of Jewish Law in the twentieth century.

Selya (2006) showed that ten of the schools surveyed taught evolution in the classroom and that eight of them suggested that this scientific theory was religiously compatible. Not surprisingly, schools that separated the sexes, a sign that a school is more religiously oriented, either did not teach evolutionary theory or criticized it as being incompatible with religion.

As part of this research, Selya (2006) also interviewed teachers and administrators at five of the schools, all of which were coeducational, with strong college preparatory programs, and which both teach evolution and stress its compatibility with religion. All of these schools share certain philosophical and/or historical roots, including a commitment to the Torah Umadda philosophy. Three of the schools were founded by prominent rabbinic figures, one of whom was Rabbi Joseph Soloveitchik who is considered to be the unofficial leader of modern Orthodoxy during much of the twentieth century.

In sum, Selya's (2006) survey seems to show that evolution is being taught in some form in the majority of modern-Orthodox day schools. However, caution should be applied to her small-scale survey, as in the United States (alone), there are 86 schools, accounting for more than 27,000 students, classified as being modern Orthodox (Schick 2009).

Although further studies, like Selya (2006), are needed, Schick's (2009) demographic studies of Jewish day schools in the United States indirectly may point to a trend of increased resistance to science subjects that are considered to be controversial, such as evolution, among Orthodox Jews. Although the numbers of students increased in modern-Orthodox schools from 1998 to 2009, the number of students that were learning in ultra-Orthodox schools increased at a much faster rate. This is due to the much higher birth rate of the ultra-Orthodox which is more than twice that of the modern Orthodox.[20] And if the school is based on an ultra-Orthodox philosophy, it is more than probable that evolution was not part of their science curriculum.

Moreover, among the modern Orthodox, there are factors that indicate some of its adherents are moving towards ultra-Orthodoxy, a process Waxman (1998) labeled "Haredization" (based on the Hebrew term for ultra-Orthodox). This phenomenon has been documented over a 20-year period by a collection of historians and social scientists.[21] A small (educational) indicator of this shift is the fact that *Torah Umesorah*, the National Society of Hebrew Day Schools, an umbrella organization that provides educational materials to Orthodox schools is increasingly distancing itself from coeducational institutions, which is one indication of increased religious practice.

There are many reasons for this shift, but the most important factor for science education is the increasing number of ultra-Orthodox Jewish teachers who are now teaching in modern-Orthodox schools (Heilman 2005; Helmreich and Shinnar 1998). As most in the modern-Orthodox world have avoided teaching, due to its lower remuneration and lack of prestige in comparison to many other professions, the modern-Orthodox school system has turned to ultra-Orthodox teachers (especially for its Jewish studies departments), which in turn affects the philosophy of the schools and their students (Heilman 2005; Helmreich and Shinnar 1998).[22]

Inside Israel, research has focused on teachers within the National Religious system, rather than the school as the unit of analysis. Such research has importance because teachers, like the schools they teach in, are one of the most critical factors influencing the balance between Judaism and science education. We say this because, as Rutledge and Mitchell (2002) have noted, teachers' attitudes and views about a subject directly influence their instructional decisions on how to teach a subject. Their research shows that teachers' background in the philosophy of science and knowledge of evolution influences their acceptance of and willingness to teach evolution. One would assume that a similar relationship exists for scientific subjects that are considered to be challenging to Judaism.

In a similar vein, Dodick et al. (2010) surveyed teachers in the National Religious school system to understand their philosophical approaches towards the interaction between Judaism and science. In total, 56 teachers were extensively surveyed using a Likert-type questionnaire developed for this research, which surveyed the

[20] As Schick (2009, p. 12) notes, "In the 1998 census, I reported that there were 3.26 children in the families of Modern Orthodox eighth graders as compared to 6.57 and 7.92 children respectively in yeshiva-world and Hassidic families."

[21] Friedman (1991), Heilman (2005), Helmreich and Shinnar (1988), Liebman (1988), Soloveitchik (1994), and Waxman (1998)

[22] Heilman (2005, p. 265), based on a personal communication with Schick, who has completed a series of demographic studies on Jewish day schools in the United States, claims that "nearly two-thirds of today's Judaica teachers in day schools come from the haredi [ultra-Orthodox] world."

teachers' approaches to the nature of science in general, geologic time, cosmology, and evolution. Eleven of the teachers were also randomly selected for interviews.

Additionally, 15 (Orthodox) scientists from the major branches of science were surveyed with the same instruments to both contrast their views with the teachers, as well as to better understand their coping strategies when confronted by scientific topics that challenge their beliefs. In the cases of both teachers and scientists, their philosophical approaches were classified according to Rosenberg's (1988) typology.

Results indicated that no single philosophical approach earned an overwhelming support from the teachers or scientists. Instead, the teachers and scientists related separately to each source of possible conflict, such as evolution, in accordance with the philosophical approach that appears to be the most fruitful for resolving a specific conflict.

The teachers did differ from the scientists in their stronger preference towards philosophical approaches which help them better integrate the domains of science and religion. Thus, the teachers favored the explanatory and complementary approaches, whereas the scientists most preferred the explanatory and parallel approaches. Possibly, the teachers favor an integrative approach because they prefer answers that avoid delivering an open, contradictory message to their students and through them to their parents and school administrations.

With regard to the scientists, tenured in academia as they are, they have the security to research issues that are both open and controversial. This also explains why some scientists adopted a conflict approach (their third most favored approach), as they acknowledge that some problems are open and (currently) unsolvable, while concurrently accepting the inherent contradictions in this situation. Unlike the teachers, however, none of the scientists adopted a limiting approach, as they saw no reason to constrain the science they practiced. This last result is important because it counters critiques (such as Nussbaum 2002) that highlight Orthodox Jewish scientists who are charged as being antiscience towards issues such as evolution. In other words, it supports the idea that there truly is a spectrum among scientists who are also Orthodox in practice.

On specific issues of conflict, geologic time was much less controversial for teachers than either cosmology or evolution. With this issue, the teachers referred directly to religious sources which implied that were either multiple creations of older worlds or that each day of creation was much longer than 24 h. The teachers' flexibility was based on the openness of classic Biblical commentators on this issue. Such commentators provide sanction for interpreting the Bible, but particularly with the age of the Earth, this sanction has greater impact because there is no direct reference within the Bible to the traditional Jewish calculation of the age of the Earth.[23]

[23] Some Jews believe that the Earth is currently 5,722 years in age (in 2012 CE). In fact, this figure, which has also influenced Christian fundamentalists' understanding of the Earth's chronology, has been calculated based on the interpolation of ages of Biblical personalities mentioned in Genesis starting from Adam's creation on the sixth day of creation (this calculation can be found in the book Seder Olam Rabbah, ascribed to the second century CE Rabbi Yossi ben Halafta). In turn, this calculation leaves the possibility of interpreting the first 6 days of creation before man's appearance as being much longer than six 24-h days (Dodick et al. 2010).

Thus, it becomes easier for teachers, who are familiar with such commentaries, to accept geologic time.

In contrast to the age of the Earth issue, approximately half of the teachers saw some conflict between the theory of evolution and Biblical creation because its random nature contradicts the belief in creation directed by the "hand of God"; moreover, some cited the fact that it also conflicted with their belief in man as the "crown of creation." It should be noted that some scientists also felt such conflicts but they were willing to live with them.

At the university level, inside Israel there is one comprehensive university that integrates "Jewish heritage" and secular studies – Bar Ilan. Its Faculty of Life Sciences provides courses in evolution, as well as integrates this subject within its various course offerings. In the United States there is a strong dichotomy between the approach of Yeshiva University, whose very motto incorporates Torah Umadda and other Orthodox institutions of higher learning. Indeed, university President Richard Joel (2003, p. 3) in *YU Review* claimed that a "moral underpinning" for science at his university was "to marry the wisdom of faith with the need to explore our universe's mysteries."

In the same issue of this magazine, biology Professor Carl Feit noted that he saw no contradiction between Judaism and biology while arguing that the evolutionary ideas could actually be used to serve to strengthen one's faith (Eisenberg 2003). In a conversation with Selya (2006), Feit notes that he includes evolution as part of his course syllabus while adding readings from the philosophy of science, philosophers, and Jewish Biblical commentaries.

In contrast, Touro College, which was founded in 1971 to "enrich the Jewish heritage" and serves a largely ultra-Orthodox student population, takes an unsympathetic view towards evolutionary biology. As a psychology instructor at Touro, Nussbaum (2002) elicited great opposition from his students for his support of teaching evolution. Moreover, the science professors at this institution routinely criticized evolution while teaching creationism.

Schools and teachers set the curricular standards which ultimately affect students; thus, to complete our understanding of the interaction of Judaism and science education, we will be looking at studies dealing with students. Of these studies, those concerned with k-12 students have emanated from Israel.

Ruach and colleagues (1996) performed a comparative survey study with 185 students evenly distributed among National Religious and secular state students in middle (grade 9) and high schools (grades 11 and 12). The middle grade high school cohort had not yet learned evolution in contrast with the high school cohort. After learning evolution the high school students from both school systems substantially increased their knowledge of evolution. However, their attitudes towards this subject were in opposition. The students of the National Religious stream increased their acceptance of creationism, whereas the Secular State students increased their acceptance of evolution.

More recently, Allouch (2010) also examined the attitudes of middle (grade 9) and high school (grade 10) students from the National Religious school system. This sample consisted of 369 students, 79 of them who were studying evolution.

The design of this study relied upon a Likert-type questionnaire (37 statements) that focused on evolution but also included a small number of statements dealing with the nature of science, cosmology, and geologic time.

Similar to the teachers sampled by Dodick et al. (2010), prior to studying the unit, the students were more accepting of a nonliteral reading of the Biblical creation time line, as well as the "Big-Bang" explanation of cosmology, than they were of evolution. Post-program, a similar result occurred. Positive attitudes connected to time and cosmology improved significantly (despite the fact that this was not the focus of their learning), whereas for the most part, their attitudes towards evolution remained at a no agreement level, even after learning the unit.

Again, similar to the teacher sample of Dodick et al. (2010), the students displayed a variety of philosophical approaches towards different issues, although based on Allouch's (2010) results, the students seem to be more conservative (religiously) in their attitudes towards these creation issues than their teachers. Nonetheless, like the student sample of Ruach and colleagues (1996), the students significantly improved their understanding of some of the issues connected to evolution (most notably Natural Selection). In sum, the results from both Ruach and colleagues (1996) and Allouch (2010) parallel the findings of Lawson and Worsnop (1992) with US students who found that a change in knowledge (about evolution) was not necessarily associated with a change in (religious) attitudes.

It must be remembered that in the case of the students in the religious stream they are exposed to far more religious learning than learning about evolution and this would likely affect their attitudes. In fact, Allouch's (2010) study showed that one of the external factors which influenced the students developing a greater acceptance of evolution was the number of curricular hours devoted to this subject. Moreover, for this group of students, part of their school success is measured by how they understand and apply their religious training; in their community, such application is seen as having great value.

It would be expected that difficulties with issues connected to evolution would also affect Orthodox students who attend university. This was addressed by Nussbaum's (2006) survey study among a sample of 176 Orthodox Jewish students at a single public college in New York City. This study provides for some rather disheartening conclusions about the state of science education among Orthodox Jews. The responses received to questionnaire probes dealing with evolution, such as "Evolution correctly explains the origin of life," and geologic time, such as "What is the age of the universe?" indicate that the subjects tended towards creationist or intelligent design perspectives. Moreover, Nussbaum's (2006) data also seems to show that the students that were science majors were even less accepting of mainstream science than those who were not science majors.

However, we should be careful in viewing this study as a summary of the attitudes of all Orthodox university students in the United States because of its methodological problems. The sample consisted of 176 subjects, surveyed at one university, with little demographic data collected concerning the subjects (such as school background). Moreover, the wording of some of the probes is to be questioned. For example, "Evolution correctly explains the origin of life" with a binary answer format would necessarily exclude theistic-evolutionist approaches.[24] Finally, interviews were not held with any of the subjects, which would have more deeply probed their philosophical approaches. Still, given the fact that Jewish Orthodox society is moving towards more ultra-Orthodox views, the results of this study do seem to reflect such societal change.

54.6 Conclusion

In looking at the relationship between Jews and science education in our modern world, we see mostly positive trends. This might be a surprising conclusion in light of what was written in the previous section; however, these trends are supported by historical, sociological, and demographic factors.

Those denominations in Judaism who have difficulties with science learning, most notably the ultra-Orthodox, and some of the modern Orthodox represent a demographic minority; in total, the Orthodox in the United States represent no more than 13 % of the Jewish population (Ament 2005).[25] Thus, over all, it is possible to say that among Jews connected to their Judaism, methods have developed, historically, in order to deal with the conflicts posed by modern science.

Indeed, it is possible to say that Reform and Conservative Jews have few or no problems with modern science education; this is why their role is downplayed in the previous section which discussed science education. There is simply no evidence that what is considered to be a challenge by some in the Orthodox world is considered to be the same in schools belonging to the Conservative and Reform movements. Thus, it would seem that science is taught in their day schools in the same ways as it is taught in the public school systems.

Moreover, it must be remembered that even among those Orthodox Jews that are challenged by scientific findings, it is *not all of science* that is considered to be a challenge, but specific sciences that touch upon issues of Biblical creation. This is the reason why our review of the science education research has not focused on science in general, but specific issues, such as evolution which are considered to be threatening to the religious sensibilities of its followers.[26] In fact, most modern-Orthodox day schools in the United States are known for the high quality of their secular studies including science education.

[24] Theistic evolution claims that God's method of creation was to design a universe in which various systems would naturally evolve.

[25] Among those of Jewish origin, who see their faith as an integral part of their lives, the Orthodox represent a higher percentage than stated; still the Orthodox do represent a minority when compared to the number of Jews belonging to movements such as the Conservative and Reform.

[26] Simply put, there are no science education studies that have examined Jewish attitudes/approaches towards science as a whole. All of the known studies focus on one or a few specific subjects such as geologic time, cosmology, and especially evolution, which (supposedly) are threats to the Jewish worldview.

It should also be remembered that although there is large demographic growth within the worldwide Orthodox sector and, especially among the ultra-Orthodox, their philosophical approach to challenging issues of science does not carry over to other school systems, largely because of their isolationist approach. Such physical and social isolation was historically adopted by the ultra-Orthodox to limit contact with and infiltration of foreign ideas (which included ideas promulgated by other Jewish denominations) that do not fit into their religious worldview (Liebman 1983; Heilman 2005). In terms of educational policy, this has meant that in Israel the ultra-Orthodox and their independent school system do not affect curricular policy within the Secular and National Religious school systems.

Outside of Israel, a similar isolationist approach has been adopted by both ultra-Orthodox and their school system·towards other Jewish denominations. This is very different from the situation in the Christian world, most notably in the United States, where fundamentalists have sometimes successfully gained elected control of school boards leading to critiquing evolutionary theory (http://www.discovery.org/a/9851) and even removing the teaching of (macro)-evolution, as occurred in Kansas in 1999 (http://www.agiweb.org/gap/legis106/evolution.html). Nonetheless, it was noted previously that there is a more subtle influence from the "Haredization" of the modern-Orthodox education system due to the increased number of ultra-Orthodox religious teachers entering that system (Heilman 2005; Helmreich and Shinnar 1998). Such sociological factors could have a stronger influence than the actual content of the science education curriculum. However, there are no studies indicating how that is specifically affecting science education.

Thus, there are challenges to the science education enterprise in the Jewish world. Many Orthodox elements see the challenges posed by the secular philosophy of science as being more of an educational threat than science per se; however, the larger issue of how various groups see secular learning in general, especially secular higher learning, is an indication of how they see science as well.

Certainly Reform and Conservative Jews attend university with no limits to what they study. So too, the modern Orthodox also attend college, although as Soloveitchik already noted in 1994 (p. 64) with "somewhat less enthusiasm" than in previous years. The ultra-Orthodox, who are known historically for their opposition to higher education, are divided in their approaches (Soloveitchik 1994). In the United States, there is recognition of the (economic) utility or even necessity of a degree, and various arrangements have been made to enable ultra-Orthodox to receive such degrees, albeit with (societal) restrictions on what is studied. In Israel, the opposition to higher education is much stronger, although due to the very weak economic condition of most ultra-Orthodox, colleges specifically designated for training this populace have been rapidly developing (Lamm 2010).

Certainly, economic incentive is a path towards increased involvement in secular learning, in general and science, specifically. And in fact it may be the only path that the ultra-Orthodox will accept in the near future. But some in the modern-Orthodox world desire a synthesis represented by a Torah Umadda philosophy because they see the inherent value in secular knowledge and learning. How is this possible?

One possible synthesis is provided by Rabbi Norman Lamm (2010) who has articulated a series of models, based on the philosophies of important rabbinic thinkers of the past. Obviously, to understand and then adopt any of these models requires a deep investment in studying the religious sources and their philosophy, as well as science and its nature.

Indeed, such an investment in learning fits well with one of the recommendations made by the Orthodox scientists interviewed by Dodick et al. (2010) when they were asked how Orthodox Jewish teachers might cope with what they felt were controversial scientific topics, such as geologic time or evolution.

It must be remembered that the Orthodox (high school) science teachers in Israel are adequately educated in science, and many received training in a religious seminary. However, such education rarely deals, systematically, with the possible philosophical conflicts between science and religion. Therefore, the scientists argued that it was important to improve the teachers' understanding of both scientific and Jewish sources that will permit them to settle their internal conflict while providing them with the tools to teach such conflicting subjects with confidence.

Indeed, Lamm (2010) shows how both in the past and today confusion has been created by the lack of understanding about the Jewish philosophical approaches to all secular learning. Moreover, he even shows how traditions that normally would never be considered to be in a Torah Umadda world (specifically, his "Hassidic" model) can be designed to create a synthesis between Judaism and science.

Although knowledge is a primary tool, the Jewish education world is a hierarchical system, especially among the Orthodox, in which teachers must answer to a series of authorities including rabbis, administrators, and parents. Thus, teachers feel more comfortable if they have experts upon whom they can rely (scientists, rabbis, and texts) which allow them to teach scientific issues that are considered to be challenging. Unlike academic scientists, teachers have less freedom to express controversial notions in science or religion, nor are they as well trained in science. Thus, their desire to have a support network of rabbis and scientists who can deal with this conflict is understandable.

This approach also has support from a previous science education study in which Colburn and Henriques (2006) interviewed a group of eight Christian clergymen for whom evolution and religion were compatible and who also believed that Scripture was not meant to be understood, literally. Based on their findings, Colburn and Henriques (2006) suggested that the science education community might find in the educated clergy an articulate ally in helping citizens better understand contentious issues surrounding science and religion.

Because they are not constrained by authority, the scientists interviewed by Dodick et al. (2010, p. 1541) instead recommended that teachers focus on two interrelated issues connected to *application* and *education* to deal with possible conflicts. *Application* is connected to how scientists see religion and science possibly integrating; by *education*, the scientists were referring to how they would like to see the conflict being taught.

Concerning application, the scientists showed a divergence in their choice of philosophies, with the two dominant approaches being the parallel and explanatory approaches. These philosophical differences fit well with application in that some of the scientists favor an approach which emphasizes the points integrating science and religion (comparable to the explanatory approach); in contrast, some of the scientists would rather avoid using science in building religious understanding (comparable to the parallel approach), as the use of science in this way resonates with a fundamentalism with which they don't identify.

Regarding "education," some of the scientists which Dodick et al. (2010) surveyed emphasized critical understanding of learning materials dealing with this conflict. Moreover, some of the scientists desired to see issues of conflict being taught pluralistically by showing students the different philosophical approaches in science and religion that deal with this conflict. This suggestion connects nicely with scientists holding either an explanatory or complementary philosophy as they connect between the domains of science and religion. As these approaches are sympathetic to the preconceived desires of the teachers to also bridge the gap between science and religion, this pluralistic approach might be easier for teachers to implement.

Such integration also has support from the literature (Jackson et al. 1995; Shipman et al. 2002; Smith and Siegel 1993). For example, Jackson and colleagues (1995, p. 605) noted that the current treatment of controversial scientific topics in schools, such as the evolution of humans, is independent of any other way (including via religion) that a student or teacher might seek answers to such topics. They argue that "Scientists and science teachers cannot continue to see themselves as participating in an epic struggle to eradicate mystical superstition and hasten the irresistible ascendancy of materialistic naturalism."

What is missing from this discussion is empirical research. As was seen, much of the education research concerning Judaism and science have focused on the attitudes of students, teachers, and scientists, mostly within the Orthodox world, towards issues of controversy, such as evolution. Future work needs to be more expansive in widening its perspective to other denominations within Judaism[27] and other branches of the sciences. Moreover, for those issues that challenge religious sensitivities much more research must be invested in testing different models of instruction, based on the philosophical approaches that have developed in Judaism. Selya's (2006) work shows that in the modern-Orthodox world, schools have already adopted a number of instructional philosophies for dealing with such controversy; however, research has not yet been conducted to determine their effectiveness. If the goal of Jewish educators is to attain some sort of balance between the world of science and Judaism, then these next steps are crucial.

---

[27] We could reference only one paper concerning the interaction between science education and Judaism from a non-Orthodox perspective. Authored by Rabbi Laurie Green (2012), who comes from the Reform movement, this policy paper argued for greater integration (similar to the explanatory approach) between religion and science studies for students belonging to the Reform movement.

References

Allouch, M. (2010). *The influence of teaching evolution on the attitudes of students in Yeshiva high schools in Israel*. Unpublished M.Sc. thesis, Jerusalem, Israel, The Hebrew University of Jerusalem.

Ament, J. (2005). American Jewish religious denominations. United Jewish Communities, Report 20, 1–19.

Aviezer, N. (1990). *In the beginning: Biblical creation and science*, Hoboken, NJ, Ktav Publishing House.

Aviezer, N. (2002). *Fossils and faith: Understanding Torah and science*, Hoboken, NJ, Ktav Publishing House.

Barbour, I. (1997). *Religion and science: Historical and contemporary issues* (A revised and expanded edition of *Religion in an age of science*), San Francisco, Harper-Collins.

Bar-Hiyya, A. (1968). *Higayon Hanefesh* (*Meditation of the Soul*), (G. Wigoder translator). New York, NY, Schocken Books.

Ben Adret, S. (2000). *Responsa* (*Sheelot ve-Teshuvot*) *and Minhat Kenaot*, Jerusalem, Israel.

Benamozegh, E. (1862). *Em la-Mikra, Matrix of Scripture,* (Vol. 1), Livorno, Italy, Benamozegh.

Benamozegh, E. (1877). *Teologia-Dogmatica ed Apologetica* (Vol. 1), Livorno, Italy, Vigo.

Ben-David, J. (1991) *Scientific growth: Essays on the social organization and ethos of science,* Berkeley, CA, University of California Press.

Ben Jacob, B. (1780). *Euclid* (translated into Hebrew), The Haag, Netherlands, L. Zosmans.

Ben Sasson, H. H., Jospe, R. & Schwartz, D. (2007). The Maimonidean Controversy. In M. Berenbaum & F. Skolnik (Eds.), Encyclopedia Judaica, 2nd Edition (pp. 745–754), Farmington Hills, MI, Gale Cengage Learning. Updated Online version: (http://www.jewishvirtuallibrary.org/jsource/judaica/ejud_0002_0013_0_13046.html)

Ben Simha, N. (1990), *Likutei Moharan*, (N. Sternholz editor), Jerusalem, Israel.

Bronowski, J. (1973). *The ascent of man*, London, British Broadcasting Corporation.

Brooke, J. H. (1998). *Science and religion: Some historical perspectives,* Cambridge, Cambridge University Press.

Brown, J. (2008-9). Rabbi Reuven Landau and the Jewish reaction to Copernican thought in nineteenth century Europe, *Torah UMadda Journal*, **15**, 112–142.

Cherry, S. (2006). Crisis management via Biblical interpretation: Fundamentalism, modern Orthodoxy and Genesis. In G. Cantor & M. Swelitz (Eds.), *Jewish tradition and the challenge of Darwinism* (pp. 166–187), Chicago, University of Chicago Press.

Colburn, A., & Henriques, L. (2006). Clergy views on evolution, creationism, science, and religion, *Journal of Research in Science Teaching*, **43** (4), 419–442.

DellaPergola, S. (2010). *World Jewish population 2010* (Number 2-2010), Storrs, CT, World Jewish Data Bank, The University of Connecticut.

Dodick, J., Dayan, A., & Orion, N. (2010). Philosophical approaches of religious Jewish science teachers towards the teaching of "controversial" topics in science, *International Journal of Science Education*, **32** (11), 1521–1548.

Efron, N.J. (2007). *Judaism and science: A historical introduction*, Westport, CO, Greenwood Press.

Eisenberg, N. (2003). Torah and Big Bang. *Yeshiva University Review*, Summer 2003, 13–15.

Ettinger, Y. (2007). Ultra-Orthodox Jewish group declares war on computers: Gerrer Hasidic sect debates use of internet; fears it will promote slander, conflict. May 20, 2007. http://www.haaretz.com/news/ultra-orthodox-jewish-group-declares-war-on-computers-1.221022.

Feinstein, M. (1982). *Igrot Moshe, Yoreh Deah* (vol. 3, Responsum 73). New York, NY, Noble Press.

Feldman, T. (1986). *Rav A. I. Kook, selected letters* (translated from Hebrew), Ma'aleh Adumim, Israel, Ma'aliot Publications of Yeshivat Birkat Moshe.

Freedman, H., & Simon, M. (1939). *Midrash Rabbah* (Standard Vilna edition, on the Pentateuch) London, Soncino Press.

Friedman, M. (1991). *The Haredi ultra-Orthodox society: Sources trends and processes*, Jerusalem, Israel, The Jerusalem Institute for Israel Studies.

Gould, S.J. (1997). Nonoverlapping magisterial, *Natural History*, **106** (March 1997), 16–22.
Gould, S.J. (1998). *Leonardo's mountain of clams and the diet of worms*, New York, NY, Harmony Books.
Gould, S.J. (1999). *Rocks of ages: Science and religion in the fullness of life*, New York, NY, Ballantine Publishing.
Green, L. (2012). The case for science education in our religious schools, *CCAR Journal: The Reform Jewish Quarterly,* Winter 2012, 186–193.
Grynbaum, M.M. (2012). Ultra-Orthodox Jews Rally to Discuss Risks of Internet. New York Times, May 21, 2012, A17. http://www.nytimes.com/2012/05/21/nyregion/ultra-orthodox-jews-hold-rally-on-internet-at-citi-field.html
Halevi, J. (1998). *Kuzari: In defense of the despised faith* (D. Korobkin translation), Northvale, NJ, Jason Aronson.
Haught, J. F. (1995). *Science and religion: From conflict to conversation,* New York, Paulist Press.
Heilman, S.C. (2005). How did fundamentalism manage to infiltrate modern Orthodoxy? (Marshall Sklare Memorial, Lecture, 35th Annual Conference of the Association for Jewish Studies, Boston, December 21, 2003), *Contemporary Jewry*, **25** (1), 273–278.
Helmreich, W.B. & Shinnar, R. (1988). Modern Orthodoxy in America: Possibilities for a Movement Under Siege. *Jerusalem Letters/Viewpoints*, Jerusalem Center for Public Affairs.
Helmreich, W. B. & Shinnar, R. (1998). Modern Orthodoxy in America: Possibilities for a movement under siege, *Jerusalem Letters/Viewpoints*, Jerusalem Center for Public Affairs, http://jcpa.org/cjc/jl-383-helmreich.htm.
Hirsch, S. R. (1992). The educational value of Judaism. In E. Bondi and D. Bechhofer (Eds.), *The Collected Writings of Rabbi Samson Raphael Hirsch* (vol. 7), New York, NY, Feldheim.
Ibn Paquda, B. (1970), *Duties of the Heart*, (M. Hyamson translation), Jerusalem, Israel, Feldheim Publishers.
Jackson, D. F., Doster, E. C., Meadows, L., & Wood, T. (1995). Hearts and minds in the science classroom: The education of a confirmed evolutionist, *Journal of Research in Science Teaching,* **32**, 585–612.
Joel, R.M. (2003). A journey worth taking. *Yeshiva University Review*, Summer 2003, 3.
Karo, J.E. (2009), *Shulhan Arukh (Code of Jewish Law)*, D. Ikhnold ed., Tel Aviv, Israel.
Kehati, P. (1991). *Mishna* (P. Kehati edition), Jerusalem, Israel, Feldheim Press.
Kook, A. I. (1938). *Orot Ha-Kodesh* (Part II), Jerusalem, Israel, Mercaz Press.
Lamm, N. (2010), *Torah Umadda* (3rd Edition). Milford, CT, Maggid Books.
Landa, J. (1991). *Torah and science*, Hoboken, N.J., Ktav Publishing House.
Lawson, A. E., & Worsnop, W. W., (1992). Learning about evolution and rejecting belief in special creation: Effects of reflective reasoning skill, prior knowledge, prior belief and religious commitment, *Journal of Research in Science Teaching*, **29** (2), 143–166.
Liebman, C.S. (1983). Extremism as a religious norm, *Journal for the Scientific Study of Religion*, **22** (1), 75–86.
Liebman, C.S. (1988). Deceptive Images: Towards a redefinition of American Judaism, New Brunswick, Transaction Books.
Leibowitz, Y. (1985). *Discussions on science and values*, Tel Aviv, Israel, Defense Ministry of Israel.
Lipsett, S. M. & Raab, E. (1995). *Jews and the new American scene*, Cambridge, MA, Harvard University Press.
Maimon, M. (1956). *Guide for the perplexed* (2nd edition) (M. Friedlander translation), New York, NY, Dover Publications.
Maimon, M. (1961). *Introduction to the Mishna, Avot, in Hakdama Le-Perush Hamishna*, (M. D. Rabinovitch, editor), Jerusalem, Israel, Mosad Rav Kook.
Maimon, M. (1987). *Mishne Torah*, (Rambam La'am edition, M. D. Rabinovitch editor). Jerusalem, Israel, Mosad Rav Kook.
McGrath, A. E. (1999). *Science and religion: An introduction*. Oxford: Blackwell.
Nieto, D. (1993). *Second Kuzari*, Brooklyn, NY.
Numbers, R. (1998). *Darwinism comes to America*, Cambridge, MA, Harvard University Press.
Nussbaum A. (2002). Creationism and geocentrism among Orthodox Jewish scientists. *National Center for Science Education Reports*, **22** (1-2), 38–43.

Nussbaum, A. (2006). Orthodox Jews & science: An empirical study of their attitudes toward evolution, the fossil record, and modern geology, *Skeptic*, **12** (3), 29–35.

Poupko, Y. (1990). *Midrash Tanhuma* (Metzuda edition), Lakewood, NJ, Metsudah Publications.

Preisler, Z. & Havlin, R. (Eds.). (1998). *Babylonian Talmud* (standard Vilna edition), Jerusalem, Israel, Ketuvim Publishers.

Robinson, I. (1983). Hayyim Selig Slominski and the diffusion of science among Russian Jewry in the nineteenth century, In Y.M. Rabkin & I. Robinson (Eds.), *The interaction of scientific and Jewish cultures in modern times* (pp. 49–64), New York, NY, E. Mellen Press.

Robinson, I., (1989). Kabbalah and science in Sefer Habrit: A modernization strategy for Orthodox Jews, *Modern Judaism,* **9**, 275–288.

Robinson, I. (2006). "Practically I am a fundamentalist": Twentieth-Century Orthodox Jews contend with evolution and its implications. In G. Cantor & M. Swelitz (Eds.), *Jewish tradition and the challenge of Darwinism* (pp. 71–88), Chicago, University of Chicago Press.

Rosenberg, S. (1988). *Science and religion in the new Jewish philosophy,* Jerusalem, Israel, The Department of Torah Culture, Israeli Ministry of Education and Culture.

Rosenbloom, N. (1996). Cosmological and astronomical discussions in the Book of the Covenant (translated from Hebrew). *Proceedings of the American Academy for Jewish Research*, **62**, 1–36.

Rozenboim, D. (Ed.). (2010). *Jerusalem Talmud.* Jerusalem, Israel.

Ruach, M., Gross, C., Peled, L., & Tamir, P. (1996). Research on the topic: Attitudes of secular and religious public high school students towards the teaching of evolution. *Journal of (Israeli) Biology Teachers*, 147, 90–96.

Rutledge, M. L., & Mitchell, M. A. (2002). Knowledge structure, acceptance and teaching of evolution, *The American Biology Teacher*, **64** (1), 21–28.

Sacks, J. (2011). *The great partnership, God, science and the search for meaning*, London, Hodder & Stoughton.

Schick, M. (2009). *A Census of Jewish Day Schools in the United States 2008-2009*, Jerusalem, Israel, The Avi Chai Foundation.

Schneerson, M. M. (1972). *Holy letters* (Section 1), Chabad Village, Israel, Organization of Young Chabad in Israel.

Schroeder, G. L. (1998). *The science of God: The convergence of scientific and biblical wisdom,* New York, Broadway Books.

Selya, R. (2006). Tora and madda? Evolution in the Jewish educational context, In G. Cantor & M. Swelitz (Eds.), *Jewish tradition and the challenge of Darwinism* (pp. 108–207), Chicago, IL, University of Chicago Press.

Shavel, C (ed.). (1963). *Kitvei Haramban* (vol. 1: Letters), Jerusalem, Israel, Mossad HaRav Kook.

Shilat, Y., ed. (1995) *Maimonides epistle on the resurrection of the dead*, Igrot HaRambam (Maimonides Letters, Vol. 1), Jerusalem.

Shipman, H. L., Brickhouse, N. W., Dagher, Z., & Letts, W. J. (2002). Changes in student views of religion and science in a college astronomy course. *Science Education*, **4**, 526–547. 2009

Shuchat, R. (1996). The Vilna Gaon and secular studies (Hebrew), *Badad*, **2**, 89–106.

Shuchat, R. (2005). Attitudes towards cosmogony and evolution among rabbinic thinkers in the nineteenth and twentieth Centuries: The resurgence of the doctrine of the sabbatical Years, *Torah UMadda Journal*, 13, 15–49.

Shuchat, R (2008). What is Jewish philosophy? From definitions to historical context, *Studia Hebraica*, **7**, 346–354.

Shuchat, R. (2008-2009). R. Isaac Halevi Herzog's attitude to scientific evolution, *Torah UMadda Journal*, **15**, 143–171.

Shuchat, R. (2009) Did the Kabbalists believe that the world was created in six days? A discussion of the duration of creation in the writings of the Kabbalists in the medieval and modern Period (Hebrew article), *BaDaD* **22**, 75–95.

Smith, M. U., & Siegel, H. (1993). Comment on "The rejection of nonscientific beliefs about life", *Journal of Research in Science Teaching*, **30**, 599–602.

Soloveitchik, H. (1994). Rupture and reconstruction: The transformation of contemporary Orthodoxy, *Tradition*, **28** (4), 64–130.

Stampfer, S. (2005). *The coming into being of the Lithuanian Yeshiva*, Jerusalem, Israel, Zalman Shazar Centre for Israeli History.
Steinberg, A. (2003). *Encyclopedia of Jewish medical ethics*, Jerusalem, Israel, Feldheim Publishers.
Swelitz, M. (2006). Responses to evolution by Reform, Conservative and Reconstructionist Rabbis in Twentieth-Century America, In G. Cantor & M. Swelitz (Eds.), *Jewish tradition and the challenge of Darwinism* (pp. 47–70), Chicago, IL, University of Chicago Press.
Touati, C. & Goldstein, B.R. (2007). Levi Ben Gershon (Gersonides), In M. Berenbaum and F. Skolnik (Eds.), Encyclopedia Judaica, 2nd Edition (pp. 698–702), Farmington Hills, MI, Gale Cengage Learning. www.bjeindy.org/encyclopedia_judaica_online
Waxman, C. I. (1998). The Haredization of American Orthodox Jewry, Jerusalem Letter/ Viewpoints, *Jerusalem Center for Public Affairs*, http://jcpa.org/cjc/jl-376-waxman.htm.
Wolfson, H. A. (1929). *Crescas' critique of Aristotle: Problems of Aristotle's physics in Jewish and Arabic philosophy*, Cambridge, MA, Harvard University Press.
Wolfson, H. A. (1976). The Philosophy of the Kalam. Cambridge, MA, Harvard University Press.
Wolowesky (1997). Teaching evolution in Yeshiva high school. *Ten Da'at: A Journal of Jewish Education*, **10** (1), 33–39. http://www.daat.ac.il/daat/english/education/evolution-1.htm
Zevin, S. (1963). *Talmudic encyclopedia*, Jerusalem, Israel, Yad HaRav Herzog.

Jeff Dodick is lecturer in the Department of Science Teaching at the Hebrew University of Jerusalem. His training is in biology and earth science (B.Sc., the University of Toronto), with specializations in paleontology and evolutionary biology (M.Sc., University of Toronto). His doctorate is in Science Education from the Weizmann Institute of Science in Rehovot, Israel. He did postdoctorate research at McGill University in Montreal, QC, Canada, and also served as a research assistant professor in the Faculty of Education and Social Policy at Northwestern University, in Evanston Illinois, United States. Broadly stated, his research interests focus on the reasoning processes that are used in sciences which investigate phenomena constrained by long spans of time, such as evolutionary biology and geology; such scientific disciplines fall under the general rubric of "historical-based sciences." In the past he has published research papers dealing with how students of different ages understand geologic time; this research interest has also extended towards testing how visitors to the Grand Canyon understand a new geoscience exhibit that explains "deep time." More recently, he has been comparing how scientists, graduate students, and high school students understand the epistemologically derived differences between experimental- and historical-based sciences.

Raphael B. Shuchat is lecturer in the Center for Basic Jewish Studies at Bar-Ilan University in Ramat Gan, Israel. His undergraduate degree in Psychology is from Empire State, SUNY Saratoga Springs, New York. He received an MA in Jewish Philosophy from the Hebrew University in Jerusalem, Israel, and a Ph.D. in Jewish Philosophy from Bar-Ilan University. He has written three books and over 30 articles. His latest book *Jewish Faith in a Changing World – A Modern Approach to Classical Jewish Philosophy* is an attempt to bring the issues of Jewish philosophy to the modern reader. His articles on the issue of science and Judaism include "Rabbinic attitudes to Scientific Cosmogony and Evolution in the nineteenth and twentieth century," "R. Isaac Halevi Herzog's Attitude to Scientific Evolution and His Dialogue with Velikovsky," "Did the Medieval Kabbalists Believe that the World was created in 6 Days?," and "From Monologue to Dialogue: Between Jewish Thought and World Culture."

Chapter 55
Challenges of Multiculturalism in Science Education: Indigenisation, Internationalisation, and *Transkulturalität*

Kai Horsthemke and Larry D. Yore

55.1 Introduction: Philosophy, Transfer, and Transformation in Education

The central purpose of this chapter is to position the indigenous–Western knowledge debate within the context of contemporary philosophical views of science and pedagogical insights into science learning and teaching. This will be done by providing a respectful, honest, and straightforward commentary on the strengths and weaknesses of including indigenous knowledge and wisdom (IKW) about nature and naturally occurring events in school science programs. The prior debates between traditionalists, modernists, and postmodernists about the issue have been unproductive. The conflict resolutions that were seeking binary judgements of *good* or *bad* and *science* or *pseudoscience* have not fully reflected the realities of world and science classrooms. These debates have neither considered the sociopolitical and social justice influences nor provided learners with opportunities to engage science without being misinformed about what they were learning—let alone its ontological assumptions and epistemological beliefs. IKW about nature and naturally occurring events is different from Western modern science (WMS), but each has personal value within the parallel worlds of interpersonal-/place-based and public/generalised knowledge systems.

Philosophy might be claimed, cautiously, to be one of the deliberative and critical resources that ought to be brought to bear on questions of accessibility and relevance and on the transfer and transformation of educational systems, knowledge, concepts, and practices—if such processes are to be justifiable, consistent, and effective.

K. Horsthemke (✉)
Wits, School of Education, University of the Witwatersrand,
Johannesburg, South Africa
e-mail: Kai.Horsthemke@wits.ac.za

L.D. Yore
University of Victoria, Victoria, BC, Canada
e-mail: lyore@uvic.ca

M.R. Matthews (ed.), *International Handbook of Research in History, Philosophy and Science Teaching*, DOI 10.1007/978-94-007-7654-8_55,

'Cautiously', because the contribution that philosophy can offer is likely to be modest—for at least two reasons. First, philosophy is only one of the deliberative and critical resources relevant to educational transfer and transformation. Second, deliberation and criticism may be necessary but are not sufficient for the justifiability, consistency, and effectiveness of the processes in question; there are vast and significant contingencies in context and practice that are likely to remain decisive (McLaughlin 2000; Schily 2009).

Over the last 50 years, pedagogical insights into this debate about science education, teaching, and learning have been based on a changing array of learning theories—behaviourism, cognitive development, cognitive psychology, and learning sciences. Based on an integration of cognitive psychology, linguistics, philosophy, and pedagogy, the most recent dominant stance has promoted a learner-activated and learner-controlled meaning-making process that combines prior knowledge, sensory input, interpersonal interactions within a sociocultural context, and intrapersonal reflections within *self*. However, this interpretation of learning might conflict with some philosophers' view of the nature of science.

A promising approach to establishing the appropriate contribution of philosophy to science education in this regard is arguably to focus on the embeddedness of philosophical considerations in many concerns and processes driving transfer and transformation (McLaughlin 2000). Many of these contain, to a greater or lesser extent, concepts, beliefs, values, assumptions, and commitments that—although they themselves may not be of a directly philosophical kind—can be subjected to philosophical scrutiny and analysis. In considering the issue in this chapter, a philosophical distinction might be drawn between worldview and critical activity. Unless seriously mentally impaired, everyone has a worldview (e.g. thoughts or ideas, views or opinions about one's own life and one's place in nature, natural events, and about the world, life, and nature in general); consequently, there exists a multitude of frequently competing and conflicting worldviews. When applied to knowledge about nature and naturally occurring events, people's views range from folklore, religious, indigenous, to scientific. Each can be personally useful, but they should not be confused, confounded, or equated (National Research Council [NRC] 1996).

However, is WMS merely one such worldview among several (Aikenhead 1996; Fakudze and Rollnick 2008)? Alternatively, does it also involve a special type of activity, a process of critical interrogation in which we are concerned with examining and evaluating the ways in which our views, opinions, thoughts, or ideas about the world, life, and nature in general are or have developed (epistemology) and are explained (ontology)? In other words, this involves critical reflection on our ontological and epistemic assumptions, the implications of our views, and the problems that have to be considered in order to reach these conclusions. Even if worldviews 'are culturally validated presuppositions about the natural world' (Aikenhead 1996, p. 4), this does not necessarily mean that that they are equally valid and congruent in terms of epistemology and ontology (Yore 2008), which does not discount their practical and personal value.

A further contribution that philosophy can make concerns the analysis of knowledge and the grounds for knowledge claims, in science education as elsewhere. The debate is between those who treat science and scientific knowledge as a 'cultural enterprise' and those who attest to the 'universality of science' (Stanley and Brickhouse 1994, p. 9). Many people would agree that some folklore about agriculture, meteorology, and navigation is laypeople's well-based expressions of fact, while other folklore in these areas is deemed myth or fancy. Religious-oriented statements about nature and naturally occurring events based on only faith may be of great personal value, but they generally lack the ontological features and epistemic means to be tested and validated using empirical evidence, standardised means, and accepted processes set by scientific communities. There is debate about whether science should be used to test one's religious beliefs, with some maintaining that religious claims about the natural world and its processes (evolution, faith healing, etc.) should be amenable to normal scientific appraisal. The more difficult issue is the consideration of IKW about nature and naturally occurring events, where claims were place based and developed over time using reasonably rigorous epistemologies and oral traditions, but the ontology allows a hybrid explanation of spiritualism and physical causality (Snively and Williams 2008).

55.2 Indigenisation and Internationalisation

The IKW–WMS deliberations need to be placed within the context of worldwide sociopolitical, economic, and social justice influences and efforts. With rapid changes in recent decades of scientific and technological advances, transnational mobility, communication and travel, economic connectivity and dependencies, and—even more recently—increasing democratisation of societies, it comes as no surprise that corresponding changes have occurred and continue to occur in education. These changes concern not only how education, its nature, and its aims are or will be conceptualised but also the very transfer and transformation of educational systems, knowledge, concepts, and practices.

The biggest challenges facing science education have arguably been accessibility and relevance to mainstream boys and girls; these challenges have become more pronounced with the increasingly multicultural nature of teaching and learning environments. How does one render accessible a discipline that has often been viewed as unnatural, difficult, and the intellectual playground of a select, gifted few? How does one instil in students a sense of relevance of science to their lives and experiences, especially if teaching and learning take place in a language and cultural context other than their home language and culture? Many people can remember the shock of entering primary school after years of the free, informal, and unstructured experiences of early childhood to the middle-classed, formal, and structured classroom. Your non-standard forms of the dominant language were not allowed; *stuff*, *ain't*, other colloquialisms, or localisms were deemed inappropriate language. *The gods bowling* as an explanation for thunder was replaced with an abstract idea about a sonic boom caused by rapid thermo-expansion of the atmosphere associated with a static electrical discharge—lightening. This culture shock would be magnified by powers of ten for nondominant language speakers and underserved or underrepresented students who entered the new culture and language environs.

There have been a variety of responses to the transformational implications of globalisation for education and, in particular, for higher education. Chief among these are the drives toward indigenisation, on the one hand, and toward internationalisation, on the other. The radical, extreme versions of these approaches reject any claim to validity or legitimacy by the rival approach. Thus, radical indigenisation involves a back-to-the-roots traditionalism and nationalism that more often than not are inspired by negative effects of the colonial experience and the need for political consolidation or cultural restoration. Globalisation is nothing new; its roots are in the historical explorations and immigrations of the fifteenth and sixteenth centuries when adventurers or refugees picked up their place-based cultures, languages, knowledge claims, and technologies and moved to a new place. It has continued at a more rapid pace today with mobility of populations leading to many free-choice immigrants—such as the two authors of this chapter. Traditionalism and nationalism are forms of sociopolitical priorities designed to conserve or restore cultural identity or forge a national identity.

The Māori Nation's successful efforts to establish the Putaiao curriculum in Aotearoa New Zealand to address colonialism (McKinley and Keegan 2008) and the First Nations' continuing efforts at linguistic and cultural restoration/conservation in Canada to address the social justice ills of residential schools (Snively and Williams 2008) are readily documented examples. Like most political and democratic endeavours based on majorities, not all people involved agree with the fundamental premises. Webb (in press) found that many parents did not want their traditional language and IKW to be the medium and focus of science instruction in South African schools; instead, they supported the English language and WMS curricular mandates with some consideration of IKW. None of the transfer and transformation issues and solutions to this point has considered how people learn—other than that the target learning content needs to be relevant to the target learners. Clearly, many people involved in this debate still adhere to the passive transmission model (i.e. knowledge is passed from teacher to learner, 'teacher directed') rather than the more contemporary interactive-constructive interpretations (i.e. shared student-teacher directed) that involve accessing, engaging, and challenging prior knowledge and supporting learners as they make meaning, store these new ideas in existing knowledge structures, or reorganize knowledge structures to accommodate discrepant ideas (Henriques 1997; NRC 2000, 2007).

Indigenisation involves what Wolfgang Welsch, a cultural theorist, has referred to as the 'return of tribes' (Welsch 2000, p. 349) and may be interpreted as a reaction against globalisation and colonisation. Given the historical, political, and socio-economic background of exploitation and oppression that motivates and explains indigenisation, the eagerness of people to return to what they perceive to be the sources of their cultural identity—their roots—is perfectly understandable. While this desire to return and to reembrace local values and indigenous traditions is not implausible, the move toward indigenisation has produced some collateral damage. Compounded by problems emanating from unhelpful immigration legislation and bouts of xenophobia (violent actions against foreigners), there has been no transfer, exchange, or mobility on the African continent comparable to that within, or produced by, European Union (EU) member states' secondary and higher education. Instead, the net result has been a marginalisation of Africa, not only of the continent as a whole but also within Africa in terms of increasing isolation of sub-Saharan African countries from each other. Indeed, these policies of 'indigenisation may exacerbate existing societal divisions and lead to new forms of intolerance and discrimination' (Andreasson 2010, p. 427). The reverse racist rhetoric and growing Zulu nationalism as an example of internal dislocation has been highlighted by Chetty (2010). Furthermore, there is the real danger that indigenisation in science education may mislead students and ill prepare them for future higher studies and careers in science, technology, and engineering.

By contrast, radical internationalisation envisages the spread of a universal, more or less monolithic, educational, and socioeconomic culture and tends to ride roughshod over local indigenous histories, values, beliefs, and cultural ways of knowing and explanatory traditions about IKW (Auf der Heyde 2005). Furthermore, such approaches may not engage students' actual prior knowledge, beliefs, and values about science and technology—thereby disenchanting more indigenous and culturally diverse students from actually learning about nature and naturally occurring events from a scientific perspective.

It is evident that neither position holds much promise for science and science teaching and learning. While the former errs in favour of increasing insularity and self-marginalisation, the latter errs in favour of dogmatic homogenisation and lack of regard for difference and diversity. More seriously still, apart from manifesting an essentialist conception of culture and identity, both perpetuate a cycle of disregard, disrespect, and intolerance, with ever-increasing ossification of the opposing fronts and polarisation.

Depending on one's sympathies, it is easy to eulogise one's preferred orientation by pitting it against a straw person that is swiftly and summarily dismissed. Consider, for example, Aikenhead's characterisation of First Nations knowledge of nature as contrasted with Western scientific knowledge:

> Aboriginal knowledge about the natural world contrasts with Western scientific knowledge in a number of ways. Aboriginal and scientific knowledge differ in their social goals: survival of a people versus the luxury of gaining knowledge for the sake of knowledge and for power over nature and other people. They differ in intellectual goals: to co-exist with mystery in nature by celebrating mystery versus to eradicate mystery by explaining it away. They differ in their association with human action: intimately and subjectively interrelated versus formally and objectively decontextualised. They differ in other ways as well: holistic First Nations perspectives with their gentle, accommodating, intuitive, and spiritual wisdom versus reductionist Western science with its aggressive, manipulative, mechanistic, and analytical explanations (Aikenhead 1997, pp. 5–6).

Elsewhere, Aikenhead listed the following attributes characterising the 'subculture of science': 'mechanistic, materialistic, reductionistic, empirical, rational, decontextualised, mathematically idealised, communal, ideological, masculine, elitist, competitive, exploitive, and violent'.[1]

---

[1] See Aikenhead (1996, pp. 9 & 10, 1997, pp. 2 & 5, 2001, pp. 11 & 12). Similar attributes of science were purported by Bishop (1998, pp. 200, 201, 210) and Witt (2007, p. 227).

It is not difficult to sympathise with the concerns that underlie advocacy of IKW projects, but it must be ensured that any contrasts of WMS and IKW about nature and naturally occurring events are based on accurate, informed views about the ontological and epistemic aspects of both knowledge systems. Loving (2002) mapped views of science on an ontological–epistemic plane, which led Yore et al. (2004) to identify three general clusters of views (traditional absolutist–realist, modern evaluativist–naïve realist, and postmodern multiplist–idealist) as anchor points for interviews with scientists that delved into their beliefs about their research enterprise. They found that although some scientists use the metalanguage of and identify with the traditional view, most scientists agree with the description of the modern view; a few scientists, mainly from the newer hybrid biosciences, lean toward the postmodern view. However, at the practical and historical perspective, people must be aware that Western knowledge, science, technology, and rationality have led to, or have had as a significant goal, the subjugation of nature; thus far, it has been devastatingly efficient.

The pursuit of nuclear energy (Fig 2005), wholesale deforestation, and the destruction of flora and fauna are arguably deplorable and irrational or, at least in hindsight, questionable applications at times and in certain places. Similarly, apart from being ethically suspect, factory farming of nonhuman animals for human consumption and, especially, vivisection are also examples of bad science (Horsthemke 2010). However, many of these examples used in defining *whose* science and knowledge systems in the modern–postmodern debates about the North–South divide may be more technological rather than strictly scientific (Good 1996; Harding 1991, 2011). Clearly, the definitions of technology as design or as applied science change these debates and claims. Wolpert stated, 'Much of modern technology is based on science, but this recent association obscures crucial differences and the failure to distinguish between science and technology has played a major role in obscuring the nature of science'(1993, p. 25). The 'central distinguishing characteristic between science and technology is a difference in goal: The goal of science is to understand the natural world and the goal of technology is to make modifications in the world to meet human needs' (NRC 1996, p. 24). However, the disparagement and belittling of indigenous peoples' practices, skills, and insights has, to a large extent, been arrogant and of questionable rationality. If honest engagement is desired in education and public awareness activities, it will require recognising cultural values, lived experiences, and prior knowledge and beliefs of those being engaged. Finally, current attempts by industrial, first-world nations to colonise or appropriate for commercial gain indigenous people's knowledge, practices, skills, and insights in the name of globalisation and worldwide development are exploitive and contemptible.

People hold many misconceptions and myths about the nature of science regarding the singularity of the scientific method, hypotheses as educated guesses, evolution of hypotheses and theories growing into laws, absolute truths based on accumulated evidence, procedural nature of inquiry, scientific enterprise can address all questions, scientists as supernatural people, all science is experiment-based, and all science claims are peer-reviewed to ensure honesty (McComas 1996). Therefore, a pro-Western/Eurocentric perspective might mislead one to characterise the differences as follows: the celebration of knowledge as intrinsically valuable versus the view of knowledge having no value in itself; reliance on a critical, rational scientific method versus superstition, magic, witchcraft beliefs, ancestor worship, and unquestioning obedience toward traditional authority; commitment to universal applications and solutions versus ethnocentrism and thoroughgoing cultural and epistemological relativism; and the value of scientific evidence versus faith and reliance on revelation. These binary positions have led to polarised controversy and winner-takes-all solution strategies, whereas in the pragmatic worlds of science education and science teaching, the greater good might lie or involve moving between these poles—two-way border crossing with mutual honour and respect (Chinn et al. 2008).

55.2.1 Knowledge Claims and Justification

If anything qualifies as science, there are certain criteria that must hold. It needs, at minimum, to involve reference to regularity, observation, description, explanation, prediction, and testable hypothesis. If it does not meet these criteria, it is not 'science', as commonly understood to be people's endeavour to search out, describe, and explain patterns of events in nature and naturally occurring phenomena where the claims are based on evidence and are generalisable and the explanations involve physical causality (Good et al. 1999). When only the epistemic beliefs and practices are considered, there is little difference between traditional IKW and WMS; but when the ontological requirements are considered, the two knowledge systems demonstrate critical differences (Yore 2008). Both indigenous and scientific epistemologies have well-established, rigorous systems and routines of observations, interpretative frameworks, and feedback loops to make, update, and revise descriptions of patterns and knowledge claims (Snively and Williams 2008). However, on the ontological dimension, indigenous explanations involve a mixture of spiritual and physical cause–effect mechanisms, while scientific explanations are limited to physical causality devoid of mysticism, magic, and spiritual causes.

With regard to scientific knowledge, one generally distinguishes between two kinds: practical knowledge and theoretical knowledge. The former denotes craft knowledge, skill, ability, practice, or custom taught or passed down from one generation to another without evidence, justification, or explanation. Some historical technologies and practices within skilled trades demonstrate such characteristics. Apart from necessarily incorporating belief, theoretical knowledge involves commitment to truth and justification (i.e. scientific evidence; Haack 2003). In other words, a person knows that something is the case if she believes that it is so and she has adequate evidence for believing that it is. Science, as well as being about inquiry, is about evidence-based empirical argumentation. Toulmin's (1958) pattern and elements (data, backings, warrants, evidence, claim, counterclaim, and rebuttal) were used frequently to describe and evaluate the quality of an argument. Adequacy, however, cannot be determined by a checklist of elements; it must be determined by the kind, degree, and context of evidence (Gott and Duggan 2003; Tytler et al. 2001; Walton 2005; Walton et al. 2008).

Different kinds of evidence pertain to the different sciences, natural as well as social. They include observation, sensory experience, oral and written testimony, and deductive and nondeductive (inductive, abductive, and analogical) reasoning. As far as the requisite degree is concerned, minimal evidence is clearly not enough, while conclusive evidence is usually not available. Normally, other than in mathematics and deductive logic, we accept evidence that is less than conclusive, that is, reasons that are nonetheless compelling within an evaluative context (argument–critique–analysis). These science and engineering practices were recently included as a central part of the new framework for science education regarding scientific inquiry and technological design, which will influence the next generation of science education standards in the United States of America (NRC 2012).

Yet, what makes evidential reasons compelling has partly, and importantly, to do with context—not only the particular scientific context but also, for example, the problem space, the environment, the cultural and social biography, and the reasoning level of the person making the knowledge claim (Aikenhead 2005). Considerations of context determine leniency or stringency, and ascription of scientific knowledge reflects the social component of knowledge, which reveals that attributions of knowledge are context sensitive (Cohen 1986). Scheffler argued that the idea of *suitability* is:

> [A] matter of appraisal, involving standards of judgment that may differ from age to age, from culture to culture, and even from person to person. The variability of such standards does not, however, imply that assessments of knowledge are arbitrary or that the would-be assessor is somehow paralyzed. He needs to assess in accord with his own best standards at the time, but he may hold his assessment subject to change, should he later have cause to revise these standards (Scheffler 1965, p. 57).

He pointed out that these standards might be applied more strictly in some cases, more approximately in others, 'thus giving rise to multiple interpretations of *knowing*' (p. 58). Therefore, the justification component permits some leeway toward a multiplist view of science where multiple interpretations of datasets are expected as the interpreters apply their lived experiences and individual interpretative frames, but the alternatives will be evaluated in the public arena of the involved science community. What counts as suitable justification in the case of a young child or person from a remote rural area—with limited opportunities, resources, or access to information—differs from that required of an older, more mature person from an industrialised, technologically advanced, privileged, and urban background. Scheffler noted the implications for education when he stated, 'As the child grows and as his prior learning takes hold, his capacity increases, allowing us to tighten the application of our standards in gauging his current performance' (p. 57). Thus, with this growth in the child's cognitive capacity, 'the same subject may thus come to be known under ever more stringent interpretations of *known*' (p. 57).

Yet, in all the various cases, the justified belief must be true and present. In the absence of truth, one cannot meaningfully speak of, or ascribe, knowledge. Scheffler suggested a subtle shift from examining beliefs to examining the *contexts* in which beliefs were advanced as knowledge-claims when he distinguished the question concerning justification of a belief from the 'question of *appraisal of the believer* …. To speak of the right to be sure is, in the present context, to appraise the *credentials* of belief from the vantage point of our own standards; it is to spell out the attitude of these standards toward specific *credentials* offered for a belief' (p. 64).

Like Scheffler, Cohen (1986) argued that the suitability of justification, or having good reasons, depends on the relevant epistemic community. He advanced his argument through an analysis of what it means to have good reasons for believing something. The concept of *defeasibility* is crucial. One's reasons for believing something are *defeasible* if there is something else that could count against them, that is, something that could defeat them or undermine their feasibility. According to Cohen, we can say that someone (e.g. a 6-year-old) has good reasons if, given her reasoning ability, it is epistemologically permissible for her to believe that something is the case. Scheffler would be inclined to apply the standards of justification more leniently in the case of the 6-year-old and more strictly in the case of the 16-year-old. Both the 6-year-old and the 16-year-old may form a justified true belief that the table they see in a darkened room is red. However, when they are informed about the presence of a red light bulb in the room (hidden from their view), the younger child will hardly be able to appreciate the significance of the presence of this *defeater*. Therefore, if she continues to cling to her belief that the table is red, we would normally credit her with sufficient justification for her knowledge claim. On the other hand, if the older child fails to see anything wrong with clinging to her belief, we would normally be more reluctant to credit her with knowledge that the table is red. What passes for sufficient or adequate justification, then, will differ, and a progression of plausible reasoning would be established.

The important point for science teachers is that what counts as a good reason depends on who is giving the reason and in what context. One of the responsibilities of a science teacher is to assess learners' knowledge *of*, *for*, and *as* learning in ways that are sensitive both to their level of understanding and to the context of assessment—accountability (of), empower learning and inform teaching (for), and stimulate learning (as). Another related responsibility is to develop the learners' grasp of the intersubjective standards of different learning areas, such as common core learning outcomes in English, social studies, science, and technical subjects (National Governors Association Center for Best Practices and Council of Chief State School Officers 2010) and cross-cutting concepts with science and engineering domains (NRC 2012). In Cohen's terms, to help learners move from a level of reasoning that provides *subjectively evident* grounds for believing something to a level that provides *intersubjectively evident* grounds for beliefs about conceptual systems and socioscientific issues.

The concept of having good reasons is not without ambiguity. A person can have subjectively good reasons (i.e. reasons that are clear and convincing to her, given her level of understanding) or intersubjectively good reasons (i.e. reasons that are clear to her and that comply with the standards of reasoning of the social group to which she belongs). How are these different applications of justification relevant to the concept of knowledge? When is an educator entitled to say that a learner knows something in the sense of knowing that in a deep way? To put the question more formally: Under what conditions may a teacher attribute knowledge to a learner?

Scheffler urged that someone possesses suitable justification when that person possesses reasons for the quality of understanding: 'In saying he knows, we are not merely ascribing true belief but asserting that he has proper credentials for such belief, the force of which he himself appreciates' (1965, p. 74). Scheffler's and Cohen's arguments imply that even if a learner has subjectively good reasons for believing something to be true, she does not have knowledge unless she also has intersubjectively good reasons for a true belief or knowledge claim. One of the tasks of effective science teaching is to assist learners to acquire the relevant concepts and intersubjective standards of justification for evidence-based empirical arguments. Here, one's sense experiences must be reliably connected with the world, one's sense organs must be intact, and one's reasoning must be correct. An analysis of good reasons indicates why reference to them is context sensitive and why neither reasoning nor our sensory experiences are infallible. Nonetheless, if they are generally reliable sources of justification, the reasons they produce might be called *intersubjectively certain*. Cohen stated, 'Reasons can be permissible grounds for belief, relative to that standard, even though they are not ideally correct'(1986, p. 575). Essentially, both the beliefs and the justifications (reasons) given for them, but not their truth, may depend on particular social and cultural contexts or circumstances. Therefore, people can acknowledge differences in cognitive resources, skills, and opportunities without thereby having to commit to epistemic relativism.

In Plato's cave parable, whatever the enlightened person knows about reality stands in stark contrast to the (majority) view that the prisoners in the cave claim to know is reality. The cave parable indicated that knowledge is ambiguous between various concepts, when each is based on a different standard. Is knowledge context dependent? Scheffler's and Cohen's arguments suggest that it may be better to say that attributions of knowledge are context sensitive. The term *context sensitive* does not offer an open invitation to or endorsement of epistemological relativism. It is important to note that, in terms of the present definition, while belief and what counts as evidence may vary from individual to individual, society to society, and culture to culture, truth does not.

The present account acknowledges that people do not have the same cognitive resources, skills, and opportunities. They do not all act or operate in the absence of constraints. Their situations are characterised by different levels of expertise, by different opportunities to access and gather information, by different levels of cognitive maturity and training, and by considerable differences in available time and deadlines. Goldman cautioned that a 'social epistemology for the real world needs to take these constraints into account' (1991, p. 233). So, when Aikenhead argues that 'the knowledge, skills, and values found in the typical secondary science curriculum have been widely criticised throughout the world for being isolated and irrelevant to everyday events that affect economic development, environmental responsibility, and cultural survival' (1997, p. 7), one might respond by acknowledging the need for science education and education in general to be anchoring in and connected to the real world and to each other.

55.2.2 Politics and the Knowledge Enterprise

A central theme of this chapter has been a social justice motive within a rigorous epistemological and ontological stance applied to IKW, WMS, science education, and science teaching. Occasionally these motives can run opposed to one another, but they need not. Worldwide there have been efforts to rationalise educational and economical energies and programs across diverse cultural and ethnic communities. In June 1999, the EU ministers of education stated in the Bologna Declaration:

> *A Europe of Knowledge* [emphasis added] is now widely recognised as an irreplaceable factor for social and human growth and as an indispensable component to consolidate and enrich the European citizenship, capable of giving its citizens the necessary competencies to face the challenges of the new millennium, together with an awareness of shared values and belonging to a common social and cultural space (European Commission 1999, p. 1).

Among the central concerns of the Bologna Declaration were the transformation of educational systems and the transfer of educational experiences and knowledge, as well as the possibility of active and meaningful engagement across historical, social, cultural, and linguistic borders. The Bologna Declaration was a pledge by each of the 29 signatory countries. In the explanation prepared by European university administrators, the following points were listed:

- [A] commitment freely taken ... to reform its *own* higher education system or systems in order to create overall convergence at European level. ...
- The Bologna process ... is not a path toward the "standardisation" or "uniformisation" of European higher education. The fundamental principles of autonomy and diversity are respected.
- The Declaration reflects a *search for a common European answer to common European problems*. The process originates from the recognition that in spite of their valuable differences, European higher education systems are facing common internal and external challenges related to the growth and diversification of higher education, the employability of graduates, the shortage of skills in key areas, the expansion of private and transnational education, etc. (Confederation of EU Rectors' Conferences and Association of European Universities 2000, p. 3).

In addition,

The Declaration specifically recognises the fundamental values and the diversity of European higher education:

- It clearly acknowledges the necessary *independence and autonomy of universities*; ...
- It stresses the need to achieve a common space for higher education within the framework of the *diversity of cultures, languages and educational systems* (p. 6).

This agreement attempted to make the transfer of university experiences, course work, and certifications possible and to expedite the movement of students and professionals among the various jurisdictions of the EU member states. These efforts within the founding EU member states have gone reasonably well, but there have been some concerns with the blending of different university systems, traditions, and conventions. The long, historical, collaborative efforts in science, technology, and mathematics in the academies and research institutions have helped the rationalisation efforts of the Bologna Declaration. The real test will be the integration of new members of the EU without long records of collaboration and similar traditions. Will this effort in Europe influence similar reform efforts in other regions of the world where knowledge systems, epistemologies, and ontologies might differ drastically, like Africa?

55.3 Transformation in Africa

Finger (2009) suggested that people should disregard the new system for the inherent value of knowledge (as contrasted with its purely instrumental value) and this would provide a similar motive and advocacy of *Africa of knowledge* as that which drives toward a *Europe of knowledge*. This pertains not only to political leaders opening tertiary institutions in recently liberated African countries in the 1950s and 1960s and potential changes to established knowledge-building institutions resulting from the Arab Spring but also and especially of contemporary theorists and academics emphasising the need for secondary and higher education to develop an African identity. Makgoba stated:

> The issue of pursuit of knowledge for its own sake and the so-called standards have … become contentious factors around the African university. … The pursuit of knowledge for its own sake has been one of the cornerstones of university education; but is there such a thing as knowledge for its own sake today? Knowledge is a human construction that by definition has a human purpose. Knowledge cannot be sterile or neutral in its conception, formulation and development. Humans are not generally renowned for their neutrality or sterility. The generation and development of knowledge is thus contextual in nature (Makgoba 1997, p. 177).

Does this mean that funding and support for curiosity-driven inquiries (science) will be threatened and replaced by this call for mission-driven inquiries (technology/engineering)?

That knowledge ascription and justification have a crucial contextual component is surely not in doubt (Horsthemke 2004), but this does not mean that the pursuit of knowledge must be described and explained in consequentialist or constructivist terms. It might be the *object* of knowledge that is and continues to be the legitimate cornerstone of secondary and higher education. 'The global competition, the involvement of industry in universities, and the social, economic and political pressures of modern society have made the [pursuit of knowledge for its own sake—curiosity-driven pursuits] obsolete. … The pursuit of knowledge and the truth with a purpose and social responsibility [mission-driven pursuits] is what universities are about' (Makgoba 1997, pp. 181–182). Surely, setting up a commission like the Truth and Reconciliation Commission also involved a noninstrumental understanding of knowledge and truth (Horsthemke 2004). If they had an exclusively instrumental function, then substituting them would be entirely permissible—say, with an amnesia drug—as long as the desired end effect or outcome was the same. With regard to the traditional roles that universities throughout the world have in society, Makgoba considered the social responsibility of knowledge systems when he stated:

> [The] preservation, the imparting and the generation of knowledge. … It is important to recognise … that the imparting of inappropriate or irrelevant education, even of the highest calibre, would … lead to a poor and ineffective product. Thus, university education has to be relevant not only to the people, but also to the culture and environment in which it is being imparted (Makgoba 1997, p. 179).

Without doubt. The trick, of course, is to avoid an education system that is impoverished as a result of excessive concerns with people's culture and user-friendliness. Makgoba's comments are equally applicable to elementary and secondary education and especially to science education and science teaching.

However, the resolution of this pedagogical issue in learning and teaching science can place politicians, philosophers, and educators at odds. Some politicians have stressed that the social justice issue overrides any considerations of the nature of science and how people learn science. Some philosophers have questioned the validity of constructivism applied to the scientific enterprise (Matthews 2000; Nola 1997; Suchting 1992), while science educators advocate an interpretation of how people learn that has a constructivist flavour (NRC 2000, 2007). Henriques (1997) investigated the conceptual mix of science and pedagogy (behaviourism, cognitive development, and learning sciences) within the sociocultural contexts of schools and found that inquiry teaching within the constructivist learning perspectives ranged from information processing, interactive constructivist, social constructivist, or radical constructivist. She considered the underlying factors (i.e. nature of science, ontological, epistemological, cognitive, pedagogical, discourse/language influences, and realities of classrooms) in these interpretations and found little support for the strong sociocultural and radical constructivist approaches, some lingering uses of behavioural-based information processing, and sizeable support for the centralist approaches and modified learning cycle (engage, explore, consolidate, assess).

55.3.1 Africanisation and Afrocentrism

Nowhere have indigenisation and internationalisation efforts been more apparent than in Africa—a continent of diversity. Africanisation (continent-wide/global perspective) and Afrocentrism (cultural/ethnic perspective) have included radical endorsements, which tend to reject any outside (e.g. colonial, Western, Northern, European, Eurocentric, etc.) influence and also segregationist forms of nationalism (such as some trends manifested in the former Soviet Union, Yugoslavia, etc.). What they arguably share, apart from an intense belief about internal homogeneity and an equally strong rejection of heterogeneity is an instrumental usage of the concept of indigeneity. Indigenisation was not only seen as an effective instrument for political persuasion, mobilisation, and justification but also as a tool in transformation, educational, socioeconomic, and cultural aspects of the larger issue and national goals. As such, it becomes symbolic and may actually produce a virulent form of the 'ethnicisation' of education, politics, and the economy (Andreasson 2008, p. 7).[2] A characteristic of this approach, one of its 'normative entanglements', is the rejection of Eurocentrism, which is linked to an express sympathy with the ethnocentrism of non-European cultures (Cesana 2000, p. 452). Yet, to respond to Eurocentrism by embracing Afrocentrism is relevantly like responding to school-ground bullying with corporal punishment or to murder with capital punishment. Motivational reasons do not amount to justification of the prescribed solution in any of these cases (Horsthemke 2006).

So much for the caricatures. There are obviously more nuanced versions that deserve correspondingly serious consideration. Thus, in the instance of indigenisation, there is an emphasis on the local that nonetheless acknowledges the significance, if not the inescapability, of the global. 'We [Africans] have to construct our own epistemological framework from which we can explore ideas and build our own knowledge. … Africans must create our own paradigm from which we can also dialogue meaningfully with Europeans' (Masehela 2004, p. 11). A vice chancellor of the University of KwaZulu-Natal in South Africa maintains:

> It is the duty of academics and scholars to internationalise, articulate, shape, develop and project the image, the values, the culture, the history and vision of the African people and their innovations through the eyes of Africans: African people should develop, write, communicate and interpret their theories, philosophies, in their own ways rather [than allow these to be] construed from foreign culture and visions (Makgoba 1997, p. 205).

Moreover, he stated, 'global economic competition is high and unless we develop a competitive high technology economy we face economic ruin, stagnation and under-development, with dire consequences for the impoverished rural and urban communities' (p. 179). While the latter insight is surely correct, Makgoba does not elaborate on the assumption that Africanisation is compatible with internationalisation or with developing a competitive high-technology economy. Furthermore, he appears to use technology and science interchangeably and does not differentiate the ontological and epistemic characteristics of science, technology, engineering, and traditional knowledge systems. Fuller deliberations are needed to establish how an Afrocentric orientation is supposed to cater for the demand, 'as we enter the era of globalisation, … to rethink ourselves anew, and bring in new ideas if we are to be a significant part of the information age and an era of knowledge industries' (Ntuli 2002, p. 66) and with the 'need to develop people and prepare young South Africans for the future and the tough world of global competition' (Makgoba 2003, p. 2). Conversely, in the instance of internationalisation, the emphasis on the global is seen as compatible with or perhaps even requiring an acknowledgement of diversity, difference, and locality/indigenousness. This sociopolitical, socioeconomic, and techno-science problem space requires fulsome and rigorous considerations

[2] For a thinly veiled endorsement of this kind of reverse racist, indeed ethnocentric orientation, also see Makgoba and Mubangizi (2010), especially the chapter on Leadership Challenges.

and deliberations regarding those involved, ensuring fair distribution of risks and benefits are appropriately distributed across the participants.

There are further, remarkable parallels between the Bologna Declaration and the call for the Africanisation of education: emphasis on the Africanisation of knowledge, teaching and learning, and finding *African answers to African problems*, the endeavour to make the African university internationally attractive and competitive, to establish international respect for Africa's rich and extraordinary cultural and scientific traditions. The major difference is that Africanisation and Afrocentrism emanate less from the political/economic precedent of the African Union and the common objectives of convergence and transnational mobility than from a shared rejection of the European education system and Eurocentrism. While the Bologna Declaration may be interpreted as a call to unity by harnessing Europe's many strengths, the emphasis in Africanisation and Afrocentrism is more on unity as a means of resisting the external economic, cultural, and political influences.

Africanisation is closely associated with educational and institutional transformation and embodies traits of both internationalisation and indigenisation in which the former link may be more controversial. Africanisation of education has a clearly *international* element (i.e. between African nations/nation states), just like Europeanisation of education has been between EU member states. Moreover, the idea of Africanisation of knowledge bears more than a fleeting resemblance to the Bologna Declaration's internationalist reference to a Europe of knowledge. Africanisation binds together a plethora of Saharan and sub-Saharan nations and states. The late Libyan head of state Muammar Gaddafi's vision of a United States of Africa, with himself as Emperor of Africa, may have been a delusional, autocratic fantasy—but at least the first part of it is shared by many people in Africa. This desire for pan-African unity is captured in the frequent appeal to communalism as a typically African value and reference to the essence, identity, and culture of Africa (note the singular).

On the other hand, there is a strong emphasis in Africanisation and Afrocentrism on indigenous, local—as contrasted with, say, global, international, European/Eurocentric—educational knowledge, practices and values. For example, there is a frequent endorsement of African mathematics as ethnomathematics, traditional knowledge about nature and naturally occurring events as ethnoscience, and African knowledge systems as indigenous knowledge systems as opposed to academic or mainstream mathematics and science and world knowledge, respectively. The African *is* the indigene: colonised, exploited, marginalised, and historically excluded from the international mainstream.

A South African report on transformation, social cohesion, and the elimination of discrimination in public higher education institutions stated that 'at the centre of epistemological transformation is curriculum reform—a reorientation away from the apartheid knowledge system, in which curriculum was used as a tool of exclusion, to a democratic curriculum that is inclusive of all human thought' (South Africa Department of Education 2008, p. 89)—later referred to as 'the Africanisation of the curriculum' (p. 91). The report contends that 'resistance to Africanisation is often advanced under the guise of a spurious argument suggesting that the debate is not about privileging Western scholarship, but rather emphasizing the universality of knowledge' (p. 91). It is 'the local context [that] must become the point of departure for knowledge-building in universities [across Africa and, indeed,] the world' (p. 92). However, not all stakeholders and scholars fully endorse an Africanised curriculum and knowledge (Horsthemke 2004; Webb in press).

55.3.2 *Cosmic Africa*

Efforts to identify and document an African perspective on knowledge about nature and naturally occurring events have taken many forms. The film *Cosmic Africa* (Rogers et al. 2003) documents the journey of South African astrophysicist Thebe Medupe in his mission to connect occidental science and astronomy to the cosmological models of some of the oldest civilisations on earth. Astronomy survives in these ancient societies despite the eroding effects of colonialism and its modern heir, globalisation. Medupe emphasises that astronomy has never just been a science in these cultures, where it is an 'intimate tapestry merging into their prayers, their lives, their dreams and their deaths'.[3] Occidental culture, on the other hand, has separated astronomy from daily experience and turned it into pure science from astrology. Medupe's mission was stated at the very beginning of the film: 'I need to discover whether my science has a place in Africa, and whether Africa has a place in my science.' His journey leads him to the Ju/'hoansi in northeastern Namibia, the Dogon in Mali, and finally to Nabta Playa in the southern Egyptian Sahara, to what is conceivably the site of the first solar observatory (see also Rogers 2007, p. 19).

During his visit to Namibia, Medupe learns not only of Ju/'hoansi reliance on the stars as to when to plant and to harvest as an astro-calendar but many of the stories connected to the sun, moon, and stars:

> One memorable night, Kxau Tami and /Kunta Boo, two elderly shamans, demonstrated how they would throw burning sticks in the direction of a very bright meteor—as they threw the sticks into the air, they uttered swear and curse words which they said would help to divert the meteor's path and thereby prevent its dangerous potential. They believe that bright shooting stars with fiery tails are invested with very powerful !nom (extreme potency) and that they have the potential to cause sickness (Rogers 2007, p. 21).

Medupe's visit to the Ju/'hoansi coincided with a total solar eclipse. He worried about whether he should tell the people about what is going to happen but decided not to; they would want to know how he knows. Instead, he sets up his equipment. When the eclipse happens, people talk about the return of winter and blame the intruder and his equipment: 'The telescope is eating up the sun.' After the eclipse and subsequent reconciliation, Medupe says, 'For the first time, I see how the stars interact with long-held beliefs to affect the way people live. My science and my Africa are beginning to come together.'

[3] Note that quotations attributed to Medupe in this and following paragraphs are taken from the film's dialogue; hence, no page or paragraph numbers are provided.

This impression was deepened with the visit to the Dogon, whose knowledge of the stars is legendary. Their daily and seasonal activities, routines, and customs are guided, for example, by the appearance of Venus (for which the 'Dogon have a number of different names ..., depending on its station in the sky'; Rogers 2007, p. 21), 'Toro Jugo—the Pleiades' (Rogers 2007, p. 20), etc. One of the elders, spiritual leader Annayé Doumbo, claimed, 'In our Dogon way, the man who makes technology is the sorcerer of the sun'. Given the harsh conditions under which they live, to the Dogon, knowing the stars can mean the difference between life and death. Does the elder know that human beings have walked on the moon? 'There is no gate to the moon', is the reply, 'it is not possible for anyone to go there, unless they are the little brother of God'.

The last leg of Medupe's journey was presented as the origin of astronomy, Egypt. However, there was no mention of the innovations and discoveries of the Maya and Aztecs, which can be taken as evidence of the lack of knowledge transmission between continents in the southern hemisphere. In the southern Egyptian desert, near the border of Sudan, he discovered what is conceivably the oldest observatory conceived and constructed by the Nabtans, nomadic pastoralists, now long dead. Predating Stonehenge in England by almost 1,000 years, it consists of stones emanating from a centre, in order to trace the rising and setting of the sun during the year, as well as the passage of the moon and stars. Medupe stated, 'The origin of astronomy, its measuring and predicting is in Africa ... Stones took the place that my computer takes now.'

It was Medupe's prime intention to create an African star chart; unfortunately, he and the research team never explored any of the tensions between IKW and WMS worldviews, knowledge claims, and explanations. They seem satisfied with just noting the different perceptions and appear to assume that there is no problem of reconciliation of myth or legend with scientific claims and explanations. At the end of the film, Medupe stated that he has come 'full circle', that his journey has served to (re)unite 'his [postmodern] science' and 'his Africa', without any attempt to account for the contradictions encountered between spirituality and astronomy.

55.4 Reflecting on Attempts to Indigenisation/Africanisation and Internationalisation

Attempts to promote specific geographic, cultural, or ethnic interpretations of knowledge and knowledge construction have encountered problems in the search for truth and wisdom within a global society. Continental landmass or geopolitical boundaries and ethnocultural groupings cannot confine knowledge, especially knowledge about nature and naturally occurring events. Efforts to indigenise or Africanise and internationalise have received philosophical, pedagogical, and practical critiques.

55.4.1 Problems with Indigenisation

Medupe's long-term goal was to develop a database and to set up a formal ethnoastronomy research group. The pertinent questions, for present purposes, were: Does the idea of ethnoastronomy make sense? What, if anything, distinguishes ethnoscience from mainstream, academic science? Is it a spiritual, contextual, or cultural element? The differences between WMS and IKW about nature and naturally occurring events may vary across science domains and between science and technology.

Jegede (1999) emphasised that the local and contextual character of his interpretation of scientific knowledge and truth, in terms of learning and teaching where a 'strong relationship that exists between the prior knowledge and sociocultural environment [of the student. It is] deemed primitive, inferior and unscientific' (p. 120) by/in the 'Western view, especially with regard to science teaching and learning' (p. 123). No wonder, the cynic might question his four fundamental features of the African belief and thought system—belief in a creator/god, belief in life after death/ reincarnation, anthropocentrism or the idea that human beings constitute the centre of the universe, and the theory of causality, which 'is the sociocultural cloak the African child takes to the science classrooms' (p. 125)—and their application to knowledge, cognition, or science. Good (2005) expressed sincere concern about mixing scientific and religious habits of mind and related ontological requirements for explanation of nature and naturally occurring events.

Aikenhead and Jegede stated:

> In the culture of Western science, students learn that the refraction of light rays by droplets of water causes rainbows; in some African cultures, a rainbow signifies a python crossing a river or the death of an important [traditional] chief. Thus, for African students, learning about rainbows in science means constructing a potentially conflicting schema in their long-term memory. Not only are the concepts different (refraction of light versus pythons fies") (Aikenhead and Jegede 1999, p. 276).

Aikenhead and Jegede appear to have confounded the ways of knowing as a learning process with the established procedures for doing and knowing science. They were actually, at best, using the contrast to differentiate epistemologies since they have provided little insights into the differences in metaphysics and underlying ontological requirements for scientific explanations. This raises the question: Which of these accounts constitutes science? Traditional African education appears to discourage critical interrogation of received knowledge, wisdom, and practice. In terms of validity and usefulness (explanation limited to physical causality void of magic, mysticism, and spiritualism; prediction and predictive power; etc.), there is no equivalence here—WMS and traditional African IKW are different! However, if Jegede's account was meant to exemplify an initial engagement strategy and 'collateral learning' (which is principally about students holding a multitude of worldviews at the same time), it might be argued that 'traditional thought' or 'explanation' (1999, p. 131) might access and engage students with traditional prior knowledge and experiences; frequently, however, it fails to involve acquisition of facts or truth and, therefore, does not constitute knowledge. For this reason, one might even consider the reference to different epistemologies/ontologies incomplete, inappropriate, or misleading.

In a related vein, le Grange mentioned the *localness* of all knowledge systems: All knowledge is local, 'located/situated and motley (messy situatedness)'. While it makes some sense to say that 'all knowledge systems have localness in common', they also share objectivity and *trans*localness (Le Grange 2004, p. 87). Le Grange would probably concur with Visvanathan's statement that 'Morality, like science, has to be invented individually' (Visvanathan 2002, p. 51). However, this view indicates a basic misconception since neither science nor morality is an individual invention. There is also a disconcerting relativism manifest in views like these. WMS attempts to make evidence-based claims, generalised explanations involving physical causality, and public evaluation where Nature is the final arbiter (Ford 2008). However, some science events and several technologies are place based, which make them suitable engagements for students in these locations.

When Māori scholar Russell Bishop refers to *Kaupapa Māori*, 'the philosophy and practice of being and acting Māori', as an 'orientation in which Māori language, culture, knowledge and values are accepted in their own right' (Bishop 1998, p. 201), this points to a fairly thoroughgoing relativism. He adds a disclaimer: 'It is also important not to ignore the impact of European colonialism by claiming that Māori culture has all the answers. Nor is this to say that *all* knowledge is *completely* [emphasis added] relative' (p. 210). The postmodernist relativist phrase 'all knowledge is completely relative' is troublesome for many philosophers of science, nature of science in education researchers, and curriculum developers as it implies that all claims are equally valid and should not be evaluated for fear of disempowering the people making the claims. Such a position might represent the slippery slope in educational thinking for WMS, public evaluation of science claims, nature as final arbiter, rigorous epistemologies, and restricted ontological assumptions. Application of these relativistic ideas will not allow biologists to differentiate among divine creation, intelligent design, and evolution explanations of changes in living organisms (including humans), which has been the focus of much legal deliberation and scholarly debate.

Relativism as a social justice stance, in particular, is problematic in that one would not be able to compare and evaluate competing knowledge claims. However, many postmodern curriculum theorists welcome this implication as a solution to the disciplinary power structure that disempowered minority scholars and silenced underrepresented voices. Onwu and Mosimege were worried about the gatekeeping mechanisms set up by WMS to determine 'what is to be included or excluded as science' (2004, p. 4). If relativism were true, for the sake of the present argument (i.e. assuming that its truth could be established nonrelatively and that this would not constitute a vicious logical inconsistency), then there would be no epistemic or veritistic grounds for choosing between the claim that rain is the result of a chain of physical cause–effect relationships involving condensing moisture that occurred as a result of evaporation driven by solar radiation and the belief that 'rain can arise at will as a result of human action' and 'the rain by-passes the farm/field of the person who stands while drinking during the ploughing season' (Onwu and Mosimege 2004, p. 7). Most disturbingly, this kind of approach would thwart all scientific inquiry into, or curiosity about, phenomena for which there already exists a traditional, folkloric account or explanation. This barrier to continuous disbelief and inquiry is similarly deterred by an absolutist–realist view of science for established ideas.

55.4.2 Problems with Internationalisation

The critique of internationalisation as a viable, defensible approach to transformation of science education—and the teaching and learning of scientific knowledge, concepts, and practices in non-Western or indigenous societies—was diverse in both articulation and geographical orientation. Bishop has argued that

> [a]ttempts to locate Kaupapa Māori research within the broad framework of international perspectives on *participatory research*, [emphasis added] indeed even to search for a methodology of participation, may defeat the very purpose of Kaupapa Māori research, which is to reduce researcher imposition in order that research meets and works within and for the interests and concerns of the research participants within their own definitions of self-determination (Bishop 1998, p. 210).

Participatory research has a distinct political action function that sets it apart from most educational research approaches. Embracing a First Nations–Aboriginal perspective and referring to transmission and transformation of science education in particular, Aikenhead stated that the 'nature of the transformation [requires] science to articulate with practice' (Aikenhead 1996, p. 29). Elsewhere, he added:

> Science education's goal of transmission runs into ethical problems in a non-Western culture where Western thought (science) is forced upon students who do not share its system of meaning and symbols … the result is not enculturation, but assimilation or "cultural imperialism"—forcing people to abandon their traditional ways of knowing and reconstruct in its place a new (scientific) way of knowing … (Aikenhead 1997, p. 11).

Similarly,

> if … science is generally at odds with a student's everyday world, … then science instruction can disrupt the student's view of the world by forcing the student to abandon or marginalise his/her indigenous way of knowing and reconstruct in its place a new (scientific) way of knowing. The result is assimilation … [which] has caused oppression throughout the world and has disempowered whole groups of people (Aikenhead 1997, p. 4).

Abrams et al. (in press) discussed 'pedagogy of hope' regarding culturally relevant science teaching involving traditional knowledge systems, mathematics, and science. They identified contemporary research findings that outline the successes and failures involved in engaging indigenous students with traditional indigenous knowledge, which have been promoted by the Alaska Native Knowledge Network (http://www.ankn.uaf.edu/) and the Indigenous Science Network Bulletin (http://members.ozemail.com.au/~mmichie/network.html).

Education can inculturate or assimilate students into a specific culture, forcing them to leave behind their home culture, or it can acculturate students into living in two cultures with free movement between these cultures. Many practicing scientists report having both religious and scientific beliefs representing their professional and personal cultures where they strategically move between these cultures as needed. Several science educators have endorsed a two-way bridge or border-crossing analogy for working with indigenous and nonindigenous students in IKW and WMS.

The notion of internationalisation, then, involves the assumption that the worldwide trend of cultures and societies is toward increasing synchronisation of local environments—presumably following the Western model. This is clearly not a wholly accurate assumption, as evidenced by the complementary development or resurgence of indigenisation and particular phenomena like Africanisation, which is about cultural identity and language restoration as the face of culture conservation. Despite its lip service to diversity, differentiation, and particularities and however benevolent its motivation and intentions, internationalisation is by its very nature ultimately unable to accommodate these differences and countercurrents, especially if and where these are at odds with its central tenets (e.g. where they are manifestations of religious fundamentalism, involve nondemocratic practices). A less favourable view considers this rival trend to be a bothersome, regressive phenomenon that is facing imminent extinction.

With regard to science education, the major challenges facing internationalisation are those of accessibility and relevance. Unfortunately, Aikenhead stated that 'the "taught" science curriculum, more often than not, provides students with a stereotype image of science: socially sterile, authoritarian, non-humanistic, positivistic, and absolute truth' (Aikenhead 1996, p. 10). The traditional absolute realist view of science found in many textbooks represents a 'real science' that is outdated with few scientists endorsing it and may have never existed (Yore et al. 2004; Ziman 2000). Aikenhead characterised:

> [school] science [as] conveying an ideology that exalted Western science over all other ways of knowing... [an] ideology [that] assumed that science was purely objective, solely empirical, immaculately rational, and thus singularly truth confirming. ... Scientism is scientific fundamentalism (science is the only valid way of knowing). ...[Science teachers, he continued] tend to harbour a strong allegiance to values associated with scientism, for instance, science is: non-humanistic, objective, purely rational and empirical, universal, impersonal, socially sterile, and unencumbered by the vulgarity of human bias, dogma, judgments, or cultural values. ... For the vast majority of students, however, enculturation into Western science is experienced as an attempt at assimilation into a foreign culture. Because students generally reject assimilation into the culture of Western science ..., they tend to become alienated from Western science in spite of it being a major global influence on our lives (Aikenhead 2001, p. 2).

Haack (2003) provided a readable and understandable defence of WMS, scientism, and cynicism within reason and common sense that questions and clarifies some of these assertions. However, based on his understanding of *culture* as 'the norms, values, beliefs, expectations and conventional actions of a group' (Aikenhead 1996, p. 7) and of science as a *subculture*, Aikenhead outlined a cultural perspective on science education and culturally responsive pedagogy founded on several tenets:

1. WMS is a cultural enterprise itself, one of many subcultures of Euro-American society,
2. People live and coexist within many subcultures identified by, for example, language, ethnicity, gender, social class, occupation, religion and geographic location,

3. People move from one subculture to another, a process called "cultural border crossing";
4. People's core cultural identities may be at odds with the culture of WMS to varying degrees,
5. Science classrooms are subcultures of the school culture,
6. Most students experience a change in culture when moving from their life-worlds into the world of school science,
7. Learning science is a cross-cultural event for these students,
8. Students are more successful if they receive help negotiating their cultural border crossings, and
9. This help can come from a teacher (culture broker) who identifies the borders to be crossed, who guides students back and forth across those borders, who gets students to make sense out of cultural conflicts that might arise, and who motivates students by drawing upon the impact WMS and technology have on the students' life-worlds. (Aikenhead 2001, p. 4)

Aikenhead's context-sensitive argument, although initially attractive, rests on some rather problematic assumptions. Thus, the claim that science, science education, teaching, resources, and learning were cultural entities remains open to counter-argument. The claim that WMS is no more than one worldview among several equally valid worldviews about nature and naturally occurring events, too, remains unsupported (Yore 2008). Each of these worldviews has personal value, but they are not based on the same ontological assumptions and epistemological beliefs (NRC 1996). Surely, science is not just 'another cultural point of view' (Aikenhead 2001, p. 6); it has been a very successful human endeavour leading to positive and negative outcomes and implications. However, most people do not want their scientific beliefs to be tested by hermeneutics of religious scriptures. Nor do they want their religious beliefs to be tested by established scientific procedures. Evolution is a theory in the scientific sense—an umbrella idea that integrates other ideas and has predictive and explanatory powers; it is not merely a theory in the common lay sense—a *crazy ass* guess or a speculation among several competing, equally valid ideas (e.g. divine creation, intelligent design).

The assertion that people's core cultural identities may be at odds with the culture of WMS to varying degrees is perhaps true—in the sense that creationists' core cultural/religious identities may be at odds with science. However, many scientists and science educators report holding religious ideas and attend houses of worship, which would indicate that they maintain parallel worlds, each with specific purposes. An example might be a religious scientist standing atop a tall building and considering suicide takes time to pray, which is not for the grand architect of the universe (his god) to discontinue gravity for a few seconds after he jumps but is most likely for the grand architect to console his family and forgive his bad deeds.

Aikenhead's understanding of the aims or purposes of science education raises serious questions, when he stated that 'if the subculture of science generally *harmonises* [emphasis added] with the student's life-world culture, science instruction will tend to support the student's view of the world' (Aikenhead 1996, p. 4).

Harmonisation is the critical point here. Does this mean sensitive and respectful engagement of prior knowledge, beliefs, and experiences—a basic axiom of contemporary learning theory—or does it mean uncritical acceptance of all opinions as equally valid? '[Teachers] should teach science embedded in a social and technological milieu that has scope and force for students' worlds, worldviews, or practical experiences (respectively); and [they] need to dismantle barriers between students and science' (1996, p. 18). He later stated:

> Most students have a chance to master and *critique aspects of Western science* [emphasis added] without losing something valuable from their own cultural way of knowing. By achieving smoother border crossings between those two cultures, students are expected to become better citizens in a society enriched by cultural differences. This is an essence of cross-cultural teaching (Aikenhead 2001, p. 16).

This invites the question whether critique of Aboriginal ways of knowing, perhaps by way of critical self-reflection, is equally encouraged. If not, then how could one speak of *successful* cross-cultural teaching and learning? How does one draw a distinction between what is scientific and what is unscientific? Moreover, how does one get students to grasp the difference—if there is one? Why, according to Aikenhead, should students learn and appropriate the content of Western science at all? These are just a few questions that would need to be addressed by a compelling critique of internationalisation and its idea of the universality of science and applicability of their knowledge about nature and naturally occurring events outside of the classroom, in future studies, and for career preparation.

There is also the issue of language. In a study that builds in part on Aikenhead's ideas of cross-cultural science education and border crossing, Fakudze and Rollnick pointed out that

> African students enter the classroom with a rich heritage of traditional beliefs that, if handled sensitively and with understanding, can play an important role in enabling learning of science. Recent developments in the understanding of how students acquire this knowledge may assist in promoting this process (Fakudze and Rollnick 2008, p. 69).

They stated that

> Most learners in Southern Africa speak one language at home and are expected to study in a different language at school. The extent of separation of these two contexts is determined by whether the school is urban or rural. ... [The] learning of science is further distanced from the home culture by its expression in either a second or foreign language, creating further logistical borders to be crossed (Fakudze and Rollnick 2008, p. 73).

Yore and Treagust (2006) pointed out that all learners of science face the 3-language problem (i.e. home, school, and science language), but nonspeakers of the language of instruction face much more distinct barriers and difficult transitions between these languages. Fakudze and Rollnick explored two significant possibilities: (a) using a discourse-based model to demonstrate 'how accessing either spoken or written mixed discourse may facilitate learners' comprehension of scientific discourse and allow a teacher to assist in its production' (Fakudze and Rollnick 2008, p. 76) and (b) how code switching is a useful strategy to 'facilitate the establishment of meaning by providing a linguistic and cultural bridge to understanding' (p. 78), that is, to assist border crossing in the science classroom. In so doing, they seek to augment Aikenhead's 'Cultural Border Crossing Hypothesis', which 'has not considered the issue of language' (p. 81):

> The use of language is an important aspect of border crossing and its management. … Where the two sides of the border are reinforced by a difference of language and the need for code switching, the gap can appear wider and more difficult to cross (p. 91).

Given that 'the issue of language … plays a very important role in the acquisition of science concepts' (p. 81), any account of internationalisation (or indigenisation, for that matter) that ignores this is likely to remain somewhat impoverished.

Guo (2008) examined traditional Chinese and indigenous cultural and language practices in Taiwanese science classrooms and found that the habits of mind of traditional Chinese philosophers tend to be intuitive, metaphorical, descriptive, and holistic in contrast to the rational, causal, analytical, and reductive ways of thinking that are emphasised in WMS. He also suggested that distinctive features of Chinese words and cultural beliefs might influence students' learning of science. Any teacher who believes in and uses an interactive–constructivist teaching approach (conceptual change, guided inquiry, etc.) needs to realise that nondominant language speakers will have much of their meaningful prior knowledge, experiences, and linguistic resources stored in their home language, which cannot be accessed, engaged, challenged, and applied in the dominant language of instruction. Furthermore, many of their knowledge-building language resources will be in their home language; therefore, these students need time in science classroom to share, negotiate, construct, argue, explain, and apply the prior knowledge and cognitive resources in peer groups using the same home language before whole-class deliberation in the languages of instruction and science (Yore 2012).

An additional problem with *both* internationalisation *and* indigenisation is that these approaches commit what might be called the fallacy of the collective singular. This is an essentialist fallacy that pervades reference to, say, German culture, European identity, Asian humility, the African university, the essence of Africa, and the like. The Bologna Declaration also seems to contain what Welsch has defined as 'the traditional concept of culture', where cultures are seen as separate and distinct 'islands' or closed 'spheres' (Welsch 2000, p. 330):

> The vitality and efficiency of any civilisation can be measured by the appeal that *its culture*[emphasis added] has for other countries. We need to ensure that the European higher education system acquires a world-wide degree of attraction equal to our extraordinary cultural and scientific traditions (European Commission 1999, pp. 2–3).

Botha (2010) provided a similar critique applied to Africa. In fact, neither internationalisation nor indigenisation appears to be able to do justice to the ways in which scientific content, principles, and practices—let alone culture and identity—are learned, developed, and transformed. It also remains unclear how these approaches could satisfactorily account for the worldwide attractiveness of 'the European' or 'the African' secondary and higher education system, respectively.

55.5 Multiculturality and Interculturality

Welsch's (2000) analysis of the traditional notion of culture was characterised by three pillars: social homogenisation, an ethnic foundation, and cultural delimitation. The problem, concisely, was that the depiction of cultures as separate, distinct islands or self-contained spheres was both unrealistic and normatively dangerous. It was unrealistic because it is descriptively and empirically weak, if not altogether mistaken. Throughout human history, there have been extensive transmission and dissemination (transsemination) among cultures and civilisations. Even during the eighteenth century for German philosopher Johann Gottfried Herder (to whom Welsch attributes this notion), there would have been few, if any, cultures completely untouched, uninfluenced, or not otherwise inspired by coexisting cultures. The idea of single cultures is also normatively dangerous because of its proximity to what might be called *culturism* (cultural racism, elitism, or exclusivism).

Given recognition of the significance of these problems, both empirical and normative, there have been two trends (not least in educational theory) in the latter half of the twentieth century to account for the ever-increasing transsemination and, importantly, to promote recognition, tolerance, and respect among human beings. Both multiculturality and interculturality seek to transcend the narrow confines of the traditional concept and to foster mutual understanding among cultures. Does either of these ideas provide a resolution to the impasse in the internationalisation–indigenisation debate?

Welsch (2000) argued that both concepts are problematic in that their very structure (one might say, more accurately, their grammar) still presupposes the very notion of the single cultures they purportedly repudiate. The idea of multiculturality emphasises the coexistence of different cultures within one and the same society. While this constitutes an improvement on the demand for social homogenisation, multiculturality is unable to address the resultant problems of this cultural plurality. It is not able to do so because of its conception of this multitude of cultures as individually homogenous. In fact, all it implies is the mere fact of coexistence—it says, or can say, very little about transsemination, whether descriptively or prescriptively. Welsch suggested it comes as no surprise that circumstances in the United States should have entailed some kind of justification of and increasing appeals to intercultural delimitation by theorists of multiculturality.[4]

The idea of interculturality does not appear to fare much better, for very similar reasons. It does go beyond emphasising mere coexistence of different cultures, by concerning itself with the issue of difficulty in cooperation and collaboration[5]—but it, too, conceptually presupposes the traditional conception of single, distinct cultures. Therefore, the problems it hopes to address must remain elusive since they arise because of the very presupposition that cultures are separate islands or self-contained spheres. The diagnosis of intercultural conflict is followed by

---

[4] Welsch cites Amy Gutmann and Will Kymlicka, among others.

[5] See Council of the European Union (2010, p. 2).

advocacy of intercultural dialogue.[6] Yet, the basic problem remains, encapsulated in the thesis of essential separateness or distinctness of the conflicting and dialoguing cultures.[7] Thus, any of the envisaged changes would ultimately be little more than cosmetic. Nevertheless, is this thesis, which constitutes not only the traditional conception of culture but also underlies the ideas of multiculturality and interculturality, *correct*? If it were, then the problems of the coexistence and cooperation/collaboration of different cultures would remain with us—and would arguably remain unsolvable.

55.6 Transkulturalität

The central goal of this chapter—to position the IKW–WMS debate within the context of contemporary philosophical views of science and pedagogical insights into science learning and teaching—remains unsatisfied since multiculturalism and interculturalism appear to be a weak solution; one last position would be transculturalism. In Africa, *Transkulturalität*, or transculturality, presents itself as a possible response to the impasse. The central thesis is that the conception espoused in the traditional view of culture, and more or less unintentionally adopted or presupposed by the views that have succeeded it, is simply false. In other words, the depiction of cultures as islands or spheres is factually incorrect and normatively deceptive. Our cultures, Welsch (2000) suggested, no longer have the purported form of homogeneity and separateness but are, instead, characterised by mixtures and permeations. Welsch described this new structure of cultures as '*trans*cultural'—insofar as the determinants of culture now *traverse* (i.e. go *through*) cultures and *cross* their traditional boundaries and insofar as the new form *transcends* (i.e. goes *beyond*) the traditional conception (Welsch 2000, p. 335).

The understanding of transculturality so explained applies both on a macro-level, pertaining to the changed and changing configuration of present-day cultures, and on a micro-level, referring to the cultural make-up and shape of individuals. The mixtures and permeations that characterise our cultures are the result of technological advances, communication and travel, economic connectivity and dependencies, and—even more recently and importantly—of the increasing democratisation of societies. Examples of these permeations include moral and social issues and states of awareness that characterise many, if not all, allegedly different cultures: the debates about human and nonhuman rights, feminist thinking, same-sex relationships, and ecological consciousness, to mention only a few. Examples from commercial interaction (*trans*actions), sport and popular culture abound—rugby has invaded Asia, Eastern Europe, and North and South America; smart phones, hand-held technologies, and electronic gaming are ubiquitous, while hip-hop can be found in

[6] See Aikenhead (2001, p. 4), described earlier, Problems with Internationalisation.

[7] See Welsch (2000, pp. 334–335).

Africa, Asia, Australia, and Europe. Welsch suggested that contemporary cultures are generally marked by 'hybridisation' (Welsch 2000, p. 337). Nonetheless, some critics would disagree with him when he claimed that the grounds for selectivity between one's own culture and foreign (or other) culture have all but disappeared and that

> there is little, if anything, that is strictly 'foreign' or 'other'; everything is within reach. By the same token, there is little, if anything, that can be called 'own': Authenticity has become folklore. It is ownness simulated for others, to whom the indigene himself has long come to belong (Welsch 2000, p. 337).

The transcultural (transmission between cultures) can be seen in the uptake of procedures and products among cultures and nation states. The Truth and Reconciliation process, underpinned as it was by a commitment to restorative justice, was historically and recognisably South African—even though it has been successfully applied and has transformed judicial thinking and practice, globally. Similarly, knowledge of the thirst- and appetite-suppressing qualities of the *!khoba* cactus (or *Hoodia gordonii*) originated with the San community, although the product has since been commercialised and is now available at pharmacies all over the world. Transculturality also operates on a micro-level (i.e. individual) where the vast majority of human beings are constituted in their cultural formations by a multitude of cultural origins, affiliations, and connections. 'We are cultural hybrids' (Welsch 2000, p. 339) or 'individuals in any cultural context are multiply situated/positioned' (López 1998, p. 227). We may have a particular national identity, but we may also have a multitude of cultural identities.

So, does transculturality yield a pertinent philosophical perspective on transmission of knowledge and practices, on the transformation of educational systems, or on the IKW–WMS issue? It *may*, but this verdict may require some additional conceptual clarification as well as more empirical substantiation. Welsch asserted that transculturality is itself a temporary diagnosis, which refers to a transition or, rather, a phase within a transition. It takes as its starting point the traditional idea of single cultures and maintains that this idea—whatever the appeal it may still hold for many—no longer applies, at least not to the vast majority of contemporary cultures. The concept of transculturality seeks to capture an understanding of a contemporary and future constitution of cultures that is no longer monocultural but transcultural. This does not mean that the concept of culture has become empty; according to Welsch, it makes good sense to speak of a coexistence of reference cultures and of new, transcultural nets or webs that emanate from these anchor points.

An objection that might be raised at this point may take the form of the *argument from entropy*—that the ever-increasing transsemination will itself logically lead to a kind of homogenisation, that the erstwhile individual (trans)cultural systems will become indistinguishable from one another, and that transculturality will level out in a kind of bland pan-cultural sameness, a global closed system. The argument is that not only that the idea of cultures will have been rendered redundant but the very notion of transculturality will also have ceased to apply. It would appear that Welsch himself has brought on this objection, by claiming that transculturality is itself a *temporary* diagnosis. However, further elucidation shows that the new reference cultures will themselves have transcultural configurations that are the reference point for the weaving of new transcultural webs. In addition, the different individual, social, geographical–environmental, and historical–political contexts will more than ensure that an entropic end state is highly unlikely to be bought about.

How does transculturality help address the central issue of IKW and WMS in teaching indigenous and nonindigenous students ideas about nature and naturally occurring events? Yore and Guo (2008) suggested that transcultural science instruction in Taiwan might be less conflicting (philosophy pedagogy) if indigenous technologies (e.g. animal traps, food preservation, fabrics, games, house construction, household tools, jewellery) were used to engage students and build social capital and trust before considering WMS ideas. Modern information communication technologies (smart phones, hand-held devices, etc.) and young people's ubiquitous engagement in electronic games and gaming might provide a common platform for exploring technological design and technologies. Thereby, historical technologies would serve as bridge into modern technology as design and engineering/technology practices (NRC 2012).

A prototypical example of transcultural science instruction in Taiwan illustrated the underlying principles and procedures for development of responsive and respectful approach to IKW and WMS (Lee et al. 2011). They reported on a case study of collaborative curriculum planning and teaching of a Grade 4 unit about the conception and measurement of time in an Amis (indigenous peoples of southeastern Taiwan) community school, which enrolled both indigenous and nonindigenous students. The planning involved Amis elders (knowledge keepers), the classroom teacher, and science educators identifying IKW about time events and devices and resource people that could be embedded into the prescribed unit of study and textbook coverage. Ideas about ritual celebrations, appearance of plants and animals, lunar phases, tides, and other cyclic events were embedded into the normal study of the Earth–moon–sun system; definitions of days, months, and years; and measures of minutes and hours. A reflective–responsive mechanism was used to monitor and adjust the teaching sequence and contents. Qualitative information indicated that the indigenous and nonindigenous students were engaged in the lessons and the teacher believed that the achievement of all students was much better than normal based on previous classes of students. Lee et al. stated:

> Although the current decline of Amis culture is difficult to halt, the "Measuring Time" module at least promotes student interest in learning and helps restore their cultural identity and pride. Although a cultural gap will continue to exist, we believe that students individually will adjust their innermost thoughts in ways that make sense to each student' (Lee et al. 2011, p. TBD).

Furthermore, indigenous and nonindigenous students will have a better appreciation of each other's cultures.

55.7 Conclusion: Philosophy of Education and the Role of the University

French philosophers Gilles Deleuze and Félix Guattari claimed, 'La philosophie ... est la discipline qui consiste à créer des concepts' [Philosophy ... is the discipline of creating concepts] (Deleuze and Guattari 1991, p. 10). Far more, one might argue, apart from its task being the creation of concepts (*if* it is that!), philosophy—including philosophy of education—addresses conceptual clarification and helps in determining the appropriateness or applicability of concepts, their interconnectedness, role in argumentation, etc. One of the most important functions of philosophy is arguably that of tireless critical interrogation—not only of concepts but also of premises, beliefs, values, assumptions, and commitments—and, by inquiring into their meaning and justification, not to mention their truth, to attempt to resolve some of the most fundamental ontological, epistemological, ethical, and educational questions.[8]

As Thomas Auf der Heyde (former dean of research, University of Johannesburg; 2005) has pointed out, universities clearly stand to benefit from globalisation—so, from an economic point of view, the question whether they are justified in embracing globalisation (e.g. of the knowledge economy) receives a quick and simple answer. The more interesting and difficult question is in what way, if any, their role as social observer and commentator, and their responsibility to critically reflect on the phenomenon of globalisation (Auf der Heyde 2005), can be made to complement the interest of the state, the universities' key stakeholders, etc. If Auf der Heyde is correct in saying that 'universities ... should also be critically appraising the issues raised by [globalisation]' (p. 41), then this is where philosophy of education arguably has its natural home. The role of philosophy consists in part in counteracting the hegemony and despotism of both homogenising (colonising) and traditional (indigenising) authority.

Acknowledgements The authors accept responsibility for the message in this chapter but would like to recognise the collective contributions that included insightful comments provided by reviewers, Cathleen Loving (Texas A&M University, USA), Christine McCarthy (University of Iowa, USA), Dawn Sutherland (University of Winnipeg, Canada), and Don Metz (University of Winnipeg, Canada), and the insights and help provided by Michael Matthews (University New South Wales, Australia) and Ron Good (Emeritus, Louisiana State University, USA). We wish to express special thanks to Sharyl Yore for the technical editing and blending two dispersant voices and languages into a reasonably understandable text. Thank you all!

References

Abrams, E., Taylor, P., & Guo, C-J. (Eds.). (in press). Pedagogies of hope: Culturally relevant teaching for indigenous learners in science and mathematics [Special issue]. *International Journal of Science and Mathematics Education.*

Aikenhead, G.S. (1996). Science education: Border crossing into the subculture of science. *Studies in Science Education*, *27*(1), 1–52.

[8] See McLaughlin (2000, pp. 444 & 448), Wimmer (2000, pp. 413–414).

Aikenhead, G.S. (1997). Toward a First Nations cross-cultural science and technology curriculum. *Science Education, 81*(2), 217–238.

Aikenhead, G.S. (2001). Integrating western and aboriginal sciences: Cross-cultural science teaching. *Research in Science Education, 31*(3), 337–355.

Aikenhead, G. S. (2005). Science-based occupations and the science curriculum: Concepts of evidence. *Science Education, 89*(2), 242–275.

Aikenhead, G.S., & Jegede, O.J. (1999). Cross-cultural science education: A cognitive explanation of a cultural phenomenon. *Journal of Research in Science Teaching, 36*(3), 269–287.

Andreasson, S. (2008). Indigenisation and transformation in southern Africa. Paper prepared for the British International Studies Association Annual Conference, University of Exeter, United Kingdom, 15-17 December. http://works.bepress.com/cgi/viewcontent.cgi?article=1010&context=stefan_andreasson (retrieved 12 May 2009)

Andreasson, S. (2010). Confronting the settler legacy: Indigenisation and transformation in South Africa and Zimbabwe. *Political Geography, 29*(8), 424–433.

Auf der Heyde, T. (2005). Globalisation, resistance and the university. *Discourse, 33*(2), 41–48. Retrieved from http://reference.sabinet.co.za/document/EJC31137

Bishop, R. (1998). Freeing ourselves from neo-colonial domination in research: A Māori approach to creating knowledge. *Qualitative Studies in Education, 11*(2), 199–219.

Botha, M.M. (2010). Compatibility between internationalizing and Africanizing higher education in South Africa. *Journal of Studies in International Education, 14*(2), 200–213.

Cesana, A. (2000). Philosophie der Interkulturalität: Problemfelder, Aufgaben, Einsichten [Philosophy of interculturality: Problem areas, tasks, insights]. In A. Cesana & D. Eggers (Eds.), *Thematischer Teil II – Zur Theoriebildung und Philosophie des Interkulturellen/ Jahrbuch Deutsch als Fremdsprache 26* [Thematic part II - On the theory and philosophy of the intercultural? Yearbook German as a foreign language 26] (pp. 435–461). Munich, Germany: Iudicium Verlag.

Chetty, N. (2010, August 27). Recounting the myths of creation, *Mail & Guardian,* pp. 5–6. Retrieved from http://mg.co.za/article/2010-08-27-recounting-the-myths-of-creation

Chinn, P. W. U., Hand, B., & Yore, L. D. (2008). Culture, language, knowledge about nature and naturally occurring events, and science literacy for all: She says, he says, they say [Special issue]. *L1—Educational Studies in Language and Literature, 8*(1), 149–171. Retrieved from http://l1.publication-archive.com/show?repository=1&article=220

Cohen, S. (1986). Knowledge and context. *Journal of Philosophy, 83*(10), 574–583.

Confederation of EU Rectors' Conferences & Association of European Universities. (2000). *The Bologna Declaration on the European space for higher education: An explanation*. Brussels, Belgium: European Commission. Retrieved from http://ec.europa.eu/education/policies/educ/bologna/bologna.pdf

Council of the European Union. (2010, May 11). *Council conclusions on the internationalisation of higher education*. Brusells, Belgium: 3013th Education, Youth, & Culture Council. Retrieved fromhttp://www.sefi.be/wp-content/uploads/114378%20Conclusiones%20Consejo%20Internacionalización.pdf

Deleuze, G., & Guattari, F. (1991). *Qu'est-ce que la philosophie?* [What is philosophy?] Paris, France: Les Éditions de Minuit.

European Commission. (1999). *The Bologna process: Towards the European higher education area* (Joint declaration of the European Ministers of Education). Brussels, Belgium: Author. Retrieved from http://ec.europa.eu/education/policies/educ/bologna/bologna_en.html

Fakudze, C., & Rollnick, M. (2008). Language, culture, ontological assumptions, epistemological beliefs, and knowledge about nature and naturally occurring events: Southern African perspective [Special issue]. *L1–Educational Studies in Language and Literature, 8*(1), 69–94. Retrieved from http://l1.publication-archive.com/public?fn=enter&repository=1&article=216

Fig, D. (2005). *Uranium road: Questioning South Africa's nuclear direction*. Johannesburg, South Africa: Jacana.

Finger, E. (2009, June 25). Hegel, hilf! [Help, Hegel!]. *Die Zeit*, p. 27. Retrieved from http://www.zeit.de/2009/27/01-Studium

Ford, M. J. (2008). Disciplinary authority and accountability in scientific practice and learning. *Science Education, 92*(3), 404–423.
Goldman, A.I. (1991). *Liaisons: Philosophy meets the cognitive and social sciences*. Cambridge, MA: MIT Press.
Good, R. G. (1996, March-April). *Trying to reach consensus on the nature of science*. Paper presented at the annual meeting of the National Association for Research in Science Teaching, St. Louis, MO.
Good, R. G. (2005). *Scientific and religious habits of mind: Irreconcilable tensions in the curriculum*. New York, NY: Peter Lang.
Good, R. G., Shymansky, J. A., & Yore, L. D. (1999). Censorship in science and science education. In E. H. Brinkley (Ed.), *Caught off guard: Teachers rethinking censorship and controversy* (pp. 101–121). Boston, MA: Allyn & Bacon.
Gott, R., & Duggan, S. (2003) *Understanding and using scientific evidence*. London, England: Sage.
Guo, C.-J. (2008). Science learning in the contexts of culture and language practices: Taiwanese perspective [Special issue]. *L1 – Educational Studies in Language and Literature, 8*(1), 95–107. Retrieved from http://l1.publication-archive.com/show?repository=1&article=217
Haack, S. (2003). *Defending science—within reason: Between scientism and cynicism*. Amherst, NY: Prometheus.
Harding, S. G. (1991). *Whose science? Whose knowledge? : Thinking from women's lives*. Ithaca, NY: Cornell University Press.
Harding, S. G. (Ed.). (2011). *The postcolonial science and technology studies reader*. Durham, NC: Duke University Press.
Henriques, L. (1997). *A study to define and verify a model of interactive-constructive elementary school science teaching* (Unpublished doctoral dissertation). University of Iowa, Iowa City, IA.
Horsthemke, K. (2004). Knowledge, education and the limits of Africanisation. *Journal of Philosophy of Education, 38*(4), 571–587.
Horsthemke, K. (2006). The idea of the African university in the 21st century: Some reflections on Afrocentrism and Afroscepticism. *South African Journal of Higher Education, 20*(4), 449–465.
Horsthemke, K. (2010). *The moral status and rights of animals*. Johannesburg, South Africa: Porcupine Press.
Jegede, O.J. (1999). Science education in nonwestern cultures: Towards a theory of collateral learning. In L.M. Semali & J.L. Kincheloe (Eds.), *What is indigenous knowledge?: Voices from the academy* (pp. 119–142). London, England: Falmer Press.
Le Grange, L. (2004). Western science and indigenous knowledge: Competing perspectives or complementary frameworks? *South African Journal of Higher Education, 18*(3), 82–91.
Lee, H., Yen, C-F., & Aikenhead, G. S. (2011). Indigenous elementary students' science instruction in Taiwan: Indigenous knowledge and western science. *Research in Science Education*. Advance online publication. doi:10.1007/s11165-011-9240-7
López, G.R. (1998). Reflections on epistemology and standpoint theories: A response to 'a Māori approach to creating knowledge'. *Qualitative Studies in Education, 11*(2), 225–231.
Loving, C. C. (2002). Nature of science activities using the scientific profile: From the Hawking-Gould dichotomy to a philosophy checklist. In W. F. McComas (Ed.), *The nature of science in science education: Rationales and strategies* (Vol. 5, pp. 137–150). Dordrecht, The Netherlands: Springer.
Makgoba, M.W. (1997). *Mokoko: The Makgoba affair*. Johannesburg, South Africa: Vivlia.
Makgoba, M. W. (2003, May 2–8). An African vision for mergers, *Mail and Guardian Supplement, Beyond Matric,* pp. 1–2.
Makgoba, M.W., & Mubangizi, J.C. (Eds.). (2010). *The creation of the University of KwaZulu-Natal: Reflections on a merger and transformation experience*. New Delhi, India: Excel.
Masehela, K. (2004, September 30). Escaping Europe's clutches. *This Day*/Opinion, p. 11.
Matthews, M.R. (2000). Constructivism in science and mathematics education. In D.C. Phillips (Ed.), *Constructivism in education: Opinions and second opinions on controversial issues* (Vol. 1 of 99th Yearbook, pp. 161–192), Chicago, IL: National Society for the Study of Education.

McComas, W. F. (1996). Ten myths of science: Reexamining what we think we know about the nature of science. *School Science and Mathematics, 96*(1), 10–16.

McKinley, E., & Keegan, P. J. (2008). Curriculum and language in Aotearoa New Zealand: From science to *Putaiao* [Special issue]. *L1–Educational Studies in Language and Literature, 8*(1), 135–147. Retrieved from http://l1.publication-archive.com/public?fn=enter&repository=1&article=219

McLaughlin, T.H. (2000). Philosophy and educational policy: Possibilities, tensions and tasks. *Journal of Education Policy, 15*(4), 441–457.

National Governors Association Center for Best Practices & Council of Chief State School Officers. (2010). *Common core state standards*. Washington, DC: Author. Retrieved from http://www.corestandards.org/

National Research Council. (1996). *The national science education standards*. Washington, DC: National Academies Press.

National Research Council. (2000). *How people learn: Brain, mind, experience, and school—Expanded edition* (J. D. Bransford, A. L. Brown, & R. R. Cocking, Eds.). Washington, DC: National Academies Press.

National Research Council. (2007). *Taking science to school: Learning and teaching science in grades K-8* (R. A. Duschl, H. A. Schweingruber, & A. W. Shouse, Eds.). Washington, DC: National Academies Press.

National Research Council. (2012). *A framework for K-12 science education: Practices, crosscutting concepts, and core ideas* (H. Quinn, H. A. Schweingruber, & T. Keller, Eds.). Washington, DC: The National Academies Press.

Nola, R. (1997), Constructivism in Science and in Science Education: A Philosophical Critique, *Science & Education 6* (1–2), 55–83. Reproduced in M.R. Matthews (ed.), *Constructivism in Science Education: A Philosophical Debate*, Kluwer Academic Publishers, Dordrecht, 1998, pp. 31–59.

Ntuli, P.P. (2002). Indigenous knowledge systems and the African renaissance: Laying a foundation for the creation of counter-hegemonic discourses. In C.A. Odora Hoppers (Ed.), *Indigenous knowledge and the integration of knowledge systems: Towards a philosophy of articulation* (pp. 53–66). Claremont, South Africa: New Africa Books.

Onwu, G., & Mosimege, M. (2004). Indigenous knowledge systems and science and technology education: A dialogue. *African Journal of Research in Mathematics, Science and Technology Education, 8*(1), 1–12.

Rogers, A. (2007). The making of *Cosmic Africa*: The research behind the film. *African Skies/Cieux Africains, 11*(July), 19–23.

Rogers, A., & Rubin, C. (Producers), & Foster, C., & Foster, D. (Directors). (2003). *Cosmic Africa* [Motion picture]. Cape Town, South Africa: Aland Pictures.

Scheffler, I. (1965). *Conditions of knowledge: An introduction to epistemology and education*. Chicago, IL: Scott, Foresman.

Schily, K. (2009). Leitwährung [Reserve currency]: Credit Point. *Die Zeit*, p. 46.

Snively, G. J., & Williams, L. B. (2008). "Coming to know": Weaving Aboriginal and western science knowledge, language, and literacy into the science classroom [Special issue]. *L1–Educational Studies in Language and Literature, 8*(1), 109–133. Retrieved from http://l1.publication-archive.com/public?fn=enter&repository=1&article=218

South Africa Department of Education. (2008, November 30). *Report of the ministerial committee on transformation and social cohesion and the elimination of discrimination in public higher education institutions*. Johannesburg, South Africa: Author. Retrieved from http://www.pmg.org.za/files/docs/090514racismreport.pdf

Stanley, W.B., & Brickhouse, N.W. (1994). Multiculturalism, universalism, and science education. *Science Education, 78*(4), 387–398.

Suchting, W.A. (1992), Constructivism Deconstructed, *Science & Education 1*(3), 223–254.

Toulmin, S. E. (1958). *The uses of argument*. Cambridge, England: Cambridge University Press.

Tytler, R., Duggan, S., & Gott, R. (2001). Dimensions of evidence, the public understanding of science and science education. *International Journal of Science Education, 23*(8), 815–832.

Visvanathan, C.S. (2002). Between pilgrimage and citizenship: The possibilities of self-restraint in science. In C.A. Odora Hoppers (Ed.), *Indigenous knowledge and the integration of knowledge systems: Towards a philosophy of articulation* (pp. 39–52). Claremont, South Africa: New Africa Books.

Walton, D. (2005). *Fundamentals of critical argumentation: Critical reasoning and argumentation.* New York, NY: Cambridge University Press.

Walton, D., Reed, C., & Macagno, F. (2008). *Argumentation schemes*. New York, NY: Cambridge University Press.

Webb, P. (in press). Xhosa indigenous knowledge: Stakeholder, value and choice. *International Journal of Science and Mathematics Education*

Welsch, W. (2000). Transkulturalität: Zwischen Globalisierung und Partikularisierung [Transculturality: Between globalization and particularization]. In A. Cesana & D. Eggers (Eds.), Thematischer Teil II – *Zur Theoriebildung und Philosophie des Interkulturellen Jahrbuch Deutsch als Fremdsprache* 26 [Thematic part II - On the theory and philosophy of the intercultural/Yearbook German as a foreign language 26] (pp. 327–351). Munich, Germany: Iudicium Verlag.

Wimmer, F. M. (2000). Kulturalität und Zentrismen im Kontext interkultureller Philosophie [Culturalityand centrisms in the context of intercultural philosophy]. In A. Cesana & D. Eggers (Eds.), *Thematischer Teil II – Zur Theoriebildung und Philosophie des Interkulturellen/ Jahrbuch Deutsch als Fremdsprache 26* [Thematic part II - On the theory and philosophy of the intercultural/Yearbook German as a foreign language 26] (pp. 413–434). Munich, Germany: Iudicium Verlag.

Witt, N. (2007). What if indigenous knowledge contradicts accepted scientific findings? – The hidden agenda: Respect, caring and passion towards aboriginal research in the context of applying western academic rules. *Educational Research and Review, 2*(3), 225–235.

Wolpert, L. (1993). *The unnatural nature of science: Why science does not make (common) sense.* Cambridge, MA: Harvard University Press.

Yore, L. D. (2008). Science literacy for all students: Language, culture, and knowledge about nature and naturally occurring events [Special issue]. *L1—Educational Studies in Language and Literature, 8*(1), 5–21. Retrieved from http://l1.publication-archive.com/show?repository=1&article=213

Yore, L. D. (2012, September). *Science and engineering language practices: Communicative, epistemic, and rhetorical functions of language in science/engineering.* Paper presented at the National Taiwan Normal University, Taipei, ROC/Taiwan.

Yore, L. D., & Guo, C-J. (2008, March). *Cultural beliefs and language practices in learning and teaching science--symposium: The intersection of the influence of schooling, culture, and nature on the motivation of Hawaiian and Taiwanese indigenous children Taiwanese perspective.* Paper presented at the annual meeting of the National Association for Research in Science Teaching, Baltimore, MD.

Yore, L. D., Hand, B. M., & Florence, M. K. (2004). Scientists' views of science, models of writing, and science writing practices. *Journal of Research in Science Teaching, 41(4)*, 338–369.

Yore, L. D., & Treagust, D. F. (2006). Current realities and future possibilities: Language and science literacy—empowering research and informing instruction [Special issue]. *International Journal of Science Education, 28*(2/3), 291–314.

Ziman, J. (2000). *Real science: What is it, and what it means.* Cambridge, England: Cambridge University Press.

Kai Horsthemke is an associate professor and current head of the Division of Studies in Education at Witwatersrand University. He was educated both in Germany and in South Africa and was awarded his Ph.D. in Applied Ethics in the Department of Philosophy, University of the Witwatersrand. A professional musician for several decades, he has been involved in stage, studio, television, and cruise ship work (in South Africa, Namibia, Mozambique, Germany, Luxembourg, England, Indian Ocean islands, Japan, Hawaii, and Colombia). He teaches philosophy of education—ethics, social and political philosophy, epistemology, philosophy of science, and logic and critical thinking, all with a strongly educational focus. He has published extensively since 2004, including a book entitled *The Moral Status and Rights of Animals* (Porcupine Press 2010). Apart from animal rights, his research interests include African philosophy (of education), indigenous knowledge (indigenous science, ethnomathematics, ethnomusicology), as well as humane and environmental education. He is currently working on a book with the provisional title *The myth of 'indigenous knowledge'*.

Larry D. Yore is a University of Victoria distinguished professor emeritus, retiring in June 2011 after 41 years. He was educated at the University of Minnesota (B.S., 1964; M.A., 1968; Ph.D., 1973). In over four decades of postsecondary teaching and research, Larry has been engaged in developing provincial science curricula, national science frameworks, national K-12 assessment projects in North America, and he has published widely in major science education journals and anthologies. He has served on or is currently a member of the editorial boards or review panels of the *Journal of Research in Science Teaching, School Science and Mathematics, Science Education, Journal of Science Teacher Education, Journal of Elementary Science Education, International Journal of Science Education, L1—Educational Studies in Language and Literacy, Science and Technology Education*, and *International Journal of Science and Mathematics Education*. Larry currently serves as associate editor and founding member (2001) of *International Journal of Science and Mathematics Education*, a SSCI listed journal. He received the 2005 Association for Science Teachers Education's Science Teacher Educator of the Year Award and 2012 National Association for Research in Science Teaching's Distinguished Contributions through Research Award. His 2012 publications include chapters entitled 'Science Literacy For All—More than a slogan, logo, or rally flag!' In K. C. D. Tan, M. Kim, and S. Hwang (Eds.), *Issues and challenges in science education research: Moving forward* (pp. 5–23) and with Carole L. Yore 'Toward convergence of metacognition, reflection, and critical thinking: Illustrations from natural and social sciences teacher education and classroom practice' In A. Zohar and J. Dori (Eds.), *Metacognition in science education: Trends in current research* (pp. 251–271); both books were published by Springer.

Chapter 56
Science, Religion, and Naturalism: Metaphysical and Methodological Incompatibilities

Martin Mahner

56.1 Introduction

In many countries, children receive both science education and religious education.[1] "Religious education" is here understood as an education under denominational auspices, however liberal. That is, students are not taught some unbiased comparative, historical, cultural, and social aspects of religion, but are expected to accept and internalize the doctrines of a particular religious belief system, usually the one their parents are affiliated to.[2] From a nonreligious perspective, this situation is unfortunate as it appears that an education emphasizing the need for empirical tests and evidence is incompatible with an education that allows for, or even encourages, the acceptance of factual beliefs without or even contrary to evidence. In other words, learning to accept statements only if there is sufficient evidence for them and learning to accept claims on sheer faith appear to be antagonistic educational goals (Mahner and Bunge 1996a; Martin 1997).

Evidently, this concern rests on the assumption that science and religion are mutually incompatible, whereby "incompatible" means that one cannot rationally accept both a scientific and a religious world view. Though common among (consistent) naturalists and secular humanists,[3] this view is of course contested by many

[1] This contribution uses material published earlier in the journal *Science & Education*, namely, from Mahner and Bunge (1996a, b) and Mahner (2012).

[2] Of course, there are approaches to teach religion in a very general sense of "spirituality," whatever that exactly means (see, e.g., Stolberg and Teece 2011). Even so the presupposition is that this spirituality comprises more than what can be obtained in a comprehensive scientific worldview.

[3] See, e.g., Clements (1990), Dawkins (2006), Dennett (2007), Edis (2007, 2008, 2009), Kanitscheider (1996), Kitcher (2004), Kurtz (2003), Mahner and Bunge (1996a, b), Martin (1997), Provine (2008), Rachels (1991), Smart (1967), Stenger (2007, 2011), and Suchting (1994).

M. Mahner (✉)
Zentrum für Wissenschaft und kritisches Denken, Rossdorf, Germany
e-mail: mahner@gwup.org

M.R. Matthews (ed.), *International Handbook of Research in History, Philosophy and Science Teaching*, DOI 10.1007/978-94-007-7654-8_56,

religionists and even some naturalistically inclined scientists and philosophers.[4] Therefore, it will be necessary to defend it. If we succeed in showing that science and religion are incompatible, it is a mere corollary that science and religious education are also incompatible.

Any argument for either the compatibility or incompatibility of science should work with a reasonably clear definition of both *science* and *religion*. However, the very existence of such definitions has been contested for a long time (see, e.g., Glennan 2009). There have been arguments to the effect that there is no reasonable demarcation between science and nonscience, in particular pseudoscience (Laudan 1983), and that, similarly, religion is so diversified that any attempt to formulate a definition that covers all religions is futile (Platvoet and Molendijk 1999). As this is not the place to review these arguments,[5] it will be helpful to narrow down what we take science and religion to be, so that we can focus on those aspects that may or may not be compatible. Before we get to this point, however, it will suffice to work with the undefined everyday usages of "science" and "religion." So we start with the question of how to avoid conflict between religion and science.

56.2 How to Avoid Conflict Between Religion and Science

Claims, theories, or world views may be in mutual conflict only if there is an at least partial overlap in their subject matter. Indeed, traditionally religions have offered general cosmologies (or metaphysics) helping to explain the major features of the world, in particular the place of humans and their relationship to the various supernatural entities allegedly populating the world alongside humans. After all, "[a]ll religions do share a feature: ostensible communication with humanlike, yet nonhuman, beings through some form of symbolic action" (Guthrie 1995, p. 197) – and this requires some factual background assumptions. While science has emerged from such a religious cosmology, it has now superseded the latter, it has significantly changed its metaphysical framework, and it has taken over the explanatory function of the old cosmologies. It appears therefore that, concerning matters of fact, religion has ceded this explanatory role to science, focusing now on other tasks. And it appears as if this concession has removed any former conflicts. While we shall see later on that this appearance is deceptive, let us first take a look at the attempts at reconceptualizing either religion or science to prevent them from being in conflict.

---

[4] See, e.g., Alston (2004), Barbour (2000), Clayton and Simpson (2008), Drees (1996), Gould (1999), Harrison (2010), Haught (1995), Peacocke (1993), Polkinghorne (1987), Rolston (1987), Ruse (2001a, 2011), Stenmark (2010), and Wentzel van Huyssteen (1998).

[5] For critiques of Laudan's view, see Mahner (2013) and Pigliucci (2013). For a comparison of concepts of religion, see Guthrie (1995). As for the demarcation of science in general, see Mahner (2007) and Thagard (2011).

56.2.1 Science and Religion Deal with Different Aspects of the World or Even with Different Realities

There are several ways to render science and religion independent so that they cannot be in conflict. One way is ontological. It splits the world into two radically different parts: a material world (nature) and a transcendent world (supernature). Whereas nature is studied by science, supernature is studied by religion.[6] A historically important example of this approach is deism, which allowed scientists to study the natural world without resorting to supernatural interventions (apart from the initial act of divine creation). However, conflict with science can only be avoided if these two worlds are causally independent: if there are causal interactions, we sooner or later face conflicting explanations. Yet if such a supernature is causally independent of the natural world, there can be no evidence for its existence so that it remains merely a conceptual possibility, moreover, one without any explanatory function as its existence or nonexistence would make no difference to our world. Such a radical split then is not very attractive to most religious believers, who usually long for a connection between themselves and the divine.

A second ontological possibility is to assume that supernatural agents are not agents in the familiar sense, but only "underlying" causes. While natural causes are (merely) "secondary causes" studied by science, God works as the "primary cause" behind the scenes. Indeed, according to some authors, God has been rather busy pulling the strings behind quantum physics and evolution, for example (Freddoso 1991; Barbour 2000; Plantinga 2011). The concepts of agency and cause involved here are best understood from the viewpoint of Scholastic metaphysics, which, though long superseded in science, is still going strong in Catholic philosophy. From a naturalist perspective, the involvement of supernatural agents "behind" natural causes is an unparsimonious and hence superfluous add-on to natural causes.[7] Assuming sustaining supernatural causes behind the web of natural causes does avoid conflict at the superficial level of the daily business of science and religion, but it does not avoid conflict at the deeper metaphysical level.

A third way of keeping science and religion separate is methodological and referential. Science and religion have different tasks, and they study different objects, or different properties or aspects of the world. For example, Rolston (1987) claims that religion is concerned with morality and meaning (not in the semantic sense of course, but in the sense of "purpose"), not material facts.[8] Following this

---

[6] It may be argued that attempts to split the world into two or more "worlds" are incoherent because, by definition, the world is everything that exists (Worrall 2004). However, if our metaphysics requires that not any old collection of causally unconnected things is itself a material thing and hence a real entity, worlds are real things (or, more precisely, systems) only inasmuch as their parts are causally connected, however weakly. Two causally unconnected universes would then be two different things, and there would be no supersystem of which they would be physical parts.

[7] For the metaphysical and epistemological problems of the idea of divine intervention, see Fales (2010).

[8] For an analysis of the various meanings of "meaning" in this context, see Martin (2002). For a critique of related noncognitive concepts of religion, see Philipse (2012).

idea, the evolutionary biologist Stephen J. Gould devoted an entire book to the "nonoverlapping magisteria" (NOMA) of science and religion (Gould 1999). Of course, if two areas have different referents, goals, and methods, they can hardly be in conflict. Dancing tango and doing science are not in conflict because they are quite different pursuits. Moreover, the NOMA approach allows for the claim that science and religion are not just compatible, but even complementary: morality and the search for purpose belong to human life just as the factual study of the world.[9]

The average religious believer, however, has to pay a high price for NOMA, because the concept of religion has to be redefined in a major way; so much so that no ordinary believer may recognize it afterwards.[10] For example, Gould considers as religious "all moral discourse on principles that might activate the ideal of universal fellowship among people" (Gould 1999, p. 62). This definition is so wide that it even applies to secular ethical discourse. So if an atheist engages in such discourse, he would be religious. While being too wide on the one hand, this definition is too narrow on the other, because it presupposes that religion has no factual content. For example, whatever people have said about the soul and the afterlife, or about the existence and properties of gods or cosmic forces, is illegitimate because it involves factual, not ethical, discourse. As a consequence, most traditional religious "truths" are excluded from the legitimate business of religion. Worse, without some factual assumptions about gods (for instance, god's will) or the order of creation (natural law doctrine) or the karma, moral values and norms cannot even be justified in a religious world view (Nowell-Smith 1967). As McCauley states:

> Religions certainly do try to make sense of our lives and of the world in which we find ourselves. The problem, though, is that that process of making sense of things inevitably involves appeals to explanations about the origins, the makeup, and the behavior of things generally and about our origins, makeup, and behavior in particular. (McCauley 2011, p. 229)

Last but not least, it can be argued that ethics cannot even be based on religion (see, e.g., Rachels 1995; Martin 2002). Thus, the identification of religion with ethics fails and hence also the "different domains" approach.

56.2.2 Religion Is Not Necessarily Bound to Supernaturalism

If science is tied to naturalism, whereas religion is based on supernaturalism, as is widely held, there is ample room for conflict at both the metaphysical level and the level of scientific explanation. But is religion really tied to a supernaturalist metaphysics? According to quite a number of scientists and philosophers, it is not. Indeed, Auguste Comte, John Dewey, Henry Wieman, Julian Huxley, Charles

[9] In science education, Sinatra and Nadelson (2011) follow this approach by postulating different epistemologies for science and religion, that is, "epistemologies that have different roles and explain different aspects of the human condition" (p. 175). Obviously, and sadly, this is an instance of epistemological relativism.

[10] See McCauley (2011), Orr (1999), and Worrall (2004).

Hardwick, and others have argued for religious naturalism.[11] Thus, "God" is redefined as the unity of our ideals, or as a cosmic process unfolding for the benefit of humans, or as the creative exchange among humans, etc. Often feelings of awe towards nature or the universe are regarded as religious feelings or as a feeling of the "sacred." For example, Einstein (1999) believed that the scientist's religiosity lies in "the amazement at the harmony of natural law."

Whether pantheism is a form of religious naturalism remains unclear. In an everyday understanding according to which the world is God, pantheism does appear to be a naturalist conception. In this sense, however, Schopenhauer's criticism applies: "to call the world 'God' is not to explain it; it is only to enrich our language with a superfluous synonym for the word 'world'" (Schopenhauer 1951, p. 40).[12] Levine (2011) rejects Schopenhauer's criticism for resting on a misunderstanding of what the pantheistic "divine unity" of the world means. However, Levine's own characterization of "unity" is fuzzy to the point of being incomprehensible, and he regards "divine" simply as experiential: whatever someone experiences as numinous is divine. Thus, the divine is turned into a subjective category.[13]

With the exception of traditional pantheism, it appears that the common motif of such non-supernaturalist approaches is to redefine "religion" in terms of either feelings or experiences, leaving no room for any factual content of religion. As Barbour (2000, p. 159) rightly remarks, religious naturalism thus simply conflicts with "most of the heritage of religious traditions."

56.2.3 Defining Religion in a Merely Functional Way

Psychologists and sociologists usually refrain from defining "religion" in a substantive way, that is, with regard to its content. Instead, they define it in terms of the functions religious beliefs, practices, and institutions have in human life and society

---

[11] See, e.g., Alston (1967), Drees (1996, 2008), and Hardwick (2003).

[12] To the consistent naturalist, the attempt to naturalize religion reduces to a game of words:

> The bogus procedure is this: When there is something that clearly does not exist, but one wishes that it did exist and wants to be able to say that it does exist, then choose something real that is similar in some respects and give it the name of the nonexistent entity. Voilà! You have now proved the existence of something that doesn't exist. Suppose one wants to prove that God exists. Find something awe-inspiring, or powerful, or infinite, or fundamental .... and call it "God". Now God exists, and the various practices with respect to that God are "religious". Unfortunately, in reality, all you've done is play with words and, thereby, pull off a shabby, unconvincing trick. (Pasquarello 2002, p. 51)

This applies not only to religious naturalism but also to the various hermeneutic approaches in modern theology, such as Paul Tillich's definition of God as "ultimate concern" (for a criticism of hermeneutic theology, see Chap. 5 in Albert 1985).

[13] Peacocke's (1993) panentheism does not seem to be a consistent naturalism, as it makes God only partly natural, so I shall not discuss his view here.

(see, e.g., Yinger 1970). That religious beliefs and practices have evolved along with humankind and that they have various functions for individuals and groups is of course nothing but a scientific description and explanation of religion.[14] It is exactly these functions that remain once the cognitive content of religion is removed for being illusory. Obviously, a naturalist, scientific view of religion cannot be in conflict with science. Yet again, the problem remains that such characterizations do not match the self-conceptions of most religions.[15]

56.2.4 The Argument from Religious Scientists

Another psychological and sociological argument to consider is the claim that science and religion cannot be in conflict because there have been many religious scientists. Indeed, quite naturally there is no shortage of historical examples, which are often used to reject the historical conflict view (see, e.g., Russell 2002). And even today the number of religious scientists is high.[16] Interestingly, average scientists tend to be more religious than elite scientists (Gross and Simmons 2009). According to the latest study of Ecklund (2010), about 64 % of elite US scientists are atheists or agnostics.

The argument from religious scientists, however, is a weak one at best. At worst, it is an *argumentum ad populum*. It would come as no surprise that a large number of people can be mistaken about something. And as we know from psychology, many people hold inconsistent beliefs. This applies also to scientists. For example, it is quite telling that most religious scientists have not used religious concepts in their scientific work (Mahner and Bunge 1996b). There are no variables referring to supernatural entities or processes in scientific theories. If someone believes in the reality of the supernatural, it is inconsistent to not make use of religious entities and methods in science. Rather, we should expect religious scientists to defend a theistic science, as Ratzsch (1996) and Plantinga (2001) consistently (though of course unsuccessfully) do. But this is rarely the case. Therefore, pace Ratzsch (2004) and others, it is not implausible to suspect the world views of religious scientists to be inconsistent.

So the problem of whether science and religion are compatible or not is not a matter of psychology and sociology but of philosophy, more precisely, of metaphysics, epistemology, and methodology. If there is no conflict at this level, then the world view of religious scientists may be consistent; otherwise, it is not.

[14] See Boyer (2001), Dennett (2007), and Guthrie (1995).

[15] Further criticism of religious functionalism in Guthrie (1995).

[16] See, e.g., Ecklund (2010), Gross and Simmons (2009), Larson and Witham (1998), and Margenau and Varghese (1992).

56.2.5 Religious Discourse Is Nonsense

According to neopositivism, metaphysical sentences including religious ones are semantically nonsensical because they are not verifiable (Ayer 1990). If religious discourse is nonsense, it can be neither compatible nor incompatible with scientific discourse. So there is no conflict with science. As a consequence, however, atheist discourse is nonsensic too: if "God exists" is nonsense, its negation "God does not exist" is also nonsense.

It is rather obvious that the neopositivist answer is not a good option for religious believers. After all, they believe that they make meaningful statements about nature or supernature or both. Indeed, the neopositivist meaning criterion of verifiability has long been abandoned: in order to verify or falsify a statement, it must be semantically meaningful in the first place, not the other way round. So we cannot keep religion away by declaring it nonsense *tout court*. However, in particular academic theology does have meaning problems, as it often resorts to an irrationalist, fuzzy discourse that helps to immunize theology from factual criticism (Albert 1985; Bartley 1984). And as we shall see later, it is not at all clear what the very term "God" exactly means.

56.2.6 Distorting Science

Removing the cognitive content of religion is not the only way to avoid conflict between science and religion. There are also attempts to remove all truth claims from science by adopting antirealist views of science, such as instrumentalism or relativism. If scientific theories are not attempts at approximating truth by stating something about how the world really is, but only more or less useful tools for systematizing or predicting empirical statements, or if scientific theories are nothing but yet another way at looking at the world, on the same par as any other, even mythical way, then science may of course peacefully coexist with religion (Byl 1985; Stenmark 2010). A less radical view is constructive empiricism, which replaces truth by empirical adequacy, but it too is a view that castrates science. Given the fact that both instrumentalism and constructive empiricism are still popular in the philosophy of science, at least much more so than relativism, it may appear bold to charge them with distorting science, but this is not the place to defend this view.[17]

56.2.7 Conclusion

As we have seen, there are many ways to construe "religion." Some of them would indeed be compatible with science. But as we have also seen, the believer has to pay a high price if he accepts them. Religious entities are either rendered causally

[17] More on this in Vollmer (1990), Psillos (1999, 2003), Worrall (2004), and Ladyman (2012).

inefficacious and hence irrelevant, or religion is emptied of any factual content so that it can no longer make any (objective) truth claims. Notwithstanding the attempts of modern theology at immunizing religion from criticism by obfuscating and subjectifying its concepts (Albert 1985), the vast majority of believers of all ages has believed that their religion does make some true factual statements about the world, in particular about humans and their relation to the divine or at least to certain spiritual entities (McCauley 2011). That is, real life religions have always included a cosmology.[18] The following characterization reflects this situation. Accordingly, religion can be seen as

> ...the belief in numinous personal or impersonal entities - gods, spirits, demons, angels, or divine powers - which have certain causal powers, and which therefore are relevant to human fate and salvation, as well as [...] an associated practice of the believers, which is adequate to make allowance for the powers of these entities and to influence them for the benefit of the believers' salvation, that is, a cult characterized by a salvation technology. (Albert 2000, p. 142, my free translation)

Both religion and science thus have an overlap in that they are epistemic enterprises. Both search for truth, partly in the same, partly in different domains. We can therefore construe both as epistemic fields.

56.3 Science and Religion as Epistemic Fields

In the following, science and religion are compared by means of a list of criteria that helps to define epistemic fields.[19] By "science" I mean factual science as opposed to formal science like logics and mathematics. Now, factual science is often called "empirical science." However, "empirical" refers to the methods of science, not to the concrete facts it studies. Science studies concrete facts (material things having certain properties and the processes they undergo) by both theoretical and empirical means. So by "fact," I do not mean *statements* about concrete facts but the referents of such factual statements. I shall ignore the question of whether there are formal or abstract facts as these do not exist in the same way as concrete facts. The following questions yield some of the criteria that help to define an epistemic field:

---

[18] This is echoed by Plantinga who has the gall to call naturalism a quasi-religion because it fulfills this world view aspect: "It offers a way of interpreting ourselves to ourselves, a way of understanding our origin and significance at the deep level of religion. It tells us where we come from, what our prospects are, what our place in the universe is, whether there is life after death, and the like. We could therefore say that it is a 'quasi-religion'" (Dennett and Plantinga 2011, p. 16f., see also Plantinga 2011). Needless to say, it is disingenuous to call a world view that has overcome religion a quasi-religion. A similar theological ploy is to compare the philosophical underpinnings of science to religious faith.

[19] Modifiyng earlier analyses by Bunge (1983), Bunge and Mahner (2004), Mahner (2007), and Mahner and Bunge (1996a).

1. Which objects does it refer to? What is the domain of facts it is concerned with?
2. What is its fund of knowledge?
3. Which background knowledge does it use in the study of its domain?
4. What are the aims of the given field?
5. Which methods does it work with?
6. Which are the philosophical background assumptions presupposed in its work? That is, what are its metaphysical, methodological, axiological, and moral foundations? Finally, which general attitude or mind-set is considered to be exemplary for those who work in the given field?

56.3.1 Science

1. The domain of factual science comprises everything existent, i.e., the whole world. Although there are certainly things that are de facto beyond scientific investigation for lack of information, there is nothing natural that could not be de jure studied scientifically. As a matter of principle, the domain of science also includes, for instance, the how and why of subjective feelings and emotions in general, as well as the origins and functions of morality and religion – fields of inquiry that are sometimes believed to be beyond scientific understanding.
2. The fund of scientific knowledge is a body of factual knowledge, in particular law statements, which grows along with research. (More on laws in Sect. 56.4.1.)
3. The background of a specific scientific field is the collection of up-to-date well-confirmed knowledge (data, hypotheses, theories) borrowed from neighboring fields. Each scientific discipline connects thus to other scientific fields. Science consists of a network of subfields or disciplines, aiming at a consilient description of the world.
4. The aims of a basic science are purely cognitive. They include, for example, the discovery of the laws of its referents, the explanation of the facts it studies, the systematization of its knowledge base (e.g., by constructing general theories), and the refinement of its methods. By contrast, the aims of technology are practical: it is concerned with design and application.
5. The *methodics* of a scientific field is the collection of its specific and general methods, where specific methods are often called "techniques." (The term "methodology" is reserved here for normative epistemology.) For example, scanning electron microscopy is a specific method, whereas the scientific method is the most general method of the sciences. Specific methods must be scrutable and objective, and we must be able to explain, at least roughly, how they work. The scientific method in general may be conceived of as consisting of the following ordered sequence of cognitive operations: Identify a problem–search for information, methods, instruments–try to solve the problem with the help of those means; if necessary, invent new means, produce new data, or design new experiments–derive the consequences of your solution (e.g., predictions)–check

the solution (e.g., try to replicate your findings by alternative means)–correct the solution if necessary in repeating the cycle–examine the impact of the solution upon the body of background knowledge and state some of the new problems it gives rise to. The structure of any scientific paper roughly reflects these steps and is thus an instance of the scientific method. Of course, there is no single specific method that could be applied to each particular case of research.

6. The philosophical background assumptions of science comprise a naturalist ontology (or metaphysics), a realist epistemology, and a system of values that is particularly characterized by the ethos of the free search for truth.[20] The value system of science includes such logical values as exactness, systemicity, and logical consistency; semantical values such as meaning definiteness (hence clarity) and maximal truth (or adequacy of ideas to facts); methodological values such as testability and the possibility of scrutinizing and justifying the very methods employed to put ideas to the test; and, finally, attitudinal and moral values such as critical thinking, open-mindedness (but not blank-mindedness), veracity, giving credit where credit is due, and more.

These philosophical assumptions are by no means generally accepted in the philosophy of science, so each of them would need further justification. Since there is no room to justify all of them here, the focus will be on the two most important aspects: the metaphysics and methodologies of science and religion. But let us take a closer look at religion first.

56.3.2 Religion

1. In addition to all religiously relevant parts of nature and society, the domain of religion comprises also supernature. Of particular interest are of course the relations of natural things (especially humans) to supernatural entities, and vice versa.
2. The fund of knowledge is a fixed or at most slowly changing collection of (mostly untestable) doctrines and beliefs, whether conveyed by means of an oral tradition or through sacred scriptures. Whatever change in religious beliefs may appear to take place is not due to research and hence newly discovered facts but is almost entirely a result of either (a) a change in the exegesis and interpretation of traditional doctrines, which, if taken literally, often are unpalatable to modern people, or (b) squabbles or even wars between rival factions in the same religious community. Hence, any substantial changes in the belief system are due to authority or external influence, not research. If genuine research takes place, such as historical investigation, this research is not accomplished by religious but scientific means even if undertaken by theologians. Accordingly, it has to be regarded as an external influence.

[20] "Ontology" is used synonymously with "metaphysics" in this paper.

3. The factual background of religion contains at best ordinary knowledge, not scientific knowledge. This is just because most religions are older than science. Some scientific knowledge may be compatible with religious doctrines up to a certain point, and some theologians may make use of scientific knowledge in certain arguments, but in the end this should not be necessary for the (alleged) truth of any religious doctrine.
4. The aims of religion are foremost practical. Moreover, they are ultimately, though mostly tacitly, a matter of self-interest in that they consist in attaining personal advantage such as salvation or eternal life (individual or cosmic). Religions are salvation technologies after all. To obey and worship the divine, or to live a virtuous life, though the explicit goal of the religious person, is, in the end, only a means to attain the blessings expected from the supernatural. All religion is ultimately anthropocentric.
5. The methodics of religion is a collection of practices, such as prayer, incantation, fasting, meditation, and other rituals that are supposed to connect human beings to the supernatural. As far as a cognitive aim is pursued, the religious person may make use of intuition, contemplation, meditation, or revelation. There is neither use for the scientific method in general nor use for specific scientific techniques.
6. The philosophical background assumptions of religion consist of a supernaturalist metaphysics, which is a collection of doctrines about the supernatural and our relations to it. Supernatural entities may be impersonal forces such as karma or more or less anthropomorphic "persons" such as gods. The epistemology of religion is usually a realist one, though religion may be consistent with any epistemology. The value system of religion seems to have only one item in common with science: the quest for truth. However, whereas the truth looked for by religionists is absolute or ultimate, scientific truth is partial or approximate. Neither exactness nor logical consistency and neither clarity nor testability are strong in religion. Moreover, it can be argued that many religious beliefs can only be upheld by disregarding such values. Otherwise, it would not be possible to cherish the mysterious or to confess *credo quia absurdum*. A religious value that is alien to science is (blind) faith, which allows the religionist to always retreat to commitment or fideism if pressed by rational analysis (Bartley 1984). Finally, religion contains an ethos of acceptance and defense of unquestionable doctrines, i.e., dogmas. As for the latter, witness Augustine's dictum, "Greater is the authority of Scripture than all the powers of the human mind," or Paul's injunction "Beware lest anyone cheat you through philosophy and empty deceit, according to the tradition of men, according to the basic principles of the world, and not according to Christ" (Col. 2: 8).

56.3.3 Conclusion

The above listed several commonalities and differences between science and religion, among them some obvious incompatibilities. Both science and religion aim at gaining knowledge about the world. Both operate from a realist perspective, and both are truth seekers. While, today, most of the world is left to science to study and explain, there is an area of overlap the closer we get to the description and explanation of the place of humans in the world. The differences concern the nonnaturalist metaphysics of religion as well as, among others, the methodological status of evidence versus faith and the role of authority versus the free search for truth. And these differences will turn out to be the major incompatibilities.

56.4 The Metaphysics of Science and Religion

Modern science emerged from a mixture of prescience, philosophy, and religion. These areas were strongly intertwined during Scholasticism, but developed apart from the early sixteenth century on (Schrader 2000; Matthews 2009). The emancipation of science from theology is thus one of the characteristic features of its development. Mainstream philosophy also emancipated itself from theology, although even today philosophy in general is still so diverse that it ranges from materialism to quasi-religious thinking and even obscurantism. Even authors who come to a rather conciliatory conclusion concerning the historical relationship between science and religion admit that the main area of conflict concerns metaphysics: "The famous episodes of conflict between science and religion are not strictly conflicts between science and religion. Rather, they are instances of a more general conflict that arises within the process of changing metaphysical frameworks" (Schrader 2000, p. 400). For example, science superseded Scholastic metaphysics, in particular the teleology inherent in its Aristotelian foundation. It also abandoned the intentional teleology of the long-respectable argument to design. Indeed, the result is that, today, the metaphysics of science is consistently naturalist, which is incompatible with any supernaturalist metaphysics, however minimally furnished it may be.

This raises the question of whether the naturalism of science is just the result of a contingent historical development or whether this historical development has just brought forward what had applied all along: naturalism as a metaphysical condition of science. This latter thesis will be defended here.

Now, metaphysical conditions or presuppositions are not exactly popular in contemporary analytic philosophy because they smack of Kantianism. Kantian apriorism is the very antithesis of the aposteriorist approach of epistemological and methodological naturalism, which is widespread in contemporary analytic philosophy. However, metaphysical presuppositions need not be apriorist in the Kantian sense. But what is meant then by "condition" or "presupposition"?

A presupposition is often understood in the sense of a statement that is entailed by a set of premises or in the sense of a necessary condition implied by some antecedent statement. Is metaphysical naturalism entailed by science in one of these senses? From a formal point of view, it is not. It is not part of a deductive argument in the sense that if we collected all the statements or theories of science and used them as premises, then metaphysical naturalism would logically follow. After all, scientific theories do not explicitly talk about anything metaphysical such as the presence or absence of supernatural entities: they simply refer to natural entities and processes only. Therefore, naturalism rather is a tacit metaphysical *supposition* of science, an ontological *postulate*. It is part of a metascientific framework or, if preferred, of the metaparadigm of science that guides the construction and evaluation of theories and that helps to explain why science works and succeeds in studying and explaining the world. As such, it is the best framework available yet, justified by its very success and its unifying and heuristic power.

56.4.1 The Metaphysical Presuppositions of Scientific Research

> …what Kant and Hume show, I think, is that limiting oneself to seeking natural causes for natural effects is not […] a metaphysical principle with no inherent grounding in science but rather a disciplinary condition of doing science, the only way to get the particular kinds of answers that science seeks within the terms of the evidentiary warrants it demands. (Loesberg 2007, p. 96f.)

A popular view among scientists maintains that science need not bother with philosophy, let alone metaphysics, at all; scientists should just apply and follow the scientific method or, if preferred, the collection of scientific methods. Somewhat more sophisticatedly, if science is ultimately about finding the truth, all that counts is evidence. Whether it confirms the natural or points to the supernatural, we should follow the evidence wherever it leads (Fishman 2009; Monton 2009). This antimetaphysical stance is importantly wrong because it rests on the assumption that both scientific methods and the evidence they produce are free of metaphysical presuppositions.

To show that there is quite a number of metaphysical postulates of science (Bunge 1983), we take a look at the three (overlapping) general empirical methods in science by means of which we gain data, which, in turn, may function as evidence: observation, measurement, and experiment. The question is whether these methods can work in a metaphysical vacuum, or whether their successful application rests upon certain metaphysical assumptions. In other words, could these methods work successfully in just any world, or can they work only in a world with a particular nature? A simple experiment chosen from a high school biology textbook will function as an example (Fig. 56.1, following Mahner 2007).

Let us focus here on the question of how much metaphysics is hidden in this simple experiment, addressing possible objections mostly in footnotes so as not to interrupt the exposition.

First, we assume that this experiment involves real entities in a real world, not just objects existing in our mind. That is, we work on the basis of ontological realism, which helps to explain not only the success but in particular the failure of scientific theories.[21]

---

[21] A very general ontological realism is probably the least controversial metaphysical presupposition of science (Bunge 1983, 2006; Alters 1997; Gauch 2009; Ladyman 2012), although there is an ongoing realism/antirealism debate in philosophy. However, this debate concerns mostly epistemological

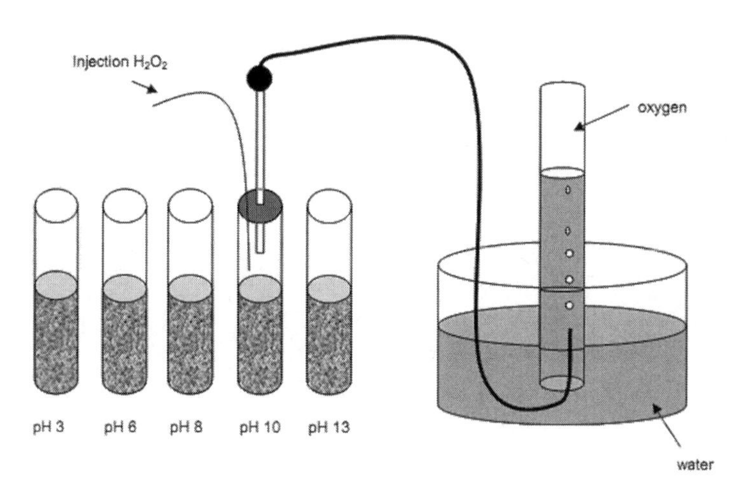

Fig. 56.1 By determining the pH optimum of the enzyme catalase, this experiment is used to demonstrate that the functioning of enzymes is pH dependent. The experimental setup is as follows. Five test tubes are halfway filled with water. We add a piece of yeast to each of them. By adding different amounts of hydrochloric acid (HCl) or caustic soda (NaOH), we arrange for a different acidity or alkalinity, respectively, in each tube, say, pH 3, pH 6, pH 8, pH 10, and pH 13. The yeast cells contain the enzyme catalase, which enables them to break down hydrogen peroxide into water and oxygen (i.e., $2\ H_2O_2 \rightarrow 2\ H_2O + O_2$). We inject a certain amount of hydrogen peroxide solution into the test tubes (e.g., by means of a syringe). Each time, we close the tube and measure the amount of gas produced after 2 min by collecting it in a measuring tube, which is connected to the given test tube by a thin rubber hose. We do not need to specify the precise amounts and conditions here, because the basic setup of this experiment will be clear anyway (from Mahner 2007; redrawn and modified from Knodel 1985, p. 39). The result of this experiment: the oxygen production is highest at pH 8 (in fact, at pH 8.5, which can only be discerned by refining the experiment)

Now that we are talking real test tubes with real yeast and real chemicals, we may ask why an experiment like this is found in a textbook. Obviously, we assume that we can repeat this experiment as many times as we see fit, and that we will obtain (roughly) the same results, provided we do not make any mistakes. The gas produced is always oxygen, neither nitrogen nor carbon dioxide. The test tubes remain test tubes, and do not spontaneously transform into chewing gum or thin air. It appears then that things and their properties remain the same under the same conditions. Certain properties of things seem to be constantly connected, so that they change together: they are covariant. In other words, certain properties of things are lawfully related.

Of course, ordinary experience already indicates that the world is lawful, but the thesis of a lawful world is not a piece of empirical knowledge: it is a necessary condition of cognition. Without things behaving regularly due to their lawful properties, no organism would be able to learn much about the world. Note that

---

problems regarding the justification of more detailed realistic claims such as the status of unobservable entities and the truth of scientific theories. Thus, someone who rejects more specific forms of realism, such as scientific realism, usually is still an ontological realist. I shall not defend ontological realism in more detail here (for such a defense, see, e.g., Vollmer 1990), because both ontological naturalists and supernaturalists share a basic realist outlook anyway.

what I am referring to here are laws in an ontological sense of lawfully related properties, not general law statements as conceptual representations of such ontic laws (Bunge 1983, 2006; Ellis 2002). This must be emphasized because the view that laws of nature are nothing but universal statements is still popular.[22]

Imagine next that we fail to obtain oxygen in our measuring tube. In this case we would look for mistakes in the setup, like a leakage in the rubber tube. We would check whether the yeast is still alive, whether we have correctly set the pH value of the water, or whether the substance we add is really a hydrogen peroxide solution. No scientist would seriously entertain the idea that somewhere in the experimental setup the gas has literally dissolved into nothing. Conversely, no scientist would assume that we can produce gas out of nothing. There is simply no point in doing experiments and "wiggling parameters" if things simply could pop out of or into nothing. Let's call this the ex-nihilo-nihil-fit principle.[23]

What initiates the production of oxygen? Oxygen does not originate spontaneously: it starts to emerge only after we add some hydrogen peroxide solution. Thus, by meddling with certain parts of the setup, we can produce a certain effect: we can

---

[22] I submit that the mainstream view of laws in the philosophy of science is inadequate. Science calls for a (neo-)essentialist view of laws, according to which "the laws of nature are immanent in the things that exist in nature, rather than imposed on them from without. Thus, [...] things behave as they do, not because they are forced or constrained by God, or even by the laws of nature, but, rather, because of the intrinsic causal powers, capacities and propensities of their basic constituents and how they are arranged" (Ellis 2002, p. 1). Thus

> not even an omnipotent God could change the laws of nature without changing the things on which they are supposed to act. Therefore, the idea that the laws of physics are contingent, and superimposed on intrinsically passive things that have identities that are independent of the laws of their behavior, is one that lies very uneasily with modern science. (Ellis 2002, p. 5)

The lawful behavior of things neither entails that we can always represent them as law statements nor that every scientific explanation is a subsumption under some law. For example, due to the enormous variation of organisms, many biologists believe that there are no laws (= law statements) in biology. But this does not entail that organisms do not behave lawfully: it is just that it often makes not much sense to try to find general, let alone universal, law statements because their reference class is rather small, holding only for some subspecies, variety, or even smaller units, for example, that is, only for those organisms sharing the same lawful properties (more on laws in biology in Mahner & Bunge 1997, Ellis 2002). Finally, even some cases of randomness are lawful because they are based on stochastic propensities such as in quantum physics. That is, there are probabilistic laws. For the neo-essentialist approach to laws adopted here, see Bunge (1977), Mahner and Bunge (1997), Bunge and Mahner (2004), and Ellis (2002).

[23] Note that "nothing" really means "nothing," not some form of radiation or some other massless form of matter. For example, what is called particle annihilation is just a transformation of a particle with mass into one or more massless particles, that is, into some form of radiation. However, it seems that the ex-nihilo-nihil-fit principle is being challenged by cosmologists, who keep entertaining the idea that the universe originated from nothing (see, e.g., Stenger 2011). In particular, according to multiverse cosmology, some primordial "nothing" keeps randomly popping out universes. But since this "nothing" has at least one property, namely, the propensity to pop out universes, it doesn't seem to be a genuine nothing which should have no properties at all and hence be unable to change.

(causally) interact with the setup. Moreover, the steps in this chain of events are ordered: their sequence is not arbitrary. That is, we must assume that causation is for real and hence an ontological category, as well as that there is a principle of antecedence: causes precede their effects in time, so that the present is determined causally or stochastically by the past, but not conversely. In other words, we need to assume not only that the experimental setup (or the world in general) is real but also that we can interact with it and that our actions can trigger orderly chains of events. Otherwise, no deliberate effect could be produced, variables could not be controlled for, etc.

If the results of our empirical methods are expected to be the results of real processes in a real world, we must rule out the possibility that the experimental setup can be causally influenced in a *direct* way solely by our thoughts or wishes (or more precisely our thinking and wishing), that is, without the interposition of motoric actions by our bodies (Broad 1949). Indeed, if the world were permeated by causally efficacious mental forces, we would have no reason to trust the reading of any measuring instrument or the results of any experiment. In other words, the data obtained through observation, measurement, or experiment could not function as evidence if they were literally the telepathic or psychokinetic product of wishful thinking. Worse, we could not even trust our own perceptions and conceptions, as they could be the result of telepathic manipulation. We may call the assumption that no such mental forces exist the "no-psi principle" (Bunge 1983, p. 106).[24] This principle must hold not only for humans but for any organism anywhere that is able to think. Neither humans nor little green aliens from another galaxy must be able to meddle, just by thinking alone, with empirical methods or our perceptional and conceptual processing of their results.

What holds for natural entities applies a fortiori to supernatural entities. We must stipulate, then, that no supernatural entity manipulates either the experimental setup or our mental (neuronal) processes or both.[25] We can even make the case that this holds not only for science but for perception and cognition in general. Indeed, this "no-supernature principle" as we may call it is also needed to avoid Cartesian skepticism. In his *Meditationes*, Descartes (1641) wrote:

> I will suppose, then, not that Deity, who is sovereignly good and the fountain of truth, but that some malignant demon, who is at once exceedingly potent and deceitful, has employed all his artifice to deceive me; I will suppose that the sky, the air, the earth, colors, figures, sounds, and all external things, are nothing better than the illusions of dreams, by means of which this being has laid snares for my credulity…. (*Meditation* 1, §12)

---

[24] The no-psi principle was one of Broad's (1949) so-called basic limiting principles of science. Being a strong believer in the paranormal, Broad maintained that this basic limiting principle had been refuted by parapsychology. However, Broad was fooled by the sloppy and partly even fraudulent parapsychological research of his time.

[25] This was already acknowledged by J. S. B. Haldane (1934), who stated that his "practice as a scientist is atheistic," that is, when he sets up an experiment, he assumes "that no god, angel, or devil is going to interfere with its course" (p. vi).

Unlike Descartes, we no longer have reason to believe that the supernatural is dominated by an all-good God, who, by his very nature, not only refrains from malicious manipulation but even functions as the guarantor of the truth of our cognition and thus our knowledge.[26] Even in traditional Christianity, there are many other supernatural entities than God, like devils, demons, and angels. Now add the many supernatural entities of other religions and finally everything we can imagine. As the fantasy and horror movie genre shows, the possible inhabitants of supernature are only limited by our imagination. If we admit the supernatural, there is no reason to rule out a priori the existence of a malicious entity that could meddle with the world including our cognitive processes. So we need to start with the postulate that no such entities exist.

Let us summarize then the metaphysical suppositions of the general empirical methods of science:

(a) Ontological realism
(b) The (ontological) lawfulness principle
(c) The ex-nihilo-nihil-fit principle
(d) The antecedence principle and an ontological conception of causation
(e) The no-psi principle
(f) The no-supernature principle

Whoever subscribes to empirical scientific methods and their function to generate evidence must also subscribe to these metaphysical principles: without them, what we are doing would not be scientific measurements or experiments but rather meaningless games. Thus, these principles are part of the ontology behind science's methodology. In a world that has these properties, science is possible.

It may be seen as problematic that the principles (c), (e), and (f) are formulated negatively. It would not be a problem, though, to reformulate the above in positive terms, for example, by offering a full-fledged metaphysical theory, elucidating the notions of property, thing, event, process, lawfulness, etc. (see Bunge 1977; Bunge and Mahner 2004; Mahner 2012). In the sense of an axiomatic definition, we could then claim that everything which works that way is natural, and that the only real existents are such natural things and events. The above negative principles, then, would simply be corollaries of such a metaphysics formulated in positive terms. However, for the sake of simplicity and convenience, I shall stick to the negative formulations.

Now, are these principles also necessary conditions, perhaps even a priori conditions? Or could the scope of at least some of these principles be somewhat restricted while science could still work successfully? In other words, are they just default principles? For example, the traditional metaphysical principle of strict causality (every event has a cause) has been shown by quantum physics to be false, because some events such as radioactive decay are spontaneous (uncaused). This is why a

[26] As Fales (2010) argues, even God may not know whether his thinking is manipulated by some evil demon. Does this require a second-order God of higher power who guarantees the truth of God's knowledge? If so, we would end up with an infinite regress of truth guarantors.

principle of strict causality is not part of the above list. And if the universe had popped out of nothing (however magical this would be), principle (c) would still hold within the universe. This suggests the possibility that some metaphysical principles could be revised. Similarly, it may be argued that even if the universe were initially created by a supernatural being, science would still be possible if there had been no further interventions since or if the number of interventions were very small. As our focus here is on metaphysical naturalism, the principles (a)–(e) will not be further addressed, so that we turn right to this possible objection to the no-supernature principle.

56.4.2 Naturalism or Noninterventionism?

We have just seen that observation, measurement, and experiment must not be subject to supernatural manipulation because they would then lose their status as empirical methods for the generation of evidence. Does this really warrant a no-supernature principle? Prima facie, it does not, at least not without further ado: it seems to warrant at most a principle of nonintervention with respect to scientific research and cognitive processes. How, then, can we justify a no-supernature principle?

To see how, it will be helpful to take a closer look at the definition of noninterventionism. It may be tempting to analyze it as a conditional statement such as "If supernatural entities exist, they do not intervene in the course of the natural world." However, this would turn nonintervention into a necessary condition for the existence of the supernatural. Indeed, by contraposition, we would obtain the absurd statement, "If supernatural entities intervene in the course of the world, they do not exist." Therefore, we better analyze "noninterventionism" as the conjunction of two statements, namely, "Supernatural entities exist really & Supernatural entities do not intervene in the course of the natural world."

This analysis shows that while at first sight noninterventionism appears to be a reasonable minimal supposition, it is in fact not, because it presupposes the existence of supernatural entities. The first statement of the above conjunction, "Supernatural entities exist," cannot be a metaphysical supposition of science because there is no reason why science should postulate the existence of something that, by not intervening in the course of the natural world, plays no part in any scientific explanation of the world.

Indeed, it is common practice in science to adopt the null hypothesis until there is evidence for an alternative substantive hypothesis. The null hypothesis usually negates that something is the case, such as that something exists or that two variables are related. Examples are the following: "Junk food is not the cause of obesity," "Men and women do not perform differently in mathematical tests," or "The Loch Ness monster does not exist." In order to prove some substantive hypothesis, its corresponding null hypothesis must be refuted empirically. The null hypothesis approach is not restricted to science: it is also adopted in modern law where a defendant is presumed innocent until proven guilty. Mutatis mutandis, the null hypothesis principle may – nay, should – be applied also in metaphysics, in particular when it comes to existential claims. For example, in the philosophy of religion, Antony Flew (1972) was the first to suggest defining "atheism" in this sense, although he did not use the scientific term "null hypothesis." An atheist, then, is not someone who positively and dogmatically denies the existence of gods, but someone who just adopts the "presumption of nonexistence" as a court of law adopts the presumption of innocence. Correspondingly, one way to conceive of metaphysical naturalism is as a metaphysical null hypothesis, stating that a supernature does not exist.[27]

Of course, there is an important difference between scientific and metaphysical null hypotheses; the latter are usually regarded as unfalsifiable by direct empirical evidence. This distinction at least was the upshot of both the neopositivists' and Popper's demarcation efforts. If we disregard the neopositivist view that metaphysics is untestable because it is nonsense, and thus accept Popper's distinction for the time being, we can say that at least some metaphysical hypotheses can be refuted (or, more cautiously, disconfirmed) *indirectly*, for example, by turning out to be incompatible with scientific practice or in being unable to explain it. For example, science could fail as a cognitive enterprise, either in its entirety or in some particular area, so that we would have to reconsider metaphysical naturalism.

In any case, the notion of a metaphysical null hypothesis implies that even metaphysical assumptions remain fallible in principle. At the same time it allows us to consider metaphysical naturalism as a necessary condition of science: if metaphysical naturalism fails, science fails too.

56.4.3 The Metaphysics of Supernaturalism

It appears that the supernatural can be characterized by simply negating most of the metaphysical principles listed in Sect. 56.4.1. Thus, a supernatural entity would be one that:

- May be able to create things out of nothing or annihilate them
- May not be subject to the antecedence principle in that it could make past events undone or change the natural sequence of events
- May not be subject to the lawfulness principle because it may be able to change the lawful properties of (natural) things or the lawful course of (natural) events
- May be able to influence (or to manipulate, if not fully control) natural things, including thinking entities and their perceptions and conceptions

[27] In his debate with Plantinga, Dennett has recently called naturalism a null hypothesis (Dennett and Plantinga 2011, p. 49). Plantinga had argued that science is compatible with theism, because science doesn't explicitly state that there is no God. This shows that Plantinga is not familiar with the concept of a null hypothesis. The same seems to apply to Flanagan (2008, p. 437), who argues against "imperialist naturalism" that we would simply not know everything that there is or is not. Yet this is exactly the reason why we have to start with naturalism as a metaphysical null hypothesis.

This is essentially what is behind the common characterization of a supernatural entity as one that has magical abilities and can thus perform miracles. Whether or not supernatural entities are subject to any supernatural laws (whatever these may be) is irrelevant here. All that matters is that, in principle, they could be able to interfere with the lawful course of natural events, hence also with our brain functions. This is why a supernaturalist ontology invites (and maybe even entails) a nonnaturalist epistemology and methodology in which special forms of cognition, such as revelation, religious experience, a *sensus divinitatis*, or whatever nonnatural ways of communication with the supernatural may obtain, are accepted as legitimate sources of knowledge and means of justification. For example, Ratzsch (1996) and Plantinga (2001) defend the idea of a "uniquely Christian science" or a "theistic science," respectively, so that there is no reason why a Christian should not make use of particular religious "methods" in science. These examples illustrate that methodology is not free of metaphysics. It comes as no surprise therefore that accommodationist scientists and philosophers, who reject metaphysical naturalism to make room for religion yet at the same time want to keep supernaturalism out of science, struggle hard to make a consistent case (see Sect. 56.5.3).

That the supernatural is characterized mostly, if not exclusively, in negative terms has been shown in more detail by Spiegelberg (1951). Even prima facie positive attributes of the supernatural turn out to be negative ones in that they are just denials of known natural characteristics. For example, "transcendence" is the negation of "immanence," that is, *not* being "located" within the confines of our spatiotemporal world. Or being a first cause is nothing but being an *uncaused* cause. And the few positive attributes such as omnipotence or omniscience are actually natural properties raised to an absolute degree. In this regard they are not fully supernatural – a statement that may require some elaboration.

Spiegelberg distinguishes two conceptions of the supernatural, quantitative and qualitative. In the former case supernatural entities are ascribed properties that differ from the natural only in degree, though often to an absolute degree. For example, a supernatural entity is more powerful than a natural entity, perhaps even all-powerful, or more knowledgeable, perhaps even omniscient.[28] The attributes of supernatural entities are then still conceived of on the basis of familiar natural properties. Thus, such conceptions are more or less anthropomorphic, which suggests that the quantitatively supernatural, if any, would still have to be spatiotemporal. By contrast, according to qualitative supernaturalism, supernatural entities are *categorically* different from natural ones, so much so that their properties are essentially mysterious, ineffable, and incomprehensible. God, then, is the *Wholly Other*, not someone or something to be understood even by the faintest analogy with anything known natural. Spiegelberg called these two types of the supernatural *overnatural* and *transnatural,* respectively (1951, p. 343). Whereas the overnatural seems to be somewhat intelligible by analogy with known natural properties, the transnatural is incomprehensible. To obtain or retain a modicum of intelligibility, conceptions of the supernatural usually combine overnatural and transnatural

[28] Despite many theological defenses, the notions of omnipotence and omniscience are incoherent (Martin 1990), so that we have reason to reject characterizations of the supernatural that employ them.

features. This allows the believer to oscillate between these two conceptions, depending on his argumentative needs. Modern theology tends to reject a merely overnatural conception of the supernatural as being too anthropomorphic and seems to prefer a more "sophisticated" conception of the supernatural in terms of the transnatural. Yet the transnatural is defined but negatively.

Spiegelberg's philosophical analysis is backed by cognitive psychology, which has shown that there is a rift between theological conceptions of religious entities and everyday religion. The latter is inevitably anthropomorphic but needs the counterintuitive features of the theological conceptions as an attention-grabbing potential.[29] It has been shown experimentally, for example, that although everyday believers know the theologically correct properties of God, they do conceive him in anthropomorphic terms when it comes to working with the concept of God in an everyday context (Barrett and Keil 1996). Whatever theology does to transnaturalize religious entities, believers will inevitably revert to overnatural concepts that better match their natural intuitive thinking. If religion is anthropomorphism, as Guthrie (1995) argues, this comes as no surprise.

Now, it may be argued that science faces a similar problem. Scientific concepts are often counterintuitive too, so ordinary people tend to stick to their more intuitive common sense understanding of the world. As McCauley (2011) shows, in this sense there is a divide between reflective attempts at cognition (science and theology) and non-reflective, popular – or as he calls them – "maturationally natural" attempts (commonsense cognition and popular religion). At the same time, however, both theology and popular religion are characterized by an unrestricted use of concepts of (intentional) agency or causality, whereas both science and commonsense cognition make a rather restricted explanatory use of intentional agents, that is, restricting it to the behavior of higher animals including humans. McCauley reminds us to not just compare science and religion simpliciter but in the correct respects. What is relevant here, then, is the metaphysical divide between science and commonsense cognition on the one hand and theology and popular religion on the other – which is the divide between naturalism and supernaturalism. Whereas the metaphysical divide, if any, between science and commonsense cognition is small, it is wider between theology and popular religion – which is the divide between the overnatural and the transnatural. In any case, both the overnatural and the transnatural are incompatible with the metaphysical naturalism of science.

56.5 Metaphysics and Methodology

56.5.1 Evidence Is No Metaphysics-Free Lunch

If metaphysical naturalism is a metaphysical presupposition of science, science should be unable to deal with anything supernatural. By contrast, if one believes that science is free of metaphysical presuppositions, the answer to the question of

[29] See Boyer and Walker (2000), Boyer (2001), and McCauley (2011).

whether the supernatural is testable is quite simply affirmative. For example, if angels descended from the sky and raised the dead or if studies on the effects of intercessory prayer yielded significant positive results, we would have empirical evidence for the supernatural and hence a valid test. (In the first case, we would have *direct* evidence, in the second case *indirect* evidence.) While many authors agree with this view,[30] others maintain that the supernatural is untestable as a matter of principle.[31] This disagreement can be explained by the distinction between the overnatural and the transnatural.

Those who maintain that the supernatural is testable seem to conceive of the supernatural as merely overnatural. That is, the supernatural is intelligible to a certain degree because its properties are not actually categorically different from natural properties: overnatural entities are more or less superpowered entities with quasi-natural properties. By contrast, those who believe that the supernatural is untestable seem to regard the supernatural as transnatural and hence as categorically different from anything known natural – which makes it both inaccessible and unintelligible and thus untestable.

But let us first take a look at the two central concepts of this debate, testability and evidence. In the broad sense, a statement, a hypothesis, a model, or a theory is empirically testable if there is empirical evidence for or against it (Bunge 1983), whereby the evidence *e* is another statement – a datum – that is relevant to the hypothesis *h* (or model or theory) in that *e* either confirms or disconfirms *h*. Now both *e* and *h* must be semantically meaningful (nonsense is untestable), and they must not be logical truths or falsities. For some evidential statement *e* to be relevant to some hypothesis *h*, *e* and *h* must share at least one referent or, if preferred, one predicate. For example, data about the crime rate in Australia in 2011 are irrelevant to quantum theory, because the data and the theory are not (partially) co-referential. Last but not least, we must demand that *e* has been acquired with the help of empirical operations that are accessible to public scrutiny, and – here enters metaphysics – both the empirical operations and our cognitive processes involved in the perception and processing (interpretation and evaluation) of the data gained by these operations must involve only lawful natural processes – that is, they must not be the result of supernatural manipulation.

So for there to be some evidence *e* about the supernatural, *e* would have to share at least one predicate with the respective hypothesis *h* referring to some supernatural entity. This, in turn, would require that the supernatural referred to in *h* possesses at least one property that can be represented by a meaningful (positive) predicate – which could only be a natural or quasi-natural property. But this is possible only if the supernatural is conceived of as *not* qualitatively or categorically different from the natural. For example, if we found reproducible significant positive effects of intercessory prayer and if these empirical data were supposed to function as evidence for a hypothesis involving a supernatural being as the cause of this effect, the

[30] See, e.g., Augustine (2001), Boudry et al. (2010, 2012), Fales (2010), Fishman (2009), Monton (2009), Stenger (2007), and Tooley (2011).

[31] See, e.g., Forrest (2000), Pennock (2000, 2001), and Spiegelberg (1951).

supernatural entity referred to would have to be able to "listen" to prayers, if only telepathically (however this would work), and understand and consider them in a way that is analogous to a human person listening to the requests of others and considering them on the basis of his or her background knowledge including a code of ethics. It is not intelligible how some solely negatively characterized transnatural entity should be able to do any of that; worse, we would not even know or understand what it means that any such entity *does* anything. For this reason, there could be evidence at most for the more or less anthropomorphically defined overnatural, so that only the overnatural may be testable in the broad sense.

Therefore, Pennock (2000, 2009) is right when he says that for the supernatural to be testable, it would have to be understood in a naturalized way, and the supernatural would have to be able to partly naturalize itself (or simply be natural to begin with) so as to interact with the natural world. If some process were actually transnatural, we could not observe it, however indirectly. Think of transubstantiation. Or think of the theological concept of continuous creation, according to which everything is constantly recreated ex nihilo by God, from moment to moment, and thus sustained in its existence. Continuity of existence is therefore just an appearance, whereas the reality behind it is a continuous transnatural intervention.

However, we still have to consider the last condition of evidence mentioned above, namely, the one that prohibits supernatural manipulation. Even if some empirical data fulfilled the formal conditions of evidence – provided the supernatural is construed as overnatural – we are still faced with the paradoxical situation that the empirical operations employed to produce such evidence presuppose the nonexistence of the very entities whose existence is supposed to be confirmed by this evidence. It may be tempting then to retreat to a principle of nonintervention with regard to our cognitive processes. But on what grounds could we defend noninterventionism? Of course, we could come up with various ad hoc assumptions. For example, the powers of the supernatural entities involved could somehow be limited, God could guarantee local noninterventionism with regard to our cognition, or God could even be the ultimate cause of our cognition and thus guarantee its correct functioning. But would it be epistemically warranted to accept any of these ad hoc contrivances, unless they are independently testable, that is, unless they are more than just logical possibilities? I don't think so. For this reason, naturalism remains the metaphysical default position of science, so that we have good reason to reject prima facie evidence for the overnatural as long as not all alternative natural explanations are exhausted.

56.5.2 Scientific Explanations Must Be Naturalist Explanations

Scientific theories are assessed (among others) with respect to their explanatory power. A scientific theory is expected to explain a certain fact or domain of facts. That is, it is supposed to tell us how something came about or how something works. In so doing, it employs law statements or reference to mechanisms (Bunge 1983;

Mahner and Bunge 1997). For example, a theory of photosynthesis informs us about the physiological processes (mechanisms) by means of which plants use light to transform carbon dioxide and water into carbohydrates and oxygen. These mechanisms are specific enough to explain what they are supposed to explain. Thus, they cannot be used to explain, for example, how birds fly or how earthquakes occur, because the respective laws and mechanisms are quite different. Do theories referring to supernatural causes or entities comply with this requirement?

They do not – even if we focus on the supernatural in the sense of the overnatural because transnatural entities devoid of positive properties are incomprehensible and hence nonexplanatory anyway. At first sight, invoking an overnatural cause to account for some fact does seem to have explanatory power. For example, intelligent design creationists claim that the theory of evolution cannot explain how certain complex organs have originated. So they invoke a supernatural entity, an intelligent designer (who allegedly need not be but is in fact considered to be God himself) who either created the organ or at least helped to accrue the given complexity. This answer appears to have explanatory power because, by analogy with human handicraft, we all understand what creating or developing artifacts is about. Yet in fact, it explains nothing because it explains too much. The problem is that an answer like "God made it the way it is" can be applied to all facts.[32] Whatever exists and whatever happens can be explained thus by reference to the will and actions of some supernatural entity. But an explanation that explains everything explains nothing.[33] Thus, supernatural explanations explain nothing because they are omni-explanatory.

The all-purpose God-did-it explanation is not something naturalists have come up with to ridicule supernaturalism. As a matter of fact, the philosophical doctrine of *occasionalism* seriously held that God is the cause of each and every event because matter is passive and cannot change or bring about anything on its own. If occasionalism, assuming 100 % supernatural causation, were true, there would be no need for natural explanations at all: a single supernatural cause would explain everything. So why do supernaturalists not adopt occasionalism? Why is science allowed to come up with natural explanations in some cases, but not in others? It seems that since the naturalist approach of science has been so successful, many supernaturalists have conceded its explanatory power and retreated to a god-of-the-gaps approach.[34]

Even some philosophers defend this view (e.g., Monton 2009), claiming that it is legitimate *in some cases* to fill an explanatory gap with a supernatural explainer, and that this would refute the charge that supernaturalist explanations are

[32] More on the problems of supernaturalist explanations in Pennock (2000), who also explores the consequences of supernaturalism for the legal system, which would have to reconsider the-devil-made-me-do-it arguments including historically superseded forms of evidence based on "higher insights" and revelations.

[33] Note that the famous "theory of everything" in theoretical physics is a misnomer, because it would not explain everything. It would just offer a unified theory of the fundamental forces of physics. But this would not even begin to explain all the emergent properties of higher-level systems.

[34] Those supernaturalists who dislike the god-of-the-gaps approach for theological reasons have retreated to a transnatural conception of the supernatural, which is immune to any empirical refutation.

omni- explanatory. However, it is doubtful that this rejoinder works. After all, supernaturalist explanations come with two proliferation problems. First, if we admit one supernatural entity into the explanatory realm of science, we are on a slippery slope to admitting as many as we fancy (Kanitscheider 1996). Christian creationists, for example, will of course tell us that the number of supernatural entities is limited by scripture. But if science admits entities from the biblical cosmos, nothing prevents it from admitting entities from other religions as well. There is no a priori reason why a Christian supernatural entity is a better explainer than a Hindu one, for example. The more supernatural explainers we get, however, the closer we get to omni-explanation again. Second, even if science were able to incorporate the overnatural into its explanations, how do we know that reference to such entities provides ultimate explanations? If science could study the overnatural, what would happen if we encountered explanatory gaps in the overnatural world too? The analogous procedure would be to resort to super-supernatural entities to fill these gaps in the first-order supernatural world, and so on, possibly ad infinitum. Just think of the famous question, "Who created the creator?".

In any case, there is another and perhaps better reason for rejecting supernatural explanantia than their omni-explanatory power. As we know nothing about the laws and mechanisms, if any, of the supernatural, we better argue that supernatural explanations explain nothing, not because they are omni-explanatory but because they are pseudo-explanatory. Indeed, to explain the unknown by means of something even more unknown and, worse, something magic and occult is an argumentative flaw known as *ignotum per ignotius* or *obscurum per obscurius*. Of course, believers in the supernatural may object that they do know something about the supernatural, for example, by reading sacred texts, by revelation, by some special form of experience, or by simply having some special insight or epistemic faculty such as a *sensus divinitatis*, as claimed, for instance, by Plantinga (2011). However, all these "methods" are no longer acceptable because they are arbitrary: just any claim could be justified by them, and they are not intersubjective.[35] For this reason, appealing to the supernatural for explanatory purposes is tantamount to saying that we do not know how a certain fact works or has come about. Supernatural explanations are therefore *argumenta ad ignorantiam*: appeals to ignorance (see also Smith 2001). Thus, they cannot, as Clarke (2009) claims, function as inferences to the best explanation: proposing a pseudo-explanation is an inference to the worst explanation.

For those who believe that filling explanatory gaps with supernatural entities is a legitimate instance of an inference to the best explanation, hypothesizing supernatural entities is analogous to postulating unobservable (or theoretical) entities in science. However, this idea faces several semantic, methodological, and ontological problems.[36]

[35] See, e.g., Mackie (1982), Martin (1990), Forrest (2000), Fales (2010), and Philipse (2012).

[36] Philipse(2012) has recently shown that the inference-to-the-best-explanation approach of natural theology faces insurmountable problems, including the failure of Bayesianism, which is also championed by radical empiricists (e.g., Fishman 2009), who believe that scientific methodology has no metaphysical presuppositions.

In science, we must be willing to endow theoretical entities with a definitive set of properties. We cannot infer a best-explaining entity whose properties may vary arbitrarily (Kanitscheider 1996). Yet this is exactly the case with concepts of supernatural entities, in particular the concept of God, which is of course the most employed concept in supernaturalist explanations. Indeed, the properties of "God" vary from theologian to theologian, from tradition to tradition, even from believer to believer, so much so that "God" in theology *A* may have properties contradictory to the ones of "God" in theology *B*. A historical example is the God of Leibniz and Newton (Kanitscheider l.c.). Whereas Leibniz's God has set up the laws of nature at the beginning so that the world has been functioning without intervention ever since, Newton's God had to intervene more or less often in the natural world in order to adjust some imperfections. Thus, both a perfectly lawful and an imperfectly lawful world can be explained by reference to God. Whatever the factual evidence, then, some concept of God can always be applied.[37]

This is not to say that "God" is meaningless in the ordinary language of a certain group, because everyone has a rough idea of what "God" means in his or her religious tradition, in particular since these traditions employ rather anthropomorphic and thus overnatural conceptions of God. But this meaning is very restricted, as it is well known that religious sects have fought each other to death over the proper meaning of "God." However, being possibly meaningful locally and in ordinary language is not enough to qualify as a legitimate scientific concept and not even as a philosophical one. As Flew (1972) put it,

> Where the question of existence concerns, for instance, a Loch Ness Monster or an Abominable Snowman, [the introduction and defense of the proposed concept] may perhaps reasonably be deemed to be more or less complete before the argument begins. But in the controversy about the existence of God this is certainly not so: not only for the quite familiar reason that the word 'God' is used – or misused – in many different ways, but also [...] because it cannot be taken for granted that even the would-be mainstream theist is operating with a legitimate concept which theoretically could have an application to an actual being.

This is important to remember because some scientists and oddly enough even some philosophers (like Monton 2009) seem to be so naive to think that the very use of the word "God" already amounts to postulating a legitimate theoretical entity with explanatory power. But it must first be ascertained that a sentence like "God caused some *x*" is more informative than "Tok caused some *x*" (Nielsen 1985).

---

[37] It may be argued that the variation in the meaning of "God" is not problematic, because scientific concepts often start out with fuzzy and variable meanings too. Think of terms like "gene" or "atom." However, the variations in the precise meanings of these concepts are adjustments guided by empirical research and theory development. These concepts could be made precise enough to even get hold of their referents: today, genes can be sequenced, and atoms can be photographed. The various concepts of God, by contrast, are not constricted and guided by empirical research, so there is no improvement in the sense of an approximation to reality. The conceptual "development" in theology is purely apologetic in that the traditional overnatural concepts of God have been transformed into transnatural ones, so that they can no longer conflict with science, or anything factual for that matter.

Assuming for the sake of the argument that it is possible to make "God" more informative than "Tok" and thus turn it into a meaningful theoretical concept and also into one whose meaning does not vary arbitrarily, an explanation referring to this God would still be arbitrary. For example, the origin of a complex organ such as the vertebrate eye may be explained by reference to some creative intervention by God. But in fact reference to any other supernatural entity would do the same explanatory work, be it a devil, an angel, a demon, or whatever. After all, we know nothing about the possible powers and intentions of such entities. So we have no empirical means for deciding among competing supernaturalist explanations (Augustine 2001). The only commonality supernatural explanantia for some fact *x* seem to share is this: some supernatural entity chose to do *x* for unknown reasons. This is hardly superior to "we do not know what caused *x*."

For all these reasons, postulating supernatural entities is *not* analogous to postulating theoretical entities in science. The semantic fuzziness, if not arbitrariness, of supernaturalist terms makes them useless as scientific concepts.

In sum, the semantic and ontological problems of supernatural concepts and statements affect both the concepts of evidence and explanation. Even if there were highly anomalous data, they would not constitute evidence for the supernatural unless there were scientifically meaningful statements about the supernatural in the first place. Until then, all we could state perhaps is that something spooky is going on, but such anomalous data could not be explained as the results of some supernatural intervention. This holds a fortiori when we are not even faced with anomalies. For example, a sentence such as "Due to its complexity, the human eye was intelligently designed by a supernatural creator" is at first sight meaningful by analogy to human design and creation. But even when applied to the merely overnatural, it is no longer clear what "intelligence," "design," and "creation" actually mean. Indeed, as Sarkar (2011) has shown, intelligent design "theorists" are unable to offer coherent and positive specifications of these concepts. This does not preclude that some overnatural concepts could be made more precise, but it shows that the road to evidence for the supernatural and the supposed benefits of its explanatory power are much rockier than the accommodationists believe.

56.5.3 Metaphysical Versus Methodological Naturalism

While it has become common knowledge that science goes together with naturalism, it is by no means commonly agreed upon what the exact nature of this relationship is. Compatibilist authors, for example, claim that science's naturalism is only a methodological naturalism, not a metaphysical one. Particularly in the philosophical context of the evolutionism/creationism controversy as well as in science education, which is concerned with *nature of science* issues, it has become common practice to distinguish methodological naturalism from metaphysical (or ontological or philosophical) naturalism and to claim that the former, not the latter, is the correct philosophical assumption of science. For the sake of convenience, let's abbreviate methodological naturalism by MN and metaphysical or ontological naturalism by ON.

Despite the popularity of MN, the characterizations of MN that we encounter in this debate are less than clear, so much so that we must guess what exactly MN is and how it differs from ON. Before substantiating this charge by taking a look at some of the most common definitions, it is important to point out first that, in this context, "MN" is used in a nonstandard way.

In philosophy, the standard meaning of "MN" is that philosophy ought to embrace the results of science and use some of its methods (weak MN) or that there is no unique philosophical method at all because only the methods of the natural sciences produce genuine knowledge (strong MN or strong scientism). In other words, weak MN states that science and philosophy are essentially continuous in that they pursue similar tasks with similar means, whereas strong MN leaves not much to do for philosophy.[38] By contrast,

> [i]n some philosophy of religion circles, 'methodological naturalism' is understood differently, as a thesis about natural scientific method itself, not about philosophical method. In this sense, 'methodological naturalism' asserts that religious commitments have no relevance within science: natural science itself requires no specific attitude to religion, and can be practised just as well by adherents of religious faiths as by atheists or agnostics. (Papineau 2007)

It is only this second meaning of "MN" that is relevant here, and it is this conception that in my view is ill-understood. The main problem is that it is unclear whether this MN actually is about scientific method rather than the metaphysics of science, in other words, whether it is a methodological (and hence an epistemological) view proper or whether it is just a covert metaphysical position, that is, a disguised form of ON. To illustrate this problem, let us take a look at some common definitions.

Pennock (2001) characterizes ON thus: "The Ontological Naturalist makes a commitment to substantive claims about what exists in nature, and then adds a closure clause stating 'and that is all there is'" (p. 84). By contrast

> [t]he Methodological Naturalist does not make a commitment directly to a picture of what exists in the world, but rather to a set of methods as a reliable way to find out about the world – typically the methods of the natural sciences, and perhaps extensions that are continuous with them – and indirectly to what these methods discover. (Pennock 2001, p. 84)

A commitment to method indicates that MN is epistemological. This is seconded by Forrest (2000), who tells us that MN is "an epistemology as well as a procedural protocol." Michael Ruse, by contrast, includes also ontological assumptions (lawfulness):

> On the one hand, one has what one might call 'metaphysical naturalism': this indeed is a materialistic, atheistic view, for it argues that the world is as we see it and that there is nothing more. On the other hand, one has a notion or a practice that can properly be called 'methodological naturalism': although this is the working philosophy of the scientist, it is in no way atheistic as such. The methodological naturalist is the person who assumes that the world runs according to unbroken law; that humans can understand the world in terms of this law; and that science involves just such understanding without any reference to extra or supernatural forces like God. Whether there are such forces or beings is another matter entirely and simply not addressed by methodological naturalism. Hence … in no sense is

[38] For further varieties of naturalism, see, e.g., De Caro and Macarthur (2008) and McMullin (2011).

> the methodological naturalist … committed to the denial of God's existence. It is simply that the methodological naturalist insists that, inasmuch as one is doing science, one avoid all theological or other religious references. (Ruse 2001b, p. 365)

Ruse's characterization reveals the main motivation behind MN: to assure the religious believer that science and religion are compatible.[39] Thus, the nonexistence of the supernatural (or rather its positive complement, ON) is not among the metaphysical presuppositions of science; it is just prohibited to refer to it. MN, then, boils down to the methodological rule, "Do not refer to anything supernatural!". The assumption of lawfulness, by contrast, is an ontological postulate. So Ruse's MN combines ontological and methodological aspects.

Even more ontological is another characterization of MN by Pennock:

> MN holds that as a principle of research we should regard the universe as a structured place that is ordered by uniform natural processes, and that scientists may not appeal to miracles or other supernatural interventions that break this presumed order. Science does not hold to MN dogmatically, but because of reasons having to do with the nature of empirical evidence. (Pennock 2009, p. 8)

Now, assumptions about the nature, structure, and workings of the world are metaphysical, not epistemological, even if most of the reasons for them are based on methodology. Moreover, Pennock's emphasis on MN as being nondogmatic indicates that in "MN" the adjective "methodological" could have a different meaning than the standard one, which is in the sense of methodology as normative epistemology, that is, the branch of epistemology concerned with the justification of beliefs and knowledge and the evaluation of methods. The standard adjective "methodological," then, classifies a position as epistemological – in contradistinction to adjectives describing some other philosophical category, such as a logical, semantical, ontological, or ethical. Another usage of "methodological," however, is in the sense of "provisional," "tentative," or "hypothetical." In this sense, "methodological" (sometimes also just "methodical") indicates either that the position in question is not regarded as an a priori truth or that it is not held dogmatically.

Consequently, there are at least two interpretations of MN:

1. MN is a genuine methodological/epistemological view, not an ontological one.
2. MN is an ontological position, namely, ON, but it is held provisionally rather than dogmatically.[40]

In the light of what was said about ON in this paper, only the second interpretation of MN is acceptable, although it would turn the name "MN" into a misnomer. The preference of "MN" over "provisional ON" could be due to the prejudice that

---

[39] That this is one of the main reasons behind MN has also been shown by Boudry et al. (2012).

[40] MN in the first sense can be held either dogmatically or provisionally. In the latter case, we may provocatively propose the name "methodological methodological naturalism," so as to point out the double meaning of "methodological." Note also that Boudry et al. (2010, 2012) distinguish intrinsic MN (in the sense of a defining feature of science) from provisional MN. The latter would be what I have just called methodological MN. Here I defend provisional ON as an intrinsic feature of science.

everything metaphysical is dogmatic. While traditional, and in particular religious, metaphysics often was dogmatic indeed, this is no longer true of a modern science-oriented metaphysics, which is fallible (Bunge 1977; Ladyman 2012). And even if modern metaphysics still were an a priori discipline, as some authors maintain (e.g., Lowe 2011), its rationalist claims would not be dogmas. For example, nobody would consider the modus ponens or the tertium quid as dogmas. This needs to be emphasized because some authors seem to confuse "a priori" with "dogmatic" (e.g., Fishman 2009, p. 814). The same would of course be true if only *some* claims of metaphysics were fallible, whereas *others* would be a priori.

If MN were indeed an epistemology, a procedural protocol, or a set of purely methodological rules, it would be a rather arbitrary choice of a protocol or of a set of rules, because it would not be backed up by a metaphysics. In a realist philosophy, being is prior to knowing. That is, the furniture and structure of the world must make cognition possible in the first place, and they must allow for the successful application of scientific methods. Hence, for a methodology to make any sense and to work successfully, there must be a metaphysics that helps to explain the functioning of this methodology. The methodology of science is therefore based on ON, just as the methodology of Plantinga's "theistic science" is based on supernaturalism.

However, if methodology cannot be separated from metaphysics, science is not religiously neutral. If science adopts ON in the sense of a metaphysical null hypothesis, it is not true that science is neutral on the existence of God, as most defenders of MN maintain (e.g., Scott 1998; Ruse 2001b; Pennock 2009). After all, the null hypothesis about some entity *x* states that *x* does not exist. Thus, science is committed to the "presumption of nonexistence" also with regard to God's existence.

56.6 Methodological and Other Conflicts

The preceding was one long argument to the metaphysical incompatibility of science and religion. It also mentioned several methodological conflicts arising from their disparate metaphysics. It may be helpful to recall them here and add a few further sources of conflict.

We have seen that the successful application of empirical scientific methods and thus the very concept of empirical evidence presuppose ON as a metaphysical null hypothesis. Whoever maintains that science can test supernatural hypotheses must find a way to resolve the paradox that any empirical test of any factual hypothesis presupposes the null assumption that supernatural entities do not exist. Most likely, an attempt at resolving this paradox will consist in some form of noninterventionism, but such an answer should not just consist in coming up with (untestable) ad hoc explanations as to why supernatural entities might refrain from such interventions: it should be a more principled approach, that is, a full theory. And, if scientific rather than philosophical, such a theory about noninterventionism should be independently testable. Yet any such test would in turn presuppose the very non-interventionist assumption….

Assuming for the sake of the argument that this paradox may be resolved, hypotheses involving supernatural entities would be empirically testable only in a limited way, namely, inasmuch as the supernatural is merely overnatural, that is, inasmuch as it has at least some natural properties. Insofar as religious convictions involve transnatural entities, they are untestable. Nonetheless, it is often claimed that even such convictions are testable. However, this often turns out to be terminological trickery, because in the context of religion, "testability" has nothing to do with empirical testability but with some alleged "experiential" or "existential" testability (Rolston 1987). Such "existential testability" is a wholly subjective notion, which is incompatible with the objective testability of science. Indeed, empirical testability undermines religion: "Because religion is an ostensible social relationship, it tends to be nonempirical, since openly testing a social relationship (…) undermines it. Testing therefore may be explicitly prohibited" (Guthrie 1995, p. 202f).

We have also seen that explanations referring to supernatural entities are either omni-explanatory or pseudo-explanatory. They are appeals to ignorance, and they may fill any explanatory gap by positing some supernatural intervention. Such "explanations," however, are arbitrary because any supernatural entity could do the same explanatory work as any other, and we may have no way to distinguish between competing supernaturalist explanations.

An important methodological incompatibility between science and religion is the latter's reliance on particular "methods" of cognition such as intuition, revelation, or religious experience.[41] Their characteristic is that they are inscrutable procedures, hence purely subjective ones. Thus, if such revelations or experiences are contradictory, there is no possibility to decide which of the alternatives is true – unless they yielded some specific factual statements that would be testable independently of the revelation or experience itself. From a methodological point of view then, they are not methods at all. However, whether such procedures are endorsed or not, religionists can always retreat to their faith when they wish to circumvent further rational and critical analysis. The difference between fundamentalist and more liberal religious views only lies at the point when such a retreat to fideism occurs (Bartley 1984; Kitcher 2004; Martin 1990).

Whereas the religionists' faith, i.e., the disregard and disrespect for evidence, is hailed as a virtue in their belief community, scientists are supposed to recognize that personal conviction or psychological certitude is no substitute for cognitive justification. The latter can only be achieved by objective evidence. Now, it may be objected that the history of science indicates that many scientists also stick to their hypotheses in an irrational manner, that they believe in them, and that they try to protect them against negative evidence. Granted. The difference, however, is that critical thinking and cognitive justification by empirical evidence belong to the ideals of the scientific community. If a particular scientist fails to comply with these ideals, he will be blamed by his peers, not praised. And if a hypothesis is not

[41] For a defense of religious experience as a valid method, see, e.g., Alston (2004). For critical analyses of the concept of religious experience, see Fales (2004, 2010), Kitcher (2004), Martin (1990), and Proudfoot (1985).

accepted by the scientific community, because there is too much negative evidence counting against it and there are perhaps better alternatives available, it will not enter the fund of scientific knowledge. By contrast, retaining one's faith even under the most averse falsifying conditions is a praiseworthy ideal in religion.

Related to faith is the role of authority in religion. While authority in religion is a methodological category, it is not so in science. Smith (2012) has recently examined the role of authority in science and religion from a mostly cognitive science viewpoint, pointing out that there are parallels between science and religion in how information is passed down from the original authority to colleagues, thence to science or religious teachers, respectively, and finally to students. Even though in science we do learn from authorities, such as colleagues, teachers, textbooks, and papers, because we cannot check every fact ourselves, and even though as individuals we do accept scientific knowledge on the basis of such authority, this is merely a matter of psychology and sociology. The real arbiter in science is evidence cum the current theoretical state of the art. This constitutes the ultimate justification. In religion, by contrast, religious doctrines are justified not by evidence, but because some authority, such as God himself or some spiritual guru, *pronounces* them as true. Justification by fiat and justification by evidence are incompatible methodologies.[42]

It may be objected that in religion "faith" does not mean "acceptance of doctrines on the basis of authority instead of evidence," but rather "trust" or "commitment." In this sense, faith is an aspect of a social relation such as trust in some other person around us. Yet such faith in persons is based on evidence: we trust our family and friends because we have some prior experience that they are trustworthy or worthy of commitment. By contrast, we have no such evidence in regard to supernatural persons, as we do not even have evidence of their very existence. So trust ($faith_2$) in such entities presupposes that we have already accepted the claim of their existence on $faith_1$ (belief without evidence). So even if there are two different concepts of faith, $faith_2$ is based on $faith_1$, so that the notion of $faith_1$ cannot be escaped. And $faith_1$ remains incompatible with science.

A different area of conflict concerns incompatible views about matters of fact. The most well-known case is the evolution/creation controversy. Liberal religionists tend to downplay such conspicuous conflicts because they are restricted to fundamentalist religions or denominations, respectively. However, fundamentalism is widespread in the USA, as well as in the Islamic world. While fundamentalism may have not much intellectual merit, it certainly is a powerful and dangerous social force. The doctrinal incompatibilities between fundamentalist religion and science are well known, so we may focus on the question of whether or not there are remaining conflicts even with respect to more liberal religion.

Apparently, there are no doctrinal conflicts left between science and liberal religion. Many scientific theories such as those in quantum physics, electromagnetism, plate tectonics, or immunology do not pose any problems for liberal religionists.

[42] Curiously, Smith (2012, p. 13) appears to realize this difference but he downplays it by saying that "in practice, however, the distinction is less stark." Yet practice is irrelevant: demarcation is first of all a matter of methodology. The cognitive and sociological similarities of learning on the basis of authority in both science and religion cannot attenuate the methodological conflict.

However, a clash between scientific theories and religious beliefs is bound to occur concerning the general cosmological views about man's (and woman's) place and status in the world, such as the evolution of *Homo sapiens*, the nature of mind, the existence of an afterlife, and the origins and social functions of religion. However liberal, religionists cannot admit that evolution has been a purely natural process (Rachels 1991; Plantinga 2011). If consistent, they must adopt at least a minimal teleological viewpoint, that is, they must posit that the evolutionary process has been guided from above and that it has a definite purpose, particularly, to establish a relationship between humans and some supernatural entity, e.g., a deity. Even if this view reduces to the claim that evolution is God's way of creation, it is at odds with evolutionary theory, because the latter makes no reference to supernatural entities and neither does any other scientific theory.

Curiously, Plantinga (2011) claims that evolutionary theory is compatible with theism, because God could have guided the process of evolution and even have caused particular mutations (see also Dennett and Plantinga 2011). Nothing in evolutionary theory would prohibit that. The supernatural is excluded only if evolutionary theory is paired off to naturalism – a union that Plantinga believes to be gratuitous. But this connection is not at all gratuitous because science is not free of metaphysics.[43]

The preceding considerations indicate that, if a religious methodology were applied in science and the scientific methodology in religion, the result would be mutual destruction. Science and religion are not only methodologically different but incompatible. The same holds for the metaphysics and the ethos of science and religion. Finally, insofar as religion makes factual statements about the world, there will also remain some doctrinal incompatibilities between religion and science. Thus, it is plainly false that, at least at a deep level, science and religion are not in conflict (O'Hear 1993). Actually, it is just at the deeper levels where the most conspicuous conflicts arise.

---

[43] Even more curiously, Plantinga (2011) argues that evolutionary theory is incompatible with metaphysical naturalism (see also Dennett and Plantinga 2011). A premise of this counterintuitive claim is that naturalists adopt an instrumentalist view of evolution, according to which natural selection favors at most cognitive faculties adequate for survival, not cognitive faculties furnishing truth, whether absolute or approximate. A purely natural evolution, then, entails that our cognitive faculties are not reliable in the sense of truth tracking. Plantinga is aware of the objection that a frog which manages to catch a fly must have correctly represented some property of its environment. But he claims that, even in the case of humans, the naturalist can talk only of appropriate behavior, not of true beliefs. He supports his case by resorting to antimaterialist arguments from the philosophy of mind, maintaining that any materialist conception of the brain and its functions, whether reductive or emergent, allows at best for appropriate behaviors, never beliefs, let alone true beliefs. True beliefs, so Plantinga's presupposition, can be had only in a nonmaterialist conception of the mind and a nonnaturalist conception of evolution. And the naturalist, who believes in the truth of naturalism, is inconsistent because naturalist evolution does not allow for the very existence of true beliefs. The real conflict, then, is between evolution and naturalism, not theism and evolution or science in general. Yet as evolutionary epistemology shows (Vollmer 2005), which is ignored by Plantinga, the evolution of cognition does lead to approximately true representations of the world (see also Dennett's reply in Dennett and Plantinga 2011). Also, naturalism requires a reconceptualization of concepts such as "knowledge" and "belief," which renders the antimaterialist argument moot (Bunge 1983).

56.7 The Conflict Between Science Education and Religious Education

> Even in his ability to be trained, man surpasses all animals. Mohammedans are trained to pray five times a day with their faces turned to Mecca and never fail to do so. Christians are trained to cross themselves, to bow, and to do other things on certain occasions. Indeed, speaking generally, religion is the chef d'œuvre of training, namely the ability to think; and so, as we know, a beginning in it cannot be made too early. There is no absurdity, however palpable, which cannot be firmly implanted in the minds of all, if only one begins to inculcate it before the early age of six by constantly repeating it to them with an air of great solemnity. For the training of man, like that of animals, is completely successful only at an early age. (Schopenhauer 1974, p. 603)

If science education is expected to inform students not only about facts but also about the philosophical background of science, it will have to address the methodological and metaphysical suppositions of science and maybe even all the world view components of science.[44] Inasmuch as religious education is also concerned with world view aspects, and we may claim that conveying a world view is even its major task, there is bound to be a conflict with the naturalist world view of science. After all, a religious education will have to state that the philosophical view of science is narrow or restricted, whereas the metaphysical and methodological outlook of religion offers so "much more" to discover. Indeed, defenders of religion argue that science and religion can be made compatible by choosing a broader metaphysics than naturalism (Barbour 2000), which entails of course that what science education teaches with respect to its philosophical foundations is inadequate.

The same holds for methodology. Religious education is likely to go against science education by allowing for exceptions concerning the acceptance of beliefs: religious beliefs need not be based on evidence, but may or even must be accepted on faith$_1$. Similarly, while it is a goal of science education to teach that it is appropriate to change one's views in the light of new evidence, religious education is prone to bringing forward a dogmatic mind-set because it teaches that unwavering faith is a good thing (Martin 1997).

Many evolutionary and developmental psychologists maintain that magical and religious thinking comes more natural because it is based on intuition rather than reflection, whereas critical or scientific thinking is something that has to be learned by keeping in check and overcoming our natural inclination towards superstitious thinking.[45]

Reinforcing our natural tendency for magical thinking by religious education thus appears to be antagonistic to the goals of science education. For example, while young children learn to master natural causality, they are at the same time exposed to religious concepts such as prayer, which teaches them that sheer wishing could

[44] See Davson-Galle (2004), Irzik and Nola (2009), Matthews (1992, 2009), and Smith and Siegel (2004).

[45] See, e.g., Guthrie (1995), Boyer and Walker (2000), Boyer (2001), Dennett (2007), McCauley (2011), and Shermer (2011). For a different view, see Subbotsky (2000) and Woolley (2000), who consider children's minds as neutral and thus to be filled with either rational or irrational cultural input.

have a physical effect. It seems that children somehow manage to put natural and imaginary causation in different "mental compartments" so as to avoid confusion (Woolley 2000). Even so, we may suspect that this compartmentalization is only partial and thus remains a steady source for ontological confusions, leading to a greater temptation to believe in various supernatural or paranormal claims and theories. Indeed, there is growing evidence that religious believers are more prone to also believing in the paranormal.[46]

If we want to raise responsible citizens who ground both their private and political decisions on scientific rather than illusory information, it is counterproductive to expose them to illusory world views. Worse, they not only learn that it is alright to accept such views as true but also to act according to those illusory beliefs. It comes as no surprise therefore that analytical thinking reduces religious belief (Gervais and Norenzayan 2012), which invites the conclusion that the converse is true too.

It may be objected here that, as the case of religious scientists shows, religion neither impedes scientific understanding nor prevents believers from choosing a career in science (see, e.g., Cobern et al. 2012). However, the empirical situation is not as straightforward.[47] Unsurprisingly, the effects of religious education very much depend on the degree of one's religiosity: the more seriously people take their religion, the worse the effects. For example, Christian fundamentalist students suffer from a lower complexity of thought and thus achieve lower educational attainment (Hunsberger et al. 1996; Sherkat 2007). In general, according to Evans (2011), religious believers take up a scientific career just as often as others. However, both scientific understanding and career choice are reduced when it comes to those scientific fields that interfere with religious belief, such as evolutionary biology and other areas that study human origins. Also, believers tend to deny scientific results if they have the impression that scientists pursue a moral agenda, for example, if scientists make recommendations for political action, as it may occur in the case of climate change. So it appears that orthodox religiosity does not lead to a general hostility towards science, unless the latter competes with central tenets of the given belief system – which shows, however, that there is a conflict with regard to a consistently scientific or else religious world view.

---

[46] See, e.g., Humphrey (1999), Goode (2000), Orenstein (2002), Hergovich et al. (2005), Lindeman and Aarnio (2007), and Eder et al. (2010). Note that the relation between religious belief and belief in the paranormal is not straightforward but depends on many variables such as level of education, gender, church attendance, and even the nature of the paranormal claims. For example, whereas astrology is mostly ruled out by Christians, creationism is not; and regular church attendance seems to prevent belief in the paranormal, presumably because the more frequent contact with official dogma protects from belief in competing paranormal or supernatural claims, respectively.

[47] For example, if science is strongly associated with technology, as in the questionnaire of Cobern et al. (2012), it may not be surprising that even orthodox believers see not much conflict between science and religion, except for ideologically contentious issues such as creationism or embryonic stem cell research. After all, even fundamentalists are glad to reap the benefits of modern technology. More importantly, personal views about the relation of science and religion, or even career choice, do not answer the de jure conflict problem concerning a consistent world view.

That a high level of religiosity may have negative effects is perhaps better seen at the social level. Comparing societal health data of the strongly religious USA with the more secular democracies of Western Europe and Japan, Paul concludes:

> In general, higher rates of belief in and worship of a creator correlate with higher rates of homicide, juvenile and early adult mortality, STD infection rates, teen pregnancy, and abortion in the prosperous democracies (…). The most theistic prosperous democracy, the U.S., […] is almost always the most dysfunctional of the developed democracies, sometimes spectacularly so, and almost always scores poorly. (Paul 2005, p. 7)

Critics have pointed out that, concerning the USA, these correlations are in most cases better explained by its higher level of social inequality rather than its high rate of religious belief (Delamontagne 2010, 2012). Using the Human Development Index (HDI), Delamontagne confirms the finding that high religiosity is accompanied by higher levels of societal dysfunction. However, he does not find a significant difference in HDI scores between moderately religious believers and nonbelievers, where the "moderately religious" are those for whom religion is "somewhat important" and the Bible is not true word by word, and who attend religious services but occasionally. Overall, higher levels of societal dysfunction in the USA are correlated with lower educational attainment, lower income, and race (Delamontagne 2012). While this may be correct sociologically, it is interesting to note from an ethical point of view that the high level of religiosity in the USA does not seem to contribute to decreasing the high level of social inequality – which casts doubt on the self-image of religion as a moral enterprise benefitting society.

It should not go unmentioned that psychological and sociological studies also report some positive effects associated with religiosity. For example, religious students tend to be more sociable, show less substance abuse problems, and tend to be more disciplined with respect to their coursework (Donahue and Nielsen 2005; Sherkat 2007). As the members of many denominations tend to form closer-knit communities, these examples may be seen as beneficial aspects of social embeddedness rather than direct effects of religious education as such. But should we not expect anyway that religious education contributes to a better morality?

Indeed, probably the major argument for religious education is that it is indispensable for moral education, in particular as science is concerned with matters of fact, not values and ethics. However, the alleged connection between religion and morality does not withstand scrutiny. First, empirical studies have shown that, overall, religious people fail to behave more morally than nonreligious people (Spilka et al. 1985; Tan 2006). For example, they neither cheat less in tests nor are they less selfish. Overall, then, religious education has no distinctly positive effect on moral behavior, which we would have to expect if the main function of religion and religious education were an ethical one.

Second, the goals of a modern moral education include acquiring the attitude and the capability of modifying one's moral principles in the light of new experience, knowledge, and insight (Martin 1991). This aim is certainly antagonistic to the religious attitude towards moral norms. If moral norms are God-given, be it by direct command or by a created natural law, they cannot be questioned or modified: they can only be obeyed or disobeyed. Finally, philosophy has amply demonstrated from Plato on that religion cannot be the basis of morality anyway.[48]

A final point, just for the fun of it, so to speak: it appears that religiosity is negatively associated to sense of humor; the more so, the more dogmatic or authoritarian believers are (Saroglou 2002). A possible objection, analogous to the argument from religious scientists, is obvious: we all know religious people with a great sense of humor. But "it is possible that religious people have a good sense of humor *despite* their religiosity; and not necessarily *because* of it" (Saroglou 2002, p. 206).

The quick upshot is this: empirical research on religiosity shows that both people and societies are the better off the less seriously they take the contents of their religious belief systems. The more literal and dogmatic the religious beliefs, the worse; the more abstract and liberal – more bluntly: the fuzzier and emptier – the better. The reason for this is, as we have seen again and again, that science and religion or, if preferred, a scientific and a religious world view are metaphysically, methodologically, and attitudinally incompatible.

References

Albert, H. (1985). *Treatise on Critical Reason*. Princeton: Princeton University Press.

Albert, H. (2000). *Kritischer Rationalismus*. Tübingen: Mohr-Siebeck.

Alston, W.P. (1967). Naturalistic Reconstructions of Religion. In P. Edwards (Ed.) *The Encyclopedia of Philosophy*, vol. 7 (pp. 145–147). London: Collier-Macmillan.

Alston, W.P. (2004). Religious Experience Justifies Religious Belief. In M.L. Peterson & Van Arragon, R.J. (Eds.) *Contemporary Debates in Philosophy of Religion* (pp. 135–145). Malden, MA: Blackwell.

Alters, B.J. (1997). Whose Nature of Science? *Journal of Research in Science Teaching* 34, 39–55.

Augustine, K. (2001). A Defense of Naturalism. http://www.infidels.org/library/modern/keith_augustine/thesis.html. Accessed 18 January 2010.

Ayer, A. (1990/1936) *Language, Truth, and Logic*. London: Penguin Books.

Barbour, I. (2000). *When Science Meets Religion. Enemies, Strangers, or Partners?* New York: HarperOne.

Barrett, J.L. & Keil, F.C. (1996). Conceptualizing a Nonnatural Entity: Anthropomorphism in God Concepts. *Cognitive Psychology* 31, 219–247.

Bartley, W.W. (1984). *The Retreat to Commitment*. LaSalle, IL: Open Court.

Boudry, M., Blancke, S. & Braeckman, J. (2010). How Not to Attack Intelligent Design Creationism: Philosophical Misconceptions About Methodological Naturalism. *Foundations of Science* 15, 227–244.

Boudry, M., Blancke, S. & Braeckman, J. (2012). Grist to the Mill of Anti-evolutionism: The Failed Strategy of Ruling the Supernatural Out of Science by Philosophical Fiat. *Science & Education*, DOI 10.1007/s11191-012-9446-8.

Boyer, P. (2001). *Religion Explained: The Evolutionary Origins of Religious Thought*. New York: Basic Books.

Boyer, P. & Walker, S. (2000). Intuitive Ontology and Cultural Input in the Acquisition of Religious Concepts. In K. Rosengren, C.N. Johnson & P.L. Harris (Eds.) *Imagining the Impossible. Magical, Scientific, and Religious Thinking in Children* (pp. 130–156). Cambridge: Cambridge University Press.

[48] See, e.g., Nowell-Smith (1967), Mackie (1982), Martin (1990, 2002), and Rachels (1995).

Broad, C.D. (1949). The Relevance of Psychical Research to Philosophy. *Philosophy* 24, 291–309.
Bunge, M. (1977). *Treatise on Basic Philosophy, vol. 3: Ontology I: The Furniture of the World.* Dordrecht: D. Reidel.
Bunge, M. (1983). *Treatise on Basic Philosophy, vol. 6: Epistemology & Methodology II: Understanding the World.* Dordrecht: D. Reidel.
Bunge, M. (2006). *Chasing Reality. Strife over Realism.* Toronto: University of Toronto Press.
Bunge, M. & Mahner, M. (2004). *Über die Natur der Dinge. Materialismus und Wissenschaft.* Stuttgart: Hirzel-Verlag.
Byl, J. (1985). Instrumentalism: A Third Option. *Journal of the American Scientific Affiliation* 37, 11–18. Online: http://www.asa3.org/ASA/PSCF/1985/JASA3-85Byl.html. Accessed August 24, 2012.
Clarke, S. (2009). Naturalism, Science and the Supernatural. *Sophia* 48, 127–142.
Clayton, P. & Simpson, Z.R. (Eds.) (2008) *The Oxford Handbook of Religion and Science*. Oxford: Oxford University Press.
Clements, T.S. (1990). *Science vs. Religion*. Buffalo, NY: Prometheus Books.
Cobern, W.W., Loving, C., Davis, E.B. & Terpstra, J. (2012). An Empirical Examination of the Warfare Metaphor with Respect to Pre-Service Elementary Teachers. *Journal of Science Education and Technology*, DOI 10.1007/s10956-012-9408-6
Davson-Galle, P. (2004). Philosophy of Science, Critical Thinking, and Science Education. *Science & Education* 13, 503–517.
Dawkins, R. (2006). *The God Delusion*. Boston: Houghton Mifflin.
De Caro, M., & Macarthur, D. (2008) Introduction: The Nature of Naturalism. In M. De Caro & D. Macarthur (Eds.) *Naturalism in Question* (pp. 1–17). Cambridge, MA: Harvard University Press.
Delamontagne, R.G. (2010). High Religiosity and Societal Dysfunction in the United States During the First Decade of the Twenty-First Century. *Evolutionary Psychology* 8, 617–657.
Delamontagne, R.G. (2012). Overgeneralization: The Achilles Heel of Apocalyptic Atheism? *Free Inquiry* 32(6), 38–41.
Dennett, D. (2007). *Breaking the Spell: Religion as a Natural Phenomenon*. London: Penguin Books.
Dennett, D. & Plantinga, A. (2011). Science and Religion. Are They Compatible? New York: Oxford University Press.
Descartes, R. (1641). *Meditationes*. http://www.wright.edu/cola/descartes/mede.html. Accessed 17 January 2010.
Donahue, M.J. & Nielsen, M.E. (2005). Religion, Attitudes, and Social Behavior. In R.F. Paloutzian & C.P. Park (Eds.) *Handbook of the Psychology of Religion and Spirituality* (pp. 274–291). New York: Guilford Press.
Drees, W. (1996). *Religion, Science, and Naturalism*. Cambridge: Cambridge University Press.
Drees, W. (2008). Religious Naturalism and Science. In P. Clayton & Z.R. Simpson (Eds.) *The Oxford Handbook of Religion and Science* (pp. 108–123). Oxford: Oxford University Press.
Ecklund, E.H. (2010). *Science vs. Religion: What Scientists Really Think.* Oxford: Oxford University Press.
Eder, E., Turic, K., Milasowsky, N., Van Adzin, K. & Hergovich, A. (2010). The Relationships Between Paranormal Belief, Creationism, Intelligent Design and Evolution at Secondary Schools in Vienna (Austria). *Science & Education*, DOI 10.1007/s11191-010-9327-y
Edis, T. (2007). *An Illusion of Harmony. Science and Religion in Islam*. Amherst, NY: Prometheus Books.
Edis, T. (2008). *Science and Nonbelief.* Amherst, NY: Prometheus Books.
Edis, T. (2009). Modern Science and Conservative Islam: An Uneasy Relationship. *Science & Education* 18, 885–903.
Ellis, B. (2002). *The Philosophy of Nature. A Guide to the New Essentialism.* Chesham, UK: Acumen.
Einstein, A. (1999). *The World As I See It*. Secaucus, NJ: Citadel Press.
Evans, J.H. (2011). Epistemological and Moral Conflict Between Religion and Science. *Journal for the Scientific Study of Religion* 50, 707–727.

Fales, E. (2004). Do Mystics See God? In M.L. Peterson & R.J. Van Arragon (Eds.) *Contemporary Debates in Philosophy of Religion* (pp. 145–158). Malden, MA: Blackwell.

Fales, E. (2010) *Divine Intervention. Metaphysical and Epistemological Puzzles*. New York: Routledge.

Fishman, Y.I. (2009). Can Science Test Supernatural Worldviews? *Science & Education* 18, 813–837.

Flanagan, O. (2008). Varieties of Naturalism. In P. Clayton & Z.R. Simpson (Eds.) *The Oxford Handbook of Religion and Science* (pp. 430–452). Oxford: Oxford University Press.

Flew, A. (1972). The Presumption of Atheism. http://www.positiveatheism.org/writ/flew01.htm. Accessed 21 February 2011. [originally published in *Canadian Journal of Philosophy* 2, 29–46]

Forrest, B. (2000). Methodological Naturalism and Philosophical Naturalism: Clarifying the Connection. http://www.infidels.org/library/modern/barbara_forrest/naturalism.html, Accessed 27 February 2011 [originally published in *Philo* 3(2), 7–29].

Freddoso, A.J. (1991). God's General Concurrence With Secondary Causes: Why Conservation Is Not Enough. In J. E. Tomberlin (Ed.) Philosophy of Religion, pp. 553–585. Atascadero, CA: Ridgeview Publishing Company.

Gauch, H.G. (2009). Science, Worldviews, and Education. *Science & Education* 18, 667–695.

Gervais, W.M. & Norenzayan, A. (2012). Analytic Thinking Promotes Religious Disbelief. *Science* 336, 493–496.

Glennan, S. (2009). Whose science and whose religion? Reflections on the relations between scientific and religious worldviews. *Science & Education*, 18, 797–812.

Goode, E. (2000). *Paranormal Beliefs. A Sociological Introduction*. Prospect Heights, IL: Waveland Press.

Gould, S.J. (1999). *Rocks of Ages: Science and Religion in the Fullness of Life*. New York: Ballantine.

Gross, N. & Simmons, S. (2009). The Religiosity of American College and University Professors. *Sociology of Religion* 70: 101–109.

Guthrie, S.E. (1995). *Faces in the Clouds. A New Theory of Religion*. New York: Oxford University Press.

Haldane, J. B. S. (1934). *Fact and Faith*. London: Watts.

Hardwick, C.D. (2003). Religious Naturalism Today. *Zygon - Journal of Religion and Science* 38, 111–116.

Harrison, P. (Ed.) (2010). *The Cambridge Companion to Science and Religion*. Cambridge: Cambridge University Press.

Haught, J.F. (1995). *Science and Religion: From Conflict to Conversation*. Mahwah, NJ: Paulist Press.

Hergovich, A., Schott, R. & Arendasy, M. (2005). Paranormal Belief and Religiosity. *Journal of Parapsychology* 69, 293–303

Humphrey, N. (1999). *Leaps of Faith. Science, Miracles, and the Search for Supernatural Consolation*. New York: Copernicus.

Hunsberger, B., Alisat, S., Pancer, S.M. & Pratt, M. (1996). Religious Fundamentalism and Religious Doubts: Content, Connections, and Complexity of Thinking. *International Journal for the Psychology of Religion* 6, 201–220.

Irzik, G. & Nola, R. (2009). Worldviews and their Relation to Science. *Science & Education* 18, 729–745.

Kanitscheider, B. (1996). *Im Innern der Natur*. Darmstadt: Wissenschaftliche Buchgesellschaft.

Kitcher, P. (2004). The Many-Sided Conflict Between Science and Religion. In W.E. Mann (Ed.) *The Blackwell Guide to the Philosophy of Religion* (pp. 266–282). Oxford: Blackwell.

Knodel, H. (Ed. 1985). *Neues Biologiepraktikum. Linder Biologie (Lehrerband)*. Stuttgart: J.B. Metzler.

Kurtz, P. (Ed.) (2003). Science and Religion: Are They Compatible? Amherst, NY: Prometheus Books.

Ladyman, J. (2012). Science, Metaphysics, and Method. *Philosophical Studies* 160, 31–51.

Larson, E.J. & Witham, L. (1998). Leading scientists still reject God. *Nature* 394, 313.

Laudan, L. (1983). The Demise of the Demarcation Problem. In R. S. Cohen & L. Laudan (Eds.)., *Physics, Philosophy, and Psychoanalysis*, (pp. 111–127). Dordrecht: D. Reidel.

Levine, M. (2011). *Pantheism*. http://plato.stanford.edu/entries/pantheism. Accessed July 29th, 2012.

Lindeman, M. & Aarnio, K. (2007). Superstitious, Magical, and Paranormal Beliefs. An Integrative Model. *Journal of Research in Personality* 41, 731–744.

Loesberg, J. (2007). Kant, Hume, Darwin, and Design: Why Intelligent Design Wasn´t Science Before Darwin and Still Isn't. *Philosophical Forum* 38, 95–123.

Lowe, E.J. (2011). The Rationality of Metaphysics. *Synthese* 178, 99–109.

Mackie, J.L. (1982). *The Miracle of Theism. Arguments for and against the Existence of God*. Oxford: Clarendon Press.

Mahner, M. (2007). Demarcating Science from Non-Science. In T.A.F. Kuipers (Ed.) *Handbook of the Philosophy of Science, vol. 1: General Philosophy of Science – Focal Issues* (pp. 515–575). Amsterdam: North Holland Publishing Company.

Mahner, M. (2012). The Role of Metaphysical Naturalism in Science. *Science & Education* 21, 1437–1459.

Mahner, M. (2013). Science and Pseudoscience: How to Demarcate after the (Alleged) Demise of the Demarcation Problem? In M. Pigliucci & M. Boudry (Eds.) *Philosophy of Pseudoscience: Reconsidering the Demarcation Problem*. Chicago: Chicago University Press.

Mahner, M. & Bunge, M. (1996a). Is Religious Education Compatible with Science Education? *Science & Education* 5, 101–123.

Mahner, M. & Bunge, M. (1996b). The Incompatibility of Science and Religion Sustained: A Reply to Our Critics. *Science & Education* 5, 189–199.

Mahner, M. & Bunge, M. (1997) *Foundations of Biophilosophy*. Berlin: Springer Verlag.

Margenau, H. & Varghese, R.A. (Eds.) (1992). *Cosmos, Bios, Theos*. LaSalle, IL: Open Court.

Martin, M. (1990). *Atheism - A Philosophical Justification*. Philadelphia: Temple University Press.

Martin, M. (1991). Science Education and Moral Education. In M. Matthews (Ed.) *History, Philosophy, and Science Education. Selected Readings* (pp. 102–113). New York: Teachers College Press.

Martin, M. (1997). Is Christian Education Compatible with Science Education? *Science & Education* 6, 239–249.

Martin, M. (2002). *Atheism, Morality, and Meaning*. Amherst, NY: Prometheus Books.

Matthews, M.R. (1992). History, Philosophy, and Science Teaching: The Present Rapprochement. *Science & Education* 1, 11–47.

Matthews, M.R (2009). Teaching the Philosophical and Worldview Components of Science. *Science & Education* 18, 697–728.

McCauley, R.N. (2011). *Why Religion Is Natural and Science Is Not*. New York: Oxford University Press.

McMullin, E. (2011). Varieties of Methodological Naturalism. In B.L. Gordon & W.A. Dembski (Eds.) *The Nature of Nature, Examining the Role of Naturalism in Science* (pp. 82–94). Wilmington, DE: ISI Books.

Monton, B. (2009). *Seeking God in Science. An Atheist Defends Intelligent Design*. Peterborough: Broadview Press.

Nielsen, K. (1985). *Philosophy & Atheism*. Buffalo, NY: Prometheus Books.

Nowell-Smith, P.H. (1967). Religion and Morality. In P. Edwards (Ed.), *The Encyclopedia of Philosophy*, vol. 7, (pp. 150–158). London: Collier-Macmillan.

O'Hear, A. (1993). Science and Religion. *British Journal for the Philosophy of Science* 44, 505–516.

Orenstein, A. (2002). Religion and Paranormal Belief. *Journal for the Scientific Study of Religion* 41, 301–311.

Orr, H.A. (1999). Gould on God. Can Religion and Science be Happily Reconciled? http://boston-review.net/BR24.5/orr.html, Accessed July 23rd, 2012.

Papineau, D. (2007). Naturalism. http://plato.stanford.edu/entries/naturalism. Accessed 16 January 2010.

Pasquarello, T. (2002). God: 12000. The Faith of a Rebeliever. *Free Inquiry* 22(4), 50–53.

Paul, G.S. (2005) Cross-National Correlations of Quantifiable Societal Health with Popular Religiosity and Secularism in the Prosperous Democracies. *Journal of Religion & Society* 7, 1–17. http://moses.creighton.edu/JRS/2005/2005-11.pdf

Peacocke, A. (1993). *Theology for a Scientific Age*. Minneapolis: Fortress Press.
Pennock. R.T. (2000). *Tower of Babel. The Evidence Against the New Creationism*. Cambridge, MA: MIT Press.
Pennock, R.T. (2001). Naturalism, Evidence, and Creationism: The Case of Phillip Johnson. In R.T. Pennock (Ed. 2001) *Intelligent Design Creationism and Its Critics* (pp. 77–97). Cambridge, MA: MIT-Press.
Pennock, R.T. (2009). Can't Philosophers Tell the Difference Between Science and Religion? Demarcation Revisited. *Synthese*, doi:10.1007/s11229-009-9547-3
Philipse, H. (2012) *God in the Age of Science? A Critique of Religious Reason*. Oxford: Oxford University Press.
Pigliucci, M. (2013). The Demarcation Problem: A (belated) Response to Laudan. In M. Pigliucci & M. Boudry (Eds.) *Philosophy of Pseudoscience: Reconsidering the Demarcation Problem*. Chicago: Chicago University Press.
Plantinga, A. (2001) Methodological Naturalism? In R.T. Pennock (Ed. 2001) *Intelligent Design Creationism and Its Critics* (pp. 339–361). Cambridge, MA: MIT-Press.
Plantinga, A. (2011). *Where the Conflict Really Lies. Science, Religion, and Naturalism*. New York: Oxford University Press.
Platvoet, J.G. & Molendijk, A.L. (Eds.) (1999). *The Pragmatics of Defining Religion*. Leiden: Brill.
Polkinghorne, J. (1987). *One World: The Interaction of Science and Theology*. Princeton: Princeton University Press.
Proudfoot, W. (1985). *Religious Experience*. Berkeley: University of California Press.
Provine, W.B. (2008). Evolution, Religion, and Science. In P. Clayton & Z.R. Simpson (Eds.) *The Oxford Handbook of Religion and Science* (pp. 667–680). Oxford: Oxford University Press.
Psillos, S. (1999). *Scientific Realism: How Science Tracks Truth*. London: Routledge.
Rachels, J. (1991). *Created from Animals. The Moral Implications of Darwinism*. Oxford: Oxford University Press.
Rachels, J. (1995). *The Elements of Moral Philosophy*. New York: McGraw-Hill.
Ratzsch, D. (1996). Tightening Some Loose Screws: Prospects for a Christian Natural Science. In J.M. van der Meer (Ed.) *Facets of Faith and Science, vol. II* (pp. 175–190). Lanham, MD: University Press of America.
Ratzsch, D. (2004). The Demise of Religion: Greatly Exaggerated Reports From the Science/Religion "Wars". In M.L. Peterson & R.J. Van Arragon (Eds.) *Contemporary Debates in Philosophy of Religion* (pp. 72–87). Malden, MA: Blackwell.
Rolston, H. (1987). *Science and Religion. A Critical Survey*. New York: Random House.
Ruse, M. (2001a). *Can a Darwinian Be a Christian? The Relationship Between Science and Religion*. Cambridge: Cambridge University Press.
Ruse, M. (2001b). Methodological Naturalism under Attack. In R.T. Pennock (Ed.) *Intelligent Design Creationism and Its Critics* (pp. 363–385). Cambridge, MA: MIT-Press.
Ruse, M. (2011). *Science and Spirituality: Making Room for Faith in the Age of Science*. Cambridge: Cambridge University Press.
Russell, C.A. (2002). The Conflict of Science and Religion. In G.B. Ferngren (Ed.) *Science and Religion: A Historical Introduction* (pp. 3–12). Baltimore, MD: Johns Hopkins University Press.
Sarkar, S. (2011). The Science Question in Intelligent Design. *Synthese* 178, 291–305.
Saroglou, V. (2002). Religion and a Sense of Humor: An a priori Incompatibility? Theoretical Considerations from a Psychological Perspective. *Humor* 15, 191–214.
Schopenhauer, A. (1951). *Essays from the Parerga and Paralipomena*. London: George Allen & Unwin.
Schopenhauer, A. (1974). Parerga and Paralipomena. Short Philosophical Essays, vol. 2 (transl. by E.F.J. Payne). Oxford: Clarendon Press.
Schrader, D. E. (2000) Theology and Physical Science. A Story of Developmental Influence at the Boundaries. In K. Rosengren, C.N. Johnson, & P.L. Harris (Eds.) *Imagining the Impossible. Magical, Scientific, and Religious Thinking in Children* (pp. 372–404). Cambridge: Cambridge University Press.
Scott, E.C. (1998). Two Kinds of Materialism. Keeping them Separate Makes Faith and Science Compatible. *Free Inquiry* 18(2), 20.

Sherkat. D.E. (2007). Religion and Higher Education: The Good, the Bad, and the Ugly. Social Science Research Council, http://religion.ssrc.org/reforum/Sherkat.pdf. Accessed 16 September 2012.
Shermer, M. (2011). *The Believing Brain*. New York: Times Books.
Sinatra, G.M. & Nadelson, L. (2011). Science and Religion: Ontologically Different Epistemologies. In R.S. Taylor & M. Ferrari (Eds.) *Epistemology and Science Education: Understanding the Evolution vs. Intelligent Design Controversy* (pp. 173–193). New York: Routledge.
Smart, J.J.C. (1967). Religion and Science. In P. Edwards (Ed.) The Encyclopedia of Philosophy, vol. 7, (pp. 158–163). London: Collier-Macmillan.
Smith, K.C. (2001). Appealing to Ignorance Behind the Cloak of Ambiguity. In R.T. Pennock (Ed. 2001) *Intelligent Design Creationism and Its Critics* (pp. 705–735). Cambridge, MA: MIT-Press.
Smith, M.U. (2012). The Role of Authority in Science and Religion with Implications for Introductory Science Teaching and Learning. *Science & Education*, DOI 10.1007/s11191-012-9469-1
Smith, M.U. & Siegel, H. (2004). Knowing, Believing, and Understanding: What Goals for Science Education? *Science & Education* 13, 553–582.
Spiegelberg, H. (1951). Supernaturalism or Naturalism: A Study in Meaning and Verifiability. *Philosophy of Science* 18, 339–368.
Spilka, B., Hood, R.W. & Gorsuch, R.L. (1985). *The Psychology of Religion. An Empirical Approach*. Englewood Cliffs, NJ: Prentice Hall.
Stenger, V. J. (2007). *God: The failed hypothesis. How science shows that God does not exist.* Amherst, NY: Prometheus Books.
Stenger, V. J. (2011). *The Fallacy of Fine-Tuning. Why the Universe Is Not Designed for Us*. Amherst, NY: Prometheus Books.
Stenmark, M. (2010). Ways of Relating Science and Religion. In P. Harrison (Ed.) *The Cambridge Companion to Science and Religion* (pp. 278–295). Cambridge, UK: Cambridge University Press.
Stolberg, T. and G. Teece (2011). *Teaching Religion and Science*. London: Routledge.
Subbotsky, E. (2000) Phenomenalistic Perception and Rational Understanding in the Mind of an Individual. In K. Rosengren, C.N. Johnson, & P.L. Harris (Eds.) *Imagining the Impossible. Magical, Scientific, and Religious Thinking in Children* (pp. 35–74). Cambridge: Cambridge University Press.
Suchting, W.A. (1994) Notes on the Cultural Significance of the Sciences. *Science & Education* 3, 1–56.
Tan, J. (2006). Religion and Social Preferences. An Experimental Study. *Economic Letters* 90, 60–67.
Thagard, P. (2011). Evolution, Creation, and the Philosophy of Science. In R.S. Taylor & M. Ferrari (Eds.) *Epistemology and Science Education: Understanding the Evolution vs. Intelligent Design Controversy* (pp. 20–37). New York: Routledge.
Tooley, M. (2011). Naturalism, Science, and Religion. In B.L. Gordon & W.A. Dembski (Eds.) *The Nature of Nature, Examining the Role of Naturalism in Science* (pp. 880–900). Wilmington, DE: ISI Books.
Vollmer, G. (1990). Against Instrumentalism. In P. Weingartner & G.J.W. Dorn (Eds. 1990) *Studies on Mario Bunge's Treatise* (pp 245–259). Amsterdam: Rodopi.
Vollmer, G. (2005). How Is It that We Can Know this World? New Arguments in Evolutionary Epistemology. In V. Hösle & C. Illies (Eds.) *Darwinism & Philosophy* (pp. 259–274). Notre Dame, IN: University of Notre Dame Press.
Wentzel van Huyssteen, J. (1998). *Duet or Duel? Theology and Science in a Postmodern World.* Harrisburg, PA: Trinity Press International.
Woolley, J.D. (2000). The Development of Beliefs About Direct Mental-Physical Causality in Imagination, Magic, and Religion. In K. Rosengren, C.N. Johnson, & P.L. Harris (Eds.) *Imagining the Impossible. Magical, Scientific, and Religious Thinking in Children* (pp. 99–129). Cambridge: Cambridge University Press
Worrall, J. (2004). Science Discredits Religion. In M.L. Peterson & R.J. Van Arragon (Eds.) *Contemporary Debates in Philosophy of Religion* (pp. 59–72). Malden, MA: Blackwell.
Yinger, J.M. (1970). *The Scientific Study of Religion*. New York: Macmillan

Martin Mahner graduated in 1985 at the Free University of Berlin (major: biology, minor: geography). After his second graduation as an A-level teacher in 1987, he obtained his Ph.D. in 1992 (major: zoology; minor: philosophy of science). In 1992, he was awarded a research grant by the Deutsche Forschungsgemeinschaft (DFG) to take up a postdoctoral fellowship with Mario Bunge at the Department of Philosophy at McGill University, Montreal (1993–1996). From 1996 to 1998, he was a free-lance instructor at three universities in Berlin, Germany (zoology, philosophy of science). Since 1999, he has been the executive director of the Center for Inquiry – Europe, near Darmstadt, Germany, a nonprofit educational organization devoted to collecting and disseminating critical information about pseudoscience. Among his publications are *Foundations of Biophilosophy* (with M. Bunge 1997, translated into German, Spanish, and Japanese) and *Über die Natur der Dinge. Materialismus und Wissenschaft* [*On the Nature of Things. Materialism and Science*] (with M. Bunge 2004). He is editor of *Scientific Realism: Selected Essays of Mario Bunge* (2001), and he contributed to the 15-volume *Lexikon der Biologie* [*Dictionary of Biology*] as well as the chapter "Demarcating Science from Non-Science" to the *Handbook of the Philosophy of Science*, vol. 1 (2007).

Part XII
Theoretical Studies: Science Education Research

Chapter 57
Methodological Issues in Science Education Research: A Perspective from the Philosophy of Science

Keith S. Taber

57.1 Introduction

Science education as an academic field occupies an interesting position as there can be something of a tension between its location within one type of academic setting (education, often considered a social science, but also drawing upon the humanities) and its strong links with the disciplines that are the target for that educational activity (i.e. the natural sciences). For one thing, education is primarily about teaching, which is a practical and professional activity, often considered to be as much a craft as a science (Adams 2011; Grimmett and MacKinnon 1992). So education as an academic discipline has strong links with education *as practised* in schools and other institutions 'of' education and seeks to learn about and inform educational activity in such formal learning contexts, as well as increasingly in various informal contexts where learners may be self-taught or learn through informal interactions that may not be primarily intended to bring about teaching.

For the purposes of this chapter, education will be considered to be centrally about the processes of teaching and learning (Pring 2000), acknowledging that the teaching may sometimes be in the form of self-direction of learning (Taber 2009a). So science education is a field that is centrally concerned with the teaching and learning of, and about, science and the scientific disciplines. Science education encompasses both teaching for the general population – science for citizenship, scientific literary – and for the preparation of future professional scientists (Aikenhead 2006; Hodson 2009; Holbrook and Rannikmae 2007; Laugksch 2000; Millar and Osborne 1998).

The nature of the work of teaching has raised interesting issues about the relationship between educational research and practice: issues about outsider versus participant research (Taber 2012b) and about the challenges of translating research

K.S. Taber (✉)
University of Cambridge, Cambridge, UK
e-mail: kst24@cam.ac.uk

M.R. Matthews (ed.), *International Handbook of Research in History, Philosophy and Science Teaching*, DOI 10.1007/978-94-007-7654-8_57,

that is often necessarily framed for the academic community into a form that can inform practitioners (de Jong 2000; Russell and Osborne 1993) – or alternatively, seeking to find ways to draw generalisable findings from local highly contextualised studies (Taber 2000).

In addition, educational research is usually undertaken with human participants and so is subject to ethical considerations that do not apply in the natural sciences (e.g. British Educational Research Association 2004). A common metaphor for science is along the lines of wrestling nature's secrets from her (Pesic 1999), but researchers are discouraged from seeing the collection of educational research data in this way. Indeed, objectifying the learner (or teacher) as a source of data, rather than considering them as a person participating in research, is usually considered inappropriate in educational work. (Ethical issues are considered further later in the chapter.)

These generic issues, which need to be faced by educational researchers wishing to influence educational practice, are often supplemented for science education researchers by questions raised by the juxtaposition of mindsets and commitments of two somewhat different disciplinary backgrounds. Education as an academic subject is often seen as a social science, and research training in education reflects this, with a major focus on the paradigmatic concerns and tensions that operate in social research. (Indeed, this is oversimplistic as some scholarship undertaken within education faculties fits within the behavioural sciences, and some educational scholarship is best seen as located in the humanities.) Yet science education, as a field, tends not to be populated exclusively (or even primarily) by social scientists choosing to focus on science education for their research topic. There is certainly a fair number of those, but many (and in at least some national contexts, most) of those who undertake research in science education have a background in natural sciences and often experience of teaching natural science subjects at school or a higher level.

Even when such a researcher takes on an identity as a social science researcher, this is often secondary to their established identity as scientist and/or science educator (Kind and Taber 2005). Yet, as will be discussed below, research training in the natural sciences is often somewhat different to research training in education and other social sciences. A further complication is that many of those undertaking research in science education who do not have a background of studying and teaching in the natural sciences actually have a background in psychology: a subject that itself straddles the disciplinary divide between natural and social sciences (Barrett 2009). This chapter will explore a range of issues about the nature of educational research (and the consequent implications for methodology), with a particular concern with whether research in science education can be considered scientific.

This focus is of interest for a number of reasons. The question of what can be encompassed within science, the demarcation question, has been of concern to some philosophers of science who have wished to distinguish science from pseudoscience (Lakatos 1970; Popper 1934/1959). The very adoption of the term 'social sciences' reflects an attempt to model disciplines like sociology on the natural sciences, leading to the questions of to what extent and in what ways are social sciences and natural sciences part of a larger 'science' (Kagan 2009). In addition are two complications referred to above: the diversity of educational scholarship and the professional identities of science educators.

Within education faculties, research and scholarship may take a wide range of forms, from completely nonempirical philosophical analyses and literary analysis of texts written for children to experimental studies carried out by educational psychologists following established 'paradigms' (which, in that context, in effect means an experiment design) and considering participants as interchangeable subjects from particular populations (normal 12-year-olds, dyslexic 9-year-olds, autistic secondary age boys, etc.). In some national contexts, science educators will be working as part of such diverse communities, whereas in other national contexts it is more common for science educators to be working as part of a science faculty – within a physics department, for example. Many science educators move into research in science education with a well-established professional identity as a scientist (or more specifically chemist, etc.) and/or science teacher and therefore bring expectations of what kind of activity 'research' is: expectations that will inform their understanding of the particular institutional context in which they are working.

57.1.1 The Programme for This Chapter

The present chapter will first consider the particular nature of educational research and the nature of the field of science education, before considering the question of whether induction into science education research reflects the process of induction into research in the natural sciences. The chapter then considers the logic of developing a research project and how in educational enquiry this involves a justifiable choice of methodology, which may have to be moderated to some extent by ethical considerations and which informs the construction of a specific research design drawing from a range of particular research techniques used in educational work.

The chapter then considers how best to understand the range of methodologies commonly applied in education in terms of the different ontological and epistemological commitments that may apply when enquiring into different kinds of research foci. This leads to the conclusion that science education is unlikely to develop the kinds of neat and somewhat self-contained research traditions often associated with research in the natural sciences, but rather that principled choices from a diverse repertoire of methods are likely to reflect the inherent nature of the research area – and that indeed such choices will be expected to shift as knowledge is developed in any particular area of research.

The chapter finally considers how despite the apparently 'aparadigmatic' nature of research in science education, conceptualising research traditions within the field in terms of Lakatosian research programmes is likely to support the research community in organising research in ways that can be seen as scientific, in terms of allowing judgements about progress and offering researchers more heuristic guidance about fruitful directions for research.

57.2 The Nature of Education and Educational Research

The philosopher of education, Richard Pring, has highlighted how educational research should focus on the core activities of education, that is, teaching and learning, suggesting that 'the distinctive focus of educational research must be upon the quality of learning and thereby of teaching' (Pring 2000, p. 27). However, Pring also acknowledged that this would go beyond the immediate classroom context to include research undertaken to 'make sense of the activities, policies and institutions' which were set up to organise learning (p. 17). That is, educational research commonly focuses on classrooms and learners but also encompasses studies exploring the policies that inform classroom teaching and how these are derived, developed and (often imperfectly) enacted. Research may also consider the governance and management of institutions such as schools, as well as the way that education, teaching and learning are understood in particular cultural contexts.

Education is the context for directed learning, and in formal educational institutions (such as schools and colleges), structures are put in place to encourage and channel learning. A key type of activity in such institutions is teaching, which I suggest is best understood as *deliberate actions intended to bring about particular learning* (Taber 2014). The conditional 'intended' is important here, because as the vast literature in science education testifies, what students learn is not necessarily what the teacher intended to teach (Duit 2009) and may sometimes be quite idiosyncratic. Not only is much learning spontaneous, in the sense of occurring without any specific intention to learn, but, as learning is a highly iterative process, it is strongly channelled by existing ways of understanding the world.

There is an interesting question of the relationship between terms such as teaching, pedagogy and Didaktik (Fischler 2011) – the latter being a common term in continental Europe, but used less in the Anglophone countries. Teaching is here used to refer to the activity, where pedagogy can either be used to refer to the general theoretical body of knowledge about how to go about teaching or to refer to a specific strategy adopted in a particular teaching context (an interesting parallel with the way 'methodology' is used both in a general abstract sense and to describe the specific strategy employed in a particular research study).

Learning is here understood as a change in the potential for behaviour, that is, a change in the behavioural repertoire of the learner (Taber 2009b), assumed to be underpinned by changes in the way the learner's experience of the world is represented in some form of cognitive structure. It is widely accepted that 'circuits' within the brain provide the material substrate that supports cognition, although the precise correlation between the synaptic level and experience of 'having an idea' is much less clear. Despite this, it seems that people do represent aspects of their experience of the external world internally and that, as Vygotsky (1934/1986) long ago noted, these representations are organised into structures rather than being like discrete 'peas in a pod'.

The nature (e.g. coherence) and extent of such structuring is a theme of empirical research in science education (Fellows 1994; Ganaras et al. 2008; Taber 2008), and it is clear that the ways individuals relate and integrate their conceptions of the natural world are often quite different from the way concepts are related in professional science or in formal science curricula. However, such conclusions are indirect inferences made in research, because an individual's cognitive structure is not directly observable (Phillips 1983). Rather, we rely on the learner's behaviour in representing their ideas in the 'public space' where it can be observed as the basis for modelling their thinking (Taber 2014). In science education, we are often concerned with developing students' knowledge and understanding (key concepts that are themselves not easy to operationalise in research), so commonly the 'behaviour' we are interested in is of the form of speech or inscriptions – such as are involved in answering a teacher's questions in class or completing a test paper or assignment.

Although some commentators consider informal learning, and the spontaneous mechanisms that support it, to be distinct from learning processes in educational contexts where there is an intention to teach particular things (Laurillard 2012), there is an increasing tendency for contexts for informal science learning (such as museums) to be planned according to pedagogic principles. So although informal science learning may often appear 'haphazard and incoherent' (Stocklmayer et al. 2010, p. 11), the educational work of museums and science centres is informed by similar debates and principles as those informing the design of curriculum and teaching in schools and colleges (Pedretti 2002).

As an academic area, education is something of a recent addition to the academy – despite scholarly periodicals such as *Science Education* (preceded by *General Science Quarterly*, which first appeared in 1916) and *School Science Review* (first appeared 1919) being long established – and this is reflected in the diverse disciplinary backgrounds of many education faculties as noted above. Traditionally education was seen as an applied subject which drew upon four 'foundation' disciplines: philosophy, psychology, sociology and history (McCulloch 2002). However, in recent decades, education has become more firmly established as an academic subject in its own right. There would seem to be a generational effect here, in that the first holders of PhDs 'in education' were necessarily supervised and advised by faculty who themselves originally trained in other disciplines; but that first generation of education PhDs could then start supervising their own research students from within the 'discipline' of education.

57.3 The Nature of Science Education in the Academy

Certainly in science education, this transition occurred within living memory of some of those currently still working in the field, so the most senior professors of science education active today undertook their own postgraduate studies in other subjects. Fensham (2004) has provided a very readable account of the origins of science education through this process. This means that early researchers in the field were trained in research methods of different disciplines, and for those who trained in the natural sciences, their expertise was often not optimal for transferring to a context exploring educational problems.

Some of the early studies in science education adopted methodology that would seem crude and naive to a postgraduate student (and are justified in terms that would be considered inadequate if submitted for publication) today, and there has been much borrowing of techniques from other fields. Such borrowing need not be a bad thing, providing the techniques concerned are not, in the process, decontextualised from the paradigmatic commitments which provided their justification as valid knowledge-seeking tools. To offer a crude analogy: a screwdriver can be used as a chisel, and a chisel can be used as a screwdriver, but in both cases, one is likely to do a poor job and risk the integrity of the tool itself. In the same manner, as will become clear below, research techniques have been designed with particular jobs in mind and may provide a messy outcome when used without due care and attention.

The major journals in science education now publish work not only on a wide range of themes but adopting a broad range of methodologies. Some of these methodologies reflect approaches common in the natural science, but others draw upon approaches less widely employed in the natural sciences, as they are intended to explore aspects of the human experience itself. Whilst the scope of the present chapter does not allow any detailed discussion of specific data collection or analysis techniques, the key considerations that have informed different paradigmatic stances will be considered below.

57.3.1 The Methodological Turn in Science Education

Early researchers in the field of science education were pioneers, and it is perhaps inevitable that some pioneering work seems crude or trite as a field becomes better established. However, recent decades have seen the development of an extensive literature focusing on educational research methodology, and a new researcher entering an educational field such as science education today can be introduced to a varied, if somewhat contested, range of reading, setting out the nature of educational research and how one should go about it. Often educational research is treated as a specialised area within social research (i.e. social science research) more generally, and sometimes it is grouped with other areas that relate to the professions (particularly areas such as social work and nursing). In effect, educational research has developed into a field of activity within education as a subject area, and research methodology has become the subject of primary journals (such as *Educational Research*er, ISSN: 0013-189X; 1935-102X) as well as being an active area of textbook publishing.

This reification of educational research into a subject for study as well as a means to carry out studies has led to much discussion about the diversity of methodological approaches used in educational studies and the relationships between them (Bassey 1992; Clark 2005). Some of these issues are explored later in this chapter. With this diversification of methodological approaches being employed in educational research, it has become increasingly expected that empirical research reports should provide not only a description of the methodology employed but also a justification of the approach used and acknowledgements of limitations inherent in the methodology or the specific design for a study.

These considerations might suggest that there seem to be two related key differences between methodology in education, including science education, and in the natural sciences:

1. Research in science education as a field draws upon a wide menu of available methodological choices, whereas research in most particular fields within the natural sciences is limited to a much more (to mix metaphors) limited palette.
2. Descriptions of research in science education often offer extensive justification of chosen methodology, commonly including explicit discussion of ontological and epistemological commitments underpinning research designs, whereas reports of research in the natural sciences often focus on specific technical details without extensive justification of the overall methodology.

This could be taken to suggest that research in science education, being unlike research in the natural sciences, should not be considered scientific in nature. To consider why these differences exist, it is useful to consider Kuhn's account of how researchers are inducted into the natural sciences.

57.4 Normal Science, Revolutionary Science and Another 'Sort of Scientific Research'

Thomas Kuhn has been highly influential in both science studies and discourse about the nature of work in the social sciences, largely based upon the reaction to his essay on *The Structure of Scientific Revolutions* (Kuhn 1970) and the adoption of the notion of working within a 'paradigm'. In that work, Kuhn argued that scientific revolutions were rare and that most scientists spent their careers doing what he termed 'normal science'. Although key aspects of this work have been much criticised (and some of this criticism will be discussed briefly below), Kuhn's description of how scientists are trained and inducted into traditions of research is especially relevant to the present chapter and is considered here to offer very fruitful insights when considering the way methodology is discussed and understood in educational research.

Kuhn described normal science as working within a paradigm or a 'disciplinary matrix' (Kuhn 1974/1977). In Kuhn's model, most science occurs within an established tradition, and these traditions are occasionally interrupted when a niggling anomaly leads to an individual (a) forming a revolutionary reconceptualisation of the field and then (b) persuading the scientific community to shift allegiance such that a new tradition is formed and the old one abandoned. For Kuhn such a revolution changes both the meaning of terms and ways of seeing and understanding the world to such an extent that those working within the new paradigm should be considered to speak a different language (such that the two paradigms become incommensurable) and in effect work in a different world (Kuhn 1996).

Aspects of this thesis have been widely discussed and critiqued (Masterman 1970; Popper 1994), especially the notion of incommensurability and the potential implication that there can be no objective way of judging progress in science – i.e. it could be argued from Kuhn's analysis that a revolution makes a field different, but not necessarily further advanced – although Kuhn himself argued that his analysis suggests judging progress is problematic rather than impossible (Kuhn 1973/1977).

57.4.1 Induction into Scientific Research

However the aspect of Kuhn's work most relevant for the present chapter is his description of how a new scientist prepares for work in, and becomes accepted within, a research field. For Kuhn, the process of becoming a professional scientist is in effect an induction into a particular tradition (or paradigm, in one of the senses in which Kuhn used the term) through a kind of intellectual apprenticeship. By the completion of this training process, the new scientist has adopted the norms associated with the disciplinary matrix that in effect defines the current state of the particular subfield in which the scientist has completed research training (Kuhn 1996). This disciplinary matrix provides the framework for scientific work 'based firmly upon a settled consensus acquired from scientific education and reinforced by subsequent life in the profession' (Kuhn 1959/1977, p. 227). Kuhn saw each such tradition as ultimately derived from a particular scientific achievement – such as Newton's work on mechanics or Lavoisier's work on chemistry, but other examples might be Darwin's work on natural selection or Crick and colleagues' work on the structure of DNA and the 'central dogma' of molecular biology. Such achievements were revolutionary enough to each initiate a new direction for scientific research; moreover, one which could provide a starting point for developing a whole new approach (Kuhn 1996).

In his work, Kuhn argued strongly that a scientist needed to demonstrate a commitment to the tradition in which he or she was working and that this was equally true for the few who would initiate scientific revolutions of their own, as it was for the majority that would work their entire careers on the 'mopping-up' work of normal science. Kuhn did not imply that such 'mopping-up' work lacked interest or excitement: it was routine in the sense of being within an established tradition and therefore had a strong 'convergent' focus, compared with the divergent nature of the 'discoveries' that initiated the occasional scientific revolutions.

Kuhn recognised the ubiquity of imprecision and anomaly in scientific work and considered that progress in science depended upon scientists being able to have enough commitment to the accepted theory in the field not to be continuously distracted by attempts to explain nonsignificant discrepancies between theoretical predictions and results (Kuhn 1961/1977). Michael Polanyi (1962/1969) also discussed how scientists need to be able to use personal judgement to ignore most of the multitude of apparent anomalies (in effect, prima facie refutations) met in the course of scientific work. Where Polanyi emphasised how such judgement depended upon tacit knowledge (which he related to the way an external reality becomes known through the complexity and subtlety of human perception/cognition), Kuhn stressed how the indoctrinating effect of scientific education could dull the ability to recognise a significant anomaly for what it was.

Whilst identification of a significant anomaly was central to initiating a scientific revolution in Kuhn' account, even the successful revolutionary has to make their argument for a paradigm shift from within the existing tradition – that is, they need to be recognised by others in the community working in the field as being a full legitimate participant (cf. Lave and Wenger 1991) in that particular scientific tradition (Kuhn 1996). This required a 'thoroughgoing commitment' to the existing tradition (Kuhn 1959/1977, p. 235).

The disciplinary matrix in which scientists work, and in which they draw upon the commitments underpinning their scientific work, supports 'relatively unproblematic...professional communication' and allows 'relative unanimity of professional judgment' that is 'comprised of ordered elements of various sorts' (Kuhn 1974/1977, p. 297). These included symbolic generalisations, models and exemplars (the latter providing the derivation of Kuhn's original choice of the term 'paradigm'). For Kuhn, the set of models used within a scientific tradition range from heuristics offering analogical insight to deeply held metaphysical commitments amounting to an ontology (Kuhn 1974/1977). Indeed, within normal science, 'research is directed to the articulation of those phenomena and theories that the paradigm already supplies' (Kuhn 1996, p. 24). Elsewhere, Kuhn refers to how researchers within a shared paradigm 'are committed to the same rules and standards for scientific practice' (Kuhn 1996, p. 11) and how paradigmatic exemplification derives from how scientific practice involves 'law, theory, application, and instrumentation together' (Kuhn 1996, p. 10).

As suggested above, Kuhn's thesis has not been universally accepted. Indeed, whilst it may be welcomed as a useful challenge to models of the nature of science that relied on the logical structure of research and oversold an assumption that in principle sufficient careful research could provide a basis for unambiguously interpreting nature, it arguably encouraged views of science that in turn underplayed the role of logical argument and interrogation of evidence in reaching consensus in science. In particular, the suggestion that normal science is somewhat routine, pedestrian and almost a matter of following algorithms (which has perhaps been taken from Kuhn, rather than offered by him) has been challenged by those who consider controversy to be a common if not constant feature of science, rather than a sign of a rare major shift (Machamer et al. 2000). Indeed, Feyerabend (1988) countered the notion of normal science by claiming that the history of science suggested there was no standard method or set of preferred approaches in science, but rather that scientists were much more pragmatic, adapting and inventing method to meet the needs of the problem at hand.

At first sight, there appears to be a wide gulf here, but perhaps such different accounts of science need not be as inconsistent as may appear initially to be the case. The basis for suggesting this (whilst acknowledging it may partly reflect a personal cognitive style of tending to prefer integration to fragmentation) links to notions of what might be termed 'grain' size in analysing the nature of science. This reading would acknowledge both (i) that controversy is certainly common in science and indeed is probably an important part of the motivation for much research (Machamer et al. 2000), but many controversies concern issues that are not linked to core ontological commitments within a research tradition (and so can be accommodated in something like Kuhn's normal science), and (ii) that innovative techniques for data collection and analysis are indeed common in the history of science when taking the 'long' view, but that again major new approaches (rather than refinements to existing techniques) are relatively rare within the day-to-day career of the working scientist, and that 'standard' techniques do become and remain established within research traditions.

From this view, criticism of Kuhnian normal science as a description of most scientific activity would *not* undermine what Kuhn has to say about the research training of individual new scientists, which generally takes place over a matter of a few years, working in one area of science and often within the context of one or two research teams and laboratories. Typically, then, according to Kuhn, a scientist is trained within a particular research tradition that leads to embracing the research community's commitment to the kinds of phenomena that fall within the scope of the field, the kind of entities that are used in explanations, a theoretical apparatus within which predictions and explanations can be developed, standard forms of representation, and accepted techniques for undertaking research studies. However, despite this characterisation of normal science, Kuhn did not claim that scientific work necessarily *had* to take this form, but rather saw this as the nature of 'mature' sciences.

Indeed, Kuhn (1996, p. 11) acknowledged that in fields that had not achieved such maturity, 'there can be a sort of scientific research without paradigms, or at least without … unequivocal and … binding' paradigms. That is, Kuhn was offering a descriptive account of science based on his historical scholarship, not a prescription for science. His description could be seen as providing demarcation criteria for mature sciences, but not for scientific enquiry per se. In 1983, Gilbert and Watts referred to how research into learning in science was in 'a pre-paradigmatic phase' as there was 'no general agreement on the aims of enquiry, the methods to be used, criteria for appraising data, the use to be made of the outcomes' (p. 61). Arguably, to some extent, this description could still be applied to science education as a field of research some 30 years later, and if we wish to see research in science education as a scientific activity, then we need to consider it as Kuhn's other, less mature, 'sort of scientific research'. However, there is an alternative argument, long recognised by Shulman (1986), for example, that suggests that research into such areas as teaching is unlikely to mature into something like Kuhnian normal science, because the degree of complexity of educational phenomena is such that no single perspective is likely to offer a full enough account of inform effective educational practice.

57.5 Characterising the Educational Research Project

In order to consider whether educational research is, is sometimes, or at least can be, a 'sort of scientific research', it is necessary to consider the nature of the work the educational researcher undertakes and to reflect on how and why this might be different from research in the natural sciences.

57.5.1 The Overall 'Shape' of a Discrete Research Study

The conceptualisation and development of an educational research project goes through two cycles during each of which there is a kind of expansion phase of exploring options and seeking sources of information, followed by a focusing (Taber 2013). Figure 57.1 uses the lemniscate as a visual metaphor to suggest that a study can be understood to ideally progress through three focal points (indicated on Fig. 57.1):

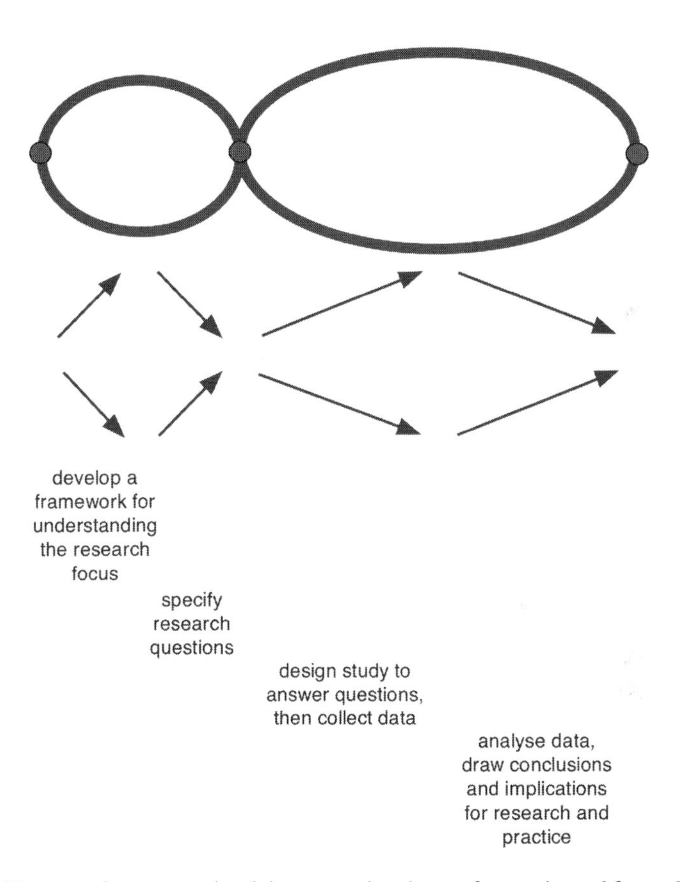

Fig. 57.1 The research process as involving successive phases of expansive and focused thinking

the initial concern or interest, the specific research questions (RQs) and the conclusions. This model assumes, for the moment, that research is largely conceptualised on a study-by-study basis, which is clearly a major simplification (Lakatos 1970).

The origin of the project is some kind of concern, issue or other focus that is seen worth investigating. The first cycle (see below) involves a process of developing the conceptual framework for the study – exploring relevant literature and reviewing previous research that may be pertinent – 'setting the scene' as it were for the new study. That is a phase that can be seen as supported by divergent thinking: allowing the recognition of relevance and forming links across diverse literature. This is followed by the framing of the particular RQ for the study. This latter step involves a focusing in on the specifics of the research (a more convergent process) and setting out how variables and constructs will be understood. Reaching this point will involve identifying any axiological commitments, the values that inform why we do research and so how we should conduct ourselves as researchers, as well as the ontological nature of what is to be researched, and so the epistemological constraints and affordances which will inform the kind of knowledge that it is possible to develop about what we are interested in.

The second 'cycle' of the project (Fig. 57.1) involves another expansive stage, where a research design is developed which can facilitate the answering of the RQ, followed by the collection of data to build up the evidence base needed to answer the RQ. This is followed by a further convergent phase where analysis 'reduces' data to results and leads to conclusions. The overall process therefore calls upon both divergent and convergent thinking: both creative and logical thought (Taber 2011).

57.5.2 Owning the Research Problem in Science and in Science Education

Formalising the process in these terms is often important in educational research because of the nature of existing literature. This reflects a difference between the common experiences of new researchers in education and those in natural sciences. A new doctoral student in one of the natural sciences will commonly be set a problem that is part of an ongoing programme within a wider research team in the laboratory and the process of identifying the relevant literature, and so conceptualising the 'gap' in existing knowledge the study is intended to 'fill' may be relatively straightforward. Indeed, it may be quite clear which techniques are to be adopted (perhaps those for which the lab is equipped with specialised apparatus) and how data will need to be analysed to produce knowledge claims acceptable to those working in the relevant field of science.

Arguably, the novice scientific researcher may be scaffolded to such an extent that they are only primarily responsible for the data collection and analysis stages, and much of the decision-making that leads up to this is largely channelled by the induction into an established way of understanding the ontology and epistemology adopted in that subfield of science. This would suggest that much of the thinking which informs such decision-making for a new researcher in education is in effect short circuited in the natural sciences. This is in line with the picture of 'normal' science (see above), described by Kuhn (1996), where the new scientist is inducted into the disciplinary matrix of the field by working through the standard paradigms. The result may often be someone who is very informed about the standard thinking and techniques in a specialised field, whilst having a much more limited knowledge of other fields within the broader discipline.

Yet the experience of a new doctoral student in education may be quite different in a number of ways. Whilst science education is now sufficiently theorised and staffed with expertise to support the natural science model outlined above, it is more likely that the research student will have greater latitude in selecting their project (if only because the apparatuses of research are less specialised and so less likely to be a constraint), and indeed within education the process of developing the project is seen as a key part of the education and training of the researcher. Moreover, whilst it remains important that doctoral supervision provides specialist support in learning about the topic area and acquiring specific skills, the student may find no single clear picture of the research area in the literature that allows an obvious conceptualisation of a 'gap' in the knowledge or a single sensible approach to an issue or problem. The state of knowledge in many educational topics would not fit Kuhn's notion of normal science, with its paradigmatic norms.

Where Kuhn suggests that the primary mode of thinking in normal science is convergent, this is often less true in educational research. Rather than being expected to 'plug' a specific 'hole' assigned by a supervisor, the educational research student is often expected to demonstrate extensive divergent thinking in accessing, evaluating and choosing between alternative potential ways of conceptualising their problem area. Within this context for undertaking research, the transition from an initial topic or issue of interest to the formation of specific RQ normally involves wide reading around a topic to appreciate and consider a range of possible ways of conceptualising the field, perhaps each based upon understanding the topic in quite distinct ways, and so suggesting different notions of how best to enquire into the subject. It is seen as the part of the student's task to develop a conceptualisation of the field and the justification for adopting (and if necessary adapting) a particular theoretical perspective (see below) for supporting the research. To caricature, the educational researcher 'owns' the research problem not because it has been 'given' (assigned) to them by the supervisor or lab director, but because they have 'built' (developed, discovered, constructed) it themselves.

Moreover, because of the lack of a clear disciplinary matrix that sets out particular tools for thinking about and doing research in the field, the research student is expected to learn about a wide range of methodologies so as to be able to comprehend and apply critical judgement to reading literature around the research topic, as it is quite likely that relevant knowledge claims in research journals will derive from a range of data collection and analytical techniques, potentially drawing upon very different (ontological and epistemological) assumptions informing different researchers' work.

The RQs themselves act as the point of transition in the flow of the study (see Fig. 57.1), and just as the RQs should reflect the thinking that has informed their formulation, they should themselves be reflected in what is to follow. A research design must address the RQs and be compatible with ontological assumptions informing the study (in terms of the nature of what is being studied) and epistemological considerations in terms of what it is reasonable to expect to be able to know about that kind of research focus. A methodology should therefore be selected (see below) which is suitable to answer the RQ, taking into account the presumed ontology of what is being studied and the kind of knowledge considered viable for such a focus; and data collection and analysis techniques are then selected which are coherent with that methodology. Data is collected (another 'expansive' stage, see Fig. 57.1) and then analysed to produce findings/results (another phase of concentration and reduction, see Fig. 57.1), developing a logical case for making new knowledge claims.

57.6 Conceptualising the Research Project

Discussions of educational research often make references to such notions as the 'theoretical perspectives' and 'conceptual frameworks' supporting particular research studies. The way these terms are understood, and will be used, in the present chapter (as unfortunately different authors do not always use a common terminology – see the comments below about phenomenography) is represented in Fig. 57.2. As well as 'theoretical perspective' and 'conceptual framework', this figure also includes three other key terms: 'research questions' (as discussed above), 'research design' and 'methodology'.

RQs are the specific questions that a research study is intended to address. These may take the form of a formal hypothesis, but in educational studies, they may instead be much more open ended, and the degree of openness will often depend upon the current state of knowledge in the topic area (as will be discussed further below).

The RQs for a particular study derive from a conceptualisation of the topic area that sets out what is already known and what is not yet known and might be worth finding out. The wording here, 'what is already known', is not intended to suggest absolute knowledge, but rather the set of knowledge claims currently considered robustly supported, and so suitable for taking as a starting point for further research. This conceptualisation, the 'conceptual framework' of a study, is often formalised in the literature review of a research report. The RQs are addressed through a 'research design' that sets out how the required data are to be collected, and how they will be analysed so as to answer the RQ. The essential logic of a research 'design' is such that it should in principle be prepared ahead of the empirical work taking place, and indeed doctoral students are commonly expected to have their research designs scrutinised and approved before commencing their 'fieldwork'. However, as in the natural sciences, research may involve false starts and

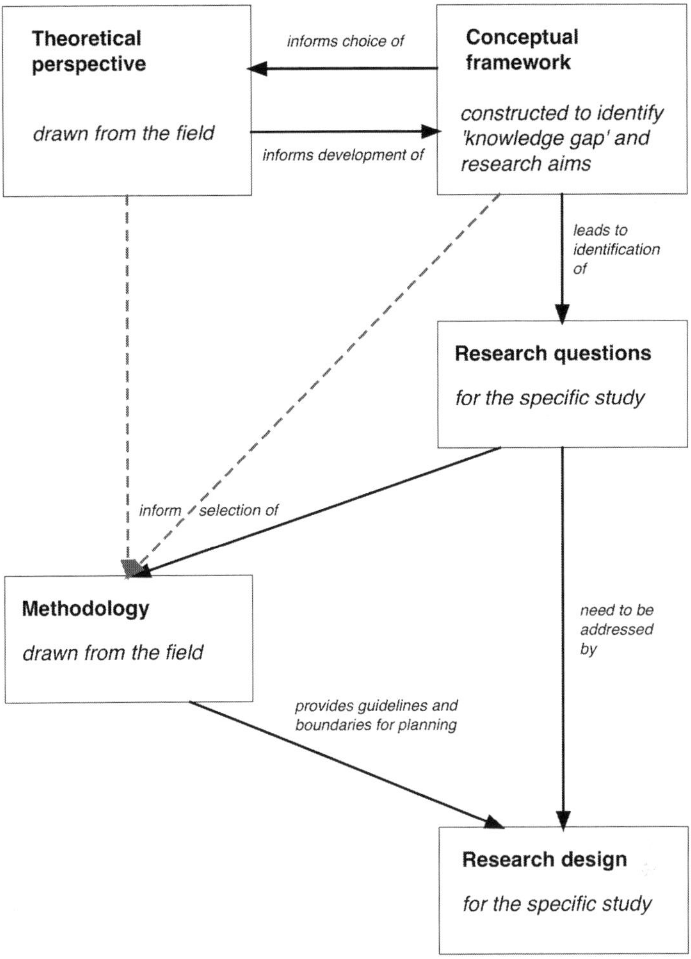

Fig. 57.2 Some key terms used to describe educational research

unproductive 'cul-de-sacs', and the design reported in published reports (and theses submitted for examination) may well – as in the natural sciences (Medawar 1963/1990) – be a rational reconstruction, in the light of experience, of what eventually 'worked'.

In some forms of educational research, the research design might be synonymous with 'experimental design', but, as is discussed below, many educational research designs are not based on experimental methods. Moreover, some research designs are 'emergent' which means that only the initial stages of data collection are firmly established before the research begins, as further detail of the research design will be informed by ongoing data collection. This is a somewhat different issue to the previous point regarding false starts (where a pre-planned approach that it was anticipated would be suitable for answering RQ is later found to be unproductive), as with an emergent design it is recognised in advance that an iterative process will be needed to refine the design.

In a grounded theory study, for example, it would be inappropriate and counter-productive to fully specify the data collection for the entire study in advance (as will be seen from the description later in the chapter, that would undermine the logic of the methodology), whereas in an experiment, it is important to specify data collection and analysis carefully in advance – although the specification that is reported in a formal account may well have been preceded by earlier versions that were abandoned as the research was developed.

Not reporting the outcome of experimental studies because those outcomes are not welcome is unethical, but not reporting studies because they are judged to have methodological failures that undermine the credibility of the results is quite appropriate (and indeed journal referees may well judge studies in these terms even when the researchers consider the procedures employed adequate). Ultimately, it is the researcher's judgement (and so their professional integrity) that has to be relied upon to discriminate between results that go unreported because they are not robustly supported and results that are robust but do not support conclusions the researcher hoped to draw. This issue is familiar enough from work in the natural sciences (Polanyi 1962/1969).

The process of shifting from a conceptual framework to specific RQ and then to a research design is clearly familiar from the natural sciences, although there it is more likely (although not always the case) that research design will imply experimental design and that research designs will be specified in detail (precisely which data to collect, precisely how it will be analysed) before any data collection begins. The argument offered in this chapter is that there are necessary (essential) differences between educational research and research in the natural sciences; but that this need not exclude research in science education from being considered 'scientific'. In particular, the process of designing and justifying research is likely in science education, more (or more often) than natural science, to require *explicit* consideration of ontological, axiological and epistemological considerations.

57.6.1 Theoretical Perspectives

Educational phenomena, teaching and learning and the social institutions intended to support teaching, can (as Shulman 1986 recognised) be very complex, and there are often alternative ways of approaching the conceptualisation of a particular research focus (such as student learning about some science topic). Discussions of educational research often make references to the 'theoretical perspective' informing a study, as something other than the 'conceptual framework' underpinning the study.

Theoretical perspectives can be thought of as well-developed theoretical positions about some aspects of a social or educational phenomenon that can act as starting points for making sense of research topics. An important point is that in science education, there is no 1:1 correspondence between theoretical perspective and specific topics. Rather there will often be several theoretical perspectives that might be relevant to a topic. These might sometimes be seen as based on competing theories, but often they might be better thought of as each illuminating some of the facets of a complex phenomenon.

There are parallels to both of these alternatives in the natural sciences. So we might consider theoretical perspectives as competing in the way that (a) the oxygen theory of combustion competed with the phlogiston theory (Thagard 1992) or (b) the notion that species have an inherent essence that makes them absolutely distinct (as might be expected if each type was originally formed by an act of special creation) is at odds with the idea that all living things derive by descent from a common ancestor (in which case species are not absolute, but current loci of relatively stable forms at a particular historical moment, contingent upon a great many particulars of past events, with temporary salience against a background of constant slow modification and shifts).

But even in the natural sciences, alternative and apparently inconsistent perspectives need not be considered to be in direct competition. An analogy here might be the way interactions between colliding molecules might be conceptualised in terms of different theoretical models. One theoretical perspective that could be applied would be an ideal gas (i.e. kinetic theory) model, where molecules can be considered to behave as spheres that undergo perfectly elastic collisions. Here the molecules are stable entities, and their collective behaviour can be used in explanatory models of bulk behaviour of the gas. Another theoretical perspective that might be applied could be to consider molecules to be complex structures including electronic orbitals with associated energy levels, some of which are occupied and others unoccupied. Here descriptions in terms of potential overlap between occupied orbitals on one molecule and unoccupied orbitals on another may form the basis for explanatory models of reaction mechanisms (at the submicroscopic level) that help explain patterns of chemical reactivity at the bulk level. In this example, we might consider that both of these perspectives are potentially valid and could contribute to a full understanding of gas properties, but that, in relation to a particular scientific problem, one will be more productive than the other. So even within the natural sciences, the application of a concept may involve selecting an appropriate tool for a particular job, from a metaphorical conceptual 'toolkit' (Taber 1995) offering alternatives that all have their own range of application. Indeed, this very feature of science appears to offer a major challenge to many learners, presumably because they often misconstrue the nature of the models presented in the curriculum (Taber 2010c).

The difference between these two types of cases would seem to be whether the different perspectives can meaningfully be considered complementary. Whilst a model of molecules as like tiny billiard balls is clearly incomplete because it does not explain chemical reactions, it remains a useful analogy for some purposes and can complement other models that explain particle behaviour under other circumstances. In other words, the apparently inconsistent models are not competing for the same 'explanatory space' in this example: one perspective explains physical properties that are commonly exhibited by gases and gas mixtures, and the other perspective can explain why chemical change sometimes occurs when gases mix.

By contrast, the oxygen theory competed with the phlogiston theory to occupy the same explanatory space – of why combustion sometimes (but not always) occurs.

Similarly, descent with modification through natural selection (Darwin 1859/1968), and the notion that organisms are members of a species because of some essence (Mayr 1987) competed (and indeed for some still compete) in the 'explanatory space' for explaining how living things on earth appear to fit into a number of specific types that (although very large) is tiny compared with the number of individual organisms on earth. Theodosius Dobzhansky (1935, p. 345) enquired whether the notion of a species was 'a purely artificial device employed for making the bewildering diversity of living beings intelligible, or corresponds to something tangible in the outside world…[that is, is] the species a part of the 'order of nature', or a part of the order-loving mind?' Indeed, it has been argued that the tendency to retain elements of essentialism long after the general tenets of Darwinian evolution were widely accepted has been a major problem in biology (Hull 1965), amounting to the kind of obstacle to scientific progress discussed by Bachelard (1940/1968).

These examples from the natural sciences give some sense of how theoretical perspectives might be drawn upon in particular research contexts. At first sight, a difference between research in the natural sciences and research in science education is that in education it may not always be so clear whether alternative theoretical perspectives are competing or potentially complementary. This difference reflects the complexity of educational phenomena (discussed further below) and is brought into focus because of the use above of historical examples from the natural science (combustion, particle theory, the origin of species) where we are judging the issue with the benefit of many decades of 'hindsight'.

A wide range of theoretical perspectives have been drawn upon in research in science education, but some illustrative examples would be the following:

Exploring college students' thinking related to the concept of field drawing upon a particular theoretical perspective of the main types of mental representations people use (Greca and Moreira 1997)

Exploring teaching and learning of cell biology in upper secondary school and drawing upon a theoretical perspective based on general system theory (Verhoeff et al. 2008)

Exploring the value of a sociocultural theoretical perspective in thinking about the learning that can occur when people visit science and technology centres (Davidsson and Jakobsson 2008)

57.7 Competing Theoretical Perspectives in Science Education

Space here only allows limited exemplification, but an example of where different theoretical perspectives have competed in science education concerns research into student thinking, understanding and learning in science. Two examples here concern flavours of 'personal constructivism' and the relationship between personal constructivism and sociocultural perspectives on learning.

A very influential theoretical perspective from developmental psychology that informed work in science education was that due to Jean Piaget and his 'genetic epistemology' (Piaget 1970/1972). Within that programme of work, Piaget developed a stage theory of cognitive development which saw particular domain general structures of thought as associated with different developmental stages and which put limits on the kind of learning possible for students at each stage (Piaget 1929/1973). Although details of the Piagetian scheme, and how it is understood to relate to education, have faced criticism (Donaldson 1978; Sutherland 1992), this has been a very influential perspective in science education (Bliss 1993, 1995). In particular, Piaget's work with its focus on structures posited in mind (Gardner 1973) contrasted with work informed by the highly influential behaviourist school (largely in the United States) that had eschewed explanations relying upon non-observable constructs such as states of mind (Watson 1924/1998, 1967).

However, in the 1970s, an alternative perspective was developed in science education that focused less on general structures of thought (the complexity of thinking available to learners) and more on their particular meaning making in different scientific topics, leading eventually to an extensive research effort (Driver et al. 1994; Duit 2009). This research explored students' own ways of thinking and talking about various natural phenomena and scientific topics, such as force, plant nutrition and heat. The aim here was less to characterise student thinking at particular levels, but to allow teachers to be aware of typical conceptions students brought to (and/or took away from) lessons, and to think about how to support students in developing understanding of the scientific models that were reflected in the school or college curriculum. This work was sometimes labelled as the alternative conceptions movement (ACM).

Both of these perspectives can be understood to be personal constructivist approaches, focused on how the individual comes to iteratively build up personal knowledge in the form of internal representations of the world (as directly experienced and as heard about second hand), but with rather different foci: one domain general (so learning in any topic is constrained by the general stage of development) and one very much on a topic-by-topic basis (where familiarity with a particular domain can lead to areas of relative expertise).

These two perspectives can certainly be seen to have competed for research attention and resources, although arguably they did not compete for the same explanatory space as they focused primarily on rather different aspects of science learning. That the ACM came to dominance within science education – although an important strand of research to inform teaching from the Piagetian perspective continued (Adey 1999) – probably said less about the perceived *validity* of the Piagetian perspective than the greater perceived *fruitfulness* of the ACM for actually informing teaching. It might also be tentatively suggested that the ACM was attractive to many of those setting out on research in science education because the terminology of early work (often conceptualised as being about identifying misconceptions) was more accessible than the rather specialised and perhaps seemingly esoteric language that had been developed within the Piagetian programme. That is not to suggest that the ACM was under-theorised, as that was not so (Driver and Erickson 1983; Gilbert and Watts 1983; Osborne and Wittrock 1985). However, as a research programme developed from within education (rather than the developmental psychology base of the Piagetian work), there was always a strong impetus to report work in terms that would make sense to classroom teachers.

More recently, much discussion and some contention in science education has been focused around the question of whether the adoption of a social constructivist perspective (Roth and Tobin 2006; Smardon 2009) should be seen as complementary to, or a potential replacement for, a personal constructivist perspective. That is, does the acknowledgement of the importance of social interaction in learning (directly through dialogue or indirectly through institutions and cultural artefacts) imply that considering learning as the personal sense making of individuals in order to construct personal knowledge in the form of mental models associated with the minds of individuals (and represented in the physical substrate of that individual's brain) is invalid (or at least, unproductive)? One view would be that the personal constructivist perspective adopts notions of knowing and knowledge that are no longer viable in terms of what is commonly claimed about how learning needs to be understood as socially situated, and how knowledge-in-action depends upon social context (Hennessy 1993).

However, from within the personal constructivist perspective, it can be argued that learning is a very complex phenomenon and that a sensible simplification for many purposes is to understand learning as due to processes that occur within the cognitive system of an individual learner who perceives their environment (in which other people and the signs of culture may be highly salient and relevant to learning); constructs internal models of it, and then acts according to the perceived reality provided by those models (acts that include making public representations of personal knowledge that can be perceived by others); and, where it seems appropriate, adjusts the internal models as indicated by feedback from the environment (including the public reactions of others to that behaviour). It is possible to adopt a more synthetic 'complementary' view of personal and social constructivism as useful perspectives that can both contribute to progressing science education (Taber 2009b): but this is by no means a consensus view in the community.[1]

---

[1] Just as there is a vast literature drawing upon and adopting (labels if not always principles) of constructivism, there has been a range of criticisms of constructivist work in science education. These include criticisms of constructivist approaches that seem to support relativist stances on scientific knowledge (Coll and Taylor 2001; Cromer 1997; Matthews 1993, 1994/2014; Scerri 2003), suggestions that constructivist teaching approaches undermine traditional ecological knowledge in indigenous communities (Bowers 2007), the theoretical basis of constructivism in education (Matthews 2002), the level of empirical support for knowledge claims (Claxton 1993; Kuiper 1994; Solomon 1992), inappropriate focus on individuals (Coll and Taylor 2001; Solomon 1987, 1993b), limited linkage between result findings and implications for teaching (Harlen 1999; Johnstone 2000; Millar 1989; Solomon 1993a), associations with unstructured 'discovery' learning approaches (Cromer 1997; Matthews 2002) and diversion of resources from more productive areas of research (Johnstone 2000; Solomon 1994). An account of these criticisms and possible rebuttals is offered elsewhere (Taber 2009b). Some of these issues reflect a wider debate in education about the nature and relative merits of constructivist and enquiry-based teaching compared with other pedagogies – especially what has been labelled as 'direct instruction' (Kirschner et al. 2006; Klahr 2010; Taber 2010a, b; Tobias and Duffy 2009).

57.8 Selecting a Methodology for a Study

The term 'methodology' when used to describe research in education – or the social sciences more widely – is distinguished from 'methods', which generally means the specific 'techniques' used to collect and analyse data. Methodologies are considered to be broader: to be principled approaches to undertaking research that can provide a framework for selecting particular component techniques. A simple analogy here is that methodology refers to an overall strategy to achieve research aims, within which specific tactics (techniques) may be employed to meet particular subgoals (Taber 2007).

Although one might refer to the specific methodology used in a particular study, methodologies tend to be considered as general-purpose approaches that can be selected according to the nature of the RQ being addressed (as suggested by the analogy with pedagogy and pedagogies above). One common methodology would be the experiment, but educational research commonly also draws on a range of other methodologies such as survey, case study, ethnography and grounded theory. It is worth reflecting briefly on the core characteristics of these common methodologies.

Experiment: The experimental 'method' is taken from work in the natural sciences and is used to test a hypothesis by controlling variables to compare two sets of conditions that differ in one accord. In practice, true experiments are seldom possible in education, for reasons discussed later in this chapter.

Survey: A survey is used to find out about the level of association of one type of element with a different type of element. So, for example, a survey could be used to find how many fume cupboards school science laboratories are typically equipped with (i.e. reporting the proportion of such laboratories having no fume cupboard, one fume cupboard, etc.). Commonly surveys are used to seek self-report information from people regarding such matters as their attitudes or behaviours. Surveys may be used to test hypotheses by comparing responses to different survey items – e.g. one could test the hypothesis that a higher proportion of male science teachers than female science teachers expect to be promoted to head of department.

Surveys may be applied within limited populations (e.g. the students in one school), but are commonly used in relation to larger populations (e.g. secondary chemistry teachers in a national context) using sampling techniques and inferential statistics to make inferences about the populations sampled. A survey that all, or nearly all, science teachers responded to could tell us whether or not a higher proportion of male science teachers than female science teachers expect to be promoted to head of department; but in practice a representative sample of modest size is likely to be sought from which inferences can be drawn about the broader population.

Case study is a methodology used to explore a particular instance in detail (Stake 2000; Yin 2003). The instance has to be identifiable as having clear boundaries and could be a lesson, the teaching of a scheme of work in a school department, assessment procedures in a university teaching department, a group visit to a museum by one class of students, etc. For example, Duit and colleagues (Duit et al. 1998) report a classroom episode where one group of students undertakes a discussion task relating to the magnetic pendulum. The authors of the report provide extensive context for making sense of the case in terms of the classroom and curriculum setting of the episode.

Although case study looks at an identifiable instance, it is normally naturalistic, exploring the case in its usual context, rather than attempting to set up a clinical setting – which would often not be viable even if considered useful, as often the case is embedded in its natural context in ways that influence its characteristics (so moving a teacher and a class from their normal setting to a special research classroom in a university, for example, is likely to change behaviours that would be exhibited in the 'natural' setting).

Sometimes (instrumental) cases are chosen because they are considered reasonably typical of a class of instances, where the complexity of what is being studied suggests that more can be learnt by detailed exploration of an instant than surveying a representative sample. Other (intrinsic) cases may be selected because they have been identified as special in some sense, and the researchers want to see if they can find out why: for example, why one teacher facilitates especially impressive learning outcomes.

Ethnography is an approach drawing upon anthropology, which attempts to make sense of a particular culture or group in its own terms, that is, to understand the meaning the individuals in that culture or group assign to certain rituals or cultural practices (Agar 2001; Hammersley and Atkinson 2007). Whilst ethnographies, that is, detailed accounts produced by ethnographic methodology, are relatively rare, if not excluded (Long 2011; Reiss 2000), in science education, studies which draw on ethnographic approaches and perspectives are quite common.

Grounded theory is a set of methods for developing theory using an inductive approach. Developed – or 'discovered' (Glaser and Strauss 1967) – in sociology, grounded theory is an approach which attempts to provide methods to assure scientific rigour when researchers attempt to understand social phenomena and existing conceptual frameworks are considered inadequate. Grounded theory relies on a number of core principles (Taber 2000), including emergent research designs that build upon 'theoretical sampling' (i.e. using the analysis of initial data to inform decisions about the next steps in data collection), 'constant comparison' (an iterative approach to analysis that requires repeated revising of data coding intended to ensure analysis that provides best fit to all the data) and 'theoretical saturation' (i.e. only ceasing data collection when further data adds nothing substantive to the theory being developed).

As this suggests, the complete grounded theory methodology is very demanding and is only viable when researchers are not under strict time pressures to complete a study. Despite this, grounded theory is commonly cited as a referent in educational studies, although often in practice such studies adopt the constant comparison method without substantive theoretical sampling or reaching theoretical saturation.

57.8.1 Other Candidates for Methodology

Sometimes *phenomenography* is considered a distinct methodology, although it is alternatively considered rather to be a particular perspective (e.g. Koballa et al. 2000), an analytical framework (Ebenezer and Erickson 1996) or even a field of enquiry (Marton 1981). Phenomenography seeks to describe, explore and characterise people's experiences.

Approaches such as lesson study and design research may also be considered as methodologies. In lesson study (Allen et al. 2004), an approach to curriculum development that has been especially popular in Japan, a group of teachers work together to plan a lesson, which is then taught by one of the group and observed by others. This allows the lesson plan to be revised, before another member of the group teaches the revised lesson, allowing a further 'trial' and opportunity for further refinement.

Whilst such approaches might seem to be more about 'development' than 'pure' research, if educational research is intended to improve teaching, then such approaches certainly cannot be excluded from consideration. Some commentators on educational research see a major distinction between 'pure' and 'applied' research (Springer 2010), but arguably all 'educational' research (as opposed to, say, psychological research into learning) should potentially have at least distal implications for informing educational activity, and the pure/applied division is not an especially helpful distinction. Arguably this presents a difference between research in science and research in education: perhaps because work exploring educational phenomena that could be considered as 'pure' would be likely to be considered not as educational research but as research in another area such as educational psychology or sociology undertaken within educational contexts. Certainly if we adopt the steer offered by Pring (2000, p. 27), then educational research is always in principle 'applied' research.

Rather, a more significant issue raised here is the role and nature of theory in research and the extent to which curriculum development and lesson design need to be seen as idiographic activities specific to the particular subject matter, curriculum setting, institution and cultural contexts, of teaching and learning. This is an issue where the science education community has not reached a strong consensus (Kortland and Klaassen 2010; Tiberghien 2012). There is an argument that the complexity of teaching and learning is such that iterative processes (such as that used in lesson study) are needed within teaching and should be institutionalised within the profession to make it a 'design science' (Laurillard 2012).

This leads to consider another methodology that is often cited in educational research, i.e. '*action research*' (McNiff 1992). Like many of the descriptors used in discussing education research, action research is understood differently by different authors, but usually means research that is carried out by practitioners to address a problem or issues in their own practice. A key feature of action research is its cyclic nature, with the practitioner-researcher implementing and evaluating an innovation intended to address the concern and then modifying the innovation as indicated by the evaluation. There is then a similarity between the action research cycle and the learning cycle (Marek 2009). The focus of action research is meant to be the improvement of the practical situation, rather than the development of generalisable theoretical knowledge, and so action research often lacks detailed documentation and formal reporting.

That said, published studies are sometimes said to be examples of action research, although generally to be considered worthy of publication, such studies are expected to demonstrate both a level of documentation, and a robustness of argument for knowledge claims, outside the typical characterisation of action research. That is, the logic of action research is that at the end of each cycle, decisions about the next cycle of action are based upon judgements 'on the balance of probability' rather than waiting to accumulate sufficient evidence to support formal knowledge claims that would be robust enough for presentation in an academic research journal.

Arguably, action research is less a methodology as such than a mode of research that is context directed, where the focus is on improving practice within a specific context, rather than developing abstract, generalisable, theoretical knowledge. This is in contrast to academic research that is theory directed, but which might collect data in a limited specific context as a methodological choice (e.g., if a case study seems most appropriate to answer RQ). From this perspective, true (context-directed) action research is unlikely to provide the basis for academic research reports, but there is no 'in principle' reason why *practitioner* research cannot contribute to the academic research literature as long as it is suitably theory directed and not exclusively concerned with addressing an immediate issue embedded within the practice context. Such practitioner research would need to apply suitable methodology to support theory-directed work (i.e. action research per se would not be such a methodology), but could still be initially motivated by a local problem or issue and may well contribute to improving practice, as well as offering a more generalisable contribution. There is therefore a good reason to avoid conflating action research with practitioner research more generally.

The methodologies described here do not exhaust the methodologies claimed in research papers. As well as variations, refined and hybrid versions of the above methodologies, there are also references to quite different methodologies. However, what counts as a distinct methodology is open to debate. It could be argued, for example, that so-called feminist methodologies, such as the feminist ethnography used in a paper reported in *Science Education* (Basu 2008), are conflating a methodology (in this case ethnography) with a theoretical perspective (here, feminism) that is informing both the choice of that methodology and how research is designed based on that strategy (cf. Fig. 57.2). A counter argument would be that the more specific feminist methodology is distinct because it is informed by a particular value position (in this case 'the importance of research having benefit for research participants and their immediate community', p. 256). Whether or not feminist ethnography should be considered a distinct methodology in its own right, ultimately what is important is that methodological choices are carefully explained and justified, and as long as that is so, readers can draw their own conclusions about the worth of knowledge claims made, and the particular labels used as descriptors are secondary

However, this example raises the important point that methodological decisions in educational research are informed by axiological as well as ontological and epistemological considerations.

57.9 Ethical Considerations and Their Methodological Consequences

All researchers should be informed by professional standards of ethics. In the natural sciences, a focus on research ethics often concerns such issues as not inventing data, not selecting results for reporting based on their level of agreement with preferred ideas and giving full acknowledgement to the work of others. These considerations also apply in educational research, of course, but there are additional ethical complications in educational work that do not tend to arise in most research in natural science. Often these issues are significant enough that methodological considerations may need to be compromised because of the ethical imperative.

57.9.1 The Good

Researchers tend to feel that research is inherently a good thing because it produces knowledge, which allows us a better understanding of some aspect of the world, and so can inform our choices. Even in the natural sciences, such a view might be challenged. Science provided knowledge to allow the development of explosives used in war as well as in engineering applications, poisons used in Nazi gas chambers and the atomic bombs dropped on Hiroshima and Nagasaki. If such applications are considered inherently evil (and few would dispute that at least in the case of the gas chambers used as instruments of genocide), then questions may be raised about the wisdom of the science that provided the technology. However, there is a common argument that knowledge in itself cannot be evil, as it can only inform human actions, where there is a moral choice to be made in how to apply such knowledge.

57.9.2 Costs and Benefits of Research

In areas such as medical science, there may be questions about the costs of the knowledge produced by research. Sometimes new treatments and procedures do more harm than good (as was the case with the use of thalidomide, which led to thousands of serious birth defects): but the medical profession is bound by an imperative to do no harm and so puts in place various safeguards to avoid harming participants in studies. Sometimes there is a recognised substantial risk, and a participant may choose to take that risk in the hope of a possible benefit. In such a situation the notion of informed consent becomes very important: that the person agrees to take the risk based on an understanding of the available knowledge about possible risks and likely benefits of participation. Sometimes participation is altruistic in the sense that the participant may be aware that there is likely to be minimal benefit personally, but that knowledge obtained may contribute to developing treatments to benefit hypothetical others at some future time.

57.9.3 Informed Consent

The medical research scenario offers a strong parallel to the situation regarding much educational research. Educational research may be carried out primarily to develop theories that might be applicable at some point in the future, and such research may potentially inconvenience teachers, learners and others who are asked to contribute through their participation now. We might hope that people would welcome a chance to contribute to the development of educational knowledge through participation in studies, but a researcher cannot require or expect this. Therefore informed consent must be obtained from participants, and the wishes of those potential participants who decline involvement must be respected, regardless of the basis of their decisions, even if this weakens or undermines a research design – such as an experimental design.

There are clearly complications with obtaining informed consent that relate to the ability of – especially young – children to understand what they are being asked to give consent to; regarding when parents as well as learners need to give consent; and about when teachers, acting in *loco parentis*, are able to give consent on behalf of students. Teachers, head teachers (or school principals), area education officers and government ministers may act as 'gatekeepers' who decide whether a proposed study can be carried out in particular classrooms and schools. They may well reject requests for research that is judged to have potential for undermining normal order and procedures.

Innovations that seem promising to researchers may be judged to make too heavy demands on potential participants; and even quite straightforward procedures such as administering simple questionnaires to classes may be considered unwelcome by busy teachers. This is often likely to lead to researchers compromising research design based on what might realistically be granted when permission is sought. Experimental designs that look to compare two different teaching and learning conditions can often apply inferential statistics, providing that the learners are randomly assigned to conditions. However, in practice, researchers are usually restricted to working with intact classes, where, at best, whole classes can be randomly assigned to treatments – a much weaker design. Indeed, sometimes the choice of the 'treatment' and 'comparison' groups depends upon which teacher is prepared to adopt some innovative practice, immediately suggesting that teacher characteristic may well be a confounding factor.

A particular issue that arises is that where some potential participants decline to be involved in a project, this may well bias any attempt at sampling. If a study seeks a representative sample and reasons for granting or declining consent link to the issues being researched, then the final sample may well be skewed.

57.9.4 Openness and Confidentiality

Another key issue that may lead to methodological compromises is the need to respect participants' desire for anonymity in research. Generally, it is considered appropriate to assure potential research participants that their data – and it has been argued that the data is *theirs* to gift to the researcher (Limerick et al. 1996) – will be kept confidential within the research team and that any reports will be written such that individuals (and often institutions) cannot be identified. This is more readily assured in some types of research than others. So reporting detailed case studies, where the expectation is to provide 'thick description' to support reader generalisation (i.e. where the reader makes a judgement about how well the reported context is similar to their own professional context), may be difficult without giving away information that would allow informants to be identified.

Indeed there are examples of research in the science education literature where the published details seem to make it very unlikely more than one person could match the description (see examples disucssed in Taber 2013). Sometimes it is suggested that it is appropriate for researchers to deliberately change some biographical or other details to assure anonymity of participants – but this clearly means providing a report which is known to be false in certain regards and puts the onus on the researchers to know what details can be changed without undermining the authenticity of the published account.

57.9.5 Member Checking and Rights to Withdraw

A further complication of respecting the rights of individuals involved in educational research is that it is often suggested that a participant should have the right to withdraw from a study *at any stage*: 'researchers must recognize the right of any participant to withdraw from the research for any or no reason, and at any time, and they must inform them of this right' (British Educational Research Association 2011, p. 6). This can clearly undermine research designs. In longitudinal studies, it is quite common to experience attrition as participants leave the study for various reasons, and this might modify the balance of participants sampled if decisions to continue participation or withdraw may be linked to issues being explored. To some extent, this might be accommodated by building-in redundancy through enrolling more participants than are required for what is considered likely to be a sufficient data set – but that may well require additional resources.

In an interview study, for example, it is normal to advise participants that they may stop the interview at any time or decline to answer any particular questions. If a sequence of interviews are planned, the participant is invited to continue their participation on each occasion, i.e. the researcher cannot expect them to abide by commitments perhaps made months before. It is commonly also suggested that whilst a study is in progress, participants should not only have the right to decline further participation at any point but also have the right to withdraw any data they have provided *earlier* in the study.

A related point concerns the right to comment on material written about a participant. In some forms of research, particularly interpretive studies claiming to report on the views, ideas and opinions of others, it is recommended that the participants should be invited to read, and comment on, any draft reports relating to their own cases – this is known as member checking. In itself this is as much a methodological as ethical safeguard, as it gives participants the chance to check the researchers' interpretations of their inputs are valid. Any feedback received from such 'member checking' should be treated as additional data that needs to be considered in drawing conclusions. Clearly at this point, there might be potential for a participant to request particular changes (if perhaps they feel their comments are not presented in a favourable light) and seek to withdraw their data from the study otherwise. This has potential to undermine the integrity of a study.

There are clearly circumstances where member checking has less value methodologically. One case would be where students' thinking is analysed in relation to canonical scientific thinking, where it is likely that a student holding an alternative conceptual framework may not be in a strong position to confirm or otherwise the worth of the analysis. In such research, there are techniques that can be adopted as part of interview procedures to ensure the validity of the interpretations being made by researchers during data collection (although it should be noted that further insight into students' thinking may emerge later during analysis): confirming responses by repeating or rephrasing questions; clarifying ideas by asking follow-up questions; paraphrasing what one believes to be the learner's argument, and seeking confirmation; returning to the same point in the same context later in the interview, to see if a consistent response is given by the learner; and approaching the same point through a different context later in the interview, to see if the learner gives consistent responses in the different contexts (Taber 1993). Member checking may also be of limited value in studies looking at shifts in participants' opinions, as participants may not retain a clear and accurate recollection of their earlier stances once their thinking has moved on.

57.9.6 Particular Challenges of Teacher Research

Ethical issues may become especially problematic for teachers and lecturers undertaking research with their own classes (and colleagues). There are a number of complications here compared with research carried out by 'external' researchers.

For one thing, the usual 'gatekeepers' who normally need to approve a study before researchers approach students about being potential participants can be bypassed. A second issue concerns obtaining informed consent, as students could feel that they are obliged to help their teacher who will often have a role in writing reports on them or grading their work (Taber 2002). Although the teacher may not seek coercion, safeguards are needed to assure students that participation is entirely voluntary and that nonparticipation carries no penalty in regard of their study. Both of these issues can somewhat be countered by recruiting a suitable senior colleague to act as a nominated person to check on the procedures being employed and informing students that they may refer any concerns to that person.

A difficult issue is to decide when research goes beyond normal teaching practice. The fully professional teacher is expected to be research informed and able to develop their teaching through classroom research (Taber 2013). Teachers are expected to innovate and to collect data so that they know how effective their teaching is. An innovative teacher, trying out new ideas to improve their teaching and collecting classroom data to evaluate their work, would *not* expect to have to seek permission from the learners in the class (and/or their parents for younger learners) nor to offer opportunities for some class members to decline to be involved in any lesson activities based on innovative approaches. Yet, in effect, this kind of evidence-based teaching practice is a form of research. This is indeed an area where it may not be clear when classroom enquiry and innovation should be considered primarily research rather than just good teaching practice.

However, what is clear is that the science education research journals contain many examples of studies based upon data collected and analysed by teachers working with their own classes, where the impression given is that the purpose of data collection was research (rather than as a normal part of teaching) and where often there is no mention of how the research was presented to learners nor whether they were invited to contribute and given the choice to decline. That is, some of these studies are written as though the authors feel that they are entitled to set exercises to collect data without consideration of the way they are using their students as data sources. Perhaps the researchers in such studies did follow appropriate ethical procedures, but if so they did not feel the need to report they had done so.

Increasingly, journals are expecting authors to make a declaration on submitting studies to the effect that appropriate ethical guidelines have been followed: although this relies on the researchers having a good understanding of the issues involved. It is suggested here that there are useful criteria that can be used to decide when evaluations of teaching innovation, or other examples of teacher research, should be considered to need informed consent from students (see Table 57.1). These concern the nature of the activity used to collect data, the purpose of the data collection and the intended use of the results (Taber 2013).

So it is suggested that researchers should (i) seek explicit consent from students they would like to be involved in studies and (ii) acknowledge that informed consent was given in research reports, when the research (a) requires input from students outside of the normal classroom/curriculum schedule and/or (b) is 'theory directed' (i.e. looks to answer general questions, where learners involved stand for learners

Table 57.1 Determining when teacher research requires informed consent from learners

	Teacher research should be considered a part of normal classroom practice when	Teacher research requires informed consent from learners when
Activity	It involves normal teaching and learning (including assessment) activities carried out within normal curriculum time	It goes beyond the normal range of teaching and learning (including assessment) activities and/or occurs outside of normal scheduled curriculum time
Purpose	It is intended to help understand better an aspect of the professional context or solve problems arising within that context, i.e. knowledge is sought to inform educational practice in the institutional setting that will benefit the learners involved in the research	It is intended to answer general theoretical questions and support the development of abstract knowledge (i.e. the students' concerned are just a convenient sample considered to represent a broader population of learners)
Dissemination	Research results will inform the teacher-researcher and may be shared will departmental or other colleagues working in the same institutional setting	It is intended that research results will be submitted for publication or disseminated through websites, conferences, networks, etc. (N.b. this would apply to research undertaken for an academic award)

generally) rather than context directed (where the research is aimed at specific issues relating to the teaching and learning in the particular research context) and/or (c) is intended for reporting and dissemination beyond the institutional context where the research is undertaken.

Following these guidelines will protect learners from being treated as research fodder and will protect researchers from suspicion of unethical practice.

57.10 Selecting Techniques in Educational Research

There is not a simple correspondence between methodology and particular techniques, but there are some clear patterns. Experiments require some form of quantification. Surveys tend to involve the use of questionnaires and/or structured observations. Case studies tend to use a range of techniques, commonly including interviews, observation and document analysis.

Interviews can be used as data collection techniques in a range of methodologies, but the *type of* interview used may change from one methodology to another (and this is also true of observational techniques). So an interview in a study employing survey methodology is likely to employ a highly structured schedule of questions (in effect, an oral questionnaire) which the interviewer is not supposed to vary (i.e. to ensure comparability between respondents), whereas in an ethnographic study, interviewing is likely to be based around a much more flexible interview schedule that allows the interviewer to probe for the participants' understandings and perceptions and to use the interactive nature of conversation (Bruner 1987) as a means to check and refine the researcher's interpretations of what they are being told. In effect these types of interview are rather *different techniques*, informed by rather different assumptions about what is methodologically appropriate in a particular study (see below). In the survey interview, it is assumed that (in principle) the interviewer could be replaced by another trained interviewer without influencing the responses of participants. Such objectivity may be more difficult to achieve in an ethnographic study where the sensitivity of the researcher to nuances in responses is much more significant.

A research design should include the ways in which data will be analysed, as well as how they will be collected, and again particular ways of analysing data are linked with particular methodologies. So, for example, formal hypotheses tested through experimental or survey approaches require the deductive use of quantitative methods applying inferential statistics, whereas grounded theory employs the 'constant comparison' method of ensuring theory is developed from data by an inductive approach. In some methodologies, it is expected that triangulation (Oancea 2005) from different data sources, or even different data collection techniques, is used to ensure the 'trustworthiness' of research (Guba and Lincoln 2005). However, this is not considered necessary when research techniques are considered to unambiguously access ontologically clear research foci (as in a well-designed experiment).

Whilst Fig. 57.2 does not show any direct link from the theoretical perspective (or the conceptual framework) to a research design, it is intended to imply that an indirect influence occurs through the RQ. The formulation of RQ involves selecting terms and phrasing that reflect, and imply, particular meanings that have been developed through the formation of the conceptual framework, informed by the theoretical perspective identified as the starting point for building an understanding of a topic.

An interesting question is to what extent the process reflected in Fig. 57. 2 would be recognised as relevant to research in the natural sciences. It is argued in this chapter that *in principle* the same kinds of consideration that apply in educational research also apply in research in the natural sciences, but much more can be taken taken-for-granted within 'normal science'.

57.11 Typologies of Educational Research Methodologies

A key analytical tool used in characterising educational research is a description of several levels at which the research can be described. Commonly three or four levels are posited that shift from a consideration of philosophical commitments underpinning the research to identification of particular techniques to collect and analyse data. For example, one commonly cited model is that used by Crotty (1998), who describes social research at four levels: (i) epistemology, (ii) theoretical perspective, (iii) methodology and (iv) methods. As one example within this scheme, *a questionnaire* (method, i.e. technique) might be used to carry out *a survey* (methodology) from a *positivistic* theoretical perspective drawing upon *objectivist* epistemology.

This is only one of the schemes recommended in textbooks on social and educational research, because it is very difficult to find a common analytical framework that readily fits all different forms of research in education. A somewhat more simplistic model (Taber 2007, 2013) posits three basic levels of analysis understood as philosophical (the level commonly called paradigm in the social sciences), strategic (methodology) and tactical (techniques).

Crotty discusses three epistemologies: objectivism, constructionism and subjectivism – depending on whether meaning is considered to be *inherent in* an object, to *arise from interactions with* an object or to be *imposed upon* an object by a subject. Often in accounts of research such as Crotty's, the impression is given to novice researchers that they are expected to adopt one of these epistemological perspectives as a way of understanding the world. Yet this would seem to imply seeing the world as comprised of objects that at some fundamental level are of the same basic nature, at least in terms of what we might aspire to know about them. Such a perspective may be contrasted with pragmatism (Biesta and Burbules 2003), which is unfortunately (and inappropriately) sometimes presented as having little time for philosophical issues, but rather simply looking for tools to do particular (research) jobs.

Neither the adoption of a blanket epistemology nor of a naive pragmatism offers a justifiable approach for educational researchers when considering methodologies to adopt for particular purposes. The position taken here is that the extent to which researchers can both (a) clarify the ontological status of foci of research; and (b) directly and unambiguously access the foci of research; varies considerably in educational work, and therefore the selection of epistemology must reflect the needs of a particular study. So, for example, it does not make sense to consider that the same assumptions will support research into the provision of Bunsen burners equipping school laboratories, student attitudes to practical work and teacher understandings of socioscientific issues.

57.11.1 Qualitative Versus Quantitative

In a book on research design, Creswell (1994) suggested that once a focus for a study was established, the next step was the choice of paradigm, and he presented two options: the quantitative (or positivist, experimental or empiricist) paradigm and the qualitative (or constructivist or naturalist) paradigm. According to Creswell, particular methodologies (or as he called them methods) were appropriate for each of these paradigms (Table 57.2).

The reference to paradigms here reflects the adoption of the term in the social sciences after the widespread influence of Thomas Kuhn's work (considered above).

Table 57.2 A typology of research methodologies, after Creswell

Quantitative methodologies	Qualitative methodologies
Experiments	Ethnographies
Surveys	Grounded theory
	Case study
	Phenomenological studies

The identification of a paradigm which is considered positivist, experimental or empiricist might seem to some to imply a more 'scientific' paradigm. However, in the present chapter, it is argued rather that a scientific approach involves a choice of methodology that is consistent with the aims of the particular study.

A major problem with the Creswell classification is the prominent use of the terms 'quantitative' and 'qualitative' as major labels, as these terms have come to be used in very different ways in educational research. One common way in which the terms quantitative research and qualitative research are understood is in terms of the type of data being collected and analysed. Certainly there is an important difference between quantitative *data*, which is suitable for certain types of analysis, and qualitative *data*, which needs to be treated with different analytical approaches.

However, even that distinction is not absolute, because there is a spectrum of approaches to the analysis of qualitative data (Robson 2002). So, for example, it may well be that interview transcripts, providing text (qualitative data), may be analysed by counting specific words or phrases to test some hypotheses (i.e. quantitative analysis). It is also common for qualitative data to be initially analysed using interpretive approaches (qualitative analysis), leading to the assignment of coding which then leads to counts of the frequencies of certain codes, which could be the basis of either descriptive statistics or, again, hypothesis testing.

However, in other studies, qualitative data may be treated in much more thematic and narrative ways, with no frequency counts or other quantification. So even when we restrict our focus to data, the quantitative-qualitative distinction is of limited value once we shift beyond the description of the data itself to its analysis. Moreover, if the focus is on the nature of the data itself, then it makes little sense to align methodologies such as case study and grounded theory, which may commonly employ both qualitative *and* quantitative data collection and analysis, under a qualitative paradigm as Creswell does.

Where the focus of qualitative and quantitative is sometimes on the type of data being analysed, the term quantitative research is also sometimes reserved for the use of hypothesis testing approaches, excluding studies that analyse quantitative data to offer purely descriptive statistics. Similarly, some authors limit the use of the term qualitative research to studies that admit the necessity of a subjective element (Piantanida and Garman 2009) and are based on an interpretative approach that does not claim objectivity in the normal scientific sense – because it is argued that some kinds of social phenomena can only be understood through the intersubjectivity formed through establishing researcher-participant rapport and that the kind of detached observer who could claim objectivity would not be able to access suitable data for the study. There are clearly many studies based on the collection and analysis of qualitative data that are not 'qualitative' research in *that* sense.

Table 57.3 Two traditions or paradigms for educational research after Gilbert and Watts (1983, p. 64)

Tradition	*Erklären* tradition: explanation is the goal	*Verstehen* tradition: understanding is the goal
Outlook	Realist – adopting an empirical-inductivist view of knowledge	Relativist – influenced by post-inductivist views of knowledge
Target	Seeking causal mechanisms	Seeking understanding as shown by the individual actors (without the overt pursuit of generalisations)
Characteristics	'Nomothetic': general laws are sought	'Idiographic': relates to the study of individuals
	'Quantitative': suitable sections of a general population are enquired into	'Qualitative': seeks to enquire into phenomena without undue regard to their typicality
	'Prescriptive': outcomes of enquiry are intended to determine future actions	'Descriptive': no overt intention of determining future actions
Approach to phenomena	Reductionist – phenomena are subdivided and the divisions selectively paid attention to	Holistic – phenomena are studied in their entirety
Methodological approaches	'Experimental': controlled situations	'Naturalistic': naturally occurring situations

57.11.2 Two Paradigms for Educational Research?

It seems clear that when used as primary descriptors without further qualification, the terms qualitative and quantitative can be ambiguous and so unhelpful. Gilbert and Watts (1983) also used the descriptors 'quantitative' and 'qualitative', inter alia, when they described two common traditions or paradigms for research that could be employed in science education. However, Gilbert and Watts offered explanations for their uses of the term, in the context of setting out two clusters of characteristics of these two traditions. Their two paradigm descriptions are summarised in Table 57.3, and several of their points will be reflected in the following treatment.

One aspect of the Gilbert and Watts scheme that needs comment is the notion of their 'paradigm 2' (*Verstehen* tradition) being a relativist one. For some commentators, any admission of relativism is seem as antiscientific, and indeed Scerri has attacked the prevalence of 'constructivist' thinking in science (and in particular chemistry) education because of its associations with relativism. Space does not allow this debate to be explored in detail here (see Scerri 2003, 2010, 2012; Taber 2006b, 2010c), except to note it is a rather different proposition to suggest (as a hypothetical example) (a) that the choice between (i) the ancient system of earth, fire, water, air and aether as elements and (ii) the modern periodic system as a basis for scientific progress is all a matter of cultural perspective (a kind of relativism difficult to justify scientifically) than it is to suggest (b) that it is important to investigate and respect learners' alternative conceptual frameworks because of their influence on the individual's *learning of* science.

As suggested above, the research focus on students' ideas in science derived from concerns with the common patterns of conceptual development and the difficulties of learning canonical science, rather than any suggestion that students' ideas offered a viable alternative basis for scientific progress. Indeed it has been noted that common alternative conceptions often share at least superficial similarities with historical scientific models and theories long abandoned (Piaget and Garcia 1989).

Often in education, we are concerned with exploring the personally constructed 'realities' (i.e. the reality as experienced) of individuals because personal sense making is at the heart of the learning process (Glasersfeld 1989). The decision to focus on such 'second-order' perspectives (Marton 1981), i.e. other people's construing of reality, *need not* imply abandoning a belief in an absolute external reality. This can be considered analogous to how the historian of science may use hermeneutic methods to understand how scientists of the past understood scientific concepts because of the value of knowledge of those personal conceptions to our understanding of the history of science, not because anyone is suggesting that such outdated ideas are as valid as current scientific thinking.

The extensive research into student understanding and thinking in science associated with 'constructivism'/the ACM was strongly informed by existing traditions of work which emphasised the importance of a person's existing ways of understanding the world as the basis for how they made sense of experience and so how that interpretation of experience informed their actions in the world (Taber 2009b). In particular, key constructivist thinkers in science (and mathematics) education were informed by the genetic epistemology (Piaget 1970/1972) of Jean Piaget (Driver and Easley 1978; Gilbert and Watts 1983; Glasersfeld 1989) and the personal construct theory (Kelly 1963) of George Kelly (Gilbert and Watts 1983; Pope and Gilbert 1983).

One significant distinction between research methodologies does closely resemble that suggested by Creswell, but is not best distinguished by the labels qualitative and quantitative. Rather, these two types of research are better characterised according to whether the research is intended to test out existing established theory through deductive methods or rather to better understand poorly theorised phenomena to aid the development of new theory (Biddle and Anderson 1986). Developing this idea suggests two clusters of characteristics of research studies, as shown in Fig. 57.3.

This perspective does not set out different methodologies as fundamentally concerned with different research enterprises, but rather reflects how in any area of scientific activity there has to initially be a period of exploring and categorising and 'making sense' of the phenomena of a field – what has been termed the 'natural history' phase (Driver and Erickson 1983) – that can lead to the kinds of theorising, and subsequently bold conjectures (Popper 1989), suitable for formal testing (see Table 57.4).

That much of educational research concerns the former, more exploratory, types of study may be partly related to the relative immaturity of educational research compared with the established natural sciences. However, there are also inherent features of education that channel much research towards the discovery pole. One of these features, noted above, concerns the inherent complexity of educational

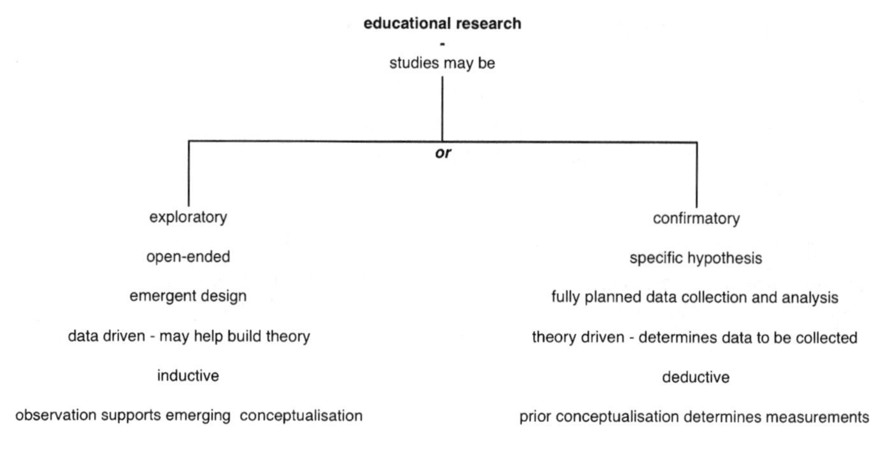

Fig. 57.3 Two main types of research in education

Table 57.4 Exploratory and confirmatory research

Paradigmatic commitment	Application	Suitable methodologies
Exploratory	In areas where no clear theoretical picture has emerged, due to limited research or complexity of phenomena	Case study Grounded theory
Confirmatory	To test hypotheses drawn from established theory	Experiments Surveys

Table 57.5 Idiographic and nomothetic research

Paradigmatic commitment	Application	Suitable methodologies
Idiographic	To enquire into educational phenomena where understanding requires detailed engagement with specific instances in their naturalistic context	Case study (to explore individual learners, classes, teachers, etc.) Ethnography (to explore cultures of identifiable groups)
Nomothetic	To enquire into aspects of educational phenomena that may be described in terms of norms and general laws	Experiments Surveys

phenomena, which are often embedded in situations from which they cannot be readily be disembodied whilst retaining their integrity.

This complicates attempts to use experimental method, as there may be myriad potential confounding factors that may be difficult to identify, let alone manipulate to control conditions, or, failing that, to measure so as to attempt to allow for during data analysis. As suggested below, this has encouraged much educational research to be focused on understanding the individual case in depth (see Table 57.5), despite

Table 57.6 Objectivist and constructivist-interpretivist research

Paradigmatic commitment	Application	Suitable methodologies
Objectivist	When dealing with issues where there is consensus ontology (the nature and demarcation of what is being studied) and clear epistemology (agreed means of learning about objects of research)	Experiments Surveys
Constructivist-interpretivist	Where exploring phenomena that are socially constructed and culturally relative or nuanced mental phenomena that can only be communicated through dialogue	Grounded theory (for understanding the central issues in social phenomena and institutions) Case study (to allow detail exploration of an individual or group) Ethnography (to provide immersion in culture to identify emic (insider) perspectives)

the problem of generalising from the individual to the wider 'population' (of teachers, of lessons, of learners, etc.).

Another key issue concerns the nature of teaching and learning as human activities. As such, there is a limit to the extent they can be seen as the subjects of objective study, because humans make personal meaning of and from their experiences, and many of the things we wish to study relate to those meanings (see Table 57.6). So whilst we might be more 'objective' when exploring class size, or curriculum content, or even whether student examination responses match specified features of canonical target knowledge; if we are interested in how a learner understands a concept, or the values they bring to science learning, or their experiences of a new teaching approach, etc., then we need to use ('constructivist'/'interpretivist') methods that can engage with and explore how others make sense of the world.

The best, though highly imperfect, apparatus we have for exploring one person's meaning making is the interpretive (meaning making) facility of another human being who can develop rapport with that first person and engage with them in some form of dialogic conversation. This affordance in some kinds of research is also linked to a serious threat to validity for those attempting to set up experimental research (i.e. in nomothetic mode). The expectations of researchers, or teachers working with them, are readily transferred to learners, and teacher enthusiasm or cynicism about some innovative approach being evaluated in a teaching-learning context can influence learners' own expectations, which in turn influence their perceptions of the innovation and so influence the learning itself. One common type of study compares learning in two 'comparable' classes where teaching by an innovative ('progressive') approach is compared with teaching through a 'traditional' approach. This immediately creates problems for making a fair comparison whether the teaching is carried out by the same teacher (will they be as equally adept and enthusiastic in both conditions?) or different teachers (who inevitably will bring different skills, and knowledge to their teaching). Added to that, the learners themselves may well react to the novelty of the innovation purely in terms of it being something different from the norm (which may well be welcomed, but could for some learners be perceived as threatening).

However we go about *collecting* data about the ideas, feelings, opinions, attitudes, etc. of others, we can only meaningfully *analyse* that data through the interpretations of other humans. This is what some commentators mean by 'qualitative' research (see above): research that relies on the intersubjectivity between researcher and study participants.

Whilst at first sight instruments such as questionnaires seem to avoid this intersubjectivity by presenting statements to be ranked or rated, the items in such instruments are only going to have validity (as statements that are both meaningful to respondents and understood by them *in the sense intended* by the researchers) when derived from previous research which explores what ideas and language will be meaningful for those surveyed – previous research which will necessarily have involved in-depth dialogic approaches (cf. Treagust 1988). Here again, the type of research which would fit under the right hand fork in Fig. 57.3 relies upon earlier rather different work that would fit under the left-hand column.

57.11.3 Mixed Methods: A Third Paradigm, a Subsuming Paradigm or a Rejection of Paradigms?

In recent years, those preferring the notions of quantitative and qualitative paradigms have admitted a 'new' paradigm known as mixed methods research: that is, research that employs a combination of quantitative and qualitative features (Creswell 2009; Creswell and Plano Clark 2007; Johnson and Onwuegbuzie 2004). Clearly if we focus on data type, there is nothing of special interest about mixed methods, as studies using quantitative and qualitative data are not themselves novel. It is less clear how a single study could at the same time employ genuinely distinct approaches such that it was at the same time objectivist/positivistic and constructivist-interpretivist if we take the former to suggest a realist ontology and an epistemology which allows claims that research offers in some sense an objective, researcher-independent, account of that reality and if the latter means accepting that the kinds of knowledge about the research foci that are possible are necessarily constructed by human beings and relative to the interpretations of a particular knower (Symonds and Gorard 2008). Given this, the claim that there is a distinct research approach known as 'mixed methods' – depending whether it refers simply to data type or something methodologically more substantial – is either fair but of no great significance, or alternatively is important but problematic (Taber 2012a).

This cynicism regarding the *label* of mixed methods derives from seeing it sometimes used in practice to describe a study's methodology simply because both quantitative and qualitative data are collected. In that situation the label is generally unhelpful as it *at best* stands in place of a more informative label for the methodology adopted and *at worse* substitutes for the choice of an actual substantive coherent methodology. That is, in practice we sometimes find the label 'mixed methods' stands in place of principled thinking about the nature of what is researched and how to best enquire into it.

However, 'mixed methods' has also been positioned as 'an approach to knowledge (theory and practice) that attempts to consider multiple viewpoints, perspectives, positions, and standpoints (always including the standpoints of qualitative and quantitative research)' (Johnson et al. 2007, p. 13). Here mixed methods research is defined as 'the type of research in which a researcher or team of researchers combines elements of qualitative and quantitative research approaches (e.g., use of qualitative and quantitative viewpoints, data collection, analysis, inference techniques) for the broad purposes of breadth and depth of understanding and corroboration' (p. 123). Clearly the discussion here is not restricted to types of data, as presumably the 'standpoints of qualitative and quantitative research' relate to ontological and epistemological issues (are we dealing with the kind of things that can be countered and/or measured?; are we enquiring into something that will require intersubjectivity as an 'instrument' to elicit data?).

The position taken in this chapter, developed further below, is that such choices cannot be established in the abstract, but need to be addressed in the context of particular studies. As a field, science education cannot be well served either by limiting data to be collected to quantitative or qualitative forms; and nor can it progress by committing to the 'standpoints of qualitative [or] quantitative research' independently of the particular questions being addressed. However, if adopting 'mixed methods' as a paradigm (Johnson and Onwuegbuzie 2004) for educational research is taken to mean that we include within our methodological repertoire a wide range of approaches, from which to select according to the need of particular studies, then this fits well with the stance adopted here. The term mixed methods is perhaps unfortunate, as this approach is less a matter of 'mixing' our methods, than of making principled choices of methodology on a case-by-case basis for each RQ we wish to address. Yet it is the very diversity of methodologies, and research techniques adopted within them, that makes this approach quite unlike the kind of 'paradigm' that Kuhn intended in characterising normal science.

57.12 The Logic of an Extraordinary 'Sort of Science': Science Education as an Aparadigmatic Scientific Field

It was suggested above that perhaps research in science education, and indeed educational research more generally, might fail to look like Kuhn's normal science in part because of the pre-paradigmatic (Gilbert and Watts 1983; Jevons 1973) nature of the field, in which case we might be reassured by Kuhn's acknowledgement that another sort of science will be found in immature fields. Alternatively, we might share Shulman's (1986) view that this is not a matter of immaturity but rather of the nature of what is being studied (e.g. social institutions and processes; often idiosyncratic personal meaning making) which makes science education unlikely to develop a clear paradigm in Kuhn's sense. That is, we might consider science education will remain 'aparadigmatic'.

The next section focuses on these key features of educational research, the complexity and diversity of the phenomena of education and the inherent complications of research with human participants. The argument here is that science education may be a relatively immature field, but that even as it matures it is unlikely to develop a structure that supports an array of relatively discrete sub-fields each with its own disciplinary matrix to support the induction of researchers into a kind of normal science. Rather, given the high level of interconnectedness between different foci of research, all of which should ultimately inform teaching, science education should aspire to be a different 'sort of science' to Kuhnian normal science.

57.12.1 The Ontological Diversity of Educational Phenomena

We have seen that (a) particular research methodologies (strategies) rest upon fundamental assumptions and may cease to make sense as research strategies when those assumptions do not apply; yet (b) this does not restrict the researcher to a limited range of the available methodological choices in any absolute sense. Research design in education then must always (explicitly) take account of ontological and epistemological issues which logically constrain what may be considered sensible methodologies to adopt for particular studies: and as the educational world does not comprise only of entities of one particular ontological status, the starting point for designing research can be quite different for different studies – even within a particular subfield of science education.

That is, there are things of interest to science education researchers that can be tightly defined, fairly objectively identified in the world and counted and measured. These types of things are open to forms of investigation (in particular, research which collects quantitative data to test hypotheses through inferential statistical techniques: experiments, surveys) that would not make sense when the 'objects' of research are instead clearly culturally relative, socially constructed entities. So methodological choices must relate to the nature of *what* one wishes to research (which will have been posited in developing the conceptual framework for the study, cf. Fig. 57.2). Consider the following potential starting points for educational RQ:

- What is the average secondary school science class size in different countries?
- Is teacher subject knowledge or extent of classroom experience more important for successful science teaching?
- How do 11-year-olds understand energy?
- What are 11-year-olds' perceptions of the difficulty of science lessons?
- What is it like to be the only young woman in an undergraduate physics class?

One immediate point to make is that all of these topics involve research into some kind of entity external to the researcher him or herself, so the commitment to undertake research would seem to clearly be an acknowledgement that there is external reality which can be considered the object of (or subject for!) study. This immediately excludes some extreme philosophical positions from usefully informing research. Indeed the commitment to undertake educational research would seem to require the adoption of some key aspects of what is sometimes considered the scientific worldview (Matthews 2009). In particular, embarking on any educational research project would seem to require at least tacit commitment to:

- The existence of some kind of external reality;
- Which has some form of permanence;
- And exhibits certain regularities;
- And which human beings are capable of learning more about.

The posing of particular RQ goes beyond this and sets out certain specific types of entities (schools, classes, teachers, 11-year-olds, understanding, lessons, etc.) as targets or foci of research. That is, even at this stage, certain ontological commitments are revealed. Sometimes these entities are linked to our theoretical perspectives, as when research seeks to investigate Piagetian developmental levels, students' mental models or their alternative conceptions.

Adopting a common epistemology meant to refer to all that we recognise in our world (Crotty 1998) would not seem a sensible starting point. One needs to start from ontology: schools, classes, teaching, understanding, perceptions, mental models, etc. may all be considered to *in some sense* exist in the world, but they are not the same kind of things, and consequently one's epistemological assumptions about them may justifiably differ. So a fairly crude positivistic stance might well be appropriate and effective in seeking to find out the average secondary school science class size in different national contexts, as it is likely to be possible to identify countries and secondary school science classes in ways most observers would find unobjectionable, and determining class size is in principle a simple counting task.

Yet 'successful science teaching' (for example) does not present itself so unproblematically as the subject of investigation: what counts as successful science teaching has shifted over time and is culturally relative, and even in a particular educational context, there will not always be agreement on the appropriate balance between different mooted aims of science teaching – let alone the most suitable indicators that might allow us to make comparisons. Still, even here, in principle we can envisage that researchers might be equipped with an observation schedule of some kind and sent to observe lessons to evaluate the success of science teaching.

Of course no matter how well the data collection and analysis was carried out in such a hypothetical observation study, a reader of the eventual research report would only give credence to the findings to the extent that they accepted the particular conception of 'successful science teaching' informing the design of the observation schedule and were satisfied that the instrument itself could provide valid indications of whether the observed teaching was indeed was indeed 'successful' in *those* terms.

Depending upon how 'successful teaching' is understood, it is entirely feasible that it could even be considered something that could be 'measured' based on quantifiable outcome measures (such as student grades or satisfaction ratings). Where successful teaching is seen simply as teaching that leads to high levels of student examination success, then coming to know where teaching is successful is relatively simple.

Yet, if instead, successful teaching is considered to be about inculcating attitudes and values, about developing relationships, and about supporting maturation, and an interactive process that necessarily involves modifying teaching objectives according to the goals, needs, motivations and personal situations of individual learners, *then* coming to identify and differentiate successful teaching is going to be more challenging, more complex and so inevitably less precise. As always the researcher's epistemology has to be informed by their particular ontological understanding. Where researchers do not agree on the nature of what they are researching (and so in effect are not researching 'the same thing'), they are unlikely to agree on how best to go about their work.

This potentially puts some areas of educational research well outside the type of 'normal science' that Kuhn (1996) characterised as the basis of most work in the natural sciences – adopting canonical definitions and instrumentation widely considered to give valid and reliable results when applied within accepted ranges of application. There is a good deal of creativity and ingenuity at work in the natural sciences, but usually applied within a fairly well-agreed understanding of the nature of what is being researched and the methods appropriate for the job. This is less often the case in educational research. Ziman (1968, p. 115) notes how an 'experienced professional scientist seldom comes into conflict with the referees of his [or her] papers…because…he [or she] has internalized the standards that the referee is trying to enforce, and has already anticipated most reasonable grounds for criticism'. However, in science education, few papers are published without significant revisions required by referees: the experienced professional science education academic may come into conflict with the referees of his or her papers much more regularly than they would wish. This does not reflect on the professionalism of science educators but on the lesser extent of shared commitments and standards for work of those writing and refereeing for particular journals.

57.12.2 Admitting the Subjective Element into Research

Similarly, the classic distinction between the object of research and the (nominally interchangeable) researcher that is an ideal of natural science is often inappropriate in educational research: so where in the natural sciences it might be reported in depersonalised terms that that a sample was ground in a pestle and mortar, in an educational research report, *we* might well report how *we* spoke to a group of students. Eliciting student understanding, for example, is likely to require some kind of co-(re)construction of meaning through interaction between researcher and learner; and when investigating pupils' perceptions of lessons, it is indeed appropriate to consider that meaning is imposed on their experiences by the students themselves rather than being inherent in the activities they take part in.

As suggested above, this does not mean giving up belief in an external reality, but, in this case part of that reality is the experiences of others, and these are not open to being measured or counted like class sizes. Indeed, the best we can hope for is to ask others to represent their (internal mental) experiences in the 'public space' we share (e.g. through talk, drawing, role-play, etc.) and then to look to make sense of these representations in terms of the mental frameworks we have developed through our own experiences in the world (Taber 2014). This type of research *requires* an interpretivist (i.e. subjectivist) approach that acknowledges the difficulties inherent in the task.

Of course as individuals all we ever really know are our own experiences in the world, and there is always something of a leap of faith involved in assuming we share understandings with others. Yet there are differences of degree. We would generally expect that training different observers to reach the same objective 'head count' when surveying class sizes is likely to be less problematic than expecting different interviewers to construct the same model of a learners' understanding of energy or to reach the same understanding of how a female student experiences being the only young woman in an undergraduate physics class.

In the case of understanding energy, we might reasonably expect that such factors as subject knowledge, teaching experience, interviewing experience and expertise and familiarity with prior research could all influence the process of the researcher constructing a model of a learner's understanding, and so the 'results' of the study. In the case of the only woman in the science class, we might consider that the gender of the interviewer could be influential: both in terms of the experiences that the researcher brings to the research as interpretive resources and possibly in terms of the extent to which the female student feels able (or willing) to access and express her experiences and feelings about them.

Such complications undermine the possibility of doing research on people's thinking and experiences that can be as objective as we expect when investigating the resistivity of an alloy or the rate of a chemical reaction. Research always depends on the interpretive resources we bring to the work, *even* in the natural sciences (Keller 1983; Sacks 1995), but in educational research there are many things we want to study where we are unable to eliminate subjectivity, because the interpretive resources relevant to the task (needed to understand another's understanding or to appreciate another's perceptions and experiences) are highly variable among potential researchers. Indeed, we might often expect that the most insightful work is likely to depend upon researchers who have very *particular* knowledge, understanding and experience: such that any objectivist notion that we can substitute another qualified researcher and expect the same result becomes highly questionable.

57.12.3 The Scientific Approach to Educational Research Is to Adopt a Meta-Methodology

The picture painted above is of a field that appropriately draws upon diverse methodologies *because* it deals with a range of different types of research foci, which vary both in how well they are understood and indeed how directly they might be known. The intrinsic variety of educational phenomena and the subsequent diversity in ontological status and epistemological commitments appropriate to particular studies suggest that a mature science education would still lack the kind of constrained disciplinary matrix Kuhn associated with normal science. So science education is not pre-paradigmatic because of its relative youth, but is aparadigmatic because of its need to make principled judgements about methodology in the context of each new research design. Some may refer to this as a mixed methods paradigm, but this seems to pervert the term paradigm to something quite incongruent with its original meaning of a pattern that one can follow to approach a certain kind of problem.

Rather, if we consider methodologies such as experiment, survey and case study as types of strategy that we select between, then science education needs to be informed by a meta-strategy, a meta-methodology that offers guidance on the selection process. We might consider Fig. 57.2 to represent the operation of this meta-strategy, and the principles outlined above – regarding how building a research design needs to be informed by an ontological and epistemological analysis of the basis for the enquiry – indicate the kind of guidance needed. Perhaps, we might see this as aspiring to working within a 'meta-paradigm', not looking to induct researchers into adopting turnkey solutions for well-defined problem areas but preparing them to confidently build research designs bespoke on a principled basis.

If we wish to consider science education as a scientific enterprise despite the need to abandon the aspiration of evolving a research paradigm for the field, we may need to look elsewhere for a demarcation criterion for what counts as science. Popper is well known for his prescription that science should proceed by a process of bold conjecture and seeking refutation (Popper 1934/1959, 1989), and this has been understood as offering such a demarcation criterion. However, in practice, it is well accepted that (i) there is no simple way to determine what counts as a falsification of the theory being tested (rather than, for example, of technical competence or some auxiliary theory of instrumentation) and (ii) in practice crucial experiments only become accepted as such in hindsight, whilst many apparent refutations are quarantined as simply anomalies to be put aside for the moment. However, an alternative perspective on the nature of scientific work, able to distinguish science from pseudoscience, was developed by Lakatos, in his 'methodology of scientific research programmes'.

57.13 Thinking of Research Within Scientific Research Programmes

As suggested above, Kuhn's ideas have been widely criticised although they remain highly influential. In particular, Karl Popper was very critical of the apparently relativist flavour of Kuhn's worked, and there was a high profile debate around aspects of Kuhn's thesis (Lakatos and Musgrove 1970). Popper rejected the 'myth' of the incommensurability between paradigms implied in Kuhn's original formulation of his work (Popper 1994). It was also argued that the account of mature sciences as each consisting of successions of individual paradigms only interrupted by occasional revolutions leading to paradigm shift was an over-generalisation (Machamer et al. 2000) and perhaps was less true in sciences other than physics.

In particular, Imre Lakatos argued that that whilst paradigm-like traditions existed in science and whilst individual scientists would tend to work within such traditions – and indeed often continue to work within them for extended periods of time – it was not unusual for several competing traditions to coexist over considerable periods within the same field of science (Lakatos 1970). Whereas in Kuhn's model this could only happen if one tradition was in the process of being supplanted by its revolutionary successor, for Lakatos it was quite possible for several alternative traditions to continue to be productive and successful in parallel. In Lakatos's terms, these would be considered as co-existing progressive research programmes.

57.13.1 *Lakatos's Notion of Scientific Research Programmes*

Lakatos's model of scientific research programmes is especially relevant to the theme of this chapter, as it offers a demarcation criterion for what can be considered a scientific tradition that can be applied well beyond the natural sciences. Lakatos's work can be considered to set out the nature of a research programme (RP) and to also offer the criteria upon which such a programme should be considered *scientific*.

A Lakatosian RP shares some features with a Kuhnian paradigm. Both are research traditions that involve an initial establishment providing the basis for considerable later development work, and both require those working within the tradition to make particular commitments. Lakatos (1970) described RP in terms of four key elements in particular that he called the hard core, the protective belt, and the positive and negative heuristics.

The heuristics give guidance on how to develop the RP. The hard core comprises of those key commitments (e.g. ontological commitments), set out at the establishment of the programme, which are essential to the nature of that programme, such that if abandoned the essence of the programme is undermined and in effect the programme has ceased. The protective belt comprises of auxiliary theories that build upon and develop what is established in the hard core, and the positive heuristic sets out how this component is developed (e.g. strategic and methodological aspects of the programme). The auxiliary theories act as 'refutable variants' of the programme in the sense that they are consistent with the hard core, but may be abandoned without risk to the programme as a whole.

Consider, as an example, how modern chemistry has made considerable progress since the establishment of a RP based around modern atomic theory. A core commitment there is that at a submicroscopic level matter can be understood to be quantised and to comprise of discrete entities, particles (or perhaps better, quanticles) which can be considered to have specific properties such that chemical behaviour as observed in the laboratory can be explained by models at the submicroscopic level. Few chemists today will direct research at testing hypotheses that are in direct contradiction with those commitments (i.e. the negative heuristic suggests such work would be counterproductive given the core commitment). Given that commitment, the development of the RP can be furthered by the positive heuristic guiding chemists in how to study the nature and properties of the discrete entities and how to build the theory relating the properties of these entities to macroscopic chemical phenomena.

Within such a programme, specific theoretical ideas will be developed in response to the positive heuristic: so now we tend to distinguish atoms, molecules, ions, etc. Particular models and concepts – a planetary model of the atom, the notion of discrete atomic orbitals, etc. – may be introduced, developed and perhaps sometimes abandoned. This does not threaten the programme itself as long as these refutable variants are consistent with the hard core and no aspect of the hard core itself is put aside. For example, the notion of the atom, and the role it plays within this system, has shifted considerably over time (Taber 2003), but this has not brought into question the core ideas that matter has structure at the submicroscopic level and that the properties of the quanta of matter at this level provide a basis for explaining chemical phenomena: rather such changes are part of the process of considering how best to model and understand the submicroscopic structures that are assumed to exist (i.e. these changes occur *within* and informed *by* the programme).

Lakatos (1970) thought that the notion of RP could apply well beyond the natural sciences – for example, psychoanalysis, Marxism and astrology could all be considered to be RP – but that a *scientific* RP remains 'progressive' in the sense that new theory adds to the protective belt (without simply explaining away difficult results) and empirical work continues to respond to and stimulate theoretical developments.

57.13.2 Research in Science Education as a Scientific Enterprise

Lakatos's work can be considered to offer a form of synthesis of the thesis of Kuhn and the antithesis of Popper. Where Kuhn's descriptive analysis lacked any means to distinguish science from non-science or good science from bad, Lakatos's ideas do offer a basis for deciding when a RP is 'progressive' and so deserves support from scientists. According to Lakatos, several RP may operate in parallel, as long as each offers evidence of being progressive. However, once a programme is clearly degenerating, it should only hold a scientist's loyalty until a new promising alternative appears.

Where Popper offers a prescription that is difficult to operationalise – as all scientific theories are formally refuted on a regular basis and all refutations can be explained away with sufficient imagination – Lakatos offers an analysis that is tolerant of individual failures, so long as the general trend within a programme clearly shows development. Gilbert and Swift (1985) characterised research in the Piagetian traditions and the ACM as co-existing Lakatosian RP in science education. Lakatos's approach not only has the potential to distinguish progressive (and so scientific) programmes from degenerating programmes but also highlights how within a genuine RP there is heuristic guidance for moving the field forward. This can be potentially very valuable to researchers (and new research students), providing research traditions *are conceptualised as* RP (in Lakatos's sense), where the features that offer heuristic value are made explicit.

Given the considerations explored above which lead educational research to draw upon such a multiplicity of methodologies, it seems unlikely that the adoption of an explicit Lakatosian perspective would allow the fields of science education to be reorganised (substantially or simply conceptually) into a number of discrete programmes with each developing the kind of disciplinary matrix Kuhn recognised in the natural science: RP in science education are likely to remain too pluralistic to seem like normal science.

However, a Lakatosian analysis can identify key commitments for particular strands of work, identify clear directions for those strands and make it easier to judge whether they are empirically or theoretically progressive at any point in time. That would certainly be valuable, both in the task of helping those in the field to appreciate the sense in which they are involved in a scientific enterprise – despite the multiplicity of theoretical perspectives and methodologies that will continue to be adopted across, and sometimes within, programmes – and in guiding researchers, journal reviewers and funding agencies in making rational choices regarding where to commit valuable and limited resources.

According to Lakatos, RP are adumbrated at the outset; and it is possible to identify elements of such programmes in science education. One tradition of work in science education (exploring the contingent nature of learning in science building on the tradition of the ACM, drawing initially on a personal constructivist perspective) has been analysed in some detail as a Lakatosian RP (Taber 2006a, 2009b). This analysis identifies a number of hard-core assumptions that were set out in seminal papers that established the programme and which have provided the taken-for-granted commitments of those taking up work in this tradition. The assumptions give rise naturally to a set of initial RQ (i.e. a basis for the positive heuristic) that have been answered (and refined) to differing extents through the development of a range of auxiliary theories and constructs that act as refutable variants of the programme. Arguably (i.e. according to this analysis) this has been a progressive programme, as it has developed its theoretical apparatus in relation to an expanding base of empirical investigations and results.

Despite this, there is clearly something of a shift away from a core aspect of the programme (the strong focus on learning as personal sense making and knowledge construction). This implies that many researchers see this tradition as having less potential for progress than alternative perspectives. This may be so, as undoubtedly as the programme has proceeded the questions to be answered have become more nuanced, and the means of answering those questions have required more effort (e.g. long-term, in-depth study of individuals rather than surveying groups of learners at one point in time).

Without a shared recognition of the heuristics of established RP, decisions about what RQ to follow up will be made by individual graduate students and researchers, with limited moderation by the community. Arguably that tends to be the way of scholarship in the humanities, but it is not how science is organised (Ziman 1968). Individuals will naturally tend to make decisions in their own interest, which is why the apparatus of a scientific enterprise (peer review for publication, funding opportunities, appointment and promotion committees) needs to be well informed about the state of a field to put the right motivations in place for individuals. The mechanisms of RP offer support for that community apparatus. The analysis of the programme of research into the contingent nature of learning in science (Taber 2009b) is certainly not beyond criticism and indeed invites alternative conceptualisations. However, it does show the feasibility of adopting Lakatos's approach as one means of seeking to take seriously the challenge of making science education a scientific enterprise.

57.14 Conclusion

In conclusion, research in science education may never resemble Kuhn's normal science, because of the complexity of educational phenomena, the difficulty of maintaining the integrity of many of those phenomena outside of naturalistic settings and the nature of teachers and learners as individuals each constructing their own understandings of the world and entitled to negotiate the basis on which they might participate in our research. It is likely that many areas of work in science education will continue to draw upon diverse theoretical perspectives and to call upon an eclectic range of methodological tools selected to meet the needs of different specific studies.

However, science education can certainly be a 'sort of science', albeit an 'extraordinary' sort of science: organised to ensure that the adoption of diverse perspectives and methodologies is informed by a meta-methodology and so always based upon rational choices deriving from a sound understanding of the current state of knowledge in the field. Given the nature of educational phenomena, the convergent channelling of Kuhnian paradigms would be too limiting and restrictive. Yet giving researchers completely 'free range' to seek their own problem and develop their own original approaches to solve it – often seen as the path to academic recognition in the humanities – is unlikely to lead to optimum progress in addressing pressing educational problems. Lakatos's notion of RP offers a middle road here, as RP provide guidance to researchers about research priorities and allow the community to take stock of progress, without the blinkers of 'the' paradigm. The way in which educational researchers are commonly trained to develop their projects, with a strong open-ended phase to creatively consider divergent options before making rational and justifiable methodological choices, can be framed (i.e. guided, but not prematurely constrained) within the heuristic guidance of a progressive RP. This would allow the principled development of research designs on a problem-by-problem basis, but guided by the heuristics of an established tradition that the research community considers to be progressive. Arguably, that offers a 'sort of science' that best suits the field of science education.

References

Adams, P. (2011). (Dis)continuity and the Coalition: primary pedagogy as craft and primary pedagogy as performance. *Educational Review, 63*(4), 467–483.

Adey, P. (1999). *The Science of Thinking, and Science For Thinking: a description of Cognitive Acceleration through Science Education (CASE)*. Geneva: International Bureau of Education (UNESCO).

Agar, M. H. (2001). Ethnography. In N. J. Smelser & P. B. Baltes (Eds.), *International Encyclopedia of the Social & Behavioral Sciences* (pp. 4857–4862).

Aikenhead, G. S. (2006). *Science Education for Everyday Life: Evidence-based practice*. New York: Teachers College Press.

Allen, D., Donham, R., & Tanner, K. (2004). Approaches to Biology Teaching and Learning: Lesson Study—Building Communities of Learning Among Educators. *Cell Biology Education, 3*, 1–7.

Bachelard, G. (1940/1968). *The Philosophy of No: a philosophy of the scientific mind*. New York: Orion Press

Barrett, L. F. (2009). The Future of Psychology: Connecting Mind to Brain. *Perspectives on Psychological Science, 4*(4), 326–339.

Bassey, M. (1992). Creating education through research. *British Educational Research Journal, 18*(1), 3–16.

Basu, S. J. (2008). Powerful learners and critical agents: The goals of five urban Caribbean youth in a conceptual physics classroom. *Science Education, 92*(2), 252–277.

Biddle, B. J., & Anderson, D. S. (1986). Theory, methods, knowledge and research on teaching. In M. C. Wittrock (Ed.), *Handbook of Research on Teaching* (3rd ed., pp. 230–252). New York: Macmillan.

Biesta, G. J. J., & Burbules, N. C. (2003). *Pragmatism and Educational Research*. Lanham, MD: Rowman & Littlefield Publishers.

Bliss, J. (1993). The relevance of Piaget to research into children's conceptions. In P. J. Black & A. M. Lucas (Eds.), *Children's Informal Ideas in Science* (pp. 20–44). London: Routledge.

Bliss, J. (1995). Piaget and after: the case of learning science. *Studies in Science Education, 25*, 139–172.

Bowers, C. A. (2007). *The False Promises of Constructivist Theories of Learning: A global and ecological critique*. New York: Peter Lang Publishing.

British Educational Research Association. (2004). *Revised ethical guidelines for educational research*. Southwell, Nottinghamshire: British Educational Research Association.

British Educational Research Association. (2011). *Ethical Guidelines for Educational Research*. London: British Educational Research Association.

Bruner, J. S. (1987). The transactional self. In J. Bruner & H. Haste (Eds.), *Making Sense: the child's construction of the world* (pp. 81–96). London: Routledge.

Clark, C. (2005). The structure of educational research. *British Educational Research Journal, 31*(3), 289–309.

Claxton, G. (1993). Minitheories: a preliminary model for learning science. In P. J. Black & A. M. Lucas (Eds.), *Children's Informal Ideas in Science* (pp. 45–61). London: Routledge.

Coll, R. K., & Taylor, T. G. N. (2001). Using constructivism to inform chemistry pedagogy. *Chemistry Education: Research & Practice in Europe, 2*(3), 215–226.

Creswell, J. W. (1994). *Research Design: quantitative and qualitative approaches*. London: Sage.

Creswell, J. W. (2009). *Research Design: Qualitative, quantitative, and mixed methods approaches* (3rd ed.). Thousand Oaks, California: Sage Publications.

Creswell, J. W., & Plano Clark, V. L. (2007). *Designing and Conducting Mixed Methods Research*. Thousand Oaks, California: Sage.

Cromer, A. (1997). *Connected knowledge: science, philosophy and education*. Oxford: Oxford University Press.

Crotty, M. (1998). *The Foundations of Social Research: Meaning and perspective in the research process*. London: Sage.

Darwin, C. (1859/1968). *The Origin of Species by Means of Natural Selection, or the preservation of favoured races in the struggle for life*. Harmondsworth, Middlesex: Penguin.

Davidsson, E., & Jakobsson, A. (2008). Staff Members' Ideas about Visitors' Learning at Science and Technology Centres. *International Journal of Science Education, 31*(1), 129–146.

de Jong, O. (2000). Crossing the borders: chemical education research and teaching practice. Chemistry Education Research Group Lecture 1999. *University Chemistry Education, 4*(1), 29–32.

Dobzhansky, T. (1935). A Critique of the Species Concept in Biology. *Philosophy of Science, 2*(3), 344–355.

Donaldson, M. (1978). *Children's Minds*. London: Fontana

Driver, R., & Easley, J. (1978). Pupils and paradigms: a review of literature related to concept development in adolescent science students. *Studies in Science Education, 5*, 61–84.

Driver, R., & Erickson, G. (1983). Theories-in-action: some theoretical and empirical issues in the study of students' conceptual frameworks in science. *Studies in Science Education, 10*, 37–60.

Driver, R., Squires, A., Rushworth, P., & Wood-Robinson, V. (1994). *Making Sense of Secondary Science: research into children's ideas*. London: Routledge.

Duit, R. (2009). *Bibliography - Students' and Teachers' Conceptions and Science Education*. Kiel: http://www.ipn.uni-kiel.de/aktuell/stcse/stcse.html.

Duit, R., Roth, W.-M., Komorek, M., & Wilbers, J. (1998). Conceptual change cum discourse analysis to understand cognition in a unit on chaotic systems: towards an integrative perspective on learning in science. *International Journal of Science Education, 20*(9), 1059–1073.

Ebenezer, J. V., & Erickson, G. L. (1996). Chemistry students' conceptions of solubility: A phenomenography. *Science Education, 80*(2), 181–201.

Fellows, N. J. (1994). A window into thinking: Using student writing to understand conceptual change in science learning. *Journal of Research in Science Teaching, 31*(9), 985–1001.

Fensham, P. J. (2004). *Defining an Identity: The evolution of science education as a field of research*. Dordrecht: Kluwer Academic Publishers.

Feyerabend, P. (1988). *Against Method* (Revised ed.). London: Verso.

Fischler, H. (2011). Didaktik - An appropriate framework for the professional work of science teachers. In D. Corrigan, J. Dillon & R. F. Gunstone (Eds.), *The Professional Knowledge Base of Science Teaching* (pp. 31–50). Dordrecht: Springer.

Ganaras, K., Dumon, A., & Larcher, C. (2008). Conceptual integration of chemical equilibrium by prospective physical sciences teachers. *Chemistry Education Research and Practice, 9*(3), 240–249.

Gardner, H. (1973). *The Quest for Mind: Piaget, Levi-Strauss and the Structuralist Movement* New York: Alfred A. Knopf,.

Gilbert, J. K., & Swift, D. J. (1985). Towards a Lakatosian analysis of the Piagetian and alternative conceptions research programs. *Science Education, 69*(5), 681–696.

Gilbert, J. K., & Watts, D. M. (1983). Concepts, misconceptions and alternative conceptions: changing perspectives in science education. *Studies in Science Education, 10*, 61–98.

Glaser, B. G., & Strauss, A. L. (1967). *The Discovery of Grounded Theory: strategies for qualitative research*. New York: Aldine de Gruyter.

Glasersfeld, E. v. (1989). Cognition, construction of knowledge, and teaching. *Synthese, 80*(1), 121–140.

Greca, I. M., & Moreira, M. A. (1997). The kinds of mental representations - models, propositions and images - used by college physics students regarding the concept of field. *International Journal of Science Education, 19*(6), 711–724.

Grimmett, P. P., & MacKinnon, A. M. (1992). Craft Knowledge and the Education of Teachers. *Review of Research in Education, 18*, 385–456.

Guba, E. G., & Lincoln, Y. S. (2005). Paradigmatic controversies, contradictions, and emerging confluences. In N. K. Denzin & Y. S. Lincoln (Eds.), *The Sage Handbook of Qualitative Research* (3rd ed., pp. 191–215). Thousand Oaks, California: Sage.

Hammersley, M., & Atkinson, P. (2007). *Ethnography: Principles in practice* (3rd ed.). London: Routledge.

Harlen, W. (1999). *Effective Teaching of Science: a review of research*. Edinburgh: Scottish Council for Research in Education.

Hennessy, S. (1993). Situated cognition and cognitive apprenticeship: implications for classroom learning. *Studies in Science Education, 22*, 1–41.

Hodson, D. (2009). *Teaching and learning about science: Language, theories, methods, history, traditions and values*. Rotterdam, The Netherlands: Sense Publishers.

Holbrook, J., & Rannikmae, M. (2007). The Nature of Science Education for Enhancing Scientific Literacy. *International Journal of Science Education, 29*(11), 1347–1362.

Hull, D. L. (1965). The Effect of Essentialism on Taxonomy - Two Thousand Years of Stasis (I). *The British Journal for the Philosophy of Science, 15*(60), 314–326.

Jevons, F. R. (1973). *Science Observed: Science as a social and intellectual activity*. London: George Allen & Unwin.

Johnson, R. B., & Onwuegbuzie, A. J. (2004). Mixed Methods Research: A Research Paradigm Whose Time Has Come. *Educational Researcher, 33*(7), 14–26.

Johnson, R. B., Onwuegbuzie, A. J., & Turner, L. A. (2007). Toward a Definition of Mixed Methods Research. *Journal of Mixed Methods Research, 1*, 112–133.

Johnstone, A. H. (2000). Teaching of Chemistry - logical or psychological? *Chemistry Education: Research and Practice in Europe, 1*(1), 9–15.

Kagan, J. (2009). *The Three Cultures: Natural sciences, social sciences, and the humanities in the 21st Century*. Cambridge: Cambridge University Press.

Kelly, G. (1963). *A Theory of Personality: The Psychology of Personal Constructs*. New York: W W Norton & Company.

Keller, E. F. (1983). A Feeling for the Organism: The Life and Work of Barbara McClintock. New York: W H Freeman and Company.

Kind, V., & Taber, K. S. (2005). *Science: Teaching School Subjects 11–19*. London: RoutledgeFalmer.

Kirschner, P. A., Sweller, J., & Clark, R. E. (2006). Why minimal guidance during instruction does not work: An analysis of the failure of constructivist, discovery, problem-based, experiential, and inquiry-based teaching. *Educational Psychologist, 41*(2), 75–86.

Klahr, D. (2010). Coming Up for Air: But is it Oxygen or Phlogiston? A Response to Taber''s Review of Constructivist Instruction: Success or Failure? . *Education Review, 13*(13), 1–6. Retrieved from http://www.edrev.info/essays/v13n13.pdf

Koballa, T., Graber, W., Coleman, D. C., & Kemp, A. C. (2000). Prospective gymnasium teachers' conceptions of chemistry learning and teaching. *International Journal of Science Education, 22*(2), 209–224.

Kortland, K., & Klaassen, K. (2010). *Designing Theory-Based Teaching-Learning Sequences for Science Education: Proceedings of the symposium in honour of Piet Lijnse at the time of his retirement as professor of physics didactics at Utrecht University*. Utrecht: CDBeta Press - Freudenthal Institute for Science and Mathematics Education.

Kuhn, T. S. (1959/1977). The essential tension: tradition and innovation in scientific research. In T. S. Kuhn (Ed.), *The Essential Tension: Selected studies in scientific tradition and change* (pp. 225–239). Chicago: University of Chicago Press.

Kuhn, T. S. (1961/1977). The function of measurement in modern physical science. In T. S. Kuhn (Ed.), *The Essential Tension: Selected studies in scientific tradition and change* (pp. 178–224). Chicago: University of Chicago Press.

Kuhn, T. S. (1970). *The Structure of Scientific Revolutions* (2nd ed.). Chicago: University of Chicago.

Kuhn, T. S. (1973/1977). Objectivity, value judgement, and theory choice. *The Essential Tension: Selected studies in scientific tradition and change* (pp. 320–339). Chicago: The University of Chicago Press.

Kuhn, T. S. (1974/1977). Second thoughts on paradigms. In T. S. Kuhn (Ed.), *The Essential Tension: Selected studies in scientific tradition and change* (pp. 293–319). Chicago: University of Chicago Press.

Kuhn, T. S. (1996). *The Structure of Scientific Revolutions* (3rd ed.). Chicago: University of Chicago.

Kuiper, J. (1994). Student ideas of science concepts: alternative frameworks? *International Journal of Science Education, 16*(3), 279–292.
Lakatos, I. (1970). Falsification and the methodology of scientific research programmes. In I. Lakatos & A. Musgrove (Eds.), *Criticism and the Growth of Knowledge* (pp. 91–196). Cambridge: Cambridge University Press.
Lakatos, I., & Musgrove, A. (Eds.). (1970). *Criticism and the Growth of Knowledge*. Cambridge: Cambridge University Press.
Laugksch, R. C. (2000). Scientific literacy: a conceptual overview. *Science Education, 84*, 71–94.
Laurillard, D. (2012). *Teaching as a Design Science: Building Pedagogical Patterns for Learning and Technology*. London: Routledge.
Lave, J., & Wenger, E. (1991). *Situated Cognition: Legitimate peripheral participation*. Cambridge: Cambridge University Press.
Limerick, B., Burgess-Limerick, T., & Grace, M. (1996). The politics of interviewing: power relations and accepting the gift. *International Journal of Qualitative Studies in Education, 9*(4), 449–460.
Long, D. E. (2011). *Evolution and religion in American Education: An ethnography*. Dordrecht: Springer.
Machamer, P., Pera, M., & Baltas, A. (2000). *Scientific Controversies: Philosophical and historical perspectives*. New York: Oxford University Press.
Marek, E. A. (2009). Genesis and evolution of the learning cycle. In W.-M. Roth & K. Tobin (Eds.), *Handbook of Research in North America* (pp. 141–156). Rotterdam, The Netherlands: Sense Publishers.
Marton, F. (1981). Phenomenography - Describing conceptions of the world around us. *Instructional Science, 10*, 177–200.
Masterman, M. (1970). The Nature of a Paradigm. In I. Lakatos & A. Musgrove (Eds.), *Criticism and the Growth of Knowledge* (pp. 59–89). Cambridge: Cambridge University Press.
Matthews, M. R. (1993). Constructivism and science education: some epistemological problems. *Journal of Science Education and Technology, 2*(1), 359–370.
Matthews, M. R. (1994/2014). *Science Teaching: The role of history and philosophy of science*. London: Routledge.
Matthews, M. R. (2002). Constructivism and Science Education: A Further Appraisal. *Journal of Science Education and Technology, 11*(2), 121–134.
Matthews, M. R. (2009). Teaching the Philosophical and Worldview Components of Science. *Science & Education, 18*(6), 697–728.
Mayr, E. (1987). The ontological status of species: scientific progress and philosophical terminology. *Biology and Philosophy, 2*(2), 145–166.
McCulloch, G. (2002). 'Disciplines Contributing to Education'? Educational Studies and the Disciplines. *British Journal of Educational Studies, 50*(1), 100–119.
McNiff, J. (1992). *Action Research: principles and practice*. London: Routledge.
Medawar, P. B. (1963/1990). Is the scientific paper a fraud? In P. B. Medawar (Ed.), *The Threat and the Glory* (pp. 228–233). New York: Harper Collins, 1990.
Millar, R. (1989). Constructive criticisms. *International Journal of Science Education, 11*(special issue), 587–596.
Millar, R., & Osborne, J. (1998). *Beyond 2000: Science education for the future*. London: King's College.
Oancea, A. (2005). Criticisms of educational research: key topics and levels of analysis. *British Educational Research Journal, 31*(2), 157–183.
Osborne, R. J., & Wittrock, M. (1985). The generative learning model and its implications for science education. *Studies in Science Education, 12*, 59–87.
Pedretti, E. (2002). T. Kuhn Meets T. Rex: Critical Conversations and New Directions in Science Centres and Science Museums. *Studies in Science Education, 37*(1), 1–41.
Pesic, P. (1999). Wrestling with Proteus: Francis Bacon and the "Torture" of Nature. *Isis, 90*(1), 81–94.

Phillips, D. C. (1983). On describing a student's cognitive structure. *Educational Psychologist, 18*(2), 59–74.
Piaget, J. (1929/1973). *The Child's Conception of The World* (J. Tomlinson & A. Tomlinson, Trans.). St. Albans: Granada.
Piaget, J. (1970/1972). *The Principles of Genetic Epistemology* (W. Mays, Trans.). London: Routledge & Kegan Paul.
Piaget, J., & Garcia, R. (1989). *Psychogenesis and the History of Science* (H. Feider, Trans.). New York: Columbia University Press.
Piantanida, M., & Garman, N. B. (2009). *The Qualitative Dissertation: A guide for students and faculty* (2nd ed.). Thousand Oaks, California: Corwin Press.
Polanyi, M. (1962/1969). The unaccountable element in science. In M. Greene (Ed.), *Knowing and Being: Essays by Michael Polanyi* (pp. 105–120). Chicago: University of Chicago.
Pope, M. L., & Gilbert, J. K. (1983). Personal experience and the construction of knowledge in science. *Science Education, 67*(2), 193–203.
Popper, K. R. (1934/1959). *The Logic of Scientific Discovery*. London: Hutchinson.
Popper, K. R. (1989). *Conjectures and Refutations: The Growth of Scientific Knowledge,* (5th ed.). London: Routledge.
Popper, K. R. (1994). The myth of the framework. In M. A. Notturno (Ed.), *The Myth of the Framework: In defence of science and rationality* (pp. 33–64). Abingdon, Oxon.: Routledge.
Pring, R. (2000). *Philosophy of Educational Research*. London: Continuum.
Reiss, M. J. (2000). *Understanding Science Lessons: Five Years of Science Teaching*. Buckingham.: Open University Press.
Robson, C. (2002). *Real World Research: A resource for social scientists and practitioner researchers* (2nd ed.). Malden, Massachusetts: Blackwell.
Roth, W.-M., & Tobin, K. (2006). Editorial: Announcing Cultural Studies Of Science Education. *Cultural Studies of Science Education, 1*(1), 1–5.
Russell, T., & Osborne, J. (1993). *Constructivist Research, Curriculum Development and Practice in Primary Classrooms: Reflections on Five Years of Activity in the Science Processes and Concept Exploration (SPACE) Project*. Paper presented at the Third International Seminar on Misconceptions in the Learning of Science and Mathematics.
Sacks, O. (1995). An Anthropologist on Mars. London: Picador.
Scerri, E. R. (2003). Philosophical confusion in chemical education research. *Journal of Chemical Education, 80*(20), 468–474.
Scerri, E. R. (2010). Comments on a recent defence of constructivism in chemical education. *Chemistry Education in New Zealand* (November), 15–18.
Scerri, E. R. (2012). Some Comments Arising from a Recent Proposal Concerning Instrumentalism and Chemical Education. *Journal of Chemical Education*.
Shulman, L. S. (1986). Paradigms and research programs in the study of teaching: A contemporary perspective. In M. C. Wittrock (Ed.), *Handbook of Research on Teaching* (3rd ed.). New York: Macmillan Publishing Company.
Smardon, R. (2009). Sociocultural and cultural-historical frameworks for science education. In W.-M. Roth & K. Tobin (Eds.), *The World of Science Education: Handbook of Research in North America* (pp. 15–25). Rotterdam, The Netherlands: Sense Publishers.
Solomon, J. (1987). Social influences on the construction of pupils' understanding of science. *Studies in Science Education, 14*(63–82).
Solomon, J. (1992). *Getting to Know about Energy - in School and Society*. London: Falmer Press.
Solomon, J. (1993a). Four frames for a field. In P. J. Black & A. M. Lucas (Eds.), *Children's Informal Ideas in Science* (pp. 1–19). London: Routledge.
Solomon, J. (1993b). The social construction of children's scientific knowledge. In P. Black & A. M. Lucas (Eds.), *Children's Informal Ideas in Science* (pp. 85–101). London: Routledge.
Solomon, J. (1994). The rise and fall of constructivism. *Studies in Science Education, 23*, 1–19.
Springer, K. (2010). *Educational Research: A contextual approach*. Hoboken, New Jersey: Wiley.
Stake, R. E. (2000). The case study method in social enquiry. In R. Gomm, M. Hammersley & P. Foster (Eds.), *Case Study Method: Key issues, key texts*. London: Sage.

Stocklmayer, S. M., Rennie, L. J., & Gilbert, J. K. (2010). The roles of the formal and informal sectors in the provision of effective science education. *Studies in Science Education, 46*(1), 1–44.

Sutherland, P. (1992). *Cognitive Development Today: Piaget and his critics*. London: Paul Chapman Publishing.

Symonds, J. E., & Gorard, S. (2008). *The Death of Mixed Methods: Research Labels and their Casualties*. Paper presented at the British Educational Research Association Annual Conference.

Taber, K. S. (1993). *Stability and lability in student conceptions: some evidence from a case study*. Paper presented at the British Educational Research Association Annual Conference. Retrieved from http://www.leeds.ac.uk/educol/documents/154054.htm

Taber, K. S. (1995). An analogy for discussing progression in learning chemistry. *School Science Review, 76* (276), 91–95.

Taber, K. S. (2000). Case studies and generalisability - grounded theory and research in science education. *International Journal of Science Education, 22*(5), 469–487.

Taber, K. S. (2002). "Intense, but it's all worth it in the end": the colearner's experience of the research process. *British Educational Research Journal, 28*(3), 435–457.

Taber, K. S. (2003). The atom in the chemistry curriculum: fundamental concept, teaching model or epistemological obstacle? *Foundations of Chemistry, 5*(1), 43–84.

Taber, K. S. (2006a). Beyond Constructivism: the Progressive Research Programme into Learning Science. *Studies in Science Education, 42*, 125–184.

Taber, K. S. (2006b). Constructivism's new clothes: the trivial, the contingent, and a progressive research programme into the learning of science. *Foundations of Chemistry, 8*(2), 189–219.

Taber, K. S. (2007). *Classroom-based research and evidence-based practice: a guide for teachers*. London: Sage.

Taber, K. S. (2008). Exploring conceptual integration in student thinking: evidence from a case study. *International Journal of Science Education, 30*(14), 1915–1943.

Taber, K. S. (2009a). Learning from experience and teaching by example: reflecting upon personal learning experience to inform teaching practice. *Journal of Cambridge Studies, 4*(1), 82–91.

Taber, K. S. (2009b). *Progressing Science Education: Constructing the scientific research programme into the contingent nature of learning science*. Dordrecht: Springer.

Taber, K. S. (2010a). Constructivism and Direct Instruction as Competing Instructional Paradigms: An Essay Review of Tobias and Duffy's Constructivist Instruction: Success or Failure? *Education Review, 13*(8), 1–44. Retrieved from http://www.edrev.info/essays/v13n8index.html

Taber, K. S. (2010b). Constructivist pedagogy is superior – it's a matter of definition. *Advanced Distributed Learning Newsletter for Educators and Educational Researchers,* (October 2010). Retrieved from http://research.adlnet.gov/newsletter/academic/201010.htm

Taber, K. S. (2010c). Straw men and false dichotomies: Overcoming philosophical confusion in chemical education. *Journal of Chemical Education, 87*(5), 552–558.

Taber, K. S. (2011). The natures of scientific thinking: creativity as the handmaiden to logic in the development of public and personal knowledge. In M. S. Khine (Ed.), *Advances in the Nature of Science Research - Concepts and Methodologies* (pp. 51–74). Dordrecht: Springer.

Taber, K. S. (2012a). Prioritising paradigms, mixing methods, and characterising the 'qualitative' in educational research. *Teacher Development, 16*(1), 125–138.

Taber, K. S. (2012b). Recognising quality in reports of chemistry education research and practice. *Chemistry Education Research and Practice, 13*(1), 4–7.

Taber, K. S. (2013). *Classroom-based research and evidence-based practice: An Introduction* (2nd ed.). London: Sage.

Taber, K. S. (2014). *Modelling learners and learning in science education: Developing representations of concepts, conceptual structure and conceptual change to inform teaching and research.*

Thagard, P. (1992). *Conceptual Revolutions*. Oxford: Princeton University Press.

Tiberghien, A. (2012). What is theoretical in the design of teaching-learning sequences. *Studies in Science Education, 48*(2), 223–228.

Tobias, S., & Duffy, T. M. (Eds.). (2009). *Constructivist Instruction: Success or failure?* New York: Routledge.

Treagust, D. F. (1988). Development and use of diagnostic tests to evaluate students' misconceptions in science. *International Journal of Science Education, 10*(2), 159–169.
Verhoeff, R. P., Waarlo, A. J., & Boersma, K. T. (2008). Systems Modelling and the Development of Coherent Understanding of Cell Biology. *International Journal of Science Education, 30*(4), 543–568.
Vygotsky, L. S. (1934/1986). *Thought and Language*. London: MIT Press.
Watson, J. B. (1924/1998). *Behaviorism*. New Brunswick, New Jersey: Transaction Publishers.
Watson, J. B. (1967). What is Behaviourism? In J. A. Dyal (Ed.), *Readings in Psychology: Understanding human behavior* (2nd ed., pp. 7–9). New York: McGraw-Hill Book Company.
Yin, R. K. (2003). *Case Study Research: Design and Methods* (3rd ed.). Thousand Oaks, California: Sage.
Ziman, J. (1968). *Public Knowledge: an essay concerning the social dimension of science*. Cambridge: Cambridge University Press.

Keith S. Taber is Reader in Science Education and Chair of the Science, Technology and Mathematics Education Academic Group, in the Faculty of Education at the University of Cambridge (UK). He is the Editor of the journal *Chemistry Education Research and Practice*. After completing a chemistry degree and qualifying as a teacher of chemistry and physics, he taught in schools and further education during which time he undertook part-time research for an M.Sc. (a case study of girls' underrepresentation in physics classes) and Ph.D. (into conceptual progression in chemistry). His main area of research relates to conceptual learning and understanding of aspects of science, including the nature of science. As well as many contributions to research journals and practitioner journals, he has authored/coauthored/edited a number of books, including *Chemical Misconceptions* (Royal Society of Chemistry, 2001); *Science Education for Gifted Learners* (as Editor, Routledge, 2007); *Progressing Science Education* (Springer, 2009); and *Teaching Secondary Chemistry*, 2nd ed (as Editor, Hodder Education, 2012).

Chapter 58
History, Philosophy, and Sociology of Science and Science-Technology-Society Traditions in Science Education: Continuities and Discontinuities

Veli-Matti Vesterinen, María-Antonia Manassero-Mas, and Ángel Vázquez-Alonso

58.1 Introduction

Since the Sputnik crisis in the late 1950s, Western science education has been continuously concerned with the provision of quality training for task-force scientists. Successive innovations did not seem to change the essential aim of traditional school science education, that is, to prepare a small minority of students to become scientists. This approach was plagued with three major omnipresent failures: students' disaffection towards scientific subjects; mythical and distorted image of science conveyed to students, and failure of school science to make students learn science in a meaningful manner, to include the transmission of a view of science which did not account for the broader, sociopolitical sphere of production and application of scientific knowledge (Aikenhead 2006). A humanistic perspective in school science education was thus developed to articulate the range of social, cultural, historical, and political dimensions of science education and to challenge traditional ideas of science as a value-free enterprise (Aikenhead 2006; Donnelly 2004).

The purpose of this chapter is to account for the innovative processes that pioneered the introduction of content related to the history, philosophy, and sociology of science and technology into science education curricula. Our aim is twofold: first, to follow the development of the movements and labels which accompanied these innovative processes, in particular, the science-technology-society

V.-M. Vesterinen (✉)
Department of Chemistry, University of Helsinki, Helsinki, Finland
e-mail: veli-matti.vesterinen@helsinki.fi

M.-A. Manassero-Mas
Department of Psychology, University of the Balearic Islands, Palma de Mallorca, Spain

Á. Vázquez-Alonso
Department of Applied Pedagogy and Educational Psychology, University of the Balearic Islands, Palma de Mallorca, Spain

M.R. Matthews (ed.), *International Handbook of Research in History, Philosophy and Science Teaching*, DOI 10.1007/978-94-007-7654-8_58,

(STS) movement, which was born in the 1970s, the ongoing debates surrounding the concept of scientific literacy, introduced in 1950, and the more recent stream of studies on the nature of science and socioscientific issues; second, to show that, beyond labels and names, there are some underlying similarities across all such movements. Similarities are especially important when the continuity of the innovative process with its corresponding movements and ideas is highlighted over the potential differences. A revisit of the history, aims, debates, implementation, and research of the science-technology-society movement is an opportunity to better understand and connect the current proposals for science education (Millar and Osborne 1998).

The scope of the approach adopted to develop such purposes has some specific limitations and rationales. The reflections presented here are restricted to precollege science education where the innovations under scrutiny may have a large impact on citizens' scientific literacy, the public understanding of science, and yet their pedagogical implementation is still scarce in traditional school science curricula. In order to comply with the space limitations, some decisions have also been made regarding the spread of cases covered in our account of continuities and discontinuities under scrutiny. The method selected consists of a qualitative comparison and evaluation of the starting point of the follow-up (STS) and the current situation of humanistic practices in science education. A choice has been made for the latter in favor of the former-mentioned trio of movements (scientific literacy, nature of science, and socioscientific issues). The selection of paradigms draws on the appreciation of both their relevance in the current science education research and their strong relationship to the history, philosophy, and sociology of science tradition, which is the underlying criterion of the search for the (dis)continuities.

Summing up, the account starts from the analysis of the STS movement, as the first relevant implementation of the history, philosophy, and sociology of science tradition in science education, and then it progresses to the selected movements, as the current relevant areas where one can find the history, philosophy, and sociology of science tradition under new forms and contexts. Finally, a conclusion of the account highlights some current debates and outstanding issues in the field.

58.2 Science-Technology-Society (STS) Movement

The science-technology-society (STS) movement emerged from the social upheavals of the 1960s and early 1970s. In the social arena, the Cold War era around the 1950s had enlightened the consciousness of humankind of the likelihood of a global nuclear holocaust, while the environmental movement of 1960s brought to the forefront the increasing damage that industrialization had caused to the environment (e.g., acid rain). During the 1960s several academics and activists such as Lewis Mumford (1967–1970) and Jacques Ellul (1964) as well as consumer activist Ralph Nader (1965) began to express doubts about the presumed benevolence of the new technologies. Because science and engineering sat at the core of those issues, the global concern over these challenges definitively pointed to the responsibilities of scientists and engineers in solving them. In response to this state of affairs, science-technology-society courses were first established in engineering colleges to educate science and engineering students about the societal impact of their work (Cutcliffe 1990). The STS movement in higher education was both an academic field and a social movement, as the scholarly pursuit of science and technology studies was intertwined with the more activist stance of teachers and political organizers (see Cozzens 1993).

Eventually the movement had some effect also on school science education, as it was expected that a better school science education would allow not only scientists and engineers but also general citizens to cope effectively with these challenges. By the 1970s, some researchers became focused on developing materials addressing the complex relationships between science, technology, and society with the purpose of improving school science education. They worked on the idea that science would become more meaningful to students when its relationships with technology were made apparent and how technology, in turn, directed society and the reverse from society to science. A school science education movement called science-technology-society (STS) was born out of these initial efforts for teaching science in the broader context of the mutual relationships between science, technology, and society.

The STS approach involved also a shift from the positivist, non-contextual philosophical stance of traditional science education towards a contextualized, post-positivist view of science (and thus acknowledging the influences of technology, society, culture, ethics, politics, etc.). In the 1960s, the traditional, positivist image of science had been challenged in the academic arena by the historical, philosophical, and sociological analysis of scientists' work, thus providing a new and different image of science. At the time, science education was strictly based on the positivist view of science, whereby knowledge of the natural sciences was kept separate from current history, philosophy, and sociology of science. The emerging epistemological reflection supported by science and technology studies was thus a real challenge to traditional views on curricula and science textbooks. Emerging from the environmental, civil rights, and consumerism movements and preceded by new views of science presented in Thomas Kuhn's (1962) *The Structure of Scientific Revolutions* and post-WWII attention to history, philosophy, and sociology of science in science instruction,[1] Jim Gallagher (1971) proposed that understanding the interrelationships of science, technology, and society should be one of the main goals of school science education. The rationale for STS science education was clearly formulated in his description of a new goal for science education:

> For future citizens in a democratic society, understanding the relationships of science, technology and society may be as important as understanding the concepts and processes of science. (Ibid., p. 337)

[1] See, e.g., Conant (1957), Holton et al. (1970), Klopfer (1963), and Klopfer and Watson (1957).

Although the STS movement in higher education began as an academic endeavor for understanding the social issues linked to developments in science and technology, over the course of the decades, it has blended together with sociological and philosophical research about the development of science and technology to constitute an interdisciplinary field of science and technology studies (Sismondo 2010). Building on the constantly developing field of science and technology studies, as well as on research on science, environmental, and citizenship education, STS has remained a major theme in many school science curriculum reforms both in North America and in Europe at least since the publication of Paul Hurd's (1975) article *Science, Technology, and Society: New Goals for Interdisciplinary Science Teaching*. In fact, during the past 40 years, STS has evolved into an umbrella term that includes a wide variety of different views about the connections between science, technology, and environment and approaches to teaching science.[2] According to Glen Aikenhead (1994, 2006), the core conceptual framework for STS in school science education now incorporates two domains of science studies: *internal sociology of science*, interested in the social interactions between scientists and their communal, epistemic, and ontological values, and *external sociology of science*, interested in the interactions of science and scientists with the larger cultural milieu (see Ziman 1984). The emphasis placed on either of these two domains and whether the issues are discussed implicitly or explicitly have varied from project to project and program to program (Aikenhead 1994).

The multifaceted and complex nature of the STS movement has meant that the movement struggled to achieve any kind of internal consolidation. Rather it became the starter for the worldwide development of many slogans (such as socioscientific issues and nature of science), which can be deemed a sign of the developmental power of the STS approach. We argue that these slogans fit the very tenets of STS education, whose complex ecology has been labeled by Aikenhead (2003) in his fortunate title "A Rose by Any Other Name."

58.2.1 Environmental Education and Education for Sustainable Development

From the beginning, the STS movement has had its roots in the planetary fear for nuclear holocaust and in the growing alarm about the environmental impact of science and technology-based artifacts (e.g., weapons and polluting chemicals). The emergence of the environmental crisis in the 1960s documented by works such as Rachel Carson's (1962) *Silent Spring* and Paul Ehrlich's (1968) *The Population Bomb* caused increasing concern from the side of the public about the responsibilities of scientists and citizens, who manage these affairs. This concern had a profound effect also on educational trends. In 1977, the Intergovernmental Conference on Environment Education set the objectives for students' awareness,

[2] For more detailed history of the evolution of STS programs, see Aikenhead (2003).

knowledge, attitude, skills, and participation and defined the following goals for environmental education:

> 1. To foster clear awareness of, and concern about, economic, social, political and ecological inter-dependence in urban and rural areas.
> 2. To provide every person with opportunities to acquire the knowledge, values, attitudes, commitment and skills needed to protect and improve the environment.
> 3. To create new patterns of behavior of individuals, groups and society as a whole, towards the environment. (UNESCO 1977, p. 26)

When the term of "sustainable development" was first introduced in the Brundtland Report by the World Commission on Environment and Development (United Nations 1987), it linked the environmental problems to issues of global equity and justice, such as income and resource distribution, poverty alleviation, and gender equality. Gradually, global environmental problems such as climate change and biodiversity reduction also replaced local problems as the main areas of concern and public debate. This reframing of environmentalism had its effects on educational trends; instead of simply referring to environmental education, UNESCO and other international organizations began to promote education for sustainable development (see, e.g., Jones et al. 2010; UNESCO 2005).

The vocabulary of sustainable development also had an effect on scientific practice. Although the political and societal processes preponderantly shaped the sustainable development movement during the late 1980s and early 1990s, by the beginning of the twenty-first century, a new field of sustainability science was emerging (see, e.g., Kates et al. 2001). Promoted by international scientific programs, scientific academies, and independent networks of scientists, sustainability science seeks to understand the fundamental character of interactions between nature, science, technology, and society as well as to "facilitate the move toward a more just and sustainable world as part of the politics of the practical" (Carter 2008, p. 176). Sustainability science is a transdisciplinary approach defined more by the problems it addresses than by the disciplines it employs and it seeks to advance both knowledge and action towards a more sustainable world (Clark 2007). It acknowledges the contextualized, post-positivist view of science promoted by the post-Kuhnian science and technology studies, recognizes that techno-scientific practice cannot stand outside the nature-society system, and seeks to promote social change (Colucci-Gray et al. 2006; Colucci-Gray et al. 2012). Thus, it seeks to involve not only scientists but also "practitioners, and citizens in setting priorities, creating new knowledge, evaluating its possible consequences, and testing it in action" (Friibergh Workshop on Sustainability Science 2000).

With its roots deeply set in the environmental literature, the STS movement shared many common characteristics with environmental education and education for sustainable development. Some science educators even advocated integrating environmental education into more socioscientific issue-driven science-technology-society-environment (STSE) education with the goal of fostering a voice of active citizenship in students (e.g., Hodson 1994, 2003; Pedretti 1997, 2003). Also education for sustainable development shares the goal of preparing the students for civic engagement. This means not only providing the students with knowledge about socioscientific issues and models for informed choices in their everyday life but also building capacity to think critically about what the experts say and participate in the process of developing and testing suitable ideas (see Vare and Scott 2007). As sustainability science is still a relatively new field of study, it is only just beginning to have impact on the research and practice of science education (see, e.g., Carter 2008; Colucci-Gray et al. 2006, 2012).

58.2.2 Socioscientific Issues (SSI) Approach

In the past four decades, STS education has become a relatively complex and diffuse field, which displays a wide variety of approaches and some huge variations in the proportion of STS instruction devoted to societal issues: STS education can range from small text boxes infused into science textbooks to highly specialized courses addressing STS issues (see, e.g., Aikenhead 1994; Pedretti and Nazir 2011). The dissatisfaction with the lack of focused and functional models for STS instruction however was the promoter for new ways forward. One of them is the science instructional proposal called socioscientific issues (SSI).

Promoters of the SSI approach claim that STS education has been missing a coherent developmental or sociological framework and thus "has been relegated to brief mentions in current school science textbooks as well as in science teacher preparation texts" (Zeidler et al. 2005, p. 359). The SSI instruction stresses the factors associated with formal reasoning (argumentation) and the moral principles underlying science-based issues, and it focuses on controversial social issues with conceptual and/or procedural links to science and technology. The socioscientific cases used in SSI instruction are usually open-ended problems, the solutions to which can be informed by scientific principles, theories, and data, though they are not fully determined by scientific or technological considerations (Sadler 2011). Through cases students become involved in social argumentation and reflection aimed at developing cognitive, critical thinking skills and affective moral development.[3]

The STS and SSI movement share the goal of better preparing learners to engage in discussions and decisions related to socially relevant issues associated with science or technology. The SSI movement seems to share similar visions, tenets, and pedagogies with STS, although it may present and argue them differently, so that its promoters claim SSI are "beyond STS" (Zeidler et al. 2005). Much like the

[3] Proponents of SSI instruction have suggested various instructional models for utilizing these socioscientific case studies to better achieve these aims (see, e.g., Sadler 2011). For example, Pedretti (2003) suggested a pedagogical model developed from Ratcliffe (1997) and which includes the following stages:

1. Option: Identify alternative courses of action for an issue.
2. Criteria: Develop suitable criteria for comparing alternative actions.
3. Information: Clarify general and scientific knowledge/evidence for criteria.
4. Survey: Evaluate pros/cons of each alternative against criteria selected.
5. Choice: Make a decision based on the analysis undertaken.
6. Review: Evaluate decision-making process identifying feasible improvements.

STSE education, SSI emphasizes informed citizenry and even positions promoting citizenship as the primary goal of science education (Sadler 2011). The advocates of SSI have argued that STS and STSE approaches with similar goals do not give sufficient attention to ethical issues to help the development of the moral and emotional development of students (e.g., Zeidler et al. 2005). SSI is seen as a "broader term that subsumes all that STS has to offer, while also considering the ethical dimensions of science, the moral reasoning of the child, and the emotional development of the student" (Zeidler et al. 2002, p. 344). The advocates of SSI instruction maintain that the STS approach does not appropriately deal with scientific and technological controversies and ethical environmental dilemmas, because it does not exploit the inherent pedagogical power of discourse in socioscientific issues, such as reasoned argumentation, explicit nature of science considerations, as well as emotive, developmental, cultural, or epistemological connections within the issues themselves.

Though promoters of SSI claim that their stress on social, developmental, argumentative, and moral issues is important enough to deserve an epistemological demarcation from the STS approach, the points of similarity between the two approaches seem to surpass the points of distinctiveness. Despite their criticism towards STS, even the advocates of SSI widely acknowledge the parenthood of STS in relation to SSI (e.g., Tal and Kedmi 2006; Zeidler et al. 2005). Besides, because understanding scientific argumentation and justification are fundamental skills in decision making about socioscientific issues, SSI movement is also closely related to the notion of nature of science, discussed in more detail in Sect. 58.4.

58.2.3 Systematizations and Evaluations of STS and SSI

The works of systematization provided by Aikenhead (1994) and Pedretti and Nazir (2011) help to understand the complexity of the STS field. Aikenhead suggested an interesting taxonomy of STS education that classifies the wide variety of STS projects into a spectrum of eight categories, from traditional science to STS science. This taxonomy expresses the relative proportion of the innovative STS elements compared to traditional science content, the way these elements are presented, and the relative weight of STS content in the educational assessment.

1. *Motivation by STS content:* Just a mention of STS content in order to make a lesson more interesting and students are not assessed on the STS content.
2. *Casual infusion of STS content:* A short non-cohesive STS content is attached onto the traditional science topic. Students are superficially assessed on the STS content.
3. *Purposeful infusion of STS content:* A series of short cohesive STS content is systematically integrated into science topics. Part of the students' assessment includes STS content.
4. *Singular discipline through STS content:* STS content organizes and sequences traditional science content. Understanding STS content is assessed, though still less than science content.

5. *Science through STS content:* STS content dictates, organizes, and sequences multidisciplinary science content. Students' understanding of STS content is assessed, though still less than science content.
6. *Science along with STS content:* STS content is the focus; relevant science content enriches learning. Students are assessed equally on the STS and science content.
7. *Infusion of science into STS content:* STS content is the focus and broad scientific principles are mentioned. Students are primarily assessed on STS content and only partially on science content.
8. *STS content:* A major technology or social issue is studied and science content is only mentioned to make links to science. Students are not assessed on pure science content.

Further, Pedretti and Nazir's (2011) review of 40 years of research on STS education identified six currents in STS education: application/design, historical, logical reasoning, value-centered, sociocultural, and socioeconomic currents that explicitly reflect the philosophical, historical, and sociological basis of STS. The first three currents appear to place more emphasis on science-oriented issues, while the other three emphasize socially oriented issues:

1. *The application/design current* focuses on the link between science and technology and solving utilitarian problems through designing new technology or modifying existing technology (technical and inquiry skills). It combines cognitive skills with pragmatic, experiential, creative work in applying scientific knowledge.
2. *The historical current* highlights science as a human endeavor through understanding of the historical and sociocultural embeddedness of scientific ideas and scientists' work. It promotes the intrinsic values of science (exciting, interesting, and necessary pursuit) through affective, creative, and reflexive approaches, where STS and nature of science overlap.
3. *The logical reasoning current* addresses controversial socioscientific issues through the interactions between science, technology, society, and environment. It develops competences on understanding multiple perspectives, critical thinking, and decision making.
4. *The value-centered current* develops moral and ethical values tied to controversial socioscientific issues. Again, it develops understanding multiple perspectives, critical thinking, and decision making on affective and moral issues.
5. *The sociocultural current* addresses science and technology as social institutions to understand its internal organization and external links to politics, economics, and culture.
6. *The socio-ecojustice current* addresses the sociopolitical aspects of science and science education to educate civic responsibility that allows citizens to act on the social and ecological, local, and global problems of the world in search for justice.

By describing value-centered and ecojustice currents, it seems clear that in their review of the STS movement, also Pedretti and Nazir (2011) hold the underlying implicit assumption of a direct parenthood between STS and SSI movements.

After years of researching and teaching through STS materials, Bennett et al. (2007) undertook a systematic evaluation of the effects of context-based and STS approaches in the teaching of secondary science. The study reviews17 experimental studies from eight different countries and the overall findings indicate improvements in attitudes and motivation towards science, while the understanding of scientific concepts and ideas is comparable to that of conventional approaches. Specifically, the review suggests that there is reasonable evidence of the following:

- Students of both genders in classes using a context-based/STS approach held significantly more positive attitudes to science than peers in classes using a traditional approach.
- A context-based/STS approach to teaching science narrowed the gap between boys and girls in their attitudes to science.
- In cases where boys enjoyed the materials significantly more than girls, this was due to the nature of the practical work in the unit. In cases where girls enjoyed context-based materials significantly more than boys, this was because of the nonpractical activities in the unit.

The review also suggests there is some evidence of the following:

- Students in classes using a context-based approach perceived significantly more often a close link between science, technology, and society and showed significantly better conceptual understanding of science than their peers in traditional classes.
- Girls in classes using a context-based/STS approach developed a significantly more positive attitude towards taking a science career compared with boys in these classes.
- Girls in classes using a context-based/STS approach showed equal conceptual understanding of science as male peers in the same classes.
- Lower-ability pupils in classes using a context-based/STS approach held significantly more positive attitudes to science and better conceptual understanding of science than lower-ability pupils in classes using a traditional approach and better attitudes than high-ability peers in the same classes.

Despite the limitations and caveats of the reviewed studies, Bennett and colleagues (2007) acknowledge that the evidence is reliable and valid in supporting the use of contexts as a starting point in science teaching: there are considerable benefits in terms of attitudes to school science and no disadvantages in the development of understanding science.

Sadler (2004) reviewed the literature on SSI to assess its relationship with significant variables of learning scientific literacy, such as skills of informal reasoning and argumentation, conceptualizations of nature of science, evaluation of information, and the development of conceptual understanding of science content. The review by Sadler (2004) does not claim that students will become better informal thinkers, capable of analyzing complex arguments and of developing mature epistemologies of science, by simply being exposed to SSI. On the contrary, the review and further studies (see, e.g., Sadler 2011) consistently suggest that improvements in students' reasoning and argumentation are quite difficult to achieve, and in fact the findings show that reasoning that takes into account scientific evidence is a highly elusive aim as students easily ignore evidence when it is not in accordance with their own claims or previous attitudes. However, SSI studies provide examples of the relative stability of students' ways of argumentation and decision making, though these processes are so deeply rooted in their identity and culture that resist taking into account other elements that go against their own personal points of view. Further, SSI studies can provide an important stimulus for working on the informal reasoning and argumentation skills, the nature of science conceptualizations, the skills of evaluation of information, and the development of conceptual understanding of science content.

58.3 Scientific and Technological Literacy

Since it was first introduced over 50 years ago and especially in the last three decades, scientific literacy has become a central educational objective of science education worldwide (Hurd 1998; Oliver et al. 2001; Dillon 2009). In fact, scientific literacy has developed into an umbrella term covering most aims of science education (DeBoer 2000; Laugksch 2000). In recent years also technological literacy has gained grounds as a similar central tenet for modern technology education (Wonacott 2001). We argue that the STS movement and its derivatives have informed discussion on both scientific literacy and technological literacy. In the following subsections, we summarize the development of these two concepts and their underlining similarities as well as their connections with history, philosophy, and sociology of science, as well as with the STS movement and its derivatives.

58.3.1 Scientific Literacy

The launch of the Sputnik I in 1957 and the following science policy crisis in the United States had a profound effect on science education. The United States as well as countless other nations saw a spur of initiatives aimed at fostering new generations of engineers and scientists, and the number of science-oriented programs mushroomed. The main focus of such programs and initiatives was on training the most gifted students to become scientists and engineers. Although a concern for the public understanding of science dates back at least to the early years of the nineteenth century and had influential proponents such as the educational reformist John Dewey, the concept of scientific literacy as a goal for science education surfaced in the period of reindustrialization after the Second World War. The concept was introduced by Paul Hurd (1958) and Richard McCurdy (1958) at the height of the Sputnik crisis. The interest in the notion of scientific literacy following the Sputnik crisis was focused on improving public understanding and support for the scientific enterprise and industrial programs (see Fitzpatrick 1960; Waterman 1960).

In the 1960s and 1970s, the concept of scientific literacy was being debated, defined, and reconceptualized countless of times.[4] Much like the STS movement, formulations of scientific literacy in the 1960s and 1970s were inspired by new academic research on science and technology studies, which viewed science and technology as socially embedded enterprises, as well as environmental and civil rights movements. Many science educators were also disappointed about the outcomes of a school science education targeted towards a minority of students interested in continuing towards university science and engineering courses rather than developing the capabilities of all students to function as responsible citizens in a world increasingly affected by science and technology. Thus, the definitions of scientific literacy shared many educational goals with the STS movement. For example, Pella and colleagues (1966) suggested that scientific literacy comprises understanding of the basic notions of science as well as understanding the ethics embedded in the scientists' work, the interrelationships of science and society, and the differences between science and technology.

In the late 1970s and 1980s, the advocates of STS education began to dominate the discussion on scientific literacy (DeBoer 2000). The National Science Teachers' Association (NSTA) position statement from 1982 entitled *Science-Technology-Society: Science Education for the 1980s* stated that the goal of school science education was "to develop scientifically literate individuals who understand how science, technology, and society influence one another and who are able to use this knowledge in their everyday decision-making" (NSTA 1982, quoted in Yager 1996, p. 4). Several other countries had similar STS-based programs striving for scientific literacy, such as *Science in Society* (Lewis 1981) and *Science in a Social Context (SISCON)* (Solomon 1983) in the United Kingdom, *Project Leerpakket Ontwikkeling Natuurkunde (PLON)* (see Eijkelhof and Lijnse 1988) in the Netherlands, and *SciencePlus* (Atlantic Science Curriculum Project 1986, 1987, 1988) in Canada. One of the reasons for the growth of emphasis on STS approaches was the recognized failure of science education reforms with a theoretical disciplinary emphasis which were implemented since after the Sputnik crisis (Matthews 1994). The urgency of the need for a change of focus for science education was supported by the publication in 1983 of *A Nation at Risk* (National Commission on Excellence in Education 1983), which documented how, in spite of the efforts following the Sputnik crisis, a vast majority of students were still not interested in science and they learned very little science. This "science literacy crisis" urged for the adoption of more contextual approaches to science education and the goal of scientific literacy for all.

The disenchantment with the results of traditional programs was not the only driving force behind the change. Influenced by civil rights and environmental movements as well as the tradition of liberal education, researchers such as Chen and Novik (1984) and Thomas and Durant (1987) saw scientific literacy as a means to promote more democratic and equal decision making. Scholars still justify the need for a scientifically literate society upon rationales that evoke many of the aims

[4] See, for instance, Agin (1974), Daugs (1970), Gabel (1976), Klopfer (1969), O'Hearn (1976), Pella (1967), Pella et al. (1966), and Shen (1975).

of the STS approach, such as socioeconomic development, cultural development, personal autonomy, usefulness for everyday life, decision making, democratic participation in public issues related to science and technology, and ethical responsibility of scientists, technicians, politicians, and citizens (see, e.g., Laugksch 2000). Such reconceptualizations of scientific literacy towards scientific citizenship have laid the ground for new approaches such as STSE and SSI. The influence has been reciprocal, as proponents of STS and its derivatives have been very active in debates on the meaning and purpose of scientific literacy. For example, Zeidler and colleagues (2005) reconceptualize SSI elements of "functional scientific literacy," identifying four areas of pedagogical importance in supporting students' cognitive and moral development: (i) nature of science issues, (ii) classroom discourse issues, (iii) cultural issues, and (iv) care-based issues.

Many advocates of the STS movement and its derivatives go beyond traditional definitions of literacy as knowledge and skills and advocate social action as the highest goal of science education (DeBoer 2000). Much like in science studies, where in the new interdisciplinary fields such as the women's studies researchers began to see themselves as activists working for a change towards more equal and societally conscious science, also in science education some researchers began to advocate a similar activism oriented towards a more equitable and democratic society. These new perspectives on the meaning and purpose of science education have been influenced by a broad array of work from science studies, feminist studies, sociocultural theory, and critical pedagogy. Based on such reflections STS tradition has actively reconceptualized science education as an instrument of social and political engagement and sociopolitical action, which is evident, for example, in the works of Wolff-Michael Roth and Jacques Désautels (2002) and Wildson dos Santos (2008).

58.3.2 Scientific Literacy as Literacy for All

Much like what we saw with the STS movement, differences that appear between the various definitions of scientific literacy proposed by various specialists and the level of substantive disagreement about its content seem to highlight the complexity of concept (Bybee 1997; Gil and Vilches 2001; Manassero and Vázquez 2001). For example, various stakeholders interpret the word "literacy" in numerous different ways. In science education research, scientific literacy is defined from a variety of perspectives: as a motto, which covers a broad international movement (Aikenhead 2003); as a metaphor that expresses the aims and objectives of science education (Bybee 1997); and as a cultural myth that indicates the ideal to pursue (Shamos 1995).

To analyze the wide variety of meanings of scientific literacy, Roberts (2007) suggests a heuristic tool featuring two extreme positions, which he calls Vision I and Vision II. Vision I corresponds to the literacy within science, that is, the decontextualized products (facts, laws, theories, etc.) and processes of science. According to Vision I school science should give pupils knowledge and skills to approach situations as a professional scientist would. Vision II refers to literacy that a student would be likely to require as a citizen acting in situations outside the scientific world, or not entirely belonging to science, although clearly related to it. In Vision II the aim of the education is enabling students to approach situations as citizens who are well informed about science. Proponents of the STS movement and its derivatives have usually been positioned more towards Vision II of science education (see, e.g., Zeidler et al. 2005). Having these two poles of the heuristic tool in mind helps to analyze the relative proportion of these possibly conflicting goals in various definitions and descriptions of scientific literacy (much in a similar manner as the Aikenhead's STS categories help to analyze the amount of STS content in a project).

Paralleling the aims of general literacy (reading and writing), scientific literacy has since the science literacy crisis of the 1980s been increasingly associated with its complement of being a literacy "for all," especially in the school years before choosing a major. The concept of scientific literacy has thus been increasingly associated with its complement "science for all," assuming they are inseparable, but without clearly specifying what they mean, thus creating some confusion and debate. Tippins et al. (1999) argue that scientific literacy and science for all are potentially two contradictory concepts; on the one hand, the idea of science for all requires that no one is excluded from science education, thus creating a need for inclusive and meaningful school science, which has relevance to all students (see Vázquez et al. 2005; Vázquez and Manassero 2007); on the other hand, scientific literacy appears to be based on a certain set of knowledge, skills, and attitudes that students must seek, for example, the contents of the *Benchmarks* (American Association for the Advancement of Science 1993) or the *NSE Standards* (National Research Council 1996). It seems that the goal of science for all requires different contents of school science for diverse learners, while scientific literacy implies the idea that all science programs must meet the same set of criteria. Students can be taught notions of science apparently needed to acquire knowledge and skills required for scientific literacy, but it may be that learning such knowledge and skills will prove to be uninteresting and of little value to the students (Manassero and Vázquez 2001). Thus, there seems to be an obvious tension between scientific literacy and science for all. This tension is at the core of the debates about the goals of science education and, in general, of all basic education that must be common and inclusive (Acevedo et al. 2003; Tippins et al. 1999) and of the demands of higher education and the benchmarking needs for preparation for a career as a scientist (Abd-El-Khalick 2012).

From recent international research in science education and current educational reform initiatives, the progressive visions of scientific literacy encompass broad conceptual frameworks that entail basic knowledge of science (scientific facts, laws, and theories) and increased emphasis on the knowledge about science, that is, understanding about the processes and methods used to develop such knowledge (scientific inquiry) as well as the history, sociology, and epistemology of scientific knowledge (nature of science). In particular, students are expected to develop some specific scientific knowledge, skills, and attitudes, which involve understanding science as a "way of knowing" (absolutely necessary, if informed decisions are to be made), decision making on scientifically based personal and societal issues that increasingly confront the students, as well as the development of a commitment to the moral and ethical dimensions of science education that include the students' social and character development. Such decisions represent a functional degree of scientific literacy as they necessarily involve careful evaluation of scientific claims by discerning connections among evidence, inferences, and conclusions through the ability to analyze, synthesize, and evaluate information. A degree of scientific literacy also entails practice and experience in developing scientific attitudes such as skepticism, open-mindedness, critical thinking, recognizing multiple forms of inquiry, accepting ambiguity, searching for data-driven knowledge, dealing sensibly with moral reasoning and ethical issues, and understanding the connections that are inherent among socioscientific issues (Zeidler 2001).

Since scientific literacy is closely linked to social, cultural, and ideological aims of education, it is virtually impossible to establish a complete model of school science curricula to better achieve it. Although the aims, purposes, and objectives might be widely shared, it might be unrealistic to expect all students to achieve the same specific objectives. As different societies and social groups interact differently with science and technology, the standard-based curricula should consider only general references that should be developed in the classroom through specific contexts. In practice, scientific literacy can be grasped in different ways and with different levels of complexity to adapt to different contexts and students. However, this contextualization should keep the general framework and the principle of equity. We maintain that the paradigm of science-technology-society (STS) is best able to guide the selection of basic content that is relevant and useful for all students and also by providing us with methodological guidelines that contextualize into practice this important educational innovation (see Acevedo et al. 2003).

58.3.3 Technological Literacy

Traditional technology education was based on the industrial model of technology, which during the past decades and in the new era of information technology became apparently outdated. In a technologically mediated world, technological literacy requires understanding about the impact of new and emerging technologies on society and the environment (Dakers 2006). The notion of technological literacy has been established as the central tenet for modern technology education (Wonacott 2001).

Much like scientific literacy calls for a science education for all and focuses on the nature of science, processes of scientific research, and interaction of science and society, technological literacy calls for a technology education for all and focuses on the nature of technology, process of technological design, and interaction of technology and society (see, e.g., ITEA 2000). Again, much like with scientific literacy, the goal of technology education is increasingly seen in promoting more democratic and equal decision-making processes and social action. One of the most comprehensive classifications of the goals of technology education was the model based on functional competencies described by Layton (1993). The functional competencies included:

1. *Technological awareness (receiver competence):* The ability to recognize and acknowledge the possibilities of technology in use
2. *Technological application (user competence):* The ability to use technology
3. *Technological capability (maker competence):* The ability to design and make artifacts
4. *Technological impact assessment (monitoring competence):* The ability to assess the personal and social implications of use of technologies
5. *Technological consciousness (paradigmatic competence):* The ability to work within a "mental set," defining what constitutes a problem, circumscribes what counts as a solution, and prescribes the criteria which technological activity is to be evaluated
6. *Technological evaluation (critic competence):* The ability to judge the worth of technological development and to step outside the "mental set" to evaluate it

In the last 20 years, the focus of technology education has moved from providing vocational skills in technology to developing critical competencies in multiple techno-literacies, which include critical computer literacy and critical multimedia literacy (Kahn and Kellner 2006). Change of focus in aims of technology education resembles the change from Vision I science education to Vision II science education (see Roberts 2007). With such new goal for technological literacy, learning and teaching technology becomes a dialogue; the teacher and students form a community with no right or wrong answers, only more or less informed interpretations (Dakers 2006).

Much like with the definitions of scientific literacy, academic science-technology-society programs have influenced definitions of technological literacy. In the 1980s, researchers in science and technology studies turned their attention to technology. MacKenzie and Wajcman's (1985) *Social Shaping of Technology* and Bijker et al. (1987) *The Social Construction of Technological Systems* paved the way for sociological accounts of technological change, much like the turn to naturalistic accounts of scientific progress revolutionized the view of scientific practice more than a decade before. Emerging sociology of technology shared theoretical and methodological lines with sociology of science and supported the unity among science and technology studies.

There are also some common misconceptions concerning technology, for instance, that science is exogenous to technology. The view that scientific discovery leads to technological innovation is so strong that it is often forgotten that there is a notable reverse flow from technology to science. Technological innovations are not just products of science – they have enormous influence also on the process of science (Stokes 1997). In fact, technology plays a huge role in the process of creating scientific knowledge as scientists create all instruments, experimental settings, and even objects of research. Direct observations of scientific phenomena usually happen at a level unattainable to our perception and phenomena are accessed through the window of technology, with instruments especially designed to refine our current scientific models (Hacking 1983). The way scientific research is done has always been and still is transformed by technological development of instrumentation (Ziman 1984). Science and technology education should take cognizance of the essential interdependence of science and technology (see, e.g., Tala 2009). We argue that instead of speaking simply about "scientific literacy" and "technological literacy," we should rather use the notion "scientific and technological literacy," which better acknowledges this bidirectional relation between science and technology (see, e.g., Fourez 1997; Holbrook 1998).

58.4 Nature of Science (NOS)

In developing scientific and technological literacy, the meta-knowledge that arises from the interdisciplinary reflections on science and technology plays an integral part. One of the central elements of STS and scientific literacy is to understand what science is, how it works, and how scientists operate. As science studies have discussed these issues and should inform teaching practice, there is a vast and complex literature on history, philosophy, and sociology of science in science education (for an overview, see, e.g., Hodson 2008, 2009). Within science education, suitable educational answers to these questions have been described by various characterizations of nature of science (NOS). In the research literature NOS is gaining ground as the most common representation of the essentials of informed and updated picture of science. NOS issues involve the most relevant features of history, philosophy, and sociology of science: what science is, how it produces valid knowledge, how it relates to technology and society, who are the scientists, how they work and relate among each other, and so on (e.g., McComas et al. 1998). The aim of producing an authentic image of science was also an important pursuit of STS approach, and as a crucial element of scientific literacy, NOS is now widely recognized to be a key concept in the curricular aims of science education all over the world.[5]

School science curricula are often filled with simplistic visions that produce mythical and deformed views of science, as, for instance, the absolutism of scientific knowledge and the stereotypical step-by-step approach of the scientific method, which hinder learning appropriate views of science and appropriate teaching of NOS in the classroom. The very question of NOS and science education arises from the inadequate images of science that school curricula and textbooks convey to students (see, e.g., Vesterinen et al. 2009; 2011). This concern was early assumed by STS movement as crucial, though modest educational aim, and today it is incorporated within definitions of scientific literacy as "learning about science" (see, e.g.,

[5] See, for instance, Adúriz-Bravo and Izquierdo-Aymerich (2009), Hodson (2003), Matthews (2004), and McComas and Olson (1998).

Abd-El-Khalick 2012; Hodson 2009). This new aim goes beyond learning science: students should learn some basic epistemological, historical, and sociological traits of science and scientists in order to better understand how science works in the current world.

Understanding how science works is an important key for appraising scientific claims, evaluating scientific arguments, and forming a personal opinion on socioscientific issues. NOS knowledge needed for addressing socioscientific issues includes things such as the ability to distinguish between science-in-the-making, where uncertainty is to be expected, and ready-made science, on which we can rely, and the ability to recognize how sociocultural, political, economic, and religious factors can impact science, as well as the reverse (Kolstø 2001). Zeidler and colleagues (2005) even describe NOS as one of the four areas of pedagogical importance to the teaching of socioscientific issues within a curriculum striving towards scientific literacy.

58.4.1 Systematizations and Evaluations of NOS

NOS is a very complex concept, partly because it evolves and changes as our understanding of science and science itself evolves and changes. NOS brings together a variety of aspects coming from different disciplines, such as history, philosophy, sociology, and psychology of science (Vázquez et al. 2001). Thus, NOS is meta-knowledge arising from the interdisciplinary reflections on science, which have been conducted by specialists from a multitude of disciplines (Vázquez et al. 2004). In fact, NOS might present so many faces that Rudolph (2000) even contends that a single NOS does not exist at all.

Although there might not be a general agreement on the exact definition of NOS, there seems to be some sort of consensus regarding the central features of NOS that should be covered in science education (see, e.g., Lederman 2007; Niaz 2008). On the whole, the educational consensus on NOS refers to basic and relevant NOS features while keeping them highly uncontroversial: what is science; the methods science uses to construct, develop, validate, and disseminate the knowledge it produces; the features, activities, and values of the scientific community; and the internal and external links of science, such as the links between science, technology, and society. The most striking aspect of the consensual NOS features, which in turn provides further evidence in support to the teaching of NOS in schools, is the strong similarity among the different lists that researchers have proposed.[6]

In spite of consensus, there remains some divergence about the relative weight the consensual aspects should have on school science. Looking at the current mainstream NOS literature, NOS is mainly depicted as epistemology or philosophy of science, while the relationships with technology and society are much less taken

[6] See, for instance, Abd-El-Khalick (2012), Lederman (2007), McComas and Olson (1998), and Osborne et al. (2003).

into consideration. However, an in-depth reading of the mainstream current research on NOS also evidences that broader STS social relationships, especially those referred to the works and status of the scientific community, are also recognized as part of the NOS field.[7] For example, the descriptions of the central features of NOS include both philosophical perspectives on science and characteristics inherent to scientific knowledge (epistemology of science), such as the empirical and tentative nature of scientific knowledge and the key distinction between theories and laws, as well as the sociological perspectives on scientific practice, such as social dimensions and cultural embeddedness of science.[8] Even though some NOS research tends to reduce NOS to a few epistemological aspects, the core conceptual framework of STS seems compatible with broader consensual NOS issues, whose enlarged frame embraces much more relevant features and keeps NOS more faithful to its multifaceted and contentious character (Matthews 2012).

The decades of research on the NOS conceptions of students and teachers allow Lederman (1992) to affirm consistently that students and teachers do not have appropriate knowledge about NOS. Research in science education to improve learning and teaching about NOS has largely documented the difficulties associated with developing sound understanding of NOS. The complexity of this task is due to the amount of interacting factors involved that prevent, limit, or facilitate teaching NOS and clarifying the effectiveness of different methods. In spite of the difficulties, it seems that some necessary, though not sufficient, conditions are curriculum development (planning, developing, and assessing) and the effectiveness of teaching in the classroom. The several contexts for developing and teaching NOS used in these studies express the complementary importance of domain-general NOS features and the diversity of domain-specific implementations. Some domain-specific contexts involved in the studies that test the effectiveness of teaching are the following: practical activities (inquiry processes), specific courses on methods and philosophy of science and technology, history of science and technology, techno-scientific issues of social interest, and impregnation of traditional science and technology with NOS contents.

An important part of implementing these contexts is the verification of the effectiveness of a variety of NOS teaching methods, which can be summarized in two basic approaches (Abd-el-Khalick 2012; Lederman 2007):

1. *Implicit teaching:* NOS contents are implicitly inserted into classroom activities, without any further planning or discussion of them, which presumably leads to NOS learning as an automatic by-product of activities.
2. *Explicit instruction:* NOS contents are made explicit in the educational activity (curricular planning and meaningful objectives, content, and evaluation), and clear reflective applications are developed in the classroom through argumentation, development of metacognition, or conceptual change.

---

[7] See, for instance, Abd-el-Khalick (2012), Leach et al. (2003), Lederman (2007), Osborne et al. (2003), Sandoval (2005), Tsai and Liu (2005), and Vesterinen et al. (2011).

[8] See, for instance, Abd-El-Khalick (2012), Lederman et al. (2002), and Osborne et al. (2003).

The review of literature on the relative impact of implicit versus explicit approaches towards addressing NOS issues shows that implicit approaches are less effective than explicit reflective approaches (Abd-El-Khalick and Lederman 2000). As an explicit and context-based approach for teaching NOS, the history, sociology, and philosophy of science form the natural setting for discussion, because each of them shows how to build scientific knowledge in the social and historical context (Hodson 2008; Lederman 2007). A specific pedagogical attention to the epistemic and social aspects of inquiry through discursive argumentation activities is also stressed in a recent review of curricular interventions on changing students' NOS conceptions (Deng et al. 2011).

58.4.2 From NOS to NOST

As mentioned earlier, some scholars tend to reduce NOS features especially to philosophical values and epistemological characteristics of scientific knowledge. Although reductionist definitions of NOS could also improve skills, such as distinguishing between good science and bad science as well as the capacity to read and evaluate scientific texts, there is a need for a larger set of skills and attitudes to address science and technology-related issues in a critical way and to reach informed decisions on socioscientific issues impacting our society and the environment. From this wider perspective, it can be argued that there is a need also for a broader conception of NOS encompassing a wider variety of features, such as how science builds, validates, disseminates, and develops knowledge; what values are involved in scientific activities; which are the characteristics of the scientific community; and how science is related to society and culture.[9] Further, even though the roles of social and societal dimensions of science are sometimes cited as central features of NOS, the role of technology is often neglected in most definitions. In order to produce an authentic image of science, there is clearly a need to embrace both the science and technology as complementary aspects of contemporary scientific activity and emphasize the techno-science dimensions of NOS (see, e.g., Tala 2009; Vesterinen et al. 2011).

To tackle these issues, the rich history of the STS movement provides a variety of pluralistic educational models of science and technology, which help students and teachers to answer questions or to critically assess coexisting controversial views, rather than pursuing indoctrination into a particular model of science and technology such as memorizing a couple of uncontroversial tenets about NOS. Even though certain teachers might favor a misbalanced or limited presentation of science, either empiricist or relativist, or avoid questioning on a debatable issue (see Clough 2007), a more pluralistic model which includes several coexisting views of science and technology and their interactions should be provided (see

[9] See, for instance, Acevedo (2008), Matthews (2012), Vázquez et al. (2004), and Vesterinen et al. (2011).

Matthews 2012). To cover the enormous complexity of techno-scientific systems in contemporary societies, we argue that the concept of NOS should be extended as nature of science and technology (NOST), which would take into better consideration issues such as the ethical and democratic values of science and technology and solving social and ecological problems through human agency and action.

The former reflection does not imply that students become historians, philosophers, or sociologists of science and technology. Rather, understanding NOST through science education has to be developmentally adapted to students and contexts, to attain modest and realistic goals (see Abd-El-Khalick 2012; Matthews 1998). For instance, the students should acknowledge the history of science, the role of the scientific community in the production of scientific knowledge, and some basic features of the philosophy of science, to become competent in informal reasoning on issues at the coupling of science, society, technology, affectivity, ethics, moral development, and civic participation.

58.5 Conclusion

This paper aimed to reflect on the course of implementation of innovative humanistic approaches to school science starting from the leading role of the science-technology-society (STS) movement ahead. It aimed to set up and to follow up the continuities and discontinuities of this evolution through the new emergent currents for teaching and learning science that become associated with the history, philosophy, and sociology of science. Although there are several other slogans and labels for this kind of humanistic science education (see, e.g., Aikenhead 2006), the chosen approaches have been selected because of their most notable influence on school science education.

The leading role of the STS movement is the cornerstone of what has been known as humanistic science education and its many derivatives, such as discussion of socioscientific issues (SSI) and nature of science (NOS). In fact, the six STS currents identified by Pedretti and Nazir (2011) draw quite explicitly on the thesis of the global resemblance among STS, SSI, and NOS approaches. Although it can be argued that STS and the aforementioned research lines (SSI and NOS) are different educational trends, it seems quite obvious too that they all share the importance of developing broad key competencies for scientific and technological literacy and are deeply rooted in the history, philosophy, and sociology of science.

The early distinction between the two fields of science and technology studies provides the inspiration for looking at the clear and direct evolution and connection of STS and NOS approaches. The "science and technology studies" interested in understanding scientific and technological practice and discourse and the "science-technology-society studies" interested in understanding social issues linked to scientific and technological development at some point appeared to be separate projects (see Sismondo 2010). Based on this division, STS and its direct derivates such as STSE and SSI were mainly inspired by research on social issues linked to science and technology. Conversely, NOS was mainly influenced by science and technology studies focused on scientific practice. These two broad fields of science and technology studies have in recent decades blended together. As part of the clearer current understanding on the global interrelatedness of science-technology-society issues, also STS, SSI, and NOS are now being perceived as related and overlapping constructs: STS and SSI approaches include the understanding of NOS features as a central educational objective, and conversely, socioscientific issues are used as an educational context to teach NOS features.

Alternatively it can be argued that if STS and the aforementioned research lines are dominated by the overall continuities among them to such a degree, then they should not deserve different labels. The amount of literature accumulated on each of these areas and the definitions made by their promoters on their unique characteristics and demarcation criteria project an interested image of epistemological differentiation rather than continuity from the original STS (i.e., the promoters of SSI insist on the moral and character education of SSI as a key differential trait from STS). A balanced position on humanistic approaches to science education should recognize the continuities as well as their progress along the differential discontinuities. In spite of the continuities, the rising of the different labels from the original multiform STS proposals has certainly contributed to deeper progress of research by reframing more specific educational problems, such as literacy and scientific competence beyond process skills, contribution to moral education beyond general impact on society, argumentation on facts and evidences beyond critical thinking, and understanding specific consensual issues on NOS beyond generic understanding about science, just to mention some of the most significant milestones of this progress.

The proponents of STS and its more recent derivates have also been active in redefining scientific literacy, which has developed into an umbrella term covering most aims of science education (e.g., DeBoer 2000; Laugksch 2000). Currently all the aforementioned approaches fit under the banner of scientific and technological literacy. For example, Hodson (2009) presents a conceptualization of scientific and technological literacy aligning with the STS efforts to innovate science education by adequately dealing with many challenges of a scientifically and technologically mediated world. Inspired by description of functional competencies of technology by Layton (1993) and competency model of science literacy by Gräber and colleagues (2002), Hodson (2009, p. 15) describes four major elements of scientific and technological literacy as a goal for school science education:

- *Learning science and technology* – acquiring and developing conceptual and theoretical knowledge in science and technology and gaining familiarity with a range of technologies
- *Learning about science and technology* – developing an understanding of the nature, methods, and language of science and technology; appreciation of the history and development of science and technology; awareness of the interactions among science, technology, society, and environment; and sensitivity to the personal, social, economic, environmental, and ethical implications of technologies

- *Doing science and technology* – engaging in and developing expertise in scientific inquiry and problem solving as well as developing competence in tackling "real-world" technological tasks and problems
- *Engaging in sociopolitical action* – acquiring (through guided participation) the capacity and commitment to take appropriate, responsible, and effective action on science-/technology-related matters of social, economic, environmental, and moral-ethical concern and the willingness to undertake roles and responsibilities in shaping public policy related to scientific and technological developments at local, regional, national, and/or global levels

The STS movement has been instrumental in setting up a basic shift of understanding of the term scientific literacy towards scientific citizenship and its association with science for all. New conceptualizations of scientific and technological literacy, such as the one presented above, demand that science teaching can no longer be bound to the transmission of traditional scientific and technological knowledge aimed to prepare scientists and engineers, which has been the indisputable and perennial aim of science education all over the world. Science education should go beyond science facts and laws to get a more holistic and useful approach for citizens, most of whom do become neither scientists nor science workers. Literacy for all assumed the original concern of STS movement to include all students into the culture of science and technology, no matter the minority they belong to, through improving their interest and attitudes to science, in order for them to become better citizens and better prepared to participate in society. It is argued that science education simply cannot stick to the scientific knowledge and the process of producing such knowledge, but the goals and capacities to be developed should have a more holistic approach and a genuine social relevance, including ethical and democratic values (Holbrook 1998). Thus, the STS proposals have also evolved to improve their contribution to science education, not just drawn to acknowledge the pitfalls but also to attain new educational challenges (such as scientific and technological literacy for all) in new educational settings and societal contexts (information, communication, and risk societies).

The STS movement and its conceptual variations originated from and have been influenced by interdisciplinary science and technology studies in higher education as well as environmental and civil rights movements. These roots are all but forgotten in the current humanistic perspectives such as NOS and SSI. New research on history, philosophy, and sociology of science still informs research and practices on science education, and the capacity and commitment to engage in sociopolitical action on matters of social and environmental concern are increasingly seen as the key competencies developed through science education (see, e.g., Hodson 2009). The humanistic perspective challenges traditional science education in at least four ways:

1. *The switch of focus from "science for scientists" to "science for all" (from Vision I to Vision II):* Science education must be inclusive, that is, committed to the aim that all students (not only those aiming to be scientists) should understand contents "of" science (concepts) and concepts "about" science (NOS). Science

education should enculturate students into their local, national, and global communities, rather than into scientific disciplines.
2. *Focus on the relevance of learning:* To achieve the education that excludes no one and to produce significant learning that is useful for students' everyday life, science education is expected to be interesting and relevant. Science education should thus be context-driven and student-centered.
3. *Focus on the additional key competencies in science and technology:* The focus of science education cannot be solely on knowledge, skills and characteristics demanded of scientists, but rather on the knowledge, skills and characteristics demanded of general citizens. Such competencies include, but are not limited to, the following skills: high order cognitive abilities (argumentation, analyzing, synthesizing and evaluating information), informed decision making, communicating and discussing openly ideas and facts, and designing and developing projects.
4. *Focus on the history, philosophy, and sociology of science contents:* For example, the task of making informed decisions on socioscientific issues demands some knowledge "about" science. The history, philosophy, and sociology of science can be used either as core curriculum content or as the context that surrounds disciplinary learning activities in science.

The current concern for effective and evidence-based teaching and learning is perhaps the most practical drive for connecting history, philosophy, and sociology of science content and curricular developments in classrooms. Although the number of evaluative studies is quite scarce within the STS movement proper, as the study of Bennett and colleagues (2007) evidences, the evaluative studies available on scientific literacy (e.g., OECD's PISA project), NOS (e.g., Lederman 2007), or SSI (e.g., Sadler 2011) are essential to broaden our view. Evaluations of humanistic approaches in science education point out that success in this field requires some basic conditions. Most of the current literature is unanimous in acknowledging that developmental, explicit, and reflective teaching is a necessary condition to successful learning, which in turn must be supported by detailed curricular planning.[10] These plans should involve explicit objectives, activities, and assessment and many reflective activities within classroom implementation where students are given opportunities to consider, discuss, construct arguments, and consolidate their understandings on broad features of science.

To meet the wide variability of demands from primary education to college education students, curricular planning should also acknowledge a developmental perspective. For instance, the controversial and interrelated character of the history, philosophy, and sociology of science issues suggests that decisions on sequencing issues across grades are not obvious. In spite of this, the developmental perspective has been scarcely discussed in NOS and SSI research, perhaps supposing implicitly that teachers are able to adequately cope with it while transposing their practical

[10] See Abd-el-Khalick (2012), Bennett et al. (2007), Lederman (2007), Matthews (2012), Sadler (2011), and Vesterinen and Aksela (2012).

knowledge in science teaching to teach humanistic content. Thus, the previous prescriptions and conditions to achieve effective teaching and learning need to be complemented with accurate developmental sequencing of issues and activities (see, e.g., Abd-el-Khalick 2012).

Summing up, this chapter traces a journey through humanistic perspectives in science education from STS and scientific literacy movements to the current SSI and NOS movements, influenced by and influencing the various conceptualizations and reconceptualizations of scientific and technological literacy. Along this journey some might perceive a kind of reductionism, as the wide variety of historical, philosophical, or sociological controversial issues in science and technology education, which are typical of humanistic approaches (see Aikenhead and Ryan 1992), have been progressively focused on a much more specific and limited list of features. For instance, much of the research and teaching experiences on SSI issues focuses on current social controversies, thus relegating to a corner the many past historical controversies, which are also essential to sound training in science. Further, much research and teaching on NOS focus on seven issues elaborated by Lederman and his colleagues (e.g., Lederman 1992, Lederman et al. 2002), thus relegating out of the scope of scientific literacy wider visions of NOS and many NOS issues. This reductionism could help researchers to better demarcate the field of study, while in education this effort might be justified on the basis of meeting the students' needs, for example, in the selection and choice of curriculum issues that could better fit the diverse populations of students (simpler pedagogical transposition, best modest goals, students' age and interests, etc.). In the long run, however, this reduction somewhat weakens the richness of the original historical, philosophical, or sociological roots of humanistic approaches to science education, and claims for overcoming this reductionist state of affairs have already been put forward (see Matthews 2012).

Coinciding with the planned school science reforms, developed and implemented in many countries over recent years, the international debate over humanistic perspectives in higher education has also been revitalized. According to Sjöström (2008), there is a need for taking into account the societal and ethical dimensions of scientific practice and taking part in public discussion about the discipline by practicing scientists; in turn, this involves a need for more emphasis on Vision II-like approaches also in the education of future scientists. Krageskov Eriksen (2002) argues that with a humanistic perspective guiding the educational planning, scientists and engineers capable of critically considering the premises of the system, "the rules of the game" – not just skilled players – could be the intent of higher education. She argues for the need of three kinds of knowledge in tertiary science education: (1) "ontological" scientific knowledge, i.e., scientific facts and theories; (2) "epistemological" knowledge, i.e., philosophical and sociological perspectives on the scientific practice; and (3) "ethical" knowledge, i.e., reflection on the role of science and science education in society. The humanistic perspective originating from the engineering colleges and academic science-technology-society programs seems to be returning back to colleges and universities via the STS school science movement.

References

Abd-El-Khalick, F. (2012). Examining the sources for our understandings about science: Enduring conflations and critical issues in research on nature of science in science education, *International Journal of Science Education,* 34(3), 353–374.

Abd-El-Khalick, F., & Lederman, N. G. (2000). Improving science teachers' conceptions of nature of science: A critical review of the literature. *International Journal of Science Education,* 22, 665–701.

Acevedo, J. A. (2008). El estado actual de la naturaleza de la ciencia en la didáctica de las ciencias [The state of art of the nature of science in science education]. *Revista Eureka sobre Enseñanza y Divulgación de las Ciencias*, 5(2), 134–169. Retrieved from http://www.apac-eureka.org/revista/Larevista.htm.

Acevedo, J. A., Vázquez, A., & Manassero, M. A. (2003). Papel de la educación CTS en una alfabetización científica y tecnológica para todas las personas [The role of STS education in scientific and technological literacy for all]. *Revista Electrónica de Enseñanza de las Ciencias*, 2(2). Retrieved from http://www.saum.uvigo.es/reec/.

Adúriz-Bravo, A., & Izquierdo-Aymerich, M. (2009). A research-informed instructional unit to teach the nature of science to pre-service science teachers. *Science & Education,* 18, 1177–1192.

Agin, M. (1974). Education for scientific literacy: A conceptual frame of reference and some applications. *Science Education,* 58, 403–415.

Aikenhead, G. S. (1994). What is STS teaching? In J. Solomon, & G. Aikenhead (Eds.), *STS education: International perspectives on reform* (pp. 47–59). New York, NY: Teachers College Press.

Aikenhead, G. S. (2003). STS education: A rose by any other name. In R. Cross (Ed.), *A vision for science education: Responding to the work of Peter J. Fensham* (pp. 59–75). New York, NY: Routledge Press.

Aikenhead, G. S. (2006). *Science education for everyday life: Evidence-based practice*. New York, NY: Teachers College, Columbia University.

Aikenhead, G. S., & Ryan, A. G. (1992). The development of a new instrument: "Views on science-technology-society" (VOSTS). *Science Education*, 76, 477–491.

American Association for the Advancement of Science (1993). *Benchmarks for science literacy: A project 2061 report*. New York, NY: Oxford University Press.

Atlantic Science Curriculum Project (1986). *SciencePlus 1*. Toronto: Harcourt Brace Jovanovich.

Atlantic Science Curriculum Project (1987). *SciencePlus 2*. Toronto: Harcourt Brace Jovanovich.

Atlantic Science Curriculum Project (1988). *SciencePlus 3*. Toronto: Harcourt Brace Jovanovich.

Bennett, J., Hogarth, S., & Lubben, F. (2007). Bringing science to life: A synthesis of the research evidence on the effects of context-based and STS approaches to science teaching. *Science Education,* 91(3), 347–370.

Bijker, W., Hughes, T., & Pinch, T. (Eds.) (1987). *The social construction of technological systems: New directions in the sociology and history of technology*. Cambridge, MA: MIT Press.

Bybee, R. (1997). *Achieving scientific literacy*. Portsmouth, NH: Heineman.

Carson, R. (1962). *Silent spring*. Boston, MA: Houghton Mifflin.

Carter, L. (2008). Sociocultural influences on science education: innovation for contemporary times. *Science Education*, 92, 165–181.

Chen, D., & Novik, R. (1984). Scientific and technological education in an information society. *Science Education,* 68, 421–426.

Clark, W. C. (2007). Sustainability science: A room of its own. *Proceedings of the National Academy of Science*, 104, 1737–1738.

Clough, M. P. (2007). Teaching the nature of science to secondary and post-secondary students: Questions rather than tenets. *The Pantaneto Forum, 25*. Retrieved from http://www.pantaneto.co.uk/issue25/front25.htm.

Colucci-Gray, L., Camino, E., Barbiero, G., & Gray, D. (2006). From scientific literacy to sustainability literacy: An ecological framework for education. *Science Education*, 90, 227–252.

Colucci-Gray, L., Perazzone, A., Dodman, M., & Camino, E. (2012). Science education for sustainability, epistemological reflections and educational practices: From natural sciences to trans-disciplinarity. *Cultural Studies of Science Education*, published online (pre-print): 21 February 2012.

Conant, J. (Ed.) (1957). *Harvard case histories in experimental science.* Cambridge, MA: Harvard University Press.

Cozzens, S. E. (1993). Whose movement? STS and social justice. *Science, Technology, & Human Values,* 18, 275–277.

Cutcliffe, S. H. (1990). The STS curriculum: What have we learned in twenty years. *Science, Technology, & Human Values,* 15(3), 360–372.

Dakers, J. (2006). Towards a philosophy *for* technology education. In Dakers, J. (Ed.), *Defining technological literacy: Towards an epistemological framework* (pp. 145–158). New York, NY: Palgrave.

Daugs, D. (1970). Scientific-literacy – re-examined. *The Science Teacher,* 37(8), 10–11.

DeBoer, G. E. (2000). Scientific literacy: Another look at its historical and contemporary meanings and its relationship to science education reform. *Journal of Research in Science Teaching,* 37, 582–601.

Deng, F., Chen, D.-T., Tsai, C.-C., & Chai, C. S. (2011). Students' views of the nature of science: A critical review of research. *Science Education*, 95, 961–999.

Dillon, J. (2009). On scientific literacy and curriculum reform. *International Journal of Environmental and Science Education,* 4, 201–213.

Donnelly, J. F. (2004). Humanizing Science Education. *Science Education*, 88, 762–784.

Eijkelhof, H. M. C., & Lijnse, P. L. (1988). The role of research and development to improve STS education: Experiences from the PLON-project. *International Journal of Science Education,* 10, 464–474.

Ellul, J. (1964). *The technological society*. New York: Alfred A. Knopf.

Ehrlich, P. R. (1968). *The population bomb*. New York, NY: Ballantine Books.

Fitzpatrick, F. (1960). *Policies for science education*. New York, NY: Teachers College press.

Fourez, G. (1997). Scientific and technological literacy as a social practice. *Social Studies of Science*, 27, 903–936.

Friibergh Workshop on Sustainability Science (2000). *Sustainability science: Statement of the Friibergh Workshop on Sustainability Science*. Retrieved from http://sustainabilityscience.org/content.html?contentid=774.

Gabel, L. L. (1976). *The development of a model to determine perceptions of scientific literacy*. Unpublished doctoral dissertation, Ohio State University.

Gallagher, J. J. (1971). A broader base for science education. *Science Education, 55*, 329–338.

Gil, D. & Vilches, A. (2001). Una alfabetización científica para el siglo XXI. Obstáculos y propuestas de actuación. *Investigación en la Escuela,* 43, 27–37.

Gräber, W., Nentwig, P., Becker, H.-J., Sumfleth, E., Pitton, A., Wollweber, K., & Jorde, D. (2002). Scientific literacy: From theory to practice. In H. Behrendt, H. Dahncke, R. Duit, W. Gräber, M. Komorek, A. Kross, & P. Reiska (Eds.), *Research in science education: Past, Present and future* (pp. 61–70). Dordrecht: Kluwer.

Hacking, I. (1983). *Representing and inventing: Introductory topics in the philosophy of natural science.* Cambridge: Cambridge University Press.

Hodson, D. (1994). Seeking directions for change: The personalization and politicization of science education. *Curriculum Studies,* 2, 71 –98.

Hodson, D. (2003). Time for action: Science education for an alternative future. *International Journal of Science Education,* 25, 645–670.

Hodson, D. (2008). *Towards scientific literacy: A teachers' guide to the history, philosophy and sociology of science*. Rotterdam: Sense Publishers.

Hodson, D. (2009). *Teaching and learning about science: Language, theories, methods, history, traditions and value*. Rotterdam: Sense Publishers.

Holbrook, J. (1998). Operationalising scientific and technological literacy: A new approach to science teaching. *Science Education International,* 9(2), 13–19

Holton, G., Rutherford, J., & Watson, F. (1970). *Project physics course.* New York, NY: Holt, Rinehart & Watson.
Hurd, P. D. (1958). Science literacy: Its meaning for American schools. *Educational Leadership, 16*, 13–16, 52.
Hurd, P. D. (1975). Science, technology and society: New goals for interdisciplinary science teaching. *The Science Teacher, 42*, 27–30.
Hurd, P. D. (1998). Scientific literacy: New minds for a changing world. *Science Education, 82*, 407–416.
ITEA (2000). *Standards for technological literacy: Content for the study of technology.* Reston, VA: International Technology Education Association.
Jones, P., Selby, D., & Sterling, S. (Eds.) (2010). *Sustainability education: Perspectives and practice across higher education.* London: Earthscan.
Kahn, R., & Kellner, D. (2006). Reconstructing technoliteracy: A multiple literacies approach. In Dakers, J. (Ed) *Defining technological literacy: towards an epistemological framework* (pp. 254–273). New York, NY: Palgrave.
Kates, R., Clark, W., Corell, R., Hall, J., Jaeger, C., Lowe, I., McCarthy, J., Schellnhuber, H-J., Bolin, B., Dickson, N., Faucheux, S., Gallopin, G., Grubler, A., Huntley, B., Jager, J., Jodha, N., Kasperson, R., Mabogunje, A., Matson, P., & Mooney, H. (2001). Sustainability science. *Science*, 292, 641–642.
Klopfer, L. E., & Watson, F. G. (1957). Historical materials and high school science teaching. *The Science Teacher, 24*, 264–293.
Klopfer, L. E. (1963). The history of science cases for high schools in the development of student understanding of science and scientists: A report on the HOSC instruction project. *Journal of Research in Science Teaching, 1*, 33–47.
Klopfer, L. E. (1969). Science education in 1991. *The School Review, 77*, 199–217.
Kolstø, S. D. (2001). Scientific literacy for citizenship: Tools for dealing with the science dimension of controversial socioscientific issues. *Science Education, 85*(3), 291–310.
Krageskov Eriksen, K. (2002). The future of tertiary chemical education: A bildung focus. *HYLE: International Journal for Philosophy of Chemistry, 8*, 35–48.
Kuhn, T. S. 1962. *The structure of scientific revolutions*. Chicago, IL: University of Chicago Press.
Laugksch, R. C. (2000). Scientific literacy: A conceptual overview. *Science Education, 84*, 71–94.
Layton, D. (1993). *Technology's challenge to science education: Cathedral, quarry or company stone?* Buckingham: Open University Press.
Leach, J., Hind, A., & Ryder, J. (2003). Designing and evaluating short teaching interventions about the epistemology of science in high school classrooms. *Science Education*, 87(6), 832–848.
Lederman, N. G. (1992). Students' and teachers' conceptions of the nature of science: A review of research. *Journal of Research in Science Teaching*, 29, 331–359.
Lederman, N. G. (2007). Nature of science: Past, present, and future. In S. K. Abell, & N. G. Lederman (Eds.), *Handbook of research on science education* (pp. 831–879). Mahwah, NJ: Lawrence Erlbaum Associates.
Lederman, N. G., Abd-El-Khalick, F., Bell, R. L., & Schwartz, R. (2002). Views of nature of science questionnaire: Toward valid and meaningful assessment of learners' conceptions of nature of science. *Journal of Research in Science Teaching, 39*, 497–521.
Lewis, J. (1981). *Science and society*. London: Heinemann and Association for Science Education.
MacKenzie, D., & Wajcman, J. (Eds.) (1985). *The social shaping of technology: How the refrigerator got its hum*. Philadelphia, PA: Open University Press.
Manassero, M. A., & Vázquez, A. (2001). Percepción de los estudiantes sobre la influencia de la ciencia escolar en la sociedad. *Bordón, 53*, 97–113.
Matthews, M. R. (1994). *Science teaching: The role of history and philosophy of science*. London: Routledge.
Matthews, M. R. (1998). In defense of modest goals when teaching about the nature of science. *Journal of Research in Science Teaching*, 35, 161–174.
Matthews, M. R. (2004). Thomas Kuhn's impact on science education: What lessons can be learned? *Science Education, 88*, 90–118.

Matthews, M. R. (2012). Changing the focus: From nature of science (NOS) to features of science (FOS). In M. S. Khine (Ed.), *Advances in Nature of Science Research. Concepts and Methodologies*, (pp. 3–26), Dordrecht: Springer.

McComas, W. F., Clough, M., & Almazroa, H. (1998). The role and character of the nature of science education. In W. F. McComas (Ed.), *The nature of science in science education: Rationales and strategies* (pp. 3–39). Dordrecht: Kluwer.

McComas, W. F., & Olson, J. K. (1998). The nature of science in international science education documents. In W. F. McComas (Ed.), *The nature of science in science education: rationales and strategies* (pp. 41–52). Dordrecht: Kluwer.

McCurdy, R. (1958). Toward a population literate in science. *The Science Teacher,* 25, 366–368, 408.

Millar, R., & Osborne, J. (Eds.) (1998). *Beyond 2000: Science education for the future*. London: Kings College.

Mumford, L. (1967–70). *The myth of the machine*. New York: Harcourt Brace Jovanovich.

Nader, R. (1965). *Unsafe at any speed: The designed-in dangers of the American automobile.* New York: Grossman.

National Commission on Excellence in Education (1983). *A nation at risk: The imperative for educational reform*. Washington, DC: US Department of Education.

National Research Council (1996). *National Science Education Standards*. Washington, DC: National Academic Press.

National Science Teachers Association (1982). *Science-technology-society: Science education for the 1980s*. Washington, DC: NSTA.

Niaz, M. (2008). What 'ideas-about-science' should be taught in school science? A chemistry teachers' perspective. *Instructional Science,* 36, 233–249.

O'Hearn, G. T. (1976). Scientific literacy and alternative futures. *Science Education,* 61, 103–114.

Oliver, J. S., Jackson, D. F., Chun, S., Kemp, A., Tippins, D. J., Leonard, R., Kang, N. H., & Rascoe, B. (2001). The concept of scientific literacy: A view of the current debate as an outgrowth of the past two centuries. *Electronic Journal of Literacy through Science, 1*.

Osborne, J., Collins, S., Ratcliffe, M., Millar, R., & Duschl, R. (2003). What 'ideas-about-science' should be taught in school science? A delphi study of the expert community. *Journal of Research in Science Education,* 40, 692–720.

Pedretti, E. (1997). Septic tank crisis: A case study of science, technology and society education in an elementary school. *International Journal of Science Education,* 19(10), 1211– 1230.

Pedretti, E. (2003). Teaching science, technology, society and environment (STSE) education: Preservice teachers' philosophical and pedagogical landscapes. In D. L. Zeidler (Ed.), *The role of moral reasoning on socioscientific issues and discourse in science education* (pp. 219–239). Dordrecht: Kluwer Academic Press.

Pedretti, E., & Nazir, J. (2011). Currents in STSE education: Mapping a complex field, 40 years on. *Science Education,* 95, 601–626.

Pella, M. O. (1967). Science literacy and the high school curriculum. *School Science and Mathematics,* 67, 346–356.

Pella, M. O., O'Hearn, G. T., & Gale, C. W. (1966). Referents to scientific literacy. *Journal of Research in Science Teaching,* 4, 199–208.

Ratcliffe, M. (1997). Pupil decision-making about socioscientific issues within the science curriculum. *International Journal of Science Education,* 19, 167–182.

Roberts, D. (2007). Scientific literacy/science literacy. In S. K. Abell, & N. G. Lederman (Eds.), *Handbook of research on science education* (pp. 729–780). Mahwah, NJ: Lawrence Erlbaum.

Roth, W.-M., & Désautels, J. (Eds.) (2002). *Science education as/for sociopolitical action.* New York: Peter Lang.

Rudolph, J. L. (2000). Reconsidering the 'nature of science' as a curriculum component. *Journal of Curriculum Studies*, 32(3), 403–419.

Sadler, T. D. (2004). Informal reasoning regarding socioscientific issues: A critical review of the research. *Journal of Research in Science Teaching,* 41, 513–536.

Sadler, T. D. (Ed.) (2011). *Socioscientific issues in the classroom: Teaching, learning and research.* Dordrecht: Springer.

Sandoval, W. A. (2005). Understanding students' practical epistemologies and their influence on learning through inquiry. *Science Education*, 89(4), 634–656.
Santos, W. L. P. dos (2008). Scientific literacy: A Freirean perspective as a radical view of humanistic science education. *Science Education*, 93, 361–382.
Shamos, M. H. (1995). *The myth of scientific literacy*. New Brunswick, NJ: Rutgers University Press.
Shen, B. S. P. (1975). Scientific literacy. *American Scientist*, 63, 265–268.
Sismondo, S. (2010). *An introduction to science and technology studies* (Second edition). Chichester: Wiley-Blackwell.
Sjöström, J. (2008). The discourse of chemistry (and beyond). *HYLE: International Journal for Philosophy of Chemistry*, 13, 83–97.
Solomon, J. (1983). *Science in social content (SISCON)*. London: Basil Blackwell and Association for Science Education.
Stokes, D. E. (1997). *Pasteur's quadrant: Basic science and technological innovation*. Washington, D.C.: Brookings Institution.
Tala, S. (2009). Unified view of science and technology education: Technoscience and technoscience education. *Science & Education*, 18, 275–298.
Tal, T., & Kedmi, Y. (2006). Teaching socioscientific issues: classroom culture and students' performances. *Cultural Studies of Science Education*, 1, 615–644.
Thomas, G., & Durant, J. (1987). Why should we promote the public understanding of science? In M. Shortland (Ed.), *Scientific literacy papers* (pp. 1–14). Oxford: Department for External Studies, University of Oxford.
Tippins, D. J., Nichols, S. E., & Kemp, A. (1999). *Cultural myths in the making: The ambiguities of science for all*. Paper presented at the Annual Meeting of the National Association for Research in Science Teaching, Boston, MA.
Tsai, C-C., & Liu, S-Y. (2005). Developing a multi-dimensional instrument for assessing students' epistemological views toward science. *International Journal of Science Education*, 27(13), 1621–1638.
UNESCO (1977). First intergovernmental conference on environmental education: Final report. Paris: UNESCO.
UNESCO (2005). United Nations decade of education for sustainable development (2005–2014): International implementation scheme. Paris: UNESCO.
United Nations (1987). *Our common future: Report of the World Commission on Environment and Development*. General Assembly Resolution 42/187, 11 December 1987.
Vare, P., & Scott, W. (2007). Learning for a change: Exploring the relationship between education and sustainable development. *Journal of Education for Sustainable Development*, 1, 191–198.
Vázquez, A., Acevedo, J. A., & Manassero, M. A. (2005). Más allá de una enseñanza de las ciencias para científicos: hacia una educación científica humanística [Beyond teaching science for scientists: Towards a humanistic science education]. *Revista Electrónica de Enseñanza de las Ciencias*, 4(2), Retrieved from http://www.saum.uvigo.es/reec/.
Vázquez, A., Acevedo, J.A., Manassero, M. A., & Acevedo, P. (2001). Cuatro paradigmas básicos sobre la naturaleza de la ciencia [Four basic paradigms about the nature of science]. *Argumentos de Razón Técnica*, 4, 135–176. Retrieved from http://www.campus-oei.org/salactsi/acevedo20.htm.
Vázquez, A., Acevedo, J. A., Manassero, M. A., & Acevedo, P. (2004). Hacia un consenso sobre la naturaleza de la ciencia en la enseñanza de las ciencias [Towards a consensus on the nature of science in science education]. In I. P. Martins, F. Paixão & R. Vieira (Org.), *Perspectivas Ciência-Tecnologia-Sociedade na Inovação da Educação em Ciência* (pp. 129–132), Aveiro (Portugal), Universidade de Aveiro.
Vázquez, A., & Manassero, M. A. (2007). *La relevancia de la educación científica [The relevance of science education]*. Palma de Mallorca: Universitat de les Illes Balears.
Vesterinen, V.-M., & Aksela, M. (2012). Design of chemistry teacher education course on nature of science. *Science & Education*, published online (pre-print): 23 June 2012.
Vesterinen V.-M., Aksela M., & Lavonen J. (2011), Quantitative analysis of representations of nature of science in Nordic upper secondary school textbooks using framework of analysis

based on philosophy of chemistry. *Science & Education*, published online (pre-print): 18 October 2011.
Vesterinen, V.-M., Aksela, M., & Sundberg, M. R. (2009). Nature of chemistry in the national frame curricula for upper secondary education in Finland, Norway and Sweden. *NorDiNa, 5*, 200–212.
Waterman, A. T. (1960). National Science Foundation: A ten-year résumé. *Science, 131*, 1341–1354.
Wonacott, M. E. (2001). Technological literacy. ERIC Digest, no. 233, retrieved from http://ericacve.org/digests.asp.
Yager, R. E. (1996). History of science/technology/society as reform in the United States. In R. E. Yager (Ed.), *Science/technology/society as reform in science education* (pp. 3–159). Albany, NY: State University of New York Press.
Zeidler, D. L. (2001). Participating in program development: Standard F. In D. Siebert, & W. McIntosh (Eds.), *College pathways to the science education standards* (pp. 18–22). Arlington, VA: National Science Teachers Press.
Zeidler, D. L., Sadler, T. D., Simmons, M. L., & Howes, E. V. (2005). Beyond STS: A research-based framework for socioscientific issues education. *Science Education*, 89, 357–377.
Zeidler, D. L., Walker, K. A., Ackett, W. A., & Simmons, M. L. (2002). Tangled up in views: Beliefs in the nature of science and responses to socioscientific dilemmas. *Science Education*, 86, 343 – 367.
Ziman, J. (1984). *An introduction to science studies: The philosophical and social aspects of science and technology*. Cambridge: Cambridge University Press.

Veli-Matti Vesterinen is researcher and teacher educator with a Ph.D. degree in chemistry education from the University of Helsinki (Finland). From 2007 he has taught on several pre- and in-service teacher education courses at the Department of Chemistry and at the Department of Teacher Education. He has authored several papers discussing nature of science and chemistry teacher education (both in Finnish and in English). The papers have been published in books and scientific journals such as *Chemistry Education Research and Practice* and *Science & Education*. He is also active in the field of informal science education. He is a cofounder of MyScience and Luova, popular science websites for teens, as well as a cofounder and editor-in-chief of the *European Journal of Young Scientists and Engineers*. For his work on popularizing science, he was one of the recipients of Finnish State Award for Public Information in 2009. Drawing insights from science and technology studies and the emerging field of philosophy of chemistry, his main research interests include scientific and technological literacy, nature of science and technology, education for sustainability, and the aims of science education.

María-Antonia Manassero-Mas is professor of social psychology in the Department of Psychology at the University of the Balearic Islands (Spain). She earned a special Degree Award in Psychology (University of Barcelona) and special Ph.D. award in Psychology (University of the Balearic Islands). Her research developed through peer-evaluated research projects focused on psychosocial processes and social psychology of science, science education, students' attitudes to science, science achievement, science-technology-society, gender differences, scientific literacy, and nature of science. She is author or coauthor of science education books (in Spanish) *Causal Attribution Applied to Educational Orientation (1995); Views on Science, Technology and Society (1998); Assessment of Science, Technology and*

Society (2001); Teaching Stress and Burnout (2003); Secondary Students' Interests on Science and Technology Curriculum (2007); The Relevance of Science Education (2007); and Science, Technology and Society in Latin America (2010), as well as many other book chapters, papers (e.g., Spanish Secondary-School Science Teachers' Beliefs About Science-Technology-Society (STS) Issues, Science & Education, 2012), and communications to congresses.

Ángel Vázquez-Alonso is educational inspector of the Balearic Islands (BI) Government and honorary researcher at the BI University (Spain), where he has taught physics, research methods, curriculum development, and educational evaluation and trained high-school teachers in the Department of Physics and Department of Education. He earned three degrees in Physics, Chemistry, and Education, and a Ph.D. on Educational Sciences, taught high-school Physics, and directed the BI Educational Quality and Evaluation Institute. His research interests have developed through peer-evaluated research projects on students' conceptions and attitudes to science, science achievement, science-technology-society, gender differences, scientific literacy, and nature of science. He is author, editor, or coauthor of some books (in Spanish) and curricular materials: *Clay* (1997); *Views on Science, Technology and Society* (1998); *Assessment of Science, Technology and Society* (2001); *Teachers: Commitment for Education* (2003); *Secondary Students' Interests on Science and Technology Curriculum* (2007); *The Relevance of Science Education* (2007); *Technology Education* (2010); and *Science, Technology and Society in Latin America* (2010), as well as book chapters, papers (e.g., Science Teachers' Thinking About the Nature of Science, *Research in Science Education*, 2012), and communications to congresses.

Chapter 59
Cultural Studies in Science Education: Philosophical Considerations

Christine L. McCarthy

59.1 Cultural Studies of Science

Contemporary science education is considered by many to be in need of a fundamental re-structuring, based upon a fundamental re-conceptualization of science. Work in the field of cultural studies of science education attempts to provide the basis for this revision. Prominent themes that are current in the CSSE literature can be traced to a number of classic mid-twentieth-century works in the philosophy and sociology of science. In this section five foundational themes and their sources are examined.

59.1.1 The Sociological Study of Science: Merton

It is uncontroversial that the activities of scientific inquiry are human activities, invariably occurring in a social context. The social context necessarily affects the historical course of scientific inquiry, and the social world is itself affected by the developing science. Sociological studies of science have examined the activities of practicing scientists and explicated the interactive relationship of science in society. Beginning in the 1930s, Robert Merton advanced the sociology of science by conceiving of science as itself a social institution. Merton's work focused on the social structure of science, the relations and interactions internal to, and, perhaps definitive of, the institution of science. Merton, often considered to be the founding father of the sociology of science, argued that science cannot be developed and cannot be maintained, unless certain cultural conditions obtain in the society (1938/1973, p. 254). Absent the favorable cultural conditions, hostility toward science can be expected to occur.

C.L. McCarthy (✉)
University of Iowa, Iowa, USA
e-mail: christine-mccarthy@uiowa.edu

M.R. Matthews (ed.), *International Handbook of Research in History, Philosophy and Science Teaching*, DOI 10.1007/978-94-007-7654-8_59,

Merton developed a conception of the ethos of scientific inquiry. This ethos is an "…affectively toned complex of values and norms which is held to be binding on the scientist…expressed in the form of prescriptions, proscriptions, preferences, and permissions" (Merton 1973, p. 269). This ethos is maintained and transmitted by the institutional structure of science. It is possible that some practices which occur in the context of science and which are taken to be scientific actually depart from the values and norms of science. To the extent that they do, nonscientific or pseudoscientific work is being produced under the guise of science. Science in such a case would suffer a collapse from within.

Merton also observed that science can be attacked from without: "science is not immune from attack, restraint, and repression" (p. 267) and "[l]ocal contagions of anti-intellectualism threaten to become epidemic" (p. 267). Advanced scientific work leads to the development of complex theories about the world[1] that cannot be understood without extensive study. Such theories are largely unintelligible to nonscientists. This incomprehension undermines cultural support for science and leaves the majority of persons "ripe for new mysticisms clothed in apparently scientific jargon" (Merton 1938/1973, p. 264). In addition, the organized and continual skepticism characteristic of science will at times threaten the interests of entrenched power holders and/or cast doubt on the truth of deeply cherished nonscientific beliefs. Science is then likely to be threatened by an organized social backlash against it.

Merton identified four institutional imperatives that constitute the ethos necessary for science. The first is an objective *universalism*. "[T]ruth-claims, whatever their source, are to be subjected to *pre-established impersonal criteria*: consonant with observation and with previously confirmed knowledge" (1942/1973, p. 270, emphasis in original). None of the personal or social attributes of the scientist are to enter into the assessment of the truth claims. Merton's universalism rests upon the objectivity of the world itself, the object of scientific inquiry. In science, Merton holds, "…[o]bjectivity precludes particularism. The circumstance that scientifically verified formulations refer…to objective sequences and correlations militates against all efforts to impose particularistic criteria of validity" (1942/1973, p. 270).

Merton explicitly rejects social, cultural, and/or racial particularism, holding that "[e]thnocentrism is not compatible with universalism" (p. 271). While accepting that the historical course of development of science is influenced by its cultural context, Merton rejects the notion that scientific knowledge could have a particular cultural, "national," or class-based content. It is possible, Merton maintains, to compare and assess the merits of inconsistent competing claims to knowledge that might arise in different cultures.

> …[T]he cultural context in any given nation or society may predispose scientists to focus on certain problems….But this is basically different from the second issue: the criteria of validity of claims to scientific knowledge are not matters of national taste and culture. Sooner or later, competing claims to validity are settled by universalistic criteria. (Merton 1942/1973, p. 271, n. 6)

[1] The term "the world" herein refers to the set of all entities and dynamic interactions that exist and occur and is synonymous with nature or reality, or simply "what is."

Merton's second institutional imperative in the ethos of science he terms *communism*. This is the assertion that the knowledge acquired by scientific inquiry belongs rightfully to the human community as a whole. "The substantive findings of science are a product of social collaboration and are assigned to the community. They constitute a common heritage…" (1942/1973, p. 273). Merton emphasizes that communal investigation is a necessary aspect of scientific inquiry.

The third imperative is *disinterestedness*. Merton remarks upon "[t]he virtual absence of fraud in the annals of science…" (p. 276) and attributes this honesty not to a superior moral integrity of individual scientists, but rather to the institutional structure of science. Inquiry, to count as scientific, must be subjected to the rigorous scrutiny of the community of qualified scientific peers. Given that scientific theories refer to an objective world, it is likely that peer scrutiny will eventually expose whatever fraudulent or incompetent practices might occur. It is when the institutional structure of science breaks down, Merton writes, that "[f]raud, chicane and irresponsible claims (quackery)" (p. 277) proliferate. "The abuse of expert authority and the creation of pseudo-sciences are called into play when the structure of control exercised by qualified compeers is rendered ineffectual" (p. 277).

The fourth institutional imperative is *organized skepticism*, which Merton considers to be "…both a methodological and an institutional mandate" (p. 277). The commitment to skepticism gives rise to conflict between the institution of science and other social institutions, e.g., religious, economic, and political institutions, when such social institutions demand an uncritical respect for and acceptance of certain doctrines. A revolt against science by the culturally powerful may be launched in response.

Merton's sociological research provides a clarification of the social structure of scientific inquiry and its relations to the wider society. Merton's approach involves empirical study of the practices and norms of the social activity of scientific inquiry. His findings go beyond simple description of characteristics found in scientific practice, to an analysis of which characteristics are necessary and definitive. Merton's approach has been largely replaced in contemporary cultural studies of science by a method of cultural criticism.

59.1.2 The Incommensurability of Scientific Theories: Kuhn

Kuhn's (1962) work, *The Structure of Scientific Revolutions,* cast doubt on the belief that the history of science has been one of cumulative progress in scientific knowledge of the dynamics of the world. Kuhn interprets the history of science as a series of revolutions that bring about fairly sudden and fundamental changes of scientific views of nature. These changes Kuhn calls paradigm shifts, defining a paradigm as a set of "universally recognized scientific achievements that for a time provide model problems and solutions to a community of practitioners" (p. viii).

In Kuhn's view, as long as a particular paradigm is in place, a calm period of normal scientific work occurs. The routine work eventually exposes anomalies, findings that don't fit within the current paradigm, which creates a crisis that can be resolved only by revolutionary change in the current paradigm. Following a paradigm shift, the world is understood in fundamentally different ways. Previously accepted theoretical entities are eliminated from the new conception of reality, and new entities, with new interactive dynamics, take their place. So fundamental is the shift in worldview that the theories belonging to the different paradigms, in Kuhn's view, are incommensurable – the terms in the old theory cannot be translated into the language of the new theory. It would seem to follow that there is no legitimate way to rationally assess the relative merits of the old and new theories or of any competing theories. If this is so, change in accepted theory would have to be driven largely by extrascientific factors that function causally, but irrationally, to determine individual and social subjective states of belief.

Kuhn's work contributed to a fundamental reevaluation of the conception of science, of scientific inquiry, and of scientific knowledge. Kuhn noted that theories about the world that are now judged to be false were once considered to be well justified and were counted as knowledge. He concluded that "…once current views of nature were, as a whole, neither less scientific nor more the product of human idiosyncrasy than those current today" (p. 2).

Kuhn appears to argue that if we now dismiss those widely accepted ancient beliefs about the world as mere myths, we should consider the currently accepted beliefs arising from contemporary scientific inquiry to be equally mythic. Given that the vast majority of belief systems in the past have been judged to be mostly wrong, it appears to follow that the majority of our current beliefs are also likely to be mostly wrong. This line of reasoning is sometimes called the pessimistic induction, pessimistic because it suggests that science proceeds by substituting one wrong system for another, without actually progressing toward true belief.

Kuhn's (1962) analysis is often taken as a repudiation of the possibility of rational bases for theory change. Yet Kuhn's analysis rests on the observation that crises of belief arise because new findings of scientific inquiry cannot be understood given the current explanatory theory. A problem is thus set, and a new theory is sought which will better explain the new observations and the old. Proposed theories are comparatively assessed and judged by their ability to resolve the current theoretical problem. The continued process of theoretical crisis and theoretical resolution of the crisis should be expected to lead to gradual improvement in the theoretical understanding of the world. The history of science should lead to optimism regarding the increasing verisimilitude, or nearness to truth, of scientific knowledge, despite any non-translatability of successive theories. The optimistic view of the growth of scientific knowledge is often dismissed in the cultural studies literature as naïve or as a piece of propaganda deliberately deployed in a political struggle for power and domination.

Kuhn's incommensurability thesis appears to support the belief that concurrent but inconsistent cultural systems of belief about the world are simply different and incommensurable paradigms. On this view, it is thought that no culturally neutral grounds can exist that would allow for reasoned choice among different cultural belief systems. Kuhn's work thus is taken to support the view, prominent in cultural studies of science, that there are no legitimate grounds for considering modern science to be of a special cognitive significance. Scientific knowledge is to be considered merely one system of belief among many others that are equally legitimate.

It is ironic, then, that Kuhn, in 1977, develops a set of five criteria for theory choice in science. The first is accuracy of fit to nature: the "consequences deducible from a theory should be in demonstrated agreement with the results of existing experiments and observations" (1988, p. 278). The second is consistency: a scientific theory should "…be consistent, not only internally or with itself, but also with other currently accepted theories applicable to related aspects of nature" (p. 278). The third is broad scope: a theory's "consequences should extend far beyond the particular observations, laws, or subtheories it was initially designed to explain" (p. 278). Fourth, a theory should be "simple, bringing order to phenomena…" (p. 278). Fifth, it should be fruitful: it should "disclose new phenomena or previously unnoted relationships" (p. 278). These five criteria, Kuhn states, "are all standard criteria for evaluating the adequacy of a theory…they provide the shared basis for theory choice" (p. 278). So, for Kuhn, the strict non-translatability, i.e., the incommensurability, of competing theories identified in his 1962 work does not in fact prevent comparative evaluation of the merits of the competing theories.

Kuhn does argue in 1977 that these and other criteria of theory choice cannot be fashioned into a complete algorithmic decision procedure, binding on all members of a scientific community. Judgment is always necessary in evaluation, and individuals who come to minority conclusions are not ipso facto unscientific. Subjective considerations might not be wholly expunged when individuals make evaluative judgments of the merits of competing theories. But there are, in Kuhn's view, rational criteria for judgment, and an open communal and critical discussion of differing individual judgments promotes judgment on rational criteria. "The criteria of choice…function not as rules, which determine choice, but as values, which influence it" (p. 285).

It is Kuhn's earlier work that has been incorporated into the foundations of the CSSE literature.

59.1.3 Cultural Relativism and Science: Barnes and Bloor

Sociologists Barnes and Bloor develop a relativistic theory of knowledge, which they claim is required by "…the scientific understanding of forms of knowledge" (1982/1994, p. 21). They make three claims. The first two are uncontroversial: "(i)…beliefs on a certain topic vary, and…(ii) which of these beliefs is found in a given context depends on, or is relative to, the circumstances of the users" (p. 22). Their third claim, the "equivalence postulate," however, is problematic. The equivalence postulate states that the truth status of a belief is irrelevant to the explanation of the "causes of credibility" of the belief. This may be true, in many cases; it is common to observe instances of confident belief about the world that would later prove false. But the observation is about the empirical cause of a subjective state of mind of a believing creature. This is not relevant to a theory of knowledge, which poses conceptual and normative questions, e.g., whether or not a particular confident state of belief ought to be counted as knowledge.

Barnes and Bloor, though, stipulate that they will be using the term "knowledge" in a specific way: "We refer to any collectively accepted system of belief as 'knowledge'" (p. 22, n. 5). This stipulation is problematic, however. To believe a proposition is simply to accept that the proposition is true and to be prepared to act as if the proposition is true. But, the subjective belief that a proposition is true, even when collective, itself provides no evidence about the actual truth or falsity of the proposition.

Barnes and Bloor cope with this difficulty by eliminating from their conception of knowledge any considerations of truth. There will be no consideration of the relative merit of different systems of collective beliefs in terms of truth, or of "nearness to truth," or *verisimilitude*.[2] Barnes and Bloor see no problem in their omission of the concept of truth, because they take use of the words "true" and "false" to merely "provide the idiom" through which individuals express their natural subjective preferences for their own cultural beliefs. Barnes and Bloor continue to use the word "knowledge," but their claims about knowledge must be understood to be claims only about collectively accepted systems of belief.

In the absence of concern about truth, any comparative assessment of the merits of different bodies of belief is seriously restricted. Despite Barnes and Bloor's redefinition of the term "knowledge," there remains need for a term that refers to systems of belief that are thought, for good reasons, to be true or at least more nearly true than competing systems. The term "knowledge" has long served this purpose in ordinary English usage.

This suggests another problem with the equivalence postulate. Finding a causal explanation of an individual or collective state of belief is not the same thing as finding *good reasons for* the state of belief. After one has reached a well-grounded explanation about *how* a particular belief was caused, it is still sensible to ask whether it is a good thing that the belief was so caused. Barnes and Bloor, however, collapse the important distinction between causes of believing and reasons for believing and assert that discovery of the causes of belief is identical to discovering reasons for the belief. Having collapsed the distinction, they maintain that the sociology of knowledge is not to be "…confined to causes *rather than* 'evidencing reasons'. Its concern is precisely with causes *as* 'evidencing reasons'" (p. 29).

Seeking causal explanations for states of affairs, seeking the conditions of occurrence of a state of affairs, is a central and legitimate aim of scientific inquiry, and scientific inquiry into the empirical causes of collective belief is interesting and important. But sociological inquiry into the facts of belief acquisition is not an inquiry into acquisition of knowledge, unless knowledge is conceived in the limited *subjective* sense of individual and/or collective belief. Further considerations regarding the concept of knowledge will be taken up in Part 3.

---

[2] Popper's term; see Sect. 59.3.3 for further discussion.

59.1.4 Science as Cultural Tyranny: Feyerabend

Feyerabend, in *How to defend society against science*, develops a broad critique of science as a human endeavor. He writes in support of the wholesale rejection of modern science, on the grounds that achievement of true belief limits freedom of thought:

> [m]y criticism of modern science is that it inhibits freedom of thought. If the reason is that it [modern science] has found the truth and now follows it, then I would say that there are better things than first finding, and then following such a monster. (Feyerabend 1974/1988, p. 37)

Feyerabend writes that scientific knowledge is merely an ideology, no different from the ideology of the Church or of Marxism. He rejects the notion that scientific inquiry is a distinctive mode of activity: "…there is no 'scientific methodology' that can distinguish science from the other ideologies. *Science is just one of the many ideologies that propel society and it should be treated as such…*" (p. 40, emphasis in original).

Feyerabend notes that there are practical educational consequences of his position and approves, writing: "Three cheers to the fundamentalists in California who succeeded in having a dogmatic formulation of the theory of evolution removed from the text books, and an account of Genesis included" (p. 41). It appears that, in Feyerabend's view, maximal freedom of choice in believing is the primary cognitive value to be achieved; having a well-warranted body of knowledge about the world is not a comparable value.

The animus against science that Feyerabend evinces in his writing seems based on a conception of the negative effect an ideology of science would have on humanity. In "Dehumanizing Humans," he speaks of the look on the face of a friend, in which, he says, the whole relationship is written. He asserts that a scientific worldview would devalue this important, albeit subjective, human experience. "This look is not an objective fact…it is not 'scientifically important' and if science takes over, not socially important either" (1996, p. 95).

In "The Disunity of Science," Feyerabend asserts that the objective ontology of materialism that grounds science forces human beings to accept the reality of a world without subjective human experiences. Classical physicists, he writes, "distinguished between the objective world of scientific laws – and this world is without change – and the subjective world of our experiences. They ascribed reality to the former and regarded the latter as an illusion" (1996, p. 39). Philosophers, certainly, have sometimes made this move, and Feyerabend's objection to it are well grounded. As long as creatures having a requisite type of complexity really exist, as they clearly do, subjective states of affairs – feelings, desires, hopes, and beliefs – will also really exist, as states of those creatures. It is possible, however, to reject problematic philosophical theses associated with science without rejecting scientific inquiry in toto.

Yet Feyerabend's belief that science itself is hostile to the most important facets of human experience has become a common one. If scientific inquiry were in fact hostile to human experience and to human values, purposes, and emotions, then a passionate antipathy to science would make sense. But neither scientific inquiry nor scientific knowledge requires a denial of the fact of the qualitative, subjective experiences of human life. Far from denying those experiences, scientific inquiry, in leading to knowledge of the dynamic relations in the world, promotes the human quest for valued qualitative experiences.

Preston (2000) quotes Feyerabend on the disunity of science, its lack of a characteristic epistemology:

> Science is not one thing, it is many; and its plurality is not coherent, it is full of conflict. There are empiricists who stick to phenomena and despise flights of fancy, there are researchers who artfully combine abstract ideas and puzzling facts, and there are storytellers who don't give a damn about the details of the evidence. They all have a place in science. (Feyerabend 1992, p. 6)

The conception of science as all inclusive, so that even storytellers who don't give a damn about evidence are to be counted as practitioners of science, is evident in much of the CSSE literature.

It may be that Feyerabend was not entirely serious in his assertions about science. The essays collected in Preston's *The worst enemy of science?* provide insight into Feyerabend's scholarly intentions, and philosophers of science who knew Feyerabend personally write that he was intentionally provocative and extreme in his assertions about science.[3] But, whatever his intentions, Feyerabend's writings are often taken seriously and at face value. The effect of his writings has been to provide support to contemporary critiques of, and rejection of, modern science as the mode of inquiry that is best able to lead to knowledge of the world.

59.1.5 Science as Cultural Domination: Harding

Harding's feminist work is widely cited in cultural studies in science education literature and exemplifies certain philosophical issues central to current critiques of science (e.g., 1986; 1998).[4] Harding describes her position as a *standpoint* theory, one that values the standpoints of those who are *other* with respect to some cultural milieu. The others with respect to science include women, nonwhite ethnic/racial groups, non-European cultures, and cultures from the past. Taking on others' standpoints, Harding maintains, enables the detection of culturally based distortions in knowledge. Harding's acknowledgement of the possibility of error in belief, and of correction of that error by critical communal discourse, is significant. In accepting this, Harding appears to accept that there is an objective order in nature, i.e., that there are regular patterns of events in nature and that such patterns can be recognized, if an inquiry process properly structured to promote unbiased comparative assessment of theories is developed. In this respect, Harding's position would appear to be close to that of Merton.

[3] See also Horgan (1993).

[4] See Grasswick (2011) for an overview of feminist work in the philosophy of science.

Yet, contra Merton, Harding argues that the claims of Western modern science to universality and objectivity should be rejected as illusions. Harding argues that all theories about the regular patterns in nature are social constructs and that all social constructs are strongly influenced by social, cultural factors and by political agendas. Given this, Harding concludes that scientific theories are neither objectively determined nor politically neutral. The bodies of collectively accepted belief developed in different cultures, which Harding takes to constitute the culture's science, should be expected to diverge in both method and content. The problem here is that Harding's conclusion relies on an assumption that any degree of influence on science by social factors negates the claim of science to objectivity. But, absolute objectivity need not be conceptually required. A greater degree of objectivity is the value to be sought in scientific inquiry. The institutional structure and norms of scientific inquiry support the goal of achieving greater objectivity.

Harding maintains that all theories about the world are underdetermined by empirical evidence. She takes this to mean that different and mutually inconsistent theories can be constructed, all of which will be equally consistent with all possible empirical evidence. The best that any scientific theory can achieve, then, is empirical adequacy, i.e., consistency with observational data. "Many socially constituted theories about nature can be *consistent* with nature's order, but none are uniquely *congruent* with it; none uniquely correspond[s] to it" (1998, p. 176).

Harding further maintains that it is impossible to comparatively evaluate different empirically adequate theories, because it is impossible to acquire empirical evidence that would definitively rule out some of the empirically adequate theories. Absent the possibility of disconfirmation through scientific inquiry, in Harding's view, as in Kuhn's (1962) work, it can only be social, cultural, political, and historical factors that determine which theories about the world come to be accepted in a culture.

But, Harding does not take into account the commitment in scientific inquiry to an ongoing search for new evidence, the study of previously unconsidered situations, that will, in creating Kuhnian crises, allow for comparative evaluation of competing theories. The fact that, at an early stage of scientific inquiry, competing theories are equally adequate to current observational evidence does not entail that they will always be. The existence of two or more competing theories constitutes an intellectual problem for the scientific community and encourages intensive scientific inquiry to resolve the problem.

Harding holds that, in practice, persons in positions of social power will make the decisions that establish culturally specific belief systems.[5] The resultant belief system, unsurprisingly, will serve to maintain the existing social power relations. Social power dynamics determine what research projects will be permitted and which will be funded. Power dynamics also operate within the scientific community, determining which observations, and which theories, will be considered authoritative in the scientific community. On Harding's view, it is in principle impossible that continued scientific inquiry might lead to results that would *objectively* decide a scientific issue.

---

[5] See Oreskes and Conway (2010) and Wagner and Steinzor (2006).

Dominance relations occur at the level of individuals, and of social strata, and also at the intercultural level. Harding observes that "Western" culture is currently globally dominant and takes this geopolitical dominance to explain why "Western" scientific theories are currently intellectually dominant. Modern scientific methods of inquiry, and the scientifically generated body of knowledge, on this view, have been imposed on other cultures, in the face of their resistance.

Harding claims not to reject in toto what she calls the "European scientific and epistemological legacy" (p. 125). She does, however, argue that European science should be updated and means by this that it should be reconceptualized as simply one local knowledge system among many others, all of which are to be conceived as of equal intellectual value.

Harding accepts that the theories resulting from scientific inquiry "may well 'work' in the sense of enabling prediction and control" (p. 132), but she does not believe that "working" provides evidence of the truth, or of the nearness to truth, of scientific theories. Harding maintains that quite different and inconsistent systems of knowledge can provide similar levels of prediction and control. She writes that "…the regularities of nature…may be explained in ways permitting extensive(though not identical) prediction and control within radically different and even conflicting, culturally local, explanatory models" (p. 132). Harding's examples, though, serve only to cast doubt on the general claim. She asserts, for example, that "[f]armers in radically different cultures can predict equally well a large range of weather patterns and their effects, just as health care workers in these cultures can predict when illness will occur and how to cure it" (p. 133).

Yet there is abundant evidence that Harding's claim about the equivalent predictive power of scientific and nonscientific belief systems is false. For example, contra Harding, predicting with any significant accuracy the path of a tornado, or a hurricane, requires the instruments of modern science. Losing the weather satellites we currently rely on would destroy the predictive advantages we currently enjoy, wherever such technology is employed. Reliance on alternative, nonscientifically warranted beliefs about health care is likely to be ineffective, except for its placebo effects.[6] While it is possible that some aspects of some alternative health care might be effective, there are no good reasons to accept any such claim in the absence of a rigorous scientific testing of the claim. It is indisputable that nonscientific peoples have developed practices that provide some measure of success in dealing with some problem situations – the fact of survival over time is evidence of this. But organized scientific inquiry such as Merton describes has led to a body of knowledge that permits successful practice in vastly more situations.

Harding writes that it was once possible to naively imagine that scientific inquiry provided a means to achieve or approach true beliefs about the world. "When the ideal results of research could be assumed to be socially neutral, truth or truth-approaching could appear to be a reasonable way to conceptualize the relationship of our best knowledge claims to the natural and social world" (p. 143). But Harding maintains that once the operations of scientific inquiry are understood to be

[6] See Bausell (2007) and Singh and Ernst (2008).

dominated by extrascientific influences, as she argues they are, the notion that scientific inquiry can lead to true beliefs, or truth-approaching beliefs, will be discarded. Like Feyerabend, Harding finds all truth claims to be inimical to social discourse. "Truth claims are a way of closing down discussion, of ending critical dialogue, of invoking authoritarian standards" (Harding, p. 145). Rather apocalyptically, Harding asserts "The achievement of truth would mark not only the end of science, but also of history" (p. 145).

But, contra Harding, even the best of scientific theories are not taken, by scientists, to have been absolutely proven to be true. The concept of truth, understood as correspondence of theory to reality, is a regulative ideal in scientific inquiry. But it is well understood within the scientific community that scientific inquiry is an ongoing process of appraisal. A scientific theory, when accepted, is always accepted tentatively, as the best theory to date, pending further inquiry that may falsify it. Nevertheless, a theory may have such strong warrant that it is judged unlikely to be falsified.[7]

Harding's position has many internal tensions. An orderly and objective reality is accepted, but to develop a body of objective knowledge of reality, involving beliefs that approach truth, is considered impossible. The concept of a universal[8] body of knowledge is replaced by the concept of discrete local knowledges, conceived in the fashion of Barnes and Bloor as collectively accepted bodies of belief. Claims to truth are considered to be means to achieve power or means to destroy free social discourse. The problem with "Western modern science" is taken to be that this mode of inquiry is *as incapable* of delivering a body of objective knowledge as any other mode of inquiry. The claims of science to do so Harding considers to be part of the efforts of the scientific community to achieve cultural domination. Conceiving the problem as one of cultural domination, Harding's solution seems to be to expose the deceptive nature of modern science and to promote acceptance of the legitimacy of whatever culturally specific belief system is found to be appealing. These positions figure prominently in the CSSE literature. Much less prominent have been the detailed criticisms levelled at Harding's position by prominent philosophers of science, many of them feminists.[9]

59.2 Cultural Studies of Science Education

Work in the cultural studies of science field lays the foundation for contemporary cultural studies of science education [CSSE]. The advent in 2006 of the journal *Cultural Studies in Science Education* is a significant development. Editors Roth

[7] See discussion of Popper, Sect. 59.3.3.

[8] Universal claims are those that apply *to the interaction in question* wherever and whenever that interaction occurs.

[9] Koertge argues that feminist epistemology "…stands in a sharply antithetical relationship to the core values of science" (1996, p. 416). See also Haack (2003) and Pinnick (2005 and 2008).

and Tobin introduce the journal in this way: "The journal encourages empirical and non-empirical research that explores science and science education as forms of culture….A requisite for all published articles is…an explicit and appropriate connection with and immersion in cultural studies" (Roth and Tobin 2006, p. 1). The editors note that the journal will include "…OP-ED pieces that present ideas radically departing from oppressive, hegemonic norms" (2006, p. 2).

It is clear from this introduction that the journal is to be intellectually selective, publishing only work that conforms to the politically and socially based critique of science explicitly established as the norm for the journal. Paradoxically, it appears that this is itself an establishment of an alternative norm that is oppressive and hegemonic, albeit on a local scale.

Work in the field of cultural studies in science education is diverse in some respects. Many studies focus on pedagogical practice, interpreting the particular needs of culturally diverse students and proposing means by which the science educator can meet those needs. Philosophical conceptions of the nature of science, reality, knowledge, and truth are explicitly addressed in some studies. A particular set of positions on these issues, consonant with the positions generally developed in the CSS literature, is an explicit expectation in CSSE literature.[10] In this section I examine selected works in the CSSE literature that exemplify fundamental philosophical commitments in the field.

59.2.1 Pedagogical Practice

Elmesky (2011) provides an example of the pedagogical focus in the field. Elmesky notes that culturally diverse students bring with them "…heterogeneous cultural ways of being …that would be typically considered as unscientific …" (2011, p. 54). Elmesky argues that the teacher ought to incorporate the students' cultural ways of being into the study of science. Taking marginalized African American students to constitute a culturally distinct group, and taking rap music to be one of their cultural ways of being, Elmesky incorporates the creation and performance of rap music into the science classroom. Elmesky provides the following rap, created by the students, as an example of culturally relevant pedagogy:

> People depend on sound to get us around almost everyday we sometimes hate it but then we love it especially when we play even the blind it helps them to find a way to see in mind sound makes vibrations which makes equations so take advantage of this information. (Elmesky 2011, p. 65)

Elmesky states with approval that the rap recognizes that sound travels through vibrations and that the phenomenon can be symbolically represented through equations. Elmesky does not provide any evidence that the exercise has helped students

[10] See also Roth and Tobin (2009).

acquire a scientific understanding of the nature of sound. Elmesky instead justifies[11] the incorporation of rapping in the science classroom as a matter of social justice, as a repudiation of the cultural domination based on class and race that she takes to be inherent in science.[12]

Elmesky's account of the enthusiasm of the students suggests that the time given to rapping may serve a positive pedagogic purpose. It appears to permit a relaxed interlude of simple fun that allows the students to return refreshed to the cognitively demanding work of learning science. But Elmesky does not conceive the rapping interlude in this way. The exercise is said to provide the marginalized students to be themselves; it overturns "...the ideologies of the dominant race and class [that] continue to govern what content is taught as well as the frameworks for pedagogical approaches to instruction" (2011, p. 54).

Elmesky does not consider the possibility that a disservice is done to the culturally marginalized students if they end up with a less rigorous, less thorough, instruction in science than culturally mainstream students. Social justice in education requires that the same high quality of instruction be provided to all students, regardless of social or economic status, or unwarranted assumptions of culturally based limitations in capacity to learn. Contra Elmesky, pedagogical methods in science education should be justified by reference to their effects on the acquisition of scientific knowledge, if social justice in education is part of the educational goal.

Meyer and Crawford's "Teaching science as a cultural way of knowing" provides another example of cultural studies work focused on pedagogical practice. The culturally diverse students in this study are Latino, African American, Native American, and English language learning students. Consistent with both Merton (1942) and Roth (2008), Meyer and Crawford conceive of science as a culture having its own rules and operations, "...a dynamic and negotiated way of knowing that is practiced by a particular community..." (2011, p. 530).

Like Harding, Meyer and Crawford regard the scientific way of knowing as just one way of knowing, among many other different but equally legitimate culturally specific ways of knowing. Scientific inquiry, they write "...is implicated in Western ways of knowing...[which is]...an already accepted cultural norm for many mainstream students" (p. 535). Meyer and Crawford state that "[w]ithout culturally relevant instructional practices aimed toward facilitating student border crossing... science instruction incurs a form of symbolic violence, where one way of knowing dominates and seeks to replace others" (p. 532).

There are good reasons to doubt that the everyday cognitive habits of *any* student are well aligned with the norms of scientific inquiry, when he or she begins to study science. The norms of scientific inquiry make it a decidedly unnatural way of thinking, which not only must be learned but takes considerable effort to learn.[13] Yet it is the commitment, in science, to continual skepticism toward currently accepted theories

[11] The term "to justify" here means "to give good reasons for" a proposition, belief, or practice and does not imply demonstrative proof.

[12] See also Emdin (2009).

[13] See also Wolpert (1992) and Cromer (1993).

that makes scientific inquiry especially well suited to contribute to the growth of the body of scientific knowledge. Nonscientific ways of thinking, lacking the commitment to organized skepticism, place value on a non-changing body of belief, maintained by indirect cultural norms that prohibit skepticism and by direct coercive force when necessary.

There is no good reason to conclude that students of certain ethnicities or students whose first language is not English would be drawn innately to nonscientific ways of thinking or would experience greater difficulty in learning science than Caucasian speakers of English. The educational challenge, for all students, is to find pedagogical methods that lead to an understanding of contemporary scientific knowledge, at least sufficient to support rational participation in social policy determination, and an appreciation of the distinctive intellectual merits of scientific ways of thinking.

59.2.2 An Ontology of Difference

Roth's (2008) work "Bricolage, metissage…" provides an example of CSSE work that goes beyond pedagogical practice, into the realm of ontology and linguistic meaning.

Roth begins with an analytic claim about language: that being different is generally understood as merely the negation of being the same. Roth argues that, when *difference* is treated as conceptually secondary, instances of identity are valued more highly than instances of difference. Whatever fails to conform to an accepted standard, Roth claims, is considered to be not merely different, but deficient, inferior.

Roth sets out to reverse the posited evaluative bias. He proposes that *difference* be conceived as the primary ontological category, as a concept that is "in and for itself" (p. 898). *Identity* would then be conceived as merely a limiting case of difference, an ideal state that is never fully achieved in actuality. Roth proposes that an ontology of difference should replace what he claims is the currently accepted ontology of purity.

Roth's empirical claim about the negative value currently accorded to difference is problematic. Contra Roth, it seems that whether sameness or difference is more highly valued depends on the type and context of the particular judgment of value. For example, in a dog show, sameness, i.e., close conformation to a breed standard, is the value sought; in a film festival, it is likely to be difference, originality, creativity, and novelty that are most highly valued.

The epistemological implications Roth sees in the proposed ontology of difference are problematic. Knowledge, Roth argues, on this view would no longer be conceived to be a single self-consistent ideal essence.[14] Instead, knowledge would be understood as a "singular plurality" that is constituted by myriad concrete instantiations of knowing. According to Roth, this means that knowledge must be

[14] Knowledge is not in fact generally conceived today in this Platonic sense, particularly not in the context of modern science; see Sect. 59.3 for further discussion.

conceived as comprised of many different and mutually inconsistent knowledges. We must, on this view, relinquish "…the idea of *one* true scientific knowledge against which all other forms of knowledge are evaluated, asked to be abandoned, and, still worse, to be 'eradicated'…" (2008, p. 903).

Roth takes his ontology of difference to mean that any cultural accepted interpretation of a term such as "knowledge" counts as a legitimate expansion of the set of all referents of the term. This might, at first, seem to be a laudably open and accepting interpretation of knowledge. The problem is that terms that are given an infinitely extendable reference lose their value as communicative tools. Once there are no limits to the reference of a particular term, there is no longer any communicative point to be served by using the term.

Contra Roth, the crucial linguistic goal is to clarify the different usages of the terms "knowledge," "science," etc., to better understand the various interconnected, or inconsistent, meanings of the terms. Roth's approach, which is to expand the extension of the terms by broadening their definition indefinitely, is linguistically counterproductive.

The difficulties with Roth's ontology of difference, with its conceptual consequences, become apparent when Roth applies his ontology to the practices of science education.

59.2.3 Hybridized Knowledge

Roth develops an account of knowledge that he terms "hybridized knowledge." Roth observes that a global cultural diaspora is under way and that diasporic persons struggle to form new, composite identities that allow them to accommodate to a new culture while retaining cherished aspects of their original culture. Roth then argues that the culture of modern science is distinctive in being fundamentally foreign to *all* human beings and that all persons are diasporic with respect to science. Science, Roth observes, demands that familiar everyday concepts be set aside in favor of concepts of science that are, in a sense, nonnatural, i.e., intuitively implausible. In this, Roth is correct. For example, the familiar notion that the sun circles the earth, rising in the morning and setting in the evening, is dismissed by modern science, because it has been shown to be false. The familiar but discarded view, according to Roth, becomes "…an affront to the legally embodied and administratively enforced culturally (scientifically) correct one-and-only way of explaining this phenomenon" (2008, p. 894).

It is, however, contrary to the scientific norm of continuous organized skepticism for science to coercively enforce belief in the way Roth states. It is possible and common for demonstrably false but familiar ways of speaking, and thinking, about the world to persist alongside scientifically well-warranted theories, when serious consequences are unlikely to arise. But, there are circumstances when the consequences of such nonscientific belief are or would be serious. When certain social policy issues are decided, it is desirable that the most current scientific knowledge be employed. For example, it is common in some cultures to require that children be vaccinated against certain communicable diseases before beginning school. The relevant scientific knowledge is so well warranted that, in the interests of public welfare, even those who disbelieve in vaccination are nonetheless generally required to comply with the vaccination policy. It is possible that this sort of social employment of scientific knowledge counts as the legal and administrative imposition to which Roth objects. There are indeed issues of social ethics that arise with respect to the social use of scientific knowledge. At the most fundamental level of analysis, the reliance upon scientific knowledge in social policy decision-making can be justified only if there is something distinctively valuable about scientific knowledge, when compared to bodies of belief otherwise generated. Such a distinctive value of scientific inquiry and knowledge is what Roth is denying.

Learning modern science on Roth's view is inherently a matter of cultural suppression and the imposition and enforcement of the privileged culture of science. Every person studying modern science, Roth holds, resists that cultural suppression and struggles to construct a hybridized version of knowledge that incorporates the science into the familiar. Roth provides an example of the problems faced by the diasporic science student. He examines a doubly diasporic situation – English-speaking students who are learning science in French, in an immersion context. Roth finds that the students employ a hybrid lingua franca that incorporates elements of both French and English. The same hybridization process is seen in the students' understanding of science concepts.

Roth reports that, after 8 years of learning French and science, neither the language nor the science of the students is satisfactory to their French and science teachers. Roth, though, chides the teachers for evaluating students on the basis of the traditional ontology of sameness. These teachers, in his view, mistakenly believe that modern science and French have ideal essential forms that should be accurately replicated by the students.

Roth evaluates the students' work, instead, in light of his new ontology of difference. He interprets the students' errors (departures from authorized usage) as positive effects of the creativity and tenacity of those who are doubly diasporic. Engaged in two foreign tasks, the students employ all of their cognitive resources, new and old. Roth argues that, in order to learn the new cultures, students must actively engage in them, prior to having achieved an accurate understanding of them. In Roth's view it would be miseducative to insist from the start on accurate replication, as this would inhibit the students' exploration of the two new cultures.

Roth argues that science teachers too often fail to accept and value such hybridized knowledge. Instead, they impose rigid norms of discourse, doing symbolic violence to the students, by disallowing their familiar culturally accepted forms of speaking, thinking, and knowing. The problem seems to be an unwarranted epistemological domination. Roth takes it to be problematic that scientific discourse is "…constituted and considered as superior to any hybrid discourse" (2008, p. 913), while familiar beliefs that originate outside the science classroom are considered to be "…a lesser form of knowing than the one to be inculcated" (p. 913).

59.2.4 Indigenous Knowledge, Indigenous Science

The concept of indigenous knowledge, and of indigenous science as the source of that knowledge, has an important place in the CSSE literature.[15] Proponents of indigenous science make the claim that long-resident peoples have, over many thousands of years, developed sets of culturally specific beliefs about the world and its dynamics. Nonliterate cultures have bodies of belief that have been preserved in strong oral traditions, which are said to have been handed down unchanged through many generations.[16] These bodies of traditional belief are considered to have successfully guided the cultures in acting in the world over vast periods of time, and given this, they are claimed to merit the term "knowledge" and to count as indigenous science.

Ogawa, a proponent of the concepts of indigenous knowledge and indigenous science, redefines "science" in a way that permits traditional beliefs to count as science. Ogawa adopts Elkana's conception of science, who states: "By science, I mean a rational (i.e., purposeful, good, directed) explanation of the physical world surrounding man" (1971, p. 1437). With that broadly inclusive definition in hand, Ogawa is able to conclude that "...every culture has its own science...its 'indigenous science'" (1995, p. 585). Given the redefinition of science, this is merely to say that each culture has its own rational explanation of the world.

The requirement of rationality might be thought to place some limit on what could count as science. But Ogawa makes it clear that, in his usage, even rationality is to be reconceived as relative to the cultural accepted beliefs. Each culture, he argues, has its own worldview, i.e., its own set of traditional beliefs, and each worldview gives rise to its own version of rationality. Having risen from traditional beliefs, each culturally specific concept of rationality invariably affirms that the culturally specific beliefs are "rational" in a culturally specific sense. Thus, the fact that a set of propositions about the world has been widely believed, over a great many years, is taken to count as good reason for continued belief, i.e., continued acceptance of the propositions as true. This interpretation clearly harks back to the work of Barnes and Bloor.

Ogawa is opposed to a doctrine he calls scientism. He first conceives scientism as "...believing uncritically in science..." (1998, p. 106), that is, believing *in* science "...without understanding science" (p. 107). Ogawa's point here is a good one: to believe in science uncritically would be epistemically problematic and would promote a tendency to accept any thesis on faith, as a doctrine never to be questioned. The solution to this problem would seem to be provision of more and better education in modern science, so that an understanding of science could replace "belief in" science. Yet this is not Ogawa's solution.

Ogawa sets out a second, and more standard, conception of scientism. Scientism, Ogawa writes, is "an ideology that identifies valid knowledge only with science"

[15] See also Deloria (1997) and Snively and Corsiglia (2001).

[16] Though, in the absence of recorded accounts, it is difficult to test and verify this claim.

(p. 106), and he rejects scientism in this sense.[17] Ogawa develops the educational implications of his anti-scientism and begins with what appears to be a paradoxical claim, the claim that multiculturalism is an inadequate response to cultural diversity. This is so, he argues, because multiculturalism is too often conceived to require mere mention of scientists of diverse ethnicities and cultures and of the contributions they have made to modern science. He considers this approach problematic, because it involves the deprecation of the nonscientific traditional belief systems of indigenous or other nonscience-based cultures.

Multiculturalism might be conceived as requiring a restructuring of science instruction intended to more effectively induct students from diverse cultures into the culture of modern science.[18] Ogawa takes this induction into modern science to be itself a problem, because in the induction process students belonging to non-Western cultures become alienated from their own cultural roots. Ogawa is objecting to modern science in the way Merton considered to be an attack from outside, which is expected when the organized skepticism of science threatens long cherished beliefs. It is troubling that this attack on science "from outside" is now advanced from within the field of science education. Contra Ogawa, to develop the means to critique, and discard, aspects of one's traditional culture is an inherently liberatory process, increasing each student's cognitive and moral agency.

Ogawa proposes to replace the multicultural approach to science education with what he terms a "multi-science" approach. In keeping with Barnes and Bloor's analysis, science is to be reconceived as relative to cultural norms and practices of thinking and knowing. Given this, science educators should "…view Western modern science as just one of many sciences, that is, science in the context of science education need not necessarily be Western modern science" (p. 584). "Scientism," in Ogawa's second sense, would thus cease to pose a cultural threat, since every cultural belief system would count equally as science, by definition. Accepting the legitimacy of indigenous science entails rejecting the notion that modern science is the only source, or the best source, of knowledge about the interactive dynamics of the world.

Ogawa considers it imperative that the indigenous sciences of students, as well as the students' unique personal sciences, be included in science curricula, along with "Western" modern science, "…as one of the main curriculum emphases" (p. 592). Learning "indigenous science" is a matter of studying the ancient traditional beliefs of whatever cultures are included, and incorporating those various beliefs into one's worldview. Content courses in science teacher education programs, Ogawa adds, would need to be revised to include study of a wide array of traditional belief systems of indigenous peoples.

Curiously, modern science, which Ogawa terms "Western" science, is not itself conceived to be an indigenous science. Ogawa observes that many in Western cultures feel no personal affinity to modern science and concludes that modern

---

[17] Aikenhead (2001) provides another example of the anti-scientism position. See also Cobern and Loving (2008).

[18] See also Southerland (2000).

science is uniquely isolated from the everyday lifeworld of even "Western" people. Modern science, Ogawa holds, is inherently alienating because of its ontology: modern science "…pertains to a Cartesian materialistic world in which humans are seen in reductionistic and mechanistic terms" (p.589).[19] It is the alienation of modern science from any particular culture, Ogawa maintains, that allows individuals in diverse non-Western cultures to do modern science. That modern science is not grounded in a specific culture, but is open to all, is taken as evidence that it is merely "…a theoretically materialistic science, …a kind of game open to anybody who will obey its rules" (p.589). The openness of science to all participants might seem to be to its credit, dispelling worries that students from non-Western backgrounds are unsuited for the study of science. But Ogawa sees the openness of modern science as a defect. Science, having destroyed the students' indigenous culture, leaves them adrift, belonging to no culture. The exclusivity of indigenous cultures seems to be, for Ogawa, a positive thing.

Western modern science, in Ogawa's view, is a culturally oppressive institution wherever it occurs. The authority of science, he states, arises only from the consensus of its scientists, and he holds that this socially powerful group imposes its views on nonscientists, i.e., the cultural majority, without their understanding or consent. "[W]estern modern science is justified only by the scientific community itself. All other institutions have been excluded from the 'inquisition' of scientific justification and are expected to accept it without objections or doubts" (1995, p. 589). According to Ogawa, in every cultural setting, including that of the "West," both science teachers and students feel modern science to be a foreign imposition.

The autonomy of science, with its organized skepticism toward long-standing traditions of belief and practice, which Merton lauded, is generally taken in the CSSE literature to be a negative thing. Science education is a matter of indoctrinating students into a destructive worldview, imposed on students against their will. Roth, for example, writes:

> One of the key concerns I have with science education is that it is little different from what religious education has been in the past. It is a form of indoctrination into a form of thinking about the world that has been shown to be detrimental to the well-being of our planet that we both constitute and that is our home, our dwelling. (Barton et al. 2009, p. 194)

It is true that the growth of the body of scientific knowledge has permitted the development of effective technologies that have enabled an enormous increase in the size of the human population. There is little doubt that science has contributed in this way to the creation of serious environmental problems of global scope. But it is the use to which scientific knowledge is put, coupled with a lack of knowledge of or a willful blindness to the consequences of that use, which creates the problems, not the content of scientific knowledge per se. Knowledge of the dynamics of the world is needed if we are to ameliorate the environmental and social problems we face. The methods of scientific inquiry, developed in the context of continuous

[19] If Ogawa is classifying Descartes as a materialist, his understanding of perhaps the most famous dualist among Western philosophers is deficient.

critical evaluation in an open community of inquirers, are means of attaining understanding of the actual dynamics of the interactions that constitute the world. The set of scientific methods is understood to be subject to expansion and revision. There is no bar to the incorporation of new methods, provided that the proposed new methods prove able to pass the tests of the organized skepticism essential to science. There is no sense in placing one's confidence in "other" ways of knowing, that is, in precisely those ways of fixing belief that have not been tested or that, having been tested, have failed.

59.3 Philosophical Considerations About Science

Conceptions of science are grounded in philosophical conceptions of the nature of reality, knowledge, and truth. Radically different interpretations of these basic philosophical concepts are clearly possible. Because radically different interpretations will lead to fundamentally different judgments about the value of practices in science education, consideration of these philosophical concepts has a practical importance.

The difficulty is that the philosophical concepts are interrelated and form an intricate network of meaning; they seem to defy separation and independent resolution. Nevertheless, an effort will be made here to examine them here in a more or less linear order.

59.3.1 Scientific Inquiry, Scientific Knowledge

Science developed as an improvement on the everyday efforts of individuals to understand the dynamics of the real situations in which they live and must act. The term "science" refers both (a) to a particular version of the human activity of inquiry and (b) to the intellectual product of that activity, a body of scientific knowledge. Popper's explication of scientific inquiry as a process of conjecture and refutation (Popper 1953, 1959, 1968, 1972) greatly clarifies the nature of science and informs this treatment.

Scientific inquiry, the activity, is the means developed by human beings to enable the discovery of the interactive dynamics of whatever exists, occurs, and interacts, i.e., the nature of reality. The aim of scientific inquiry is to develop explanatory theories about the world that are true of the world, that truly state the dynamic interactions that constitute the world. Scientific inquiry is an ongoing process of self-criticism, directed at the body of scientific knowledge and at the processes of scientific inquiry. The emphasis on continual self-criticism and improvement permits the hope that scientific knowledge, the product of scientific inquiry, will be increasingly determined by the interactive dynamic systems that constitute the world. There is, however, no guarantee that the process will lead to true statements about the world. Even if a statement or set of statements is in fact true of the dynamics in question, it is, by virtue of the nature of the scientific inquiry process, impossible to definitively, absolutely prove that it is true. Judgments about the truth of scientific statements are the all that can be had. Continual openness to the possibility of error is a necessary feature of scientific inquiry. Theoretical structures are always held open to improvement or replacement. Incompatible theories are taken to be in competition.

New judgments, based on new evidence, and/or on new analyses of existing evidence, are continually being made about the relative merit of competing theories. For this reason, the best-warranted scientific theories to date, are generally the more interesting and are the best bet when seeking to guide action. There are no guarantees of successful action, however; actions may fail even when guided by a true theory.

Scientific inquiry is a communal human activity. It requires a group of people acting cooperatively in pursuit of the same goal, the production of an objective body of knowledge of the world, and sharing a distinctive set of norms, methods, and institutions. Science requires a community and a community having a particular and distinctive culture that promotes the general aim, the search for truth. Maintenance of an open, sustained, cooperative, and critical discussion among qualified persons about the theories under evaluation is a necessary condition of scientific inquiry. The institutional structures of science are necessary for that maintenance.

Maintaining a self-critical attitude toward one's existing beliefs is made considerably easier by the public and interactive critical norms of science. The institutions of scientific inquiry, when they are working well, serve as guardians of the constraints of objectivity. Scheffler states that "[t]hese controls [on belief], embodied in and transmitted by the institutions of science, represent the fundamental rules of its game" (1967, p. 2). Ross et al. also consider the institutional arrangements of science to be the key factor in the demarcation of science. "…Science is, according to us, demarcated from non-science solely by institutional norms: requirements for rigorous peer review…requirements governing representational rigour with respect to both theoretical claims and accounts of observations and experiments…" (2007, p. 28). Despite human cognitive shortcomings, we can "achieve significant epistemological feats by collaborating and by creating strong institutional filters on errors" (p. 28).

Critical discussion is a necessary prelude to the experimental testing of competing theories. Focused on the logical merits of the theories, this discussion determines which theories should be experimentally tested and how those tests should be structured. Critical discussion is also required after the experimental testing, to assess the results of the testing. The rigor of communal critical discussion cannot be achieved by a single individual nor by a few individuals of like mind. Each human being is naturally subject to various biases introduced by personal history, personal desires, fears, enthusiasms, etc. The requirement that critical discussion be open and communal provides a corrective influence, as the various biases of the participants are counterbalanced. In any effort to critically evaluate one's own thinking, the cooperation of others, who think otherwise, is invaluable. But, the scope of the discussion is not limitless. Competence with respect to the history and current state of the relevant body of scientific knowledge is a practical necessity. Scientific inquiry does not require the consideration of the groundless assertions of ancient mythologies.

Scientific inquiry requires the use of well-warranted[20] methods and instruments of objective inquiry, and participation in the social institutions of science, that promote adherence to the norms of science. The history of scientific inquiry is a history of innovation and continued improvement, not only in scientific knowledge but also in the methods, instruments, and institutions of inquiry process itself.

Scientific inquiry, in Popper's conception, proceeds by a process of "conjecture and refutation." The generation of new ideas, conjectures, or, simply, guesses about the world is a creative human act, and this is the necessary first step in scientific inquiry. The source of a new conjecture is not in itself important; conjectures about the world might arise from prescientific myths, from inborn dispositional beliefs, or from dreams, etc. But not all conjectures are equally worth testing. The better conjectures will be bold, i.e., will have high informational content, and will make claims about the world that contradict current theories.

Scientific method cannot be accurately characterized as a simple, and yet universal, set of procedural steps. But, it is not the case that any activity, nor even any inquiry activity, counts as scientific. Over time, diverse inquiry practices that serve particularly well in severely testing theories about the world are identified, formalized, and adopted, tentatively, as methods of scientific inquiry. The methods of scientific inquiry form a vast and highly varied set, which is always subject to testing, revision, expansion, and improvement. Judgments about the quality of methods used in an inquiry are made continually and can only be made by scientific peers who have the requisite technical methodological knowledge.

A body of scientific knowledge is the intellectual product of scientific inquiry. Explanatory theories about the world, and the critical discussions of them, constitute the intelligible content of scientific knowledge. It is necessary for further scientific inquiry that the contents of the body of scientific knowledge be formulated linguistically and/or mathematically and that these formulations be set out in a physical form that permits open public scrutiny and evaluation. All of the contents of scientific knowledge remain open to continued public appraisal, and the best-warranted theory at one time may, on reappraisal, be demoted or may be judged to be false.

The contents of knowledge, belonging to Popper's *third world*, i.e., the world of the intelligible, are objective, i.e., real objects, real existents in their own right. Though the intelligible contents of the physical body of knowledge are the products of human activity, these intelligible objects are largely autonomous, in that they have, in their own right, properties and relations that are unaffected by human thoughts about them.

---

[20] The term "well-warranted" refers here to those theories, practices, means, etc., that have been subjected to rigorous testing and critical evaluation and have been judged to have done well in standing up to the scientific scrutiny, following Dewey's (1929) usage.

The current body of well-warranted scientific knowledge must at the least be internally consistent, if it is to have any prospect of being true of the world. Discovery of inconsistencies in the current theories about the world presents a scientific problem, to be solved by further scientific inquiry. A distinction is needed between current scientific theories, the possible truth of which is still being investigated, and theories that have been refuted or that have failed to succeed in competing with other theories. The latter remain in the body of scientific knowledge but are consigned to the archives, available for further scrutiny, but not considered in the scientific community to be worth further scrutiny.

A question is often raised: Is the content of the current body of scientific knowledge dependent on culture or on the world? The alternatives seem to be mutually exclusive, so it is not surprising that conflict has arisen over this question. Gieryn describes the conflict as a science war and places the combatants into two camps. Those in the "science studies" camp consider scientific knowledge to be culturally relative, simply a body of collectively accepted beliefs. Those in the "science defenders" camp consider scientific knowledge to be universal, equally applicable across cultures, a body of statements about the world that, if not true, are at least near to the truth (1999, p. 344).

There is no doubt that scientific inquiry is a historical process and that the events in that history are contingent, not necessary. Given this, the contingent events of human social life, and cultural circumstances, can and do affect the historical course of development of scientific knowledge. Given this, it is possible that activities of bona fide scientific inquiry could, in different and *communicatively isolated* human cultures, be developing differently over time. But it is a mistake to conclude, as cultural studies of science theorists generally do, that scientific inquiry and the intellectual product thereof, scientific knowledge, is therefore dependent upon the specifics of cultural history.

It is possible that members of two mutually isolated cultures could each independently develop a genuine culture of scientific inquiry, having the norms, practices, and institutions necessary for the practice of science and that these cultures would engage independently in scientific inquiry into the same natural phenomenon. It might happen that various explanatory theories are proposed in each isolated culture and that the competing theories are subjected to severe experimental testing and open and critical communal discussion in each culture. It is possible that judgments about the best theory of the entities and dynamics of the world in one culture would be substantially different from and inconsistent with the judgments of the other culture. The science developed in each culture is thus dependent, historically, on the unique course of events in each culture.

But the moment the isolated cultures come into communicative contact, the incompatible scientific theories come into competition. Assuming that it is the same phenomenon that is being investigated, there can be only one set of statements that truly states the dynamic interactions that constitute the interactive system in question. The presence of two or more different theories, each intended to be that set of true statements, simply sets out a new scientific problem. The situation would indicate to scientists in each previously isolated scientific community the need for further scientific inquiry to assess the relative merits of the competing theories. The theories are historically dependent on culture, but the truth, or nearness to truth, of the competing theories is not relative to culture.

The competing theories may persist side by side for a long time, pending evidence that refutes one or another of the theories. Perhaps each theory will give way to a new theory, one that is a better warranted and offers a better explanation of the phenomenon. But, in science, the appearance of two competing theories about the same phenomenon always presents a scientific problem to be solved. Refusal to engage in that reassessment of theories required by the norms of science is a clear indication that the supposed science of one of the cultures is not, after all, genuinely scientific.

It is rational to prefer, as a basis of action, the theory that has best survived severe empirical tests and communal critical discussion (Popper 1972/1995, p. 22). But acceptance of a scientific theory always remains tentative, no matter how successfully it survives severe empirical testing and critical evaluation.

59.3.1.1 Falsifiability

It is impossible to speak sensibly about science unless science can be conceptually distinguished from nonscience, from pseudoscience, and from speculative metaphysics masquerading as science. Popper distinguishes scientific theories from other theories that "…though posing as science, had in fact more in common with primitive myths than with science" (1953/1988, p. 20). It is clear from the CSSE literature that the demarcation problem is still a live question and one that has practical consequences both in education and in society at large.

Popper argues that the key to the demarcation problem is the falsifiability of scientific theories. To be scientific, a theory must be "…*incompatible with certain possible results of observation*" (p. 22, emphasis in original). According to Popper, "[a] theory which is not refutable by any conceivable event is non-scientific. Irrefutability is not a virtue of theory (as people often think) but a vice" (p. 22).

The initial conjectures about the world, which will be influenced by the contingencies of observation and inductive inference, must be formulated as testable hypotheses that specify what events must occur, when specified actions are taken under specified conditions – those actions under those conditions constitute the experimental test. Observing the events expected to occur provides a measure of confirmation of the theory; repeated confirmations are the basis of inductive inference that the theory is likely to be true or near true.

Inductive reasoning fails, however, as a method of scientific reasoning, because the scope of every scientific theory must go beyond the evidence obtained from any finite set of confirming instances (Popper 1953, 1959, 1972). In contrast, observation of events predicted by theory to be *impossible* serves logically to show the theory to be refuted. If the observation is judged in the scientific community to be a good one, if it is replicable, and judged not to be the result of error, the rejection or revision of the refuted theory is logically required.

Popper's falsifiability criterion for science does not constitute a decision algorithm to be mechanically applied to particular theories. It is not the case, in theory or in practice, that a single failed prediction is sufficient to falsify a theory. It is unlikely that events observed will perfectly match the events predicted based on a hypothesis intended. The events in the world that constitute the context of the experiment will affect the observation process in unanticipated ways. Calculations from the theory of the observations that would falsify the theory will include assumptions that are not well grounded and be incorrect. Judgments about the meaning of particular experimental results are an ineluctable part of the process of scientific inquiry.

59.3.1.2 Science and Pseudoscience

The ability to distinguish science from nonscience and from pseudoscience, and good scientific inquiries and theories from poor, is of the utmost cultural importance. Goldacre gives a detailed account of the prominence of pseudoscience in contemporary medical practice, in the form of complementary and alternative medicine [CAM]. Goldacre identifies a number of factors that work to undermine public understanding of science, even in cultures that one might expect to be most thoroughly scientific, and draws a discouraging conclusion. Addressing purveyors of pseudoscience, and opponents of genuine science, Goldacre says, "You win":

> …you collectively have almost full-spectrum dominance. Your ideas—bogus though they may be—have immense superficial plausibility, they can be expressed rapidly, they are endlessly repeated, and they are believed by enough people for you to make very comfortable livings and to have enormous cultural influence. You win. (Goldacre 2010, pp. 253 and 254)

Goldacre, however, sets out to counter the trend and explains crucial aspects of scientific reasoning that differentiate science from pseudoscience. Bausell, also focusing on the plausibility to many of complementary and alternate medicine, identifies numerous impediments to reasoning which plague human beings: "the family of logical, psychological, and physiological impediments to connecting cause and effect that necessitates the conduct of scientific research in the first place" (2007, p. 35).

Oreskes and Conway, in *Merchants of Doubt* (2010), document current efforts in the United States to keep well-established scientific knowledge from affecting governmental policy decisions. One effective tactic is to call for *more* science, for sound science, for critical reassessment of scientific findings. Such calls might seem to be the epitome of dedication to the methods of science. The problem is that well-warranted scientific knowledge is often countered and effectively undermined by well-funded disinformation campaigns that promote pseudoscience as science and that denounce genuine science as "junk science."

The proliferation of pseudoscience is a major problem in contemporary society.[21] CSSE theorists contribute to the problem in consistently fail to distinguish science,

[21] Also see Binns (2011), Gauchet (2012), Kitcher (1998), Morrison (2011), Wagner and Steinzor (2006), and Zimmerman (1995).

not only from pseudoscience but even from nonscience; even the possibility of drawing, the conceptual distinction is denied. This is the direct logical consequence of the CSSE conception that all collective systems of belief count equally as knowledge and that all methods of achieving collective belief count as but different sorts of science. In collapsing the distinction between scientific inquiry and other sorts of inquiry, CSSE theorists undercut educational efforts to bring students to understand the nature of science.

59.3.1.3 Objectivity

Objectivity is an epistemic virtue characteristic of science. Individual or collective objectivity is achieved to the degree that believing is influenced by objective factors, i.e., the controlling force of the dynamics of the world, and the constraints of logical consistency. Scheffler states the relation of science and objectivity:

> …*de facto* science articulates, in a self-conscious and methodologically explicit manner, the demands of objectivity over a staggering range of issues of natural fact, subjecting these issues continuously to the joint tests of theoretical coherence and observational fidelity. (Scheffler 1967, p. 13)

Daston and Galison (2010) trace the rise of objectivity as an epistemic virtue associated with scientific inquiry to the 1800s. Images of natural objects captured by mechanical devices, e.g., photographic plates, were considered to be objective because they would "preserve the phenomena" from the imaginative subjective impressions, intuitions, and biases of the inquirer. They conceive objectivity in this way: "To be objective is to aspire to knowledge that bears no trace of the knower – knowledge unmarked by prejudice or skill, fantasy or judgment, wishing or striving" (p. 17). Prior to the adoption of the ideal of objectivity, they claim, it was common practice for observers to discard discordant observations as defective and, guided by personal intuitions of truth and essential form, to seek out the examples that would seem to verify the favored theory.

To achieve an ideal state of perfect objectivity is in practice impossible and is sometimes conceived as an unnatural cognitive state for human beings.[22] But the employment of instruments that preserve the phenomena, coupled with public participation in studying the preserved artifacts, promoted the emergence of objectivity as an ideal or regulative idea for scientific inquiry. Objectivity, as an epistemic virtue, requires commitment to acquiring beliefs about the world that conform, as much as possible, to the actual states and dynamics of the world. In scientific inquiry, the goal is to design interactions, observations, that serve as tests of hypotheses about the world. Scientific institutions and norms are consciously designed to promote the achievement of objectivity.[23]

---

[22] See Wolpert (1992) and Cromer (1993).

[23] See Longino (1990), for another view.

Wolpert argues that scientific modes of thinking about the world are radically different from commonsense thinking about and conceptions of the world. Commonsense ideas of motion, for example, are generally not correct, and the explanations of motion found in modern physics are strongly counterintuitive. "Generally…the way in which nature has been put together and the laws that govern its behavior bear no relation to everyday life" (1992, p. 6). The need to commit to objectivity as an ideal guiding belief formation is one factor that makes the scientific mode of thought so different from everyday casual patterns of thought. Objectivity requires a willingness to be critical of one's current beliefs, even those that are comfortable, traditional and/or cherished. "Being objective is crucial in science when it comes to judging whether [one's] subjective views are correct or not. One has to be prepared to change one's views in the face of evidence, objective information" (Wolpert 1992, p.18).

Cultural studies theorists of science education are on solid ground when they observe that many students experience considerable difficulty in the study of science. To achieve an understanding of scientific inquiry and scientific knowledge does require the elimination of erroneous beliefs. Some students may indeed refuse the option to learn scientific knowledge and methods of inquiry, opting instead to maintain nonscientific traditional beliefs about the world. But the value of the scientific inquiry as the means of gaining knowledge of the world remains.

Ross, Ladyman, and Spurrett speak out "in defense of scientism." They argue that scientific inquiry is in fact the only means by which well-warranted and hence likely-to-be-true beliefs about the world can be acquired:

> …science just *is* our set of institutional error filters for the job of discovering the objective character of the world—that and no more but also that *and no less*—science respects no domain restrictions and will admit no epistemological rivals. (Ladyman and Ross 2007, p. 28, emphasis in original)

The only way to make a plausible case for there being some *other* way of knowing the world is to reconceive "knowing" so that the term refers to something other than having an accurate understanding of the world. It would be better, communicatively, to choose a different term.

Knowing the world scientifically, while highly valuable, is by no means the *only* valuable state worth experiencing. Esthetically appreciating the world is an eminently valuable state and a significant part of the full human experience. Rejoicing in the world, valuing the world, feeling a passionate connection to and responsibility for the world, and having fun in the world – these are all valuable and worthy states to experience. Imaginative stories about natural events and their meanings can be profoundly moving, emotionally, and can convey important social and personal values and thus have value. But none of these states or activities count conceptually as knowing. To conceptually distinguish knowing states, and efforts to know, from other states and other efforts does not in itself diminish the value of the other-than-knowing aspects of life. The animus against scientism, and against science, seems to rest on an unrecognized assumption that only that which counts as knowing, or as the means to knowing, counts as valuable. Having accepted that assumption, it seems important to insist that all valuable aspects of life must count as "ways of knowing," albeit very different ways. The better resolution, though, is to reject the false assumption that the sole experience that is of value in life is knowing.

59.3.2 The Object of Scientific Inquiry

59.3.2.1 Reality

Questions about the nature of reality are questions of ontology. Some of these questions are of ancient origin, e.g., is reality made up of material substance, or mental substance, or both? Reality might be conceived dualistically, allowing for both physical reals and nonphysical mental of spiritual reals. Alternatively, reality might be conceived monistically, allowing for only physical reals, e.g., chairs, dogs, planets, stars, electrons, and photons, or, alternatively, for only nonphysical reals, e.g., spirit, mind, and consciousness.

Today the question might be whether the matter and energy of contemporary physics exhaust the realm of reality or whether there is something else, in addition to the physical, that is also real, such as the mental, the spiritual, the intelligible, and the divine, or perhaps, consciousness. To be other-than-physical, the posited reals would need to be unconstrained by physical dynamics.

Popper divides the real into three distinctive worlds. Popper's first world is that of the *physical*, including the dynamic interactions thereof; the second world is that of the *mental*, of subjective states of belief, of knowing; and the third world is that of the *intelligible*, of statements, arguments, theories, and scientific knowledge.

In Popper's view, science involves the real things of all three worlds. The first world, the physical world, is clearly involved: physical human actions that have observable physical effects constitute empirical tests of theories. The second world, the mental world, is involved: human believing, human thinking, is required in the generation of hypotheses and in the critical discussion leading to evaluative judgment.[24]

The third world, the intelligible world is involved: linguistic formulations of hypotheses, theories, and arguments are required throughout the inquiry process, and theories and the critical discussions of them constitute the body of scientific knowledge that is the product of scientific inquiry. It is the casting of thoughts into linguistic form, itself a physical form, that produces intelligible objects and makes possible the ongoing public, critical examination that is a necessary characteristic of science.

Scientific inquiry begins with theorizing about the interactive dynamics of the physical world. This is not to say that only physics can be scientific – systems of dynamic interaction can be studied at any level, and an ecosystem is as legitimate an object of scientific study as a system of subatomic particles. Dynamic states of

[24] To call this the "mental" world accords well with common usage but can introduce unnecessary confusion, should the term "mental" suggest the traditional ontological dualism, with "mind" existing as a nonmaterial mental substance.

affairs of organic beings are as open to scientific inquiry as those of inorganic beings. Scientific inquiry into the dynamic states of conscious organic beings is equally feasible. Having a thought is not fundamentally different from having an urge or having a cold – all are complex dynamic states of affairs of organic beings. The mental world that Popper posits appears to be a part of the physical world.

To make sense of the practices of scientific inquiry, an assumption of metaphysical realism is necessary.[25] Metaphysical realism is the thesis that the world is objective, meaning that the dynamic interactions and resultant states of the world are as they are, regardless of any thought, wish, or belief about them.[26] There is a complication, though. Thoughts, wishes, beliefs, etc., are themselves objectively real; like all real thing/events, in a particular sense, these are mind independent. That is, the existence of a particular mental state is unaffected by thoughts, wishes, and beliefs about its being or having been in existence, except when there is an ordinary physical causal connection. This is not to deny that organismic states of affairs, including thoughts, beliefs, and desires, are, potentially, causally interactive with other objective real things and events. Causal interactivity is the very hallmark of the real, and if internal cognitive states were *not* causally interactive, that would be reason to doubt their physical reality (see Khlentzos 2004). It is to deny that the entities and interactions that constitute the world are dependent upon the operations of a postulated non-physical mind.

Scientific realism builds on the thesis of metaphysical realism. Scientific realism is the thesis that the theoretical entities in use in well-warranted scientific theories are to be understood as objectively real, and that statements about those entities are to be understood to be true, or false. Ellis sets out a well-developed version of scientific realist ontology. According to Ellis, the scientific realist accepts "that an ontology adequate for science must include theoretical entities of various kinds, and that it is reasonable to accept such an ontology as the foundation for a general theory of what there is" (2009, p. 23).

Ellis employs an argument to the best explanation in support of his scientific realism: "…if the world behaves *as if* entities of the kinds postulated by science exist, then the best explanation of this fact is that they really do exist" (2009, p. 24). He distinguishes descriptive theories, those that posit real entities, from theories that explain by developing idealized models of interactions. The idealizations employed in the latter, e.g., a frictionless surface, are not intended to be understood as really existing. The scientific realist has no need to take idealizations to actually exist.

In contrast, Ellis argues, the entities postulated by an accepted "causal process theory" should always be understood realistically. If theory A "is agreed to be the best causal account that can be given of the occurrence of some event E, and A is a satisfactory theory, then the entities postulated in A as the *causes* of E must also be thought to exist" (p. 31). The argument for real existence of theoretical entities is strengthened when the posited entities are found to have additional causal effects,

[25] Although it could be that metaphysical realism is false and that the practices of scientific inquiry actually do not make sense.

[26] See Devitt (1984/1991), Leplin (1984), and Melnyk (2003).

beyond those initially predicted. Ellis, like Khlenzos, holds that "causal connectivity is what characterized real things…real things should have a range of different properties, and so be capable of participating in various causal processes" (p. 32).

Ellis observes that the metaphysics of scientific realism quite naturally accords with scientific practices of inquiry, for the simple reason that this metaphysics "has been developed out of science, specifically to accommodate the developments that have occurred in this area" (p. 115). Given this, it is unsurprising that scientific realism "makes good sense of the nature and structure of scientific knowledge" (p. 115). But this ontology, like any ontology, is not intended only to be used in the context that gave rise to it. Ellis argues that scientific realism is likely to provide a sound basis for understanding in other areas, such as mathematics, and moral and political philosophy.

The scientific realist position sketched out here is a controversial one. Many philosophers of science, and scientists, adopt instead an instrumental conception of science. Instrumentalism in this context is the thesis that the statements that constitute scientific theories are not to be understood as truth-functional claims about reality. Instead, theories are to be understood as convenient, useful mathematical devices that happen to work well when making predictions about observable events. Instrumentalists argue that it is *not* the task of physics to arrive at true statements about things that really exist, particularly when the posited entities are unobservable. Ellis's separation of descriptive theories from theories specifically intended not to be descriptive seems to accomplish the conceptual work necessary to end the dispute.

Real events in the world affect human well-being, for better or worse, and these events are notoriously unresponsive to organismic states of wishing, desiring, believing, etc. Human action in the interactive world is required; that action, to be sufficiently successful in meeting human needs, must be guided by beliefs that are true or near to true. Instinctive dispositions to act in certain ways under certain conditions must have been sufficient at some time in our evolutionary history. But such dispositions no longer suffice. Beliefs must be carefully chosen, formulated, and tested for efficacy in guiding action successfully. The best of the tested beliefs are considered to be known.

Scientific inquiry is often still motivated by the pragmatic need to know, though it is often motivated today by the simple desire to know. In either case, the general aim of scientific inquiry is to seek truth, and its more achievable goal is to produce explanatory theories that enable humans to understand the world and to act more or less successfully in the world. A genuine explanatory theory must make true or nearly true statements about real entities and their real causal interactions.

59.3.3 Knowledge and Knowing

The term "knowledge" is used in a great many ways.[27] It is often used to refer to a particular subset of beliefs, those beliefs that are coupled with a strong feeling of

[27] See, for example, Sosa (2011).

certitude. Popper calls this the subjective sense of knowledge (1972/1995). Knowledge, in this subjective sense, is an organismic state of affairs of the individual; when the state of knowing is widely shared among a group of individuals, it is a cultural state of affairs. In either case, subjective knowing is an ordinary physical state of a very complex entity. Popper places these states of affairs in his second world, the world of mental entities, populated by knowings, as well as believings, desirings, wishings, fearings, etc.

A claim to *know* serves a linguistic function, that of providing a particular assurance. It conveys the speaker's claim that the proposition said to be known is not merely believed to be true, but is believed to be true for good and compelling reasons. When a person asserts that he *knows* that a storm is approaching, the intention is to assure the listener that this proposition is very well warranted, that it is true or near to truth, and that it ought to be relied upon as a guide to action.

States of individual or collective knowing can be divided into two sorts, based upon the quality of the inquiry process that led to the state of knowing. On one end of the continuum, there are casual inquiry processes, including minimal inquiries barely worthy of the name. On the other end, there are the carefully designed ongoing programs of concerted inquiry that count as scientific inquiry.

A casual inquiry process is often considered to be good enough to warrant the assertion "I *know*," particularly in everyday situations in which the stakes are thought to be low. Yet even at this low level of inquiry, there are standards. A body of casually acquired subjective knowledge is expected to be self-consistent. Revision and improvement of the body of knowledge is expected, by an ongoing assessment of the warrant of the beliefs, and the elimination of error. Confidently believed propositions which have been tested and have failed the test should lose the status of subjective knowledge.

Popper develops the concept of objective knowledge and distinguishes this from subjective knowing (1968/1995). Objective knowledge, in Popper's view, belongs to his third world, "…the world of intelligibles, or of *ideas in the objective sense*; it is the world of possible objects of thought: the world of theories in themselves, and their logical relations; of arguments in themselves; and of problem situations in themselves" (1978/1995, p. 154). These intelligible objects are man-made. But, having been made, these objects have an existence that is largely autonomous of the world of subjective thinking, believing, and knowing. The existence of objective knowledge does not require the existence of individual or collective subjective states of believing.

The objective body of scientific knowledge is distinctive in that it is the product of intentionally devised and tested methods of scientific inquiry. The goal of scientific inquiry is to generate theories about the complex dynamics that constitute the world that are true or near truth. The body of scientific propositions is gradually improved by the elimination of error, made possible by the process of severe testing and critical communal evaluation. The availability of this objective knowledge permits the gradual improvement of individual and collective subjective states of knowing.

There is a pragmatic advantage to employing the set of propositions that have best survived the scientific inquiry process to guide one's action. One will be better able to predict the effects of interactions in the world and to effectively act in the world. The severe testing and critical comparative communal assessment of competing theoretical explanations are the means to achieve this advantage. The norms of scientific inquiry and the institutional structures that permit and promote adherence to those norms are uniquely suited to developing knowledge in this subjective sense.

59.3.4 Truth

An important concept, truth, has been left for last. What is meant by the term "true"? The ordinary language usage of the term "truth" is explained well by the correspondence theory of truth. On this theory, a proposition about the world is "true" if and only if the relevant state of the world actually *is* the way it is stated to be. A state of the world is the truth-maker for every true proposition. Every true statement corresponds, in this sense, to some state of the world; every false statement fails to so correspond.

The correspondence theory of truth requires, clearly, that there be a world, but makes no assumptions about, nor has implications for, specific theories about the world. Any ontological theory can be matched with the correspondence theory of truth. When truth is understood as correspondence, the truth of a proposition provides a good explanation for the success of action taken on the basis of belief in that proposition.[28] Truth, understood as correspondence of the statement to the actual state of affairs, is the ideal goal of scientific inquiry.

Popper conceives truth as correspondence to reality, i.e., to the facts, to what is. Popper's conception of science requires this concept of truth. In developing the demarcation criteria for empirical science, Popper writes:

> …I was very far from suggesting that we give up the search for truth: our critical discussions of theories are dominated by the idea of finding a true (and powerful) explanatory theory; and *we do justify our preferences by an appeal to the idea of truth*: truth plays the role of a regulative idea. *We test for truth*, by eliminating falsehood. (Popper 1972/1995, pp. 29, 30)

The correspondence interpretation of truth is commonly accepted among philosophers; Vision (2004) develops a strong defense of a correspondence. But this interpretation is often considered by cultural theorists and postmodern relativists to be fatally flawed. The theory fails, it is often said, because there is no way for the human organism to have *direct access* to the actual states of the world, and there is thus no way to make the required comparison between (a) statements *about* the world and (b) the world itself. Absent direct access, it seems that on the

[28] But note that true belief is neither necessary nor sufficient for successful action.

correspondence theory of truth we could never determine, with certainty, the truth or falsity of any of our statements.

There are several problems with the criticism. The argument at best only shows that without direct access to the world, humans could not *find out* which statements are true and which false. But, it would be possible to could keep the correspondence theory of truth and accept that humans are doomed to ignorance of truth. But this is not necessary – it would be possible to accept the claim that humans do not direct access to the world, but do have sufficient access to the world to make critical judgments, based on good reasons, about the truth of our theories.[29] Or, one could accept the claim that causal interactivity of humans in the world provides the only form of direct access that is needed.

If one rejects the correspondence conception of truth, one might opt instead for an anaphoric theory of truth, conceiving truth as a pro-sentence forming operator.[30] Or, one might adopt a coherence theory of truth. On the coherence conception, each statement in a set of statements is considered to be true, provided only that the set as a whole is internally self-consistent. In a coherence theory of truth, the truth-maker for a statement is not the state of the world, but rather the internal coherence of the set of statements. The problem with this conception of truth is that, on this view, different and mutually inconsistent sets of statements would all be true, provided that each set is itself internally consistent.

Alternatively, one could choose to eliminate the concept of truth from knowledge entirely. Scientific theories could be understood to be, not true, or nearly true of the world, but instead "empirically adequate." In this case, the only claim made with respect to the theory is that it adequately accounts for the data at hand. An empirically adequate theory is often considered to be replaceable by any of an infinite set of different theories, each of which also accounts for the data. More radically, theories in science might be conceived to be nothing more than convenient fictions, contemporary mythologies developed communally by adherents to the culture of science, having no advantage over the different mythologies developed in other cultures. On this view the sole criterion for knowledge is *viability*, i.e., the mere survival of belief in the proposition. This interpretation of the relation of truth to knowledge is the one that is often in evidence in the CSSE literature.

59.4 Conclusion

Reality, knowledge, truth, and science – these interconnected concepts must form a coherent set, if any of them are to make sense. The concept of knowledge requires that there be some object of knowledge, a reality, to be known, and which can be known, at least in part, if knowledge is to be possible. Truth, understood as

[29] See also Lynch (2005).

[30] See Grover et al. (1975), pp. 73–125.

correspondence of propositions to reality, is a regulative ideal for knowledge, an ideal goal to be increasingly approached. There being no "book of correct answers," judgments must be made, continually, as to the quality of the body of knowledge developed to date, with the intention to improve upon the body of knowledge. Given this process, it should be expected that the best knowledge of the day will not be as good, i.e., neither as extensive nor as close to truth, as the body of knowledge that will be developed. But changes in the body of knowledge constitute improvements in understanding; they are not mere fluctuations arising from whim or fashion.

Scientific inquiry differs markedly from other means of belief acquisition in its explicit commitment to maintaining a continued attitude of critical skepticism toward the propositions of the current body of knowledge. Ongoing public testing of that which is currently accepted as knowledge is *de rigueur*, a necessary condition of science. Open, critical discussion of all aspects of the process of inquiry and the results of the inquiry is also necessary. Judgments that arise from that discussion are held open to question, and progress is conceived as the ongoing elimination of error. Intentionally seeking for error in the current body of knowledge is the goal of the inquiry.

Being fully real interactants, human beings have no shortage of access to reality, though understanding the interactions we experience is no easy matter. Having a need for such understanding, for knowledge, and also a passion for knowledge, human beings have developed a complex social institutional structure, science, that promotes the effort to understand the interactive dynamics of the world, to know. The social institution of science does not work perfectly in promoting knowledge of the world – this should go without saying. But this constitutes a social policy problem to be addressed, not an indictment against the social institution of science *tout court*. To have an organized social system dedicated to the testing of propositions about the world and open to criticism and improvement is better than not to have such a system. To make use of the scientific knowledge that emerges to guide individual belief and collective social policy is a good thing, better than to make use of beliefs otherwise produced. These are value judgments, not certainties, but they appear to be well grounded. If so, one of the principal educational projects must be to convey to new generations the nature and value of the social institution of science.

Matters that should be of major concern for science educators arise in the CSSE literature, traceable to the philosophical positions underlying much of that literature.

The chief concern is that, in the CSSE literature, scientific inquiry is not clearly distinguished from other sorts of activities nor is it held to be distinguishable. In the absence of this conceptual distinction, no method of generating propositions and of fixing belief in them can be legitimately excluded from the science classroom. The phrase "other ways of knowing" is ordinarily conceived to refer to ways of generating individual or collective systems of belief (subjective knowing) that are *other* than scientific. As such, they can be easily seen to be out of order in the science classroom. So, for example, a close reading of some form of sacred scripture, even though it would lead to subjective states of knowing, would be justifiably excluded. Close study of time-honored oral traditions about the world would, by the same token, be properly excluded.

But in the CSSE literature, purported other ways of knowing are reconceived and are taken not to be other than science, after all. Instead, they are taken to be forms of science, albeit radically different forms (recall Roth's ontology of difference and the "singular plurality" he employs in dissolving conceptual distinctions). Given this, ancient traditional stories about the world, as well as religious, ideological, and political doctrines, must be counted as forms of science, in the new, all-inclusive sense.

Explicit advocacy of this revision in meaning is common in the CSSE literature and is entailed when the philosophical groundwork of the CSSE literature is accepted. The loss of the concept of science as a distinctive form of inquiry generating a distinctive body of knowledge would constitute an enormous cultural change. The resultant cultural endorsement of nonscientific, and hence poorly warranted, belief systems about the world would have negative consequences. There are, first, negative effects on the young persons who are effectively denied access to modern scientific knowledge. This is, too clearly, a problem already, as educational systems fail to provide excellent science education to many students. But to intentionally link an inferior education in genuine science to the ethnicity and/or cultural background of the students, on the grounds that, for them, the study of modern science is a form of cultural suppression, considerably increases the ethical problem.

Second, there are likely to be negative social consequences, should the CSSE reinterpretation/destruction of the concept of science become fully accepted and standard in science education. Disapprobation of science, of scientific methods of inquiry and the resultant scientific knowledge, is already common in many societies across the globe – public discourse in the United States provides many examples. This is of concern because, in a representative democracy, widespread public disdain for science is easily transmitted into public policy.

References

Aikenhead, G. (2001). Integrating western and aboriginal sciences: Cross-cultural science teaching. *Research in Science Education*, 31(3), 337–355.

Barnes, B. & Bloor, D. (1982/1994). "Relativism, rationalism and the sociology of knowledge", in Hollis M. and S. Lukes (eds). *Rationality and relativism*. Cambridge, Massachusetts, The MIT Press.

Barton A.C., Tonso, K.L. & Roth, W. (2009). Diversity of knowledges and contradictions: A metalogue, in Roth, W., ed.. *Science education from people for people: Taking a stand(point)*. New York: Routledge.

Bausell, R.B. (2007). *Snake oil science: The truth about complementary and alternative medicine*. Oxford: Oxford University Press.

Binns, I.C. (2011). Battle over science in Louisiana. *Reports of the National Center for Science Education* (USA) 31.6, 2.1.

Cobern, W.W. & Loving, C.C. (2008). An essay for educators: Epistemological realism really is common sense. *Science & Education* 17:425–447.

Cromer, A. (1993). *Uncommon sense: The heretical nature of science*. Oxford: Oxford University Press.

Daston, L. & Galison, P. (2010). *Objectivity*. New York: Zone Books.

Deloria, V. (1997). *Red earth, white lies: Native Americans and the myth of scientific fact.* Golden Colorado: Fulcrum Publishing.
Devitt, M. (1984/1991). *Realism and Truth.* Princeton: Princeton University Press.
Dewey, J. (1929). *Experience and Nature*, 2nd ed. Chicago: Open Court Publishing.
Ellis, B. (2009). *The metaphysics of Scientific Realism.* Montral & Kingston: McGill-Queen's University Press.
Elkana, Y. (1971) The problem of knowledge. *Studium Generale*, 24, 1426–1439.
Elmesky, R. (2011). Rap as a roadway: creating creolized forms of science in an era of cultural globalization. *Cultural Studies of Science Education*, 6, 49–76.
Emdin, C. (2009). Reality pedagogy: Hip hop culture and the urban science classroom, in *Science Education from People for People: Taking a Stand(point)*, ed. Roth, W.M. New York: Routledge.
Feyerabend, P. (1974/1988). "How to defend society against science", in *Introductory Readings in the Philosophy of Science*, ed. Klemke, E.D., Hollinger, R., & Kline, A.D. (1988) Buffalo N.Y.: Prometheus Books.
Feyerabend, P. (1992). Nature as a work of art, in *Common knowledge*, quoted in Preston, J, Munevar, G. & Lamb, D., eds. (2000). *The worst enemy of science? Essays in memory of Paul Feyerabend.* New York, Oxford University Press.
Feyerabend, P. (1996/2011). *The Tyranny of Science.* Cambridge: Polity Press.
Gauchet, G. (2012). Politicization of science in the public sphere: A study of public trust in the United States, 1974 to 2010. *American Sociological Review* 77(2) 167–187.
Gieryn, T.F. (1999). *Cultural boundaries of science: Credibility on the line.* Chicago: University of Chicago Press.
Goldacre, B. (2010). *Bad science: Quacks, hacks, and Big Pharma Flacks.* New York: Faber and Faber.
Grasswick, H. E., ed. (2011). *Feminist Epistemology and Philosophy of Science: Power in Knowledge.* Dordrecht: Springer.
Grover, D., Camp, J. & Belnap, N. (1975). A prosentential theory of truth. *Philosophical Studies* 27 p. 73–125.
Haack, S.: 2003, 'Knowledge and Propaganda: Reflections of an Old Feminist'. In C.L. Pinnick, N. Koertge & R.F. Almeder (eds.) *Scrutinizing Feminist Epistemology: An Examination of Gender in Science*, Rutgers University Press, New Brunswick, pp. 7–19.
Harding, S. (1986/1993). *The science question in feminism.* Ithaca, N.Y.: Cornell University Press.
Harding, S. (1998). *Is science multi-cultural? Postcolonialisms, feminisms, and epistemologies.* Indianapolis: Indiana University Press.
Horgan, J. (1993). The worst enemy of science. *Scientific American* 269 p. 36–37.
Khlentzos, D. (2004). *Naturalistic realism and the realist challenge.* Cambridge, Mass.: MIT Press.
Kitcher, P. (1998). A plea for science studies, in Koertge, N., ed. (1998). *A House Built on Sand: Exposing Postmodernist Myths About Science.* Oxford: Oxford University Press.
Koertge, N.: 1996, 'Feminist Epistemology: Stalking an Un-Dead Horse'. In P.R. Gross, N. Levitt & M.W. Lewis (eds.), *The Flight from Science and Reason*, New York Academy of Sciences, New York, pp. 413–419.
Kuhn, T. (1962/1970) *The structure of scientific revolutions.* Chicago: University of Chicago Press.
Kuhn, T. (1977/1988). Objectivity, value judgment, and theory choice, in *Introductory Readings in Philosophy of Science.* Eds. Klemke, E.D., Hollinger, R. & Kline, A.D. Buffalo NY: Prometheus Books.
Ladyman, J. & Ross, D., eds. (2007). *Everything must go: Metaphysics naturalized.* Oxford: Oxford University Press.
Leplin, J. (Ed.) (1984). *Scientific Realism.* Berkeley: University of California Press.
Longino, H.E. (1990). *Science as Social Knowledge: Values and Objectivity in Scientific Inquiry.* Princeton, N.J.: Princeton University Press.
Lynch, M.P. (2005). *True to life: Why truth matters.* Cambridge Mass: MIT Press.
Melnyk, A. (2003). *A physicalist manifesto: Thoroughly modern materialism.* Cambridge, MA: Cambridge University Press.

Merton, R.K. (1938). Science and the social order. *Philosophy of Science* 5 p. 321–327. (Reprinted in Merton (1973).
Merton, R.K. (1942). Science and technology in a democratic order. *Journal of Legal and Political Sociology* 1:115–126. (Reprinted in Merton (1973) as The normative structure of science.)
Merton, R.K. (1973). *The Sociology of Science: Theoretical and Empirical Investigations.* Chicago: The University of Chicago Press.
Meyer, X. & Crawford, B.A. (2011). Teaching science as a cultural way of knowing: merging authentic inquiry, nature of science, and multicultural strategies. *Cultural Studies of Science Education* 6:525–547.
Morrison, D. (2011). Science denialism: Evolution and climate change. *Reports of the National Center for Science Education (USA).* Vol. 31 Issue 5 pp. 17–27.
Ogawa, M. (1995). Science education in a multiscience perspective. *Science Education*, 79(5), 583–593.
Ogawa, M. (1998). Under the noble flag of developing scientific and technological literacy. *Studies in Science Education, 31:102–111.*
Oreskes, N & Conway, E.M. (2010). *Merchants of doubt.* New York: Bloomsbury Press.
Pinnick, C.L.: 2005, 'The Failed Feminist Challenge to "Fundamental Epistemology"', *Science & Education* **14**(2), 103–116.
Pinnick, C.L.: 2008, 'Science Education for Women: Situated Cognition, Feminist Standpoint Theory, and the Status of Women in Science', *Science & Education* **17**(10), 1055–1063.
Popper, K. R. (1953/1988). "Science: conjectures and refutations" in. E.D Klemke, R. Hollinger, & A.D. Kline (Eds.), *Introductory Readings in the Philosophy of Science* (pp. 19–27) Buffalo N.Y.: Prometheus Books.
Popper, K.R. (1959/2002). *The logic of scientific discovery.* New York: Routledge.
Popper, K.R. (1968/1995). Epistemology without a knowing subject, in Popper, K.R., (1972/1989) *Objective knowledge: An evolutionary approach (Revised Edition)* Oxford: Clarendon Press.
Popper, K.R. (1972/1989). *Objective knowledge: An evolutionary approach (Revised Edition)* Oxford: Clarendon Press.
Preston, J, Munevar, G. & Lamb, D., eds. (2000). *The worst enemy of science? Essays in memory of Paul Feyerabend.* New York, Oxford University Press.
Ross, D., Ladyman, J. & Spurrett, D. (2007) *In defense of scientism*, in Ross, D. & Ladyman, J., Eds., *Everything must go: Metaphysics naturalized.* Oxford: Oxford University Press, pp. 1–65.
Roth, W. (2008). Bricolage, metissage, hybridity, heterogeneity, diaspora: concepts for thinking science education in the 21st century. *Cultural Studies of Science Education*, 3, 891–916.
Roth, W. & Tobin, K. (2009). *The world of science education: Handbook of research in North America.* Rotterdam: Sense Publishers.
Roth, W. & Tobin, K. (2006). Announcing Cultural Studies of Science Education, *Cultural Studies of Science Education*, 1, 1–5.
Scheffler, I. (1967). *Science and subjectivity.* Indianapolis: The Bobb-Merrill Company, Inc.
Singh, S. & Ernst, E. (2008). *Trick or treatment: The undeniable facts about alternative medicine.* New York: W.W. Norton & Company.
Snively, G. & Corsiglia, J. (2001). "Discovering indigenous science: implications for science education," *Science Education*, 85, 6–34.
Sosa, E. (2011). *Knowing full well.* Princeton: Princeton University Press.
Southerland, S.A. (2000). Epistemic universalism and the shortcomings of curricular multicultural science education. *Science & Education* 9: 289–307.
Vision, G. (2004) *Veritas: The correspondence theory and its critics.* Cambridge MA: MIT Press.
Wagner, W. & Steinzor, R. (2006) *Rescuing science from politics: Regulation and the distortion of scientific research.* Cambridge: Cambridge University Press.
Wolpert, L. (1992). *The unnatural nature of science: Why science does not make (common) sense.* Cambridge Massachusetts: Harvard University Press.
Zimmerman, M. (1995). *Science, nonscience, and nonsense: Approaching environmental literacy.* Baltimore: The Johns Hopkins University Press.

Christine L. McCarthy is associate professor in the Schools, Culture, and Society program at the University of Iowa, in Iowa City, Iowa, USA. She received an M.A. degree in Environmental, Population, and Organismic Biology from the University of Colorado and a Ph.D. in Education from The Ohio State University. Her research interests include philosophy of science, philosophy of mind, epistemology, critical thinking, ethics, and the philosophy of John Dewey. She is a coeditor of *The Sage Handbook of Philosophy of Education* (Sage Publications 2010) and author of the chapter *Concepts of Mind* therein.

Chapter 60
Science Education in the Historical Study of the Sciences

Kathryn M. Olesko

60.1 The Historiography of Science Education

Historical scholarship since the 1930s has demonstrated that science education is not merely a minor subfield of historical investigation somewhat akin to institutional history, but is in fact central to understanding the contours of scientific practice, the formation of scientific personae, and ability of the scientific community to reproduce and survive. The historiography of science education to date has highlighted the ways in which educational settings sustain clusters of values, mental habits, and material practices that make possible the epistemological and social dimensions of science, including the transmission and popularization of scientific knowledge; the conduct of teaching and research; the training of recruits; and the public's views on science, including its social, political, cultural, and economic functions and the image of the natural world it conveys.

What occurs *inside* educational settings has much to do with what is *outside* them. The values, habits, and practices of scientific practitioners acquired in training are sometimes drawn from culture at large, as they are when craft or technical practices are adapted to the study of nature. Conversely, the values and habits cultivated in science instruction are part of the socialization of the pupil, and thus, science education participates in the construction of the individual, society, the state, and civil society. In addition, norms of social interaction in educational – and by extension, professional or workplace – settings have been shown to be as important as knowledge transmission in the course of training scientists or educating pupils at all levels of instruction. Crossings between "outside" and "inside" or between science and society provide a way to understand the mutual integration of science and culture, including national goals. Studies of science education have thus

K.M. Olesko (✉)
Department of History, Georgetown University, Washington, DC 20057-1035, USA
e-mail: oleskok@georgetown.edu

M.R. Matthews (ed.), *International Handbook of Research in History, Philosophy and Science Teaching*, DOI 10.1007/978-94-007-7654-8_60,

demonstrated that the vitality of the sciences and their practices has as much to do with their internal robustness as with their linkages to broader historical contexts, including daily life.

The history of science has reached a point where science pedagogy now has a secure place in understanding the nature of science. Simply put, science cannot exist without institutional and intellectual forms of disseminating knowledge and educating students and practitioners. Yet the historical study of science pedagogy transcends concerns for disciplinary reproduction in the sciences. The histories of science education are now many, and major review articles on the topic have become more common.[1] Historical approaches to the topic, however, are bifurcated into historians of science who view science pedagogy largely (but not entirely) as a problem in disciplinary creation and reproduction and historians of education who view science pedagogy and science popularization more broadly as a means of transferring value from institutional science to the public at large for the purpose of securing social stability, economic well-being, cultural hegemony, or political power (Rudolph 2008).[2] School science, popular science, university science, laboratory science, industrial science, and government science are some of the most salient sites of the many types of science pedagogy that not only sustain the scientific enterprise but also present the public with value-laden options of how to live their lives.[3]

While current scholarship takes into account the wide variety of institutional spaces in which the transmission of scientific know-how, intellectual and manual, occurs, much remains to be done. This essay treats scholarship by historians of science who have studied science education at either institutions of higher learning or sites of professional scientific activity (e.g., postdoctoral training). To a large degree, these studies have focused on the training of practitioners, but they have also considered the broader social, cultural, economic, and political functions of science education in producing secondary school teachers, administrative bureaucrats, and engineers or in realizing the ideological goals of dominant elites, such as the German notion of *Bildung* or the American Cold War ideal of a national security state. After a brief historiographical review, this essay examines four principal loci of historical investigation: scientific textbooks; science pedagogy, or how science is taught and learned; pedagogical practices in the generational reproduction of scientists; and finally the political, social, and economic dimensions of science education.

[1] For overviews of the literature, see Macleod and Moseley (1978), McCulloch (1998), Mody and Kaiser (2007), Olesko (2006), Rudolph (2008), and Simon (2008).

[2] Concerning the transfer of values to the public, Rudolph (2008, p. 65) perceptively argues that the exchange goes both ways and that the boundary between scientific values and nonscientific ones is a zone of conflict worthy of historical investigation. Rudolph's review of the literature on science education and the lay public is exemplary (2008, pp. 69–75).

[3] For representative variety of settings, see Daum (2002), Dennis (1994), Geiger (1998), Holmes (1989), Kohlstedt (2010), Leslie (1993), Nyhart (2002), Olesko (1988), Olesko (1989), Pauly (1991), Rudolph (2002), and Schubring (1989). Studies of science instruction in primary education, secondary education, and the public sphere deserve their own dedicated historiographical reviews along the lines of Rudolph (2008).

It concludes by reflecting on how the emerging area of scholarship known as the history of the senses can be incorporated into the history of science education.[4]

60.1.1 The Early Twentieth Century

Before the 1930s dry-as-dust histories of educational institutions, dating from the late nineteenth to early twentieth centuries, had valorized the training of scientists in the industrialized world without casting a critical eye on the pedagogical process itself. With largely descriptive surveys underpinned by tables and statistics, these studies helped to create founder myths and institutional shrines within specific disciplines that subsequently proved difficult to displace in the historiography of the sciences. These myths helped to entrench a logical positivist historiography by viewing education through the lens of the progress of research. That approach faded in the 1930s when sociologists of knowledge, struck by the contrast between the liberal, rational conception of the individual promised by the Enlightenment and the conformities pressed upon the masses by totalitarian states, began to unpack the relationship between reason, behavior and social norms, and identity formation (Elias 1939; Fleck 1935; Schutz 1932).

Among this generation of sociologists of knowledge, Ludwik Fleck had particularly perspicacious insights into the nature of science learning in the context of what he called the "genesis and development of a scientific fact" – the general idea that facts are not discovered, but are rather made in a process that involved intellectual decisions, institutional practices, and social judgments that are all learned in training. Science education in his view created the mental and social frameworks necessary for the cohesiveness of a scientific community and for the creation and acceptance of new ideas. Education also established links to the past via the "syllabus of formal education" (Fleck 1979, p. 20). Fleck thus embedded the educational processes of socialization and training in broader contexts, claiming that "In science, just as in art and life only that which is true to culture is true to nature" (Fleck 1979, p. 35). Most relevant to this essay, Fleck believed that "initiation into science was based on special methods of teaching" (Fleck 1979, p. 112). But his views on science went largely unnoticed until the translation of his work into English in 1979. By then whatever he could have offered the historical analysis of science education was eclipsed by the popularity of Thomas S. Kuhn's *Structure of Scientific Revolutions* (1962) which, Kuhn later revealed, may in any event have had its origins in Fleck's work. Kuhn, however, quickly forgot he had read Fleck and could later only surmise his indebtedness to him (Kuhn 1979, pp. vii–ix).

[4] I thank Michael Matthews, editor of this volume and of *Science & Education*, for permission to reproduce and paraphrase parts of Olesko (2006) in this essay.

60.1.2 The Later Twentieth Century

In the aftermath of the Third Reich and the ideological realignment of postwar educational systems into the Cold War intellectual factories for defense, studies of the social system of science fell into two distinct phases, both of which shaped perceptions of the historical significance of science education. The first, from the end of World War II to roughly the beginning of the tumultuous social and political movements of the 1960s, was marked by an ideological capitulation to a system that placed great faith in science and technology as guarantors of the strength of the nation state, whatever its political orientation. Science education became one means among many for bolstering national security and tipping the global balance of power, as had occurred in the United States, the Soviet Union, and Great Britain and other nations that became members of the nuclear club. It also became a sine qua non for developing states that aspired to become modern. Key concepts defining the social system of science originating in this period tended to follow politics and shielded science from a deeper examination of certain features of its internal operation, including the question of how science was learned in the first place. A prominent example is Polanyi's notion of "tacit knowledge" which rendered ineffable some of the techniques of science as well as the methods of how scientists were trained (Polanyi 1958).

60.1.2.1 The 1960s and 1970s

The 1960s marked the beginning of the second phase when methodological changes in the history of science lifted the veil of secrecy that had hitherto concealed aspects of scientific work, revealing more clearly the interweaving of scientific and social practices. From historians as diverse as Michel Foucault (1966, 1975), Thomas Kuhn (1962) and Jerome Ravetz (1971) came a matrix of fruitful questions about the role of science education in the practical work of science as well as in discipline formation and maintenance.

By viewing scientific education as a process of near totalitarian indoctrination, Kuhn highlighted the powerful role of science pedagogy in transmitting paradigmatic problems, solutions, skills, and other guidelines for scientific practice. Practical activities, including instruction and knowledge production, were united in what Kuhn called normal science, his epithet for everyday scientific practices and beliefs. In his view the external world intervened in scientific practice only during periods of crisis that evolved into paradigm shifts when methods and skills metamorphosed in response to cognitive dissonance (Kuhn 1962).

More sensitive to the nuances of science pedagogy than Kuhn, Ravetz prioritized the social dimensions of instruction over intellectual ones. Training in how to make the kinds of sound judgments that avoided the pitfalls of scientific research (i.e., unsolvable problems and the dead ends of fruitless research trajectories) attracted his attention more than the content of knowledge or the means of its transmission.

Yet Ravetz was also deeply indebted to Polanyi and could not abandon the notion that skills were tacitly learned under the guidance of a master scientific instructor much in the same way that craftsmen learned trade skills. By definition skill learning could not be the object of historical investigation because it was ineffable. Ravetz viewed teaching as an intensely personal process, one so personal that were the precepts of scientific practice made explicit, learning the craft work of science would be irreparably damaged. Despite his insights, his impact on the historical study of science education has remained limited (Ravetz 1971).

Historians of science education may still genuflect to Kuhn, but it was Foucault who most invigorated theoretical discussions of history of science education. His intentionally ambiguous use of the word "discipline" – as conceptual organization but also corporeal training *and* character development – united the social, moral, and intellectual normalizing functions of education (Foucault 1975). Foucault was persistently critical of historians of science for their inability to grasp what was at stake in the construction of scientific regimes. For him the notion of "discipline" encompassed a plethora of minor procedures with major repercussions. Enforced by institutions of higher learning and the legal apparatus, disciplining *made* the modern individual and hence was constitutive of the formation of both modern society and the modern state. In three particular components of disciplining, Foucault discovered, too, the social processes at work in the pedagogic formation of modern scientific disciplines: hierarchical observation, normalizing judgment, and the examination or test (Foucault 1975). Although Foucault's views were not uniformly adopted, historians of science echoed his point of view in their study of systems of examination (Clark 2006, pp. 93–140; Macleod 1982) and in their affirmation of the centrality of teaching to launching and sustaining the disciplines (Pyenson 1978, p. 94). In other respects, however, the views of Kuhn and Foucault were often at odds with what more empirically based studies have demonstrated (Simon 2008, p. 105).

60.1.2.2 The 1980s and Beyond

A third conceptual phase, the focus of this essay, began in the last decades of the twentieth century. This phase was characterized by a deeper examination of the empirical record of science education in local, national, regional, and global contexts; a methodological pluralism that circumscribed the interpretive power of theoretical studies of science education (based nearly exclusively on Kuhn and Foucault) and expanded the role of historical contingencies in the shaping of science and its pedagogical practices; and a recognition that while science education was a subject in its own right, it was also an important site for understanding not only the larger structure and operation of the entire scientific enterprise but also more broadly in the construction of modernity. Consequently the historical study of science education became a window on the larger political, economic, and social environments of which science was a part. Due to the dominance of the military-industrial-university complex in the post-World War II period, the focus of historical studies of science education was largely, but not exclusively, upon the physical sciences.

Historiographical developments since the 1960s have refined the methodologies used to study the trio discipline, pedagogy, and practice. While not abandoning institutional contexts, new approaches have nonetheless gone beyond them. An important fruit of this effort has been the detailed historical examination of the training of neophyte scientific practitioners, which in turn has led to a recasting of how disciplinary history unfolds. Yet the historical significance of pedagogical experiences goes beyond the admittedly artificial confines of disciplinary history to include social, political, cultural, and economic history. These larger contexts have shown how widespread and necessary the framework of support and approbation was (and still is) for science education, dispelling the idea that science education is a self-driven enterprise.

60.2 Scientific Textbooks

The study of scientific textbooks was among the earliest genres in the history of science education. It still remains the most popular. Textbooks are enticing as historical objects of investigation because they present neatly packaged compilations and arrangements of scientific knowledge suited for instruction. They also confound historical investigation because they represent a selective history of their subjects. These contradictory traits led Kuhn (1962) to view them as little more than static moments or paradigms in the history of normal science and so as constraining in their effect upon students. Fleck (1935, 1979), however, created a dynamic conceptual framework that illuminated their role in discipline formation. He viewed textbooks as part of an intellectual continuum, occupying a position between journal and vademecum (handbook) science and popular science. As an intellectual hybrid, textbooks both initiated students to scientific ways of thinking and preserved some contact with ordinary knowledge.

Recent scholarship has cautioned against defining the textbook genre too narrowly, as an organized distillation of the results of research and in contradistinction to scientific popularization. The boundaries between different representations of knowledge now appear more fluid, and the distinction between genres less clear. At the most general level, textbooks are indispensable sources for capturing how thousands of students (and not merely future scientific practitioners) are exposed to science and what image of science they are likely to form. In the mid-1980s, sociologists of knowledge reinforced the association between textbooks and discipline formation by defining disciplines as "knowledge assembled to be taught" (Stichweh 1984, p. 7). Textbooks now are considered integral to understanding not only traditional topics of historical investigation, such as the development of ideas, epistemological choices and debates, the taxonomy of skill-based learning, and even the social dynamics of science such as priority disputes, but also the shifting relationship between science and society and the transnational nature of science (Simon 2011; Vicedo 2012, p. 83).

60.2.1 Textbooks and the History of the Disciplines

A defining feature of historical scholarship on scientific textbooks is its emphasis on discipline formation. Chemistry textbooks have attracted particularly sustained attention in this regard. Hannaway (1975) pioneered this branch of the historical study of science pedagogy in his study of Andreas Libavius's *Alchymia* of 1597. Regarded as the first chemistry textbook, *Alchymia* organized knowledge and united knowledge with practical skills; proffered plans for a "chemical house" or laboratory where hands-on learning would take place; and, in Hannaway's view, offered an alternative to the secretive nature of Paracelsus's alchemy by creating open chemical knowledge. By teasing out *Alchymia*'s long-standing usefulness and popularity across the century after its publication, Hannaway argued that *Alchymia* made vital contributions to intellectual dialogue on the nature of chemistry – quite the opposite of the deadening routine that Kuhn had identified with textbooks.

Historians have since qualified Hannaway's ambitious claims without dismantling its position as a turning point in the history of chemistry textbooks. *Alchymia* spread Paracelsian techniques by incorporating some of them into chemistry – thereby uniting the practical arts with science and academic forms of argument – and so to a limited degree became a textbook that was suited for both university instruction and the needs of the practical arts. According to Powers (2012), Herman Boerhaave (1668–1738) completed the transformation begun by Libavius. Boerhaave took a didactic form of chemistry based on some skills and operations, but lacking in concepts suited for examining the properties of chemical species, and combined it with elements of alchemy, chemically based medicine, and experimental natural philosophy – all of which he believed could fill in the conceptual gaps of a didactic chemistry. Furthermore, according to Powers, the instrumental practices of these latter three subjects (practices Libavius did not fully address) were crucial in shaping the practical side of chemical instruction. The result was Boerhaave's *Elementa Chemiae* which, in 40 editions between 1722 and 1791, set a pedagogical and research agenda for chemistry and defined chemistry as both an academic discipline and a practical art years before Antoine Lavoisier. Powers noted, however, that the assimilation of *techne* into teaching at the University of Leiden was not easily done, but once accomplished, chemical instruction assumed a dual nature as both theoretically and instrumentally based, with each side influencing the other. Thus, both Libavius and Boerhaave used science pedagogy as a platform for defining chemistry as a discipline.

Bensuade-Vincent in her review of textbooks from the chemical revolution (1990) argued that textbooks not only serve as snapshots of a discipline, but they are also essential for understanding the formation of schools, and so they function as tools of training, professionalization, and standardization (Bensuade-Vincent 1990). In this vein Hall (2005) has demonstrated that Lev Landau's and Evgenii Lifschitz's *Course of Theoretical Physics* played a decisive social role in the 1930s and later in shaping a Soviet research school in theoretical physics by framing problems and techniques for solving them that later carried over into research practice. In this way Soviet theoretical physicists could differentiate themselves from other schools, such as Arnold Sommerfeld's (whose German school was also created through a distinctive pedagogy and a defining textbook, *Atomic Structure and Spectral Lines,* which went through several editions during the crucial phase of quantum mechanics in the 1920s). Hence, although some textbooks defined transnational scientific communities, these Soviet and German cases indicate that the social and intellectual training of scientists could very well result in more localized sets of practices.[5]

In some quarters it has become commonplace to define and even to identify a discipline in terms of how it is taught or even represented in textbooks (Simon 2011). Certainly the creative role of textbooks in *helping* to create the disciplines cannot be denied. As textbooks are widely translated, reach transnational audiences, and become the foundation of national examinations in the sciences, the urge to associate them closely with discipline formation is compelling (Simon 2008, 2011). Especially when the creative processes at work in textbook construction, revision, and translation are considered, the ability of textbooks not only to *define* disciplines but also to *reshape* them is incontrovertible. Textbooks are remarkably fluid intellectual products (Bensuade-Vincent et al. 2003; García-Belmar et al. 2005).

Yet there are limitations to this perspective. Chief among is the danger of viewing the evolution of a textbook as teleological – as inevitably and directly reaching the terminus ad quem of a "discipline." That approach creates a deterministic pathway of analysis that could obscure the historical significance of a textbook that goes off the beaten path. Textbooks can be transnational, but they are also historically contingent in both creation and use. They can be universal, but they are also sites of conflict and competition. Arguments over which textbooks to use (or even to create) in science education are instances where there are competing views of reality, interpretation, and method coming to terms with one another. Such arguments could also be indicative of a struggle for scant resources (as when representatives of different approaches compete for the same clientele) or a struggle for prestige (as when scientists define their allegiances through the use of a particular textbook in teaching). These and other adaptations to or constraints of context limit the universal and transnational nature of textbooks. And context, in turn, modulates the degree to which a textbook does or does not contribute to discipline building.

The persistence of local scientific practices (especially industrial ones of relevance to the sciences, such as chemical technologies) that resist incorporation into textbooks, for instance, forestalls their broader recognition and acceptance and makes their adaptation elsewhere difficult if not impossible (Lundgren 2006). Other countertendencies to discipline building include the production of textbooks that challenge what later become dominant approaches (say alternatives to Newtonian physics in the eighteenth and early-nineteenth centuries, including Romantic nature philosophy) (Lind 1993, pp. 278–314). Examining only those textbooks which fed into the dominant tradition would be to represent falsely what historical reality was at the time. Most textbooks also fail to address some of the investigative techniques and skills of scientific practice which are incorporated, instead, into laboratory manuals (Olesko 2005). A textbook may be a partial map to a discipline, but it is not the discipline as a whole.

[5] As also demonstrated by Kaiser (2005a), Olesko (1991), and Warwick (2003).

60.2.2 Textbooks and Their Historical Contexts

Textbooks can also be viewed as focal points for many of the historical contingencies that shape both scientific practice and the roles of science and the scientist in society and so carry historical significances that transcend that genre. Their physical dimensions, for instance, are not boundaries that mark the "inside" and "outside" of science but rather can be likened to porous filters that permit the intermixing of several different cultural elements and so have been studied as a part of culture writ large. Recent scholarship has exposed the connections between textbook culture and the constitution of the public sphere; teased out the relationship between textbook production and social structure; and, most importantly, provided strong evidence that the decisive century in textbook culture may not be the nineteenth, when textbook culture matured, but the eighteenth, when textbook culture was just beginning.

A particularly productive locus of scholarship on scientific textbooks has been the team of Bernadette Bensaude-Vincent, Antonio García-Belmar, and José Ramón Bertomeu-Sánchez.[6] Their collective results are the most comprehensive, thorough, and innovative studies to date of the textbook culture in any of the sciences. To their credit they have viewed textbooks as active agents of culture, but not necessarily as carriers or even creators of disciplinary knowledge as early works in the genre, such as Hannaway's (1975), argued. They view textbook writing as a negotiation between author, public, press, and state (García-Belmar and Bertomeu-Sánchez 2004). The richness of their collective findings is in large part of the result of their ability to assemble international teams of scholars whose combined linguistic abilities enable them to examine cultures less well known and to achieve results attainable only through careful comparative histories. Of special note is the team's decision to examine the scientific periphery, including such places as Portugal, Hungary, and the Greek-speaking areas of the Ottoman Empire. Just as earlier works on science pedagogy during the Cold War adapted to a culture of secrecy and national security, this team's work on textbooks shows the impact of ongoing European integration.

Although their collective approach is largely empirical, their findings nonetheless mesh with earlier theoretical writings on science pedagogy. Of relevance to their project is Fleck's depiction of the historical role of publishing in sustaining science pedagogy where published knowledge becomes a "part of the social forces which form concepts and create habits of thought" determining "what cannot be thought in any other way" (Fleck 1979, p. 37). His account of the viability of scientific knowledge necessitates a reading public that takes an active part in the public sphere where discussions concerning the relevance and interpretation of scientific knowledge occur. So when Antoine Lavoisier's chemistry entered Portugal by way of Vicente Coelho Seabra's *Elementos de Chimica* around 1790, the absence of a local chemical community and a weak public sphere, constrained by the inquisition

---

[6] A partial list of their projects includes Bensuade-Vincent (2006), Bensaude-Vincent et al. (2002), Bensuade-Vincent et al. (2003), García-Belmar & Bertomeu-Sánchez (2004), García-Belmar et al. (2005), and Lundgren and Bensuade-Vincent (2000).

despite the expansion of print culture under Maria I (1777–1792), were reasons why Seabra's textbook was not adopted (Carniero et al. 2006).

Likewise in Russia the cumulative effect of the Church's monopoly on printing was to stunt the growth of a healthy public sphere where the free exchange of information could take place, thereby also restricting the growth of scientific communities (Gouzevitch 2006). In the Greek-speaking regions of the Ottoman Empire along the western end of the Mediterranean, the dominating presence of merchant elites meant that practical knowledge, conversions (weights and measures, coinage, and the like), and navigational issues were more important than Isaac Newton's *Principia*, so the former dominated textbooks in the physical sciences (Petrou 2006; Patiniotis 2006). Yet in each of these cases, the limited audience reached by textbooks did not diminish their roles in creating conditions conducive to the future growth of the public sphere: to wit, they promoted the standardization of language, vocabulary, scientific idiom, and alphabet that would eventually promote a larger reading public and audience for the sciences.

Publication patterns in scientific textbooks thus help in understanding the social structure and technical and scientific interests of the region over which they are found. The strong elite merchant class in the Ottoman Empire accounts for Greek translations of textbooks on practical geometry, geography, and commerce (all were useful for trade) and the relative paucity of textbooks on physics and chemistry, which carried little of significance for merchants. Conversely, as Patiniotis (2006) has observed, the absence of social support can doom a branch of knowledge. Textbook distribution reflects the balance of power among elites, as it did in the Ottoman Empire where the laws of the marketplace were more important than the laws of nature.

Characterized by discipline building, university history, the reform and extension of the secondary school, and the professionalization of the career of the scientist, the nineteenth century is often considered the defining moment in the modern social and institutional forms of science education. Recent studies of scientific textbooks demonstrate, however, that the eighteenth century may actually have more to offer us in terms of *why* (rather than *how*) these changes took place. As Patiniotis (2006) has pointed out, the word *textbook* was coined in the eighteenth century. The protracted shift from Aristotelian scholarship to more recent knowledge, as took place in Portugal under the *estrangeirados* during a period of enlightened educational reform, suggests that the intellectual dynamics of textbook organization in the eighteenth century may have been more problematic and difficult than they were in the nineteenth. Likewise the rapid intellectual shift in those areas under Napoleonic rule, such as northern Italy in 1796–1797 where the new French chemistry was established by law under public educational reform acts (Seligardi 2006), calls to mind the popular and social support required to make the shift permanent.

60.3 Science Pedagogy

Yet textbooks have their shortcomings as historical sources: they cannot reveal what went on the classroom, and they provide little information on how students learned and what their experiences meant to them. Since the late twentieth century, historians of science have turned to other types of documents in an attempt to understand the behind-the-scenes activity of teaching and learning in the sciences. Lecture notes, problem sets, student notebooks, examinations, laboratory exercises, instructional instrumentation, and multiple varieties of unpublished, duplicated materials have become privileged ways of reconstructing what went on in the seminar, lecture hall, and practicum. When supplemented by complementary materials, some published and some not (such as personal correspondence; diaries; autobiographies; laboratory notebooks or simply notebooks; and published versions of lectures, often straight from raw notes), the resulting historical scholarship reached even beyond a deeper understanding of science instruction to reveal how dependent all aspects of science as a human activity were upon educational processes. From primary education to the professional level of postdoctoral fellowships, apprenticeships, and the acculturation of mature researchers to new institutional settings, pedagogy played a part.

To confine science education to the transmission of knowledge or to the internal practices of the scientific community, then, is to mischaracterize the historical roles of science pedagogy. Science education has played a role in forming value systems; the scientific self (mentally, bodily, behaviorally, sensory, affectively, emotionally); social norms, including where in the social hierarchy different kinds of sciences fell; gender relations, both in- and outside the sciences; the power relations that determined the relative position of science and scientists vis-à-vis the state, society, and the economy; the cultural function of the sciences; and, finally, the role and perception of rationality in modernity. Science pedagogy thus has become the fulcrum which rests some of the most important dimensions of modernity. With what regard science education was held and why, as well as how much support that it garnered from the state and society, have become key historical questions in the study of both local manifestations and larger systems of science instruction.[7]

60.3.1 The Pedagogical Dimensions of Science Instruction

An early focus in the study of science pedagogy was the introductory science course offered in colleges and universities. Although it goes without saying that introductory courses had to be carefully framed to both attract and retain recruits in the sciences, only slowly did historians realize that their constitution demanded historical explanation. Geison (1978), Holmes (1989), and Olesko (1991) in their studies of, respectively, the physiologist Michael Forster at Cambridge, the chemist Justus Liebig at Giessen, and the physicist Franz Neumann at Königsberg are three early examples of how student needs shaped the tenor and texture of introductory courses. The pedagogical strategies of these scientists were instrumental not only in

[7]Major studies that contributed to the broader significance of science pedagogy include Clark (2006), Gooday (1990, 2005), Gusterson (2005), Hentschel (2002), Josefowicz (2005), Kaiser (2005a, b), Olesko (1991, 2005), Pyenson (1983), Rossiter (1982, 1995, 2012), Schubring (1989), Traweek (1988, 2005), and Warwick (2003).

accommodating their student clientele but also in preparing them for advanced exercises and eventually research.

The instructional successes in each of these cases were dependent upon intimate knowledge of their students' prior preparation, a judicious integration of the techniques of research into teaching, and a willingness to deploy pedagogical techniques that both worked and accommodated student needs. Effective teaching also depended upon coordinating the introductory science course with secondary school science instruction. Foster's evolutionary approach to biology and physiology challenged the former anatomical bias in English physiology; Neumann's instrumental use of mechanics brought astronomical techniques into the core of physics teaching; and Neumann's and Liebig's emphasis on instrumental and error analysis promoted more rigorous standards of precision in physical and chemical investigations (Geison 1978; Holmes 1989; Olesko 1991).

Using the introductory science course as representative of science teaching and learning, though, is a bit like claiming that a textbook represents what is taught and learned. In both cases access to what actually went on in the classroom is limited. The sources available to Olesko in her study of Neumann's seminar, though, overcame that limitation. With seminar reports, seminar exercises, correspondence, lecture notes, and student problem sets, she was able to render how both teaching and learning transpired in the seminar. The results were unexpected. Rather than inculcating only the mathematical techniques of theoretical physics, Neumann concentrated instead on teaching his students the methods of an exact experimental physics: to wit, the determination of both the constant and accidental (random) errors of an experiment, the latter by the method of least squares. Bessel's exemplary seconds-pendulum investigation, undertaken for the determination of an official unit of length in Prussia, served as a model for the precision-measuring exercises of the seminar (Olesko 1991) (Fig. 60.1).

The cumulative effects of doing these exercises were transformative for students. Their investigations demonstrated how they acquired what Fleck called the professional habits needed to become a "trained person" (Fleck 1979, pp. 89–90). But something more happened. The emphasis on the precision and reliability of their data, the determination of constant and accidental errors, and the marginalization of techniques of approximation meant that there was an "epistemological and technical concern for certainty that at times bordered on obsession" (Olesko 1991, p. 17). That obsession, which Olesko called the "ethos of exactitude," failed to sensitize students to when the quest for epistemological certainty should end. The ethos became an ethic in the sense that it "guided professional actions and decisions by providing the ways and means of separating right from wrong, truth from error, and the even the called from the damned. It helped to define professional identities, structure investigative strategies, and identify significant problems" (Olesko 1991, p. 450). While this ethos thus played a determinative role in shaping the professional behavior of Königsberg seminar students, it also created psychological limitations that were often crippling: the quest for absolute precision was in the end an illusion, one that sometimes prevented them from seeing more pragmatic, and quicker, solutions to the problem at hand.

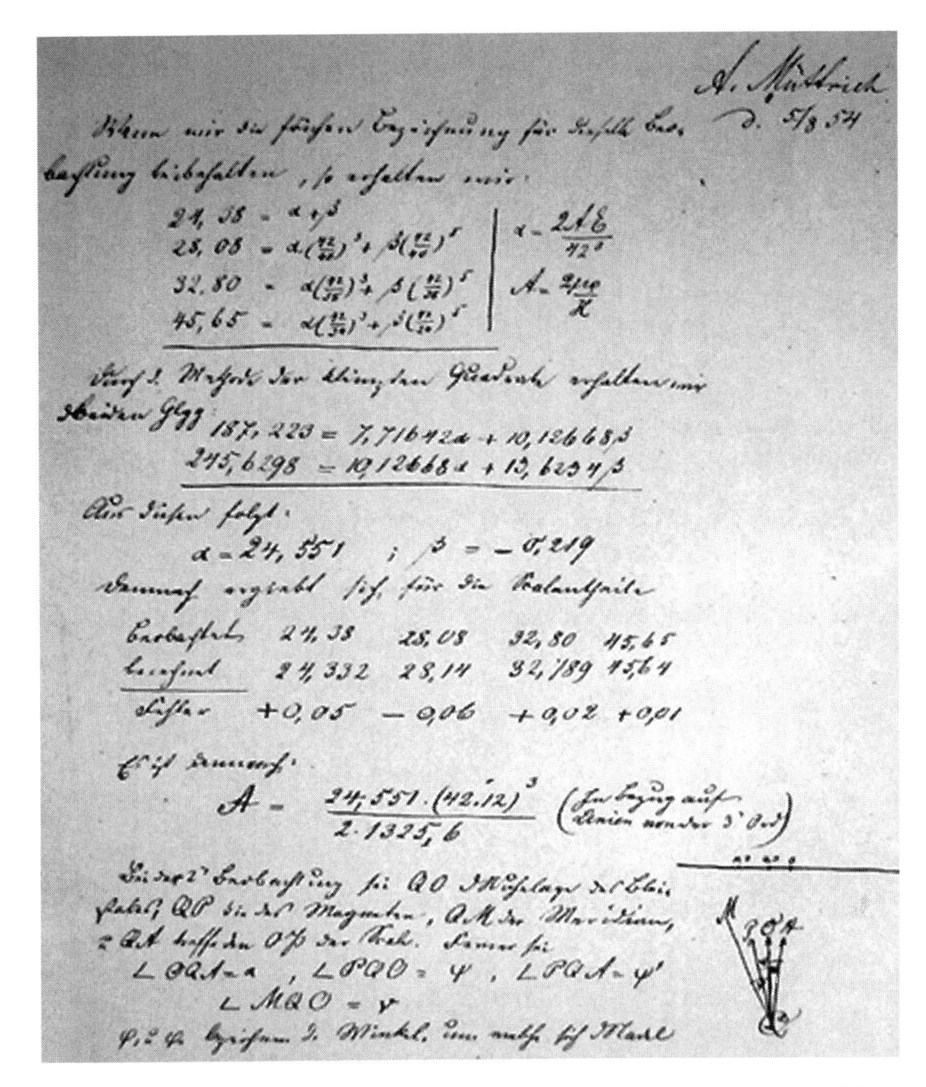

Fig. 60.1 Gottlieb Anton Müttrich (1833–1904), notebook from the physical division of Franz Neumann's seminar at the University of Königsberg, 1854. In his determination of the horizontal component of the earth's magnetism, Müttrich applies the method of least squares, a hallmark technique of the seminar (Source: Arbeiten der physik. Abteilung des mathem. Physikalischen Seminars der Königl. Universität in Königsberg 1854–55. Heft 1 [21945/55]. Abt. Va Rep. 11 Planck 1836/26. Max Planck Gesellschaft Archiv, Berlin)

Like the Königsberg case, other detailed studies of science pedagogy have demonstrated how intensely local some practices were. Warwick's history of the Cambridge Mathematical Tripos, an examination on analytical mathematical methods rooted in Newtonian mechanics, rested on actual tests (but not on the students' answers, which would have revealed how students performed) and other sources that illuminated the process of learning, including the notes of the coaches who offered preparatory training for the test. He concluded that coaches developed such distinctive solutions to problems that when they were applied outside of the Tripos setting, the Cambridge connection was immediately recognized. These techniques were designed to enable the virtuoso performance necessary for scoring high enough on the examination to attain the coveted rank of Wrangler. But at the same time, they restricted analytical solutions to closed algebraic expressions and eliminated infinite series or approximate solutions. The ability to engage in research was not the goal of instruction, yet the impact of these techniques upon practice in physics was profound and long lasting. Of note, James Clerk Maxwell's *Treatise on Electricity and Magnetism* (1873) was not a response to the British Association for the Advancement of Science's study of a suitable electrical metrology (as had so often been assumed), but rather an attempt to resolve pedagogical issues left unsettled when the Tripos incorporated electromagnetic theory in 1868 (Warwick 2003).

The maintenance of the Cambridge coaching system relied on forms of sociability that not only mitigated some of the intense pressures of the examination but that also guaranteed the type of intellectual self-identification associated with a scientific school: face-to-face interaction, bonding with the coach, and small-group learning. This sociability was certainly similar to that attained at Königsberg, but the results were different. Analytical virtuosity was the goal at Cambridge; in Königsberg, competency to pass the state examination for secondary school teachers. At Cambridge the Tripos was for undergraduates, was not in service of a profession, and was part of an intensely local culture. At Königsberg, by contrast, the state examination was for graduate students, was designed to certify the suitability of students who wished to teach mathematics or science in secondary schools, and was administered by academics for the entire state.

Similar to the nineteenth-century examples of Cambridge and Königsberg was the twentieth-century implementation of the newly created Feynman diagrams as a quick way to train physicists, the largest group in the postwar glut of science students. Feynman diagrams were in this sense created to accommodate a particular student clientele. This example demonstrates how a technique that began as a *pedagogical* device ended up as a *standard* tool for solving particular kinds of problems in quantum electrodynamics. In other words, a pedagogical device became a practice not only in the field for which it was created but also in nuclear physics, particle physics, and various forms of experimental physics. Moreover, this new calculational and visual tool "transform[ed] the way physicists saw the world" and eased the conceptual difficulties in teaching quantum electrodynamics (Kaiser 2005c, p. 4). Although the population that used Feynman diagrams was composed mostly of graduate students, the physicists who found them useful constituted a community that recognized the diagram's ability to solve certain problems quickly. Feynman diagrams are thus an example of a pedagogical innovation that was created to accommodate a large student clientele but also became a means to ease the computational tasks in a growing field of science (Kaiser 2005a; Kaiser 2005c) (Fig. 60.2).

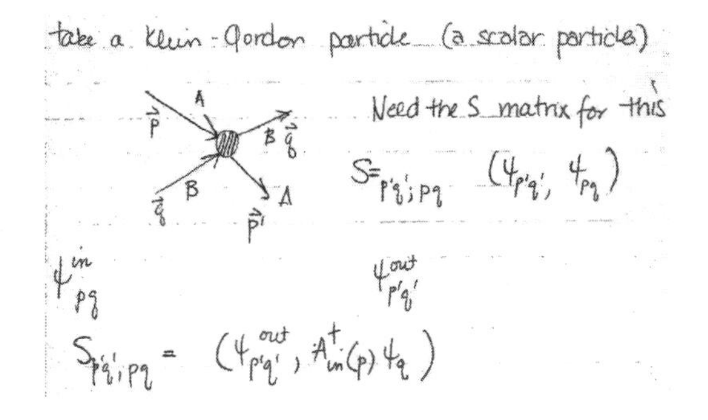

Fig. 60.2 Feynman diagram (Source: Kathryn M. Olesko, Notes for PHYS 490: Quantum Electrodynamics, Cornell University, Spring Semester 1973.Taught by Howard Tarko. Author's personal possession)

60.3.2 Science Pedagogy as Learning by Doing

While much of the historical literature on science pedagogy has focused on how science is *taught,* a small but growing body of scholarship has examined how science is *learned.* The methodological challenges of studying the latter are considerable, for the historian must find sources – notebooks, correspondence, and the like – that reveal the experiences, values, and attitudes of students as they make the transition from neophyte to practitioner. How brightly historians have been able to shed light on what transpired in exercises has depended upon available sources, not only written records of laboratory exercises but also instruments used for them. Success has been mixed, and much has to be inferred. Holmes' (1989) study of the relationship between teaching and research in Justus Liebig's Giessen chemistry laboratory relied on traces of laboratory teaching in either Liebig's publications or those of his students, and hence, his findings were necessarily incomplete. Liebig's concerted efforts to transform chemistry instruction through the introduction of the components of research procedures as smaller manageable exercises can only be inferred indirectly.

To varying degrees historians have been able to ascertain the exact exercises assigned to students and to assess their ability to complete them, but largely only for the case of physics. In the United States, Great Britain, and Germany, laboratory instruction began between the 1860s and 1880s, although, in Germany, smaller private instrument collections enabled hands-on learning decades earlier. But here too the results are skewed toward what documentary evidence is available. What is known about British laboratory practices also comes from comments in scientific publications. Far better reconstructed from printed sources are the reasons why such instruction succeeded in the first place and how that instruction was sustained. In Britain the factors contributing to the introduction of precise measuring methods into teaching laboratories between 1865 and 1885 were the development of precise measuring methods in the committees of the British Association for the Advancement of Science (e.g., for electrical standards), the inauguration of a student laboratory at Glasgow by William Thomson in 1855, and the example of professional physicists using precise measurements. Precision in measurement as a part of instruction was legitimated by the presence of a type of liberal education that emphasized rational and accurate reasoning, especially for future teachers; by the need to demarcate scientific methods from craft-based procedures; and by the association of precision measurement with economic production, especially in the telegraphic industry (Gooday 1990).

Industrial connections and lofty ideological goals were less in evidence in the United States when student laboratory instruction started after 1850. Here findings have relied on manuscript sources, laboratory manuals, and the printed record. Laboratory exercises became especially popular after the publication of Edward C. Pickering's *Elements of Physical Manipulation* (1873–1876), a manual adopted by most universities and colleges having the necessary space and instruments for such instruction (Kremer 2011). Laboratory instruction and instrument production were robust and flexible enough in America to accommodate student exercises in the new field of spectroscopy, which relied on precision gratings of sufficient resolution to give sufficiently differentiated visual results for instructional purposes and to do so at affordable cost (Hentschel 2002).

The development of laboratory instruction and the construction of university laboratories in Germany arose in response to student needs around 1870, although private collections afforded the opportunity to offer exercises earlier in the century, especially in Germany's numerous science seminars (Cahan 1985; Olesko 1991; Schubring 1989). For the German case, the archival record is rich and rewarding. Not only do historians have access to student notebooks, student exercises, lecture notes, and annual reports on teaching; they also have, in some cases, notebooks depicting the genesis of laboratory exercises. Such is the case for the most well-known and popular of laboratory manuals in physics, Friedrich Kohlrausch's *Leitfaden der praktischen Physik* (1870), which by 1996 went through 24 editions. Kohlrausch, who became an assistant to the physicist Wilhelm Weber at the University of Göttingen in 1866, worked for 4 years exploring which physical exercises worked best especially for beginning students. He left behind meticulous records of his experiences with exercises, as well as of student responses to them. Historical documents of this type, while rare, provide unsurpassed insight into how hands-on learning took shape, as well as student reactions to it (Olesko 2005) (Fig. 60.3).

60.4 Generational Reproduction

Generational reproduction is a complex issue in science pedagogy because it straddles traditional and nontraditional pedagogical settings. The reproduction of scientists is in one sense the direct result of the efficacy of science pedagogy. Yet that reproduction is also dependent upon robust pedagogical practices at the postgraduate institutions. At the simplest level, handbooks – compilations, distillations, and novel organization presentations of "what everyone knows" – are examples of higher-level pedagogies that sustain scientific practice in professional settings (Gordin 2005). At the next level, bureaucracies like standards institutions have to develop and deploy pedagogy simply to accomplish their mission. For instance, at Germany's Imperial

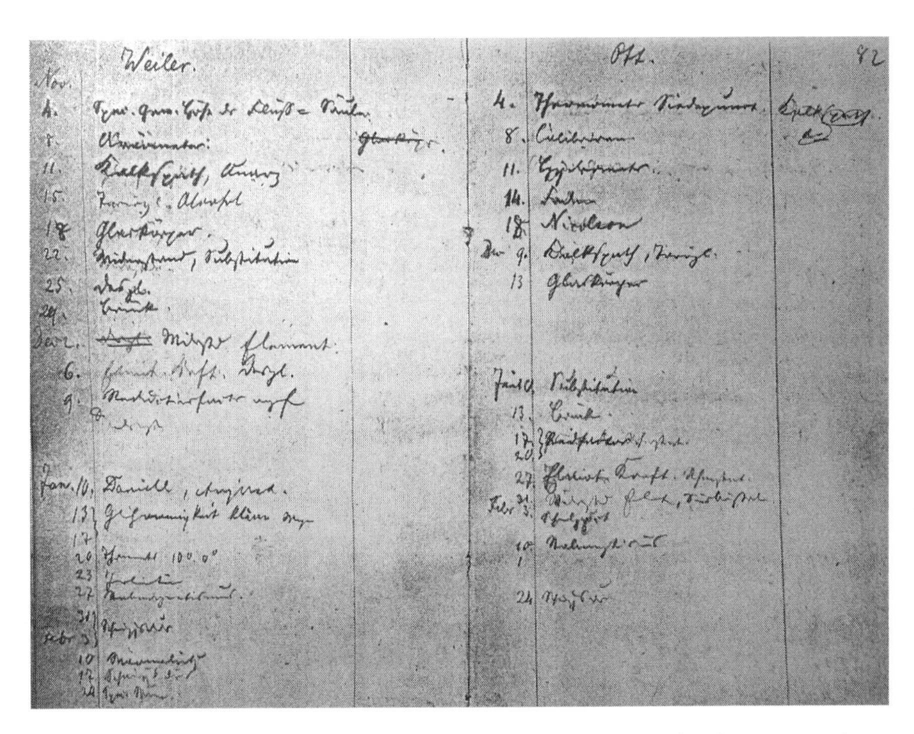

Fig. 60.3 Friedrich Kohlrausch's journal of laboratory exercises assigned to two students, November 1871–February 1872 (Source: Friedrich Kohlrausch Nachlass, Tagebuch Nr. 2504, Deutsches Museum Archiv, München)

Institute of Physics and Technology (established 1887), young physicists fresh from their doctorate had to acculturate themselves by learning the institutional norms of a bureaucracy whose purpose was both fundamental (as in standards determination) and novel (as in measuring black body radiation) (Cahan 1989). Indeed standards institutions around the world rely on higher forms of pedagogy not only for their own practitioners at home but also in order to normalize metrologies across the globe.

Such strategic interventions of science pedagogy have become apparent especially in instances of scientific disputes over the interpretation of data or when analytical representations fail to mesh. As Gooday has shown, pursuing solutions in the manner of the Mathematical Tripos could persist years after taking the examination, resulting in conflict with other professional norms. That's what happened to John Hopkinson who, in posing a solution to a particular electromagnetic problem using Cambridge techniques, clashed with a well-entrenched engineering graphical tradition. In the end Hopkinson accommodated the analytical and practical-graphical traditions, but his story is one that underscores the persistence of science pedagogy in making sense of the world (Gooday 2005, p. 142).

A special case of the strategic role of science pedagogy is found in the realm of nuclear weapons scientists. From 1945 to 1963 when the Limited Test Ban Treaty was approved and nuclear bomb testing went underground, nuclear weapons scientists enjoyed what Gusterson has called the "charismatic" era characterized by high levels of innovation and guidance from physicists whose experience with testing was indispensable for training new recruits in the ways and means of aboveground testing. In the 1970s and 1980s, however, routinization set in, with the result that innovation slowed, bureaucratic hardening occurred, and individual contributions to the effort were small. By the 1993 ban on all testing, experienced nuclear scientists retired; a new generation of scientists came on board to maintain devices they could not test in reality, and virtual computerized testing replaced real-life experiences with the bomb. Less and less knowledge and know-how about nuclear bombs were passed down generation to generation, resulting in an "involuted pedagogy of diminishing returns" (Gusterson 2005). In other words, the absence of real-life exercises (in bomb testing) means that the teachers (older nuclear scientists) could not train students (newer nuclear scientists) in how to use a test as a feedback mechanism to improve a nuclear weapon. In this case, generational reproduction did not so much as fail as wither away.

Yet perhaps there is no more important issue in the realm of generation reproduction than why women are so poorly represented among the practitioners of certain sciences, especially the so-called "hard" sciences. The gender implications and consequences of science pedagogy are critical problems of its history that beg for deeper analysis. As Rossiter (1982, 1995, 2012) has argued for the American case, women's gains in the scientific professions after initial marginalization and continued second-class status after World War II were ones that took place in the safe haven of women's colleges, through activism and organization, by piggybacking on the women's movement, and eventually favorable federal legislation. At the same time, however, educational benefits like the G.I. Bill of 1944 (and later amendments), the National Defense Education Act of 1958, the National Defense Student Loan program, and other Cold War measures to improve American standing in the sciences resulted in the further masculinization of science education at coed institutions.

In both science education and professional settings where postdoctoral training and professional grooming took place, institutionalized science pedagogy did more injustice than good for women scientists through its perpetuation and legitimation of sexism and other discriminatory practices. In addition systems of scientific training produced a gendered hierarchy of fields where the most impervious to allowing women entry were the hard core sciences. Traweek (1988) has demonstrated how training in high-energy particle physics promulgated gendered norms that worked against the incorporation of women. Over the long term, then, science pedagogy replicated the classical gender hierarchy of modernity.

60.5 The Historical Contexts of Science Education

As a disciplinary practice that often finds itself nestled closely to other branches of science and technology studies, the history of science often neglects, ironically, larger historical contexts as a venue for understanding the past. The result for the history of science education is a tendency to view key elements as static categories: discipline, pedagogy, practice, persona, textbook, and other units of analysis tend to acquire universal dimensions faster than they are understood as categories shaped by historical contingencies that change them over time. As a category of *historical* analysis, science pedagogy thus must be viewed from frameworks larger than either disciplinary or institutional history. The problem is to determine how large that framework should be and what factors are important within it.

For instance, the long-term transition from Aristotelianism to natural philosophy can only be understood by looking at what transpired in educational institutions, but to fully understand that transition, other factors such as the intellectual predilections and activities of religious orders have to be taken into account. Key agents in bringing about that transition were the Jesuits who, through teaching and textbook writing, were instrumental in institutionalizing newer frameworks for learning such as Cartesianism, Newtonianism, and, by the eighteenth century, hands-on learning (Brockliss 2006; Feingold 2003). At a later time, Boerhaave's *Elementa Chemiae* took shape within and absorbed the values of the local context of Leiden's religious, medical, and commercial cultures (Powers 2012). Great Britain's social transformation in the wake of industrialization played directly into Forster's innovations, which were implemented when Cambridge education became accessible to a broader socioeconomic clientele (Geison 1978).

In nineteenth-century Germany where mathematics had political value before it had economic currency, intimate forms of seminar instruction instilled in secondary school science teachers a belief in the powerful role of pure mathematics in interpreting physical reality, a perspective their students carried with them to the university (Pyenson 1977, 1979, 1983). Liebig and Neumann trained students for whom state qualifying examinations for secondary school teaching offered the possibility of upward social mobility and greater economic security (Holmes 1989; Olesko 1991).

Foucault thought that the problem of determining the relations of physics "with the political and economic structure of society" was to pose "an excessively complicated question" (Rabinow 1984, p. 51). Studies of physics pedagogy have nonetheless demonstrated a tightly woven connection between abstract knowledge and social norms and values. Warwick turned his study of the Cambridge Mathematical Tripos into a revealing window on Victorian culture by demonstrating how both mind and body were implicated in scientific and mathematical training. Coaching for the Tripos built character and cultivated the values of the Victorian gentleman. Public events surrounding the Tripos were filled with stress and sweat, ritual, and, for the highest-scoring Wranglers, an earned social status associated with merit (Warwick 2003).

Finally, a historically contextualized view of study of science pedagogy offers an unparalleled opportunity to examine the political dimensions, broadly conceived, of science education. Foucault is widely cited for his advocacy of viewing education as a political process: teachers, who controlled classroom disorder and reported on individual performance, were a strategic professional group whose members were the architects of power relations that both defined and disciplined the individual (Foucault 1975). But this focus on disciplining the subject has tended to ignore the degree to which individual agency was circumscribed by the systems and arrangements that make successful science education possible. Consider the nuclear scientists studied by Gusterson (1996). They learned while working in a nuclear weapons laboratory to create divided selves: a self that during the day created and maintained weapons of mass destruction and a self that on evenings and weekends cordoned off the workaday world in secrecy and silence. The history of science pedagogy is thus not only about understanding the transmission of knowledge and generational reproduction: it is more importantly about pedagogy as a moral and political practice where the examination of textbooks, pedagogical techniques, and institutions is part of understanding the structure of power (Giroux 2011), gender relations (Traweek 1988), civil society (Nyhart 2002), and other dimensions of extra-scientific contexts.

60.6 On the Horizon: The History of the Senses

Intellectual flexibility is a prime desideratum for the future of studies of science education: first, in order to make connections to new areas in historical scholarship and, second, in order to begin to analyze what is emerging as the next phase of science education in the early twenty-first century. Two developments – one historiographical and three contextual – loom large as challenges in writing the history of science pedagogy: the history of the senses, the emergence of massive open online courses (MOOCs), the corporatization of the university, and the growing number of technical professionals who bypass formal modes of science instruction en route to positions in the information technology and other economic sectors relying on scientific and technical knowledge. The controversies erupting over the latter three issues are fascinating (especially in the policy realm) and certainly worthy of study; but it is still too early to discern how they fit overall in the history of science education.

Nevertheless, these changes in the form and manner of science education at the beginning of the twenty-first century are designed to assist students where they need help most: in the mastery of foundational concepts. Scientists and policy makers argue that in the "learning science revolutions," training the eye is essential: "Visual representations are crucial to conceptualizing and communicating science, but students often have difficulty interpreting the models, simulations and graphs that are key to attaining a true understanding of science domains (Singer and Bonvillian 2013, p. 1359)." It seems appropriate then to conclude this essay with an examination of how the history of the senses can be incorporated into the history of science education as a tool of analysis as science instruction takes its next turn.

60.6.1 Integrating the History of the Senses

To a degree historians of science have taken the senses, especially vision, into account in their examination of science education. Most of these studies have focused on instruction in the life sciences, but the recognition that new printing techniques in the nineteenth century transformed textbooks has renewed the interest of historians of science in the role of vision more broadly in science instruction.[8] In addition to vision, hearing and touch are central to science learning, yet these have scarcely been studied and perhaps with good cause. Ideological frameworks, for one, make it difficult to isolate the historical roles of the senses. Karl Marx, among others, held that because the senses were alienated from the individual under capitalism, their history was impossible to write. Practical concerns too have impeded an examination of the senses in history. General historians have acknowledged over and over again the difficulties in writing the history of the senses even as they have maintained that cultural conditioning, which varies over time and across space, determines how individuals and groups deploy their senses (Jay 2011).

Science education is not only one of the strongest contributors to that cultural conditioning: science also cannot exist without sensory training, which in turn is a foundation for scientific judgment. Sharpening the senses to the point of achieving a disciplined focus (of several types) is a process that takes place both in science education and the practice of science. How science instruction enabled students to achieve focus is only beginning to be understood. Boerhaave, for instance, considered it essential to train students in the management of sensory data and for that purpose drew upon more general medieval pedagogical methods that fostered concentrated logical thinking. The new public course on instruments that he introduced in 1718 deliberately linked empirical information (the student's sensory perceptions) to chemical theory, trained students to interpret phenomena according to the instruments that measured their qualities (as in using Fahrenheit's thermometer to measure warmth), and educated the senses by disciplining them. His course on instruments thus complemented his course on chemical theory where the objective was to train reasoning processes (Powers 2012). Yet even as science education transformed the senses, the senses have a history of their own outside scientific contexts.

A transition from aural culture to an ocular one occurred in the passage from the eighteenth to the nineteenth centuries, opening the way for what both contemporaries and historians have called *Anschauungsunterricht* – a type of instruction that enables students both to visualize things and to interpret visual images. This passage entailed the cultivation of more impersonal forms of perception when abstract forms of representation replaced mimetic ones as the "culture of the diagram" replaced copying nature (Bender and Marrinan 2010). Moreover, visual learning expanded in the nineteenth century with the introduction of photographs, charts, spectroscopy, graphs, and X-rays. These instrument-mediated images revealed patterns, as in spectroscopy, that were typical of some aspect of nature (the wavelength patterns of elements) but also mysterious as to what they signified beyond a characteristic pattern. Spectral patterns were difficult to interpret, and so the student's perceptual apparatus had to be formally trained (Hentschel 2002, pp. 368–385). In the twentieth century, image-based science exploded to include electron microscopy, moving

[8] See Anderson and Dietrich (2012), Bucchi (1998), Dolan (1998), Hentschel (2002), and Lawrence (1993).

images, and digital imagery. Concomitantly, images transformed textbooks to the point where "visual literacy" became essential both for science learning and as preparation for scientific research (Anderson and Dietrich 2012, p. 2).

60.6.2 Fleck and the Senses in Science Education

How might historians of science education take into account the history of the senses? Fleck's work (1979) could with profit be used here. By isolating three elements of learning that reshape (and so educate) the prospective knower – experience, cognition, and sensation – Fleck offers a way to view science pedagogy as a process that transforms science students into something they are not. The first, experience, concerns the formation of scientific behaviors like the acquisition of skills through observation and experiment and the ability to think scientifically, both of which Fleck claims "cannot be regulated by formal logic" (Fleck 1979, p. 10). What is seen in the form of "words and ideas," he warns, is merely the "phonetic and mental equivalents of the experience coinciding with them." They are merely symbols (p. 27).

Fleck challenges us to view the past of science education differently by replacing our rapt concern for the transmission of knowledge with a fresh look at the behavioral and psychological transformations of the science learner. Experience, sensation, and cognition are all socialized by training, a process he describes as a transformation of the senses: the "slow and laborious revelation and awareness of what 'one actually sees' or *the gaining of experience*" (Fleck 1979, p. 89). Experience thus reshapes not only our minds but also our bodies. Sharpened vision – the ability to identify phenomena, for instance – is indicative of a state of "readiness for directed perception" (p. 92). In a similar fashion, he interprets cognition as a social activity ("the most socially conditioned activity of man"), making knowledge "the paramount social creation" (p. 42). Cognition can, in fact, only be understood according to Fleck as a deeply historical and contextual process that renders the mind nearly one with the beliefs of others around it. So associations between knowledge and value (say when sickness is linked to sin) can only be explained through the lens of cultural history.

Taken together, experience, sensation, and cognition form the core of the professional habits that a scientist exercises day in and day out. They are the foundation of a "collective psychology" (p. 89) transmitted through education which keeps a scientist within the cognitive framework of his or her community. The main characteristic of a thought style is that through it a trained scientist progresses nearly automatically from a vague perception to a stylized and visual one "with corresponding mental and objective assimilation of what has been so perceived" (p. 95).

What makes Fleck's analysis of scientific training useful for the historical study of science pedagogy is its ability to account not only for *learning* science but also for *becoming* a scientist, a process that entails both mental and sensory transformations. Although the strength of a thought collective depends on the existence of active science pedagogies that can carry the thought style from one generation to the next (Fleck 1979, p. 39), Fleck believed that education, although a constraint that both compelled the learner to see only in a certain way, was also pliable enough to allow for the recognition of experiences that resisted their automatic inclusion in a community thought collective. In this way the learner could also become the creative scientist. Indeed he argued that the inability to recognize resistances was the mark of the "inexperienced individual" who "merely learns but does not discern" (p. 95).

References

Anderson, N. & Dietrich, M. (2012). *The Educated Eye: Visual Culture and Pedagogy in the Life Sciences*. Hanover, N.H.: Dartmouth College Press.

Bender, J. & Marrinan, M. (2010). *The Culture of the Diagram*. Stanford, Calif.: Stanford University Press.

Bensuade-Vincent, B. (1990). A View of the Chemical Revolution through Contemporary Textbooks: Lavoisier, Fourcroy, and Chaptal. *British Journal for the History of Science*, 23, 435–460.

Bensaude-Vincent, B. (2006). Textbooks on the Map of Science Studies. *Science & Education*, 15, 667–670.

Bensaude-Vincent, B., García-Belmar, A. & Bertomeu-Sánchez, J. (2002). Looking for an Order of Things: Textbooks and Classifications in Nineteenth Century France. *Ambix*, 49, 227–251.

Bensaude-Vincent, B., García-Belmar, A. & Bertomeu-Sánchez, J. (2003). *La naissance d'une science des manuels (1789–1852)*. Paris: Editions des Archives Contemporaine.

Brockliss, L. (2006). The Moment of No Return: The University of Paris and the Death of Aristotelianism. *Science & Education*, 15, 259–278.

Bucchi, M. (1998). Images of Sciences in the Classroom: Wallcharts and Science Education, 1850–1920. *British Journal for the History of Science*, 31, 161–184.

Cahan, D. (1985). The Institutional Revolution in German Physics, 1865–1914. *Historical Studies in the Physical Sciences*, 15(2), 1–66.

Cahan, D. (1989). *An Institute for an Empire: The Physikalish-Technische Reichsanstalt 1871–1918*. Cambridge: Cambridge University Press.

Carniero, A., Diogo, M.. & Simöes, A. (2006). Communicating the New Chemistry in 18th-century Portugal: Seabra's *Elementos de Chimica*. *Science & Education*, 15, 671–692.

Clark, W. (2006). *Academic Charisma and the Origins of the Research University*. Chicago: University of Chicago Press.

Daum, A. (2002). Science, Politics, and Religion: Humboldtian Thinking and the Transformation of Civil Society in Germany, 1830–1870. *Osiris*, 17, 171–209.

Dennis, M. (1994). Our First Line of Defense—Two University Laboratories in the Postwar State. *Isis*, 85, 427–455.

Dolan, B. (1998). Pedagogy through Print: James Sowerby, John Mawe, and the Problem of Colour in Early Nineteenth-Century Natural History Illustration. *British Journal for the History of Science*, 31, 275–304.

Elias, N. (1939). *Über den Prozess der Zivilisation: Soziogenetische und psychogenetische Untersuchungen*, 2 vols. Basel: Haus zum Falken.

Feingold, M. (2003). *Jesuit Science and the Republic of Letters*. Cambridge, Mass.: MIT Press.

Fleck, L. (1935). *Entstehung und Entwicklung einer wissenschaftlichen Tatsache: Einführung in die Lehre vom Denkstil und Denkkollektiv*. Basel: Benno Schwabe & Co.

Fleck, L. (1979). *Genesis and Development of a Scientific Fact*. Chicago: University of Chicago Press.

Foucault, M. (1966). *Les mots et les choses: un archéologie des sciences humaines*. Paris: Gallimard.

Foucault, M. (1975). *Surveiller et punir: Naissance de la prison.* Paris: Gallimard.
García-Belmar, A. & Bertomeu-Sánchez, J. R. (2004). Atoms in French Chemistry Textbooks during the First Half of the Nineteenth Century. *Nuncius,* 19, 77–119.
García-Belmar, A., Bertomeu-Sánchez, J., & Bensuade-Vincent, B. (2005). The Power of Didactic Writings: French Chemistry Textbooks of the Nineteenth-Century. In D. Kaiser (Ed.) (2005b), pp. 219–251.
Geiger, R. (1998). The Rise and Fall of Useful Knowledge: Higher Education for Science, Agriculture and the Mechanic Arts, 1850–1875. *History of Higher Education Annual,* 18, 47–66.
Geison, G. (1978). *Michael Forster and the Cambridge School of Physiology: The Scientific Enterprise in Late-Victorian Society.* Princeton: Princeton University Press.
Giroux, H. (2011). *On Critical Pedagogy.* New York: Continuum.
Gooday, G. (1990). Precision Measurement and the Genesis of Physics Teaching Laboratories in Victorian Britain. *British Journal for the History of Science,* 23, 25–52.
Gooday, G. (2005). Fear, Shunning, and Valuelessness: Controversy over the Use of "Cambridge" Mathematics in Late Victorian Electro-Technology. In D. Kaiser (Ed.) (2005b), pp. 111–149.
Gordin, M. (2005). Beilstein Unbound: The Pedagogical Unraveling of a Man and his *Handbuch.* In D. Kaiser (Ed.) (2005b), pp. 11–40.
Gouzevitch, I. (2006). The Editorial Policy as a Mirror of Petrine Reforms: Textbooks and their Translators in Early 18th-Century Russia. *Science & Education,* 15, 841–862.
Gusterson, H. (1996). *Nuclear Rites: A Weapons Laboratory at the End of the Cold War.* Berkeley: University of California Press.
Gusterson, H. (2005). A Pedagogy of Diminishing Returns: Scientific Innovation across Three Generations of Nuclear Weapons Science. In D. Kaiser (Ed.) (2005b), pp. 75–107.
Hall, K. (2005). "Think Less about Foundations": A Short Course on Landau and Lifshitz's *Course of Theoretical Physics.* In D. Kaiser (Ed.) (2005b), pp. 253–286.
Hannaway, O. (1975). *The Chemists and the Word: The Didactic Origins of Chemistry.* Baltimore: Johns Hopkins University Press.
Hentschel, K. (2002). *Mapping the Spectrum: Techniques of Visual Representation in Research and Teaching.* Oxford: Oxford University Press.
Holmes, F. (1989). The Complementary of Teaching and Research in Liebig's Laboratory. *Osiris,* 5, 121–194.
Jay, M. (2011). In the Realm of the Senses: An Introduction. *American Historical Review,* 116, 307–315.
Josefowicz, D. (2005). Experience, Pedagogy, and the Study of Terrestrial Magnetism. *Perspectives on Science,* 13, 452–494.
Kaiser, D. (2005a). *Drawing Theories Apart: The Dispersion of Feynman Diagrams in Postwar Physics.* Chicago: University of Chicago Press.
Kaiser, D. (Ed.) (2005b). *Pedagogy and the Practice of Science: Historical and Contemporary Perspectives.* Cambridge, Mass.: MIT Press.
Kaiser, D. (2005c). Making Tools Travel. Pedagogy and the Transfer of Skills in Postwar Theoretical Physics. In D. Kaiser (Ed.) (2005b), pp. 41–74.
Kohlstedt, S. (2010). *Teaching Children Science: Hands-On Nature Study in North America, 1890–1930.* Chicago: University of Chicago Press.
Kremer, R. (2011). Reforming American Physics Pedagogy in the 1880s: Introducing 'Learning by Doing' via Student Laboratory Exercises. In Herring, D. & Wittje, R. (Eds.) *Learning by Doing: Experiments and Instruments in the History of Science Teaching.* Stuttgart: Franz Steiner Verlag, pp. 243–280.
Kuhn, T. (1962). *The Structure of Scientific Revolutions.* Chicago: University of Chicago Press.
Kuhn, T. (1979). Forward. In L. Fleck (1979), pp. vii-xii.
Lawrence, S. (1993). Educating the Senses: Students, Teachers, and Medical Rhetoric in Eighteenth-century London. In Bynum, W. & Porter, R. *Medicine and the Five Senses.* Cambridge: Cambridge University Press, pp. 154–178.
Leslie, S. (1993). *The Cold War and American Science: The Military-Industrial-Academic Complex at MIT and Stanford.* N.Y., Columbia University Press.

Lind, G. (1993). *Physik im Lehrbuch, 1700–1850: Zur Geschichte der Physik und ihrer Didaktik in Deutschland.* Berlin: Springer Verlag.
Lundgren, A. & Bensuade-Vincent, B. (Eds.) (2000). *Communicating Chemistry: Textbooks and their Audiences, 1789–1939.* Canton, Mass.: Science History Publications.
Lundgren, A. (2006). The Transfer of Chemical Knowledge: The Case of Chemical Technology and its Textbooks. *Science & Education,* 15, 761–778.
Macleod, R. (Ed.). (1982). *Days of Judgement: Science, Examinations, and the Organization of Knowledge in Victorian England.* Driffield, N. Humberside: Studies in Education.
Macleod, R. & Moseley, R. (1978). Breadth, Depth and Excellence: Sources and Problems in the History of University Science Education in England, 1850–1914. *Studies in Science Education,* 5, 85–106.
McCulloch, G. (1998). Historical Studies in Science Education. *Studies in Science Education,* 31, 31–54.
Mody, C. & Kaiser, D. (2007). Scientific Training and the Creation of Scientific Knowledge. In Hackett, E., et al. (Eds.) (2007). *The Handbook of Science and Technology Studies.* Cambridge, Mass.: MIT Press, pp. 377–402.
Nyhart, L. (2002). Teaching Community via Biology in Late-Nineteenth Century Germany. *Osiris,* 17, 141–170.
Olesko, K. (1988). Michelson and the Reform of Physics Instruction at the Naval Academy in the 1870s. In Goldberg, S. and Stuewer, R. (Eds.), *The Michelson Era in American Science, 1870–1930.* New York: American Institute of Physics, pp. 111–132.
Olesko, K. (1989). Physics Instruction in Prussian Secondary Schools before 1859. *Osiris* 5, 92–118.
Olesko, K. (1991). *Physics as a Calling: Discipline and Practice in the Königsberg Seminar for Physics.* Ithaca, N. Y.: Cornell University Press.
Olesko, K. (2005). The Foundations of a Canon: Kohlrausch's *Practical Physics.* In D. Kaiser (Ed.) (2005b), pp. 323–356.
Olesko, K. (2006). Science Pedagogy as a Category of Historical Analysis: Past, Present, and Future. *Science & Education,* 15, 863–880.
Patiniotis, M. (2006). Textbooks at the Crossroads: Scientific and Philosophical Textbooks in 18th Century Greek Education. *Science & Education,* 15, 801–822.
Pauly, P. (1991). The Development of High School Biology. *Isis,* 92, 662–688.
Petrou, G. (2006). Translation Studies in the History of Science: The Greek Textbooks of the 18th Century. *Science & Education,* 15, 823–840.
Polanyi, M. (1958). *Personal Knowledge: Toward a Post-Critical Philosophy.* Chicago: University of Chicago Press.
Powers, J. (2012). *Inventing Chemistry: Herman Boerhaave and the Reform of the Chemical Arts* Chicago: University of Chicago Press.
Pyenson, L. (1977). Educating Physicists in Germany circa 1900. *Social Studies of Science,* 7, 329–66.
Pyenson, L (1978). The Incomplete Transmission of a European Image: Physics at Greater Buenos Aires and Montreal, 1820–1920. *Proceedings of the American Philosophical Society,* 122, 92–114.
Pyenson, L. (1979). Mathematics, Education, and the Göttingen Approach to Physical Reality, 1890–1914. *Europa,* 2, 91–127.
Pyenson, L. (1983). *Neohumanism and the Persistence of Pure Mathematics in Wilhelmian Germany.* Philadelphia, American Philosophical Society.
Rabinow, P. (Ed.) (1984). *The Foucault Reader.* New York: Pantheon Books.
Ravetz, J. (1971). *Scientific Knowledge and its Social Problems.* Oxford: Clarendon Press.
Rossiter, M. (1982). *Women Scientists in America: Struggles and Strategies to 1940.* Baltimore: Johns Hopkins University Press.
Rossiter, M. (1995). *Women Scientists in America: Before Affirmative Action, 1940–1972.* Baltimore: Johns Hopkins University Press.
Rossiter, M. (2012). *Women Scientists in America: Forging a New World Since 1972.* Baltimore: Johns Hopkins University Press.

Rudolph, J. (2002). *Scientists in the Classroom: The Cold War Reconstruction of American Science Education*. New York: Palgrave.
Rudolph, J. (2008). Historical Writing on Science Education: A View of the Landscape. *Studies in Science Education*, 44, 63–82.
Schubring, G. (1989). The Rise and Decline of the Bonn Natural Sciences Seminar. *Osiris,* 5, 56–93.
Schutz, A. (1932). *Die sinnhafte Aufbau der sozialen Welt: Eine Einleitung in die verstehende Soziologie.* Wien: J. Springer.
Seligardi, R. (2006). Views of Chemistry and Chemical Theories: A Comparison between Two University Textbooks in the Bolognese Context at the Beginning of the 19th Century. *Science & Education,* 15, 713–737.
Simon, J. (2008). Communicating Science and Pedagogy. In J. Simon et al. (Eds.) (2008), *Beyond Borders: Fresh Perspectives in History of Science*. Newcastle, U.K.: Cambridge Scholars Publishing, pp. 101–112.
Simon, J. (2011). *Communicating Physics: The Production, Circulation, and Appropriation of Ganot's Textbooks in France and England, 1851–1887.* London: Pickering and Chatto.
Singer, S. & Bonvillian, W. (2013). Two Revolutions in Learning. *Science,* 339 (22 March 2013), 1359.
Stichweh, R. (1984). *Zur Entstehung des modernen Systems wissenschaftlicher Disziplinen: Physik in Deutschland, 1740–1890.* Frankfurt am Main: Suhrkamp.
Traweek, S. (1988). *Beamtimes and Lifetimes: The World of High Energy Physicists.* Cambridge, Mass.: Harvard University Press.
Traweek, S. (2005). Generating High-Energy Physics in Japan: Moral Imperatives of a Future Pluperfect. In D. Kaiser (Ed.) (2005b), pp. 357–392.
Vicedo, M. (2012). Introduction: The Secret Lives of Textbooks. *Isis,*103, 83–88.
Warwick, A. (2003). *Masters of Theory: Cambridge and the Rise of Mathematical Physics.* Chicago: University of Chicago Press.

Kathryn M. Olesko is an associate professor in the Department of History and the Program in Science, Technology, and International Affairs at Georgetown University where she has also held several administrative positions. She majored in physics and mathematics at Cornell University and then took her doctorate there in the history of science. She is the author of *Physics as a Calling* (1991), the forthcoming *Why Prussians Measured: The Myth of Precision, 1648–1947,* and articles on the history of science. For 11 years she was the editor of the History of Science Society's annual thematic volume, *Osiris*, which she transformed into a forum for mediating the history of science and "mainstream" history. Her interest in the history of science education, which spans her entire career, has focused especially on studies of identity formation in the sciences and on the moral, professional, and epistemological roles of measurement in both science education and professional practice. She is now engaged in the history of measurement and an interdisciplinary historical study of water management in Prussia. Her honors include several fellowships from the National Endowment for the Humanities and the National Science Foundation, short- and long-term appointments as fellow at the Max Planck Institute for the History of Science in Berlin, and the Dibner Distinguished Fellowship at The Huntington Library. In 1998 she was elected a fellow of the American Association for the Advancement of Science for her "contributions to scholarship and teaching in the history of science and for leadership in the History of Science Society and the AAAS."

Part XIII
Regional Studies

Chapter 61
Nature of Science in the Science Curriculum and in Teacher Education Programs in the United States

William F. McComas

61.1 Introduction

In the past 50 years, advocacy for and scholarship in nature of science (NOS)/history and philosophy of science (HPS) in school science has grown from a few scattered references in the science education literature to a veritable flood of support for, interest in, and research in the field.

A single definition of NOS shared by the majority of science educators would be difficult to find, but many would agree that NOS is the area of study in which students learn how science functions, how knowledge is generated and tested, and how scientists do what scientists do. McComas and colleagues (1998, p. 4) suggest that:

> The nature of science is a fertile hybrid arena which blends aspects of various social studies of science including the history, sociology, and philosophy of science combined with research from the cognitive sciences such as psychology into a rich description of what science is, how it works, how scientists operate as a social group and how society itself both directs and reacts to scientific endeavors. The intersection of the various social studies of science is where the richest view of science is revealed for those who have but a single opportunity [such as the case in school settings] to take in the scenery.

Just as there is a lack of complete agreement on the definition of nature of science, the name itself has engendered some debate. Some scholars suggest that it would be best to call this domain nature of science studies, history and philosophy of science, ideas about science, nature of sciences, nature of scientific knowledge, and views on the nature of science, among others. Of course each of these labels has some advantage over the others and is usually preferred by one group or another, typically for philosophical reasons. However, given the extent of scholarship associated with the NOS label and references to it, that will be the term used throughout this chapter.

W.F. McComas (✉)
University of Arkansas, Fayetteville, AR 72701, USA
e-mail: mccomas@uark.edu

M.R. Matthews (ed.), *International Handbook of Research in History, Philosophy and Science Teaching*, DOI 10.1007/978-94-007-7654-8_61,

61.2 The Context of Education in the United States: An Overview

To fully appreciate this report on NOS in the United States, one must recognize the unique way in which education is organized and governed in the nation. At the founding of the republic in the mid-1700s, the original 13 governing entities coalesced into a nation. This new nation had two axes of command and control; the federal (or national) government reserved some powers for itself (defense and diplomacy as examples), and other responsibilities (education, for instance) remained in control of the states. In the United States, education is frequently mentioned as the quintessential *states' rights* issue. In many ways the individual US states such as New York and California function like independent nations; this is particularly true with respect to education.

What has evolved in the United States is a blend of laws, policies, governing traditions, and educational systems that have much in common but leave the ultimate control for schooling to the state rather than federal government. This is a unique situation, with Germany and perhaps only a few other nations sharing such a decentralized system. Each state, therefore, has full responsibility for teacher licensure, achievement testing, the establishment and maintenance of an education bureaucracy, school funding, and the development of educational goals and standards.

The US federal government, generally through the Department of Education (equivalent to the Ministry of Education in other countries), provides some guidance and encourages specific policies by commissioning studies and rewarding states through monetary support (or the threat to withhold such support). The US Congress is also involved in education with its mandate to produce a periodic "report card" on the education situation across the nation through the National Assessment of Educational Progress (NAEP). While it may be useful to talk about education in the United States, overarching statements about the nation as a whole are difficult to make with any assurance; we must look widely and infer liberally.

However, as we will see, the first decades of the twenty-first century have seen a gradual loosening of the states' formally tight grip on education policy with new sets of goals for school science under development by broad groups with representatives from the science and education communities with funding from public and private entities. Soon, virtually all of the US states will adopt what is called *The Next Generation Science Standards (NGSS)*. This is a major change in the educational governance in the United States with vast implications for assessment, teacher preparation, curriculum development, and classroom practice.

61.3 NOS in the Schools of the United States: Some Historical Perspective

There is no single moment when NOS and related ideas entered the educational area, but more than 150 years ago, the Duke of Argyll in his Presidential Address to the British Association for the Advancement of Science stated that "What we want in the teaching of the young is not so much the mere results as the methods and, above all, the history of science" (in Matthews 1994, p. 11). This may be among the first suggestions that an element of what is now called *nature of science* should be part of the school science curriculum.

A century later this view crossed the Atlantic Ocean to the United States, where the Educational Policies Commission *Report on Education for All American Youth* raised the promise of the use of NOS (1944, p. 132) by stating:

> These scientists are thought of as living men [sic], facing difficult problems to which they do not know the answers, and confronting many obstacles rooted in ignorance and prejudice. In imagination, the students watch the scientists at work, and look particularly for the methods which they use in attacking their problems . . .

In 1946, James Bryan Conant, educator, scientist, and president of Harvard University, delivered his famous Terry Lectures at Yale and stated that students must understand the tactics and strategies of science, an obvious reference to NOS. He later expressed the view that "some understanding of science by those in positions of authority and responsibility as well as by those who shape opinion is therefore of importance for the national welfare" (1947, p. 4). Conant (1951) later expanded on these ideas by suggesting that "every American citizen ... would be well advised to try to understand both science and the scientist as best he can" (p. 3). These are among the earliest and most clearly stated rationales for the inclusion of nature of science as an essential part of science literacy in the United States. Even if the term *nature of science* was not widely used, it is clear that is what these early advocates mind.

The 1957 Soviet launch of Sputnik and the perceived threat to US superiority in science and technology gave rise to what has been called the *Golden Age of Science Education*. During the period following Sputnik extending through the 1960s, various US government agencies funded a large number of projects targeting the improvement of science and mathematics education with study groups and a staggering number of curriculum development projects. These were all designed to bolster the nature and effectiveness of science and mathematics teaching in the nation (DeBoer 1991).

By 1960, the National Society for the Study of Education argued even more clearly for the inclusion of NOS in school science:

> There are two major aims of science-teaching; one is knowledge, and the other is enterprise. From science courses, pupils should acquire a useful command of science concepts and principles. Science is more than a collection of isolated and assorted facts . . . A student should learn something about the character of scientific knowledge, how it has been developed, and how it is used. (in Hurd 1960, p. 34)

An additional justification for the inclusion of the nature of science in science class comes from science educator Joseph Schwab (1964), who correctly observed that science is taught as an "unmitigated rhetoric of conclusions in which the current and temporal constructions of scientific knowledge are conveyed as empirical, literal, and irrevocable truths" (p. 24). Many of the science curriculum projects of the 1960s – sometimes called alphabet soup projects because of their letter-rich acronyms (such as S-APA, ESS, CHEM Study) – were designed to shift science instruction away from a focus on "what do scientists know" (i.e., content) to an examination "how do scientists know" (i.e., process). Interestingly, several of the science curriculum projects funded by the government as a response to the perceived Soviet threat were expressly designed with NOS elements so that students would have the opportunity to understand how to "do" science in the real world. Of course, this is hardly a surprise since the expressed purpose of these new curricula was to encourage more students to become scientists and engineers.

As the 1960s became the 1970s, several authors reminded science educators of the importance of the nature of science by using that term expressly. Robinson (1968) in *The Nature of Science and Science Teaching* discussed the nature of physical reality including probability, certainty, and causality and concluded with by considering the interplay between science instruction and the nature of science. In *Concepts of Science Education: A Philosophical Analysis*, Martin (1972) reiterated some of these arguments in support of including NOS in science instruction by advocating the use of inquiry learning along with discussion of the nature of explanation and the character of observation in the science classroom.

NOS studies gained traction in the final decades of the twentieth century. There were contributions to NOS studies from increasing numbers of scholars, focused publications in the field and increased advocacy and understanding for the place of the history and philosophy of science in the science classroom. At this same time Duschl (1985) reminded the science education community that the way science was increasingly represented in classrooms was often at odds with a modern view of how science functions.

Next, various organizations in the United States such as the National Research Council (NRC) and the American Association for the Advancement of Science (AAAS) expressed interest in the nature of science. AAAS released an important report defining what literate individuals should know about science, called *The Liberal Art of Science: Agenda for Action* (AAAS 1990). It featured an entire chapter with recommendations for what sort of NOS topics ought to be included in school science. These included science values and ways of knowing, methods of collecting, analyzing and classifying data, the nature of explanations in science, and the limits on scientific understanding.

Without NOS, students will very likely continue to see science only in its "final form," a label coined by Duschl (1990) describing the situation in which students learn only the conclusions of science with little opportunity to experience how these scientific discoveries were made. "Final form science" provides such a shallow view of the scientific enterprise that students are unable to use the methods of science for themselves or to gauge the scientific worth of ideas proposed by others.

61.4 Why Nature of Science Matters: The United States Context

At this point it is important to recognize a special challenge with respect to NOS in US classrooms. Certainly, NOS can help students understand and appreciate the inner workings and limitations of science as a way of knowing. With that goal in mind, it would be hard to imagine than anyone would argue with its inclusion in science class, particularly if NOS knowledge assists science learners become better decision makers on scientific matters.

However, this is precisely why NOS is needed and why some might rather omit it. If students can judge the worth of scientific evidence and conclusions on their own, they will be far less likely simply to accept what others tell them. This has become an important issue with respect to topics perceived as controversial by some in the United States. such as evolution and, more recently, climate change. If students understand how science functions, they would quickly realize that issues like these are not just matters of opinion. In fact, they are matters of science not politics, not religious doctrine, and most certainly not just opinion. If the scientific evidence demands reaching a particular conclusion, then that is the most reasonable conclusion to accept even if it is unpalatable for some external reason like religion, politics, or preference.

As a case in point, consider the case of biological evolution and those who reject it because of their motivation by a particular worldview (Moore et al. 2009). Evolution denial has been active for more than a century; since education was governed locally for much of the history of the United States, evolution was simply not taught as too heretical. Following the famous 1925 Scopes trial in the southern state of Tennessee, there was a national debate about the teaching of evolution that generally left evolution out of most science classrooms. Space does not permit a full account of the battles in both courthouses and the court of public opinion, but even the last major skirmish which occurred probably will not end the attacks. In 2005, the Dover Area School District in Pennsylvania lost an expensive and foolish fight to defend their policy requiring the teaching of intelligent design along with evolution as an "alternative" in biology class (Humes 2007).

It is not clear if those who deny evolution and climate change are sophisticated enough to understand that when students have the tools to think for themselves they will either accept or reject scientific ideas based on the merits of the ideas themselves rather than with reference to some preformed ideology. Perhaps the long tradition of democracy in the United States has caused some to think that we can and should vote on everything including the validity of scientific ideas. If students and their parents fail to understand the guiding principles of science, they may think that all knowledge from any source is essentially a matter of opinion, and therefore, equally valid. This relativistic approach has given rise to the "science wars," which are a subset of the greater "culture wars" that have raged in some decades. Such "wars" result either when one group feels excluded by the knowledge generation methods and traditions of another or when a group assumes the position that no source of knowledge – including science – produces more secure results than any other (Parsons 2003; Brown 2001). Students are the causalities of these "wars" when important elements of the curriculum are eliminated or minimized or even mocked. This happened with evolution and is occurring with climate change.

The battle against evolution and the growing rejection of global warming have shared roots in a lack of understanding of how science functions. A firm appreciation of the limits of science should enable those who question evolution to recognize that its acceptance does nothing to negate most religious beliefs. Those who reject climate change on ideological grounds will be much less likely do to so if they understand that science forms conclusions not on the opinions of a few but on the preponderance of the evidence as analyzed and interpreted by the community of scientists worldwide. Only through knowledge of the nature of science can students understand how science produces and validates knowledge. Given the continuing tensions about the process and products of science in the United States, knowledge of NOS is arguably more important here than in almost any other nation. With that thought in mind, we can turn our attention to what aspects of the nature of science should inform science teaching and learning.

61.5 Nature of Science in US Science Education Standards Documents

Many involved with science education in the United States would likely agree with a decision to call the recent decades as the *Age of Standards* in science education. In a relatively short time two documents were designed to guide the teaching and learning of science at a national level. Of course, some states already had their own documents, but the advent of the national standards movement gave rise to the development of the first major sets of science standards. Interestingly, two documents from two different groups – with some overlap in committee membership – appeared almost simultaneously.

The first of these came from the American Association for the Advancement of Science (AAAS)-sponsored *Project 2061* (1989) (designed to reform science teaching by the time Halley's Comet returns in 2061) and the related *Benchmarks for Science Literacy* (AAAS 1993). Just 3 years later, the National Research Council released its own proposal for the content of school science boldly titled the *National Science Education Standards* or *NSES* (NRC 1996). In spite of a high degree of similarity between them, the *Standards* made the recommendation that science should be taught through inquiry, while *Benchmarks* is generally silent on how science should be taught. Of course, given the issue of educational governance in the United States, neither set of guidelines could be imposed and, in fact, no state fully adopted all of the recommendations.

A detailed comparison of *NSES* and *Benchmarks* (McComas and Olson 1998) revealed that both documents provided many targeted recommendations regarding the nature of science for K–12 science instruction. Both documents generally describe how science functions and builds new knowledge by mentioning the common characteristics of science including the use of careful measurement, observation, experiment, peer review, and potential for replication. Science is labeled as tentative, calling theories and laws as related but distinct aspects of science; science is portrayed as a collaborative and social human endeavor affected by the social and historical milieu within a domain of creativity. Neither document provided as complete a description of science for science teaching as recommended by Lederman (2002, 2007), McComas (1998, 2008), and Osborne and colleagues (2003). However, the strong NOS message in these two documents affirmed the nature of science as a necessary part of the science teaching enterprise that could no longer be ignored.

So, the debate about *whether* to teach NOS has ended leaving open only the question about *what* NOS elements should be taught. The response to this challenge may best be achieved by using the consensus approach which considers the common views of various science education experts. Some, such as van Dijk (2011), reject the consensus approach for making this determination partially on the grounds that the final picture that emerges is not a complete view of NOS. Perhaps, but this is no more or less true than the consensus about what to teach that resulted from an examination of traditional science topics such as chemical reactions, photosynthesis, and the rock cycle. Anything we teach students about science and its nature should be accurate with respect to the needs and abilities of the target audience. However, the goal in defining what NOS to teach in schools is predicated on what students need to appreciate the scientific enterprise, what they can understand, and what the curriculum can support, not in providing an obsessively complete view of the nature of science as known by historians and philosophers of science.

One particularly fruitful approach about what NOS to teach is found in an analysis of science standards from each of the US states. These documents can be seen as quasi-independent opinions (of course, many of those who crafted these documents reviewed the foundation same literature) on NOS that allow us to reach some common ground on what elements of the nature of science should best inform school science teaching.

61.6 Nature of Science in the US State Science Standards

The US state science teaching standards have traditionally been in flux both due to the ongoing cycle of revision and updating. This change will continue because most of the state science standards will disappear as the new standards are adopted. Therefore, an analysis of the existing standards just before the adoption of the *Next Generation Science Standards* can provide both an interesting historical review and a rich picture of NOS instruction in the United States with great potential for reaching agreement on the nature of science in schools.

McComas and colleagues (2012) reviewed the then-current science content standards of the 50 US states and the District of Columbia (N=51) to determine what NOS content was included. The investigation was focused on a search for the appearance of any of 12 elements of the nature of science[1] most commonly recommended by experts (Al-Shamrani (2008). These are called key aspects of NOS and detailed as a part of Table 61.1.

[1]The 12 key NOS elements were used to guide the search, but researchers were attentive to and noted all instances of NOS-related language found in thousands of lines of text in 51 documents. In these documents, more than 3400 instances of NOS were located and categorized.

Table 61.1 The percentage of US state standards documents in which specific key aspects of the nature of science appears at least once (N=51)

Key aspects of NOS found in a review of all US state standards documents in the spring of 2012 (N=51)	Percentage of state standards documents in which specific aspects of NOS appear (%)
Science is based on empirical evidence	96
Cooperation exists in science	90
Scientific conclusions have a degree of tentativeness	90
Distinction between observation and inference	86
Role of experiments in science	86
Distinction between science and technology	84
Science is socially and culturally embedded	76
There is no stepwise scientific method	71
Science has a subjective element	61
Distinction between law and theory	55
Science cannot answer all questions (i.e., there are limits to science)	49
Creativity has a role in science	25

Each appearance of NOS was noted to (a) identify which NOS elements appear, (b) gauge where they appear with respect to educational level (elementary, middle, and secondary), (c) determine how many of the key elements are included (this is called completeness), (d) measure the distribution or comprehensiveness of NOS elements with respect to their inclusion in and across grade levels, and (e) rank the state standards documents based on a combined measure of completeness and comprehensiveness. In brief this analysis provides an in-depth look at NOS in US public schools just before the introduction of new science instructional standards.

61.6.1 Nature of Science in the US State Standards

Table 61.1 reports the specific key NOS elements that appear in the science standards of each state; data provided next will provide much more detail about where the NOS element appears with respect to grade level and with what frequency. These data reveal which key NOS elements (empiricism, tentativeness in science, cooperation and collaboration in science, the distinction between observations and inferences, and the role of experiments) are included within the standards of most states. From this review, creativity, the idea that science cannot answer all questions, and the distinction between law and theory are shown to be included less frequently than other notions in the state science standards.

Table 61.2 The US states listed with the number of key NOS aspects included in their science content standards along with a "letter grade" associated with that number of key aspects of NOS

Score	Number of key aspects of NOS included	State
A+	12	NE, NH, NC, OH
A	11	FL, IN, KY, ME, MI, MN, MO, NV, NJ, NY, OR, TN
B	10	AK, GA, KS, LA, MA, MS, NM, ND, DC
C	9	ID, IL, MD, VA, WA, WV
D	6–8	AL, AZ, AR, CA, CO, HI, MT, OK, PA, SD, UT, VT, WI

61.6.2 Which US State Standards Contain the Most NOS Elements?

The data in Table 61.2 show how likely it is that a state document would include a complete range of all recommended NOS elements. Sixteen states rate highly in terms of completeness; some states (Nebraska, New Hampshire, North Carolina, and Ohio) have all 12 key NOS elements in their science content standards.

61.6.3 The Nature of Science Recommendations for Grade Level in the US State Standards

The empirical aspect of science, the role of cooperation, the distinction between observation and inference, and the distinction between science and technology are the most likely NOS elements to be found across grades K–12. Creativity and the distinction between theory and law are introduced only at the higher grade levels (Table 61.3).

61.6.4 Combining Completeness and Comprehensiveness: How Do the States Rank for NOS Inclusion?

Using a weighted system by which states earn "high marks" for having the most NOS elements included most frequently *across* grade levels, we ranked each state based on how NOS is featured in its science teaching standards. This ranking system is based on determining a top score in which all of the NOS elements would appear in the standards at every grade level. However, achieving this top score is not anticipated or even advocated since some NOS elements might be inappropriate for younger learners. Therefore, the final analysis was a norm-referenced scale that reveals what states thought was possible rather than for the researchers to suggest

Table 61.3 The percentage of state documents in which each key aspect of the nature of science appears generally (N = 51) and appears at a particular grade-level span

Key aspects of NOS found in US state standards	Percentage of documents in which the NOS element appears	Percentage of state documents in which a NOS element (sub-domain) appears at a particular grade level span as a measure of comprehensiveness			
	(%)	Grades K–2 (%)	3–5 (%)	6–8 (%)	9–12 (%)
Science is based on empirical evidence	96	59	88	78	80
Cooperation exists in science	90	63	71	80	84
There is a tentative element in science	90	24	41	73	67
Distinction between observation and inference	86	51	65	55	57
Role of experiments	86	27	69	76	71
Distinction between science and technology	84	59	73	73	67
Science is socially and culturally embedded	76	35	51	55	63
There is no stepwise scientific method	71	22	43	57	57
Science is somewhat subjective	61	14	35	47	43
Distinction between law and theory	55	4	16	33	47
Science cannot answer all questions (i.e., there are limits to science)	49	39	39	35	37
Creativity play a role in science	25	6	10	16	14

what was ideal. Space limitations preclude including the results for each state, but Tables 61.4 and 61.5 provides the norm-referenced ranking of those states that scored highest when considering both completeness and comprehensiveness.

Not unexpectedly, no state suggested that all of the key NOS elements should be included at every age/grade level. If we treat the 50 states as independent experiments in writing NOS-related standards, we can learn much about consensus and innovation. For instance, it is possible to note that some of the NOS elements are more likely to be recommended for younger students, while other NOS content is generally reserved for those in the upper grades. This is an important consideration in the design of science curricular.

Not surprisingly, there is a huge variance in terms of NOS inclusion when comparing the fifty-one US science content standards. The authors of this study found that many states include both a complete picture of recommended NOS content and do so across grade levels. This strategy of returning to content repeatedly at higher

Table 61.4 States within the top 10 ranks with respect to completeness and comprehensiveness taken together (Note: some states' ranks were tied with others for their rank so the list does not extend to 51)

Highest rank with respect to NOS completeness and comprehensiveness in the state standards	State
1	OH
2	FL, NH, OR
3	MO
4	NC
5	KY, NJ
6	NE
7	MI, NY
8	MA
9	MN
10	LA

Table 61.5 States within the lowest 10 ranks with respect to completeness and comprehensiveness taken together (Note: some states' ranks were tied with others for their rank so the list does not extend to 51)

Lowest rank with respect to NOS completeness and comprehensiveness in the state standards	State
31	AZ
32	CA
33	WI, AR
34	DE, WY
35	TX
36	AL, CO
37	CT, SD
38	SC
39	RI
40	IA

levels of abstraction and/or complexity across grade levels is an excellent application of the spiral curriculum in instructional design. Some of the recommended NOS elements are complex enough that students would learn little from a one-time encounter. NOS is so integral to an understanding of the scientific enterprise that including some elements only at one grade level or only at a particular time in the science curriculum would reduce the likelihood that students would understand the concept and appreciate its significance. So, those states that expect students to engage NOS content early and often are to be congratulated.

However, as with everything else of importance in the educational enterprise, actual NOS instruction in classrooms is in the hands of teachers. No matter how robust the state standards are, such standards are embraced and interpreted by the 7.2 million teachers in 15,000 school districts across the nation (US Census Bureau 2011). What NOS instruction looks like in one classroom may be quite distinct from such instruction in another classroom, even in the same school.

This analysis of the multiplicity of guidelines produced in the last decade demonstrates that NOS is important and that there is a consensus among science educators and policymakers toward what elements of the nature of science are most appropriate to inform school science instruction (see Table 61.1). Furthermore, the analysis has also provided evidence of which particular NOS aspects are most appropriately included at each grade-level span (Table 61.3). Such data are of vital importance in designing future science standards guidelines.

61.7 Nature of Science: An Emerging Consensus View

Figure 61.1 is a graphical organizer of a consensus view of the nature of science designed to inform school science. This plan is based on a review of the prevailing experts, a review of the fifty-one US state documents (McComas et al. 2012) and earlier suggestions from McComas (2008).

Here, the nine recommended elements of the nature of science are arranged in three suites of related items: the tools and products of science, science knowledge and its limits, and the human elements of science. Nine domains or key aspects of NOS (empirical basis, shared methods, subjectivity, creativity, and others) are not suggested as complete descriptions. Also, when elucidated and discussed, it will be obvious that some of these nine are more extensive in scope and content than others. For instance, the NOS element "science has shared methods" relates to a range of issues including the ideas that there is no single step-by-step method and that scientists share techniques such as good record keeping and use of induction, deduction, and inference. On the other side of the figure, we find the distinction between science and technology noted. This is a relatively discrete idea stating that students should understand the distinction between the roles and processes of science and those of technology and engineering. Such distinctions are more important than ever given the inclusion of engineering in the *Next Generation Science Standards*. The point to keep in mind is that the "size" of each other nine recommendations is

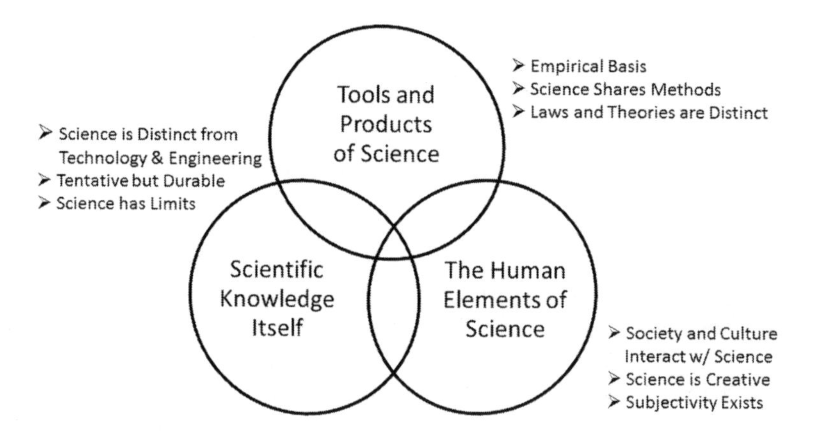

Fig. 61.1 A graphical representation of nine basic NOS elements frequently recommended for use in K-12 science instruction, arranged into suites of related items (Adapted from McComas (2008))

not the same nor is the level of complexity of the underlying idea. Textbook authors, curriculum designers, and, of course, teachers will ultimately be responsible for how these ideas are integrated into classroom instruction.

Figure 61.1 is nothing more than a potentially useful way to illustrate the range and relationship of the NOS elements most commonly recommended to inform and enliven science teaching in schools. There should be no implication that these elements are simply to be memorized on the first day of school and quietly forgotten. Rather, these elements must be explicitly explained, illustrated with examples, mentioned in context with discussion of science content, and assessed to gauge student understanding.

61.8 The Next Steps in US Science Instruction: The Framework for K-12 Science Education and the Next Generation Science Standards

One of the challenges with respect to providing an overview of the nature of science in the US context is that change in the educational landscape is constant. This is particularly true now that many of the US states are poised to adopt shared new science instructional standards, presumably abandoning their specific state learning goals. Colleagues in mathematics and English/language arts education have already faced this situation. Many of the US states have accepted what is called the "Common Core" (*Common Core State Standards for English Language Arts & Literacy in History/Social Studies, Science, and Technical Subjects*), a set of shared instructional standards in these disciplines (Council of Chief State School Officers 2010; Rothman 2011). Also, students are to develop appropriate levels of writing and reading literacy in the content area of science.

The latest revolution in science instruction in the United States began with the release of a new document designed to guide science teaching at the precollege level titled *A Framework for K-12 Science Education: Practices, Crosscutting Concepts, and Core Ideas* (National Research Council 2012). The recommendations in the *Framework* have already been transformed into various draft versions of the *Next Generation Science Standards (NGSS)* which, in turn, have been reviewed by working groups across the nation. As is often the case when it comes to shared initiatives, a few states have decided not to endorse the *Next Generation Science Standards*, so some diversity in the nature of the US science curriculum will continue to exist. However, the vast majority of US states plan to adopt the *NGSS* recommendations thus shifting the balance of power in education, in a curricular sense, quite dramatically.

61.8.1 A Framework for K-12 Science Education

Since the *Framework* (NRC 2012) represents a major new source of thinking about science teaching in the United States, it will be examined in detail with respect to its

Table 61.6 The three dimensions of the Framework for K-12 Science education: practices, crosscutting concepts, and core ideas (NRC 2012, p. 3). One example from each disciplinary core idea is provided as an illustration, but there are many more

1. Scientific and engineering practices
   1. Asking questions (for science) and defining problems (for engineering)
   2. Developing and using models
   3. Planning and carrying out investigations
   4. Analyzing and interpreting data
   5. Using mathematics and computational thinking
   6. Constructing explanations (for science) and designing solutions (for engineering)
   7. Engaging in argument from evidence
   8. Obtaining, evaluating, and communicating information
2. Crosscutting concepts
   1. Patterns
   2. Cause and effect: mechanism and explanation
   3. Scale, proportion, and quantity
   4. Systems and system models
   5. Energy and matter: flows, cycles, and conservation
   6. Structure and function
   7. Stability and change
3. Disciplinary core ideas
   Physical science (such as matter and its interactions)
   Life sciences (such as from molecules to organisms; structures and processes)
   Earth and space science (such as Earth's place in the universe)
   Engineering, technology, and applications of science (such as engineering design)

inclusion of aspects of the nature of science. The Framework is a narrative result of discussions from expert groups empaneled to ensure accurate science content and provide a set of broad expectations for science and engineering education in the K-12 learning environment. The *Framework* provides recommendations in three domains including science and engineering practices, crosscutting concepts, and core ideas in life, physical, and Earth/space sciences. Although engineering and technology were mentioned in previous standards documents, engineering has risen to a status almost equal to that of science itself. In fact, the *Framework* is endorsed by the presidents of the National Academy of Sciences *and* the National Academy of Engineering.

Since these three dimensions are inherent in the design of the *Framework* and to the new standards, it is useful to include them here (Table 61.6). Some of these dimensions do relate to elements of the nature of science, but it is important to note that NOS itself does not appear explicitly even though "how science works" could certainly be considered a crosscutting concept.

A number of techniques were used to analyze the *Framework* for its inclusion of NOS elements including the application of the new computer search tool (Jiang 2012) specifically developed to look for inclusions of specific content in large digital files. In addition, the search examined the index in the *Framework and* performed keyword searches within the *Framework*, accompanied by a reading of the *Framework* to establish context.

The computer-assisted search found that most of the major recommended ideas in the nature of science appear in the *Framework*. Of course this is encouraging. However, a closer examination of the document reveals that although NOS is stated and implied at various places in the text, but not explicitly recommended anywhere. The key NOS concepts are scattered throughout the document rather than being located in one section, and only a few appear in the index. Even the term "nature of science" only appears in the index (science, nature of) in a discussion of recommendations for improvement to the *Framework*. This is curious considering the following strong statement:

> any [science] education that focuses predominately on the detailed products on the detailed products of scientific labor—the facts of science—without developing an understanding of how those facts were established or that ignores the many important applications of science in the world misrepresents science . . (NRC 2012, p. 43)

As a way of considering what the *Framework* contains with respect to the nature of science, these next few sections will examine how the NOS concepts of creativity, scientific method, inference, law/theory, tentativeness, limits of science, subjectivity, and the social elements of science are treated in the *Framework*. It is particularly noteworthy to see how widely scattered the references are and how diligently one would have to search the entire *Framework* and infer in order to derive a useful recommendation for nature of science.

61.8.2 Creativity in the Framework

Most experts agree that creativity is an important part of science, and the *Framework* mentions this fact in the following places: "They [students] should come to appreciate that science and the current scientific understanding of the world are the result of many hundreds of years of creative human endeavor" (NRC 2012, p. 9). "Not only is such an approach alienating to young people, but it can also leave them with just fragments of knowledge and little sense of the creative achievements of science, its inherent logic and consistency, and its universality" (NRC 2012, p. 10). "...the insights thus gained help them [students] recognize that the work of scientists and engineers is a creative endeavor—one that has deeply affected the world they live in" (NRC 2012, pp. 42–43). "...construction of explanations or designs using reasoning, creative thinking, and models" (NRC 2012, p. 44), and "the creative process of developing a new design to solve a problem is a central element of engineering" (NRC 2012, p. 206).

61.8.3 Scientific Method in the Framework

The notion that there is no single stepwise scientific method appears in the *Framework* in a variety of places, but there is no single description of what this means, nor is the term found in the index. Consider the following statements: "Second, a focus on practices (in the plural) avoids the mistaken impression that there is one distinctive approach common to all science – a single "scientific method" – or that uncertainty is a universal attribute of science" (NRC 2012, p. 44). "For example, the notion that there is a single scientific method of observation, hypothesis, deduction, and conclusion—a myth perpetuated to this day by many textbooks—is fundamentally wrong" (NRC 2012, p. 78). "Thus the picture of scientific reasoning is richer, more complex, and more diverse than the image of a linear and unitary scientific method would suggest" (NRC 2012, p. 78) and "…and not as rote procedures or a ritualized 'scientific method'" (NRC 2012, p. 254). All of these statements accurately represent the issue related to "scientific method" correctly and appropriately, but the new Framework focuses on science and engineering practices, and the way in which those practices are discussed together in the document could be a source of new misunderstandings about this issue.

61.8.4 Law/Theory in the Framework

For decades, educators have been concerned that students do not understand the relationship between hypothesis, theory, and law. The *Framework* has a multitude of references to the term "hypothesis," several to "theory," but no direct references to the role of law in science. There are the following implied references to "law," but it is not clear why the term itself is missing.

With respect to laws, the *Framework* offers the following: "Repeating patterns in nature, or events that occur together with regularity, are clues that scientists can use to start exploring causal, or cause-and-effect, relationships, which pervade all the disciplines of science and at all scales" (NRC 2012, p. 87). The following statement comes close to providing a definition of "law," but unfortunately it is not referenced in the index:

> One assumption of all science and engineering is that there is a limited and universal set of fundamental physical interactions that underlie all known forces and hence are a root part of any causal chain, whether in natural or designed systems. Such "universality" means that the physical laws underlying all processes are the same everywhere and at all times; they depend on gravity, electromagnetism, or weak and strong nuclear interactions. (NRC 2012, p. 88)

There is a very well-constructed discussion of hypotheses and laws in the *Framework*, a section of which is shown below. Unfortunately, the obvious and necessary link to "law" is absent.

> Theories are not mere guesses, and they are especially valued because they provide explanations for multiple instances. In science, the term "hypothesis" is also used differently than it is in everyday language. A scientific hypothesis is neither a scientific theory nor a guess; it is a plausible explanation for an observed phenomenon that can predict what will happen in a given situation. (NRC 2012, p. 67)

61.8.5 Tentativeness in the Framework

Finally, we consider the idea of tentativeness in the *Framework*. In addition, we find a few statements about this issue such as "any new idea [in science] is tentative ..." (p. 79) and "any new idea is initially tentative, but over time, as it survives repeated testing, it can acquire the status of a fact—a piece of knowledge that is unquestioned and uncontested, such as the existence of atoms" (p. 94). Unfortunately, there are problems with both of these statements. In the first case, there is no corresponding statement about how long-lasting (or durable) science knowledge is in practice even if it is tentative. In the second case, the statement is incorrect because it implies that once a "fact" is established, its tentative character is somehow eliminated. Philosophers of science would remind us that all knowledge in science is tentative.

61.8.6 Limits of Science and Subjectivity in the Framework

These important issues are neglected in the *Framework*. There are many mentions of the targets of science and engineering but there is no explicit statement about what science and engineering cannot do. Unless one has a strong background in the philosophy of science, it would not be intuitively obvious that science does not address all problems in all domains. This is a particularly troublesome issue since many students fail to understand that many of their religious notions are not in the domain of science and therefore are not subject to attack by science. In the case of evolution, for instance, students would be well served by knowing that the scientific explanation of biological change through time does not demand that they give up a belief in some metaphysical component to such change.

There are vague implications in the *Framework* that science has subjective elements, but the word appears nowhere in the document. This is unfortunate, because an examination of the history of science provides many examples of how scientists have personally pushed and rejected ideas and interpreted evidence even contrary to the views of others. Sometimes this subjective element is exactly what was necessary for an idea finally to be accepted by the scientific community (i.e., Milliken's discovery of the charge on the electron). Giving students opportunities to recognized that science has a subjective element is valuable in portraying science accurately and situating science as a human pursuit.

61.8.7 Science as a Social Activity in the Framework

There are a number of references in the *Framework* to the social aspect of science. Consider the following as a good description of this aspect:

> ... science is fundamentally a social enterprise, and scientific knowledge advances through collaboration and in the context of a social system with well-developed norms. Individual scientists may do much of their work independently or they may collaborate closely with colleagues. Thus, new ideas can be the product of one mind or many working together. (NRC 2012, p. 26)

The only objection to this phrase relates to its placement. As is the case with many of the NOS elements in the Framework, this one is not included along with other nature of science ideas so that readers could see this as part of a broader description of NOS in science education.

61.8.7.1 The Framework and NOS: Conclusions and Concerns

The *Framework* does contain some meaningful suggestions for the nature of science and even includes a section called "understanding how scientists work" with the following powerful statement reproduced here in full:

> The idea of science as a set of practices has emerged from the work of historians, philosophers, psychologists, and sociologists over the past 60 years. This work illuminates how science is actually done, both in the short term (e.g., studies of activity in a particular laboratory or program) and historically (studies of laboratory notebooks, published texts, eyewitness accounts). Seeing science as a set of practices shows that theory development, reasoning, and testing are components of a larger ensemble of activities that includes networks of participants and institutions, specialized ways of talking and writing, the development of models to represent systems or phenomena, the making of predictive inferences, construction of appropriate instrumentation, and testing of hypotheses by experiment or observation.
>
> Our view is that this perspective is an improvement over previous approaches in several ways. First, it minimizes the tendency to reduce scientific practice to a single set of procedures, such as identifying and controlling variables, classifying entities, and identifying sources of error. This tendency overemphasizes experimental investigation at the expense of other practices, such as modeling, critique, and communication. In addition, when such procedures are taught in isolation from science content, they become the aims of instruction in and of themselves rather than a means of developing a deeper understanding of the concepts and purposes of science. (NRC 2012, p. 43)

Unfortunately, what the *Framework* fails to do is to clearly and explicitly feature exactly what NOS elements are recommended and provide some definitions and descriptions in one place to serve as a useful guide for those hoping to use the document to frame instruction in this important area. The NOS suggestions are implied, too widely scattered and much too implicit to be of use to educators not already familiar with the nature of science. To profit from the *Framework*, educators would have to work very hard and infer much to derive a useful list of NOS goals, a task that would be difficult without considerable expertise in this area. It is ironic that the document reports that "Many of those who provided comments thought that the "nature of science" needed to be made an explicit topic or idea. They noted that it would not emerge simply through engaging with practices" (NRC 2012, p. 334). It is curious that the authors had to be informed of this clear fact.

61.8.7.2 The Next Generation Science Standards and the Nature of Science

Following the release of the *Framework*, committees began the development of the *Next Generation Science Standards (NGSS)* with a March of 2013 target date for

Table 61.7 An example of the first of the content standards from the next generation science standards (NRC 2013) showing the link to the nature of science associated with that standard

Section title: K. forces and interactions: pushes and pulls	
Specific goals:	K-PS2-1. Plan and conduct an investigation to compare the effects of different strengths or different directions of pushes and pulls on the motion of an object K-PS2-2. Analyze data to determine if a design solution works as intended to change the speed and direction of an object with a push or pull
Science and engineering practices:	Planning and carrying out investigations Analyzing and interpreting data
Disciplinary core ideas:	PS2.A: Forces and motion PS2.B: Types of interactions PS3.C: Relationship between energy and forces ETS1.A: Defining and engineering problem
Crosscutting idea: Patterns and cause and effect	
Nature of science: Scientific investigations use a variety of methods	

completion. What makes this document different from others is that almost all of the states agreed to use this single set of guidelines to guide science teaching. Another major change is that the NGSS would be based on the science and engineering practices, core ideas, and crosscutting concepts (see Table 61.6) in addition to a blending of science with engineering. The goal of these new standards is that all high school graduates would have enough knowledge of science and engineering to participate in debates about science issues, make informed choices as consumers, and be prepared to enter science-related careers.

Various groups and individuals reviewed the *Framework* and agreed that it failed to adequately illustrate and define the various elements of the nature of science that should inform science teaching. What resulted from this review is a much more robust and complete vision for NOS in the NGSS with eight specific NOS domains discussed in an appendix and included throughout the document. Each domain of grade-level content is linked to the other design elements, but now the nature of science is explicitly included. Consider the following example from the NGSS illustrated in Table 61.7.

Unlike in the *Framework* where the nature of science elements is highly scattered, those in the *Next Generation Science Standards* (NGSS) are found together in a single appendix. There are eight NOS categories each with specific learning goals divided into recommendations for K–2, 3–5, middle school, and high school learners. Table 61.8 includes all of these NOS recommendations along with a description of that learning goal. The authors also indicate that some of the NOS goals are most closely associated with science and engineering practices and some with the crosscutting concepts. While it is not clear why this is important, perhaps the authors hope that teachers will be able to teach aspects of NOS while teaching some of the other recommended content.

Table 61.8 Categories of nature of science found in Appendix H of the next generation science standards (NRC 2013). Additional detail provided has been extracted from the specific details associated with the learning goals stated for each of the grade level (K–2, 3–5, middle, and high school). Those goals thought to be most associated with science and engineering practices are labeled Pr, and those most closely linked to crosscutting concepts are labeled Cr

Scientific investigations use a variety of methods (Pr); this implies that there is no step-by-step method, but this is not stated, that students are to distinguish between science and nonscience, and that scientists share values such as objectivity and open-mindedness
Scientific knowledge is based on empirical evidence (Pr); the grade-level expectations for this NOS element are much the same and self-evident that science requires evidence
Scientific knowledge is open to revision in light of new evidence (Pr); this implies that science is tentative but this is not stated explicitly
Science models, laws, mechanisms, and theories explain natural phenomena (Pr); theory and law are said to be individual and unique elements of science. Useful and accurate definitions are provided
Science is a way of knowing (Cr); additional information relates to the unique elements of science (such as empiricism and skepticism) and the view that scientific modes of inquiry can be used by sciences and nonscientists and that contributions to science come from various types of people
Scientific knowledge assumes an order and consistency in natural systems (Cr); this notion relates to an expectation that nature is orderly and develops this idea across the grade levels
Science is a human endeavor (Cr); all types of people have contributed to science and continue to make contributions and that science influences and is influenced by society. Other issues mentioned include creativity, some discussion of science, and its link to technology
Science addresses questions about the natural and material world (Cr); the NOS category includes discussion of the target of science, the notion that science cannot answer all questions, and the role of science in decision-making

The way in which these eight NOS ideas are linked to the specific learning objectives seems more of an afterthought than something strategic, but this reviewer is pleased to see NOS represented so prominently. In the example provided earlier (Table 61.7), please note that the NGSS considers that when young students learn about "forces and interactions," this would be a good time to demonstrate that "scientific investigations use a variety of methods."

61.8.8 Next Generation Science Standards: Conclusions and Critique

There is no doubt that the authors of the *NGSS* responded appropriately to the comments made by those who reviewed the *Framework*. The nature of science goals is more explicitly stated and organized. However, there are still some issues regarding NOS of some concern. These include the perception that some of the ideal NOS elements are missing, the nature of the narrative justifying NOS in the NGSS Appendix, the placement of NOS in the *NGSS*, and the potential challenges associated with potential conflating of science and engineering in the minds of students.

With respect to what is missing, the most glaring omission is an explicit statement indicating that science advances both through organized means and through somewhat subjective ones. It would be reasonable to include this in the category of "science is a human endeavor." Creativity should be mentioned more consistently. Currently, the ideas of "creativity and imagination" are noted only as grades 3–5 and high school goals within the human dimension category. Of course, it is reassuring to see this characteristic of science listed at all, but there is little reason for its lack of prominence. It would be useful to discuss the relationship of tentativeness and durability across grade levels rather than saving such discussion only for high school students. The *NGSS* standards are silent on the distinction between inference and observation, an issue that could be easily corrected in either of the first two categories, "variety of methods" or "empiricism."

Since the *Standards* include engineering goals along with those for science, one would have expected to see a NOS goal statement distinguishing science from technology and engineering. Formally, suggestions to include this NOS element had not been embraced by the science education community, but must now rise in importance because of the curious design of the *NGSS*. The *NGSS* and its blend of science and engineering present the danger that students may consider science and engineering as similar, therefore confuse the two. These two disciplines are not the same; they have distinct goals and distinct methods, and although there are advantages in blending scientific ideas with their application through engineering, educators must work very hard to distinguish one from the other.

In conclusion, two other concerns about NOS in the NGSS come to mind. First, the rationales, research, and proposed instructional mechanism for NOS provided in Appendix H are remarkably shallow. The brief section discussing the development of NOS is incomplete; there has been significant work done in recent decades to justify the inclusion of NOS but that has not been included in any strategy or complete fashion. For instance, the use of history as a means by which NOS may be included in the classroom is fine, but there are many other ways to integrate nature of science and science teaching, yet the *NGSS* offer no other suggestions. The two charts listing the key NOS elements are very well designed, but they offer a learning progression for teaching aspects of NOS across the grade levels that does not seem to have been based on a review of the research. Finally, the positioning of the NOS links at the bottom of each page of science content goals makes it appear that NOS was an afterthought. All those who recognize that NOS is fundamental to students' understanding of the scientific process will have to redouble efforts to ensure that reality prevails as the national embraces these new science standards.

61.9 Standards and the Nature of Science: Challenges and Conclusions

In the United States there is no mechanism yet in place to guarantee that any particular science content, including NOS, is taught in each classroom in an appropriate and complete manner. Having well-reasoned standards is a very good step, but issues of assessment, teacher preparation, and targeted curriculum approaches to teaching the nature of science are also important.

61.9.1 Assessment and the Nature of Science

Through national legislation known as *No Child Left Behind*, the US federal government has mandated that some science content acquisition be measured by end-of-course (EOC) evaluations. However, presently there is no single measure applied nationally and few of the state-developed instruments measure much that could be considered related to the nature of science. This situation may change with the widespread adoption of the *Next Generation Science Standards.* When the majority of US states are using the same science teaching standards, it will become increasingly cost-effective to produce assessment tools for use widely. Such assessments become a powerful "policy lever" to encourage teachers to include any specific content in instruction. Of course, this content must include the nature of science.

Although there is now an increasing consensus on what aspects of NOS should be communicated in school settings, the problem of assessment has not yet been fully addressed. We simply do not have valid, reliable, and complete measures of NOS that can be administered to large groups of students and scored quickly in a cost-effective manner. We are also still grappling with important considerations such as how contextualized any measures of NOS must be to be valid. In other words, must NOS items in a biology end-of-course test be different markedly from those used for physics students? Another challenge has been to develop assessment tools that provide some detail about what students know of the various NOS subdomains such as those indicated in Table 61.7.

Therefore, some effort must be expended to develop valid and reliable NOS assessment instruments with useful and robust subscales to help determine what students understand about the nature of science and make sure that such items are included on all end-of-course examinations. Completing both of these tasks will encourage educators to include NOS in instruction, while the results of these assessments will permit educators and researchers to visualize what specific elements of NOS are being communicated effectively in schools.

61.10 Science Teacher Preparation and the Nature of Science

Another force that has potential to unite the states with respect to science teaching relates to preparation of science teachers. Many states have adopted the teacher education standards associated with one of the national accreditation organizations (such as the National Council for the Accreditation of Teacher Education or NCATE), a nongovernmental group. Therefore, by default and design, the teacher education programs governed by NCATE have common elements because of the guidelines imposed on them by these accreditation boards.

This issue becomes somewhat complicated because NCATE refers instructions that prepare teachers to the science content standards developed by the National Science Teachers Association (NSTA). The NCATE database shows that there is at least one academic institution within each state in the United States accredited by NCATE. So, for those science teacher education programs that use NCATE/NSTA, the 2003 standards are quite detailed with respect to the nature of science.

Strand 1 of the NSTA Standards (NSTA 2003) (science content) requires that teachers have implicit knowledge of nature of science under the subheading "concepts and principles" and perhaps "unifying concepts." Explicit knowledge is required under Standard 2 (nature of science), which includes both history of science and nature of science. The rationales provided for NOS inclusion are strong and include many of the NOS elements mentioned frequently by science educators including issues such as tentativeness, empiricism, subjectivity, creativity, the use of inference, the lack of a step-by-step scientific method, and the fact that science is embedded in culture. As well, we find recommendations for teaching NOS in an explicit fashion and the use of history to augment NOS lessons.

Beyond this general reference to NOS and NCATE/NSTA, the nature of science is specifically referenced in the competencies only for the states of Texas, Oklahoma, Florida, Missouri, and Massachusetts. As an example, the *Preparation Manual for the Texas Examinations of Educator Standards in* Domain I for Life Science states:

> The science teacher understands the process of scientific inquiry and its role in science instruction. The science teacher understands the history and nature of science. The science teacher understands how science affects the daily lives of students and how science interacts with and influences personal and societal decisions. The science teacher knows unifying concepts and processes that are common to all sciences. (Texas Examination of Educator Standards 2006, p. 6)

Competency 2 further describes what the beginning teacher should know about nature of science, the process of scientific inquiry, and the unifying concepts that are common to all sciences.

The required competencies in the biological sciences for Oklahoma list two areas with explicit reference to nature of science. Competency 2 requires that students "understand the nature of science including the historical and contemporary contexts of biological study" (p. 2–2). Competency 3 is to "understand the process of scientific inquiry and the role of observation, experimentation, and communication in explaining natural phenomena" (Oklahoma Subject Area Tests – Study Guide, p. 2–3). Florida (Florida Department of Education 2011) provides the most comprehensive inclusion of nature of science within the teacher licensure requirements in listing nature of science within all science disciplines including biology, physics, chemistry, and Earth-space science.

The Missouri Department of Elementary and Secondary Education Certification Requirements for Secondary Education (2009) require two semester hours of study in the history/philosophy of science and technology for all secondary science certifications. Under the regulations for educator licensure and preparation approval (603 CMR 7.00), Massachusetts (2012) lists subject matter knowledge requirements for teacher licensure that include methods of research in the sciences, history and philosophy of science, and principles and procedures of scientific inquiry (which many now include as part of the nature of science).

The licensure requirements for the majority of other states demand that a prospective teacher have a degree in the content area plus passing the Praxis II test.

Based on the *Biology: Content Knowledge* informational guide available from the developer of the test, the Educational Testing Service (ETS), we find approximately 12–22 multiple-choice questions out of the total 150 questions that relate specifically to NOS knowledge. These are all listed under the heading *Basic Principles of Science.* The nature of science questions relate to processes involved in scientific inquiry such as making observations, formulating and testing hypotheses, identifying experimental variables and controls, drawing scientific conclusions, and using scientific sources and communicating findings appropriately. Questions also relate to distinguishing differences among facts, hypotheses, theories, and laws; the testable nature of hypotheses; formulation of theories based on accumulated data; and the durability of laws. Finally, questions are also included that deal with the idea that scientific ideas change over time. These explicit questions about the nature of science comprise approximately 8–15 % of the Praxis II content knowledge test for biology, for example.

This is encouraging; it is clear that those thinking about science teacher preparation have come to value the nature of science, and there is some expectation that teachers who complete accredited programs would have some basic knowledge of NOS. Of course, there is no way of knowing if these teachers value and include NOS content in their classes nor do we know if these teachers preparation programs include a discussion of curriculum strategies for teaching the nature of science.

61.11 US Textbooks and Curriculum Innovations to Support NOS Instruction

One way that nature of science could become a more visible and integral part of the science curriculum is through its focused inclusion in science textbooks. If texts were to feature strong NOS elements – particularly if NOS is included throughout the book – there is a good chance that teachers and students would give more attention to this important topic.

However, in the highly competitive world of textbook publishing in the United States and the long-standing principle of permitting most schools to choose which science books to purchase, all textbooks authors will have to focus more clearly on the nature of science. This is more likely now given the widespread adoption of the *Next Generation Science Standards* and their reasonable focus on the nature of science. The challenge is not just to ensure that science textbooks include an accurate treatment of the nature of science linked to the traditional science content, but such books must integrate NOS throughout the text in a robust fashion.

The hope is that new texts will avoid the current tradition in which whatever NOS content included is relegated to the opening chapter with almost no additional NOS content woven into the subsequent chapters that feature standard science content. This is unfortunate, because many teachers likely move quickly through or skip that first chapter and therefore miss opportunities to share NOS content with students. An additional challenge is for teachers to have access to curriculum models to support engaging instruction in the nature of science. As we will see, although some models have been developed, almost none have entered common use.

61.11.1 Curriculum Innovations and the Nature of Science

Even though explicit and wide-ranging NOS lessons are all absent from current US textbooks, there have been a few important curriculum projects designed to support NOS instruction. The first example of a project targeting NOS instruction draws on the history of science and a set of ancillary study units that apply the case approach. The next two come as a direct result of the US government funding designed to reinvigorate and redirect science and mathematics education through new curriculum models following the launch of Sputnik in the late 1950s. These examples include a classic "alphabet soup" curriculum project called *Science – A Process Approach*; a textbook, *Project Physics*, with multiple NOS inclusions; and several media-centered approaches. There are other NOS examples found in a review of the history of curriculum development in the United States, but sadly none made long-lasting impacts on the inclusion of NOS in the typical science classroom. However, all of these stand as important testaments to the potential a NOS-focused science curriculum and should be studied carefully by those wishing to reestablish the nature of science in science instruction.

61.11.2 The Case History Approach: Conant and Klopfer

A focus on NOS in these curriculum innovations was offered in the previous decade by J.B. Conant an influential scientist, government official, and president of Harvard who stated "...it is my contention that science can best be understood by laymen through close study of a few relatively simple case histories..." (Conant 1947, p. 1). With this simple statement, the use of history as a dominant approach to NOS instruction was born.

Conant's suggestion for the use of history of science in science instruction resulted in what is the most noteworthy example of the case approach, *The Harvard Case Studies in Experimental Science,* which grew out of a general education course beginning at Harvard in 1948 and ultimately published in two volumes (Conant and Nash 1957). The seven cases focused on issues such as Robert Boyle and pneumatics, phlogiston theory, temperature and heat, atomic molecular theory, plants, spontaneous generation, and the historical development of the concept of electric charge.

Later, Conant's student and later fellow Harvard professor, Leo Klopfer along with Cooley (1963), adapted the case study approach for use in high schools with *the History of Science Cases (HOSC)*. Each of these units included the exploration of a major scientific idea through the examination of excerpts of historical documents and experimentation carried out either by students themselves or as a demonstration by the teacher (Lind 1979). The nine titles proposed or developed for HOSC were each represented by individual guides for teachers and their students. The units included exploration of cells, the chemistry of carbon dioxide, Fraunhofer lines, electricity and life (with the intriguing title *Frogs and Batteries*), the discovery of halogen elements, air pressure, plant reproduction, atomic theory, and the speed of light.

The overarching goals for HOSC were to show students the methods used by scientists; the means by which science advances and the conditions under which it flourishes; the personalities and human qualities of science; the interplay of social, economic, technological, and psychological factors with the progress of science and the importance to science of accurate and accessible records; constantly improved instruments; and free communication between scientists (Klopfer 1964).

61.11.3 S-APA: Science: A Process Approach

This project was developed starting in 1962 by the AAAS Commission on Science Education with funding from the National Science Foundation based on the learning theories of psychologist Robert Gagne. The basic premise of S-APA was that the "first and central purpose of science education is to awaken in the child, whether or not he will become a professional scientists, a sense of the joy, the excitement, and the intellectual power of science" (AAAS 1967, p. 1). The basic notion was to give students a variety of hands-on experiences based on a hierarchy of skills (called process skills) that scientists were seen to use regularly.

The processes that the students experienced included observation, classifying, using numbers, measuring, using space/time relationships, communicating, predicting, and inferring (called the basic processes for the primary grades). These were joined by integrated process for upper grade students and included defining operationally, forming a hypothesis, interpreting data, controlling variables, and experimenting.

It should be clear that a number of important NOS ideas were contained among the processes even though S-APA was not designed expressly to communicate such goals. Unfortunately, S-APA like so many of these curriculum projects was never widely used. It quickly faded from the scene because of its uniqueness, its lack of focus on science content, and its decontextualized nature (Finley 1983). However, many of the individual lessons provide wonderful lesson opportunities for teachers today who want a hands-on approach to the nature of science.

61.11.4 Project Physics: NOS in a Textbook Setting

One of the most important text-based projects to feature NOS was *Project Physics* (also called *Harvard Project Physics*) authored by Gerald Holton, F. James

Rutherford, and Fletcher Watson. This book was remarkable in its desire to merge the traditional content of physics with the human side of science. In the introduction to the 1981 edition, the authors write that one of their goals was for students to see "physics as the wonderfully many-sided human activity that it really is. This meant presenting the subject in historical and culture perspective..." (p. iii). The book meets this challenge on almost every page by prominently discussing the people who contributed to modern physics.

As an example, consider the prologue which begins "It is January 1934 in the city of Paris. A husband and wife are at work in a university laboratory. They are exposing a piece of ordinary aluminum to a stream of tiny changed bits of matter called alpha particles" (Holton et al. 1981, p. 1). It is impossible to find another science text that so deliberately shares the human side of science with students in such an explicit fashion. It the pages of *Project Physics*, readers could encounter all of the greats and near-greats of physics while learning the traditional science content.

With both Conant's case study approach and the humanistic physics of Holton, Rutherford, and Watson, we see innovative approaches to the inclusion of nature of science content in science curriculum. Sadly, there have been very few other such products, and even these were only marginally successful. Project Physics had a much longer life; it was produced in several editions, and by far more teachers and students than the case study materials or *S-APA*, but in the end, neither approach can be found in use today.

61.11.5 Using Media to Teach the Nature of Science

There have been some excellent ancillary materials produced such as the *Mechanical Universe Project* created and hosted by David Goodstein of the California Institute of Technology in 1985 and *Mindworks* in 1994 by Barbara Becker (Becker et al. 1995). Both of these initiatives produced high-quality video reenactments of important scientists and their discoveries. The good news is that *Mechanical Universe* is available on line for streaming by teachers who would like to use the reenactments to enliven their presentations of science content and teach some important NOS lessons at the same time.

In fairness, many former curriculum projects and current texts do include some discussion of NOS, typically in the opening chapter, and often feature the names and dates of the most famous scientists, typically in side bar material. Unfortunately, NOS in this fashion is too easy to ignore by teachers and students. The sad reality is that in spite of a number of notable innovations including those mentioned here, the US science curriculum today does not look much different than it did before the widespread interest in and rationales for the inclusion of the nature of science.

61.12 Conclusion

The United States is a unique environment for education with its decentralized educational bureaucracy, a multitude of special interest groups involved in policymaking, free-market choice of textbooks, and individual state control over the curriculum and end-of-course (or exit) assessments. However, this review provides some reasons to be encouraged. We have seen the development of some interesting curriculum models to support NOS instruction, and there is increasing and widespread interest on the part of the science education community that NOS is an important learning goal. In addition, science educators have provided strong rationales for the inclusion of NOS and have defined a set of NOS sub-domains appropriate for use in school settings. *The Next Generation Science Standards* feature most of the recommended elements of NOS in a prominent fashion using a spiral curriculum approach to ensure that NOS is included each year. Perhaps the states will respond to these new standards by requiring that preservice teachers have significant experience with the nature of science. This suggestion for improvement cannot be overstated. Those who teach science certainly will have learned much science content while earning their degrees, but the same claim may not be made regarding teachers' understanding of how science functions. As a case in point, only science teachers repeat the inaccurate notion that there is a scientific method; scientists know better. However, science teachers learned this faulty idea in the pages of the science textbooks since they learned very little about the structure and processes of science while digesting the diet of "final form" science in their prior school and university experiences. Perhaps if those learning science were permitted to look "behind the curtain" and gain an understanding of the nature of science, those who become teachers could carry those lessons into their classrooms.

On the other hand, there are reasons to be concerned. Currently, NOS is not a topic of any prominence in most US science textbooks and is certainly not included in texts in a systematic and explicit fashion. It is unclear what new teachers need to know and be able to do with respect to NOS leading to concern that they may simply ignore this content. Only with strong curriculum models aligned to the NOS content in the *NGSS* and shared examinations that include a strong NOS component will the promise offered by robust inclusion of NOS become a reality.

The inescapable conclusion is that the history and philosophy of science must be a strong part of science teacher preparation, must be included in standards in a clear and robust fashion, must be featured in an engaging fashion in textbooks, and must be assessed on science exit examinations. With these challenges in mind, there is hope that the nature of science will take its place along with traditional science content as a vital part of the science learning experience for the next generation of US science learners.

Acknowledgments The author appreciates the many helpful suggestions offered by the anonymous reviewers. In addition the author acknowledges the extraordinary work of colleagues Drs. Carole Lee and Sophia Sweeney who helped to prepare the state-by-state NOS comparisons, Dr. Feng Jiang who assisted with the analysis of the new *Framework* using a specially designed computer model to discover incidences of NOS in text, and Dr. Lisa Wood who made important contributions to the section on teacher preparation standards as they relate to the inclusion of the nature of science. Needless to say, all errors, interpretations, opinions, and conclusions are the responsibility of the author.

References

Al-Shamrani, S. M. (2008). Context, accuracy and level of inclusion of nature of science concepts in current high school physics textbooks. Unpublished doctoral dissertation. University of Arkansas, Fayetteville, AR.

American Association for the Advancement of Science (AAAS) (1989). *Project 2061: Science for all Americans,* Washington, D.C.: Author.

American Association for the Advancement of Science (AAAS) (1990). *The liberal art of science: Agenda for action.* Washington, DC: Author.

American Association for the Advancement of Science (AAAS) (1993). *Benchmarks for science literacy.* New York, Oxford University Press.

American Association for the Advancement of Science (AAAS) (1967). *Science-A process approach, Part A: Description of the program.* Washington, D.C.: Author.

Becker, B. J., Younger-Flores, K., and Wandersee, J. H. (1995). MindWorks: Making scientific concepts come alive. In F. Finley, D. Allchin, D. Rhees, and S. Fifield (Eds.), *Proceedings of the Third International History, Philosophy, and Science Teaching* (vol. 1, pp. 115–125). Minneapolis, Minnesota: The University of Minnesota.

Brown, J. R. (2001). Who rules in science? An opinioned guide to the wars. Cambridge, MA: Harvard University Press.

Conant, J. B. (1947). *On understanding science.* New Haven, CT: Yale University Press.

Conant, J.B. (1951). Science and common sense. New Haven, CT: Yale University Press.

Conant, J.B. and Nash, L.K. (1957). Harvard case histories in experimental science (Vols I and II). Cambridge, MA: Harvard University Press.

Council of Chief State School Officers and National Governors Association. (2010). *Common Core State Standards for English Language Arts & Literacy in History/Social Studies, Science, and Technical Subjects.*

DeBoer, G. (1991). *A history of ideas in science education: Implications for practice,* New York, Teachers College Press.

Duschl, R. A. (1985). Science Education and Philosophy of Science Twenty-Five Years of Mutually Exclusive Development. *School Science and Mathematics, 85* (7) 541–555.

Duschl, R. A. (1990). *Restructuring science education: The importance of theories and their development* New York, Teachers College Press.

Educational Policies Commission (1944). Education for all American youth. Washington, DC: National Education Association.

Florida Department of Education. (2011). Competencies and skills required for teacher certification in Florida, sixteenth edition.

Finley, F. (1983). Science processes. *Journal of research in science teaching, 20*(1), 47–54.

Klopfer, L. (1964). The use of case histories in science teaching. *School science and mathematics, 64,* 660–666.

Klopfer, L.E. & Cooley, W.W. (1963). The history of science cases for high schools in the development of student understanding of science and scientists. *Journal of Research in Science Teaching 1*, 33–47.

Holton, G., Rutherford, F.J. and Watson, F.G. (1981). *Project physics.* New York: Holt, Rinehart and Winston.

Humes, E. (2007). *Monkey girl: evolution, education, religion, and the battle for America's soul.* New York: HarperCollins Publishers.

Hurd, P.DeH. (1960). 'Summary', in N.B. Henry (ed.) *Rethinking Science Education: The fifty-ninth year-book of the National Society for the Study of Education,* Chicago, IL, University of Chicago Press, 33–38.

Jiang, F. (2012). The Inclusion of the Nature of Science and Its Elements in Recent Popular Science Writing for Adults and Young Adults. Unpublished doctoral dissertation. University of Arkansas, Fayetteville, AR.

Lederman, N. G. (2007). Nature of science: Past, present, and future. In S. K. Abell & N. G. Lederman (Eds.), Handbook of research on science education (pp. 831–879). Mahwah, NJ: Erlbaum.

Lederman, N.G. (2002). The state of science education: Subject matter without context. Electronic Journal of Science Education [on line], 3(20), December http://wolfweb.unr.edu/homepage/jcannon/ejse/lederman.html

Lind, G. (1979). The history of science cases (HOSC): Nine units of instruction in the history of science. *European journal of science education, 1*(3), 293–99.

Martin, M. (1972). Concepts of science education. Glenville, IL: Scott-Foresman.

Massachusetts Department of Elementary and Secondary Education (2012). Regulations for Educator Licensure and Preparation Program Approval. Last amended June 26, 2012

Matthews, M. R. (1994). Science teaching: The role of history and philosophy of science. New York: Routledge.

McComas, W. F. (2008). Proposals for core nature of science content in popular books on the history and philosophy of science: Lessons for science education. In Lee, Y. J. & Tan, A. L. (Eds.) *Science education at the nexus of theory and practice*. Rotterdam: Sense Publishers.

McComas, W. F. (1998). The principal elements of the nature of science: Dispelling the myths of science. In W. F. McComas (Ed.) *Nature of science in science education: rationales and strategies* (pp. 53–70). Kluwer (Springer) Academic Publishers.

McComas, W. F., Clough, M. P., and Al-mazroa, H. (1998). A review of the role and character of the nature of science in science education. In W. F. McComas (Ed.). The nature of science in science education: rationales and strategies. Kluwer (Springer) Academic Publishers.

McComas, W.F., Lee, C. & Sweeney, S. (2012, April). A Review of the U.S. State Science Standards with Respect to the Nature of Science. Annual Meeting of the National Association for Research in Science Teaching. Indianapolis, IN.

McComas, W.F. & Olson, J.K. (1998). The nature of science in international science education standards documents. In W.F. McComas (Ed.) *The nature of science in science education*, 41–52. Dordrecht, Netherlands: Kluwer (Springer) Publishers.

Missouri Department of Elementary and Secondary education Certification Requirements for Secondary Education. (2009). Code of State regulations 5 CSR 20–400

Moore, R., Decker, M., & Cotner, S. (2009). *Chronology of the evolution-creation controversy.* Santa Barbara, CA: Greenwood Publishing Company.

National Science Teachers Association Teacher Education Standards (2003). (http://www.nsta.org/pd/ncate/docs/NSTAstandards2003.pdf). Accessed 10 May, 2012.

National Research Council (NRC) (1996). *National science education standards.* Washington, D.C., National Academy Press.

National Research Council (NRC) (2012). *A framework for K-12 science education: Practices, crosscutting concepts and core ideas.* Washington, D.C., National Academy Press.

National Research Council (NRC) (2013). *Next generation science standards.*http://www.nextgenscience.org/next-generation-science-standards. Accessed 30 May, 2013

Oklahoma Subject Area Tests – Study Guide, 010 Biological Sciences. (n.d.) Oklahoma Commission for Teacher Preparation, OK-SG-FLD010-03

Osborne, J., Collins, S., Ratcliffe, M., Millar, R., & Duschl, R. (2003). What "ideas-about-science" should be taught in school science? A Delphi study of the expert community. *Journal of Research in Science Teaching, 40*, 692–720.

Parsons, K. (2003). *The science wars: Debating scientific knowledge and technology.* Amherst NY: Prometheus Books.

Robinson, J.T. (1968). The nature of science in science teaching. Belmont, CA: Wadsworth Publishing Company.

Rothman, R. (2011). *Something in common: The common core standards and the next chapter in American education*. Cambridge MA: Harvard Education Press.

Schwab, J.J. (1964). 'The teaching of science as enquiry', in J. J. Schwab & P. F. Brandwein (eds.), *The teaching of science*, Cambridge, MA, Harvard University Press, 31–102.

Texas Examinations of Education Standards (TEES) (2006). Preparation manual: 136 Science 8–12. http://geft.tamu.edu/science_test.pdf. Accessed 8 June, 2012.

U.S. Census Bureau (June 22, 2011). http://www.census.gov/newsroom/releases/archives/ facts_for_features_special_editions/cb11-ff15.html. Accessed 3 March, 2012.

Van Dijk, E. M. (2011). Portraying real science in science communication. Science education, 95(6), 1086–1100.

William F. McComas is the inaugural holder of the Parks Family Endowed Professorship in Science Education at the University of Arkansas where he directs the Project to Advance Science Education (PASE). This follows a career as a biology and physical science teacher in suburban Philadelphia and professorship at University of Southern California. He has earned B.S. degrees in Biology and Secondary Education, M.A. degrees in Biology and Physical Science, and the Ph.D. in Science Education from the University of Iowa. McComas has served in leadership roles with the National Science Teachers Association, the International History, Philosophy and Science Teaching Group, the National Association of Biology Teachers, and the Association for Science Teacher Education. McComas is a recipient of the Outstanding Evolution Educator and Research in Biology Teaching awards, the Ohaus award for Innovations in College Science Teaching, and the ASTE Outstanding Science Teacher Educator award. He has written and edited several books including The Nature of Science in Science Education: Rationales and Strategies, numerous book chapters, and articles and has given more than 100 keynote speeches, workshops, and presentations at public events and professional meetings in the United States and in more than a dozen other countries. Most recently, he was a Fulbright Fellow at the *Centre for the Advancement of Science and Mathematics Teaching and Learning* at Dublin City University, Ireland.

Chapter 62
The History and Philosophy of Science in Science Curricula and Teacher Education in Canada

Don Metz

62.1 Introduction

In Canada, the contributions to the field of history and philosophy of science (HPS) and the role and influence of HPS and science education have a rather recent, but significant, history. In the last few decades, we find influential philosophers such as Mario Bunge, Ian Hacking, and Paul Thagard; noted historians of science such as Stillman Drake; and influential science educators like Derek Hodson, Jacques Désautels, Glen Aikenhead, Arthur Stinner, Stephen Norris, and Kieran Egan (and many more, too numerous to mention) providing a diverse and extensive collection of books, essays, journals, lectures, conference keynotes, and graduate students in the field of HPS.

Indeed, many of these Canadian academics have enjoyed a wide range of international success and recognition. Mario Bunge, still active in his 90s, alone has contributed over 80 books and 400 journal articles in his tenure including his oft-referenced eight-volume *Treatise on Basic Philosophy* (1974–1989). Ian Hacking, well known for his defense of entity realism, wrote the introduction to Feyerabend's (1975) "*Against Method*," and Stillman Drake is universally considered to be the authority on Galileo, publishing over 130 books and articles on the great master (Buchwald and Swerdlow 1993, p. 663). More recently, Paul Thagard has emerged as a leader in the field of philosophy and cognitive science (Thagard 2010[1]). In the field of science education, Derek Hodson gave us the influential "Towards a Philosophically More Valid Science Curriculum" (Hodson 1988) and recently published *Towards Scientific Literacy: A Teachers' Guide to the History, Philosophy and Sociology of Science* (Hodson 2008). Nadeau and Désautels (1984) advanced an

[1] Note that Thagard (2010) is just the most recent of many books in philosophy and cognitive science by Paul Thagard.

D. Metz (✉)
Faculty Of Education, University Of Winnipeg, Winnipeg, MB, Canada
e-mail: d.metz@uwinnipeg.ca

M.R. Matthews (ed.), *International Handbook of Research in History, Philosophy and Science Teaching*, DOI 10.1007/978-94-007-7654-8_62,

epistemological perspective, and Glen Aikenhead provided an often-used research instrument on the Views on Science, Technology, and Science (Aikenhead and Ryan 1992) and has worked extensively in the last number of years arguing for a cultural component in science education. Arthur Stinner's Large Context Problems are widespread and cited by many (Stinner 1995), and Stinner's dramatizations are well known at the conferences of the International History and Philosophy of Science and Science Teaching organization. Additionally, in terms of scientific literacy and its connections to literacy, we find significant contributions from Stephen Norris (Norris et al. 2005), and in terms of narrative and story structuring, Kieran Egan has argued that language and reality develop with a constant interaction using intellectual tools of metaphor, imagery, and binary structuring (Polito 2005).

It is always somewhat risky to highlight so few examples of HPS in your own country when many more contributions to conferences (three of the ten IHPST conferences have been hosted in Canada), journals (editors like Ian Winchester), books (Roth and Desautels 2002), countless journal articles, and contributions in the field of curriculum can be found. However, to the reader, I provide these brief references to provide a context in which I can lay the foundations of a review of HPS in science curricula and teacher education in Canada.

Education in the Canadian political system is a provincial responsibility. Given a geographically expansive country, a diversity of cultures, and ten provinces and three territories, the role of history and philosophy of science (HPS) in science education varies considerably across the nation. In this chapter, I will explore a wide range of aspects of HPS and the teaching of science in Canada. Some of these aspects will include the evolution of ideas surrounding Science, Technology, and Society (STS) and a move towards scientific literacy and "Science for All." These features include the historical underpinnings of the history and philosophy of science and science curricula, the influence HPS has had on national and provincial curriculum from the process/product debates of the 1960s to the national conversations surrounding the 1993 Victoria Declaration (CMEC 1999), and the 1997 Pan-Canadian protocol (CMEC 1997). Additionally, I will address the inclusion of HPS in teacher education programs (including some best practices) and some recent developments. Finally, some time will be taken to examine the status of local indigenous knowledge, specifically its integration in science curriculum in the province of Saskatchewan.

62.2 Historical Perspectives

An early reference to the inclusion of HPS in curriculum in Canada can be found in a memorandum that was signed in 1962 by many prominent mathematicians in the United States and Canada.[2] Among other things the memorandum called for as a fundamental principle a genetic method:

> **Genetic method.** "It is of great advantage to the student of any subject to read the original memoirs on that subject, for science is always most completely assimilated when it is in

[2] On the Mathematics Curriculum of the High School, The Mathematics Teacher of March 1962, American Mathematical Monthly of March 1962

> then ascent state." wrote James Clerk Maxwell. There were some inspired teachers, such as Ernst Mach, who in order to explain an idea referred to its genesis and retraced the historical formation of the idea. This may suggest a general principle: The best way to guide the mental development of the individual is to let him retrace the mental development of its great lines, of course, and not the thousand errors of detail. (Ahlfors et al. 1962, p. 426)

Two Canadians signed the memorandum, the prominent geometer H. S. M. Coxeter, from the University of Toronto, and a relatively new Canadian Alexander Wittenberg, from Laval University. Coxeter was very active in the Upper Canada Branch of the Canadian Society for the History and Philosophy of Science for many years. As well, the Canadian Mathematical Bulletin reported in 1963 that Wittenberg was appointed as a professor of mathematics at York University in Toronto. At this time, York was in the process of developing courses to relate mathematics to the humanities and social sciences as well as the natural sciences. Wittenberg was to initiate this venture by developing a first year course in mathematics and philosophy. Hayo Siemsen (2011), tracing the influences of Ernest Mach in science education, notes the strong connections of Wittenberg to such notable professors as Pólya and Nevanlinna, and he argues convincingly about the influence that Mach had on science education (quite an obvious influence on Wittenberg it seems from the inclusion of the genetic method in the aforementioned memorandum). Siemens further suggests that is was Wittenberg who provided Freudenthal with the initial ideas leading to the PISA study (see Wittenberg 1965, 1968; Siemsen and Siemsen 2008, 2009). Wittenberg could then be considered to be an early pioneer in Canada promoting the inclusion of HPS. However, Wittenberg unfortunately died at an early age and published very little. The reality is that he had very little influence and is essentially unknown in the country because of his early demise. The genetic method, a historical approach to knowledge, had no influence in curriculum developments and remained for the most part and academic argument.

In most developed countries, the science curriculum, often referred to as "school science" (Duschl 1994), generally provides a wide breadth of coverage intended to prepare students for postsecondary science courses acting as a "pipeline" to future studies (Millar and Osborne 1998). Glen Aikenhead (2003), tracing the origins of school science in the Western tradition, notes that traditional science curriculum "assumes that 'science' in 'school science' has the same meaning as it has in, for example, 'the American Association for the Advancement of Science'" (p. 3). That is, the science curriculum is essentially a nineteenth-century curriculum in scope and intent.

The Canadian curriculum for the first half of the twentieth century generally followed this model with a great emphasis on the training of scientists and engineers in the 1960s and 1970s fueled by international competition and the space race. In reflection, Rutledge (1973) notes "The concern was also expressed that too little emphasis was being given to the philosophy and processes of science and that meaningful laboratory work was not an important part of most school science" (p. 600). However, emerging from these concerns, a large number of curriculum resources were produced that mostly focused on the processes of science. Many of these resources such as Science: A Process Approach (AAAS 1967), Science Curriculum Improvement Study (SCIS 1972), Elementary Science Study (EDC 1969), Physical Science Study Committee (PSSC 1960), Chemical Education

Material Study (CEMS 1960), Biological Sciences Curriculum Study (BSCS 2006), Earth Science Curriculum Project (Heller 1964), and Introductory Physical Science Program (ESI) found their way into Canadian science curricula. As elsewhere, the need for scientists was considered essential, and although it was generally accepted that all students needed to be "scientific literate," this literacy was oriented towards investigations, observation, learning of abstract concepts, and the acquisition of knowledge as though the student had discovered it. In general, weaknesses identified in these approaches were a failure to meet the needs of the general student population, readability levels were usually above the grade level, and most experiments and investigations were close ended rather than open ended (Baptiste and Turner 1972). Others argued that "there was no basis in school science on which to develop a critique of the role science plays in society or an appreciation of its strengths and limitations" (Fensham 1973, cited from Fensham 1997, Reconsidering Science Learning, p. 22).

With the advent of curricula that focused on the preparation of scientists, criticism began to emerge at that time that questioned why the only students worthy of curricular attention were those who might become professionals. Questions concerning "science for whom?", the balance of content and process, the relevance of the curriculum, and the emphasis on the transmission model of instruction were beginning to be asked by many educators and scientists themselves.

As an effort to address the relevance of science curriculum, the early development of a humanistic alternative to school science could be found in Science, Technology, and Society (STS) programs formally initiated in the late 1960s, in the United States, United Kingdom, Australia, and the Netherlands (Aikenhead 2003; Solomon and Aikenhead 1994). Aikenhead argued that "These university academic programs responded to perceived crises in responsibility related to, for instance, nuclear arms, nuclear energy, many types of environmental degradation, population explosion, and emerging biotechnologies. Thus, social responsibility for both scientist and citizen formed one of the major conceptions on a humanistic perspective in school science" (p. 8).

The dichotomy of purposes of science education, that is, science for future scientists versus a more humanistic approach for scientific literacy, helped fuel debates about the purposes of science education in the country. For example, the introduction of an STS orientation of high school science courses was not well received by university faculty. In some provinces, STS science courses were not approved for university entrance.

An important step in examining the various viewpoints in science education was initiated in the 1980s by the Science Council of Canada (SCC). The SCC was created by the federal government in 1966 to advise the government on science and technology policy (Millin and Steed 2012). The SCC members were broadly representative of the Canadian scientific community, in both the academic and private sectors. Consisting of a panel of 30 experts from the natural and social sciences and from business and finance, they produced many published works which were qualified by the council for reliability and methodology. In some cases, these reports provided recommended actions to governments or other institutions. However, the reports were most often intended to create a climate for policy debate rather supporting specific guidelines for government action.

In order to foster dialogue in science education, the SCC undertook a series of four discussion papers for a study of Canadian science education in the early 1980s. As noted by Hugh Munby (1982), one of the objectives was "to stimulate active deliberation concerning future options for science education," suggesting that the Science Council had no collective view on a direction for science education in Canada and in order to develop such a view, it was soliciting a variety of viewpoints.

The position papers were "A Canadian Context for Science Education" by James Page (1980) who argued that science education could contribute to "improved national awareness." A second paper by Glen Aikenhead (1980), "Science in Social Issues: Implications for Teaching," supported the view that science should be taught such that students learn about social and political issues. A third paper, "An Engineer's View of Science Education," by Donald George (1981) advocated for the equal treatment of the engineer's intellectual processes, and the final paper of the series, "What is Scientific Thinking?," by Hugh Munby (1982) examined what it meant to think critically and scientifically. The SCC's work culminated with an extension research program which included examination of the science curriculum guidelines in every province and territory, an analysis of common textbooks, survey of teachers and students, and case studies of science teaching across the country. Additionally, 11 conferences attended by ministry officials, school board members, teachers, students, and representatives from university, labor, and industry helped produce a consensus which endorsed a position of "Science for All" (SCC 1984).

The SCC report 36, "Science for Every Student: Educating Canadians for Tomorrow's World," established a goal of scientific literacy for all with four broad aims:

To encourage full participation in a technological society
To enable further study in science and technology
To facilitate entry to the world of work
To promote intellectual and moral development of individuals

Among the 47 recommendations, we find a recommendation to incorporate a science-technology-society emphasis in science courses at all levels and a recommendation to "ensure that courses at all levels present a valid representation of the nature of science and scientific activity through the judicious use of examples from the history of science" (p. 48).

While these recommendations perhaps fostered some debate, there is little evidence that they influenced provincial or national curriculum at the time (except perhaps by the individuals locally). Arguably, it was the slow process of curriculum development and not the debate that was most influential. Curriculum development in Canada is usually marked by major overhauls in 20-year (approximately) intervals. At this time, any significant changes would have to wait.

Around the same time as the SCC report, The National Science Teachers' Association adopted Science, Technology, and Society (STS) as its official position for science education (Yager 1993, p. 145). Many other countries also developed resources to address teaching of STS. For example, in the United Kingdom, we find Science in a Social Context (Solomon 1983), and in Canada, SciencePlus (ASCP 1988) was developed in Atlantic Canada and used across various provinces. As ideas around STS, STSE (adding the Environment), and socioscientific issues began to evolve, the Canadian context, especially in terms of the intended curriculum, began to advance towards a more national viewpoint. Aikenhead (2000), following the work of Rosenthal (1989) and Ziman (1984), illustrates two types of social issues in STS science that began to envelope the views of most science educators:

1. Social issues *external* to the scientific community ("science and society" topics, e.g., energy conservation, population growth, or pollution)
2. Social aspects of science – issues *internal* to the scientific community (the sociology, epistemology, and history of science, e.g., the cold fusion controversy, the nature of scientific theories, or how the concept of gravity was invented)

Aikenhead also noted that these views would be flexible enough to embrace the different goals and content found the curricula across different provinces. Ultimately, these dual issues began served as a major influence in the development of the Pan-Canadian Framework's STSE emphasis.

Today, many Canadian educators continue work on these STSE issues, in both external[3] and internal aspects of STS science.[4]

62.3 Pan-Canadian Science Framework

Generally, provincial curricula in Canada have been developed through the years independently by each province with varying degrees of cooperation. No national curriculum existed in any form. As a result, a considerable amount of disparity and unnecessary duplication of effort existed among the provinces. Moreover, families moving from one part of the country to another were often caught in a curriculum mismatch with children repeating some content while completely missing other topics.

Recognizing that with increased mobility of Canadians, a more coordinated effort would help harmonize the curriculum nationally and provide some economies of scale, the Provincial Ministers of Education (Canadian Ministers of Education of Canada (CMEC)) agreed on the Victoria Declaration in 1993 identifying education as lifelong learning and highlighting the need for national curriculum compatibility (Milford et al. 2010). The CMEC established the Pan-Canadian Science Project as the first national curricular effort, culminating with the development of the *Common Framework of Science Learning Outcomes* (CMEC 1997). The framework was intended as a guide for provinces to develop their own curriculum based on the framework, giving some degree of

[3] For example, see Yore (2011), Pedretti (1997), Pedretti et al. (2008), Pedretti and Hodson (1995), Sammel and Zandvliet (2003), and Bencze (2010).

[4] For example, see Norris and Phillips (2003), Stinner (2003), and Klassen (2006).

commonality between the provincial jurisdictions. In reference to the development of the framework, Aikenhead (2000) writes:

> In keeping with Canadian culture, the *Common Framework of Science Learning Outcomes* (the *Framework*) evolved through negotiation and compromise among provincial bureaucrats, advised by interested parties (stakeholders) in each province. This political process, however, did not meet the standards of curriculum policy development held by the Canadian science education academic community. (p. 51)

Aikenhead, who was instrumental in the previous debates more than a decade earlier on the nature of science education promoted by the Science Council of Canada (SCC), contrasted the processes followed by the CMEC and SCC. He noted that the SCC conducted educational studies "with the highest of scholarly standards," while the CMEC allowed the provincial civil servants, many with little expertise or research knowledge in field of study they were addressing, to complete the framework. Many science educators, originally recruited as consultants intended to provide input at the national level, were relieved of their responsibilities when some provinces did not want to fund equal participation. In other words, participation was reduced to the lowest common denominator.

However, as Aikenhead also noted, some of the conclusions of the SCC science education study in the early 1980s did find their way into the CMEC's bureaucratic negotiations and development of the framework. Additionally, other international documents such as the US National Research Council's *Standards* (NRC 1996) and Project 2061 (AAAS 1989) were significant influences. By the late 1990s the framework, with a major emphasis on STSE goals, became a set of guidelines for science curricula across the country.

A critique of the framework, which in the author's opinion suffers greatly as compendium of fragmented outcomes, is beyond the scope of this paper. However, within the framework, many outcomes with respect to the history and nature of science begin to emerge in Canadian science education for the first time and merit mention here. The framework takes the position that the promotion of students' scientific literacy requires students' understanding of HPS and the nature of science (NOS) for rational and scientific decision making in an everyday context.

As a vision for scientific literacy, four foundation statements were established for this framework:

1. Science, Technology, Society, and the Environment (STSE)
2. Skills
3. Knowledge
4. Attitudes

It is within the STSE foundation that we find explicit mention of the nature of science (although it is always coupled with technology, i.e., "the nature of science and technology"). The STSE foundation focuses on three major dimensions: the nature of science and technology, the relationships between science and technology, and the social and environmental contexts of science and technology. Table 62.1 highlights the HPS and NOS attributes that are found in the description of the STSE foundation.

Table 62.1 HPS and NOS attributes

Nature of science and technology
Science is a human and social activity with unique characteristics and a long history that has involved many men and women from many societies. Science is also a way of learning about the universe based on curiosity, creativity, imagination, intuition, exploration, observation, replication of experiments, interpretation of evidence, and debate over the evidence and its interpretations. Scientific activity provides a conceptual and theoretical base that is used in predicting, interpreting, and explaining natural and human-made phenomena. Many historians, sociologists, and philosophers of science argue that there is no set procedure for conducting a scientific investigation. Rather, they see science as driven by a combination of theories, knowledge, experimentation, and processes anchored in the physical world. Theories of science are continually being tested, modified, and improved as new knowledge and theories supersede existing ones. Scientific debate on new observations and hypotheses that challenge accepted knowledge involves many participants with diverse backgrounds. This highly complex interplay, which has occurred throughout history, is fuelled by theoretical discussions, experimentation, social, cultural, economic, and political influences, personal biases, and the need for peer recognition and acceptance
While it is true that some of our understanding of the world is the result of revolutionary scientific developments, much of our understanding of the world results from a steady and gradual accumulation of knowledge
Social and environmental contexts of science and technology
The history of science highlights the nature of the scientific enterprise. Above all, the historical context serves as a reminder of the ways in which cultural and intellectual traditions have influenced the questions and methodologies of science and how science in turn has influenced the wider world of ideas

The framework itself is organized into general learning outcomes (GLOs), which are established for each foundation. General learning outcomes are broad statements of what students are expected to learn and be able to do. Table 62.2 shows the GLOs for the STSE foundation, many of which address HPS, STSE, and NOS outcomes.

In turn, the GLOs are delineated into several specific learning outcomes (SLOs). The GLOs and SLOs are then linked to specific learning outcomes in the knowledge foundation (the SLOs are not shown here). Each GLO in Table 62.2 has a similar path that can be traced to knowledge outcomes. Sometimes the GLO is represented in more than one outcome, and sometimes it may not be represented at all. One should note that the curriculum remains organized in the traditional way in terms of knowledge outcomes. While the inclusion of the nature of science outcomes should be seen as encouraging progress, the outcomes are scattered across a wide range of knowledge outcomes. In conclusion, we can say that in terms of promoting HPS in the science curriculum, we see for the first time in Canada a set of general and specific learning outcomes. While this is a very positive development, these outcomes are fragmented and dispersed without any clear connections to any overall perspective on the nature of science. In other words, we now have HPS and NOS outcomes, but there is no coherent approach to teaching of the nature of science.

Table 62.2 General learning outcomes (GLO) for the STSE foundation

It is expected that students will...
109
Describe various processes used in science and technology that enable us to understand natural phenomena and develop technological solution
110
Describe the development of science and technology over time
111
Explain how science and technology interact with and advance one another
112
Illustrate how the needs of individuals, society, and the environment influence and are influenced by scientific and technological endeavors
113
Analyze social issues related to the applications and limitations of science and technology, and explain decisions in terms of advantages and disadvantages for sustainability, considering a few perspectives
114
Describe and explain disciplinary and interdisciplinary processes used to enable us to understand natural phenomena and develop technological solutions
115
Distinguish between science and technology in terms of their respective goals, products, and values, and describe the development of scientific theories and technologies over time
116
Analyze and explain how science and technology interact with and advance one another
117
Analyze how individuals, society, and the environment are interdependent with scientific and technological endeavors
118
Evaluate social issues related to the applications and limitations of science and technology, and explain decisions in terms of advantages and disadvantages for sustainability, considering a variety of perspectives

62.4 HPS Outcomes in Provincial Curriculum: Ontario

We also must remember at this point that the framework is intended as a guideline and each province may (or may not) adapt the curriculum in their own way. Indeed, this is exactly what has happened. For example, the Ontario (Canada's largest province) curriculum organizes expectations for the grades 11 and 12 science courses in five strands that are intended to "where possible" align with the topics set out in the *Pan-Canadian Common Framework of Science Learning* Outcome (CMEC 1997). Table 62.3 shows how an HPS outcome can be traced to the Ontario provincial curriculum implementation from the vision present in the framework to the implementation aligned with specific knowledge outcomes. The GLO (general learning outcome) (GLO – 114), "Describe and explain disciplinary and interdisciplinary processes used to enable us to understand natural phenomena and develop technological solutions," has nine specific learning outcomes in the framework

Table 62.3 Tracing an HPS outcome to provincial curriculum

Vision			
Science is a human and social activity with unique characteristics and a long history that has involved many men and women from many societies			
General learning outcome (GLO – 114)			
Describe and explain disciplinary and interdisciplinary processes used to enable us to understand natural phenomena and develop technological solutions			
Specific Learning Outcome (SLO – 114.x)			
114-1			
Explain how a paradigm shift can change scientific worldviews			
114-2			
Explain the roles of evidence, theories, and paradigms in the development of scientific knowledge			
114-3			
Evaluate the role of continued testing in the development and improvement of technologies			
114-4			
Identify various constraints that result in trade-offs during the development and improvement of technologies			
114-5			
Describe the importance of peer review in the development of scientific knowledge			
114-6			
Relate personal activities and various scientific and technological endeavors to specific science disciplines and interdisciplinary studies			
114-7			
Compare processes used in science with those used in technology			
114-8			
Describe the usefulness of scientific nomenclature systems			
114-9			
Explain the importance of communicating the results of a scientific or technological endeavor, using appropriate language and conventions			
Life science	**Chemistry**	**Physics**	**Earth science**
114-2	*114-2*	*114-2*	*114-2*
Explain the roles of evidence, theories, and paradigms in the development of scientific knowledge (e.g., explain how the cloning of a sheep in 1997 affected the scientific theory of differentiation)	Explain the roles of evidence, theories, and paradigms in the development of scientific knowledge (e.g., explain how bonding theory can help one understand certain colligative properties)	Explain the roles of evidence, theories, and paradigms in the development of scientific knowledge (e.g., explain the role of evidence and theories in the concept of fields)	Explain the roles of evidence, theories, and paradigms in the development of scientific knowledge (e.g., describe the historical development of theories to explain the origin of the universe)

document as shown in Table 62.3. I show how one of these, 114-2, "explain the roles of evidence, theories, and paradigms in the development of scientific knowledge," can be found in a knowledge outcome for each science course, biology, chemistry, earth science, and physics, at the secondary level (from the framework). Table 62.4

Table 62.4 HPS at the provincial level (GLO 114-2)

Life Sciences	The Ontario expectation is virtually exactly the same as the framework
Earth Science	The Ontario intent is the same but the wording is different, i.e., "describe origin and evolution of the Earth and other objects in the solar system"
Chemistry	There is no corresponding outcome in Ontario
Physics	The Ontario intent is the same but specific reference to models and evidential examples are provided

describes how the outcome appears in the Ontario curriculum. It is possible, although somewhat challenging and time-consuming, to trace each GLO and SLO to a knowledge outcome.

Thus, we can see that for the most part the nature of science outcomes as described in the framework is making its way into the provincial curriculum documents. Although some of the outcomes may be omitted, we can find many outcomes that have actually been enhanced, and even new outcomes emerge as they may relate to local interests. For example, our physics outcome becomes in Ontario:

> explain how the concept of a field developed into a general scientific model, and describe how it affected scientific thinking (e.g., explain how field theory helped scientists understand, on a macro scale, the motion of celestial bodies and, on a micro scale, the motion of particles in electromagnetic fields). (Ontario 2000, p. 108)

And in terms of local interest, we can find outcomes such as:

> describe advances in Canadian research on atomic and molecular theory (e.g., the work of Richard Bader at McMaster University in developing electron-density maps for small molecules; the work of R.J. LeRoy at the University of Waterloo in developing the mathematical technique for determining the radius of molecules called the LeRoy Radius). (Ontario 2000, p. 65)

We must see the adoption of these types of outcomes as a positive step in promoting an understanding of HPS in science curriculum and as a step towards achieving many of the claims of the HPS community of the benefits of using an HPS approach in science education (Matthews 1994; Winchester 1989). However, the inclusion of the HPS outcomes still remains mostly ad hoc and fragmented without any internal consistency in the science programs. Such consistency can be found in independent courses in philosophy that are offered through such programs as the International Baccalaureates (IB) and as stand-alone courses such as teaching philosophy in high school. These courses, worthwhile and generally taught with some expertise, are offered as electives and usually found as "gifted and talented" offerings in some jurisdictions.

This emergence of a coherent curriculum in philosophy (albeit more general), the growing acceptance of philosophy as a teaching major or minor in more provincial jurisdictions, and the organization of teachers associations such as the Ontario Philosophy Teacher's Association should be seen as positive force in the future for the continued development of the inclusion of HPS in curriculum in general and in science education specifically. Arguably, Canadians who support an HPS perspective are pleased with the progress, albeit flawed in many ways. However, we should take our pleasure with some caution.

62.5 HPS Outcomes in Provincial Curriculum: Saskatchewan

Contrasting the Ontario use of the Pan-Canadian Framework is the recent development of the Saskatchewan science curriculum. The province of Saskatchewan is in Western Canada and has a population of just over one million persons with approximately 16 % Indigenous and Métis people. The vision for science education in the province is stated in the Saskatchewan Learning Science 9 Curriculum Guide:

> The aim of K-12 science education is to enable all Saskatchewan students to develop scientific literacy. Scientific literacy today embraces Euro-Canadian and Indigenous heritages, both of which have developed an empirical and rational knowledge of nature. A Euro-Canadian way of knowing about the natural and constructed world is called science, while First Nations and Métis ways of knowing nature are found within the broader category of Indigenous knowledge (Hounjet et al. 2011). (Sask Science 9, p. 5)

The Saskatchewan curriculum delineates the foundations of the Pan-Canadian Framework from a cultural perspective. For example, traditional and local knowledge is included alongside of life, physical, earth, and space science knowledge. Scientific knowledge is represented as a set of understandings, interpretations, and meanings that are part of cultural complexes that encompass language, naming and classification systems, resource use practices, ritual, and worldview. Indigenous knowledge is represented as a set of understandings, interpretations, and meanings that belong to a cultural complex that encompasses language, naming and classification systems, resource use practices, ritual, spirituality, and worldview (note: the difference is spirituality).

The Saskatchewan curriculum emphasizes cultural perspectives throughout all of their science documents stating that:

> Students should recognize and respect that all cultures develop knowledge systems to describe and explain nature. Two knowledge systems emphasized in this curriculum are First Nations and Métis cultures (Indigenous knowledge) and Euro-Canadian cultures (science). In their own way, both of these knowledge systems convey an understanding of the natural and constructed worlds, and they create or borrow from other cultures' technologies to resolve practical problems. Both knowledge systems are systematic, rational, empirical, dynamically changeable, and culturally specific. (SASK Science 9, p. 20)

The cultural perspective is carried through to the general and specific learning outcomes often in terms of identifying and/or comparing and contrasting cultural perspectives. For example, in grade 9 science, we find curriculum outcome RE9.4 "Analyze the process of human reproduction, including the influence of reproductive and contraceptive technologies" (p. 32). An indicator of achievement is given as "Acknowledge differing cultural perspectives, including First Nations and Métis perspectives, regarding the sacredness, interconnectedness, and beginning of human life" (p. 32). The achievement of these outcomes is also carried through the recommended textbook, Saskatchewan Science 9, developed especially for this curriculum. First Nations and Métis content, perspectives, and ways of knowing are an integral part of the Saskatchewan science textbook. For example, the social and cultural perspective in the section on contraception states, "All cultures and religions have moral and ethical beliefs surrounding reproduction and contraception. There are also different beliefs about when exactly a fertilized egg is considered to be a person. For this reason, some methods of contraception are considered to be the ending of a human life" (p. 103). In another section on chemical change, a cultural perspective is given as "Some medicine women would go on a fast for several days during which time they would meditate on the medicines and plants they would mix to create the desired effects. Certain families have spiritual understanding to mix medicines for heart disease, diabetes, hepatitis, and other illnesses" (p. 140). The textbook does not just pay "lip service" to cultural perspectives, but integrates differing viewpoints, Elder's wisdom, and traditional knowledge throughout all units. It should be noted that in no way does the textbook shy away from an open and clear discussion of controversial topics in science such as contraception. Given the recent publication of the textbook series, it remains to be seen how it will be accepted by students, teachers, parents, and other stakeholders.

As widely noted in educational circles, we have an intended curriculum and an implemented curriculum. Teachers remain the final filter in the delivery of any curriculum, and their professional development and preservice preparation is an important factor in the actual achievement of these outcomes in the classroom. This point reflects the practical orientation of the teaching field and what they value or at least what they believe schools and stakeholders value. Interestingly, in a study conducted by the SCC in 1980, a survey of over 4,000 science teachers in Canada ranked the importance of objectives in science education. Out of 14 different objectives, the number one was "understanding scientific facts, concepts and laws" and number 14 was "understanding the history and philosophy of science." Many teachers hold preconceptions about teaching science from their own experiences, from their preservice education experiences, and from their own teaching experiences (Duffee and Aikenhead 1992). Aikenhead suggests that "a simple in-service intervention by itself holds little promise for altering a teacher's acceptance of STS science" and argues that in addition to changing deep-seated values and images of teaching science, teachers must add new methods to their repertoire of instructional strategies. However, his recommendation that new routines of instruction are best learned from fellow teachers means we must provide meaningful preservice instruction to develop these teachers. In the next section, I'll outline a couple of exemplars oriented towards developing new "routines of instruction" in preservice teachers.

62.6 Teacher Education

Teacher education and certification in Canada is governed by each provincial or territorial agencies through regulations established by colleges of teachers and/or ministries of education. At the university level, teacher education programs must be designed to meet these specific requirements in terms of required coursework and practical teaching experience. Teacher education programs in Canada generally have two forms: an integrated program or an after-degree program.[5] In the integrated program students enter directly into a faculty of education and take subject area courses at the same time as they take education courses. Typically, the majority of the education courses can be found in the final 2 years of the program; at this time students complete an in-school practicum which varies in duration from province to province. In after-degree programs, students first complete a degree and then enter a 12-month or a 24-month education program (depending on provincial requirements). In both the integrated and after-degree programs, students typically finish with two degrees, a subject area general bachelor degree and a bachelor of education degree (although some programs may vary).

In terms of course offerings, students in education programs will take courses in educational psychology, teaching methodology, literacy, aboriginal education, special education, plus curriculum, instruction, and assessment in their teaching specialties. Teacher programs, in both the integrated and after-degree format, can be very crowded places in terms of compulsory requirements that vary from province to province. Some education professors across the country who maintain an interest in HPS may be found to include the nature of science as a lecture or topic in their methods courses. However, it is unusual to find a course offering in the history and philosophy of science and science teaching at the undergraduate level.[6] Therefore, I offer here two exemplars in teaching the history and philosophy of science and science teaching that is required at the undergraduate level. The first is a course for secondary preservice science teachers who have a major or minor specialty in the sciences. The second exemplar is a course for nonspecialist middle-year teachers.

62.6.1 HPS for Secondary Science Teachers

The history and philosophy of science and science teaching course, developed and taught by the author, is offered at the University of Winnipeg and is compulsory for all students with a double science major/minor. The course can also be taken as an elective by other interested students. The course has essentially three components, philosophical perspectives in science, teaching strategies for HPS, and a historical case study (Stinner et al. 2003). Using a seminar format, students examine, chapter by chapter, John Losee's (1993) book, *Introduction to the History and Philosophy of Science*. Losee contends that the philosopher of science seeks answers to such questions as:

What characteristics distinguish scientific inquiry from other types of investigation? What procedures should scientists follow in investigating nature?

[5] There are many variations of these programs across the country, including specialized access programs.

[6] It is possible from time to time to find graduate courses that relate specifically to HPS themes. For example, Derek Hodson offered a graduate course, Curriculum Making in Science: Considerations in the History, Philosophy, and Sociology of Science.

What conditions must be satisfied for a scientific explanation to be correct?
What is the cognitive status of scientific laws and principles?

Providing a commentary on original works from Aristotle to Van Frassen (and everyone in between), Losee's approach is appropriate for the undergraduate student and provides an insight into the major questions in the philosophy of science. Following the grounding in the philosophy of science, a number of teaching strategies are examined including the implementation of historical experiments, storytelling, debates, and a variety of differentiated instruction strategies (Metz et al. 2007). Students complete their own case study that traces the historical development of a scientific idea and design a plan to implement instruction in a science class. Case studies over the years have included all of the major ideas in science such as the development of the models of the atom and the cell and the history of pasteurization. However, many creative, unique, and interesting case studies have also been developed that include the history of the pacemaker, birth control, the story of blood, play doh, and climate change.

62.6.2 HPS for Nonspecialist Middle-Year Science Teachers

In teacher education programs in Canada, preservice middle-year programs typically follow a set of standard courses in content area pedagogy. As generalists, these teachers are faced with teaching a broad range of subjects with a core of language arts, social studies, mathematics, and science. Students, who enter these programs, overwhelmingly, have a background in the humanities. Very few students have a background in mathematics and science, and their self-efficacy in teaching these subjects is generally low.

As a model program incorporating HPS, I will outline here a middle-year science methods course initiated by Brian Lewthwaite at the University of Manitoba (Lewthwaite 2012). Recognizing that their teacher candidates viewed science as primarily a body of knowledge and not a process of inquiry, they use the nature of science as a pedagogical framework to assist their students in developing a more authentic view of science:

> More importantly, we believe it has the potential to assist teacher candidates in developing a positive self-image of themselves as teachers of science because their role is focused more on an understanding of the pedagogical processes to be employed in science instruction rather than, simply, their content knowledge. (Lewthwaite 2012)

In their course, suitable topics are identified for students to teach in their practicum experience that follows the course. The course instructor would then present a historical account of the development of understanding associated with the topic under consideration in a 1-h presentation. Following the presentation, the students were asked to consider the nature of science (NOS) attributes embedded within the historical context. Next, the student teachers, working in pairs, planned three linked authentic science lessons that addressed the topic to be taught. The students were required to use their understanding of the nature of science as a rationale for the task orientation taken in the lesson. Following the course, students participate in a 5-week teaching practicum where they co-teach their lessons. In a post-teaching reflection, the students are required to identify at least five nature of science characteristics that they believed informed the development of their lessons and their teaching of science. Additionally, the students critiqued their own lessons and reported on how these attributes resulted in positive learning experiences for their students.

Initial results of this innovative program have been very positive (Lewthwaite 2012). Their students speak of transformative experiences viewing science as a human activity, focusing on educational goals that are social and personal and an instructional emphasis that is more process oriented. The authors make a strong claim that NOS must not just be an ad hoc component of a science methods course for middle-year teachers but that

> NOS needs to be the foundation upon which teacher education science courses should be premised, especially in using the human story of science as vignettes for teacher candidates to come to an appreciation of what science is. (Lewthwaite 2012)

62.7 Conclusion

Canada, as a country, is geographically very expansive and culturally highly diverse. In many ways, it is difficult for outsiders to imagine how the nation brings such a diverse set of ideas together into a national perspective. In reality, we never do. Within the most recent round of curriculum renewal, provinces interpret and implement the national standards in their own, sometimes unique, ways.

It is clear from our review of HPS perspectives in Canadian curriculum and teacher education that we have made significant progress in enabling the inclusion of HPS perspectives in science education today. A great deal of attention is being given, especially in the academic community, to STSE and sustainable initiatives. Additionally, cultural perspectives, especially as noted in the province of Saskatchewan, are being integrated in a significant manner. However, it is also evident that more work needs to be done to prepare teachers to teach STS, HPS, and cultural perspectives such that the intended curriculum more closely matches the implemented. Most elementary and middle-year school teachers still have little science background, and most teachers still give low priority towards teaching the more subtle aspects of HPS. Unfortunately, knowledge content is still the emphasis in science education today dominated by teacher-directed instruction and motivated in a large part by assessment techniques. As we move forward, we are thankful for the appearance of HPS in national documents and initiating, at least in part, a national discussion.

Further studies are needed to demonstrate the effectiveness of HPS perspectives and the extent to which it can be a guiding principle as advocated by Lewthwaite instead of merely an "add-on" to traditional science instruction. Probably, the most interesting period of curriculum reform lies ahead.

References

Ahlfors, L. V. et al., 1962. On the Mathematics Curriculum of the High School. *American Mathematical Monthly*, 69, 425–426. The Memorandum was signed by 65 Mathematicians.

American Association for the Advancement of Science (AAAS), 1989. *Project 2061: Science for all Americans*. Washington, DC: American Association for the Advancement of Science.

Aikenhead, G.S., 1980. *Science in Social Issues: Implications for Teaching*, Science Council of Canada, Ottawa.

Aikenhead, G., *2000*. STS Science in Canada: From Policy to Student Evaluation, In: David, D. & Chubin, D., eds., *Science, Technology, and Society Education A Sourcebook on Research and Practice Series: Innovations in Science Education and Technology*, Vol. 6, Springer.

Aikenhead, G., 2003. Review of Research on Humanistic Perspectives in Science Curricula, A paper presented at the *European Science Education Research Association ESERA 2003 Conference*, Noordwijkerhout, The Netherlands, August 19–23.

Aikenhead, G.S. & Ryan, 1992. The Development of a New Instrument: "Views on Science Technology Society", VOSTS , *Science Education*, 1992.

American Association for the Advancement of Science (AAAS), 1967. *Science: A Process Approach*, Washington, DC.

Atlantic Science Curriculum Project (ASCP), 1988. *SciencePlus3*, Harcourt, Brace, Jananovich, Canada, Inc.

Baptiste, H. P. & Turner, J., 1972. A Study of the Implementation of the Elementary Science Study Program on a System-Wide Basis in the Penn-Harris-Madison Schools, Paper presented at the *National Science Teachers Association Annual Meeting*, New York City, April 1972.

Bencze, J.L., 2010. Promoting student-led science and technology projects in elementary teacher education: Entry into core pedagogical practices through technological design. *International Journal of Technology and Design Education*, 20, 1, 43–62.

Biological Sciences Curriculum Study (BSCS), 2006. BSCS *Science: An Inquiry Approach*, Dubuque, IA: Kendall/Hunt.

Buchwald, J. & Swerdlow, N., 1993. Eloge: Stillman Drake, 24 December 1910-6 October 1993, *Isis,* 85, 4, 663–666.

Bunge, M., 1974–1983. Treatise on Basic Philosophy, Volumes 1–6, Dordrecht, Boston, USA.

Chemical Education Material Study (CEMS), 1960. *Chemistry*, Pimentel, G. C., ed., W.H. Freeman & Co., San Francisco.

Council of Ministers of Education (CMEC), 1997. *Common Framework of Science Learning Outcomes K to 12: Pan-Canadian Protocol for Collaboration on School Curriculum*, Council of Ministers of Education, Toronto, ON.

Council of Ministers of Education (CMEC), 1999. *The Victoria Declaration,* Council of Ministers of Education, Toronto, ON.

Duschl, R. A., 1994. Research on the history and philosophy of science. In D. L. Gabel Ed., *Handbook of research on science teaching and learning* pp. 443–465 . New York: MacMillan.

Duffee, L., & Aikenhead, G.S., 1992. Curriculum change, student evaluation, and teacher practical knowledge. *Science Education*, 76, 5, 493–506.

Education Development Center (EDC). *Elementary Science Study*, McGraw-Hill, 1969.

Fensham, P. J., 1973. Reflections on the social content of science education materials, *Science Education Research* 3, 1, 143–50.

Fensham, P., 1997. School science and its problems with scientific literacy. In Scanlon, E., Murphy, P., Thomas, J. & Whitelegg, E., eds. , *Reconsidering Science Learning*. London: Routledge.

Feyerabend, P., 1975. *Against Method*, 4th ed., New York, NY: Verso Books, 2010.

George, D., 1981. *An Engineer's View of Science Education*, Science Council of Canada, Ottawa.

Heller, R., 1964. The Earth Science Curriculum Project – A Report of Progress, *Journal of Reasearch in Science Teaching*, v2, pp. 330–334.

Hodson, D., 1988, Toward a Philosophically More Valid Science Curriculum, *Science Education* 72, 1, 19–40.

Hodson, D., 2008. *Towards Scientific Literacy: A Teachers' Guide to the History, Philosophy and Sociology of Science* Sense Publishers, Rotterdam, The Netherlands.

Hounjet, C., Kvamme, B., Mohr, P., Phillipchuk, K., & View, T., 2011. *Saskatchewan Science 9*, Pearson Canada, Inc., Toronto.

Klassen, S., 2006. A Theoretical Framework for Contextual Science Teaching. *Interchange 37, 1–2,* 31–62.

Lewthwaite, B., 2012. Revising Teacher Candidates' Views of Science and Self: Can Accounts from the History of Science Help?, *International Journal of Science and Environmetal Education.,* in press.

Losee J., 1993. *Introduction to the History and Philosophy of Science*, Oxford University Press.

Matthews, M.R., 1994. *Science teaching: The role of history and philosophy of science. New York: Routledge.*

Metz, D., Klassen, S., McMillan, B., Clough, M., & Olson, J., 2007. Building a foundation for the use of historical narratives, *Science & Education, 16, 3–5,* 313.

Milford, T., Jagger, S., Yore, L. & Anderson, J., 2010. National Influences on Science Education Reform in Canada, *Canadian Journal of Science, Mathematics and Technology Education*, 10, 4, 370–381.

Millar, R., & Osborne, J. eds., 1998. *Beyond 2000: Science education for the future.* London: King's College, School of Education.

Millin, L. & Steed, G., 2012. *Science Council of Canada*, in *The Canadian Encyclopedia*, downloaded from http://thecanadianencyclopedia.com.

Munby, H., 1982. *What is Scientific Thinking?,* Science Council of Canada, Ottawa.

Norris, S., Guilbert, S., Smith, L., Hakimelahi, S., Phillips, L., 2005. A Theoretical Framework for Narrative Explanation in Science, *Science Education,* 89, 4, 535–563.

Nadeau, R. & Destautels, J., 1984. *Epistemology and the Teaching of Science*, Science Council of Canada, Ottawa.

Norris, S. & Phillips, L., 2003. How literacy in its fundamental sense is central to scientific literacy, *Science Education*, 87, 2, 224–240.

NRC National Research Council, 1996. *National science education standards*. Washington, DC: National Academy Press.

Ontario Ministry of Education, 2000. *The Ontario Curriculum, Grades 11 And 12: Science*, downloaded from http://www.edu.gov.on.ca.

Page, J., 1980. *A Canadian Context for Science Education*, Science Council of Canada, Ottawa.

Pedretti, E. ,1997. Septic tank crisis: A case study of science, technology and society education in an elementary school. *International Journal of Science Education*, 19, 1211–1230.

Pedretti, E., & Hodson, D., 1995. From rhetoric to action: Implementing STS education through action research. *Journal of Research in Science Teaching*, 32, 463–485.

Pedretti, E., Bencze, L., Hewitt, J., Romkey, L, & Jivraj, A., 2008. Promoting issues-based STSE perspectives in science teacher education: Problems of identity and ideology. *Science & Education*, 17, 8/9, 941–960.

Physical Science Study Committee (PSSC), 1960. *PSSC Physics: Teacher's Resource Book and Guide*, D. C. Heath.

Polito, T., 2005. Educational Theory as Theory of Culture: A Vichian perspective on the educational theories of John Dewey and Kieran Egan, *Educational Philosophy and Theory, 37, 4, 475–494.*

Rosenthal, D.B., 1989. Two approaches to STS education. *Science Education*, 73, 5, 581–589.

Roth, M., & Desautels, J., 2002. *Science Education as/for Sociopolitical Action*, Peter Lang Publishing, Inc.

Rutledge, J. A., 1973. What Has Happened to the New Science Curriculum?, *Educational Leadership*, 30, 7, 600-03, Apr 1973.

Sammel, A., & Zandvliet, D.B., 2003. Science Reform or Science Conform: Problematic Epistemological Assumptions with/in Canadian Science Reform Efforts. *Canadian Journal of Science, Mathematics and Technology Education.* 2003 October, University of Toronto Press.

Science Council of Canada, 1984. *Science For Every Student: Educating Canadians for Tomorrow's World.* Report 36. Ottawa: Science Council of Canada, 1984.
Science Curriculum Improvement Study (SCIS), 1972. *Sample Guide.* Rand McNally & Company, Chicago, Illinois.
Siemsen, K.H. & Siemsen, H., 2008. Ideas of Ernst Mach Teaching Science. In: Gh. Asachi University, Cetex, Iasi, *5th Int. Seminar on Quality Management in Higher Education QM 2006*, Tulcea, Romania, June 2008.
Siemsen, H. & Siemsen, K.H., 2009. Resettling the Thoughts of Ernst Mach and the Vienna Circle to Europe – The cases of Finland and Germany, *Science & Education* 18, 3, 299–323.
Siemsen, H., 2011. Ernst Mach and the Epistemological Ideas Specific for Finnish Science Education. *Science & Education,* 20, 245–291.
Solomon, J., 1983. *Science in a Social Context*, United Kingdom, Basil Blackwell and the Association for Science Education.
Solomon, J., & Aikenhead, G., eds., 1994. *STS education: International perspectives on reform.* New York: Teachers College Press.
Stinner, A., MacMillan, B., Metz, D., Klassen, S. & Jilek, J., 2003. The Renewal of Case Studies in Science Education, *Science & Education* 12, 7, 645–670.
Stinner, A., 1995. Contextual Settings, Science Stories, and Large Context Problems: Toward a More Humanistic Science Education, *Science Education* 79, 5, 555–581.
Stinner, A., 2003. Scientific Method, Imagination, and the Teaching of Science. *Physics in Canada.* 59, 6, 335–346.
Thagard, P., 2010. *The Brain and the Meaning of Life*, Princeton University Press.
Winchester, I., 1989. History of science and science teaching, *Interchange, 20,* 2, 1–2.
Wittenberg, A.I., 1968. *The Prime Imperatives: Priorities in Education.* Toronto: Clarke, Irwin & Company.
Wittenberg, A.I., 1965. Priorities and Responsibilities in the Reform of Mathematical Education: An Essay in Educational Metatheory, *L'Enseignement Mathématique* 11, pp. 287–308.
Yore, L. D., 2011. Foundations of scientific, mathematical, and technological Literacies—Common themes and theoretical frameworks. In L. D. Yore, E. Van der Flier-Keller, D. W. Blades, T. W. Pelton, & D. B. Zandvliet Eds., Pacific CRYSTAL centre for science, mathematics, and technology literacy: Lessons learned pp. 23–44 . Rotterdam, The Netherlands: Sense.
Yager, R.E., 1993. Science-Technology-Society as Reform, *School Science and Mathematics,* 93, 3, 145–151.
Ziman, J., 1984. *An introduction to science studies: The philosophical and social aspects of science and technology.* Cambridge: Cambridge University Press.

Don Metz is an associate professor in the Faculty of Education at the University of Winnipeg. He holds a B.Sc. in Physics and an M.Ed. and Ph.D. in science education. He has been extensively involved with curriculum writing in Canada, and he has been very active in the professional development of in-service teachers for many years. He has presented at numerous conferences and published papers in a wide variety of education journals including *Teacher Education Journal, Science Education, Science & Education, The Physics Teacher*, and the *Canadian Journal of Environmental Education.* His current interests include student-generated questions and academic engagement in science, contextual teaching, the use of historical narratives and the interrupted story form, sustainability, and international practicum for preservice teachers.

Chapter 63
History and Philosophy of Science and the Teaching of Science in England

John L. Taylor and Andrew Hunt

63.1 The Years Leading Up to the National Curriculum

Interest in teaching about the history of science and aspects of its philosophy in schools dates back, in the UK, at least as far as the mid-nineteenth century. In his exploration of the claims made for the teaching of the history of science, Edgar Jenkins (1990) quotes from the presidential address at the British Association for the Advancement of Science meeting in Glasgow in 1855. The Duke of Argyll said that what was wanted in the teaching of the young, was 'not so much the mere results, as the methods, and above all, the history of science', if education were to be 'well-conducted to the great ends in view'.

In a review, Michael Matthews (1994/2014) traces the weak and uneven tradition of incorporating the history of science in science education and takes up the story at the start of the twentieth century when, in 1918, a committee chaired by J J Thomson issued a report called *Natural Science in Education* which argued that:

> It is desirable…to introduce into the teaching some account of the main achievements of science and of the methods by which they have been obtained. There should be more of the spirit, and less of the valley of dry bones…One way of doing this is by lessons on the history of science (Cited in Brock (1989, page 31)).

The report went on to say that:

> some knowledge of the history and philosophy of science should form part of the intellectual equipment of every science teacher in a secondary school.

As Mathews has shown, these recommendations were included in the 'Science for All' curriculum that was developed after the First World War. Also in the

J.L. Taylor (✉)
Rugby School, Rugby CV22 5EH, UK
e-mail: jlt@rugbyschool.net

A. Hunt
National STEM Centre, York, UK

M.R. Matthews (ed.), *International Handbook of Research in History, Philosophy and Science Teaching*, DOI 10.1007/978-94-007-7654-8_63,

interwar years, Percy Nunn, the philosopher of science, Richard Gregory and other historically minded educationalists argued the case for history. Popular science textbooks incorporating these ideas were written by a number of authors including the chemist, E. J. Holmyard. Holmyard argued for a historical approach, not just on motivational or instrumentalist grounds but on cognitive grounds: teaching a topic historically was the only way that the nature of scientific truth could be conveyed.

After the Second World War, the history of science gradually diminished in importance in school science. Science education was subject-focussed and directed towards external examinations controlled by the universities. By modern standards, this was a small-scale business. In 1952 only about 20,000 candidates, aged 16, took O-level Biology with about 15,000 taking O-levels in each of physics and chemistry (ASE 1979). These examinations were taken by a subset of the students attending selective grammar and independent schools which catered for about 25 % of the age cohort.

At that time, most young people attended secondary modern schools and left at the age of 15 with no qualifications in science. However, during the late 1950s and early 1960s, some local authorities began to move away from the selective system and establish comprehensive schools. By the late 1960s, comprehensive secondary education had become a central part of government policy. In the same decade, new regional examination boards were set up to provide the Certificate of Secondary Education in all subjects, including science, for young people who were not able to take O-levels.

Then, in 1972, the school leaving age was raised to 16. From that time on, most students were expected to take CSE or O-level examinations before leaving school or staying on for further education. As a result of these developments, more and more young people chose to study some science throughout their time in secondary school. Subsequently, in 1989, the study of science became a compulsory part of the curriculum with the total number of students gaining science qualifications at the age of 16 exceeding 600,000 per year by the twenty-first century.

The tradition established in selective schools continued to be very influential. In this tradition, school science was regarded as a fixed body of knowledge, related to and derived from real science, which young people need to acquire in order to understand the world they live in and which they must master in order to take their studies of science further (ASE 1979). The persistence of this general approach meant that teaching about the history and nature of science did not enter the mainstream of science education for all in England and Wales until the national curriculum was introduced in 1989 (Department of Education and Science 1989).

The first version of the curriculum was detailed and complex. It was divided into 17 attainment targets (ATs). One of the significant innovations was the seventeenth attainment target, AT17 ‘The nature of science’. The intention of AT17 was that students should study historical case studies to develop their understanding of how scientific theories arise and change through time. They were also expected to explore how the nature and application of scientific ideas are affected by the social and cultural context in which they develop.

None of the ideas in AT17 were new, but for the first time it was compulsory that they should feature in the science courses of most learners. The feasibility of teaching these ideas, and the resources needed for the teaching, had been investigated on a relatively small scale in a series of innovative curriculum projects during the previous three decades starting with the Nuffield projects of the 1960s.

63.2 Nuffield O-Level Sciences in the 1960s

By the mid-1950s, it had become clear to teachers and politicians that secondary science education was in need of major reform. There was dissatisfaction with the emphasis on factual recall in examinations. Chemistry courses focussed on the preparation and properties of elements and compounds, whilst in physics the content was largely confined to the classical themes of nineteenth century science such as electricity and magnetism, heat, light and sound. Biology, as a combination of botany and zoology, was not fully established as a subject for boys as well as girls (Hunt 2011).

Pressure for change arose particularly from activists in the Science Master's Association and Association of Women Science Teachers (organisations which merged to form the Association for Science Education in 1963). Leading teachers from selective schools had built up a consensus that there was an urgent need to modernise both the content of the science curriculum and how it was taught. Much work had been done in the late 1950s to draw up fresh syllabuses.

The problem was that at that time there was no statutory, centralised control of the curriculum. The government did not expect to intervene directly in curriculum matters. As a result, there was no established way to bring about reform and no obvious source of resources to fund the development work. However, after a period of negotiation, the Nuffield Foundation[1] decided to make a very large grant towards the cost of a long-term programme to improve the teaching of school science and mathematics. This was the beginning of a major era of curriculum development

The work began with the Nuffield Science courses to O-level. These were very influential, but their main focus was on practical work carried out by students as the starting point for introducing scientific concepts. Nevertheless these courses did not ignore the tradition of using episodes from the history of science as a context for teaching science. One of the aims of Nuffield Biology,[2] for example, was to present the subject as part of human endeavour showing, with the help of historical examples, that biological knowledge is the product of scientists working in many different parts of the world (Nuffield Foundation 1966a). The course aimed to demonstrate that the science of biology is based not only on observation and experimentation but

[1] The Nuffield Foundation website has an account of the charity's involvement in the science curriculum over 50 years: http://www.nuffieldfoundation.org/curriculum-projects

[2] All the original Nuffield Biology publications are available from the National STEM Centre eLibrary at http://stem.org.uk/cxhs

also on questioning, the formulation of hypotheses, testing of hypotheses and, above all, on communication between people.

Nuffield Chemistry[3] aimed to encourage students to be 'scientific about a problem' (Nuffield Foundation 1966b). This was based largely on the project team's perception of what 'being scientific' means to a scientist: the application and the personal commitment involved, the importance of the disciplined guess or 'hunch' as well as logical argument, the feeling of exploration and the readiness to make apparently unwarranted jumps whilst knowing how to check their validity.

The Nuffield Chemistry project produced an extensive series of Background Books for students which were intended to amplify and extend work done in class and to stimulate the interest of pupils in wider aspects of their study. Many of these short readers were historical featuring the work of famous scientists from the early nineteenth century to the mid-twentieth century. *The Way of Discovery*, one of the Stage III Background Books, began with a short statement about the origins of scientific knowledge as an introduction to a series of personal accounts based on interviews with 14 Nobel Prize-winning scientists.

However, there were no explicit learning outcomes about the nature of science. Understanding of the methods of scientific enquiry was treated as a tacit aspect of science learning (Millar 1997). There was no significant credit in the O-level examinations for this kind of learning.

Nuffield Physics,[4] like the other Nuffield courses, was committed to presenting the subject as a connected fabric of knowledge with a focus on key concepts (Nuffield Foundation 1966c). Under the influence of Eric Rogers, the project team wanted students to acquire the feeling of doing science, of being a scientist – 'a scientist for the day'. As with the other courses, the teaching approach was based on a guided introduction to physics concepts related to hands-on practical work by learners. Even so, the history of physics was of particular importance in the fifth year of the course which traced the development of theories about the solar system from ancient Greek models to the grand Newtonian synthesis (Nuffield Foundation 1966d). However, as the Teachers' Guide to Year V shows, the main focus of the teaching was on Newton's laws of motion, circular motion and other related concepts.

Paul Stevens was one teacher who explored the contradictions between the 'discovery learning' encouraged by the Nuffield courses and the ultimate aim of teachers to induct their students into the accepted scientific explanations (Stevens 1978). He argued that there was a logical conflict between 'discovering' and 'learning' which could produce some undesirable psychological results. In his view the fragmentary Nuffield philosophy, in so far as it could be abstracted from the publications, was bound to influence teachers into misleading students about the nature of science.

[3] All the original Nuffield Chemistry publications, including the Background Books, are available from the National STEM Centre eLibrary at http://stem.org.uk/cxew

[4] All the original Nuffield Physics publications are available from the National STEM Centre eLibrary at http://stem.org.uk/cxqc

Jonathan Osborne (2007) has also pointed out that Eric Rogers' vision of what it means to be a 'scientist for a day' was a very narrow one based predominantly on the exact sciences of physics and chemistry and a hypothetico-deductive methodology. Osborne also gives examples to show why it was wrong to think that the learning of science and the doing of science could be regarded as one and the same thing. Whilst the practice of science is the search for answers concerning unanswered questions that we have about the material world, the task of science education is different. Its role is to construct in the young student a deep understanding of a body of existing knowledge.

63.3 Integrated and Process Science

Another course devised for more able learners was the Schools Council Integrated Science Project (SCISP),[5] known as *Patterns*, which was designed as a 3-year course to O-level for 13–16-year-olds (Schools Council 1973). This was developed in the late 1960 and early 1970s with the aim of providing a broad and balanced science curriculum in about a fifth of the available curriculum time in schools. That is the time that was typically allocated to two of the three traditional science subjects.

The rationale for integration was that all scientists use the same general approach to their work: the 'art' of science that is common to all areas of the subject. In *Patterns* this 'art' was expressed in terms of 'pattern-seeking' and 'problem-solving'. Throughout the scheme, students were expected to search for generalisations which could be expressed as patterns. These patterns were then used to solve scientific problems.

The American psychologist Robert Gagné was a major influence on SCISP. His thinking shaped the pedagogy and also helped to justify the notion that the content of science did not matter. His view was that knowledge was evolving so rapidly that anything learnt today might be redundant tomorrow. The enduring features of science were its processes. Hence, what young scientists needed to learn were processes: measurement, observation, hypothesis generation, experimental design and so on (Adey 2001).

The SCISP project team pointed out that they were not presenting their approach as *the* 'scientific method'. However, in later projects such as *Warwick Process Science* (Screen 1987) and *Science in Process* (Inner London Education Authority 1987), there was even more emphasis on a common set of scientific processes such as observing, classifying, inferring, investigating, predicting, hypothesising and evaluating. In these courses there was a clear implication that these activities in school reflected the ways in which scientists work across a wide range of scientific disciplines.

Unlike SCISP, these later integrated and process-based science courses, published in the 1980s, were planned in order to offer something to students across the ability

[5] All the original Schools Council Integrated Science Project publications are available from the National STEM Centre eLibrary at http://stem.org.uk/cxug

spectrum in the expectation that the most able students would find challenge in working at processes and meet the demand for problem-solving with sophistication, whilst the less able students would find the work stimulating without experiencing failure and consequent disillusionment with science.

However, this particular view about the nature of science was later criticised by Paul Black (1986) in his presidential address to the Association for Science Education. He warned against the temptation for curriculum planners to invent their own theories of knowledge, of philosophy and of culture. He pointed out that the philosophy of biology has its own agenda, distinct from the philosophy of physics, dealing, for example, with questions of reductionism, the general problem of teleology, the notion of organicism and whether or not biology can claim to have its own distinctive laws. One of the principles for curriculum design in science, he suggested, is that teaching should acknowledge the diversity and spectrum of differences across the sciences.

Robin Millar and Rosalind Driver (1987) in an article called *Beyond Process* presented a detailed critique of the assumptions that were being used at the time, more or less implicitly, as a basis for the case for 'process science' and which were, in their view, in danger of luring science educators along a misguided path. In examining the 'processes of science' from the perspective of the philosophy of science, they argued that there is no evidence to support the view that there is a unique and distinctive method of science based on a set of generalisable processes.

63.4 Science, Technology and Society (STS) Courses

In the mid-1970s science teachers recognised a fresh challenge. Despite the development work of the previous decade, young people were not opting for science. The term 'flight from science' had become current since the publication of the 1968 Dainton Report which had shown that the number of students studying science in the upper secondary school was declining. Some of the teachers who had taken leading roles in Nuffield projects, including John Lewis, a physics teacher in a selective independent school, surmised that they had enjoyed themselves too much investigating the intellectual niceties of theoretical science, when they should have been more concerned with the interplay between science and society (Lewis 1987). However, at the time, most science teachers felt ill-equipped to present a broader view.

In response the Association for Science Education gave support to two curriculum development projects in England that produced Science, Technology and Society (STS) courses for school students (Aikenhead 1994; Hunt 1994). One of the STS course was 'Science in Society', which was led by John Lewis (1981).[6] This course was strongly influenced by the 'Club of Rome' report on the limits to global

[6]The *Science in Society* Teacher's Guide and Readers are available from the National STEM Centre eLibrary at http://stem.org.uk/cx9s

growth (Meadows 1972). The other was 'SISCON in Schools'[7] headed by Joan Solomon (1983), then a teacher in a state school. Her team drew inspiration from SISCON – Science In a Social CONtext – that had been inaugurated in 1971 for students in universities and polytechnics.

Both these courses were able to innovate and experiment with new themes because they were studied by post-16 students in the slot in the timetable for general studies intended to broaden the curriculum for those taking two or three specialised academic A-level courses. Teachers had a great deal of freedom in their choice of courses to run as part of general studies because they did not count for much outside schools.

The place of the science in STS courses was a matter for debate (Solomon 1988). Research had suggested that students were often resistant to accepting and applying scientific knowledge in the context of social issues because they had difficulties in reconciling the uncertainties of the issues with what they perceived as science's claims to 'truth' (Aikenhead 1987; Fleming 1986). This led to the view that in STS courses it was important that students should learn something about the nature of scientific theories. Proponents of STS courses argued that students needed to appreciate the role of imagination in the development of theories, the importance of modelling and the relationships between data and explanations. In addition it was suggested that students needed to appreciate something about the provisional and uncertain nature of scientific theories.

Science in Society Reader J called *The Nature of Science* featured essays by a range of authors. One essay contrasted Baconian induction with the views of Karl Popper on the scientific method. Another essay in the reader outlined the ideas of Thomas Kuhn on how scientific ideas develop. Other essays dealt with the role of imagination in science and with the interfaces between science and religion. Reader K *Science and Social Development* opened with an essay by Sir Desmond Lee included a discussion of the extent to which Greek speculation about the nature of the physical world could be regarded as scientific in the modern sense of the term.

The nature of science was explored in the SISCON reader called *How can we be sure?* which included chapters about observations and generalisations and about the birth of scientific theories. This reader also included a historical section about changes in scientific explanations such as changes in theories about light and about burning.

In his review of the rationale for STS courses, John Ziman (1994) discussed the diversity of approaches to STS. In his view, the fundamental purposes of STS education were genuinely and properly diverse and incoherent. He identified and commented on seven different approaches drawn from STS initiatives around the world including the historical approach and the philosophical approach both of which had featured to a limited extent in the English STS programmes.

Commenting on the historical approach, Ziman noted that the history of science had long been regarded as the most natural medium for humanising science education

[7] The *Science In a Social CONtext* Teachers' Guide and a series of Readers are available from the National STEM Centre eLibrary at http://stem.org.uk/cx9z

but that this was not without serious disadvantages. One disadvantage was that since the actual history of science is peculiarly deep, subtle and complicated, for most students the history of science is just too academic and remote. This leads to a related disadvantage, namely, that elementary accounts of the history of science in school end up as historically inaccurate celebrations of scientific progress which attribute heroic qualities to a few exceptional individuals. This echoes Stephen Brush's concern that the history of science in schools all too often leads to fictionalised idealizations and conveys a view of history that implies that science is a steady and cumulative progression towards the pinnacle of modern achievements (Brush 1974). Despite these warnings, teachers and curriculum developers continued to feature historical case studies whilst generally ignoring the new historical interpretations.

The two pioneering post-16 STS courses were only adopted by a small minority of students for study after the end of their compulsory period of education. Nevertheless, they proved influential because the resources they published helped to introduce teachers to ways of linking their science teaching to the world outside school or college. The projects also featured new teaching approaches for engaging the interest of their students in issues related to the nature of science and its applications (Lewis 1987). Curriculum developers who had been involved in these projects were soon taking leading roles in the Science, Technology in Society (SATIS) projects. These were very much more widely adopted and later were major influences on the first version of the national curriculum.

63.5 The SATIS Projects

Michael Young (1971), when discussing the stratified and specialised nature of the traditional school curriculum, contrasted the closed, narrow, high-status curriculum suited to potential future science specialist (such as the Nuffield O-level schemes) with open, broad, low-status programmes for the rest. He argued that under this regime it was inescapable that most of those who succeeded in school science would be systematically denied the opportunity to grasp science as an integral and unavoidable part of social life, whilst the rest, the failures, would leave school to become members of a scientifically illiterate public.

By the early 1980s this view was becoming more widely shared as shown by the first UK government statement about the science curriculum which made the case for broader, less specialised science programmes in school (Department of Education and Science 1985). This statement defended the proposed changes by claiming that science courses could challenge able students not by the task of accumulating greater stores of scientific knowledge but by the application of scientific principles to the real world, by the opportunity to investigate and solve problems and by the necessity of bringing scientific method to bear on assignments where the answer cannot be predicted in advance.

In line with the official guidance, significant changes in science education took place. In many schools, the option to study one, two or three of the separate sciences was replaced by compulsory science courses occupying 20 % of curriculum time from the age of 14 to the age of 16. These programmes of 'balanced science' were variously integrated, co-ordinated or combined, but, like SCISP, they had the common aim of making sure that all young people continued to study elements of biology, chemistry and physics as well as other aspects of science such as earth science and astronomy throughout their secondary education. This could be a challenge for teachers who were no longer working with students who had chosen to study their specialist subject.

Also at this time, the old system of public examinations for 16-year-olds (based on O-levels and the CSE) was being replaced by the new General Certificate of Secondary Education (GCSE) for all young people in schools. For the first time national criteria were introduced to control the new GCSE syllabuses. Amongst other things, the criteria for science required that not less than 15 % of the assessment of science should be related to the technological applications of the subject with their social, economic and environmental implications.

In this context, the Association for Science Education (ASE) launched the first Science and Technology in Society (SATIS) project in 1984 with the support of a charitable trust and a range of industrial companies (Hunt 1988). John Holman was appointed as director having played a key role in the *Science in Society* project. The small team recruited to the project was made up mainly of teachers. Despite the imposition of the new GSCE criteria, teachers still had the expectation that they could shape the interpretation of the new courses in schools.

The team decided to develop a bank of short resource units. Each unit related to a major topic in science syllabuses whilst exploring important social and technological applications and issues. Almost all of the units were written by a larger number of contributing teachers, often in co-operation with experts from industry, higher education, agriculture or the public services. In this way the project brought new stories, case studies and examples into science education. Between 1986 and 1991 the project published 120 units each requiring a lesson or two of class time.[8]

The SATIS units were not developed according to a predetermined plan, but, in response to the coverage of GCSE syllabuses and examinations, they clustered into themes such as materials, energy resources, the environment, health and ethical issues. None of the units in this project dealt with the nature of science because this did not feature in the GCSE courses at that time. However, there were some historical units thanks to contributions from the historian, Anthony Travis (1993), a former science teacher, whose research was focussed on the origins of the chemical industry, notably the dyestuffs industry.

As a result, the project published units about the work of Fritz Haber (number 207), the discovery of Perkin's mauve (number 510), the quest for medical 'magic bullets' by Paul Ehrlich (number 805) and the ingenious contributions of Carl Bosch to the commercialisation the Haber process (number 810). These units did not

[8] All the SATIS units together with a General Guide for Teachers and an update of the first 100 units are available from the eLibrary of the National STEM Centre (Accessed in March 2012 from http://stem.org.uk/cx9n)

present a particular view of the nature of science but touched on the human side of scientific discovery and the social implications of technological changes.

The units were presented in the form of short teachers' notes with student worksheets which schools were free to copy. ASE was able to publish the units cheaply at just the right time with the result that they were very widely adopted. This form of publishing became very popular in the UK and other projects adopted a similar pattern. Between 1995 and 1999, for example, ASE produced a series of SATIS-style units to celebrate key centenaries in the history of science.[9]

Following the success of the first SATIS project, the ASE decided to extend the work to older students. The new project, SATIS 16–19, was similar to SATIS in many ways: it was funded on a similar basis, teachers determined the policy and did the writing, and the intention was to produce a bank of varied resource materials to enrich existing courses. The project team set out to support both the general education of all students as well as to provide resources to supplement specialist science programmes.

In addition to the 100 SATIS 16–19 units[10] published between 1990 and 1992, the project also published three readers written by Joan Solomon which were designed to provide a rationale for planning the STS component in programmes of general education. The first of the readers, *What is Science?*, encouraged students to reflect on the nature of scientific theories and to consider where imaginative scientific theories come from. Examples from the history of science were included to show how scientific theories change and to explore the interplay between science and society. The sections of this reader were cross-referenced to related SATIS 16–19 units, allowing students to examine selected topics in more depth. The related units included *The retrial of Galileo* (unit 1), *Two games and the nature of science* (unit 26), *Patterns in the sky* (unit 51), *Why do we grow old?* (unit 61) and *Science and religion – friends or foes?* (unit 77).

An important aspect of these two SATIS projects was the way that they helped to legitimise activities which had not previously been common in science lessons. The units provided teachers with models for lessons involving discussion techniques, role plays, analytical reading, data analysis and problem-solving. These approaches to teaching and learning were essential in broadening school science to encompass historical topics and studies of the nature of science.

Following their experience with STS and SATIS projects, John Holman and Joan Solomon went on to make significant contributions to the development of the first national curriculum for England and Wales. John Holman was a member of the core working group that drafted the new curriculum, and Joan Solomon acted as adviser for the development of AT17.

---

[9] See the four units in the collection called 'Celebrating Centenaries of Famous Discoveries' in the eLibrary of the National STEM Centre (Accessed in March 2012 from http://stem.org.uk/cxg6)

[10] All the SATIS 16–19 units together with the three readers, *What is Science*?, *What is Technology?* and *How does Society decide?*, are available from the eLibrary of the National STEM Centre (Accessed in March 2012 from http://stem.org.uk/cxas)

63.6 The Nature of Science in the National Curriculum

The first version of the national curriculum for England and Wales was introduced in 1989. It was divided into 17 strands, called attainment targets (Department of Education and Science 1989). Two of these attainment targets conveyed generic messages about science: AT1 Exploration of Science and AT17 The Nature of Science. Jim Donnelly (2001) has discussed the tensions between AT1 which drew on a broadly empiricist and inductivist tradition in science education and AT17 which was strongly influenced by STS thinking with its emphasis on the social and cultural dimensions of science.

Professor Jeff Thompson was chair of the working group that drafted the science curriculum. He had been active in the Association for Science Education and 10 years earlier had chaired the ASE group that wrote the consultative document *Alternatives for Science Education* (ASE 1979). This report had aimed to anticipate possible developments in the structure of the science curriculum by describing three possible curriculum models. All three models included aspects of the history and philosophy of science. Thus the chair and several members of the national curriculum group had an interest in introducing AT17 despite the challenge that this would present to many science teachers.

Stephen Pumfrey's critique (1991) of this first version of the national curriculum concentrated on AT17. He pointed out that a curious feature of this attainment target was that it nowhere gave an explicit answer to the question 'What is the nature of science?' He suggested that these nine ideas were the key ingredients of what the answer implied by AT17 appeared to be:

1. Meaningful observation is not possible without a pre-existing expectation.
2. Nature does not yield evidence simple enough to allow one unambiguous interpretation.
3. Scientific theories are not inductions, but hypotheses which go imaginatively and necessarily beyond observations.
4. Scientific theories cannot be proved.
5. Scientific knowledge is not static and convergent, but changing and open-ended.
6. Shared training is an essential component of scientific agreement.
7. Scientific reasoning is not itself compelling without appeal to social, moral, spiritual and cultural resources.
8. Scientists do not draw incontestable deductions, but make complex expert judgements.
9. Disagreement is always possible.

In exploring these ideas, students were expected to learn about stories from the history of science. Pumfrey highlighted tensions implicit within AT17 which, on the one hand, recognised that there have been different values and forms of natural knowledge in different times and contexts whilst, on the other hand, expecting historical case studies to illustrate the norms of scientific practice today.

Pumfrey illustrated his arguments with reference to teaching resources developed under the leadership of Joan Solomon. She had not only contributed to the formulation of AT17 but also had been one of the people who led initiatives to support its implementation in schools. Her first publications were short readers telling stories about developments in the history of science written for 11–14-year-olds to illustrate aspects of the nature of science.[11] Next she worked with groups of teachers in schools to devise a varied collection of teaching and learning activities covering all the aspects of AT17. These lesson ideas were published in two volumes called *Exploring the Nature of Science*: one for 11–14-year-olds (Key Stage 3) (Solomon 1991) and one for 14–16-year-olds (Key Stage 4)[12] (Solomon undated).

Many of the activities in *Exploring the Nature of Science* were based on historical case studies. The introduction to the Key Stage 3 book shows that the authors were sensitive to the tensions identified by Stephen Pumfrey by stating that:

> The greatest difficulty with teaching history is understanding just what scientists in another age meant by the terms they used – like 'atoms', 'heat' or even '…made of water'. In none of these cases were the meanings exactly as they are today. The subtlety of historical research lies in trying to understand the thinking of scholars who were influenced by philosophies that few of us have even encountered. Inevitably we are forced to simplify, both through lack of time, and because our basic intention is not to lecture on the history of science, but to teach children about the nature of science (Solomon 1991, Introduction – *no page numbers*).

The Nuffield-Chelsea Curriculum Trust[13] had also responded to AT17 with a publication called *Investigating the Nature of Science.*[14] The file was devised by a team from King's College London. It aimed to support teaching of all aspects of AT17 to 11- to 16-year-olds. The resource provided detailed lesson ideas covering 12 topics explored with the help of a variety of methods that included discussion, role-play, experimental work, models and guided reading (Honey 1990).

This file contained an introduction to explain the rationale of AT17. This began with a discussion of the reasons for covering the nature of science in a science course. The authors pointed out that a view of the nature of science is implicit in any science course and suggested that it is desirable to articulate and clarify what this view is so that, for example, students understand better the relationships between experiments and theory.

The introduction to the file went on to explain how misconceptions about science can arise in the minds of learners both from messages in the media and from the

---

[11] The *Nature of Science* readers were published by the Association for Science Education (1989–1990), and most of them are now available from the eLibrary of the National STEM Centre (accessed in March 2012 from http://stem.org.uk/cx9r)

[12] The *Exploring the Nature of Science* publications are available from the eLibrary of the National STEM Centre at http://stem.org.uk/cxfb

[13] From the late 1970s to the early 1990s, the Nuffield Foundation's support for curriculum development in science, maths and technology education was the responsibility of the Nuffield-Chelsea Curriculum Trust based at Chelsea College.

[14] *Investigating the Nature of Science* is available from the National STEM Centre eLibrary at http://stem.org.uk/rx3a7

experience of school science. Finally the authors reviewed some differing views about the nature of science before suggesting ways of making the history and philosophy of science part of the school curriculum.

63.7 The Short Life of AT17

The first version of the science national curriculum was so detailed and complex that a revision began very soon after it was launched in schools. Given the many demands made by the new curriculum, and the lack of expertise and limited resources for teaching AT17, there was little opposition when, in 1991, this attainment target was merged with the first attainment target (AT1) that covered investigative practical science (Department of Education and Science 1991).

It is not surprising that few teachers regretted the passing of AT17 in view of the argument put forward by Martin Monk and Jonathan Osborne (1997). They suggest that the failure of the history of science to contribute to the mainstream of science teaching arises because teachers have no confidence that a historical context adds anything to their students' knowledge and skills. Within the classroom, teachers' dominant concerns are the development of the student's knowledge and understanding of the content of science; that is with 'what we know' rather than 'how we know'. Many teachers also lack the requisite knowledge of either the history of science or the nature of science to explore any of these issues appropriately. Teachers are heavily influenced by the demands of external assessments. The early national curriculum tests did little to support the teaching of AT17.

63.8 Rethinking the Purposes of Science Education: Beyond 2000

The first major review of the national curriculum as a whole was carried out by Sir Ron Dearing, a former senior civil servant who had recently been chief executive of the Post Office Ltd. His report (Dearing 1993), published in 1993, argued that the curriculum had become an unwieldy structure which was virtually impossible to implement. As a result a further revision was carried out to cut down the content and to eliminate overlap between subjects. Coverage of the nature of science was reduced in emphasis in the resulting 1995 version of the science curriculum (Department of Education and Science 1995). The consequence was that there was little or no attention given to teaching about the history of science and the nature of science by most teachers in pre-16 courses for the next 10 years or so.

Following the Dearing review, the government announced a 5-year moratorium on curriculum change. This provided an opportunity for reflection which was taken up by Rosalind Driver and Jonathan Osborne of King's College, London. A grant from the Nuffield Foundation provided the support for a seminar series to 'consider and review the form of science education required to prepare young people for life in our society in the next century'.

The series of six invitation seminars were held between 1997 and 1998. In addition there were two open seminars to widen the discussion to include teachers and others with an interest in the future of science education. The *Beyond 2000* report (Millar and Osborne 1998) from the seminar series made the case for finding better ways to meet the two main purposes of science education for 14–16-year-olds:

- To develop the 'scientific literacy' of all students in preparation for adult and working life
- To provide the foundations for more advanced courses in science

The report proposed that the same curriculum could not serve both purposes, especially in the final 2 years of compulsory education. It recommended that the structure of the science curriculum should differentiate more explicitly between those elements designed to enhance 'scientific literacy', and those designed as the early stages of a specialist training in science, so that the requirement for the latter would not distort the former.

The report outlined the aims and nature of a 'scientific literacy' course. Alongside the aim of introducing young people to the major 'explanatory stories' of science about life and living things, matter, the Universe and the made world, a key aim of such a course was that it should give young people an understanding of how scientific inquiry is conducted. The report summarised the rationale for teaching 'about science' in this paragraph:

> In order to understand the major 'explanatory stories' of science, and to use this understanding in interpreting everyday decisions and media reports, young people also require an understanding of the scientific approach to inquiry. Only then can they appreciate both the power, and the limitations, of different kinds of scientific knowledge claims. They also need to be aware of the difficulties of obtaining reliable and valid data. Science issues are often about the presence or absence of links and correlations between factors and variables, often of a statistical and probabilistic kind, rather than directly causal – so young people need an understanding of these ideas, and practice in reasoning about such situations. Often the plausibility of a claimed link depends on seeing a mechanism which might be responsible – and here again an understanding of the major 'explanatory stories' of science is needed. Finally, young people need some understanding of the social processes internal to science itself, which are used to test and scrutinise knowledge claims before they can become widely accepted – in order to appreciate their importance, but also to recognise the ways in which external social factors can influence them (Millar and Osborne 1998, pp. 19–20).

The *Beyond 2000* report set out in some detail the ideas-about-science that it recommended should feature in a scientific literacy course. The authors of the report warned that these ideas-about-science represented a significant expansion of the range and depth of treatment that such issues currently demanded in the existing curriculum. As a result, they pointed out that any development based on the recommendations needed to be undertaken in collaboration with science teachers so that their introduction could be a managed process and not a sudden, and possibly unwelcome, event.

63.9 Science for Public Understanding

Following his review of the national curriculum, Sir Ron Dearing was asked to review post-16 qualifications in England, Wales and Northern Ireland including A-levels. Included in his report was a recommendation for a new Advanced Subsidiary (AS) exam to be taken after 1 year post-16 to encourage students to broaden their studies by studying up to five subjects in place of the traditional three (Department for Education and Employment 1996). This would allow students to begin their post-16 programme with, say, five AS-levels – and then choose three of these to continue to A-level. They could obtain an AS-level qualification in the two that they took for just 1 year. It was suggested that the two additional AS courses taken might complement or contrast with the main areas of the students' specialist interests.

This change to the post-16 academic curriculum threw up an opportunity to reassess existing STS courses and make a fresh start. As before, there was considerable freedom allowed for the style and content of courses intended to broaden the curriculum. These courses were not subject to the detailed national guidelines that applied to mainstream advanced courses. One of the examining boards[15] invited Robin Millar to take the lead in developing a new course (Millar 2000). He had published an article outlining his thinking which set out ideas that he was also contributing as co-author of the report from the *Beyond 2000* seminars (Millar 1996).

In this way, the AS *Science for Public Understanding* course became a test-bed for trying out a model of a scientific literacy course as described in the *Beyond 2000* report. The syllabus[16] for this new course gave the same weighting to ideas-about-science as it did to explanatory theories (Millar and Hunt 2002). The course was presented as a series of topics which provided the contexts for teaching about science explanations and about the nature of science. The topics were divided equally between issues in the life sciences and issues in the physical sciences.[17]

Some of the topics were treated historically. One example was 'Understanding health and disease' that, in part, used the work of people such as Snow, Semmelweis and Pasteur to explore ideas-about-science including the distinction between correlation and cause, the origins of scientific explanations and how the scientific community resolves the conflicts between competing theories.[18] Other topics treated

---

[15] The invitation came from the Northern Examinations and Assessment Board (NEAB) which later merged with other examining bodies to form the awarding organisation that is now called AQA.

[16] Over the period covered by this chapter the older term 'syllabus' has been replaced by the term 'specifications'. Specifications are more explicit about aims, content, assessment model and grade criteria.

[17] From 2007, *Science for Public Understanding* became a full A-level subject with the name *Science in Society*. The AS specification (syllabus) for *Science in Society* is very much the same as its precursor and can be downloaded from the AQA website together with past examination papers (accessed in March 2012 from http://www.aqa.org.uk/qualifications/a-level/science/science-in-society/science-in-society-key-materials)

[18] For more details of the approach, see the revised version of the AS course textbook (Hunt 2008).

from a historical perspective were 'Understanding who we are' which included study of the theory of evolution and its origins and 'Understanding where we are' which traced the development of theories of the solar system and the Universe. The other topics in the course also introduced and applied ideas-about-science when dealing with contemporary issues, including ethical issues. The textbook included an outline of some ethical frameworks to help students discuss the issues.

The teaching approach adopted for the course drew on the experience of earlier STS courses. *Science for Public Understanding* aimed to help students understand more about the nature of science and to provide them with the skills needed to participate as citizens in debates on topical science. The course therefore expected students to develop their abilities across a wider range of concepts and skills than in most traditional science classes.[19]

The timetable for introducing the new AS courses made it possible to pilot AS *Science for Public Understanding*. The first cohorts of students took pilot examinations in June 1999 and June 2000. By this time it was becoming clear that there would be an opportunity to test out the recommendations of the *Beyond 2000* report in a set of novel GCSE courses supported by a large-scale curriculum development project that came to be called *Twenty First Century Science*. With this in mind, the Nuffield Foundation commissioned a team from King's College London to make a study of the teaching and learning of science explanations and the ideas-about-science in *Science for Public Understanding*.

The research was carried out between 2001 and 2002. The findings were published in a report called *Breaking the mould?* (Osborne et al. 2002a). The data gathered by the research team suggested that the course had been successful in achieving its first aim: to sustain and develop students' enjoyment of, and interest in, science. The overwhelming majority of students said that the course was both enjoyable and interesting. Furthermore, the *Science for Public Understanding* course had managed to attract students who would not otherwise have studied science post-16. It was notable that nearly 60 % of all the students were female, which the researchers suggested was a significant achievement for a course where 50 % of the content was physical science.

However, the report showed that the new course made demands on teachers' pedagogic techniques such as the skill to run and organise effective discussions that engage all students in thinking critically about socioscientific issues. The researchers pointed out that students need to be explicitly taught, not only how to evaluate media reports about science critically but also how to construct effective arguments which are reliant on evidence rather than personal or group opinion. The report stated that the available guidance on teaching methods was inadequate.[20]

The research demonstrated that changing the culture that forms and moulds teachers is much harder than simply changing the curriculum. To bring about changes requires

[19] Schemes of work and lesson activities originally devised for AS *Science for Public Understanding* are available as revised and updated versions from the AS section of the *Science in Society* website: http://www.nuffieldfoundation.org/science-society (accessed March 2012).

[20] The *Science for Public Understanding* (now *Science in Society*) website was developed in response to this criticism (http://www.nuffieldfoundation.org/science-society)

considerable support, effort and time. The researchers concluded that the course had begun that process and planted the seeds of a different way to teach and engage students but that enabling it to take root would require continued endeavour.

Within a year of the launch of the new AS courses, as part of Curriculum 2000, it became clear that, despite expectations, most students would opt for four rather than five AS courses. As a result the interest in 1-year AS courses, such as *Science for Public Understanding*, was less than expected, but the numbers of candidates for the exams were significantly larger than had been the case for its STS predecessors at this level.

63.10 Evidence-Based Practice in Science Education (EPSE)

Between January 2000 and June 2003, a research network, funded by the Economic and Social Research Council (ESRC), carried out four interrelated projects to improve the interface between science education researchers and teachers. The aim of the programme was to develop and evaluate several examples of evidence-based practice in science education.

The lead researchers had all been involved in the *Beyond 2000* seminar series. One of the projects set out to examine the curriculum implications of the recommendation that the aim of compulsory science education should be to develop 'scientific literacy' by providing a basic philosophical understanding of the nature of science, the function and role of data in scientific argument and how the scientific community functions (Osborne et al. 2006). This project was begun in response to the notion that a shift towards science education for citizenship implied a broadening in the range of stakeholders with a legitimate interest in determining the goals of the school science curriculum.

Accepting that the nature of science is seen by contemporary scholars as a contested domain, and given the lack of empirical evidence of consensus, the research team decided to try to determine empirically the extent of agreement amongst a group of experts about those aspects of the nature of science that should be an essential feature of the school science curriculum. The method chosen for eliciting the view of the expert community was a Delphi study (Osborne et al. 2003). In this way the researchers sought to establish consensus amongst a group made up of five representative individuals from each of these groups: leading scientists, historians, philosophers and sociologists of science, science educators, those involved in science communication and primary and secondary teachers of science.

Using three rounds of linked questionnaires, where the responses were sifted and successively commented on, the researchers determined the level of agreement amongst the experts about which ideas-about-science were so important that they should be included in school science. At the end they arrived at a strong consensus about nine common themes that are summarised in Table 63.1 (Osborne et al. 2002b).

Table 63.1 Findings of a Delphi study summarising the views of experts about those aspects of the nature of science that should be an essential feature of the school science curriculum (Osborne et al. 2002b, p. 30)

Nature of scientific knowledge

Science and certainty

Students should appreciate why much scientific knowledge, particularly that taught in school science, is well established and beyond reasonable doubt and why other scientific knowledge is more open to legitimate doubt. It should be explained that current scientific knowledge is the best we have but may be subject to change in the future, given new evidence or new interpretations of old evidence

Historical development of scientific knowledge

Students should be taught some of the historical background to the development of scientific knowledge

Methods of science

Scientific methods and critical testing

Students should be taught that science uses the experimental method to test ideas, and, in particular, about certain basic techniques such as the use of controls. It should be made clear that the outcome of a single experiment is rarely sufficient to establish a knowledge claim

Analysis and interpretation of data

Students should be taught that the practice of science involves skilful analysis and interpretation of data. Scientific knowledge claims do not emerge simply from the data but through a process of interpretation and theory building that can require sophisticated skills. It is possible for scientists legitimately to come to different interpretations of the same data and, therefore, to disagree

Hypothesis and prediction

Students should be taught that scientists develop hypotheses and predictions about natural phenomena. This process is essential to the development of new knowledge claims

Diversity of scientific thinking

Students should be taught that science uses a range of methods and approaches and that there is no one scientific method or approach

Creativity

Students should appreciate that science is an activity that involves creativity and imagination as much as many other human activities and that some scientific ideas are enormous intellectual achievements. Scientists, as much as any other profession, are passionate and involved humans whose work relies on inspiration and imagination

Science and questioning

Students should be taught that an important aspect of the work of a scientist is the continual and cyclical process of asking questions and seeking answers, which then leads to new questions. This process leads to the emergence of new scientific theories and techniques which are then tested empirically

Institutions and social practices in science

Co-operation and collaboration in development of scientific knowledge

Students should be taught that scientific work is a communal and competitive activity. Whilst individuals may make significant contributions, scientific work is often carried out in groups, frequently of a multidisciplinary and international nature. New knowledge claims are generally shared and, to be accepted by the community, must survive a process of critical peer review

63.11 School Science and the Changing World of the Twenty-First Century

A new version of the whole national curriculum for England was published in 2000, but there was little change to the science curriculum. The main change was a modification of the section about scientific enquiry to incorporate some new content covering ‘ideas and evidence’.

Influenced by the *Beyond 2000* report, the government asked the Qualifications and Curriculum Authority (QCA) to begin a project which it called ‘Keeping School Science in Step with the Changing World of the twenty-first Century’. The QCA was responsible for curriculum in England at that time. By now it had become very difficult for school teachers to take the initiative and shape the core curriculum through their active membership of the Association for Science Education. Politicians and regulators were in charge.

QCA commissioned researchers to investigate three issues: what students would need to become scientifically literate citizens, what should constitute a curriculum to meet those needs and how students’ learning in a new and different science curriculum could be assessed (QCA 2006).

The first study involved consulting groups with an interest in school science education about the features of the proposed curriculum. The consultation was carried out by a group of researchers from the Centre for Studies in Science and Mathematics Education, University of Leeds (Leach 2002). Secondary science teachers in focus groups were presented with an outline of a curriculum, broadly in line with the thinking in the *Beyond 200*0 report, with these features:

- It views pupils as potential users and consumers of science, rather than as potential producers of scientific knowledge.
- It aims to give pupils a sense of the cultural significance of science.
- It covers less of the traditional conceptual content of science, allowing time to cover key areas in significantly more depth.
- It gives more attention to the ways in which science works, emphasising ‘how we know what we know’.
- It introduces scientific disciplines that predict risk, such as epidemiology, and reduces the amount of time spent on the ‘traditional’ school science disciplines of physics, chemistry and biology.

The findings from this study suggested that secondary science teachers were generally dissatisfied with the existing curriculum and supportive of change; however, they did not share a common vision of what a science curriculum for all students might look like. The researchers concluded that any future attempts to change the focus of the science curriculum would have to take seriously the need for such a shared vision to emerge within the science teaching profession if the changes were to be successful. Consulting more widely, the study identified many ‘critical voices’ in relation to the meaning and feasibility of the goals of scientific literacy.

The second study was carried out by researchers from King’s College London and the University of Southampton (Osborne and Ratcliffe 2002). The aim was to explore appropriate methods of assessment for a new curriculum featuring ideas and evidence, in a curriculum intended to develop understanding of the nature and limitations of scientific endeavour, through historical and contemporary contexts. The challenge was to find assessment tasks which would encourage teachers to explore not only which scientific ideas are believed but why they are believed. The researchers were looking for assessment items covering the relationships between the claims, data and warrants for trust in scientific ideas.

The research team studied examples of assessment in this field from across the world. In their interim report the researchers stated that it was their general impression that, internationally, the assessment of the nature and processes of science was an underdeveloped field. They reported that they had found relatively few sources that had created a significant body of items for testing understanding of the nature of science, the analysis and interpretation of data and the processes of science. Nevertheless they were able to use their findings to assemble four written tests and four teacher-assessment tasks that were tried out in schools.

From the analysis of the test results, the research team concluded that reliable and valid items for testing pupils' understanding of the processes and practices of science in contemporary or historic contexts could be developed. The researchers argued that many of the items and the teacher-assessment tasks offered authentic contexts for assessment. However, they pointed out that teachers considered the amount of reading and the language level to be off-putting, but that such comments reflected a tension between what was currently being taught and the comparative 'novelty' of the content and emphases of the items being trialled. The researchers identified unresolved issues and areas in which further work was needed.

The third study was carried out by the University of York Science Education Group (UYSEG) in collaboration with the Association for Science Education and the Nuffield Curriculum Centre. The aim was to devise a curriculum model that would meet the two overarching purposes of science education identified in the *Beyond 2000* report: the development of scientific literacy for all and preparation for post-compulsory science study for some.

The curriculum model proposed that all students would complete a 'core' course (UYSEG 2001). This course would provide 'a broad, qualitative grasp of the major science explanations' and also include insights into the nature of science and its relation to social and ethical issues. In addition, most students would also opt for one of two additional science courses offering either traditional science content or a focus on the applications of science within everyday and work-related contexts (*Twenty First Century Science* project team 2003).

QCA acted on the recommendations of the third study and commissioned the OCR awarding organisation[21] to produce a suite of pilot GCSE qualifications to match the curriculum model. QCA used the findings of the second project to inform

[21] There are three awarding organisations in England responsible for GCSE and A-level specifications (syllabuses) and examinations. The three were formed by the merger of a number of previous examination boards. The three are AQA, Edexcel and OCR. Specifications and examinations have to conform with national criteria that were formerly produced by QCA but are now controlled by a regulator called Ofqual.

the national testing of science for younger pupils, but there was a lack of thorough development work to devise appropriate methods of assessment of ideas-about-science for GCSE courses. The recommendations of the first study were not followed up (Ryder and Banner 2011).

The University of York Science Education Group and the Nuffield Curriculum Centre set up *Twenty First Century Science* as a large-scale curriculum development project to provide the teaching resources and support needed by the schools taking part in the trials of the new GCSE courses commissioned by QCA and run by OCR[22] (Millar 2006). The project was funded by grants from three charitable foundations including the Nuffield Foundation whose trustees supported the recommendations of the *Beyond 2000* report.

Development of the resources began in 2002 in preparation for a pilot that ran in nearly 80 schools from 2003. QCA carried out early, small-scale evaluation studies of the pilot and concluded, well before the end of the pilot, that the findings were sufficiently positive to justify reworking the national curriculum and GCSE science criteria broadly in line with the model being trialled (QCA 2005). A larger-scale evaluation of the pilot was commissioned by the charitable trusts funding the development project, but the findings were published too late to influence the national developments (Burden et al. 2007).

63.12 A New National Curriculum Featuring 'How Science Works'

Drawing on all the work done since 2000, the QCA introduced a new version of the national curriculum in 2004.[23] This was divided into two major strands: 'knowledge and understanding' and 'how science works'. The curriculum model was a response to the notion that science education should not only communicate a body of knowledge but also convey an understanding of how that knowledge has been, and continues to be, developed. This required that the curriculum place greater emphasis on the nature of science and the way scientists and the scientific community as a whole operate.

There were four main sections of the 'how science works' strand (Toplis 2011):

- Data, evidence, theories and explanations
- Practical and inquiry skills
- Communication skills
- Applications and implications of science

[22] The latest version of the Twenty First Century Science GCSE Science specification can be downloaded from the OCR website at http://www.ocr.org.uk/qualifications/type/gcse_2011/tfcs/ (accessed November 2011).

[23] This version of the national curriculum for students aged 14–16 will probably continue to apply until 2014. It is available from the Department for Education website: http://www.education.gov.uk/schools/teachingandlearning/curriculum/secondary/b00198831/science/ks4/programme (accessed June 2012).

Table 63.2 How science works in AQA and OCR-A specifications

Concepts of evidence in AQA specifications (the thinking behind the doing)	Ideas-about-science in Twenty First Century Science
Observation as a stimulus to investigation	Data and their limitations
Designing an investigation	Correlation and cause
Making measurements	Developing explanations
Presenting data	The scientific community
Using data to draw conclusions	Risk
Societal aspects of scientific evidence	Making decisions about science and technology
Limitations of scientific evidence	

However, the national curriculum has never provided enough information to allow regulators to set a detailed framework for GCSE specifications. To give awarding organisations more specific guidance, the curriculum authorities produce GCSE subject criteria. The new criteria published for the 2004 curriculum offered considerable flexibility with the result that the awarding organisations came up with very different interpretations of 'how science works'. Two sets of specifications were based on fully developed rationales derived from research and scholarship. These are compared in Table 63.2. Most of the 600,000 or so students that take GCSE Science each year follow one or other of these popular courses.

AQA[24] adopted a rationale for teaching about the methods of science based on the work of Gott and Roberts (2008) at Durham University.[25] This approach focussed on the procedural understanding and understanding of concepts of evidence that are needed to carry out and interpret science investigations.

The OCR Twenty First Century Science specifications[26] were updated from the pilot versions and used the rationale for teaching ideas about science that informed the *Beyond 2000* report based on the work of Robin Millar and his collaborators at the universities of York, King's College London, Southampton and Leeds. The approach in the core science specification was designed to develop the scientific literacy of all young people. The thinking was that a course based on this rationale would help to develop the knowledge and understanding that are most useful in interpreting and evaluating the sorts of science-based information and claims that everyone encounters in their adult and working lives (Millar 2013).

[24] Details of AQA GCSE Science specifications are available from the website of the awarding organisation: http://web.aqa.org.uk/qual/newgcses/science.php (accessed June 2012).

[25] See the Research Report: Background to Published Papers by Richard Gott and Ros Roberts at http://www.dur.ac.uk/education/research/current_research/maths/msm/understanding_scientific_evidence/ (accessed June 2012).

[26] The Nuffield Foundation website provides details of the Twenty First Century Science courses (rationale, published resources, assessment methods and support for teachers). See http://www.nuffieldfoundation.org/twenty-first-century-science (accessed March 2012).

63.13 Assessing 'How Science Works'

At that time there was no significant investment in developing appropriate methods of assessing 'how science works' in written examination papers. Bringing in this new emphasis to the curriculum made fully explicit ideas that had previously been implicit. There was no accumulated expertise in assessing the ideas, which meant that the examiners for all the awarding organisations had much to learn. Consequently, in the early years the assessments had unsatisfactory features as shown by a SCORE[27] (2009) report on GCSE examinations.

The problems with the assessment of 'how science works' were investigated in more detail by research commissioned by SCORE in 2010. The research team reported that they had found wide variation in the breadth and depth of the treatment in the different GCSE course specifications and examinations (Hunt 2010).

One clear conclusion was that the societal (STS) aspects of this area of the curriculum were being given very substantial, but rather trivial, emphasis compared with the treatment of ideas related to the methods of science and the nature of scientific explanations.

The consequence was that the assessment practices were not fit for purpose and so did not have the confidence and support of the community including teachers. Test items in written examinations failed to show that they assessed knowledge and understanding that every young person needs. As a result many people failed to appreciate that the teaching of 'how science works' could be based on rigorous concepts and challenging learning goals.

The findings of SCORE and others led to a further revision of GCSE specifications and assessments with the aim of bringing greater clarity and consistency to the assessment of ideas related to the methods and nature of science.

63.14 Argumentation and the IDEAS Project

The introduction of 'how science works' emphasises the importance of educating students about how we know and why we see science as a distinctive and valuable way of knowing. This means that students need to explore reasons why accepted theories have become established and why alternative ideas are thought to be wrong (Simon 2011).

Shirley Simon, with Jonathan Osborne and Sibel Erduran at King's College London decided to study the implications of teaching science based on the view that what lies at the heart of science is the commitment to evidence as the rational basis for argument and justification. The group worked with teachers in a research project

[27] SCORE is a partnership of organisations which aims to improve science education in UK schools and colleges. The organisations: Association for Science Education, Institute of Physics, Royal Society, Royal Society of Chemistry and the Society of Biology.

called *Enhancing the Quality of Argument in School Science* (Osborne et al. 2004a). Influenced by their findings, they then developed the publications of the IDEAS project (Osborne et al. 2004b).

The team produced the pack to support the professional development of teachers. They did so because they believed that, in presenting scientific ideas and their supporting evidence to school students, it was essential to consider the arguments for the scientific ideas and other competing theories. One reason for this was that the research evidence suggested that the opportunity to consider why the wrong idea is wrong is as important as understanding the justification for the scientific idea. A second reason was that engaging in argument provides students with a better insight into the nature of scientific enquiry and the work of scientists.

Activities in the pack explored what is meant by the term 'argument' in science and why it is a significant feature of science, bearing in mind that the everyday meaning of the word 'argument' is not the one that is being used in this context. Teachers were introduced to Ron Giere's (1991) epistemological framework and asked to discuss its relevance to sciences ranging from cosmology and geology to biochemistry and physics.

The pack also introduced teachers to the ideas of the philosopher Stephen Toulmin (1958). Activities in the pack gave participants opportunities to use Toulmin's model to analyse and construct scientific arguments. The arguments were mainly related to evidence and explanations. Debate about socioscientific issues was included but was not the main focus of the activities.

63.15 How Science Works in Post-16 Specialist A-Level Science Courses

Following the changes to the national curriculum, the QCA, added a section on 'how science works' to the national criteria for A-level sciences when they were revised for post-16, specialist science courses starting in 2008 (Ofqual republished annually). Section 63.7 of the criteria includes the requirement that science A-levels should enable students to appreciate the tentative nature of scientific knowledge, consider ethical issues in the treatment of humans, other organisms and the environment and appreciate the role of the scientific community in validating new knowledge and ensuring integrity.

A review of the new A-level specifications based on these criteria shows that the new section covering how science works has had a limited impact on most courses.[28] However, the Advancing Physics[29] specification is of particular interest because it has a coherent rationale for 'how science works' fully integrated into the course

[28] This was found, for example, during an unpublished review of specifications carried out by one of the authors for SCORE in 2012.

[29] Information about the course, the published resources and the assessment can be found on this website: http://www.advancingphysics.org/ (accessed in June 2012).

design, content, assessable learning outcomes and scheme of assessment. This rationale covers relatively few ideas but in greater depth and provides a distinctive justification for the inclusion of 'how science works' in advanced courses. Students who follow this course are, for example, assessed on their ability to:

- Identify and discuss ways in which interplay between experimental evidence and theoretical predictions have led to changes in scientific understanding of the physical world
- Use computers to create and manipulate simple models of physical systems and to evaluate the strengths and weaknesses of the use of computer models in analysis of physical systems
- Identify and describe the nature and use of mathematical models
- Identify and describe changes in established scientific views with time

Some other advanced courses take a context-led approach which provides opportunities to feature case studies in the history of science. One example is Salters-Advanced Chemistry,[30] which uses aspects of the history of chemistry to illustrate conceptual and technological developments in some of the topics. Storylines in the course with a strong historical dimension include 'The polymer revolution', 'What's in a medicine' and 'Colour by design'.

Another context-led approach is Salters-Nuffield Advanced Biology.[31] This is a course that introduces students to ethical principles that enable them to analyse and discuss biological issues (Reiss 2008). Students learn about four ethical frameworks: rights and duties, utilitarianism, autonomy and virtue ethics. Implicit in the approach, as Michael Reiss explains, is the notion that one can be confident about the validity and worth of an ethical argument if three criteria are met (Reiss 1999). First, if the arguments that lead to the particular conclusion are convincingly supported by reason. Secondly, if the arguments are conducted within a well-established ethical framework. Thirdly, if a reasonable degree of consensus exists about the validity of the conclusions, arising from a process of genuine debate.

63.16 Teaching About the Nature of Science at A-Level

Research and development work to explore post-16 students' understanding of aspects of 'how science works' has been limited. However, starting in 1999, a team at the University of Leeds carried out a small-scale project, called *Teaching about*

---

[30] Information about the course, the published resources and the assessment can be found on this website: http://www.york.ac.uk/org/seg/salters/chemistry/index.html (accessed in March 2012).

[31] Information about the course, the published resources and the assessment can be found on this website: http://www.nuffieldfoundation.org/salters-nuffield-advanced-biology (accessed in March 2012).

Science during which they designed teaching resources[32] to develop students' understandings about aspects of the nature of science in post-16 A-level science courses. They evaluated the impact of the resources and identified areas of knowledge and expertise that act as barriers to teachers in using the materials to promote student learning about the nature of science (Leach et al. 2003).

The focus of the project was on epistemological aspects of AS-/A-level syllabuses. The lessons were designed to address misconceptions which research had shown to be commonly held by students. Students have misconceptions about the nature of theoretical explanations in science. They tend to believe that theoretical models emerge directly from data and that all features of a theoretical model correspond directly to features in the real world. They often fail to recognise the conjectural and tentative nature of many scientific explanations and that scientific explanations are often expressed in terms of theoretical entities which are not 'there to be seen' in the data. Three lessons were designed to address these misconceptions: A *Electromagnetism,* B *Cell membranes* and C *Continental drift.*

Other misconceptions relate to the assessment of the quality of scientific data. Students tend to see examination of the quality of scientific evidence as simply a matter of making a judgement about whether the scientists involved had made any mistakes. They often fail to recognise the inherent uncertainty of measurements and have little idea of how scientists deal with this uncertainty. Few students use ideas about the validity and repeatability of evidence in evaluating its quality or recognise the significance of examining the spread of a set of data. Lesson D *Chemical data* and lesson E *Mobile phones* were designed to address these misconceptions.

A third set of misconceptions relate to the purposes of scientific investigations. Students tend to see scientific investigation as a process of careful description. For such students collecting a 'good' set of data is the end of the data interpretation process. They often fail to recognise that many investigations involve the testing of ideas. The need to interpret the data in terms of scientific ideas is not recognised. Lesson F *Purposes of Science* was designed to address these misconceptions.

The technical report on the project (Hind et al. 2001) concluded that overall the interventions provided by the lessons had been successful in broadening the profile of views about the nature of science which some of the students in the sample drew upon in response to a specific context. However, there were also a number of students for whom these single interventions had little effect judging by the evidence of their evaluative probes. Furthermore, observations and interviews with teachers highlighted the difficulties for teachers in teaching about the nature of science, something that for many teachers is unfamiliar.

---

[32] The lessons developed and trialled by the Teaching about Science project can be downloaded from the Nuffield Foundation website: http://www.nuffieldfoundation.org/teaching-about-science (accessed in March 2012).

63.17 An Alternative Approach to Teaching the History and Philosophy of Science

63.17.1 Origins of the Perspectives on Science Course

The Perspectives on Science course has its origins in conversations about the need for a qualification which would give post-16 UK students an opportunity to explore and develop their own ideas concerning topics related to the history and philosophy of science.[33] The course development process was initiated by Becky Parker, John Taylor (both science teachers) and Elizabeth Swinbank from the University of York Science Education Group. The rationale for developing the qualification was that students enjoy discussions in which issues relating to the epistemology and metaphysics of science are raised but that there is often little scope for exploring these during science lessons due to lack of time (Taylor 2012). Moreover, it was felt that providing students with an opportunity to explore 'the human face of science' by a study of the history of science would help them to develop a more realistic understanding of the nature of science and also help to break down the divide between 'scientific' and 'humanities' styles of thinking. It was felt too that study of the history and philosophy of science had value in helping to develop students' skills in critical thinking (Swinbank and Taylor 2007).

63.17.2 The Perspectives *Approach to Teaching the History of Science*

In certain respects, the *Perspectives on Science* course was similar in aim and design to the *Science for Public Understanding* course described above. Both developments exploited the opportunity afforded by changes in the post-16 national curriculum, specifically the introduction of the 'AS-level'. Both were developed in a context in which there was growing emphasis on the importance of developing the scientific literacy of students (Millar and Osborne 1998) and both therefore aimed to help students to develop their ideas about science.

However, the *Perspectives on Science* approach was distinctive in a number of respects. Firstly, it was not developed as a science qualification but as a programme in the history and philosophy of science. Whilst the course materials drew on scientific cases studies, the programme of study was designed to teach students how to apply the methods of historical and philosophical analysis to scientific material, rather than to teach them more science. Moreover, the qualification aimed to give a central place to pedagogical techniques, such as classroom discussion and debate,

[33] See Bycroft (2010) for further discussion of the development of the course.

more commonly associated with humanities subjects, and to teach students how to produce extended research dissertations.

Secondly, the course developers deliberately chose a different approach to the question of how ideas about science should be taught than that embodied in the *Science for Public Understanding* course. Whilst the *Science for Public Understanding* course developers selected a list of ideas about science which were taught as part of the prescribed course content, the *Perspectives on Science* chose not to use any such list. The course was designed specifically to allow students the opportunity to develop their own ideas. There was in fact no prescribed course content. Instead, the programme of study was constructed in such as way as to teach students the skills they needed to begin thinking for themselves about historical and philosophical questions relating to science. The aim of the course was to teach students *how* to think in these ways, not *what* to think. The course materials did use a series of case studies, in which issues from the history and philosophy of science were introduced, but these were selected because they were felt to provide good contexts for the skills of historical and philosophical research and analysis, rather than because they were thought to contain essential subject knowledge.[34]

A third distinctive feature of the *Perspectives on Science* was the mode of assessment. The course was unique amongst post-16 UK AS-level qualifications in that assessment was entirely by means of a student research project and oral presentation and not by means of a written exam. This mode of assessment was felt to be the most suitable, given that the aim of the course was to encourage students to develop their own ideas through processes of research and argumentative discussion. Students were allowed to make a free choice of topic for their dissertation, although guidelines existed to lead them towards research questions which were well focussed, with links to research literature and with scope for analytic thought and argumentative engagement.

As Edgar Jenkins (1994) points out, the common feature of most attempts to include insights from the history and philosophy of science in school curricula is that the teaching is assumed to be essentially supportive of the goals of science education itself; it is not expected to challenge its traditional purposes. This is where *Perspectives on Science* opened up a new line of development, allowing teachers and students to respond directly, and in its own terms, to scholarship in the history and philosophy of science. One student, for example, chose to write a project in which she argued that witch-hunting was a scientific activity (witch-hunters based their conclusions on evidence and carried out trials). The historical component of her project was informed by documentary research involving a visit to a county records office to read records of witch trials, and, philosophically, she drew on Feyerabend's radical critique of the notion of scientific method. Another student chose to write about the extent to which the problem of the incommensurability of rival paradigms undermines the idea of objective progress in science.

[34] The student and teacher guides for the course provide further details (Perspectives Project Team 2007a, b).

These examples illustrate the way in which students had a degree of freedom which was unusual in courses which are part of a national qualification framework to engage with scholarly arguments and challenge conventional conceptions of the nature of science.[35]

The point made in Ziman (1994) about the tendency of courses in the history and philosophy of science to endorse oversimplified, Whiggish interpretations of the history of science was addressed by the use of carefully constructed case studies which embody good historiographical practices, including contextualization of scientific developments, exploration of rival conceptual schemes and the exploration of historically complex narratives, such as that of the developing understanding of oxygen, or the recent controversy about cold fusion, which act to counter the idea that the history of science is a tale of steady progress towards an agreed-upon truth.[36] Students addressing projects with historical dimensions were expected to apply these techniques in writing literature reviews, which showed critical awareness of issues such as source reliability and objectivity, and manifested awareness of the wider context within which the developments they discuss took place. The successful application of these techniques of historical analysis formed one strand in the assessment criteria for the dissertation.

63.17.3 *The* Perspectives *Approach to Teaching the Philosophy of Science*

The approach to the philosophy of science was determined by the fact that the aim of the course was to equip students with the knowledge and skills to begin to develop their own philosophical ideas about topics related to science. This was achieved in the programme of study by using a sequence of lessons designed to help students develop skills in critical thinking and conceptual analysis. In most cases, the lessons involved little didactic instruction, although some teaching of common philosophical frameworks took place. Topic areas addressed include science and religion, the nature of scientific truth, genetics, animal welfare, medical ethics, the mind, artificial intelligence, free will and determinism, pseudoscience and the paranormal. Case

[35] Examples such as these may reawaken the concern expressed – with tongue-in-cheek – by Brush (1974) about the risk that teaching students the history of science will have the damaging consequence of subverting a convergent realist interpretation. But as Brush himself concludes, it can be argued that helping students to develop a more realistic picture of how scientists behave will have 'redeeming social significance'. As Matthews (1994/2014, p. 7) notes, HPS programmes can humanise the sciences, and there is evidence that this makes science programmes more attractive to students, particularly girls. It was this thought which informed the approach of the *Perspectives on Science* course. In this connection it is worth pointing out that both of the projects exemplified above were written by girls, one of whom chose to change from studying History at Cambridge to studying History and Philosophy of Science because she had found it such an enjoyable subject.

[36] The history of science section of the *Perspectives on Science* course was written by Peter Ellis, with John Cartwright contributing a section on the discovery of oxygen.

study material from these subject areas was used as a stimulus for classroom discussion and debate, the aim of these discussions being to engage student interest, as well as to help strengthen students' ability to engage in reasoned argument. To this end, teaching about argument structure was integrated into the programme of discussion of philosophical topics.

Inevitably, a short course in the philosophy of science can only provide an introduction to a select number of the central debates in the field. The written dissertation provided the context for greater depth of study of a specific research question with an historical and/or philosophical dimension. So, for example, a student who found discussion of philosophical questions linked to genetics particularly interesting might choose to study the question of whether autism has a genetic cause, and as part of this, to explore in greater depth the way the concept of causation functions in such a setting.

63.17.4 Teaching and Teachers

In 2007/2008, a team from the Institute of Education at the University of London carried out a research study of *Perspectives on Science* (Levinson et al. 2008, 2012). The researchers noted that discussion topics mostly tended to involve ethical questions, which tend to be very effective in helping students engage in discussion as the substantive knowledge needed is quite accessible. They noted that diversity, passion and extreme viewpoints amongst students can be productively harnessed. They also noted that there were far fewer projects which focussed on the history of science (a point further discussed in Bycroft 2010) and urged that teachers should try to address more challenging philosophical and historical issues as students' discussion skills develop.

The research study of the *Perspectives on* Science course echoed others (Monk and Osborne 1997) in noting teachers' problematic lack of knowledge of the philosophy of science. This is an area which teachers who felt that they would like further training identified as the one in which they were most in need of additional support.

63.17.5 Implications for Science Education

The *Perspectives on Science* course ran as an AS pilot between 2004 and 2008. In the final pilot year, 31 UK schools taught the course. As a small-scale programme, the question arises as to whether it constitutes another instance of the observed failure of programmes in the history and philosophy of science to contribute to mainstream science education (Monk and Osborne 1997).

Despite its small size, the *Perspectives on Science* course has some claim to have had an influence on wider educational developments and to have the potential to contribute to mainstream science education. This is mainly because of the part it has played in shaping the *Extended Project Qualification*, which was launched nationally in 2008 and which by 2011 attracted almost 25,000 student entries. This qualification provides post-16 students with an opportunity to carry out a major research project on a topic of their own choice. The *Perspectives on Science* course served as a model for the dissertation unit of the *Extended Project* offered by the Edexcel awarding body. A number of elements of the *Extended Project* dissertation model, namely, an emphasis on the writing of an historically focussed literature review, with critical evaluation of source objectivity and reliability and consideration of context, use of philosophical reasoning in addressing issues which are not susceptible to empirical resolution and exploration of ethical issues arising from scientific and technological developments, derived directly from the approach piloted in the *Perspectives on Science* course.

The *Extended Project* also offers a Field Study/Investigation unit, and this unit provides an opportunity for in-depth scientific investigation, with the construction of hypotheses, data collection and analysis which involves exploration of the extent to which data supports hypotheses. Significantly for the purposes of teachers wishing to integrate elements of the history and philosophy of science within their science teaching, *Extended Project* investigations also require students to show awareness of the wider context of their research, and this can lead them in the direction of historical study or exploration of social, economic or ethical issues related to their investigation. So, for example, one student carried out an investigation with the title 'Global warming: Anthropogenic or Astronomical', in which he researched the recent history of one aspect of debate about the causes of climate change then went on to experimentally test the hypothesis that cloud cover could be affected by cosmic ray flux. Another student carried out an investigation under the title 'What is the most effective method of producing aspirin and is it justifiably described as a 'wonder drug'?', in which she researched the history of the development of aspirin, explored the controversy about who first synthesised acetylsalicylic acid, examined the way aspirin is presented in the media, discussed problems of access to aspirin in less economically developed countries and then carried out a laboratory comparison of the yield of two different synthetic pathways to a precursor of aspirin.

Currently, the number of students who are writing *Extended Projects* in which scientific questions are explored in this contextual manner, with consideration of ethical, historical and philosophical aspects, or with links to empirical work, is not large. But it is significant that this work is beginning to take place as part of mainstream science education, at least at post-16 stage. Monk and Osborne (1997) noted that science teachers will begin to engage with the history and philosophy of science only insofar as it can be shown that this engagement contributes to their students' examinable knowledge and skills. As a result of the development of the *Extended Project,* informed by the *Perspectives on Science* course, there exists a national qualification, the assessment criteria of which have been shaped to be conducive to gaining a deeper understanding of scientific topics, as well as others, by means of historical and philosophical enquiry.

63.18 Conclusion

Despite 50 years of exploration and over 20 years of the national curriculum's treatment of the history of science and the nature of science, there is still no consensus in England about the ideas-about-science that should feature in the curriculum. This is shown, for example, by the public debate that followed the introduction of 'how science works' into the curriculum (Perks 2006). Over the period covered by this chapter, it is notable how long it has taken to bring about systemic change.

Despite the slow pace of change, much has been learned about the conditions needed to introduce teaching about the history and nature of science into mainstream science education. One lesson is that it is important that new courses lead to recognised qualifications even during the pioneering phase. It is also important that methods of assessment are devised that reward teaching and learning in line with new aims. A significant reason why it has taken so long to embed teaching about the nature of science into the everyday thinking and practice of science teachers is that it has proved difficult to specify the intended learning outcomes in language that is widely understood and accepted. This has the consequence that it has been difficult to devise assessment items for examinations that encourage good practice in schools.

Another barrier to change has been that teaching about the nature of science can be a formidable challenge to those teachers whose own education and training has not helped them to reflect on the history and philosophy of science themselves. At the very least, dissemination of ideas and teaching methods beyond an initial group of enthusiasts depends on the production of high-quality resources in print and other media. In England the production of such resources, especially in the early stages, when their commercial value was in doubt, has been heavily dependent on support from number of charitable foundations. Opportunities for professional development are important too, but they have generally not been provided on a large enough scale to change the general culture of science teaching.

The English experience shows that it is crucial that the course content and teaching methods are rooted in the practical realities of school classrooms. All the early developments described in this chapter grew out of the interests of teachers. Subsequently some of these teachers moved into curriculum development and into universities, without losing interest in the field. Thus grew up a strong and enduring partnership between practitioners in schools and academics in higher education. The footnotes and bibliography for this chapter reflect the fruitfulness of this collaborative approach with their mixture of references to a great variety of tried-and-tested resources as well as references to research papers and scholarly articles.

An extraordinary feature of the science curriculum in England since the introduction of the first national curriculum in 1989 has been the rapidity of change. New versions of the science curriculum for students aged 14–16 were published in 1989, 1991, 1995, 2000 and 2004. The curriculum for those aged 11–14 was also revised in 2006. Further change is underway following the election of a new government in 2010. This government has abolished the curriculum authority, QCA, so that changes to the curriculum are now under the direct oversight of ministers. This is a complete reversal of the situation in the 1960s when politicians kept out of 'the secret garden'[37] of the curriculum.

In its first statements about the new science curriculum, the Department for Education[38] has stated that one of the aims should be that students develop their understanding of the nature, processes and methods of science. However, the approach is likely to change and the term 'how science works' will almost certainly disappear from official documents. An early draft of the curriculum suggests that in the future students should develop their understanding of these aspects of science through practical activity.

The development of many of the initiatives discussed here derived from the desire to create curriculum space for discussions of questions about the history and philosophy of science to be taken further. The *Perspectives on Science* approach pioneered the use of such discussions as a catalyst for independent student enquiry and project work. It has recently been argued (Taylor 2012) that teaching itself should be seen as a philosophical activity and that, to counter the tendency of assessment to determine pedagogy, teachers should focus more on equipping students to think critically and enquire independently in all areas of study. It has also been argued that conceptual understanding, which is fundamental to all learning, presupposes a grasp of the historical and philosophical matrix within which scientific knowledge exists (Blackburn 2010). If these arguments can be sustained, in the teeth of an educational culture dominated by the pressures of high-stakes assessment and accountability measures, it may yet come to be recognised that ideas emerging from the history and philosophy of science teaching community can be beneficially applied to mainstream science education.

Acknowledgments We are grateful for advice from Peter Ellis, Edgar Jenkins (University of Leeds, UK), Robin Millar (University of York, UK), Jonathan Osborne (Stanford University, USA), Michael Reiss (Institute of Education, University of London, UK) and Jim Ryder (University of Leeds, UK). We are also grateful to Ralph Levinson, Michael Hand and Ruth Amos at the Institute of Education, University of London, UK, for sharing their research findings ahead of publication.

References

Adey, P. (2001). 160 years of science education: an uncertain link between theory and practice *School Science Review*, 82 (300), 41–46

Aikenhead, G. (1987). Student beliefs about Science Technology Society: four different modes of assessment, *Science Education*. (71) 145–161.

[37] The term 'secret garden' was popularised in connection with a speech by the Labour Prime Minister Jim Callaghan in 1976. He started a debate that led in time to the national curriculum. He argued that it should not be teachers alone that determine the curriculum but that parents, learned and professional bodies, representatives of higher education and both sides of industry, together with the government, all have an important part to play in formulating and expressing the purposes of education and standards.

[38] The latest information about the national curriculum is published on the Department for Education website: http://www.education.gov.uk/schools/teachingandlearning/curriculum/nationalcurriculum

Aikenhead, G. (1994) What is STS Science Teaching? in Solomon, J. & Aikenhead, G. eds. *STS Education. International Perspectives on Reform* (pp 47–59). New York: Teachers College Press.

ASE (1979), *Alternatives for Science Education*, Association for Science Education

Black, P. (1986). Integrated or co-ordinated science? *School Science Review* 67 (241), 699–681.

Blackburn, S. (2010). Science, history and philosophy. In Derham, P. and Worton, M. (eds) *Liberating Learning; Widening Participation*, University of Buckingham Press, 49–53

Brock, W. H. (1989). History of Science in British Schools: Past, Present & Future. in Shortland, M. & Warwick, A. eds. *Teaching the History of Science* (pp 30–41). Oxford: Basil Blackwell.

Brush, S.G. (1974) Should the History of Science Be Rated X? *Science*, vol 183 pp. 1164–1172

Burden, J., Campbell, P., Hunt, A. & Millar, R. (2007). *Evaluation of the twenty first century science pilot* (Evaluation Report). Retrieved June, 2012, from http://www.nuffieldfoundation.org/pilot-evaluation)

Bycroft, M. (2010) Perspectives on Science *Newsletter of the History of Science Society,* 39(1), 16–19

Dearing, R. (1993). *The National Curriculum and its assessment: final report*. London, SCAA.

Department of Education and Science (1985) *Science 5–16: a statement of policy*, HMSO, London.

Department of Education and Science. (1989). *Science in the National Curriculum.* Her Majesty's Stationary Office. (Accessed in March 2012 from the National STEM Centre eLibrary at http://stem.org.uk/rx4rb.)

Department of Education and Science. (1991). *Science in the National Curriculum.* Her Majesty's Stationary Office. (Accessed in March 2012 from the National STEM Centre eLibrary at http://stem.org.uk/rx4ur)

Department of Education and Science. (1995). *Science in the National Curriculum.* Her Majesty's Stationary Office. (Accessed in March 2012 from the National STEM Centre eLibrary at http://stem.org.uk/rx4uu)

Department for Education and Employment/Welsh Office/Department of Education for Northern Ireland. (1996). *Review of Qualifications for 16–19 Year Olds (The Dearing Review).* London/Cardiff/Belfast.

Donnelly, J. (2001). Contested Terrain: The Nature of Science in the National Curriculum for England and Wales. *International Journal of Science Education* 23, 181–96.

Fleming, R. (1986). Adolescent reasoning in socioscientific issues. *Journal of Research in Science Teaching.* 23 (8), 677–688.

Giere, R. (1991). *Understanding Scientific Reasoning (3rd ed.).* Fort Worth, TX: Holt, Rinehart and Winston.

Gott, R. & Roberts, R. (2008) *Research Report: Background to Published Papers* (Accessed in June 2012 from http://www.dur.ac.uk/education/research/current_research/maths/msm/understanding_scientific_evidence/)

Hind, A., Leach, J. & Ryder, J. (2001). *Teaching about the nature of scientific knowledge and investigation on AS/A level science courses*, Technical Report from the Centre for Studies in Science and Mathematics Education, University of Leeds. (Retrieved in November 2011 from http://www.nuffieldfoundation.org/project-team)

Honey, J. ed. (1990). *Investigating the Nature of Science*. Published for the Nuffield-Chelsea Curriculum Trust by Longman Group UK Books. (Accessed in March 2012 from the National STEM Centre eLibrary at http://stem.org.uk/rx3a7.)

Hunt, A. (1988). SATIS approaches to STS. *International Journal of Science Education.* 10 (4) 409–420.

Hunt, A. (1994) STS Teaching in England. In Solomon, J. & Aikenhead, G. (eds.) *STS Education. International Perspectives on Reform* (pp 68–74). New York: Teachers College Press.

Hunt, A. ed. (2010). *Ideas and Evidence in Science: lessons from assessment*, SCORE (Accessed March 2012 from http://www.score-education.org/policy/qualifications-and-assessment/how-science-works)

Hunt, A. ed (2008). *AS Science in Society*. Heinemann/Pearson Education, Harlow UK.

Hunt, A. (2011). Five decades of innovation and change. In Hollins, M. (ed) ASE Guide to Secondary Science Education, Association for Science Education, pp 12–22

Inner London Education Authority. (1987). *Science in Process*. Heinemann Educational Books.

Jenkins, E. W. (1990). The history of science in British schools: retrospect and prospect. *International Journal of Science Education*, 12 (3), 274–281.

Jenkins, E. W. (1994). HPS and school science education: remediation or reconstruction?. *International Journal of Science Education*. 16 (6), 613–623

Leach, J. (2002). Teachers' views on the future of the secondary science curriculum. *School Science Review*, 83(204), 43–50.

Leach, J., Hind, A. & Ryder, J. (2003). Designing and evaluating short teaching interventions about the epistemology of science in high school classrooms. *Science Education, 87*(3), 831–848.

Levinson, R., Hand, M. & Amos, R. (2008). *A Research Study of the Perspectives on Science AS Level Course*. London: Institute of Education, University of London

Levinson, R., Hand, M. & Amos, R. (2012) What constitutes high quality discussion skills in science? Research from the Perspectives on Science course. *School Science Review*, 93 (344), pp. 114–120.

Lewis, J. (1981). *Science in Society*. (Teacher's Guide and Readers). Association for Science Education/Heinemann Educational Books. (Accessed in March 2012 from the National STEM Centre eLibrary at http://stem.org.uk/cx9s.)

Lewis, J. (1987) Teaching the relevance of science for society. In Lewis, J. & P. J. Kelly (eds) *Science and technology education and future human needs*, Pergamon London pp 19–24.

Matthews, M. (1994/2014). *Science teaching: the role of history and philosophy of science*. London: Routledge

Meadows, D. H. (1972). *The Limits to Growth: A Report for the Club of Rome's Project on the Predicament of Mankind*, Earth Island

Millar, R. & Driver, R. (1987). Beyond processes. *Studies in Science Education*, (14) 33–62.

Millar, R. (1996). Towards a science curriculum for public understanding. *School Science Review*. 77 (280), 7–18.

Millar, R. (1997). *Student's Understanding of the Procedures of Scientific Enquiry* in Connecting Research in Physics Education with Teacher Education, International Commission on Physics Education. (Accessed March 2012 from http://www.physics.ohio-state.edu/~jossem/ICPE/C4.html)

Millar, R. & Osborne, J. (1998). *Beyond 2000: Science education for the future*. (Accessed in March 2012, from http://www.nuffieldfoundation.org/beyond-2000-science-education-future.)

Millar, R. (2000). Science for public understanding: Developing a new course for 16–18 year old students. In R.T. Cross & P.J. Fensham (Eds.), *Science and the citizen for educators and the public*. Melbourne: Arena Publications, pp. 201–214.

Millar, R. & Hunt, A. (2002). Science for Public Understanding: a different way to teach and learn science. *School Science Review*, 83 (304), 35–42

Millar, R. (2006). Twenty first century science: Insights from the design and implementation of a scientific literacy approach in school science. *International Journal of Science Education*, 28 (13), 1499–1521.

Millar, R. (2013). Improving science education: Why assessment matters. In Gunstone, R., Corrigan, D., & Jones, A. (Eds.). *Valuing assessment in science education: Pedagogy, curriculum, policy*. Dordrecht: Springer.

Monk, M., & Osborne, J. (1997). Placing the History and Philosophy of Science on the Curriculum: a model for the development of pedagogy. *Science Education*. 81(4), pp 405–424.

Nuffield Foundation. (1966a). *Nuffield Biology Teachers' Guide I*. Longmans/Penguin Books (Accessed in March 2012 from the National STEM Centre eLibrary at http://stem.org.uk/cxhs.)

Nuffield Foundation. (1966b). *Nuffield Chemistry Introduction and Guide*. Longmans/Penguin Books. (Accessed in March 2012 from the National STEM Centre eLibrary at http://stem.org.uk/cxf3)

Nuffield Foundation. (1966c). *Nuffield Physics Teachers' Guide III*. Longmans/Penguin Books. (Accessed in March 2012 from the National STEM Centre eLibrary at http://stem.org.uk/cxqc)

Nuffield Foundation. (1966d). *Nuffield Physics Teachers' Guide V*. Longmans/Penguin Books. (Accessed in March 2012 from the National STEM Centre eLibrary at http://stem.org.uk/cxqc)

Ofqual (republished annually) GCE AS and A level subject criteria for science (Accessed in March 2012 from http://www.ofqual.gov.uk/downloads/category/191-gce-as-and-a-level-subject-criteria)

Osborne, J., Duschl, R. & Fairbrother, R. (2002a). *Breaking the mould? Teaching Science for Public Understanding*. Nuffield Foundation (Accessed in March 2012 from the eLibrary of the National STEM Centre at http://stem.org.uk/rx38b.)

Osborne, J., Ratcliffe, M., Bartholomew, H., Collins, S. & Duschl, R. (2002b). EPSE Project 3: Teaching pupils 'ideas-about-science'. *School Science Review*, 84 (307), 29–33.

Osborne, J. & Ratcliffe, M. (2002). Developing effective methods of assessing 'Ideas and Evidence' *School Science Review*, 82 (305), 113–123

Osborne, J., Ratcliffe M., Collins S., Millar R. & Duschl R. (2003). What 'ideas-about-science' should be taught in school science? A Delphi study of the expert community. *Journal of Research in Science Teaching*, 40 (7), 692–720.

Osborne, J., Eduran, S. & Simon, S. (2004). Enhancing the quality of argument in school science. *Journal of Research in Science*, 41(10), 994–1020.

Osborne, J., Erduran, S. & Simon, S. (2004). *IDEAS pack* (in-service training materials, a resources file and disc with video clips of argument lessons), King's College London.

Osborne, J., Ratcliffe, M., Collins, S. & Duschl, R. (2006). Specifying Curriculum Goals. In Millar, R., Leach, J., Osborne, J. & Ratcliffe, M. *Improving Subject Teaching: Lessons from research in science education*, Routledge, pp 27–43.

Osborne, J. (2007). In *Praise of Armchair Science Education*, speech at the Annual Conference of the National Association for Research in Science Teaching (Accessed in June 2012 at http://www.kcl.ac.uk/content/1/c6/01/29/36/joconference.)

Perks, D. (ed) (2006) *What is Science Education For?* Institute of Ideas

Perspectives on Science Project Team. (2007a). *Perspectives on Science Student Book*, Heinemann, Oxford.

Perspectives on Science Project Team. (2007b). *Perspectives on Science Teacher Resource File,* Heinemann, Oxford.

Pumfrey, S. (1991). History of science in the National Science Curriculum: a critical review of resources and their aims. *British Journal of the History of Science*. 24, 61–78.

QCA (Qualifications and Curriculum Authority). (2005). Evaluation and analysis of the 21st Century science pilot GCSEs. London: QCA.

QCA (Qualifications and Curriculum Authority) (2006). *Written evidence from QCA to the House of Lords Science and Technology Select Committee* (Science Teaching in Schools Report). (Accessed in March 2012, from http://www.publications.parliament.uk/pa/ld200506/ldselect/ldsctech/257/257we29.htm.)

Reiss, M.J. (1999). Teaching ethics in science. *Studies in Science Education*, 34, 115–140

Reiss, M.J. (2008). The use of ethical frameworks by students following a new science course for 16–18 year-olds. *Science & Education*, 17, 889–902 http://ioe-ac.academia.edu/MichaelReiss/Papers/482897/The_use_of_ethical_frameworks_by_students_following_a_new_science_course_for_16-18_year-olds)

Ryder, J. & Banner, I. (2011). Multiple Aims in the Development of a Major Reform of the National Curriculum for Science in England. *International Journal of Science Education*, 33 (5), 709–725.

Schools Council. (1973). *Patterns Teachers' Handbook*. Longman/Penguin Books. (Accessed in March 2012 from the National STEM Centre eLibrary at http://stem.org.uk/cxug)

SCORE (2009) *GCSE Science 2008 Examinations report*. (Accessed in June 2012 from: http://www.score-education.org/policy/qualifications-and-assessment/gcse-science)

Screen, P. (1987). *Warwick Process Science*. Ashford Press Publishing.

Simon, S. (2011) *Argumentation*. In Toplis, R. (ed) *How Science Works, Exploring effective pedagogies and practice,* Routledge, London and New York

Solomon., J. (1983). *Science In a Social CONtext.* (Teachers' Guide and a series of Readers), Association for Science Education/Blackwell. (Accessed in March 2012 from the National STEM Centre eLibrary at http://stem.org.uk/cx9z)

Solomon, J. (1988). Science, technology and society courses: tools for thinking about social issues. In *International Journal of Science Education,* 10 (4) 379–387.

Solomon, J. (1991). *Exploring the Nature of Science at Key Stage 3.* Blackie and Son Ltd. (Accessed in March 2012 from the eLibrary of the National STEM Centre at http://stem.org.uk/cxfb)

Solomon, J. (undated). *Exploring the Nature of Science at Key Stage 4.* Association for Science Education. (Accessed in March 2012 from the eLibrary of the National STEM Centre at http://stem.org.uk/cxfb)

Stevens P. (1978). On the Nuffield Philosophy of Science. *Journal of Philosophy of Education.* 12, 99–111.

Swinbank, E. & Taylor, J.L. (2007). Designing a course that promotes debate: the Perspectives on Science (PoS) AS level model. *School Science Review*, 88(324) 41–48.

Taylor, J.L. (2012). *Think Again: A Philosophical Approach to Teaching*, Continuum, London and New York

Toplis, R. (2011) *How did we get her? Some background to How Science Works in the school curriculum.* In Toplis, R. (ed) *How Science Works, Exploring effective pedagogies and practice,* Routledge, London and New York

Toulmin, S. (1958). *The Uses of Argument.* Cambridge: Cambridge University Press.

Travis, A. (1993). *The Rainbow Makers.* Associated University Press.

Twenty First Century Science project team. (2003). *21st Century Science* – a new flexible model for GCSE science. *School Science Review*, 85(310), 27–34.

UYSEG (University of York Science Education Group). (2001). *QCA Key Stage 4 Curriculum Models Project. Final Report.* London: Qualifications and Curriculum Authority.

Young, M. (1971). An Approach to the Study of Curricula as Socially Organized Knowledge. In Young,

Ziman, J (1994) The Rationale of STS Education is in the Approach. In Solomon, J. & Aikenhead, G (eds) *STS Education: International Perspectives on Reform*, Teachers College Press, New York pp. 21–31

Dr. John L. Taylor is Head of Philosophy and Director of Critical Skills at Rugby School. He studied Physics and Philosophy at Balliol College, Oxford before going on to do research in Philosophy and teach Philosophy of Science at the university. In 1999 he moved to Rugby School to teach physics. He has directed the Perspectives on Science project. He is a Chief Examiner for the Extended Project and a Visiting Fellow of the Institute of Education.

Andrew Hunt is science consultant to the eLibrary of the National STEM Centre. He studied Natural Sciences at Cambridge before teaching in four contrasting schools over 20 years. During that time he was also a contributor, as writer and examiner, to the development and assessment of the Nuffield Chemistry courses. Subsequently he took on leading roles in a series of curriculum development projects for the Nuffield Foundation, the Association for Science Education and the University of York Science Education Group. He was the first director of the Nuffield Curriculum Centre until 2007 where he was responsible for oversight of major projects in science, technology and mathematics. He is author and editor of a variety of chemistry textbooks and multimedia resources. He is chair of examiners for the AQA A-level Science in Society course. In 1988, Andrew was awarded the Royal Society of Chemistry Award in Chemical Education. He received an honorary doctorate at the University of York in 2009.

Chapter 64
Incorporation of HPS/NOS Content in School and Teacher Education Programmes in Europe

Liborio Dibattista and Francesca Morgese

64.1 Government Policies and Recommendations for the Teaching of Science with the Historical-Critical Approach

64.1.1 Europe

The crisis of scientific vocations has been a pressing subject in the agenda of developed countries, and the relationship between the *crisis of scientific education and social, political and economic development* has been widely acknowledged. The question had already arisen in the White Paper *Teaching and Learning: Towards the Learning Society* (European Commission 1995): facing rapid changes, the widening of the exchanges to a world dimension, the rise of the information society and rapid progress in science and technology. There was a *paradoxical* reaction by European citizens that would see as a *threat* and with *irrational fear* the scientific and technological innovations. These should instead have been considered as instruments for the acquisition of new competences and competitiveness on the job market.

The 1995 European Commission document hoped for the involvement of initiatives and actions for the diffusion of culture and scientific and technological information that might emphasise the value of science and technology for the progress of humanity. Promoting *general culture* and *scientific culture* in this case was meant as a support for the society of knowledge: 'Clearly this does not mean turning everyone into a scientific expert but enabling them to fulfil an enlightened role in making choices which affect their environment and to understand in broad terms the social implications of debate between experts. There is similarly a need to make everyone

L. Dibattista (✉) • F. Morgese
Centro Interuniversitario Seminario di Storia della Scienza,
Università degli Studi "Aldo Moro", Bari 70100, Italy
e-mail: liborio.dibattista@uniba.it; francesmorg@yahoo.it

M.R. Matthews (ed.), *International Handbook of Research in History, Philosophy and Science Teaching*, DOI 10.1007/978-94-007-7654-8_64,

capable of making considered decisions as consumers' (p. 11). Traditional science teaching, aiming at the mastery of a strictly logic order, of the deductive system, of abstract notions among which mathematics dominate, seems to paralyse and to make a passive subject of the learner, suffocating his imagination.

The finger was pointed at the tendency of teaching to 'present[ing] the world [...] as a completed construction', and member states were invited to promote initiatives for the introduction of the history of science and technology as part of science education.

The debate on education *tout court*, and particularly on scientific and technological education, was dealt with later by the commissions of a number of European Councils. The one in Lisbon in 2000 (European Council Lisbon 2000) identified for the decade 2000–2010 the goal of making European economy *knowledge based*: 'the most competitive and dynamic knowledge-based economy in the world, capable of sustainable economic growth with more and better jobs and greater social cohesion', confirming the relationship between education, training/growth and European economic development. After Lisbon, the new 'noticeable' issue has been to promote the *candidateship of scientific and technical studies*, one of the objectives discussed in the European Council of Stockholm 2001 where it was asserted that European competitiveness needs a number of mathematics and science experts and it is dangerous that these studies are deserted by European youths, who show a negative attitude towards them and an inferior learning to expectations (European Council Stockholm 2001).

The need to *attract more students to technical and scientific studies* has become from that moment one of the *leitmotif* of the European documents: in Barcelona (European Council Barcelona 2002), for example, they proposed a 'general renewal of pedagogy' and the use of 'development strategies aiming at the performance of schools in encouraging pupils to study natural science, technology and mathematics and in teaching these subjects'. In Brussels (European Council Brussels 2003), the objective to be achieved was an increase of 15 % in the number of graduates in mathematics, science and technology. Subsequent European Councils would monitor and record the implementation of the Lisbon strategy and the achievement of the objectives proposed in the preceding years.

In this panorama are included the huge survey campaigns on the level of learning of European students, to which Italy too has been participating for at least 30 years (INVALSI 2008).

In short, these documents strongly stressed *education/economic development* and the need to for scientific culture to *renew its methods* and *teaching practices*. On closer look, however, there is not a precise indication with regard to the strategies to follow in order to achieve such renewal.

The first document to express more practical insight on the subject was the 2004 report released by the European Commission, *Europe needs more scientists!* (European Commission 2004), a series of recommendations aimed at increasing European human resources in the fields of science and technology. For the first time, a finger is pointed extremely clearly at the 'perception' of science that students form during their school years and which seems to be one of the most significant reasons for young people's difficulty in imagining themselves working in the future in a scientific career. School, by continuing to insist on 'counter-intuitive concepts and abstract ideas with no relevance to their daily lives', continues to immerse students in a context void of any real and deep understanding of science. Member states are invited to enrich their science curricula with wide-ranging interdisciplinary relationships between science, technology and other disciplines, with historical considerations which are neither stereotyped nor anecdotal and with wide-ranging and significant reflections on the nature of science (p. 138).

The 2005 survey called *Special Eurobarometer: Europeans, Science and Technology* had pointed out that 50 % of the European citizens interviewed agreed with the statement: 'science classes at school are not sufficiently appealing' (European Commission 2005, p. 99, 102), so it is the way science is taught in schools that turns off interest towards scientific subjects. The survey devoted a special section to the analysis of *scientific studies and the mobilisation of young people* to monitor European citizens' awareness about the role of science in society and to understand the causes for the loss of interest in scientific studies. What the survey showed was a *lack of understanding* of scientific and technological matters and a general feeling of *distance* from such issues.

Similar conclusions emerge from a survey made in 2006 (OECD 2006; European Commission 2006) on the evolution of young people's *interest* in science and technology: once again, formal scientific education is taken as the cause for the lack of natural curiosity towards science, this being even more serious since 'student choices are mostly determined by their image of S&T professions, the content of S&T *curricula* and the quality of teaching'.

These discussions and statements on scientific education have produced many of the research projects and teaching activities that will be illustrated in the second part of this chapter, one of which will be the description of the ROSE project.

The international[1] comparative project ROSE (the Relevance of Science Education) collected and analysed data on the factors that influence young people's perception of science and technology in scholastic and extra-scholastic contexts and reasons for the choices students make about whether or not they continue to study science. The target population is students towards the end of secondary school (age 15) and the research instrument is a questionnaire. The objective of ROSE is to collect enough information to develop proposals that will affect policy decisions in terms of scholastic curricula, reduce the extent of differences between the actual science and technology curriculum and that which the students would like to learn about and reinforce the relevance, attractiveness and quality of S&T education (Schreiner and Sjøberg 2004).

The reports of numerous countries which participated in the project[2] and a general summary which identifies the main results of the project are currently available (Sjøberg and Schreiner 2010). These reports show widespread interest on

[1] About 40 countries have taken part or are taking part in ROSE. http://roseproject.no/.

[2] http://roseproject.no/publications/english-pub.html; http://roseproject.no/publications/other languages.html. For Italy, see Neresini et al. (2010).

the part of 15-year-olds in the subjects of science and technology, especially for those issues with more easily perceivable links to daily life and future work and professional possibilities. At the same time, knowledge of the importance of science and technology for society seems to be widespread among youth. However, the judgment of the relevance and attractiveness of the scientific disciplines studied at school[3] is more negative in the more highly developed nations than it is in those with a lower (HDI).[4]

Generally, in European countries the scientific subjects studied at school are seen to be important but difficult to understand, and this perception deters students from imagining themselves working in scientific professions in the future: '[…] school science fails in many ways' from the inability to involve students to the inability to arouse their curiosity, from shortcomings in making occupational possibilities perceivable to shortcomings in helping students to appreciate nature (Sjøberg and Schreiner 2010, p. 11). Additionally, it can be seen that students give a low rating to the interest of the contents present in texts and in the curricula. Low to average approval was found for knowledge about famous scientists and their lives, while high values were found for knowledge of the applicative 'context' of science and technology. An important result of the project is that 'there seems to be a need to 'humanize' school science, to show that science is part of history and culture' (Sjøberg and Schreiner 2010, p. 30).

A key document in the field is the so-called Rocard report (European Commission 2007) which finds new strategies to be implemented in teaching through the identification and promotion of *inquiry-based science education* (IBSE) and *problem-based learning* (PBL). This document promotes non-rote teaching based on abstract information and a teaching model based on the *processes of science* and on how it is practised, on concepts *transmitted through concrete experience*, on rich laboratory work, but most of all, an experience in which *learning proceeds through problem-solving*.

Since *problem-solving* implies the need to gather information, to identify the possible solutions, to evaluate in a critical way all the alternatives and show the conclusions, it would allow the students to engage in *active learning*, through the *building* of their own scientific knowledge. Teaching should rely on *concrete examples* as a way of providing access to understanding of the science.

Therefore, school being the most appropriate educational institution for providing training in scientific culture and career counselling, we reckon it is the place where these initiatives should be carried out.

We must clarify that the European documents have just an advisory value for member states and do not have compulsory effects on educational policies.

A detailed report prepared by the Eurydice European Unit on behalf of the Directorate-General for Education and Culture della European Commission (Eurydice 2011) contains a comparative investigation of regulations and official

[3] Sixteen questions in the ROSE questionnaire focused on the assessment of this area (*My science classes*).

[4] Human Development Index.

recommendations on the subject of science teaching in Europe. Significantly, the title of executive summary is 'Countries support many individual programmes, but overall strategies are rare'. In fact:

> Few European countries have developed a broad strategic framework to raise the profile of science in education and wider society. However, a wide range of initiatives have been implemented in many countries. The impact of these various activities is nevertheless difficult to measure ... Most European countries recommend that science should be taught in context. Usually this involves teaching science in relation to contemporary societal issues ... The more abstract issues relating to scientific method, the 'nature of science' or the production of scientific knowledge are more often linked to the curricula for separate science subjects which are usually taught in the later school years in most European countries. (Eurydice 2011, p. 9)

Moreover,

> Context-based science teaching emphasises the philosophical, historical or societal aspects of science and technology, as well as connecting scientific understanding with students' everyday experiences. This approach is considered by some researchers to increase students' motivation to engage in scientific studies, and possibly lead to improved scientific achievement and increased uptake. The science-technology-society approach requires science to be embedded into its social and cultural context. From a sociological perspective, this includes examining and questioning the values implicit in scientific practices and knowledge, looking at the social conditions as well as the consequences of scientific knowledge and its changes, and studying the structure and process of scientific activity. From a historical perspective, changes in the development of science and scientific ideas are studied. From a philosophical perspective, context-based science teaching raises questions regarding the nature of scientific inquiry and evaluates the grounds of its validity. It also recognises science as a 'human endeavour' where imagination and creativity play a role.
>
> Embedding science into its social/cultural context is considered important when teaching because the development of scientific knowledge may be viewed as a social practice which is dependent on the political, social, historical and cultural realities of the time. The process involves examining/questioning the values implicit in scientific practices and knowledge, looking at the social conditions as well as the consequences of scientific knowledge and its changes, and also studying the structure and process of scientific activity. At primary level, this approach is recommended in approximately half of European education systems. At lower secondary level, embedding science into its social and cultural context is suggested in 27 educational systems. The history of science is recommended in less than half of European education systems at primary level. At lower secondary level, the history of human thought about the natural world (from its beginnings in prehistoric times to the present) is suggested in more than half the European countries. The least common contextual dimension in science teaching at ISCED[5] 1 and 2 is the philosophy of science. Only about one third of European education systems at primary level and about a half of countries at lower secondary level suggest addressing questions regarding the nature or validity of scientific inquiry.(ibidem, pp. 65–66)

Figure 64.1 from the Eurydice report (reproduced after this paragraph) illustrates in which countries and at what level of education the *curriculum* for science teaching in primary and lower secondary education refers to science in context, either in terms of the history of science or contemporary societal issues or both.

[5] ISCED: International Standard Classification of Education by UNESCO. (1) Primary education, (2) lower secondary, (3) upper secondary.

◆◆◆

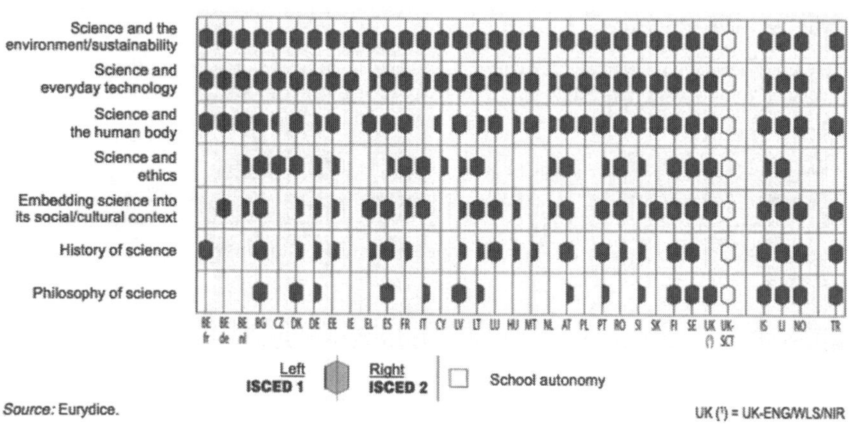

Explanatory note

At ISCED 2 a particular contextual issue is marked as 'recommended' if it is recommended in an integrated science course or in at least one of the three individual science subjects – biology, chemistry or physics. When a certain issue is *not* recommended in *all* science subjects, the subjects are mentioned below.

Science and everyday technology – **Greece** and **Lithuania**: chemistry and biology. **Poland**: physics.
Science and human body - (biology not considered – see text above)). **Denmark, Hungary** and **Slovakia**: chemistry. **Greece**: physics.
Science and ethics – **Slovenia**: biology and chemistry. **Denmark, Spain, France, Cyprus** and **Latvia**: biology.
Social/cultural context of science – **Austria**: physics and biology. **Denmark**: biology.
History of science – **Estonia**: chemistry and physics. **Austria**: biology and chemistry.
Philosophy of science – **Austria**: biology.

Country specific notes

United Kingdom (ENG/WLS/NIR): History of science only in England and Northern Ireland.
United Kingdom (SCT): No recommendations are made in steering documents. However, strong emphasis is placed on interdisciplinary learning within contextualised frameworks and all of the above could be included in teaching and learning.

◆◆◆

Fig. 64.1 Contextual issues to be addressed in science classes, as recommended in steering documents (ISCED 1 and 2), 2010/11

An update of this framework can be found at https://webgate.ec.europa.eu/fpfis/mwikis/eurydice/index.php?title=Home, where the 'Eurypedia (European Encyclopedia on National Educations Systems)' provides current information on national education systems of 38 European countries. The same report provides some news about improvements to science teacher education. In particular, it emphasises the relevance of improvement of teachers' views of the nature of science (NOS):

> In a professional development programme focusing on scientific modelling, teachers improved their views of NOS and inquiry as they expanded their definitions of science from a knowledge-based orientation to a process-based one. Furthermore, an informed understanding of the NOS can be enhanced by the use of meta-cognitive strategies (Abd-El-Khalick and Akerson 2009) and it seems that pre-service teachers who receive explicit instruction in the nature of science as a stand-alone topic are more able to apply their understanding of the nature of science appropriately to novel situations and issues than teachers learning within the context of a case such as for instance, climate change. (p. 104)

Many European countries are conducting programmes and projects for improving science teachers' skills. The survey SITEP (Survey on Initial Teacher Education

Programmes in Mathematics and Science) conducted by the Eurydice unit at the EACEA, which closed in late 2011, demonstrated no significant changes in teachers' approach. As a matter of fact, 'the most important competence addressed in teacher education is the knowledge and ability to teach the official mathematics/science curriculum'. However 'creating a rich spectrum of teaching situations, or applying various teaching techniques, is usually a part of a specific course in both generalist and specialist teacher education programmes. Applying collaborative or project-based learning and inquiry- or problem-based learning is frequently addressed in both types of teacher education programmes'.

The uniformity of the curricula in terms of competence, learning and, as a consequence, of certification and evaluation of the courses of study is something that Europe aspires to, as also shown by the publication of *The European Framework for Key Competencies*[6] (European Parliament 2006) and the *European Qualification Framework* – EQF[7] (European Parliament 2008) – which are two fundamental documents from the viewpoint of EU cooperation in education and lifelong learning. The European viewpoint is to promote collaboration between member states in view of the consolidation of a common European base of key skills and the facilitation of international exchanges and labour mobility. Nevertheless, we are still far from the attainment in Europe of uniformity in terms of science curricula at the level of primary and secondary instruction and also of teacher training.

An important European initiative which intends to enhance the range of quality of research in science education in Europe is ESERA (European Science Education Research Association),[8] formed at the European Conference on Research in Science Education held in Leeds, England, in April 1995. ESERA is an association of European science educators which aims to provide a forum for collaboration in science education research between European countries and to relate research to the policy and practice of science education in Europe. Through the organisation of conferences and the publication of reports, ESERA aims to highlight the major issues facing formal secondary science education, identify similarities and differences between countries and make a series of recommendations for improvement in key areas.

The latest report published, *Science Education in Europe: Critical Reflections* (Osborne and Dillon 2008), contains the conclusions of science educators from nine European countries. There are brief recommendations that analyse the current situation in science education in Europe and lay out the desired prospects for the future. Perhaps the most striking of these is the first recommendation which predicts the obligatory teaching of the nature of science (NOS) in secondary schools, while courses in individual scientific disciplines would become electives available to students who wish to prepare for careers in science. This is exactly the opposite of the current situation in European secondary schools. The reasoning behind this recommendation lies in the perception of standard science teaching as being both of

[6] http://ec.europa.eu/education/lifelong-learning-policy/key_en.htm

[7] http://ec.europa.eu/education/lifelong-learning-policy/eqf_en.htm

[8] http://lsg.ucy.ac.cy/esera/index.html

little use and unappealing, based on the rote learning of abstract concepts, at a time when there is a need to provide students with basic knowledge of how science works and skills that will prepare them to address the problems of modern life and be informed citizens. On the other hand, the type of preprofessional training provided by a curriculum of single scientific disciplines ends up distancing youth from the prospect of undertaking a career in science.

The other recommendations underline aspects of science in context, the need for more female science teachers in the schools and the importance of early intervention since studies have demonstrated that student interest in science already begins to decline after age 14. Additionally, it is hoped that European governments will invest in the recruitment and support of a highly qualified teaching force, to be recognised both economically and socially, and realise the importance of modifying the current systems for evaluating abilities, knowledge and skills, which at this time focus too heavily on content performance tests such as the PISA. *Developing and extending the ways in which science is taught is essential for improving student engagement. Transforming teacher practice across the EU is a long-term project and will require significant and sustained investment in continuous professional development.* Finally, in the opinion of the report's authors, the best way to obtain this kind of science education for students lies in the study of the history of science and reflections on the epistemology and ethics of scientific enterprise.

In May 2011, the *Scientix Conference*, the conference of the community for science education in Europe, was held in Brussels. *Scientix* is a project of *European Schoolnet*, a network of 31 European Ministries of Education, supported by the European Commission. *Scientix* gathers and coordinates information on activities and results in the field of science instruction, including projects financed by the sixth and seventh European Union Framework Programmes, the initiatives of the *Lifelong Learning Programme* of the DGEC (*Directorate-General for Education and Culture*) and national initiatives. The key ideas set forward in Brussels highlighted how great the need is for scientific literacy in Europe, how important it is for educational policies to be coordinated with the needs of the workplace, how it is necessary to develop creative curricula and innovative teaching practices and how all of this is fundamentally related to the recognition (formal, intellectual, economic and political) of the key role played by teachers in European schools. John Holman, at that time director of the British *National Science Learning Centre*, mentioned in the welcome address, among other topics, the issue of how fundamental it is that teaching innovation be conveyed through curricula able to transmit, in addition to scientific contents, a clear idea of the *nature of science*, hence of the epistemology that structures scientific knowledge. If the magic formula is *inquiry-based science learning*, the questions and problems that produce 'science' as their answer are best learned by studying their history and philosophy. However, after this splendid inaugural declaration, epistemology and the history of science disappeared from the presentations at the Belgian conference.

In conclusion, in European policies on science education we can report a decrease in the emphasis given to the historical-philosophical approach during the last

15 years. As a matter of fact, in the final findings of the 1998 Strasbourg conference, conducted by Claude Debru, about the *History of Science and Technology in Education and Training in Europe*, it was expected that:

> all European students in science, technology or medicine should be strongly encouraged to attend a special course in the history of science, technology or medicine at least once during their studies. This teaching should be delivered at undergraduate level, as a element of general culture ... Compulsory courses in history of science and technology should be part of the training received by science teachers at universities, so that they can convey a more dynamic view of science in their subsequent teaching at secondary schools...Students of history, philosophy and the social sciences should receive a specialized course in the history of science and technology at least once during their curriculum. The course should be compulsory in the training of future school teachers ... The teaching of the history of science and technology should be systematically introduced in institutes devoted to the training of secondary-school teachers. (Debru 1998)

Eight years later, during the 2nd International Conference of the European Society for the History of Science, the round table discussion conducted by Claude Debru on the same argument showed no substantial changes; on the contrary:

> We should encourage the interest of DG (Directorate General) Research which lost interest in the Humanities in recent years after some signs of interest for master programmes ... One relevant aspect lies in the intensified economic pressure on universities. This results in a challenge towards several academic disciplines that can have neither many students nor many third-party funds. Unfortunately, among these disciplines is also the history of science (P. Heering). (Kokowsky 2006)

Thirteen years later, sadly we cannot but repeat the account from Pietro Corsi in Strasbourg: 'History of Science: Star of Research, Cinderella of Education' (Corsi 2000).

64.1.2 Italy

The European guidelines have been acknowledged by the member states through initiatives, research and interventions that have contributed to bringing up the question of *scientific culture* in ministerial agendas and have engaged with public opinion, thus better defining the outlines and the content of the problem. In Italy, the fundamental question has been: *which factors are responsible for the falling interest in science and hamper the desire of the young in taking up a career in science?*

A common answer is that sciences per se do not arouse interest, since they are difficult and demanding disciplines (whether this is true or just a stereotype, prompted by an established cultural tradition, is the object of much research) and that a scientific degree, with the commitment and hard work employed in the process, does not necessarily turn into a well-paid and lasting profession.

However, these views do not coincide with those who work in science museums and science centres that are now quite widespread, including in Italy. They witness growing curiosity, interest and enthusiasm in the students who take part in educational visits as extra-school activities (Rodari 2008). This is an interest that seems to arise 'only outside the classes and the departments' (Crespi et al. 2005; Gouthier and Manzoli 2008, p. 143), and it has been predicted that it is rather 'a positive involvement that will gradually fade away with the passing of the years' (Cigada 2007).

These external contexts to the school world show that the problem lies in the *teaching methods*, in the *content*, in its *tools* and in the *methodologies* that are employed in the teaching process of scientific disciplines and also in the *image of science* that formal education conveys. Moreover, the formal organisation of school, compared with the non-formal approach of science centres, is an additional issue. The question seems to be the *lack of correspondence* between practised science and taught science in school. By *lack of correspondence* we mean to say that the way that science is taught and learnt leaves out the reconstruction of the problems that led scientists to carry out their research and the reasons that led them on the path that they undertook. We would like here to cite an expression of Dario Antiseri: 'there is an urgent need, in the teaching of sciences, to make students *stumble* onto the problems' (Antiseri 1977, pp. 111–112). If science is to be intended as *construction*, why are the students denied the access to the way in which this construction happened in the past and still happens now?

Some important experiences show that the active involvement of students in building, elaborating and communicating scientific paths transforms itself into motivation and learning. As an example of this, we can indicate an initiative called *Scienza under 18* (Science under 18), a regional council project, promoted by the Regional School Office of Lombardia (Italy), that has worked on the valorisation of the didactic experimentation and public communication of science by the students themselves for about a decade.[9]

In Italy, an important step, in accordance with the European and Italian debate, has been the creation in 2006 of the *Interministerial Work Group for the Development of the Scientific and Technologic Culture*, recently re-established as *Committee for the development of scientific and technological culture* (July 2010).[10] This institution has the role of drawing up guidelines to support the diffusion of scientific culture and the improvement of the quality and efficacy of education in the field. The documents produced make the same assumption that *science* is *culture*, a *way of knowledge*, and that as such it has to be explicated in the *curriculum*. The renewal of teaching proposed there does not only imply an in-depth study of the content matter of science but also of the discourse on science and scientific work, aiming at the student's building and personal re-elaboration of the knowledge (Documento di lavoro 2007, pp. 3–4) in synergy with the professional development of teachers, for whom a lifelong training is necessary.

The case against superficial factual knowledge teaching leads to the identification of key objectives to be achieved: to update *teaching methods*, which, in fact, are not introducing young students to experimental inquiry; to lead them to the pleasure

[9] http://www.scienza-under-18.org. Accessed 23 January 2011.

[10] http://archivio.pubblica.istruzione.it/argomenti/gst/ and http://www.istruzione.it/web/ministero/scienza-e-tecnologia. Accessed 23 January 2011.

of discovery and to a *taste for problem-solving*; to enhance systematic laboratory practice, to facilitate the development of activities like observation, measuring, handling and building; to diversify and enrich *didactic tools*, being too often only represented by *low-quality textbooks* (Fierli 2004); and to introduce a *humanised approach* with special attention given to *historic contextualisation*, 'to be considered as the understanding of the way and the time in which the concepts have been dealt with and the way the discoveries have been made'.

The aim is to make a *historical approach* to scientific disciplines and their connection to the *humanities* highly recommended; to situate the birth of concepts, of theories and of inventions in their social and cultural context; and to highlight the role of science and technology in the history of mankind.

In short, the finger is pointed against an idea of science presented without an epistemological, historical, dynamical and sociocultural approach, in which the description of the results has the priority over science making, in which scientific theories are presented as an absolute thing, to be learned in rote fashion, in the form of statements and principles, without any interdisciplinary approach. In addition, the new approach stresses the importance of the formative power of scientific culture, something that has long been neglected by the philosophic tradition and in Italian teaching practice (Morgese 2008).

However, though considered a priority in the Italian ministerial courses of study, these remain only hints and programme statements, since no specific document on the historical contextualisation of the *humanistic approach* and on the *connection between the sciences and humanities* has been issued (Morgese 2010b).

We recall the workshop, organised by the *Work Group* called *Per un nuovo liceo 'scientifico' nel XXI secolo. Fondazione culturale e rilevanza sociale* (1–2 April 2009) (Towards a new 'scientific' senior high school for the XXI century. Cultural foundation and social relevance), which constituted one of the preparatory tasks to the secondary school reform currently in force in Italy. The workshop moved in this direction, with the aim of promoting a debate among those in schools, universities and research on the cultural structure and the curriculum of senior high schools that award the science diploma,[11] which are the secondary schools with the greatest number of accepted applicants in Italy. The main issue was the reflection on the cultural and formative value of the science curriculum that it must be based on the '*holism of the project*, on the correct integration between the humanist-social component and the scientific one … not as a cold addition of the two disciplines but as a unified course' (Berlinguer 2009) in order to allow, for the different disciplines, the promotion of an integrated knowledge and the acquisition of scientific culture as the heritage of a community, not just for those who will take up a career in that field (Gouthier 2009a, p. 3, b).

Finally, though only as an initial recognition, in Italy, the importance of the issue was acknowledged through the introduction of references to this matter in the recent

[11] Diploma is the qualification you get at the end of five years in high school; there are two types of Lyceum in Italy: *Liceo Classico* (Diploma in Classical Studies) and *Liceo Scientifico* (Diploma in Sciences).

National Guidelines concerning the specific learning objectives of high school courses. Specifically, it is prescribed for mathematics, physics and the natural sciences that the theories studied should to be set in the historical, philosophical, scientific, social, economic and technological environment within which they developed (see below).

Starting in the 2010–2011 school year, a *reform of the secondary level of education* was implemented in Italy.[12]

The objective of the new system is to revitalise the quality of secondary school education, intended as the ability to provide the student with 'the cultural and methodological tools for an in-depth understanding of reality, so that he can, in a rational, creative, active and critical way, deal with situations, phenomena and problems and acquire knowledge, abilities and skills in line with his capacities and personal choices and suitable for the continuation of higher level studies, for integration into social life and the world of work' (Regulation, Art. 2, Comma 2, cited in MIUR 2010a).

The reform was accompanied by the publication of ministerial reference documents. Two that are essential for a clear understanding of the new system are *profilo educativo, culturale e professionale (PECUP) dello student/educational, cultural and professional student profile (PECUP)* (MIUR 2010c) and *indicazioni nazionali/national indications* for high school courses of study (MIUR 2010b). The PECUP defines the student profile at the conclusion of the high school programme in terms of knowledge, abilities and skills. The document calls for a complete evaluation of all aspects of the student's scholastic work, in particular including 'the study of disciplines from a systematic, historical and critical perspective'. The centrality of the historical and critical perspective can be clearly inferred from the ministerial auspices. Firstly, these aspects are referred to as the result of common learning to be cultivated across the board in all six types of high school in the educational system. In particular, it is explicitly called for in the humanistic historical area,[13] in which the profile calls for the ability 'to contextualize scientific thought, the history of its discoveries and the development of technological inventions in the greater sphere of the history of ideas'. However, it is revealing that this proficiency is absent from the scientific, mathematical and technological area.

Thus, it can be inferred that the promotion of the development of this proficiency is basically entrusted to teachers of history and philosophy and not to teachers of scientific and mathematical disciplines. Secondly, the historical perspective is constantly referred to as the result of learning, even in the distinct high school courses of study. At the end of each high school programme, the student must be able appropriately and knowledgeably to place specific cultural products within their historical and cultural contexts.

---

[12] This reform reordered the secondary school system into six types of academically oriented high schools and divided the professional institutes into two sectors, for a total of six courses of study, and the technical institutes into two sectors, for a total of 11 courses of study.

[13] One of the five areas that the results of cross-disciplinary learning are divided into at the academically oriented high schools. The others are the methodological area; the logical-argumentative area; the linguistic and communicative area; and the scientific, mathematical and technological area

The national indications are the PECUP's guidelines for each discipline: teachers look to this document when creating their course outline so that their students can reach the objectives for learning and the acquisition of skills provided for in high school education. Therefore, the national indications basically focus attention on the disciplines to be studied, analysing them from two points of view: (1) from the point of view of the *competencies*[14] expected at the end of the course of study and (2) from the point of view of the ongoing *specific learning objectives*, aimed at the achievement of competencies, structured in disciplinary units relative to each 2-year period and to the fifth year. At the same time, a few core disciplines are identified: Italian language and literature, foreign language and literature, mathematics, history and sciences.

The guiding principles of the national indications are the following:

1. The *unity of knowledge*, with no separation of the 'notion' from its transformation into a skill. It is explicitly stated that 'Knowing is not a mechanical process, it implies the discovery of something that enters through the sensory process of a person who 'sees,' 'realizes,' 'tests,' and 'verifies,' in order to understand'.
2. *Interdisciplinarity*: the need to build a dialogue between the various disciplines for a coherent and homogeneous profile of cultural processes. The indications take on the task of highlighting the fundamental points of convergence, the historical moments and the conceptual connections that require the joint intervention of more than one discipline to be understood to their true extent.

As in the case of the PECUP, the national indications also make abundant reference to the historical dimension of knowledge, understood as 'reference to a given context' and to the need to promote that dimension in all disciplines of study to ensure that students obtain critical and mindful knowledge.[15]

---

[14] The recommendations of the European Parliament and Council, 23 April 2008, in the European Qualifications Framework for Lifelong Learning, define this competency as 'Proven capacity to use personal, social and/or methodological knowledge, abilities and capacities in situations of work or study and in personal and professional development'.

[15] For mathematics, the student must know how to set the various mathematical theories studied in the *historical context in which they were developed*, understand their conceptual significance and have acquired a *historical-critical vision of the relationships between the main themes of mathematical thought and the philosophical, scientific and technological context*. It is explicitly stated that 'This articulation of subjects and approaches will be the basis for *establishing connections and comparisons* in concepts and methods with other disciplines like physics, natural sciences, philosophy and history'. For physics, at the conclusion of the course of study, the student will have learned the basic concepts of physics, acquiring knowledge of the *cultural value of the discipline and of its historical and epistemological evolution*. For the natural sciences, the subject matter should be proposed following the historical and conceptual development of the single disciplines, both temporally and as per *their links with the entire cultural, social, economic and technological context of the periods in which they were developed*. These links must be made explicit, through underscoring the reciprocal influences in the various spheres of thought and culture. A student completing the humanities high school course of study must also know how to contextualise scientific thought within the humanistic dimension. A student completing the scientific high school course of study must be aware, especially in physics, of the connection between the development of the knowledge of physics and the historical and philosophical context in which it developed.

The introduction in Italy of the reform was accompanied by a national convention[16] and by a series of seminars[17] aimed at teachers and principals, which included presentations by members of the commission for national indications and by representatives of the academic, productive and research fields. The relationships regarding the role and nature of science education in the reform show that the writers of the national indications were aware of the cultural value of the teaching of scientific disciplines with a historical, contextual and epistemological approach, referred to particularly in the presentations by Tommaso Ruggieri (2010), Giorgio Bolondi (2010), Nicola Vittorio (2010a, b) and Andrea Battistini (2010),[18] all of which are available on the website dedicated to the reform. What is missing, however, is any explicit reference to the results of national and international research on the use of the history and philosophy of science in teaching. Also missing are materials in this field of research which could be accessible for teachers on the web portal dedicated to the reform. Therefore, it seems that there is a gap between the national indications and the results of research.

Nonetheless, we would like to focus the reader's attention on one point: within the total hours for scientific disciplines, which are already few, there is no specific course time scheduled for the introduction of contents regarding the history and philosophy of science.

Moreover, the ability of the teacher to design interdisciplinary courses of study and to create connections between mathematics, the scientific disciplines and history and philosophy is entrusted to the free choice, competence and sensibility of the teachers themselves. Therefore, it is not part of their training, nor is it indicated as a practice that must necessarily be part of the teaching profession. The history of science is practically absent from the training of teachers, both pre-service and in-service.

Ministerial Decree 249/2010 (MIUR 2010d) redefined the initial training of Italian teachers at all levels. Therefore, starting in the 2011–2012 academic year, Italian universities instituted degree courses and Active Internship Training (TFA) courses for graduate students preparing to teach in the schools. The scientific field 'history of science and technology' is almost completely missing and is actually completely left out of the course of study for future teachers of history and philosophy at academically oriented high schools. The history of science and technology is taught to teachers who will teach philosophy, psychology and educational sciences in human sciences high schools, while it is an elective course (and with fewer course

[16] National Convention *I nuovi Licei: l'avventura della Conoscenza*, Rome, 11 October 2010. Organising agency: Fondazione per la scuola della Compagnia di san Paolo, http://www.fondazionescuola.it/magnoliaPublic/iniziative/nuovi-licei/presentazione.html.

[17] Available at the previous link.

[18] Giorgio Bolondi, professor of geometry, faculty of economics and business, Università degli Studi di Bologna, and Nicola Vittorio, professor of astronomy and astrophysics, department of physics, Università degli Studi di Tor Vergata, Rome. They are both members of the commission for the national indications http://www.indire.it/lucabas/lkmw_file/licei2010/Decreto%20n.%2026%20del%20%2011%20marzo%202010.pdf.

hours than the mandatory courses) for future teachers of mathematics, sciences and technology in junior high schools (Bernardi 2011).

This appears to be a step back from what was required for the initial training of teachers prior to Ministerial Decree 249/2010. Previously, starting in 1998, teacher training was entrusted to specialised teacher training schools for secondary education (SSIS), a 2-year postgraduate training course required of anyone wishing to become qualified as a high school teacher. In the SSIS programme, study of the history of science and technology was obligatory for teachers of mathematics and scientific disciplines.[19] This experience was interrupted after nine cycles: in 2009, the activity of the SSIS was brought to a stop (Anceschi and Scaglioni 2010).

We wonder, then, how it is actually possible for teachers to teach science with an historical approach when this dimension is absent from their training. Moreover, a lot of questions arise: the guidelines are not based on the results of national or international research on the use of history and philosophy of science in science teaching, and there is a lack of availability of reference materials available for teachers. Finally, the time devoted to science teaching is far from enough to make possible an approach where history and philosophy of science can play an important role.

64.2 The Historical-Critical Approach: Concrete Experiences in Europe and Italy

In this section, we illustrate a few initiatives, projects and teaching activities fielded in Europe and Italy with the aim of carrying out the European recommendations outlined in the previous section of this chapter, and we illustrate the profile of some associations and research institutions dealing with education projects that have HPS/NOS as content.

64.2.1 Europe

In Europe, it is possible to identify a few recent experimental proposals that blend scientific education with the historical-critical approach.

[19] The SSIS Apulia was a special case in that history of science and technology was a cross-discipline course for the teachers of all subjects (Dibattista and Morgese 2011). The aim of the course of 'history of science and technology' in SSIS Apulia was the didactic application of the discipline through the training of both science and humanities teachers. The result of the ten-year experience was the publication of a book, with the collaboration of some trainees of the various SSIS courses, that displays a series of actual proposals, usable case studies for the teaching of science in an interdisciplinary way and is based on the historical-critical approach (Dibattista 2008).

HIPST (*History and Philosophy in Science Teaching*)[20] has organised the collaboration of international research groups in order to produce and develop case studies for teaching and learning science with the historical-critical method oriented towards the discovery of NOS and the inclusion of the production of scientific knowledge in authentic contexts.

The theoretical assumption of the point of departure is that scientific concepts are more easily understood if presented in the historical context of their discovery, rather than presented in the decontextualised and systematic manner typical of the traditional didactic approach. The tools used were the production of teaching materials, in the form of case studies; documentation by the teachers involved in the study; and the formation of a solid network of science teachers, researchers and institutions which disseminate scientific culture, with the objective of collaboration in synergy for the implementation of the project and its follow-up (Höttecke and Riess 2009; Höttecke and Henke 2010, 2011).

In the initial phase of the project (1–16 months), a collection was made of materials related to the current teaching practices and activities in each partner country in the area of science teaching and learning, so as to establish a starting point on which to build the following project phases. The second phase of the project (16–28 months) was dedicated to the creation of a corpus of case studies on the basis of the national needs which had emerged in the previous phase, for example, on the basis of their adaptation to the study programmes in force in each nation. The corpus, translated into the various languages of the partner countries, was distributed and put into practice. The last phase of the project (29–30 months) was the period of the transfer of know-how, the fine tuning and distribution of the materials produced and their evaluation. All of the material produced was made accessible online.[21]

The Catalan Society for the History of Science and Technology (*Societat Catalana d'Història de la Ciència i de la Tècnica* – SCHCT) created a pilot course in the history of science for in-service teachers sponsored by the Department of Education (Departament d'Educació)[22] in Catalonia. The courses,[23] taught online with the Moodle learning management system, were held during 2009–2010 (Science and Technology through History) and 2010–2011 (the History of Mathematics and Science for Secondary Education) academic years, and more

---

[20] The project, financed as part of the Settimo programma quadro (FP7, *The Seventh Framework Programme*, 2007–2013), had a duration of 30 months and concluded in July 2010, http://hipst.eled.auth.gr/. Project participants included ten partners from seven European nations and Israel.

[21] Available on the platform hipstwiki: http://hipstwiki.wetpaint.com/. The section of the corpus of *case studies* developed over the course of the project is at page http://hipstwiki.wetpaint.com/page/hipst+developed+cases.

[22] http://www20.gencat.cat/portal/site/ensenyament

[23] The course was presented at the symposium *La Història de la Ciència i de la Tècnica en l'Ensenyament i en la Formació del Professorat* dell'VIII *Congreso Internacional sobre Investigación en la Didáctica de las Ciencias* – Enseñanza de las Ciencias en un mundo en transformación, Barcelona, 7–10 September 2009, http://ice.uab.cat/congresos2009/eprints/cd_congres/propostes_htm/htm/inici.htm.

than 50 % of those enrolled completed the courses. The same introductory module was used to present the history of science and its use in teaching for the three modules specific to the teaching of mathematics, physics and biology. The greatest achievement of the initiative was that the teachers learned about the effectiveness of making use of primary sources, such as the writings of Galileo, Newton and Darwin. The greatest difficulty lay in the fact that the course was perceived as a university course in the teaching of the history of science, rather than as a tool developed specifically for school teaching (Grapì 2009, 2011; Massa and Romero 2009). The initiative is part of a change which took place in Catalonia between 2007 and 2008 with the adoption of a new curriculum for secondary education[24] which places at the centre of the teaching-learning of science the discovery of NOS, the historical context of scientific knowledge and the relationships between science, technology and society, and that, as a consequence, makes specific teacher training necessary. A similar situation can be found in the new study programme for mathematics, in which the knowledge of a few notions of the historical genesis of key events in mathematics is included (Grapì 2009).

Nevertheless, the history of science is not currently a mandatory discipline in the training of future teachers, and curricula in the history of physics, chemistry, biology or mathematics are optional. The CAP (*Certificado de Aptitud Pedagogica* – the certification for the qualification to teach in secondary school), in force until 2009, included the possibility to attend one or two sessions dedicated to the history of a particular scientific discipline. The current *Master en Formacion de Profesorado* includes in the Especialidad de Matemáticas only an obligatory module on 'History of Mathematics' and one on 'Mathematics, Society and Culture'.

The current Spanish curriculum, issued by the Education Law (2006), includes HPS/NOS contents in compulsory and high school education through two ways: on the one hand, through the definition of the key competence in science as a cross-curricular keystone for the curricula of all science subjects and, on the other hand, through the design of common specific content for science subjects. The specific common contents designed for the curricula of science subjects involve a miscellaneous set of issues that can be grouped into the following categories: searching for information and using information technologies (ICTs); compliance with safety and operating standards in the laboratory; autonomy and creativity; processes of scientific inquiry; and science, technology and society.

Though the former categories include some aspects related to HPS/NOS issues, the latter display the core of HPS/NOS contents for science subjects. These include the recognition of the role of scientific and technological knowledge on social development and people's lives; the recognition of the importance of scientific knowledge to make decisions about objects and about yourself; the assessment of the contributions of the natural sciences to meet the needs of human beings and improve the conditions of their existence; the appreciation and enjoyment of natural and cultural diversity; and the contribution to their conservation, protection and improvement, the recognition of the relations among science (physics, chemistry,

[24] Ley Orgànica 2/2006, http://noticias.juridicas.com/base_datos/Admin/lo2-2006.html.

biology and geology) to technology, society and the environment and the potential applications and implications of their study and learning. Further, the institutionalisation of a specific subject, called Science for the Contemporary World, provides the longest and most systematic list of HPS/NOS content. The subject is compulsory for both science and nonscience students in high school (eleventh grade). It focuses on scientific and technological literacy by addressing current issues that are important for the general citizen.

The situation in higher education degrees is quite complex at the moment; on the one hand, similarly to Italy, it greatly depends on the research interests at each university; on the other hand, it is now changing due to changes to curricula to adapt them to Bologna guidelines, so that until the process is complete, little can be said about it. Finally, the initial training of teachers is also influenced by the development of the Bologna process. In the case of primary teachers, as science is for them just one subject among many other subjects to learn, the HPS/NOS contents are scarce. In the case of secondary and high school teachers (a single group in Spain from the view of initial training), the master's degree must prepare teachers to teach the science curricula, and thus, as the HPS/NOS curricular issues detailed above are compulsory, it must follow that HPS/NOS issues should be part of the master's training.

The ATLAS (Active Teaching and Learning Approaches in Science) group of the School of Primary Education at the Aristotle University of Thessaloniki has created a web tool (atlaswiki) that allows its teacher-users to find (and propose) relevant materials for the creation of teaching units, starting with the history of science (Koulountzos and Seroglou 2007).

Moreover, the Institute of Neohellenic Research of the National Hellenic Research Foundation in collaboration with the Laboratory of Science Education, Epistemology and Educational Technology (ASEL) of the University of Athens, Greece, started the History, Philosophy and Didactics of Science Programme (HPDST) (http://www.hpdst.gr) that is active in publishing journals and monographs of historical and scientific interest and in the organisation of symposia and conferences and has launched the project Hephaestus (Hellenic Philosophy, History and Environmental Science Teaching Under Scrutiny), funded under the FP7, currently ongoing and aiming to improve the activity of the programme in the field of history of science, educational activities and dissemination of scientific information and results.

More actions funded by the European Community through FP7 and are relevant to this review are as follows:

SONSEU (Science on Stage Europe), a European initiative designed to encourage teachers from across Europe to share good practice in science teaching, which publishes *Science in School*, a European journal for science teachers which often proposes papers about history and philosophy of science

MATERIAL SCIENCE, a partnership of six European Universities from Cyprus, Finland, Greece, Italy and Spain for the design and implementation of research-based ICT-enhanced modules on material properties. The project has a strand of educational resources based on the HPS (insights from the history of science and technology).

In France, the group Patrimoine, Histoire des Sciences et des Techniques (PaHST) at the University of Brest, in addition to the study of the scientific and industrial patrimony, is also active in the use of the history of science and epistemology in the teaching of the sciences. The group took part in the project 'Mind the Gap' in the cluster of projects financed by FP7 which attempt to meet the needs highlighted by the aforementioned Rocard report. In 2010, it organised a European workshop, History of Science and Technology Resources and Methods for Inquiry-Based Science Teaching (IBST), and it manages a block of courses at the University of Brest dedicated to HST, IBST and cultural mediation in science (Laubé 2011).

Still in France, the HPM[25] – History and Pedagogy of Mathematics – an international study group on the relations between the history and pedagogy of mathematics affiliated to the *International Commission on Mathematical Instruction* (ICMI), should be noted. By combining the *history* of mathematics with the *teaching and learning* of mathematics, the group aims at stressing the conception of mathematics as a living science, a science with a long history, a vivid present and an as yet unforeseen future. Among the group's activities of particular interest are the satellite meetings at the *International Congress on Mathematical Education* (ICME), organised by ICMI every 4 years with the aim of disseminating and exchanging considerations on and practices in the use of the history and epistemology of mathematics in teaching.

Among the HPM's recent activities, of particular interest was the organisation in Nantes on 4–6 July 2011 of the international conference *European Perspectives in the Use of History in Mathematics Education*, dealing with the developments of research in the field of the use of history in mathematics education, with particular reference to Europe.

Another activity worthy of note is the organisation of the *European Summer University on the History and Epistemology in Mathematics Education* (ESU). These are conferences which are moments of reflection and international exchange on the use of history and epistemology in mathematics education and bring together a network of teachers, researchers in education and historians. The proceedings are a landmark in the evolution of this approach. A high point of the ESU conferences is the teachers' participation in the activities in close collaboration with the results of research. The initiative of organising a Summer University (SU) on History and Epistemology in Mathematics Education belongs to the French Mathematics Education community in the early 1980s.

In France, the network of IREMs[26] – Instituts de Recherche sur l'Enseignement des Mathématiques – organises conferences and seminars that focus on reflection, research and training teachers in how to integrate the historical approach in teaching mathematics. The research group also produces teaching materials (reprints of original sources, reports on classroom activities, teaching plans) which are the result of

[25] For the history of the founding group in 2000, see the document available at http://www.clab.edc.uoc.gr/HPM/HPMhistory.PDF. The group's website is at http://www.clab.edc.uoc.gr/HPM/INDEX.HTM. The HPM Newsletter is at http://grouphpm.wordpress.com/.

[26] http://www.univ-irem.fr/spip.php.

the hands-on classroom experience of teachers who participate in the IREM network. Ongoing activities worthy of mention include the organisation of the international conference *La didactique des mathématiques: approches et enjeux. Hommages à Michèle Artigue,*[27] 31 May–2 June 2012, which included among its topics of discussion consideration of the practices and research in the history and epistemology of mathematics in the teaching of mathematics (Plenary Lecture *Epistemology, History and Didactics*; Atelier *Epistemology and Didactics*).

A group of five European universities (Kapodistrian University of Athens, University of Pavia, University of Oldenburg, University of Cyprus, University of Thessaloniki) has created, as part of the actions of the European Union's Comenius 2.1, the STeT, Science Teacher e-Training project. This project, financed in 2006 as the continuation of a similar programme – the MAP project of 2004 – has created a series of tools which use case studies from the history of science to provide in-service science teachers with innovative materials to help them reconceptualise their views in some important teaching and learning aspects of science education and gradually transform their teaching practice. Currently, the Kapodistrian University of Athens coordinates five other partners (University of Flensburg, University of Brest, Polish Association of Teachers, Diamantopoulos School and the University of Winnipeg) in a new Comenius multilateral project named *Storytelling@Teaching Model* (S@TM) that started in 2011 and will end in 2013. This aims to enhance the professional development of science teachers by the use of case stories from the history of science, building a digital resource kit based on the storytelling teaching method.

The Scientific and Technological Research Council of Turkey (TÜBİTAK) is the leading agency for managing, funding and conducting research in Turkey. Thanks to funding provided by TÜBİTAK's 1001 programme, which funds scientific and technological research projects, the University of Marmara carried out a project for the development of teaching materials specifically developed for the teaching of modules on the history of science in secondary schools and tested its effectiveness on a sample of teachers of scientific disciplines (Irez et al. 2011).

One of the most complex projects for the integration of the history and philosophy of science into the coursework of preuniversity level students is the Perspectives on Science course which has currently been assessed as an Extended Project Qualification, an accepted qualification for university entrance in the UK. Developed by the Centre for Innovation and Research in Science Education, Department of Education, University of York, the course has the main goal of providing students with, in addition to the typical subject matter of the history, philosophy and ethics of science, the development of critical thinking skills typical of the epistemological approach. The goal is not only to prepare them for study in science after they leave school but, more generally, to foster an inquisitive, rational approach to life in general. For this reason, the course does not require acquisition of specific subject matter, but after an introductory phase in which the students learn how to use source materials and develop skills in philosophical and ethical argumentation and logical

[27] http://www.colloqueartigue2012.fr/

reasoning, they must analyse case studies in the history of science and produce an individual research project. This dissertation takes the place of a final examination and must be defended orally. The teaching of the course and the evaluation of the final dissertation is carried out by the teachers from the institutions which have adopted the qualification[28]; this introduces the question of the training of teachers for a type of didactics based on the contextual and cultural dimension of science. This issue was addressed through the creation of teaching materials[29] and an *in-service training* programme.[30] The objectives of the course are to foster the mental skills that make it possible for the students to address science's 'big questions', to develop research and argumentation skills and to be open to ethical debate (Taylor and Swinbank 2011).

Moreover, the University of York, in partnership with the Nuffield Foundation, developed the Twenty First Century Science qualification, a suite of General Certificate of Secondary Education (GCSE) courses that 'meet[s] the needs, through flexible options, of those who will go on to be professional scientists and of those who will not'. The materials are designed to achieve scientific literacy through an understanding of ideas about science and science explanation. Basically, this is the realisation of the wishes from the Nuffield report that we have previously mentioned.

Still in the UK, mention should be made of the HIMED (History of Mathematics in Education) conferences, run by the Education Section of the British Society of History of Mathematics (BSHM). These were established in 1990 and promote the use of history in mathematics education (Fauvel and van Maanen 2000).

Finally, in the catalogue of the initiatives of scientific instruction collected by the STENCIL project (Science Teaching European Network for Creativity and Innovation in Learning (http://www.stencil-science.eu/)), it is possible to find numerous localised activities in European schools which use the history of science as a tool for teaching scientific disciplines. For brevity's sake, we will only point out here *Maths in Wonderland* in Romania, *Maths to Play* and *History, Maths, History of Mathematics* in Italy and *Energy is our Future* in Belgium.

---

[28] In 2008, approximately 30 secondary schools participated in the project; in 2011, more than 700 institutes in the UK are taking advantage of this opportunity.

[29] A *Student Book* (Perspectives on Science Project Team 2007a) and a *Teacher Book* (Perspectives on Science Project Team 2007b) have been produced. Both are organised by *case studies* related to scientific problems and questions which are explored in their historical, epistemological and ethical dimensions and offer ample study materials. For example, the *Student Book* is organised as follows: Part 1: *Researching the History of Science,* Part 2: *Discussing Ethical Issues in Science,* Part 3: *Thinking Philosophically about Science*, and Part 4: *Carrying out a Research Project.*

[30] These are courses carried out during the school year both residentially and not. They introduce the teachers, of both scientific and humanistic disciplines, to the historical, epistemological and contextual approach to science, to the strategies for developing an active and dialogue-based method of teaching, to the development of *case studies* and to the writing of and *coaching* for the final dissertation.

64.2.2 *Italy*

In highlighting the existing situations, we distinguish between university research groups and professional teachers' associations.

64.2.2.1 University Research Groups

The university research groups listed carry out theoretical research on the methodologies for science teaching-learning, also developing operative projects for teacher training and orientation, and for the production of didactic materials.

University of Pavia: Group of History and Didactics

Historically, the oldest and better established group in Italy devoted to this matter is the Group of History and Didactics of the University of Pavia, which has worked with the aim of introducing the history of science in school curricula, especially through conferences and publications (Bevilacqua et al. 2001; Bevilacqua and Fregonese 2000–2003).

The activity of the group is focused on the identification of tools and methodologies that can contribute to the improvement of the teaching-learning of physics and the issues related to the initial and in-service training of junior and senior high school teachers, with an eye to developing innovative approaches to the teaching of physics, including the historical developments in the field of physics and the use of new technologies.

The *Group of History and Didactics* of the University of Pavia participated in the PRIN F21[31] project, *Percorsi di Formazione in Fisica per il 21° secolo/Physics Training Courses for the twenty first Century*, carried out in 2006 by University of Naples 'Federico II' under the scientific direction of Prof. P. Guidoni and in collaboration with various university groups of Italian researchers.

Within the F21 PRIN, the Pavia group was involved in the production of *teaching-learning sequences* (TLSs) for teachers of scientific disciplines, set in the wider panorama of research on TLSs. The TLSs implement the historical approach in the teaching of friction, in the belief that the role of the history of science is particularly effective and justifiable in this specific subject because it helps teachers:

> to clearly position recent developments, which have opened new areas and issues of research. A short historical overview, in addition to looking back at the characters and episodes of the past, serves to draw attention to recent events and future prospects, working together to

[31] PRIN is the acronym which indicates 'research programmes of considerable national interest'. An overview of PRIN F21 is available at http://www.ricercaitaliana.it/prin/dettaglio_prin-2004020419.htm. For the part of the programme regarding the Pavia group, under the scientific direction of Prof. Paolo Mascheretti, see http://fisicavolta.unipv.it/didattica/SeqAttr/xxx.html and http://www.ricercaitaliana.it/prin/unita_op-2004020419_006.htm.

show how this is a subject of current interest and study. It also makes it possible to provide simplified but effective insight into the issues regarding the subject, while learning about its complexity, linked to the diversity of materials and situations, and the theoretical uncertainties, revealed in interpretative controversies which have not yet been entirely resolved.

Additionally, the Pavia group has produced *the Pavia Project Physics – Gateway for the Circulation of Scientific Historical Culture* (http://ppp.unipv.it). The research carried out by the group is divided into three areas:

- Science history and philosophy: to position research on scientific knowledge within its cultural, institutional and social contexts and the context of the philosophies of nature that scientists explore and to more correctly contextualise the products of science
- Science education: research and experimentation with constructivist methodologies of teaching-learning which stimulate students' ability to formulate and resolve problems and to be active creators of their own scientific culture
- Digital technologies: the construction of hypermedia with differentiated approaches for different levels which facilitate the proliferation and personalisation of learning paths in physics

The products created by the group include:

- The analysis of *case studies* of the history of physics from Galileo to the modern day, with particular interdisciplinary focus on the relationships between scientific concepts and philosophical, religious and epistemological concepts
- Restoration and appreciation of primary sources, including the work of identifying and cataloguing collections of scientific tools and library collections
- Development and testing of learning activities for various scholastic levels on the basic concepts of physics, also through the use of ICT tools and multimedia, along with the production and testing of teaching modules for training physics teachers
- The creation of websites and teaching hypertexts with simulations of scientific experiences and the use of two-dimensional and three-dimensional presentations and animations to illustrate the theoretical principles of physics or how tools used in this field work
- A series of books, including essays, studies, catalogues of collections and teaching guides, also available on CD-Rom, exhibits, teleconferences and television programmes

These projects and products were tested as part of the course in physics, chemistry and the natural sciences for high school teachers.

University of Rome 'La Sapienza': Dipartimento di Fisica

Other important contributions in this direction are the activities of the research group of Dipartimento di Fisica of University of Rome 'La Sapienza' about the history of thermodynamics (Tarsitani and Vicentini 1991). Of particular interest is the didactic and research activity of Carlo Tarsitani, professor of the foundations of physics, regarding the history and philosophy of physics (developments in the field of physics in the nineteenth and twentieth centuries, history of quantum physics, conceptual foundations and the philosophical implication of quantum mechanics) and didactics of physics (the study of the conditions that could make possible the effective teaching of the physics of the twentieth century, in the final years of secondary school) (Tarsitani 2009).

University of Bologna: Physics Department

The Bologna group is identified around the didactic and research activity of Silvio Bergia, Grimellini Tomasini and Olivia Levrini. Here, we intend to focus on their considerations regarding the history and philosophy of science as effective tools in the teaching of physics in the SSIS programme (the *teacher training specialising course for high school teachers*) of Bologna. The materials developed for the didactic activity of the SSIS of Bologna concern space-time physics (from classical mechanics to the basic ideas of general relativity) and represent the results of a process of educational reconstruction, in which subjects' aspects are integrated with historical-epistemological and cognitive considerations with the following criteria: privileging the quality of knowledge rather than the quantity of notions to be transmitted, addressing topics and questions of twentieth century physics on the basis of a 'modern teaching' of classical physics, and fostering an image of physics as a 'cultural product' characterised by a coexistence of different interpretations of the same formalism and the interconnections with other cultural fields (Grimellini Tomasini and Levrini 2003).

University of Udine

The second level interuniversity master's degree in 'Didactic Innovation in Physics and Orientation' (M-IDIFO3)[32] (De Ambrosis and Levrini 2010), with headquarters at the University of Udine, is part of the Scientific Degree Programme (PLS) and is one of the most important Italian programmes for the orientation of teachers in the didactics of physics. A part of the didactic programme is dedicated to historical content (20 h of the history of cosmology from antiquity to Einstein and 30 h of laboratory work on the historical evolution of the concept of time).

University of Bari

We want to highlight two further innovative experiences in Italy regarding the teaching of science with an historical approach. They are both projects funded by the Italian Ministry of Education, University and Research (MIUR) in the sphere of funding designated for the dissemination of scientific culture (Law 6/2000) and

[32] Coordinated by M. Michelini, Udine, as part of the Scientific Degree Project.

were both designed and directed by the Centro Interdipartimentale Seminario di Storia della Scienza dell'Università degli Studi di Bari 'Aldo Moro'/Centre for the Interdepartmental Seminar on the History of Science at the 'Aldo Moro', University of Bari. The seminar has, for many years, worked at training teachers, both pre-service and in-service, in the use of the historical and philosophical approach to science education.

The first project, *La storia della scienza va a scuola/The History of Science Goes to School*, was conducted during the 2009–2010 academic year and was a practical experience in introducing the history and philosophy of science into junior and senior high school classrooms in Apulia (Italy). The many schools which participated made use of the historical-scientific teaching modules through the case study approach. In the first phase, the participating teachers were trained by university tutors in how this particular teaching approach works. In the second phase, the teachers taught the modules in their classes, and, finally, these modules were presented at a concluding conference. The effectiveness of the project was also evaluated through questionnaires created specifically for this purpose.

Over 20 in-service teachers of scientific and humanistic subjects and over 400 students participated in the project.

The positive results of the research were the following: efficacy in communicating the scientific subject matter, learner (and teacher) openness to issues regarding the nature of science, the fact that the students gained a more comprehensive view of science and great student enthusiasm for publicly demonstrating the work they had done. Same critical points were the extra time and work that the teachers were required to devote to this project, on the one hand, to prepare the modules and, on the other, because of the lack of ad hoc teaching aids (Dibattista 2010). The project's products are 15 interdisciplinary didactic proposals in the form of case studies described by the teachers in a collective volume (Dibattista 2010), each one including an illustrative analytical file: the disciplines involved in the study case, the type of students it was aimed at, the prerequisites, the cognitive and metacognitive objectives, the methods and tools used, the timeframe, the proposals for verification and a bibliography.

The second project, *Il Racconto della Scienza – Digital Storytelling in Classe/The Story of Science – Digital Storytelling in the Classroom*, was conducted during the 2011–2012 academic year and was the result of a competition held in the junior and senior high schools of the Apulia region. Participants were asked to create multimedia products using the technique of Digital Storytelling to narrate a historical-scientific episode or the story of a scientist. Objectives were to promote the introduction of innovative approaches to teaching scientific disciplines, based on the history and philosophy of science, and solicit the production of highly personalised digital learning environments, created by the users themselves, starting from the specific didactic needs of each group-class and centred on narrative practice. Given the innovative nature of the methodology required to create the products, the teachers in charge of the classes entered in the competition completed a training programme on the history of science and audiovisual technologies, carried out at the University of Bari. Nineteen senior high schools and 13 junior high schools from the Puglia region participated. Over 40 teachers of scientific and humanistic disciplines and over 800 students participated in the project.[33] Twenty-four *Digital Storytelling* courses with historical-scientific content were produced by the schools. The project evaluations, carried out through the administration of questionnaires to teachers and students ex ante and ex post, are currently being processed. The project's final publication will include the description of the creation process of the winning Digital Storytelling courses, starting from the historical-scientific case study chosen as the topic; how the Digital Storytelling course fits into the framework of studies on digital learning environments; and the evaluation of the project, starting from the results of the questionnaires.

A recent project, *Performascienza. Laboratori teatrali di storia della scienza a scuola (Performascienza. Theatre Workshops of History of Science at School)*, involved historians of science and pedagogists of the University of Bari 'Aldo Moro' in the promotion of theatre workshops on history of science case studies in the junior and senior high schools of Bari and in the provinces. The project was carried out by *Scienz@ppeal Association*[34] of Bari and took place in 5 months of the school year 2009–2010. The project applied the dramatisation of *case studies* involving a wide range of actions, such as monitoring the scientific imagination of the teachers and of the students, the teaching practices of the science teachers and the receptiveness of the schools and of the territory in the diffusion of scientific culture. Finally, it involved an evaluation of the efficacy of the case study methodology, through the means of narration and drama, in building an interdisciplinary and complex *scientific literacy*. The products of the project are three videos which illustrate the process of carrying out the project in the three participating schools and a final publication (Morgese and Vinci 2010) which contains the narration of the case studies realised through theatre in the schools, the evaluation of the experience by the tutor teachers in each school, the evaluation of the project on the basis of the results of the questionnaires and a discussion of how the project fits into the framework of studies on the historical approach in science teaching.

64.2.2.2 Professional Associations of Teachers

Teacher associations listed below primarily perform guidance and training of teachers.

ANISN

The *Associazione Nazionale degli Insegnanti di Scienze Naturali/National Association of Natural Sciences Teachers* (ANISN)[35] is addressing scientific education using the historical-critical method.

---

[33] The winners received their awards on 16 December 2011 at the Bari Cittadella Mediterranea della Scienza at an event attended by over 350 students and their teachers. The list of the digital storytelling winners and the reason for which they were chosen are available at www.scienzappeal.com.

[34] www.scienzappeal.com

[35] http://www.anisn.it/

ANISN is an association of teachers, scientists and enthusiasts founded in 1979 with the aim of promoting and increasing the professionalism of natural science teachers. Today, it is an authoritative organ that interfaces with institutions to promote the quality of science teaching in Italy and the exploitation of the best teaching practices. Periodically, the association organises conferences aimed particularly at teachers and publishes information about its activities through reports, newsletters and its website, which contains a wealth of information organised in various sections. One of these sections is dedicated to the history of science[36] and its use in teaching, in the belief that many topics in the natural sciences can be effectively addressed in the classroom through the historical approach and that the history of science is of great educational value 'since it makes clear how provisional the scientific models that man has created over the years are and points out the intersections that have always existed between science and other fields of knowledge'. Additionally, it 'makes it possible to define course outlines, presenting material in a progression mirroring that which occurred in history'. It is possible to download hypertexts containing syllabuses in the natural sciences.

To make explicit ANISN's contribution to the debate on scientific training at school, we would like to cite here the report *La visione della scienza costruita nella scuola/The Vision of Science Created at School* (ANISN 2007).[37] The report is the result of a study which revealed an alarming situation in the scientific disciplines: the way students learn these disciplines at school distances them from science since it does not make its meaning and importance explicit, but, on the contrary, they are presented in an authoritarian, difficult, boring, selective form, too specifically aimed at the mechanical application of strategies for the resolution of problems. The only exception seems to be the natural sciences, which are more able than others to explain their cognitive role in young people's education.

SCI-DDC

The *Società Chimica Italiana – Divisione di didattica della chimica/Italian Chemical Society – Chemistry Teaching Division* (SCI-DDC)[38] is pursuing a project to include historical-teaching modules in the core university curriculum of the undergraduate programme in chemical sciences. The experimental phase of the programme was carried out in the 2007–2008 academic year at the University of Basilicata. During the following year, the pilot course ('chemistry and its evolution') was taught at the University of Camerino with the participation of university and high school teachers and students in the province of Macerata. The topics covered in the pilot programme were the development of electrochemistry, nuclear chemistry, organic chemistry, inorganic chemistry, toxicology and physical chemistry and

[36] http://www.anisn.it/storia_scienza.php

[37] This is a study on the perception of mathematics, physics, chemistry and natural sciences carried out through the administration of a questionnaire to 1,488 senior high school students in Italy.

[38] http://www.didichim.org/node/61

discussions with the students about the effectiveness of the historical approach to teaching chemistry. The initiative is an integral part of the Project Piano Lauree Scientifiche/Scientific Degree Project[39] in chemistry.

This proposal stems from the belief that the illustration of the historical depth of chemistry can notably improve the effectiveness of its teaching by showing where its concepts and practices come from and what their value is in terms of the work, organisation, time and passion of scientists in this field and critically focusing students' attention on concepts considered to be fully understood but which, instead, are unresolved.

The modules, taught by experts in history and the teaching of chemistry and science, are preceded by an introductory session on the historiography of the scientific disciplines and elements of the epistemology of the sciences. The subject matter modules are conducted as lessons logically inserted into the programme of host courses. The list of introductory lessons and subject matter modules is rich and varied. The former range from the epistemology of experimental cognitive science procedures to the historiography of science and chemistry. The latter illustrate the basic conceptual core of the subject, while taking an in-depth look at its historical development as related to the history of scientists and the social context of the scientific research.

Universities interested in this programme can sign up for a cycle of historical-teaching modules to be used on their campus.[40]

AIF

The activities of the AIF (*Associazione per l'Insegnamento della Fisica/Association for the Teaching of Physics*)[41] are also worthy of note. The AIF is a teacher's association founded in 1962 in Turin with the goal of popularising and promoting research in physics and the teaching of physics at all levels of school, from elementary school through to university. The association publishes and distributes scientific and teaching publications, organises teacher-training courses and conferences recognised by the MIUR (the Ministry of University Instruction and Research) and organises annual student competitions in physics. It is one of the implementing bodies (together with ANISN, SCI-DDC, the Leonardo da Vinci National Museum of Sciences and Technology Foundation in Milan and the City of Science in Naples) of the ISS (Insegnare Scienze Sperimentali/Teach Experimental Sciences) programme designed to monitor and carry out training initiatives for teachers in service under the form of research action for the improvement of the teaching-learning of the experimental sciences, with particular attention to teaching methodology.[42]

---

[39] http://www.progettolaureescientifiche.eu/il-piano-lauree-scientifiche

[40] The complete list of modules can be found at http://www.didichim.org/download/Moduli%20storico%20didattici%20AA%202009-2010.pdf.

[41] http://www.aif.it/

[42] http://archivio.pubblica.istruzione.it/argomenti/gst/iss.shtml

The association is organised in work groups: the history of physics work group,[43] founded in 1985, studies issues regarding the history of physics in terms of their teaching value. Part of the group's work includes the organisation of a winter teacher-training school with the collaboration of experts in the field, a seminar on the history of physics at the national AIF conference, refresher courses for teachers in the history and teaching of physics. The goal of the initiatives for teachers regarding history is, first of all, to help them to reflect critically on the historical developments in the field of physics, to point out the interactions between the various scientific disciplines and to promote the value of teaching the history of physics within a general physics course through the possibility of increasing historical knowledge about the development of physics theories. Furthermore, they aim to recognise and enhance the cultural and social value of science in its historical dimension, analyse the characteristics of historical research, reflect on the sources and social and cultural context of reference, improve knowledge of primary and secondary sources and analyse the available teaching materials.

64.3 Conclusion

Despite the excellent quality of the projects, research and the practical experiences carried out in Europe and Italy, a series of criticisms must be reported:

(a) The European and Italian documents that we have mentioned do not always clearly insist on the use of the history of science. Rather, they call expressly for a study of the IBSE or PBL scientific disciplines.
(b) Even where the document is explicit, this does not automatically translate into curricular content, given the 'indicative' nature of said documents.
(c) The same thing happens with academic research or field experiences: they are not included in the formal programming of teaching in the schools. This is due to the lack of a connection between universities, teacher associations and ministries for public instruction. In fact, most of the time research and field experiences do not involve those responsible for didactic policy and programming on the national level. Nor do the didactic materials produced in these studies have a level of formalisation high enough for their inclusion in formal curricular programming.
(d) While the introduction of HPS in teaching aims to combine knowledge and interdisciplinarity, in most cases teachers show a natural reticence to abandon their mono-disciplinary structure. Planning multi- or interdisciplinary teaching units requires the collaboration of other teachers who must dedicate a certain number of hours to this type of planning; these hours are currently not included in the school organisation. Additionally, the materials produced in the practical experience of carrying out interdisciplinary units, even if published, remain outside the circuit of manuals and textbooks. It is well known that 90 % of teachers' scholastic programming is based on the latter.

[43] http://www.lfns.it/STORIA/

Overcoming the difficulties listed here would require:

(1) A closer relationship between academic research, field experiences and instructional policies, involving the appropriate authorities as early on as in the planning of the research programmes.
(2) A high level of formalisation and promulgation of the teaching materials and learning units produced during the research and experimentation which would allow for their adoption in obligatory formal instruction.
(3) In any case, many studies have shown that modifying the curriculum is not very effective unless the teachers are motivated: 'It may be much more important to give teachers new frameworks for understanding what to count as learning than it is to give them new activities or curricula' (Langer and Applebee 1987, p. 87).
(4) The creation of institutional spaces in faculty meetings, specifically aimed at the planning of interdisciplinary courses.

References

Abd-El-Khalick, A., Akerson, V. (2009). The Influence of Metacognitive Training on Preservice Elementary Teachers' Conceptions of Nature of Science. *International Journal of Science Education,* 31(16), pp. 2161–2184

Anceschi, A., Scaglioni, R. (2010), *Formazione iniziale degli insegnanti in Italia: tra passato e futuro*, Napoli, Liguori.

ANISN (2007). *La visione della scienza costruita nella scuola. Indagine sull'immagine della Scienza che hanno gli studenti della scuola secondaria superiore*. Year XVI – special issue – January. Napoles: Loffredo Editore.

Antiseri, D. (1977). *Epistemologia e didattica delle scienze*. Rome: Armando Editore.

Battistini, A. (2010). *Letteratura e scienza*. Paper presented at the seminar *Nuovi Licei: l'avventura della conoscenza. Liceo Scientifico e Scienze Applicate*, Bologna 30 November 2010, Fondazione Ducati.

Berlinguer, L. (2009). Intervista, *Tutto Scienze*, 1 April, http://archivio.pubblica.istruzione.it/argomenti/gst/allegati/articolo_tutto_scienze.pdf. Accessed 23 January 2011.

Bernardi, W. (2011), *La storia della scienza nella formazione degli insegnanti*, http://www.storiadellascienza.net/app/download/5122294963/La+storia+della+scienza+nel+DM.+249+sulla+formazione+degli+insegnanti.doc?t=1304074491.

Bevilacqua, F. and Fregonese, L. (2000–2003). *Nuova Voltiana, Studies on Volta and his Times*. Pavia: Hoepli.

Bevilacqua, F. et al. (2001). Science Education and Culture: The Contribution of History and Philosophy of Science. London: Springer.

Bolondi, G. (2010). *Il percorso di Matematica e il rapporto con le altre discipline*. Paper presented at the seminar *Nuovi Licei: l'avventura della conoscenza. Liceo Scientifico e Scienze Applicate*, Bologna 30 November 2010, Fondazione Ducati, http://www.fondazionescuola.it/magnoliaPublic/iniziative/nuovi-licei/Liceo-Scientifico.html.

Cigada, F. (2007), La scienza degli studenti. In F. Cigada et al. (Eds.). *Il sapere scientifico della scuola* (pp. 140–41). Milano: FrancoAngeli.

Corsi, P. (2000). History of Science: Star of Research, Cinderella of Education. In C. Debru (Ed.), *History of Science and Technology in Education and Training in Europe* (pp. 213–220). Luxembourg: Official Publications of the European Communities.

Crespi, M. et al. (2005). L'immagine della scienza nei bambini e negli adolescenti: il ruolo dei musei. In N. Pitrelli and G. Sturloni (Eds.). *La stella nova, Atti del III Convegno annuale sulla comunicazione della scienza* (pp. 43–52). Milano: Polimetrica.

De Ambrosis, A. and Levrini, O. (2010). How physics teachers approach innovation: An empirical study for reconstructing the appropriation path in the case of special relativity. *Phys. Rev. ST_PER* 6.020107, pp. 020107-1–020107-11.

Debru, C. (1998), *History of Science and Technology in Education and Training in Europe, Euroscientia Conferences*, Strasbourg, 25–26 June 1998, European Commission, DG RTD, http://bookshop.europa.eu/en/history-of-science-and-technology-in-education-and-training-in-europe-pbCG4696013/.

Dibattista, L. (2008), Ed. *Gli spaghetti di Mendel e altri racconti. Lezioni di Storia della Scienza per i Licei*, Bari: Cacucci.

Dibattista, L. (2010), Ed. La Storia della Scienza va a scuola, Atti del Workshop, Bari 11 dicembre 2009. Bari: Adda.

Dibattista, L. and Morgese, F. (2011), History of Science Teaching to Teaches. The Italian Experience of the SSIS, *Science & Culture: Promise, Challenge and Demand, 11th International IHPST and 6th Greek History, Philosophy and Science Teaching Joint Conference*, Book of Proceedings, 1–5 July 2011, Thessaloniki, Greece, pp. 204–08.

Documento di lavoro (2007), http://archivio.pubblica.istruzione.it/argomenti/gst/index.shtml. Accessed 23 January 2011.

European Commission (1995). *White Paper Teaching and Learning: Towards the Learning Society,* http://europa.eu/documents/comm/white_papers/pdf/com95_590_en.pdf. Accessed 23 January 2011.

European Commission (2005). *Special Eurobarometer. Europeans, Science and Technology*. http://ec.europa.eu/public_opinion/archives/ebs/ebs_224_report_en.pdf. Accessed 23 January 2011.

European Commission (2006). *Science Teaching in Schools In Europe. Policies and Research,* Eurydice, The information network on education in Europe, DGEC. http://www.indire.it/lucabas/lkmw_file/eurydice///Science_teaching_EN.pdf

European Council Lisbon (2000). *Presidency conclusions.* http://www.consilium.europa.eu/ueDocs/cms_Data/docs/pressData/en/ec/00100-r1.en0.htm. Accessed 23 January 2011.

European Commission (2004). *Europe needs more scientists! Increasing human resources for science and technology in Europe*, http://ec.europa.eu/research/conferences/2004/sciprof/pdf/final_en.pdf.

European Commission (2007). *Science Education Now: A Renewed Pedagogy for the Future of Europe.* http://ec.europa.eu/research/science-society/document_library/pdf_06/report-rocard-onscience-education_en.pdf. Accessed 23 January 2011.

European Council Barcelona (2002) *Detailed work programme on the follow-up of the objectives of Education and training systems in Europe.* http://eur-lex.europa.eu/pri/en/oj/dat/2002/c_142/c_14220020614en00010022.pdf. Accessed 23 January 2011.

European Council Brussels (2003). *Council Conclusions on reference levels of European average performance in education and training (Benchmarks*). http://www.cedefop.europa.eu/en/files/benchmarks.pdf. Accessed 23 January 2011.

European Council Stockholm (2001). *Report "The concrete future objectives of education and training systems"*. http://ec.europa.eu/education/policies/2010/doc/rep_fut_obj_en.pdf. Accessed 23 January 2011.

European Parliament (2006). *Recommendation of the European Parliament and of the Council on key competences for lifelong learning.* http://eur-lex.europa.eu/LexUriServ/LexUriServ.do?uri=OJ:L:2006:394:0010:0018:EN:PDF.

European Parliament (2008). *Recommendation of the European Parliament and of the Council of 23 April 2008 on the establishment of the European Qualifications Framework for lifelong learning [Official Journal C 111, 6.5.2008]*. http://eur-lex.europa.eu/LexUriServ/LexUriServ.do?uri=OJ:C:2008:111:0001:0007:EN:PDF.

Eurydice (2011). *Science Teaching in Schools in Europe. Policies and Research.* http://eacea.ec.europa.eu/education/eurydice/thematic_studies_en.php.

Fauvel, J. and van Maanen, J. (2000), Ed. *History in Mathematics Education: The ICMI Study.* Dordercht, The Netherlands: Kluwer Academic Publishers, p. 104.

Fierli, M. et al. (2004). *Le immagini e le pratiche della scienza nei libri di testo della scuola primaria e della scuola secondaria di I grado.* Roma: ZadigRoma.

Gouthier, D. (2009a). Immagini della Matematica. Matematica per immagini. In B. D'Amore e S. Sbaragli (Eds.), *Pratiche matematiche e didattiche in aula, Atti del Convegno Castel San Pietro Terme, 6-7-8 novembre 2009*, Bologna: Pitagora Editrice, http://www.danielegouthier.it/home/wp-content/uploads/2009/10/gouthierCSPT2009.pdf accessed 23 January 2011.

Gouthier, D. (2009b). Il ruolo dell'immagine della matematica nella scelta degli studi e delle professioni scientifiche. In O. Robutti e M. Miranda (Eds.) *Atti del III Convegno Nazionale di Didattica della Fisica e della Matematica* (pp. 319–24). Torino: Provincia di Torino, http://www.danielegouthier.it/home/wp-content/uploads/2009/10/gouthierDiFiMa2007.pdf. Accessed 23 January 2011.

Gouthier, D. and Manzoli, F. (2008). *Il solito Albert e la piccola Dolly. La scienza dei bambini e dei ragazzi.* Milano: Springer.

Grapì, P. (2009). Projecte d'un curs telemàtic d'història de la ciència i de la tècnica per a la formació del professorat. *Enseñanza de las Ciencias*, Número Extra VIII Congreso Internacional sobre Investigación en Didáctica de las Ciencias, Barcelona, pp. 3704–3708, http://ensciencias.uab.es/congreso09/numeroextra/art-3704-3708.pdf.

Grapì, P. (2011), An On line Corse of History of Science and Mathematics for In-Service Teachers, Science & Culture: Promise, Challenge and Demand, 11th International IHPST and 6th Greek History, Philosophy and Science Teaching Joint Conference, Book of Proceedings, 1–5 July 2011, Thessaloniki, Greece, pp. 290–95.

Grimellini Tomasini, N. and Levrini, O. (2003). History and Philosophy of Physics as Tools for Preservice Teachers Education. Paper presented at the *III Girep Seminar*, Udine (Italy), http://www.fisica.uniud.it/URDF/girepseminar2003/abstracts/pdf/grimellini2.pdf.

Höttecke, D. and Riess, F. (2009). Developing and Implementing Case Studies for Teaching Science with the Help of History and Philosophy. Paper presented at *Tenth International History, Philosophy and Science Teaching Conference,* South Bend, USA, June 24–28 2009.

Höttecke, D. and Henke, A. (2010). Looking back into the Future: Lessons from HIPST about Implementing History and Philosophy in Science Teaching. Paper presented at *the History and Philosophy in Science Teaching Conference,* University of Kaiserslautern, Germany, March 11–14, 2010.

Höttecke, D. and Henke, A. (2011). Constructing HPS-Based Case Studies for Teaching and Learning Science. Paper presented at the *11th International IHPST and 6th Greek History, Philosophy and Science Teaching Joint Conference*, 1–5 July 2011, Thessaloniki, Greece.

INVALSI (2008). *Le competenze in Scienze Lettura e Matematica degli studenti quindicenni. Rapporto nazionale PISA 2006.* Roma: Armando.

Irez, S. et al. (2011), A Tipology of Science Teachers' Aims in Incorporating the History of Science in Science Teaching, *Science & Culture: Promise, Challenge and Demand, 11th International IHPST and 6th Greek History, Philosophy and Science Teaching Joint Conference*, Book of Proceedings, 1–5 July 2011, Thessaloniki, Greece, pp. 359–65.

Kokowsky, M. (2006) *The Global and the Local: The History of Science and the Cultural Integration of Europe. Proceedings of the 2nd ICESHS (Cracow Poland, September 6–9, 2006).* Round Table: History of Science in education and training in Europe: What new prospects? pp. 93–104

Koulountzos, V. and Seroglou, F. (2007), Designing a Web-based Learning Environment: The Case of ATLAS. Paper presented at IMICT 2007 Conference *Informatics, Mathematics and ICT: a golden triangle*, 27–29 June 2007, Boston.).

Langer, J.A. & Applebee, A. N. (1987) *How writing shapes thinking: A study of teaching and learning.* (Research Report, Number 22). National Council of Teacher of English, Urbana, IL.

Laubé, S. (2011), History and Cultural Mediation in Science and Technology versus an Example of Teacher Educating at the University of Brest (France), *Science & Culture: Promise, Challenge and Demand, 11th International IHPST and 6th Greek History, Philosophy and Science Teaching Joint Conference*, Book of Proceedings, 1–5 July 2011, Thessaloniki, Greece, pp. 428–34.

Massa, M. and Romero, F. (2009). La formació històrica per a l'ensenyament de les matemàtiques. *Enseñanza de las Ciencias*, Número Extra VIII Congreso Internacional sobre Investigación en Didáctica de las Ciencias, Barcelona, pp. 3709–3712, http://ensciencias.uab.es/congreso09/numeroextra/art-3709-3712.pdf.

MIUR (2010a), *Guida alla nuova scuola secondaria superiore*, http://www.istruzione.it/getOM?idfileentry=217468

MIUR (2010b), *Indicazioni nazionali riguardanti gli obiettivi specifici di apprendimento concernenti le attività e gli insegnamenti compresi nei piani degli studi previsti per i percorsi liceali*, http://www.indire.it/lucabas/lkmw_file/licei2010///indicazioni_nuovo_impaginato/_decreto_indicazioni_nazionali.pdf

MIUR (2010c), *Il profilo culturale, educativo e professionale dei Licei*, Allegato A al Regolamento dei Licei, http://archivio.pubblica.istruzione.it/riforma_superiori/nuovesuperiori/doc/Allegato_A_definitivo_02012010.pdf

MIUR (2010d), *Regolamento concernente: «Definizione della disciplina dei requisiti e delle modalita' della formazione iniziale degli insegnanti della scuola dell'infanzia, della scuola primaria e della scuola secondaria di primo e secondo grado...*, http://www.miur.it/Documenti/universita/Offerta_formativa/Formazione_iniziale_insegnanti_corsi_uni/DM_10_092010_n.249.pdf

Morgese, F. (2008). L'istruzione scientifica nella società della conoscenza. In L. Dibattista (Ed.). *Gli spaghetti di Mendel e altri racconti. Lezioni di Storia della Scienza per i Licei* (pp. 23–40). Bari: Cacucci.

Morgese, F. (2010b). La didattica della scienza attraverso la storia della scienza: politiche, esperienze, metodologie. In F. Morgese and V. Vinci V. (Eds.), *Performascienza. Laboratori teatrali di storia della scienza a scuola* (pp. 15–49). Milano: Franco Angeli.

Morgese, F. and Vinci V. (2010), Eds., *Performascienza. Laboratori teatrali di storia della scienza a scuola*. Milano: Franco Angeli.

Neresini, F. et al. (2010). Scienza e nuove generazioni. I risultati dell'indagine internazionale ROSE. Vicenza: Edizioni Observa – Science in Society.

OECD, 2006, *Evolution of Student Interest in Science and Technology Studies. Policy Report*, p. 2, http://www.oecd.org/dataoecd/16/30/36645825.pdf. Accessed 23 January 2011.

Osborne, J. and Dillon, J. (2008), *Science Education in Europe: Critical Reflections,* A Report to the Nuffield Foundation, King's College London, http://www.nuffieldfoundation.org/sites/default/files/Sci_Ed_in_Europe_Report_Final.pdf.

PoS Project Team (2007a). *Perspectives on Science. Student Book.* Oxford: Heinemann.

PoS Project Team (2007b). *Perspectives on Science. Teacher Resource File*. Oxford: Heinemann.

Rodari, P. (2008). Il museo, i giovani e la scienza. In D. Gouthier D. and F. Manzoli (Eds.). *Il solito Albert e la piccola Dolly. La scienza dei bambini e dei ragazzi* (pp. 123–38). Milano: Springer.

Ruggieri, T. (2010). *Matematica e innovazione*. Paper presented at the national conference *Nuovi Licei: l'avventura della conoscenza*, Rome 11 October 2010, Aula Magna Università Luiss, http://www.fondazionescuola.it/magnoliaPublic/iniziative/nuovi-licei/presentazione.html.

Schreiner, C. and Sjøberg, S. (2004). ROSE: The Relevance of Science Education. Department of Teacher Education and School Development, University of Oslo, http://roseproject.no/key-documents/framework.html.

Sjøberg, S. and Schreiner, C. (2010). The ROSE project. An overview and key findings. University of Oslo, http://roseproject.no/network/countries/norway/eng/nor-Sjoberg-Schreiner-overview-2010.pdf.

Tarsitani, C. e Vicentini, M. (1991). *Calore, energia, entropia. Le basi concettuali della termodinamica e il loro sviluppo storico.* Milano: FrancoAngeli.

Tarsitani, C. (2009). *Dalla fisica classica alla fisica quantistica. Riflessioni sul rinnovamento dell'insegnamento della Fisica*. Rome: Editori Riuniti University Press.

Vittorio, N. (2010a). *Il nuovo percorso di Fisica e le scienze*. Paper presented at the seminar *Nuovi Licei: l'avventura della conoscenza. Liceo Scientifico e Scienze Applicate*, Bologna 30 November 2010, Fondazione Ducati, http://www.fondazionescuola.it/magnoliaPublic/iniziative/nuovi-licei/Liceo-Scientifico.html.

Vittorio, N. (2010b). *L'insegnamento delle discipline scientifiche*. Video from the seminar *Nuovi Licei: l'avventura della conoscenza. Liceo Scienze Umane*. Turin, 12 April 2011, Sala Convegni, Piazza Bernini, http://www.fondazionescuola.it/magnoliaPublic/iniziative/nuovi-licei/LiceoScienzeUmane.html.

Taylor, J.L. and Swinbank, E. (2011), Perspectives on Science: a Qualification in the History, Philosophy and Ethics of Science for 16–19 year old studente, *Science & Culture: Promise, Challenge and Demand, 11th International IHPST and 6th Greek History, Philosophy and Science Teaching Joint Conference*, Book of Proceedings, 1–5 July 2011, Thessaloniki, Greece, pp. 730–34.

Liborio Dibattista is a physician with a specialisation in hygiene. He also holds a master's degree in philosophy from the faculty of humanities and philosophy at Università degli Studi of Bari and a Ph.D. in the history of science. He has been lecturing history of science courses on history of biology and medicine at the faculty of humanities and philosophy – Università degli Studi di Bari – since 1997. He is professor of the history of science and technique at Scuola Interateneo di Specializzazione per la Formazione degli Insegnanti di Scuola Secondaria (University Specialisation Course for High School Teachers' Training). He currently works as university professor of history of medicine for high-level philosophy courses at Università degli Studi of Bari.

His research interests include the history of neurophysiology, with special attention to the mind-body problem. From a methodological point of view, Dibattista proposes an original model of scientific investigation of the historiographical texts based on computational linguistics. He is involved in promoting the teaching of science based on the history of science and, therefore, in a humanistic vision of training. He is a member of the History of Science Society, of Società Italiana di Storia della Scienza and of Società Italiana di Storia della Medicina. He is scientific director of a number of experimental projects for the application of history and philosophy of science to science education and teacher training in Italy.

He is the author of several contributions to journals in the history of science and computational linguistics and the following monographs: *Storia della Scienza e didattica delle discipline scientifiche* (Armando, Roma 2004), *J.-M. Charcot e la lingua della neurologia* (Cacucci, Bari 2003) and *Il Movimento Immobile. La fisiologia di E.J. Marey e C.E. François-Franck* (Olschki, Firenze, 2010). He is editor of *Gli Spaghetti di Mendel. Lezioni di Storia della Scienza per i Licei* (Cacucci, Bari 2008) and of *Storia della Scienza e Linguistica Computazionale. Sconfinamenti possibili* (FrancoAngeli, Milano 2009).

Francesca Morgese holds a master's degree in humanities from the faculty of humanities and philosophy at Università degli Studi 'Aldo Moro' of Bari and a Ph.D. in history of science; she is a connoisseur on history of science since 2009. She teaches classics in senior high schools.

Her research activities concern these following themes: European and Italian history of education and history of science education, the application of historical narrative in science learning, the relationship between science and the humanities and science communication in school. She is an experienced teacher in several projects PON-FSE of narrative practices and history of science in secondary schools. She has participated with scientific communications at several national and international conferences with papers on the history of medicine and the teaching of science. She is legal representative of the Scienz@ppeal Association of Bari and responsible for projects of dissemination of scientific culture in which there is communication between educational agencies and educational and research institutions of Apulia, first schools and university. She is a member of Società Italiana di Storia della Scienza (SISS).

Her main publication on teaching of science and science education is Morgese F. and Vinci V. (Eds), *Performascienza. Laboratori teatrali di storia della scienza a scuola* (FrancoAngeli, Milano 2010), and several other contributions in the field have appeared in edited books and journals.

Chapter 65
History in Bosnia and Herzegovina Physics Textbooks for Primary School: Historical Accuracy and Cognitive Adequacy

Josip Slisko and Zalkida Hadzibegovic

65.1 Introduction

Although the destiny of contemporary societies highly depends on sciences, the students' interest in becoming professional scientists is in alarming decline (Osborne et al. 2003). In Germany, for instance, the number of high school students who are interested in studying physics is significantly lower than the number of students that would like to spend their lives doing math, geography, art, and even politics (Hannover and Kessels 2004).

It is circular reasoning to say that for those students who somehow fell in love with physics, the science is always their first priority, being both most interesting and everywhere present. However, if one is looking for more convincing personal and social reasons why students have negative attitudes towards physics and science, then the answers point at the quality of their prior education and their interest in the subject (Mamlok-Naaman 2011), the academic plans (Crawley and Black 1992), the authenticity of science experiences in science education (Eijck and Roth 2009), the preferences of each student by nature or by the influence of teachers and/or parents, and even how they view scientists (Lee 1998). Recently, there were a few very informative review articles of the research literature related to the factors that affect positively or negatively students' attitudes towards science (Osborne et al. 2003; Krapp and Prenzel 2011).

J. Slisko (✉)
Benemérita Universidad Autónoma da Puebla, Centro Historico, Mexico
e-mail: jslisko@fcfm.buap.mx

Z. Hadzibegovic
Department of Physics, University of Sarajevo, Sarajevo, Bosnia and Herzegovina
e-mail: zalkidah@yahoo.com

M.R. Matthews (ed.), *International Handbook of Research in History, Philosophy and Science Teaching*, DOI 10.1007/978-94-007-7654-8_65,

From these studies and their theoretical frameworks, one can conclude that the proper way to make physics and science intellectually and emotionally more attractive to young people is to choose a proper mix of teaching goals and methodology, along with an adequate usage of textbooks and other supporting materials. In other words, the learning goals, the content of textbooks, and the way physics is taught should play an important role in changing the negative attitude of a large number of students toward physics and the learning of physics. Nevertheless, one must always keep in mind that the process of building students' motivations, from a psychological point of view, is a rather complex theoretical and experimental issue (Lavigne and Vallerand 2010).

For some decades, science and physics educators (Klopfer 1969; Russell 1981; Matthews 1994/2014; Irwin 2000) have given multiple arguments why positive change in attitude towards science might be achieved by increasing the presence of history of science in physics education, especially in the content of textbooks and learning tasks. It was known that historical information might help to predict students' conceptual difficulties, with similar meaning to both the students and the earlier scientists (Wandersee 1986), and to design adequate learning sequences (Monk and Osborne 1997). The strength of these arguments was increased by the results of classroom-based studies (Klopfer and Cooley 1963; Solomon et al. 1992; Solbes and Traver 2003).

For changing students' views of science and scientists, it is convenient to use original materials from the history of science contained in books, museum collections about famous physicists, and multimedia materials based on the history of science. Unfortunately, the traditional (and frequently, superficial) way of using the history of science in science teaching and learning has typically involved brief biographical sketches of scientists, fragmented notes on inventions, and photos or drawings of scientists and their works, all mainly serving a decorative role.

The contribution of the history of science to science teaching and learning began in the second half of the twentieth century, both by historical studies whose results might be useful in teaching (Conant 1957) and by physics textbooks which had a strong historical flavor (Holton 1952). The culmination of that initial interaction between history and physics teaching was the Harvard Project Physics (Rutherford et al. 1970; Holton 2003). In the course textbook, historical episodes and information were not presented continuously but at places where they might foster learning and positive attitude towards science. Regarding the use of science history in teaching and learning school science, one can find now a great variety of considerations, didactical proposals, and experimental results that are very promising (Kokkotas et al. 2010).

Teaching that makes use of historical knowledge for designing learning tasks and cares about their real implementations produces good results related, for example, to the nature of science (Abd-El-Khalick and Lederman 2000; Galili and Hazan 2001; Abd-El-Khalick 2002), attitudes towards science (Mamlok-Naaman et al. 2005), conceptual learning in optics (Galili and Hazan 2000), or notion of experimentation styles in electrostatics (Heering 2000). In general, any use of historically oriented material in science and physics courses should be carefully analyzed in a broader perspective, determined by cognitive, metacognitive, and emotional aims of physics education (Seroglou and Koumaras 2001). History of science is also useful in better articulation of teaching (Binnie 2001; Wang and Marsh 2002) and prospective teachers' education (Riess 2000).

A main issue for textbook authors and analysts is the quantity and form of historical aspects in physics that will be included in a textbook. Laurinda Leite (2002) proposed a useful model for potential authors or educators who want to use historical episodes in their classes. According to that model, textbook analysis should discover how the history of physics is incorporated into the textbook body, including (1) variety, way, organization, and usage of historical information as one group of analysis elements and (2) accuracy of historical information, context in which the historical information appear, textbook consistency, connection between learning activity, and history of physics and references as other groups of analysis elements. The chosen historical episodes could be set out in a context useful for all learners (compulsory part) and additionally in parts reserved for only those who want to know more or, in other words, for talented pupils.

The methodology of using episodes from the history of science in physics textbooks should be based on four basic requirements:

1. Historical accuracy (use of historical episodes according to original articles, letters, notes, and patents constructed by scientists, as authentic historical documents)
2. Cognitive adequacy (concordance of the episode content with the taught topic and desired cognitive skills of pupils)
3. Motivational potential (potential of the historical information to increase pupils' interest in physics)
4. Didactical tools for the learning of the chosen content that are included in science curriculum

65.2 Characteristics of Primary School Physics Textbooks in Bosnia and Herzegovina

65.2.1 General Information on Analyzed Textbooks and Their Use of History

Before recent curriculum and policy changes (introducing 9 years long primary schooling), primary school education in Bosnia and Herzegovina was carried out in eight grades (corresponding to pupils' ages between six or seven years and 14 or 15 years). Due to the specific administrative organization, there are 13 different curricula in Bosnia and Herzegovina brought by 13 regional ministries of education (two at entity level, one at district level, and 10 at canton/county level). However, one can suppose that there are no limitations and restrictions for the authors in relation to the different curricula when organizing textbooks.

Among many physics textbooks used in different regions of Bosnia and Herzegovina, we analyzed only those written by domestic authors or published by

Table 65.1 Basic data on analyzed textbooks

Author	Textbook title	Year of publication	Total number of pages	Acronym
Esad Kulenovic	Physics for 7th grade of primary school	2006	197	A7
Esad Kulenovic	Physics for 8th grade of primary school	2006	214	A8
Nada Gabela, Hasnija Muratovic	Physics VII for 7th grade of primary school	2004	132	B7
Nada Gabela, Hasnija Muratovic	Physics VIII for 8th grade of primary school	2004	116	B8
Aziza Skoko, Kasim Imamovic	Physics 7 Textbook for 7th grade of primary school	2005	202	C7
Aziza Skoko, Kasim Imamovic	Physics 8 Textbook for 8th grade of primary school	2004	180	C8
Hedija Boskailo-Sikalo	Physics 7 Textbook for 7th grade of primary school	2004	205	D7
Hedija Boskailo-Sikalo	Physics 8 Textbook for 8th grade of primary school	2005	151	D8

domestic publishers. We carried out an analysis of four textbooks for 7th grade and four textbooks for 8th grade pupils used in primary school curricula in one of the Bosnia and Herzegovina entities – the Federation of Bosnia and Herzegovina (leaving out the physics textbooks in Republika Srpska) in the school year 2010/2011. Basic information about the analyzed textbooks is given in Table 65.1.

Textbooks A7 and A8, authored by Kulenovic, are now in their sixth edition. Textbooks B7 and B8, written by Gabela and Muratovic, are now in their second edition. The remaining four textbooks C7, C8, D7, and D8 are in their first edition.

It is important to stress that the primary school physical science syllabus in BiH does not explicitly require the usage of historical content to support physics learning. The presence of such information depends exclusively on authors' and editors' decisions.

Therefore, it is not surprising that textbook authors used various forms and amounts of historical elements. A quantitative analysis of textbooks, following the checklist by Laurinda Leite (2002), revealed a total of 15 different ways of using historical episodes of physics. These ways are presented in Table 65.2.

The largest number of historical elements was found in the C8 and C7 textbooks (25 % and 21.5 %, respectively). Notes about scientists had mostly a decorative role. Heterogeneity of historical information was noticed in textbooks C and D, whose authors belong to a younger generation of authors.

Table 65.2 Total number of contents in the textbooks for the seventh and eighth grade that take into account the criteria established in the checklist

	Count by grade level		
Historical content/information	7th	8th	Sum
Independent sections on the history of physics in some chapters	6	5	11
Historical notes integrated in the text	50	52	102
Note on the name of the units under the name of the scientist	11	9	20
Short biographical sketches of scientists (up to three lines of text)	17	43	60
Longer biographical sketch of the scientist	28	50	78
Scientists' pictures or drawings	47	65	112
Prominent scientific contribution (discovery, theory, law...)	43	26	69
Original documents/texts	13	1	14
Photograph or drawing of the invention or the experiments	12	19	31
Photograph of laboratory	3	5	8
Painting of historical episodes	5	3	8
Comical drawings of historical episodes	5	0	5
Legends as supposedly historical episodes	6	1	7
Other representations of historical episodes (stamps, banknotes...)	0	5	5
Historical episodes for those who want to know more	0	1	1
Overall historical content used	246	285	531

The most frequent historical elements were in a pictorial form integrated into the body text. Only one author gave references for such depictions. Notes, photos, and drawings of the same scientists (Galileo, Newton, Tesla, and Einstein) were present in the total sample of the analyzed textbooks. Historical information was generally intended to be used by all the pupils, whereas only one author planned to use historical data in a particular section titled "If you want to know more" (textbook A8).

Talented pupils could benefit from historical information that could lead to the formation of an idea, designing and carrying out an experiment, and formulating and applying a law or a theory. There were no instructions on how to obtain additional information about famous scientists, most influential experiments, theories, or original text, although such an information is relatively easy to be found today due to the Internet.

The comparison of the textbooks in relation to their inner consistency of historical elements showed that textbooks A and B are homogeneous, whereas C and D are heterogeneous in their presentations. Only textbook D provides references related to the materials used during their preparation. This can be seen as a generation difference because younger authors use more historical elements and materials.

Although there are important differences in the amount and type of historical episodes, the use of history of physics is generally reduced to superficial biographical notes that do not introduce pupils to the process of knowledge production in physics. Representation of the historical contents used in the analyzed textbooks in terms of percentages is given graphically in Fig. 65.1.

Fig. 65.1 Percentages of the overall content on history of science by analyzed textbooks

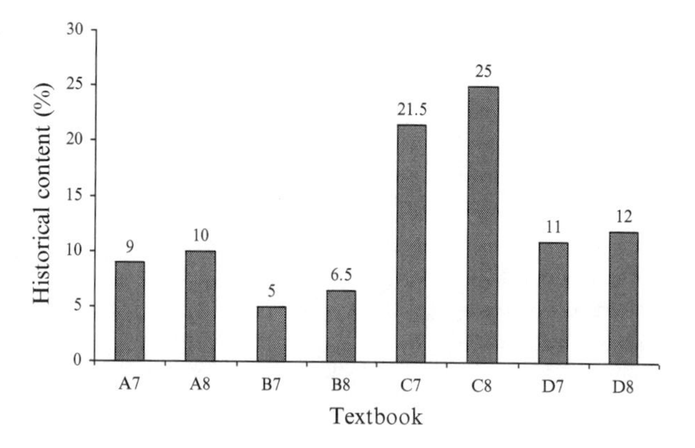

According to Fig. 65.1, it can seen that textbooks C7 and C8 contained remarkably higher amount of historical content compared to others, whereas textbook B7 had the smallest amount of historical content.

65.2.2 Some Specific Examples of History Usage in the Analyzed Textbooks

The elements of the history of science in the eight textbooks are used in a rather fragmented and superficial way. Mostly, they are biographical short texts describing the roles of physicists or of their inventions in very general terms. Frequently, texts are accompanied by physicists' pictures or other types of visual representation data, such as drawings, paintings, stamps, and banknotes.

The most important finding is that there are no explicit and meaningful learning tasks based on the history of science in these textbooks. The following presentation of the elements that appear most frequently will serve as an illustration of these not-for-learning-science uses of the history of physics in analyzed textbooks.

65.2.2.1 Biographical Portraits of Scientists

Sixty (11 %) short biographical sketches of scientists (up to three lines of text) have been counted. Basically, they provide the following information: birth and death year, nationality or country of origin, and the domain in which the scientist contributed by invention, law, or theory. Pictures or drawings that can be easily found on the Internet often accompany them.

Biographical information about scientists is usually included in a special position in the textbook page and always within a covered theme that is associated with the work of the scientists. There were three basic forms of presenting the life of physicists. The first is used in 32 % of the biographical notes, where scientists are presented in only one sentence associated with a picture or a drawing situated in a separate box containing a minimal amount of information: name and surname of the scientist, birth and death year, nationality or citizenship, and belonging to a scientific discipline (a physicist, astronomer, mathematician, chemist, or philosopher).

In the second form, textbook authors provide biographical notes about scientists ranging between two and five sentences (11 %), adding together data about their discoveries or inventions, or about their importance for civilization. In the third form, more informative data about scientists (57 %) are presented such as some short stories from their lives and works, ranging from 200 to 450 words. Such biographical stories about Newton, Galileo, Tesla, and Franklin can be found in the textbooks written by Skoko and Imamovic and by Boskailo-Sikalo.

65.2.2.2 Notes on Eponymous Names of Fundamental or Derivative Physical Units

In many instances, the history of science is used to inform about the origin of the names of units. From a total of 20 examples, the following two examples are presented (Boxes 65.1 and 65.2):

Box 65.1 The Origin of the Unit Name "watt" (D7, p. 150)

The unit of power is named the watt. Its symbol is W. This name was given in honor of the engineer Watt.

Box 65.2 The Origin of the Unit Name "coulomb" (B8, p. 9)

In the international system (SI) the unit of electric charge is the coulomb (symbol C). The unit is named after the French physicist Charles Coulomb who discovered the law of interaction of charges.

However, there was no information regarding neither the units that were used before the introduction of the International System of Units (SI) nor the reasons for introducing these new units. That curricular stand might be understood for those former units, like gauss, which are mainly out of use today. Nevertheless, some information about the usage of the mile, mile/hour, inch, or Fahrenheit degree as matter-of-fact units in some parts of the world (England, the USA) might be in place because students live and act in a globally connected world. This omission causes difficulties on students' understanding of Internet information and, further, could cause problems in their intercultural identification and recognition.

Fig. 65.2 Visual representation of Newton's famous prism experiment (D8, p. 149)

65.2.2.3 Short Historical Information About Experiments

Some important physics experiments, like Newton's prism experiment, are presented as a sentence-long information accompanying a picture (Fig. 65.2).

The text under this picture reads:

> Newton got a color spectrum by passing white light through a glass prism.

The picture used in the text is identical to the one available on-line at http://www.biographyonline.net/scientists/isaac-newton.html (accessed 10 November 2012).

This information is useless for the students' learning of the nature of light since Newton's main contribution was not the color spectrum but demonstrating via many cleverly designed experiments that the Aristotelian theory of light was wrong.

65.2.2.4 "Legendary History" of Physics

The "legendary history" of Physics is also presented in the analyzed textbooks. The following is one common example about Galileo and the Leaning Tower (Fig. 65.3):

> According to the legend, Galileo Galilei tried to calculate free fall time, observing a body falling from the Leaning Tower of Pisa. At that epoch, the time could not be calculated precisely as it happens nowadays with chronometers. Something like that was impossible. He discovered the law of uniformly accelerated motion by letting the ball rolling down an inclined plane as we indicated earlier. (C7, p. 102)

Fig. 65.3 The Leaning Tower of Pisa and Cathedral (C7, p.102) Available at URL: http://www.7wonders.org/europe/italy/pisa/leaning-tower/ (accessed 10 November 2012)

In this extract the authors incorrectly describe both the legend and the real experiment carried out by Galilei. The legend says that Galilei disapproved Aristotle's idea that "heavier bodies fall faster" by dropping two cannon balls of different weight, which hit the ground at the same time. So, "calculation of time" is not part of the legend.

Galilei's "inclined plane experiment" (Galilei 1954) was a highly discussed theme in the history of science. Repeating Mersenne's critique, Koyre (1968) claimed that Galilei neither carried out this experiment nor other experiments he described in his published books. Nevertheless, thanks to Drake's ground-breaking investigations of Galilei's handwritten notes (Drake 1973, 1975), it is now generally believed that Galilei did actually carry out his experiments.

Taking into account Galilei's description of the theory that became the basis for the "inclined plane experiment" (Galilei 1954), it is necessary to stress that Galilei didn't design that experiment to discover the law of uniformly accelerated motion, as it is claimed in the text above, but to check out if that motion is a motion with constant acceleration. For such a motion, Galilei knew the theoretical relation between distance covered and time elapsed: the covered distance is directly proportional to the square of time elapsed.

Galilei's worldview led him to believe in such a possibility for free fall, and he wanted to have experimental evidence. He repeated the experiment increasing progressively the inclinations. The data he got were the expected ones according to the theoretical model (distance covered directly proportional to the time squared). This made him to infer that very likely free fall (motion down the plane with 90° inclination) is also a uniformly accelerated motion.

Even more erroneous treatment of Galilei's "thought experiment" regarding the logical possibility of forceless motion has been found in the analyzed textbooks. This thought experiment was given in the form of Socratic conversation between Salviati and Simplicio in "Dialogue concerning the two chief world systems" (Galilei 1967). Salviati led Simplicio to accept that (a) a ball rolling down an inclined plan increases its speed; (b) the same ball rolling up an inclined plane decreases its speed. After that Salviati formulated a very disturbing question for Simplicio (a person advocating the Aristotelian view of motion):

> Now tell me what would happen to the same movable body placed upon a surface with no slope upward or downward. (Galilei 1967, p. 147)

Two different treatments of this "thought experiment" by Galilei are presented below. The first example is extracted from the A7 textbook and the second from the D7 one (Boxes 65.3 and 65.4). The illustration showed in Fig. 65.4 is one similar to the A7 and D7 illustrations:

Box 65.3 Textual Description of the Experiment (A7, pp. 93–94)

When a small ball is rolling down an inclined plane of a certain height, it will continue to move uniformly in a horizontal surface, and then climb up another inclined plane to the same height from which it started. In the absence of friction, this climbing will be independent of the distance traveled horizontally and the slope of the inclined plane along which the ball climbs. On this basis, Galileo concluded that the body moving at a certain speed on a horizontal surface in the absence of friction and other resistant forces will continue to move uniformly forever. This conclusion is known as Galilei's principle of inertia.

Box 65.4 Textual Description of the Experiment (D7, p. 121)

Galileo Galilei observed the movement of the ball rolling down a double inclined plane in which the slope of the right part can be changed. If we reduce the slope of the right inclined part, and the ball rolls down from the same position, then the ball rolls up the right inclined part to the same height but the distance it must go along the slope is greater. The maximum distance is achieved when the ball moves in a horizontal surface at the right side.

Every body persists in its state of rest or uniform motion in a straight line unless it is compelled to change that state by forces impressed on it.

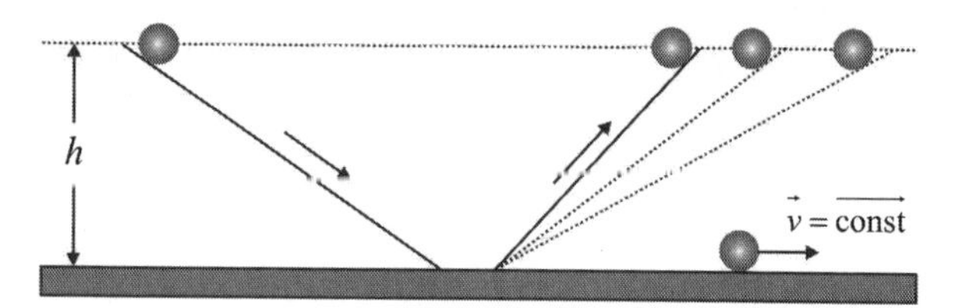

Fig. 65.4 An illustration of Galilei's "thought experiment" on inertial motion (A7, p. 94, and D7, p. 121)

We describe Galilei's experiment and his conclusion...(A7, p. 93)

As it can be seen, both treatments erroneously suggest that this "thought experiment" was a real experiment, carried out by Galilei.

Erroneous and superficially simplified presentations of physics history in textbooks, especially of those important episodes that are connected to research-based historical facts, are not an exotic syndrome of the authors from Bosnia and Herzegovina. On the contrary, it is a rather global phenomenon, as recent Niaz's books (2008, 2009, 2010) show convincingly. This unsatisfactory situation is caused, on the one hand, by the weak quality control in physics teaching and textbook writing (Slisko and Hadzibegovic 2011), which makes possible the invention of (erroneous) historical information due to the ignorance of the authors, reviewers, and users. This invented information enters and stays in the textbooks. On the other hand, there is an inclination of the research community towards the "backwards written history," which is common in the way "normal science" is presented in physics textbooks (Brackenridge 1989).

Backwards written history refers to special classes of falsified history in which historical episodes are presented in a way that leaves the impression that modern concepts and procedures were used when, in fact, they didn't exist in that particular time. So, it is not a physics history as it really was, but a physics history as it might have been but was not. One paradigmatic example of *backwards written history* is the known (erroneous) claim in the physics textbooks that Cavendish measured the value of the gravitational constant, although his research question and reasons were different, and the very idea of the gravitational constant did not yet exist (Slisko and Hadzibegovic 2011). A misleading interpretation of the Greek atomism as an anticipation of the modern scientific atomic theory is also common in philosophical literature (Chalmers 2009).

Among 531 examples of use of historical information detected in the analyzed physics textbooks for primary school in Bosnia and Herzegovina (BiH), one example was chosen, found in the D8 textbook, to explore how primary school students (pupils) make sense of superficial and incomplete presentations of an historical episode. Up to date, this is the first study of this kind related to the uses of history in BiH physics textbooks.

Our objective is to show by some initial data that such presentations are not cognitively adequate for pupils. Namely, the pupils as sense-making persons try to provide the missing information to establish the story coherence. This process, as it will be shown later, is potentially damaging for the pupils' learning because the missing information they provide may take forms that neither correspond to the historical facts nor is physically possible.

65.3 Chosen Historical Episode for Research with Primary School Students: The Measurement of the Speed of Sound in Water

The chosen historical episode is the measurement of the speed of sound in water, carried out by Jean-Daniel Colladon in 1826 at Lake Geneva (Colladon 1893). That measurement was very important for checking whether the theoretical formula for the speed of sound in combination with the measured value of the water

Fig. 65.5 The boat with underwater bell and mechanism for sending light signal

compressibility would predict a correct value of the speed of sound in water. That research was planned and successfully accomplished by Colladon and Sturm. For that particular research they received the first award of the French Academy of Science. Although these contextual details would be very important in analyzing textbook presentations of the Lake Geneva Experiment at higher educational levels, we will not deal with them in the analyzed accounts for younger students (grade VIII in Bosnia and Herzegovina and grade IX in the UK).

65.3.1 Design and Results of the Original Experiment

The basic ideas of the experiment design at Lake Geneva were as follows: An assistant of Colladon (Sturm was then in Paris and did not take part in the experiment!), being in one boat, sent simultaneously an underwater sound signal and a light signal in air. The sound signal was produced by striking an underwater church bell with a hammer. A complicated mechanism simultaneously ignited the gunpowder and struck the bell on the first boat (Fig. 65.5).

Colladon himself was in the second boat with a long horn immersed in the water, attached to his left hand (Fig. 65.6). The immersed end of the horn had an elastic membrane that would vibrate when reached by the underwater signal, making possible for Colladon to hear the sound of the bell. He activated a chronometer (stopwatch) using his right hand when he saw the light signal coming from the first boat and stopped it when he heard the sound of the bell that came through the water.

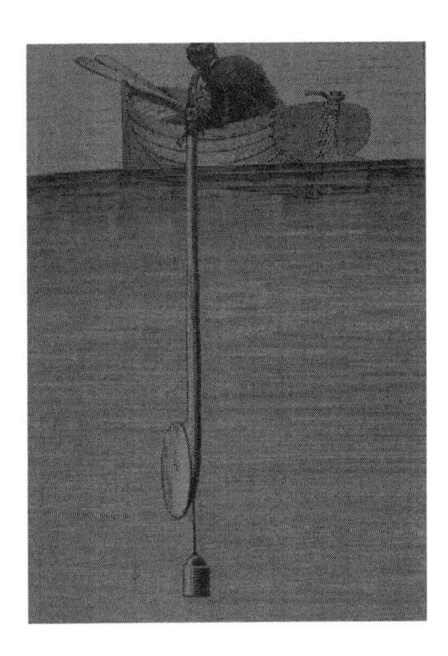

Fig. 65.6 The boat with underwater horn is used to hear the sound produced by bell

The distance between the two boats was carefully measured by triangulation methods and it was found to be 13,478 m. The mean time the sound needed to travel through water between the bell and the horn was 9.4 s. These values result in a value of 1,435 m/s for the speed of sound in water.

65.3.2 *A Textbook Presentation of the Lake Geneva Experiment in Bosnia and Herzegovina*

Only one textbook (D8) presented the experiment carried out by Colladon at Lake Geneva, combining verbal and visual information. The textual part is as follows:

> A physicist and his assistant were in two boats 1,500 m away from each other. A bell on a rope was immersed in the water from the first boat with the assistant in it. His task was to hit the immersed bell with a hammer and simultaneously send a light signal. The physicist with a long horn was in the second boat. One end of the horn was immersed in water, while the other end of the horn was held by the physicist near his ear. The moment he saw the light signal, the physicist switched the chronometer on and measured the difference of time between the moment he saw the light and the moment he heard the sound. The delay of the sound signal was one second. The physicist concluded that the sound needed 1 s to travel 1,500 m through water, and that the speed of sound in water was 1,500 m/s. (D8, p. 102)

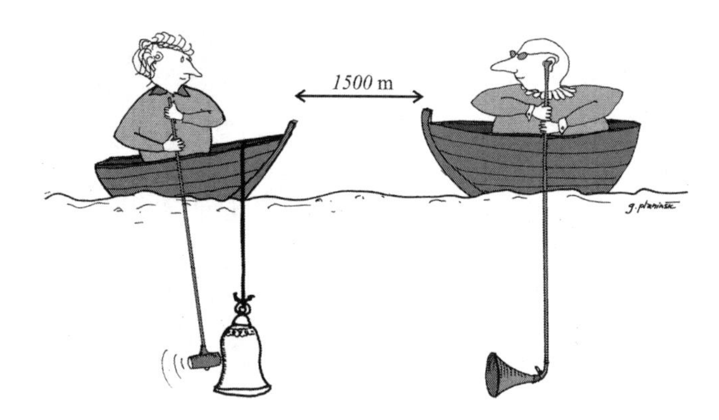

Fig. 65.7 The drawing of Colladon's experiment, similar to the picture in D8, p. 102

The visual representation of the experiment is a drawing of the situation (Fig. 65.7)

This example is analyzed from two perspectives. The first one is its historical accuracy or, in other words, whether the description of the particular episode is precise enough. It is easy to see that the author presents false experimental data, giving for distance and time the values of 1,500 m and 1 s, respectively.

The second, even more important one is the cognitive perspective: Is the presentation structured in a cognitively adequate way giving thus pupils an opportunity to learn how physics in its experimental domain works, without inducing students to accept some erroneous ideas?

Comparing verbal and visual descriptions, it is possible to detect incoherence between them, which would be a learning obstacle for all students who try to comprehend fully the way the experiment was carried out. Namely, the drawing gives students no idea of how the light signal was sent as, according to the drawing, assistant's both hands are employed to hit the underwater bell by the hammer. The other, potentially enigmatic part for students, is how the physicist can switch on and off the chronometer if, again, both of his hands are used to hold the horn under water.

So, the above textbook presentation of the historical experiment in which the speed of sound in water was measured is:

1. Historically inaccurate (the values of the distance and time are arbitrarily invented)
2. Depersonalized (Colladon became an anonymous physicist)
3. Cognitively inadequate (neither the text nor the drawing gives students an opportunity to comprehend fully how the experiment was carried out)

65.3.3 Another Textbook Presentation of the Lake Geneva Experiment

To put this particular textbook presentation in a broader international perspective, it is instructive to look at another presentation of the same experiment which comes from the UK:

Newton's work predicted that sound should travel faster in water than in air.

This was proved by an experiment on Lake Geneva in 1827. An underwater bell was rung at the same time as some gunpowder was lit. On another boat 14 km (9 miles) away, the flash was seen (at night) and the sound was heard through the water by a large ear trumpet dipping into the water (Johnson et al. 2001, p. 82, italics added).

The follow-up tasks for the students are:

1. Think about the Lake Geneva Experiment, and sketch what you think the apparatus looked like on each boat.
2. With a distance of 14 km the sound took 10 s. What was the speed of sound in water? (Johnson et al. 2001, p. 82, Question 3)

Regarding historical accuracy, the UK textbook authors present rounded values of the distance and time that are close enough to the original ones. The year of the experiment is not the correct one and corresponds to the year the results were initially published.

It is very good that the authors asked students to calculate the speed of sound in water instead of giving them its value (task b).

It is also a good idea to suggest students to sketch how the experiment was carried out (task a), although it might be a very demanding drawing task for students due to the fact that the verbal description is really incomplete and hardly can lead any student towards necessary technical details of the design (see Figs. 65.5 and 65.6). In addition, the impersonal form of the narrative about the experiment might be a serious obstacle for students' sense making and learning.

Being inspired by the drawing task (1) in the UK textbook presentation, in our research, described in the next section, we explored how students visualize and make sense of incomplete verbal and visual information of the chosen textbook presentation of the historical experiment for measuring the speed of sound in water.

To make our discussion of the process followed by the students as they try to understand a given text more understandable, it is necessary to remind readers of the ideas of Kintsch (1998) and Kintsch and van Dijk (1978) regarding text comprehension. They claim that two basic steps are important in order to understand a text: the construction of the text base and the construction of the corresponding situation model. The text base is drawn from the propositions of the text, and it expresses its semantic content, both globally and locally. The situation model is constructed by integrating the textual content in the reader's knowledge schemes. Any text for which the reader is unable to construct a correct situation model is not understandable. If the text is taken from a textbook, then such a text is not cognitively adequate for students learning.

The research procedures to find out details of the situation models students construct for a textbook presentation might be word based or drawing based. As already indicated, the second option was selected. This option is used in research on students' science learning (Benson et al. 1993; Edelson 2001; Köse 2008; Shepardson et al. 2011), students' comprehension of scientific texts (Schwamborn et al. 2010; Leopold and Leutner 2012), and students' images of scientists (Farland-Smith 2012) or mathematicians (Picker and Berry 2000).

Up to date, as far as we know, this is the first research using pupil's drawings as a tool to assess comprehension of textbook presentation.

65.4 Basic Research Description: Participating Pupils, Worksheet Design, and Evaluation

65.4.1 *Participating Pupils*

A group of 151 pupils participated in this research. The locations of the schools and pupils' gender characteristics are given in Table 65.3.

Locations of the PS1–PS3 were situated in Sarajevo Canton, whereas location PS4 was situated in Central Bosnia Canton. It is worthy to mention that these two Cantons were randomly selected and have different education policies.

Pupils were given text worksheets (Box 65.5) containing only written information on the Lake Geneva Experiment from the textbook D8 and four tasks. The original picture showed in the D8 textbook was not attached.

65.4.2 *Worksheet Design and Evaluation*

The first task was to make a drawing based on the information given in the text worksheet. The second task was to explain in textual form all the difficulties they encountered in the first task. After that, the students were given the original drawing from the D8 textbook, and they were told to complete the third and fourth task.

Table 65.3 Distribution of pupils by school, number of class unit, and gender

School	N (Classes)	Female	Male	N (Pupils)
PS1	1	11	11	22
PS2	1	13	6	19
PS3	3	22	28	50
PS4	3	30	30	60
Total	8	76	75	151

Notes: *N* number, *PS* primary school

Box 65.5 The Content of the Worksheet

(WS)Pupil's Code:

Speed of Sound in Water

The following text describes the experiment used to measure the speed of sound in water in the nineteenth century:

A physicist and his assistant were in two boats 1,500 m away from each other. A bell on a rope was immersed in water from the first boat with the assistant in it. His task was to hit the immersed bell with a hammer and simultaneously to send a light signal. The physicist with a long horn was in the second boat. One end of the horn was immersed in water, while the other end of the horn was held by the physicist near his ear. In the moment he saw the light signal, the physicist switched the chronometer on and measured the time difference between the moment he saw the light and the moment he heard the sound. The delay of sound signal was one second. The physicist concluded that the sound needed one second to travel 1,500 m through water, and that the speed of the sound in water was 1,500 m/s.

Questions/Tasks

1. Insert into the box below your own drawing according to the given text.

2. If any part of the given text is not presented in your drawing, then describe that part by your own words. Explain why it is important to understand this experiment, and why it might be difficult to draw it.
   Answer:

3. Describe your feelings and thoughts about your own drawing and the drawing from the textbook that has been shown to you.
   Answer:

4. Express your opinion here about your today learning experience.
   Answer:

The third task asked the students to give feedback on their thoughts and emotions after the presentation of the original drawing. The feedback was given verbally in a classroom in the presence of one of the researchers (Z. H.). Finally, the students were asked to complete the fourth task by expressing their attitudes towards active and passive physics learning and to compare this classroom experiment with the standard education they received previously.

The first task was used to assess the ability of pupils to construct an adequate situation model according to Kintsch's terminology.

To track pupils' achievements, a scoring scheme was developed specifically to evaluate pupil's understanding of the given text. According to this, each of the following scoring rubric elements was graded with one point (seven points in total):

1. Any drawing according to in-class reading (given text)
2. Drawing similar to the D8 (the same number and choices of drawing elements)
3. Device for sending light signal from the first boat
4. Instrument/device for time measurement placed in the second boat
5. Sound wave representation in sinusoidal shape
6. Data of distance (1,500 m)
7. Velocity value (1,500 m/s).

65.5 Results and Analysis

The statistical data based on the results of the 151 pupils who responded to the questions and solved the tasks in the WS were analyzed. Achieved scores within different groups of pupils (male or female and pupils from different cantons) were tested for normality of distribution with Kolmogorov-Smirnov and Shapiro-Wilk tests. Since the distribution was not normal, the scores were expressed as median and compared within different groups with Mann-Whitney test. Pupils reached the median value of four points.

There were no significant differences in scores between pupils from different counties (Mann-Whitney U: 2602.5, $p = 0.62$) or between male and female pupils (Mann-Whitney U: 2,740, $p = 0.68$). However, the minimum score achieved by the male pupils was 2, whereas the female pupils had a minimum score of 0 (Table 65.4).

A maximum score of seven was accomplished by five (3.3 %) pupils. Pupils were classified into five groups according to the number of achieved points as follows:

Group I: 0–3 points
Group II: 4 points
Group III: 5 points
Group IV: 6 points
Group V: 7 points.

The distribution of the pupils among the above mentioned groups is shown in Fig. 65.8.

Pupils' results that scored according to the seven expected drawing elements (items) are presented in Table 65.5.

The most important result of this research was the opportunity for each pupil to actively participate in drawing and after-drawing discussions. According to their

Table 65.4 Basic statistical data

	Male, N=75	Female, N=76	Total, N=151
Median	4	4	4
Range	5	7	7
Minimum	2	0	0
Maximum	7	7	7
Achieved/total points (%)	61	62	61

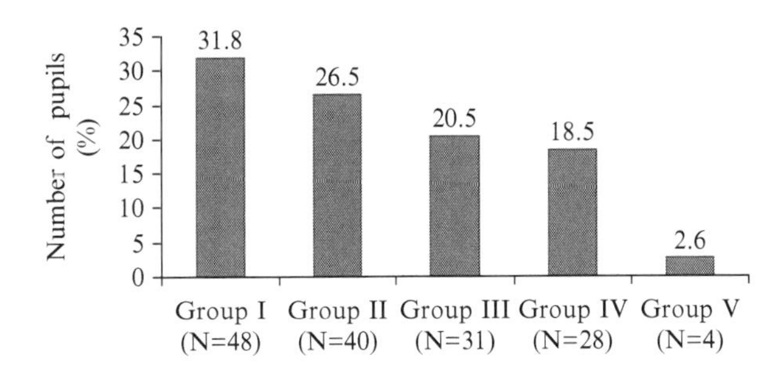

Fig. 65.8 Distribution of pupils' groups by achieved points

Table 65.5 Percentage of pupils that scored in each item

Expected drawing item	Grading point	Frequency (%)
Any drawing according to the text	1	99.3
Drawing identical to the D8	1	28.5
Device for sending light signal from the first boat	1	11.3
Instrument/device for time measurement placed in the second boat	1	15.2
Sound wave representation in sinusoidal shape	1	58.9
Data of distance (1,500 m)	1	79.5
Velocity value (1,500 m/s)	1	18.5

teachers' testimonies, many of them find common activities in physics classrooms to be boring and uninteresting because of their passive roles.

Surprisingly, the number of pupils who used sound wave representation in sinusoidal shape was high (58.9 %). This is interesting because in the primary school curriculum, they do not meet graphical representation of the sine function or analytical representation of wave phenomenon. Obviously, this is an influence from their out-of-school experiences.

Among the 151 pupils, 79.5 % considered the sound velocity value as an important detail that they needed to include in their drawings. However, it was discovered in a posterior class discussion that around 60 % of them believed that the sound velocity in air is greater than in water.

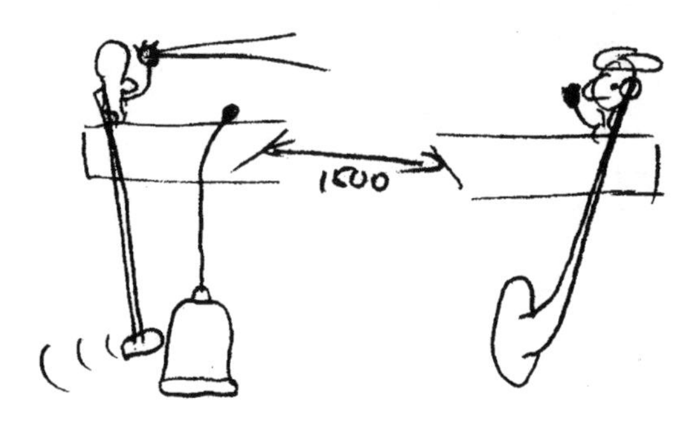

Fig. 65.9 A pupil's drawing with both light-sending and time-measuring devices

Data analysis is separated into four parts.

Part A
Basically correct drawings were given by 39 % of pupils, with *both light signal sending and time-measuring devices* being inserted into drawings that were similar to the one from the D8 (Fig. 65.9). Although their devices do not correspond to the actual devices used in the experiment, it is interesting that four out of 10 students were able to construct an adequate situation model from the given text. So, despite that some of the primary school students (from 14 to 15 years old) feel bored by the common physics learning activities, they grasped the experimental situation in a more accurate way than the artist who created the textbook drawing and the editor and reviewers who approved it!

Part B
Some pupils (11 %) revealed an abovementioned "backwards written history" approach to the reconstruction of the historical episode by drawing contemporary devices, such as lasers or digital watches. They can be hardly blamed for this "error" because that was their solution to the problem of cognitively inadequate narrative, which is unclear about the light-sending procedure or mentions time-measuring instruments that might be unknown to pupils (chronometer).

Part C
Only eight pupils (5 %) answered the second WS question. Such a poor participation can be understood within the context of classroom culture: these pupils are rarely asked to express and describe in their own words their thinking as well as the learning obstacles they encounter. This explains why it was preferred to use the drawing mode instead of the verbal mode to explore the students' comprehension of the text.

Nevertheless, what these eight pupils wrote is very informative. Five of them indicated that they did not know how to present sound after the light signal was sent. Two students wrote down that they found difficult to draw a "broken" thing (hammer, bell, horn). It is a nice example of how complex a drawing task might be for students who would like to apply their prior knowledge from optics. One student wrote that he did not know how to represent the situation of the physicist in the second boat seeing the light signal.

Part D

The students finally had the chance to observe the textbook's drawing and make comparison between the textbook's and their own drawing. This resulted to a series of different comments. According to the comments and expressed emotions, the pupils can be divided into three groups: satisfied, frustrated, and indifferent.

The first group (G1) consists of 78 satisfied students (52 %) who expressed positive emotions and gave affirmative comments after comparing their drawings with the textbook drawing shown to them by the researcher (Z. H.). The most prevalent keywords they used were "happy, satisfied, and pleased."

The second group (G2) consists of 10 frustrated pupils (7 %) who had negative comments and emotions about their drawings after the comparison with the textbook drawing. They used keywords like "disappointed, sad, frustrated, and embarrassed."

The third group (G3) consists of 63 indifferent pupils (41 %) who had no written comments or opinion presented in the WS.

In G1 four different subgroups of pupils can be distinguished who used different descriptions of their drawings as follows:

G1-a: "My drawing is *similar* to the drawing from the textbook" was stated by 61 pupils (72 %).

G1-b: "My drawing is *the same* as the drawing in the textbook" was stated by 7 pupils (8 %).

G1-c: "My drawing is *different* from the drawing in the textbook" was stated by 11 pupils (13 %).

G1-d: "My drawing is *better (richer)* than the drawing from the textbook" was stated by 4 pupils (5 %).

It is interesting to note that in the G1-a subgroup, there are pupils whose drawings include the same elements as the drawing from the textbook (case of 28 of pupils). Twenty pupils drew both signal devices, while the drawing from the textbook does not show any signal. Seven pupils used a light source (lamp, laser, sun-mirror) in their drawing, and six pupils had the time-measuring device in the boat with the physicist.

In the G1-b subgroup some pupils' drawings actually were different. Two pupils used two assistants in the boat; others drew a signaling device in the physicist's boat.

In the G3-c subgroup the pupils explicitly stated why their drawings were better or richer than the drawing from the textbook:

> My drawing is better because I have a lamp and a timepiece which is not included in the original drawing.
>
> My drawing is a better one. In the drawing from the textbook the light is not shown, the sound visualization is not found and the distance between the two boats is poorly represented.

Pupils from G2 believed that their drawings differed from the textbooks' drawing because they did not draw the physicist and his assistant (boats without people, for example). This is a very important detail, because both persons were mentioned

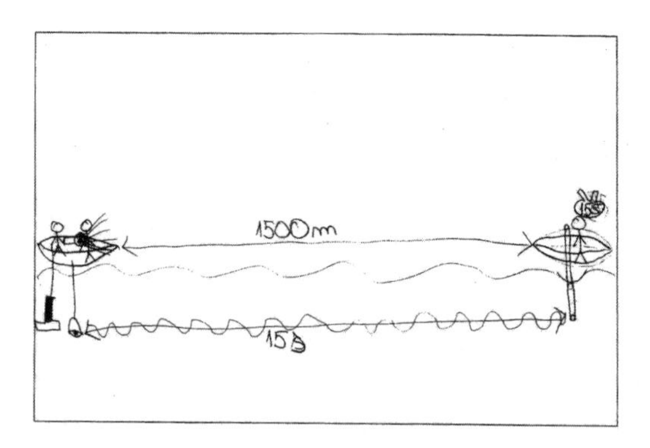

Fig. 65.10 Two experimenters in the first boat

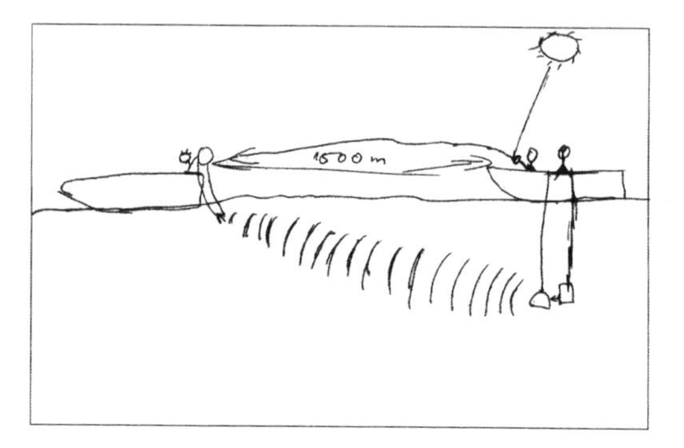

Fig. 65.11 Two experimenters in the first boat, one sending light signal by a mirror that reflects a ray from the sun

explicitly in the text. So, if some pupils omit in-text-mentioned persons in their drawings, then it is very likely that this situation model without persons will be found more frequently in pupil/student-generated drawings for the narrative given in the UK textbook (Johnson et al. 2001) that does not mention the persons who carried out the experiment.

65.5.1 Some Comments on Selected Pupils' Drawings

Pupils' sense making of the textbook verbal information about the Lake Geneva Experiment is a subtle process. It can be derived from pupils' drawings, which are their visual representations of the corresponding situation model.

In order to solve the problem how the two simultaneous events (i.e., hitting the bell and sending the light signal) were carried out, some pupils added the second experimenter in the first boat. One experimenter hits the bell and the other sends the light signal using a light source. Two different approaches were used to show the way the signal was sent. According to the first one, the signal was sent from a light source (Fig. 65.10). According to the second, the light signal consists of the solar light beam reflected by a mirror (Fig. 65.11).

A further difference appears in the second boat. Although in both cases the experimenter used the tube to receive the sound signal through the water, in the first case no

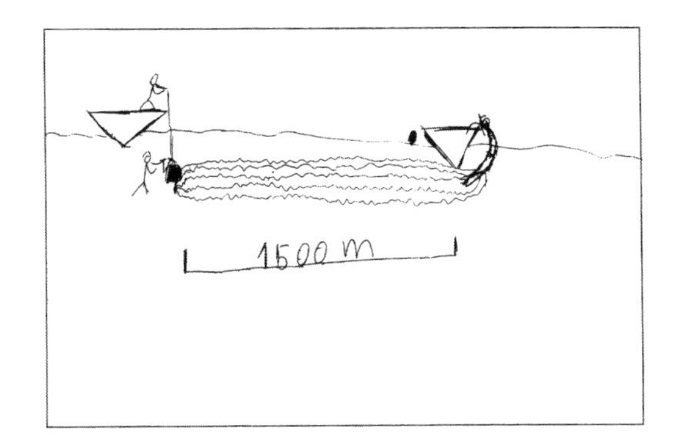

Fig. 65.12 One experimenter in the first boat and the second underwater

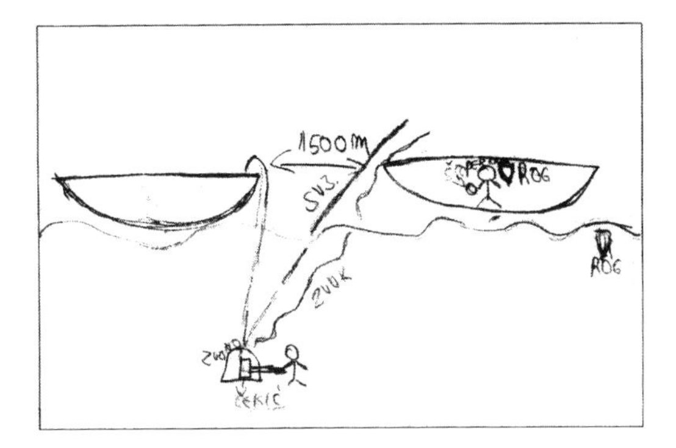

Fig. 65.13 An experimenter with two busy hands in the second boat

time-measuring device is presented, while in the second a watch is floating in the air near the experimenter. Obviously this pupil did not know the way the device operated.

Two experimenters are imagined in other drawings, too (Fig. 65.12). In the first boat one experimenter stands holding the rope with the bell, while the other experimenter is under the water hitting the bell. Obviously, for this pupil the only way to hit an underwater bell is to have somebody under the water to do it.

The same idea of hitting one underwater bell was adopted by another pupil (Fig. 65.13). Curiously, this pupil was able to imagine that the physicist in the second boat could use the right hand to hold the chronometer and the left hand to operate the tube.

Another pupil applied the two-busy-hands idea for the experimenter in the first boat (Fig. 65.14). The first experimenter uses his right hand to send the light signal from a battery lamp and the left hand to hit the underwater bell by a hammer. Interestingly, this pupil did not apply the same idea for the second experimenter who holds the tube but not the digital chronometer.

One pupil drew both experimenters with two busy hands (Fig. 65.15). The experimenter in the first boat uses one hand to hit the underwater bell and the other hand to hold a mirror to send the light signal (in the form of reflected solar rays).

The experimenter in the second boat holds with both hands the timepiece, while the tube is attached to his left ear.

The mirror as light-sending device appeared in another drawing, too (Fig. 65.16). This pupil suggested a pendulum as time-measuring device.

Fig. 65.14 Experimenter with two busy hands in the first boat

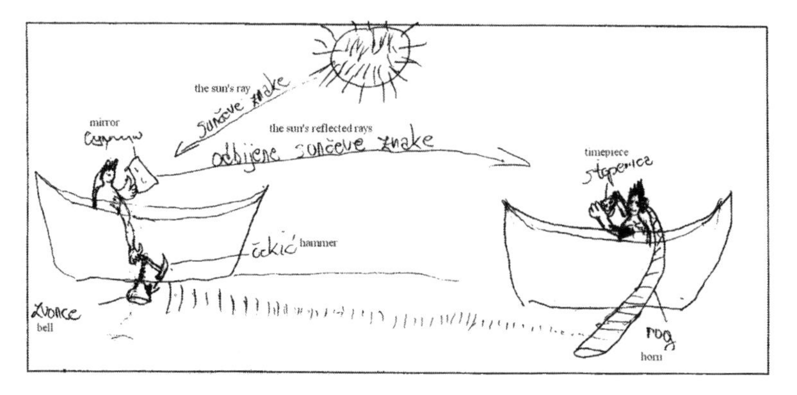

Fig. 65.15 Two experimenters with two busy hands

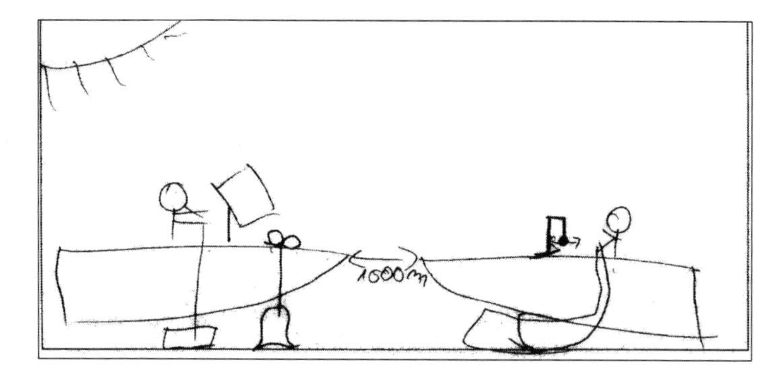

Fig. 65.16 Mirror as a light-sending device and a pendulum as time-measuring device

65.6 Conclusion

A great part of the historical information found in physics textbooks for primary school in Bosnia and Herzegovina is useless for learning about the nature of science because it is too short and rather incomplete. In the case of the Lake Geneva Experiment, a longer description of the historical episode was both historically erroneous and cognitively inadequate.

Cognitive adequacy was studied in relation to the textbook verbal description of the Lake Geneva Experiment aiming to measure the speed of sound in water. The task for the pupils was to make sense of the corresponding text by drawing the situation model concerning the specific experiment. In Kintsch's theory of text comprehension, the quality of the constructed situation model reveals the level of understanding.

Although 40 % of the pupils were able to draw an acceptable situation model with both light-sending and time-measuring devices that are absent in the textbook drawing, many pupils experienced difficulties in imagining how the experiment was done. Obviously, the text describing the experiment does not contain enough verbal clues to help these pupils construct a correct visual representation of the experimental situation. For them the textbook presentation of the experiment is incomplete and, in consequence, is not cognitively adequate.

In their sense-making efforts, some pupils suggested solutions (like underwater experimenter hitting the bell) that are erroneous. This fact shows that incomplete and superficial historical information is potentially misleading for pupils' learning.

Almost all pupils actively participated in the drawing activity, but very few provided written comments on the difficulties they experienced in thinking about how the experiment was done. The alarming absence of written pupils' notes about their thinking is very likely related to the classroom culture in which they rarely are asked to express what they think.

In the post-drawing discussion (with Z. H.), pupils commented that they enjoyed this novel activity (i.e., first draw the experimental situation and then compare their drawings with the textbook drawings) very much. It seems that most pupils prefer to talk about their ideas rather than to write about them, which is maybe a result of a rigid and more traditional education system in Bosnia and Herzegovina.

Up to date, this was the first study of the pupils' sense making of an incomplete textbook presentation of the Lake Geneva Experiment. Consequently, there are possibilities for future research. Some interesting questions are:

Would a "group drawing activity," following individual drawing attempts, improve the sense-making process of the same incomplete text?

Would a more complete verbal description of the historical episodes lead to better quality pupils' drawings?

Would impersonal account of the experiment, like the one in the UK textbook, be more difficult for pupils' drawing?

Acknowledgements We are very grateful to Prof. Gorazd Planinšič (University of Ljubljana) for his artwork in Fig. 65.7. Our thanks also go to Prof. Dewey I. Dykstra, Jr. (Boise State University), Prof. Chris Moore (Coastal Carolina University), and Dr Ioannis Papadopoulos (School of Primary Education, Aristotle University of Thessaloniki) for their kind help in improving the English.

References

Abd-El-Khalick, F. (2002). Rutherford's enlarged: A content-embedded activity to teach about the nature of science. *Physics Education* 37, 64–68.

Abd-El-Khalick, F. & Lederman N.G. (2000). The influence of history of science courses on students' views of nature of science. *Journal of Research in Science Teaching, 37*(10), 1057–1095.

Benson, D.L., Wittrock, M.C. & Baur, M.E. (1993). Students' preconceptions of the nature of gases. *Journal of Research in Science Teaching* 30(6*),* 587–597.

Binnie, A. (2001). Using the History of Electricity and Magnetism to Enhance Teaching. *Science & Education* 10(4), 379–389.

Brackenridge, J.B. (1989). Education in science, history of science, and the textbook - Necessary vs. sufficient conditions. *Interchange* 20(2), 71–80.

Chalmers, A. (2009). *Boston Studies in the Philosophy of Science: The Scientist's Atom and the Philosopher's Stone – How Science Succeeded and Philosophy Failed to Gain Knowledge of Atoms.* Dordrecht, Heidelberg, London, New York: Springer.

Colladon, J.D. (1893). *Souvenirs et Memoires*. Geneva: Albert-Schuchardt.

Conant, J.B. (1957). *Harvard case histories in experimental science*. Cambridge, MA: Harvard University Press.

Crawley, F.E. & Black, C.B. (1992). Causal modeling of secondary science students' intentions to enroll in physics. *Journal of Research in Science Teaching* 29(6), 585–599.

Drake, S. (1973). Galileo's experimental confirmation of horizontal inertia: Unpublished manuscripts (Galileo gleanings XXII), *Isis*, 64(3), 291–305.

Drake, S. (1975). The Role of Music in Galileo's Experiments. *Scientific American* 232, 98–104.

Edelson, D.C. (2001). Learning-for-Use: A framework for the design of technology-supported inquiry activities. *Journal of Research in Science Teaching* 38(3), 355–385.

Eijck, van M. & Roth, W.M. (2009). Authentic science experiences as a vehicle to change students' orientations toward science and scientific career choices: Learning from the path followed by Brad. *Cultural Studies of Science Education* 4(3), 611–638.

Farland-Smith, D. (2012). Development and Field Test of the Modified Draw-a- Scientist Test and the Draw-a-Scientist Rubric. *School Science and Mathematics* 112(2), 109–116.

Galilei, G. (1954). *Dialogues concerning two new sciences*. New York: Dover.

Galilei, G. (1967). *Dialogue concerning the two chief world systems*. Berkeley: University of California Press.

Galili, I. & Hazan, A. (2000). The influence of an historically oriented course on students' content knowledge in optics evaluated by means of facets- schemes analysis. *American Journal of Physics* 68, S13–S15.

Galili, I. & Hazan, A. (2001). The Effect of a History-Based Course in Optics on Students' Views about Science. *Science & Education* 10(1-2), 7–32.

Hannover, B. & Kessels, U. (2004). Self-to-prototype matching as a strategy for making academic choices. Why high school students do not like math and science. *Learning and Instruction* 14(1), 51–67.

Heering, P. (2000). Getting Shocks: Teaching Secondary School Physics Through History. *Science & Education* 9(4), 363–373.

Holton, G. (1952), *Introduction to concepts and theories in physical science*. Reading, MA: Addison – Wesley.

Holton, G. (2003). The Project Physics Course, Then and Now. *Science & Education* 12(8), 779–786.
Irwin, A.R. (2000). Historical case studies: Teaching the nature of science in context. *Science Education* 84(1), 5–26.
Johnson, K., Adamson, S. & Williams, G. (2001). *Spotlight Science 9. Second Edition*. Cheltenham: Nelson Thornes.
Kintsch, W. & Dijk, van T.A. (1978). Toward a Model of Text Comprehension and Production. *Psychological Review* 85(5), 363–394.
Kintsch, W. (1998). *Comprehension: A Paradigm for Cognition*. Cambridge, MA: Cambridge University Press.
Klopfer, L.E. & Cooley, W.W. (1963). The history of science cases for high schools in the development of student understanding of science and scientists: A report on the HOSG instruction project. *Journal of Research in Science Teaching* 1(1), 33–47.
Klopfer, L.E. (1969). The teaching of science and the history of science. *Journal of Research in Science Teaching* 6(1), 87–95.
Kokkotas, P.V., Malamitsa, K.S. & Rizaki, A.A. (Editors). (2010). *Adapting Historical Knowledge Production to the Classroom*. Rotterdam: Sense Publishers.
Koyre, A. (1968). *Metaphysics and Measurement. Essays in Scientific Revolution*. Cambridge: Harvard University Press. Chapters I, II, III and IV.
Köse, S. (2008). Diagnosing Student Misconceptions: Using Drawings as a Research Method. *World Applied Sciences Journal* 3(2), 283–293. http://idosi.org/wasj/wasj3%282%29/20.pdf. Accessed 8 December 2012.
Krapp, A. & Prenzel, M. (2011). Research on Interest in Science: Theories, methods, and findings. *International Journal of Science Education* 33(1), 27–50.
Lavigne, G.L. & Vallerand, R.J. (2010). The Dynamic Processes of Influence Between Contextual and Situational Motivation: A Test of the Hierarchical Model in a Science Education Setting, *Journal of Applied Social Psychology* 40(9), 2343–2359.
Lee, J.D. (1998). Which Kids Can "Become" Scientists? Effects of Gender, Self-Concepts, and Perceptions of Scientists. *Social Psychology Quarterly* 61(3), 199–219.
Leite, L. (2002). History of Science in Science Education: Development and Validation of a Checklist for Analyzing the Historical Content of Science Textbooks. *Science & Education*, 11(4), 333–359.
Leopold, C. & Leutner, D. (2012). Science text comprehension: Drawing, main idea selection, and summarizing as learning strategies. *Learning and Instruction* 22(1), 16–26.
Mamlok-Naaman, R. (2011). How can we motivate high school students to study science? *Science Education International,* 22(1), 5–17. http://stwww.weizmann.ac.il/menu/staff/Rachel_Mamlok/. Accessed 8 December 2012.
Mamlok-Naaman, R., Ben-Zvi, R., Hofstein, A., Menis, J. & Erduran, S. (2005). Learning Science through a Historical Approach: Does It Affect the Attitudes of Non-Science-Oriented Students towards Science? *International Journal of Science and Mathematics Education* 3(3), 485–507.
Matthews, M.R. (1994/2014). *Science teaching. The role of history and philosophy of science*. New York, London: Routledge.
Monk, M. & Osborne, J. (1997). Placing the history and philosophy of science on the curriculum: A model for the development of pedagogy. *Science Education* 81(4), 405–424.
Niaz, M. (2008). *Physical science textbooks: history and philosophy of science*. Hauppauge, NY: Nova Science Publishers.
Niaz, M. (2009). *Critical Appraisal of Physical Science as a Human Enterprise: Dynamics of Scientific Progress*. New York: Springer
Niaz, M. (2010). *Innovating Science Teacher Education: A History and Philosophy of Science Perspective*. New York: Routledge.
Osborne, J., Simon, S. & Collins, S. (2003). Attitudes towards science: A review of the literature and its implications, *International Journal of Science Education 25*(9), 1049–1079.
Picker, S.H. & Berry, J.S. (2000). Investigating pupils' images of mathematicians, *Educational Studies in Mathematics*, 43(1), 65–94.

Riess, F. (2000). History of Physics in Science Teacher Training in Oldenburg. *Science & Education* 9(4), 399–402.
Russell, T.L. (1981). What history of science, how much, and why? *Science Education* 65(1), 51–64.
Rutherford, F.J., Holton, G. & Watson, F.G. (1970). *The Project Physics Course Text*. New York: Holt, Rinehart & Winston.
Schwamborn, A., Mayer, R.E., Thillmann, H., Leopold, C. & Leutner, D. (2010). Drawing as a generative activity and drawing as a prognostic activity. *Journal of Educational Psychology* 102(4), 872–879.
Seroglou, F. & Koumaras, K. (2001). The Contribution of the History of Physics in Physics Education: A Review. *Science & Education* 10(1–2), 153–172.
Shepardson, D.P., Choi, S., Niyogi, D. & Charusombat, U. (2011). Seventh grade students' mental models of the greenhouse effect. *Environmental Education Research* 17(1), 1–17.
Slisko, J. & Hadzibegovic, Z. (2011). Cavendish Experiment in Physics Textbooks: Why do Authors Continue to Repeat a Denounced Error? *European Journal of Physics Education* 2(3), 20–32. http://ejpe.erciyes.edu.tr/index.php/EJPE/article/view/41/30. Accessed December 8 2012.
Solbes, J. & Traver, M. (2003). Against a Negative Image of Science: History of Science and the Teaching of Physics and Chemistry. *Science & Education* 12(7), 703–717.
Solomon, J., Duveen, J., Scot, L. & McCarthy, S. (1992). Teaching about the nature of science through history: Action research in the classroom. *Journal of Research in Science Teaching* 29(4), 409–421.
Wandersee, J.H. (1986). Can the history of science help science educators anticipate students' misconceptions? *Journal of Research in Science Teaching* 23(7), 581–597.
Wang, H.A. & Marsh, D.D. (2002). Science Instruction with a Humanistic Twist: Teachers' Perception and Practice in Using the History of Science in Their Classrooms. *Science & Education* 11(2), 169–189.

Josip Slisko (B.Sc. in physics, M.Sc. in philosophy of physics, Ph.D. in philosophical sciences) teaches courses on physics and mathematics education at Facultad de Ciencias Físico Matemáticas of the Benemérita Universidad Autónoma de Puebla. His research interests include students' explanatory and predictive models of physical phenomena, students' strategies for solving untraditional physics and mathematics problems, presentation of knowledge in textbooks, and design of active learning sequences which promote cognitive, metacognitive, and emotional development of students. Josip Slisko is author or coauthor of 80 journal articles and 12 physics textbooks. Since 1994 he joined the Sistema Nacional de Investigadores (National System of Researchers), a program of the Mexican government through which scientific production of one is evaluated and a stipend is paid to recognized scientists. He is member of American Association of Physics Teachers and serves in editorial boards of various educational journals (Latin American Journal of Physics Education, European Journal of Physics Education, Metodicki ogledi and Eureka). Since 1993, every last week in May, he is the president of the committee that organizes an international workshop called New Trends in Physics Teaching.

Zalkida Hadzibegovic (B.Sc. in physics, M.Sc. in history and philosophy of science, Ph.D. in physics) teaches courses on general physics, astronomy, and philosophy and history of science at the University of Sarajevo, Bosnia and Herzegovina. Zalkida Hadzibegovic created and implemented a new course curriculum in the Second Bologna Study Cycle in the field of philosophy of science for prospective physics teachers and researchers and has been offering a few master work themes which will begin research related to the role of history and philosophy of science in Bosnian elementary and high school education. She established in 2009 the first science education research group (Sarajevo Chemistry and Physics Education Research Group) which conducts research on integrated chemistry, physics, mathematics, and history and philosophy of science knowledge and focused on active learning approach. Her research interests include the applications of history and philosophy of science in science education and students' scientific literacy and creativity at the university and high school level especially for talented students. Zalkida Hadzibegovic is author or coauthor of several physics textbooks. She published a few articles related to different topics of physics education participating as author or coauthor at the GIREP (Groupe International de Recherche sur l'Enseignement de la Physique/International Research Group on Physics Teaching) conferences, ESERA (European Science Education Research Association), BPU (Balkan Physical Union), and WCPE (The World Conference on Physics Education) conferences, and at the SEAC (European Society of Astronomy in Culture) conferences in the field of astronomy.

Chapter 66
One Country, Two Systems: Nature of Science Education in Mainland China and Hong Kong

Siu Ling Wong, Zhi Hong Wan, and Ka Lok Cheng

66.1 Introduction

An understanding of NOS has been widely recognised as an essential component of scientific literacy and has been included as a curriculum goal in science curriculum standards documents in many developed regions of the Western world.[1] Driver, Leach, Millar and Scott (1996) have crystallised from the literature five arguments in support of developing students' NOS understandings:

1. Utilitarian – understanding NOS is necessary for making sense of the science and managing the technological objects and processes in everyday life.
2. Democratic – understanding NOS is necessary for making sense of socioscientific issues and participation in decision-making process.
3. Cultural – understanding NOS is necessary for appreciation of science as a major element of contemporary culture.
4. Moral – understanding NOS can help develop awareness of the norms of the scientific community that embodies moral commitments which are of general value.
5. Science learning – understanding NOS can support successful learning of science content.

[1] See, for example, American Association for the Advancement of Science (AAAS) 1990; Council of Ministers of Education, Canada (CMEC) 1997; Department for Education (DfE), England 1995; and National Research Council (NRC) (1996).

S.L. Wong • K.L. Cheng
Faculty of Education, The University of Hong Kong, Pokfulam Road, Hong Kong SAR, China
e-mail: aslwong@hku.hk; chengkla@hku.hk

Z.H. Wan (✉)
Department of Curriculum and Instruction, The Hong Kong Institute of Education, Lo Ping Road, Hong Kong SAR, China
e-mail: zhwanhku@gmail.com; wanzh@ied.edu.hk

M.R. Matthews (ed.), *International Handbook of Research in History, Philosophy and Science Teaching*, DOI 10.1007/978-94-007-7654-8_66,

Although NOS has been one of the extensively researched areas in science education, there is not a unified way to define the term NOS in the literature. Some authors (Abd-El-Khalick and Lederman 2000; Lederman 2007) delimit NOS to aspects related to epistemology of science. In this chapter, we adopt a broader meaning of NOS in common with definitions adopted by Clough (2006), Irzik and Nola (2011), Osborne and his colleagues (2003) and Wong and Hodson (2009, 2010). The phrase NOS used in their work encompasses the nature of scientific inquiry, the nature of the scientific knowledge it generates, how scientists work as a social group and how science impacts, and is impacted by, the social context in which it is located.

Science education in both the mainland China and Hong Kong has undergone major curriculum reform since the beginning of the twenty-first century. One of the common new aims introduced in curriculum standards documents is the development of students' appreciation of nature of science (NOS).

Prompted by the soaring economy in recent years, which has brought tremendous changes in people's lives, the mainland Chinese government started to look for strategies to sustain long-term development of the country. These strategies include reforming education that can nurture and prepare the future generations for its development. Within science education in mainland China, there is a transition from a more elite to a more 'science for all' curriculum with an emphasis on the promotion of scientific literacy (Wei and Thomas 2005). NOS has hence started to become an important topic in some science curriculum reform documents (e.g. Ministry of Education [MOE] 2001a, b), Chinese academic articles,[2] as well as textbooks for training science teachers (e.g. Liu 2004; Yu 2002; Zhang 2004).

Science education in Hong Kong has also undergone considerable changes since the implementation of the revised junior secondary science curriculum (grades 7–9) (Curriculum Development Council [CDC] 1998). It was the first local science curriculum that embraced certain NOS features, e.g. being 'able to appreciate and understand the evolutionary nature of scientific knowledge' (CDC 1998, p. 3) was stated as one of its broad curriculum aims. In the first topic, 'What is science?', teachers are expected to discuss with students some features about science, e.g. its scope and limitations and some typical features about scientific investigations, e.g. fair testing, control of variables, predictions, hypothesis, inferences and conclusions. Such an emphasis on NOS was reinforced in the revised secondary 4 and 5 (grades 10 and 11) physics, chemistry and biology curricula (CDC 2002). Scientific investigation continued to be an important component, while the scope of NOS was slightly extended to include recognition of the usefulness and limitations of science as well as the interactions between science, technology and society (STS). In the recently implemented senior secondary curricula of the science subjects (CDC-HKEAA 2007), there is a further leap forward along the direction of earlier curriculum reforms in the curriculum and assessment guides. The importance of

[2] See, for example, Chen and Pang (2005), Liang (2007), Xiang (2002), and Yuan (2009).

promoting students' understanding of NOS is explicitly spelt out together with its perceived benefits to students.

Although understanding of NOS has become a key curriculum aim in the science curriculums in both mainland China and Hong Kong, under the policy of the 'One Country, Two Systems', Hong Kong has retained its autonomy to decide on its own policies. There have been some distinctive differences of NOS education in both places as reflected in the (1) official curriculums, (2) textbooks, (3) teacher training and (4) school teachers' implementation of NOS teaching. This chapter provides an overview of the situations about NOS education in both places for these four areas.

66.2 NOS as Portrayed in Official Science Curricula

In a recent study, Cheng and Wong (in print) have examined the NOS ideas included in the two recent official senior physics curriculum documents used in mainland China (MOE 2004, thereafter known as '*CHN-Standards*') and Hong Kong (CDC-HKEAA 2007, '*HK-Guide*'). Ten NOS ideas are identified as listed below:

1. Laws as generalisations and theories as explanations of the generalisations
2. Creative elements of the scientific processes
3. Tentative and developmental nature of science
4. Distinction and relationship between science and technology
5. Theory-laden nature of scientific processes
6. Empirical nature of scientific knowledge
7. Different ways of performing scientific investigations
8. Interactions between science, technology and the society
9. Moral and ethical dimensions of science
10. Scientists as a community

Among them, some are similarly represented in both Hong Kong and the mainland China, including 'Laws as generalisations and theories as explanations of the generalisations', 'Tentative and developmental nature of science', 'Empirical nature of scientific knowledge' and 'Scientists as a community'. However, some of these NOS ideas, including 'Different ways of performing scientific investigations', 'Interactions between science, technology and the society' and 'Moral and ethical dimensions of science', are presented with significant differences.

66.2.1 Methods of Scientific Investigations

There is a dedicated chapter on scientific investigations in *CHN-Standards* in which the 'components of scientific investigations' are listed in a table as shown in Table 66.1. The seven components are stunningly similar to *the* stepwise scientific method as one of the myths about NOS highlighted by McComas (1998).

Table 66.1 Components of scientific investigations as presented in *CHN-Standards* (pp. 10–11)

Components of scientific investigations	Basic skill requirements for scientific inquiries and physics experiments
Question formulation	Able to discover physics-related problems
	Express these problems clearly as physics problems
	Aware of the significance of problem discovery and question formulation
Speculation and hypothesis formation	Propose problem-solving methods and solutions to problems
	Predict the results of physics experiments
	Aware of the importance of speculation and hypothesis
Experiment planning and design	Develop plan according to the experimental objectives and conditions
	Attempt to select the appropriate experimental methods, set-ups and instruments
	Consider the experimental variables and their control
	Aware of the role of planning
Experimentation and data collection	Collect data using a variety of methods
	Perform experiments according to guidelines and able to use common instruments
	Faithful recording of experimental data and aware of the meaning of the duplicated collection of experimental data
	Being safety-conscious
	Aware of the importance of objective collection of experimental data
Analysis and reasoning	Analysis of experimental data
	Attempt to develop conclusion according to the observations and data
	Explain and describe the experimental data
	Aware of the importance of analysis and reasoning to experiments
Evaluation	Attempt to analyse the differences between hypotheses and experimental results
	Attend to the unresolved issues in the inquiries and discover new problems
	Improve the inquiry plan with reference to the experience gained
	Aware of the significance of evaluation
Exchange and co-operation	Able to write reports for the experimental inquiries
	Upholding principles while respecting others during co-operation
	Be co-operative
	Aware of the importance of exchange and co-operation

In the *HK-Guide*, 'scientific investigation' is a subsection under Skills and Processes,[3] where the intended learning targets that could be acquired through conducting scientific investigations are listed. Students are expected to:

- Ask relevant questions
- Propose hypotheses for scientific phenomena and devise methods to test them
- Identify dependent and independent variables in investigations
- Devise plans and procedures to carry out investigations
- Select appropriate methods and apparatus to carry out investigations
- Observe and record experimental observations accurately and honestly
- Organise and analyse data and infer from observations and experimental results
- Use graphical techniques appropriately to display experimental results and to convey concepts
- Produce reports on investigations, draw conclusions and make further predictions
- Evaluate experimental results and identify factors affecting their quality and reliability
- Propose plans for further investigations, if appropriate

(HK-Guide, pp. 9–10)

It is noteworthy that among all nine subsections under Skills and Processes, 'practical work' shares a few similar learning targets as 'scientific investigations', for example, students are expected to:

- Devise and plan experiments
- Select appropriate apparatus and materials for an experiment
- Interpret observations and experimental data
- Evaluate experimental methods and suggest possible improvements

(HK-Guide, p. 10)

By listing the expected learning targets related to Skills and Processes expected to be achieved through scientific investigation instead of spelling out the 'Components' of Scientific Investigations as presented in *CHN-Standards*, the *HK-Guide* might give a lesser impression of a rigid stepwise method of doing scientific investigations.

66.2.2 Interactions Between Science, Technology and Society

While the specifications on the discussion of the positive aspect of the technological applications of science are found in both documents as expected, *CHN-Standards* attends to the societal development brought by technological advances to a larger

[3] The learning targets of the Hong Kong physics curriculum are categorised into three domains: Knowledge and Understanding, Skills and Processes and Values and Attitudes.

extent. Students are required to appreciate 'the roles of the widespread use of heat engines in bringing about the changes towards science, *social development and the mode of living*'. (p. 22, emphasis added), while the specification in the *HK-Guide* requires students to understand 'how major breakthroughs in scientific and technological development that eventually affect society are associated with new understanding of fundamental physics' (p. 49).

The *HK-Guide* also pays more attention to the controversies that can be caused by new technologies, for example, in the topic of Wave Motion, students are expected to develop understanding of the 'controversial issues about the effects of microwave radiation on the health of the general public through the use of mobile phones' (p. 38). Also, in the topic of Radioactivity and Nuclear Energy, students are 'to be aware of different points of view in society on controversial issues and appreciate the need to respect others' points of view even when disagreeing; and to adopt a scientific attitude when facing controversial issues, such as the use of nuclear energy' (p. 51).

On the other hand, the *CHN-Standards* only generally states that students should be 'aware of the social problems that are brought about by the technological applications of Physics' (p. 2); it is less emphasised on, if not silent, about controversial issues related to use of technologies. For example, in the topics of Energy Sources and Development of Society, it is suggested that students 'investigate the common pollutants causing air pollution' (p. 23) and 'understand the development and application of nuclear technology as well as its outlook' (p. 23). Yet contested views about the use of nuclear energy are not brought up.

The lack of suggested learning and teaching activities for discussion of controversial socioscientific issues in the *CHN-Standards* is strikingly consistent with the absence of the democratic argument as a reason for teaching NOS as stated by the Chinese science teacher educators reported by Wan and colleagues (2011). As the authors said,

> [T]he absence may be explained in terms of the political context in China… a socialist country governed by Chinese Communist Party. In such a less decentralized system, general public has relatively little voice in the public decision on social issues. (p. 1118)

66.2.3 Moral and Ethical Dimensions of Science

The only relevant content related to the moral and ethical dimensions of science in the *CHN-Standards* is about the positive image of scientists. The *CHN-Standards* suggests textbooks to 'use vivid information to exhibit their scientific spirit and their determined devotion to science' (p. 58). Scientists are to be honoured for their respectable character and selfless commitment to the development of science. The *CHN-Standards* also explicitly spells out the intention of using examples to illustrate such scientific spirit in order facilitate character building of learners.

In comparison, the perspective adopted by *HK-Guide* is more balanced and reflective. Students are suggested to discuss 'the roles and responsibility of scientists and the related ethics in releasing the power of nature' (p. 52) and address to the 'moral issues of using various mass destruction weapons in war' (p. 52). Unlike *CHN-Standards*, scientists as portrayed in the *HK-Guide* are not simply 'heroes' to be appreciated; the moral and ethical consequences of the technologies resulted from their endeavour are to be examined.

66.3 NOS as Portrayed in School Science Textbooks

Two sets of corresponding physics textbooks from each city, including the one published by People's Education Press (Centre for Research and Development on Physics Curriculum Resources of PEP 2004, 'PEP') and the one by Guangdong Education Press (Editorial Group for Physics Textbooks, Guangdong 2004, 'GEP') from Guangzhou, and the set published by Oxford University Press (Wong and Pang 2009, 'OUP') and by Longman (Tong et al. 2009, 'Longman') in Hong Kong, were studied.

Almost all NOS ideas that are commonly found in the official curriculums in the West (McComas and Olson 1998) are embedded in the textbooks used in both Hong Kong and the mainland China. It seems that the absence of corresponding NOS ideas in the official curriculums has not resulted in their absence from the textbooks. For instance, the text segments that could be deployed to exhibit the 'theory-laden nature of scientific processes', absent in *CHN-Standards*, could be found in the historical episodes included in the textbooks used in the mainland China. Similarly, while the 'creative elements of the scientific processes' are not included in *HK-Guide*, the use of analogical thinking in scientific process could be found in a small dose in the textbooks of Hong Kong. Considerable differences of NOS ideas presented in the textbooks of both sites are again from the three aspects described above.

66.3.1 Methods of Scientific Investigations

The implied message on the universality of scientific method is translated into concrete form in the textbooks of the mainland China. In PEP, a shortened sequence of the processes listed as 'key components of scientific investigation' (pp. 10–11, *CHN-Standards*) is diagrammatically presented as the common method to be used in science (Fig. 66.1). To illustrate such a sequence, the work of Galileo in the study of free-falling object is reconstructed and 'moulded' into the fixed sequence of steps shown in the diagram.

However, the textbooks in Hong Kong, unlike their mainland counterpart, do not produce a similar list in their texts for a less than impressive reason though: Both OUP and Longman mostly consist of 'cookbook'-type practical activities with the key purposes of facilitating students to learn the relevant target concepts and to practise experimental skills, which will be examined through written and

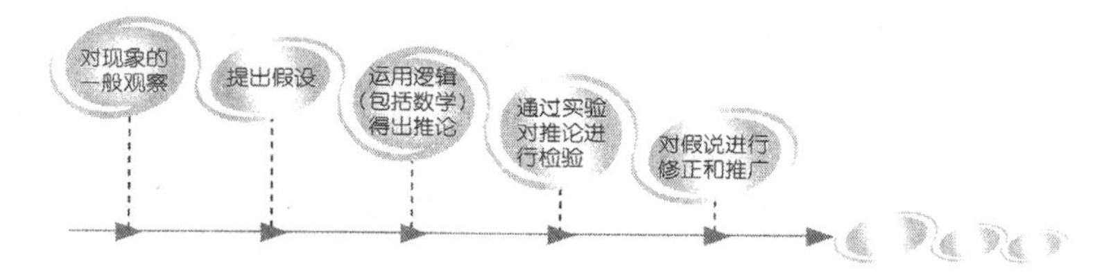

Fig. 66.1 Picture in one of the textbooks used in the mainland China (PEP 2004, Physics-1 p. 48) showing the steps of 'the scientific method' (The phrases in the picture, from left to right, mean 'General observation of phenomena', 'Hypothesizing', 'Obtaining corollary using logic (including mathematics)', 'Evaluation of the corollary through experiment' and 'Amendment and extension of the hypothesis')

practical examinations, respectively. The full detailed procedures of each practical activity are provided in the experimental workbooks, whereas the textbooks only provide an abridge version for reminding students the key features of the practical activity. The practical activities are more like heavily guided scientific investigations. Students can perhaps tell from the variation of the procedures of each practical activity that scientific investigations do not consist of a fixed set of stepwise procedures.

66.3.2 Interactions Between Science, Technology and the Society

The focus on the role of scientific and technological advances to the societal development in *CHN-Standards* is translated into textbooks in the mainland China. For example, the societal impact of the Industrial Revolution including that 'the venue of production changed from dispersed family production [houses] for agriculture and handcrafts to factories for centralized production', 'self-sufficient natural economy had evolved into market economy in which production and consumption are separated' and 'the development of cities facilitated the advancements of science, technology, education and cultural industries. Human society [were going] into the era of industrial civilization' (PEP 2004, Elective 1–2, p. 77) can be found in the textbooks.

The particular attention paid to the statements on 'the rise of manufacturing and service industries' and 'rise of market economy and capitalism' may be the result of the influence of the Marxist idea on historical periodisation, as similar influence of Marxism on the science teacher educators in the mainland China can also be found in Wan and colleagues (2012) and in the previous section of this chapter.

While textbooks are required to cover on the societal problems brought about by the products of science, the negative consequences of these applications are downplayed in PEP. For instance, safety problems stemmed from the use of nuclear energy are simply stated to be issues that 'are continuously being attended to'

(PEP 2004, Elective 1–1, p. 58), while two major nuclear meltdowns (the Three Mile Island accident and Chernobyl disaster) are just touched on (pp. 59–60) without any discussion of the related controversy.

However, GEP is not equally silent on socioscientific controversies, for instance, in the nuclear pollution problems brought by the Chernobyl disaster and the worries of the people on the use of nuclear power, which are natural results, are being mentioned (GEP 2004, Elective 2–3, p. 89). However, it is still attempting to contain students' worries by providing information like 'the nuclear reactors use low-enriched uranium... As such, reactors will not explode like nuclear bombs.' (GEP 2004, Elective 1–2, p. 50) to ensure students will not overreact towards the possible nuclear hazards.

On the other hand, for the societal implications of science and technology, the controversy about the use of nuclear energy is included in the OUP, and this corresponds well to the specifications in *HK-Guide*. In its discussion, the authors deploy a cartoon showing two demonstrations organised by pro- and anti-nuclear camps to represent the conflicting views presented on both sides (Radioactivity p. 90). Besides being told that there may be controversies, students are guided to study them. For instance, in the discussion of the possible harmful effects brought by the overhead power cables, students are required 'to find out ... action taken by pressure groups and the response of electricity companies' (Electricity p. 280). The socioscientific conflicts among different stakeholders and groups are being investigated.

However, in *Longman*, the impacts of science and technology on the society are superficially covered. Moreover, socioscientific controversies are mostly replaced by scientific findings on the possible health effects, as in the case of the overhead power cables. Socioscientific controversies are reduced to scientific colloquia:

> Some researches indicate that exposure to changing electric and magnetic fields from transmission lines may incur a number of health problems. These include leukaemia, cancer, miscarriage, clinical depression, etc. However, these conclusions are solely based on statistical studies ... The detailed mechanism is still unclear. (*Longman* Vol. 4, p. 440)

66.3.3 Moral and Ethical Dimensions of Science

The particular specification on the need for the textbooks to portray scientists as good models of students results in the narratives in *PEP* that describe scientists as extraordinarily moral such as the despise of wealth and title as presented through the example of Faraday and having the 'courage' in making the difficult decision to suggest the use of atomic bombs (which is unlikely to be presented as such in the curriculum artifacts of the West):

> Some scientists, especially those escaped from the Fascist persecution, had a premonition about the threat of atomic bombs especially after they had heard that Germany had accelerated the research on chain reactions... On July 1939, nuclear physicist Szilard and others found Einstein and wished he could use his status to urge United States to produce atomic bombs before Germany had done so... (PEP 2004, Elective 3–5, pp. 84, 85)

Similar narratives of scientists' advocacy on the use of atomic bombs to counter the possible threats can be found in the chapter of atomic physics in *GEP*. However, a more reflective stance is taken. The text initiates a discussion on whether scientists should be free from the moral consequences of their work:

> Should people develop nuclear technology if both the pros and cons are taken into account? Some say scientists should care about science *per se*, while the impacts of science and technology to the society should be under the care of sociologists. Do you agree? (*GEP* Elective 3–5 p. 88)

The textbooks in Hong Kong also invite students to reflect on the moral responsibility of scientists. The deaths, injuries and sicknesses resulted from the use of atomic bombs in World War II are mentioned, and rhetoric that attempts to identify scientists' responsibility of using atomic weapons was used (*OUP*, Radioactivity p. 96).

In *Longman*, the moral and ethical aspects are pointed out but not discussed in depth. For example, while 'international race of nano-weapons' and the 'technology of changing and even constructing DNA' are said to cause much impact on ethical values (Atomic World, p. 149) and nuclear energy is said to be able to 'destroy our civilization if it is misused' (Radioactivity, p. 108), no further discussion is following these brief mentions.

From the comparison of the official curricula in both sites, it can be seen that there are differences between mainland China and Hong Kong regarding their inclusion of NOS ideas in the official curricula. There are also significant differences in terms of the inclusion and the focus of the NOS ideas in the textbooks. It is obvious that some messages (overt and covert) in the official curricula could (and sometimes fail to) influence the corresponding textbooks. Given the important status of textbooks in the science classrooms in both mainland China and Hong Kong, a follow-up study in identifying how other factors, on top of the official curricula, play their roles in shaping the NOS ideas found in the textbooks is being conducted.

66.4 Preparing Science Teachers to Teach NOS in Mainland China

Although NOS has appeared in the revised science curriculum documents in mainland China, the training is mostly dependent on the initiative of science teacher educators in Normal (teacher training) Universities. There are no large-scale teacher professional development projects funded by the government in developing in-service teachers to achieve a critical number of teachers in implementing the new curriculum goals. It appears that the current priority for science education reform of the Chinese government is more concerned with promoting inquiry-based science teaching and learning. Nevertheless, as reported in recent studies (Wan 2010; Wan et al. 2011, 2012), there are a number of Chinese science teacher educators teaching NOS to prospective science teachers in their courses for the bachelor degree of science education.

The study by Wan and colleagues (2012) revealed the influence of Marxism on these educators' conceptions. Although, during the interviews in this study, the factors influencing Chinese science teacher educators' views of NOS had not been intentionally asked, three science teacher educators explicitly stated that their NOS views are influenced by Marxism, especially dialectical materialism, which had been taught since their school time:

> My understanding of NOS is influenced by my early education of dialectical materialism. I had been taught of Marxism when I was in the school. It is common for the people grown up under the Red Flag[4] like me. Although it is too politicized in China, I feel most of its arguments are right, especially for those on natural philosophy...When I started to read the literature on NOS, I made use of my knowledge of dialectical materialism to understand it. (Wan et al. 2012, p. 16)

In addition to such general arguments, the influence of Marxism on Chinese science teacher educators' conceptions in NOS content to be taught is reflected in some specific aspects of the findings in the study.

66.4.1 Realist Views of Mind and Natural World

It is rather rare to find realist statements in the Western literature on NOS instruction. Even when it is found in *Science for All American* (American Association for the Advancement of Science [AAAS], 1990), one of few found by the authors, the tone adopted to present it is very assumptive:

> The world is understandable...Science *presume*s that the things and events in the universe occur in consistent patterns that are comprehensible through careful, systematic study. Scientists believe that through the use of the intellect, and with the aid of instruments that extend the sense, people can discover patterns in all of nature. ... Scientist also *assumes* that the universe is, as its name implies, a vast single system in which the basic rules are everywhere the same. Knowledge gained from studying one part of the universe is applicable to the parts. (AAAS 1990, p. 2, authors' emphasis)

Wan et al. (2012) found that realist understandings of mind and natural world were considered by more than half of Chinese science teacher educators as scientific worldview and suggested by them as NOS content to be taught:

> One aspect of NOS is some basic thoughts guiding scientific investigation, including basic understanding of matter, motion...These scientific worldviews included understandings that ... *the world is material*, all matters in world are *connected*, such connections are universal, *matters are in constant motion*, there are rules underlying such motion, and all these connection and rules are knowable to human being. (p. 14)

[4] Red Flag means the flag of Chinese communist party. Actually, red is the colour of all communist parties. The people grown up under the Red Flag refer to the people that are under influence of communist party.

The statement cited above reflects several core elaborated realist arguments:

(a) The existence of an external world that is independent of the observer
(b) The universality and constancy of connection in the world
(c) The possibility of our mind to know the external world and connections within it

These three statements are the prerequisites to arrive at the final and core argument of realism, i.e. the existence of the corresponding relationship between scientific knowledge and natural world. Without the existence of an independent material world, it is meaningless to discuss the relationship between scientific knowledge and natural world. The major form of the scientific knowledge is the universal and constant connection between variables, so if it is believed that there is no such universal and constant connection existing in the real world, the natural conclusion will be that such corresponding relationship cannot possibly exist between scientific knowledge and natural world. Even with the conditions of (a) and (b), if we do not admit that we can, through the use of the intellect and with the aid of instruments that extend the sense, discover such connection in the world, such a corresponding relationship is still problematic.

Indeed the choice of the words and the tone adopted by the Chinese science teacher educators are rather affirmative. This tendency is consistent with Marxist tradition, where realist epistemology is believed by Marxists as the right philosophy or worldview to guide scientists. Actually, on the basis of their materialist ontology, Marxists hold the realist epistemology (Curtis 1970; Farr 1991; Murray 1990), which admits the knowability of the material world. In Stalin's formulation: 'The world and all its laws are fully knowable … there are no things in the world which are unknowable, but only things which are as yet not known' (Stalin 1985, p. 18). He was elaborating Engel's contention that:

> natural scientists may adopt whatever attitude they please, they will still be under the domination of philosophy of science. It is only a question whether they want to be dominated by a bad, fashionable philosophy or by a form of theoretical thought with rest on acquaintance with the history of thought and its achievement. (Engels 1976, p. 558)

For Marxists, such theoretical thought includes the realist epistemology described before.

Although the core meanings of the excerpts are similar to the popular realist statements, there is still some difference as reflected in the emphasised words. First, emphasising that the world is material may reflect a materialist tendency, which is not necessarily held by realists – many realists grant the existence of nonmaterial things (fields, etc.). Second, the concepts of connection and change, which are not commonly included in the popular realist statements, are obviously emphasised in the above excerpts. However, the concepts of connection and change are the basic points of dialectics in Marxism, so some popular Chinese textbooks of Marxism integrate them with materialism and realist arguments to make a summary of Marxism principles. Hence although the possibility may not be totally excluded that the appearance of this NOS element and its high frequency is the result of the influence of other philosophical theories, the above clues indicate that it might more probably be the outcome of the influence of Marxism.

66.4.2 Truth-Approaching Nature of Scientific Knowledge and Science as the Pursuit of Truth

There are also other two realist NOS elements found in the study, i.e. truth-approaching nature of scientific knowledge and science as the pursuit of truth, whose meanings are very similar. The first believes that 'scientific knowledge as *relative truth*…Although it may never arrive at the state of *absolute truth*, its development is a process of continuous progression towards the truth in the objective world' (Wan et al. 2012, p. 16). The idea that science cannot obtain final truth but nevertheless can progressively obtain better approximations to the truth was enunciated by Karl Popper and called 'verisimilitude'. At the same time, the second argues that 'the goal of scientific investigation is to pursue truth, to reflect the real picture of the objective world…Science is unlike art. Art aims to pursue aesthetics, which allows departure from the fact and reality. But science investigation does not allow it' (Wan et al. 2012, p. 15). Although they were not suggested as frequently as the scientific worldview, if we count together the number of educators who have suggested them, there will be a total of ten. However, these two elements are hardly found in most recent science education curriculum in the West and academic publications on NOS studies. Such a feature in Chinese educators' conception is consistent with Marxist realist epistemology introduced before.

It can be also found that the terms *relative truth* and *absolute truth* were used as a pair by Chinese science teacher educators to describe truth-approaching nature of scientific knowledge. Truth is an important issue in many other branches of philosophy, but it is uncommon to couple these two terms to discuss the nature of knowledge in their literature. On the contrary, since being used by Lenin in his work *Materialism and Empirio-criticism* (Lenin 1977), this pair of terms has been popularly used among Marxists:

> Human thought then by its nature is capable of giving, and does give, absolute truth, which is compounded of a sum-total of relative truths. Each step in the development of science adds new grains to the sum of absolute truth, but the limits of the truth of each scientific proposition are relative, now expanding, now shrinking with the growth of knowledge. (Lenin 1977, p. 135)

Similar statements can be found in almost all the Chinese books on Marxism. Therefore, the appearance of this pair of terms is another indication of the influence of Marxism on Chinese science teacher educators' conception.

66.4.3 Logical Thinking in Scientific Investigation

Logical thinking in scientific investigation was considered by 15 Chinese science teacher educators as important for a scientific worldview and suggested by them as NOS content to be taught. It was depicted by a chemistry teacher educator:

> Induction and deduction are both important logical methods in scientific investigation. Deduction is from the general theory or concepts to the specific conclusion or facts…On the contrary, induction is the reasoning from the specific conclusion or facts to the general theory or concepts…I feel human understanding of the world is an endless process from the specific to the general, and then to higher level of the specific. It is a spiral process. (Wan et al. 2012, p. 12)

On the contrary, this NOS element is uncommon in the Western literature on NOS instruction. And even it is found, little elaboration is made. The caution about elaborating this element in the West may be due to the controversies between inductivism and deductivism in the philosophy of science. However, as reflected in the data, Chinese science teacher educators did not seem to concern conflicts between them. On the contrary, deduction and induction were considered as something like a unity of the opposite. This point is rather similar to Marxist view. 'Induction and deduction belong together…Instead of one-sidedly raising one to the heavens at the cost of the other, one should seek to apply each of them in its place, and that each completes the others' (Engels 1976, p. 519).

66.4.4 Empirical Basis of Scientific Investigation, General Process of Scientific Investigation and Progressive Nature of Scientific Knowledge

As indicated in the paper, except realist understanding of mind and natural world, the other four NOS elements suggested by more than a half of the educators in this study as NOS content to be taught are all NOS elements related to empiricism, i.e. empirical basis of scientific investigation, logics in scientific investigation, general process of scientific investigation and progressive nature of scientific knowledge.

The empirical basis of scientific investigation emphasises the role played by observation and experiment in the process of scientific investigation. As stated by one educator, 'empirical method is one of major features of scientific investigation since whether the results or arguments in science will be accepted or not will depend on the result of testing through observation or experiment' (Wan et al. 2012, p. 10). The origin of the empirical nature of scientific investigation can be traced back to the empiricism in the philosophy of science, emphasising the crucial role of human senses in the generation of scientific knowledge. Just relying on the empirical data cannot give a full explanation of the development of the scientific knowledge. In order to establish the validity of the scientific knowledge, it is necessary to provide a method to bridge between empirical data and scientific knowledge. The empiricist philosophers of science suggested that logic would serve such a purpose. Therefore, logics in the scientific investigation are also an empiricist NOS element.

General process of scientific investigation means a set of elements of conducting scientific research. It usually 'starts with a question, proceeds then with proposing some hypotheses or arguments to the question, testing such hypotheses or arguments through observation or experiment, drawing some conclusions and at the end communicating such conclusion so as to convince people of the conclusions' (Wan et al. 2012, p. 13). The discussion on general process or method of science can find its origin in empiricist philosophy of science. Both inductive empiricists and deductive empiricists were concerned to provide a scientific method. For inductive empiricists like Francis Bacon (1562–1626), the scientific method generally consists of four steps: observation and experimentation, classification, generalisation and testing. But deductive empiricists, like Karl Popper (1902–1994), think that the only logic that science requires is deductive logic. The method of science advocated by Popper is known as hypothetic-deductive method, which consists of the following steps (detailed description of Popper's methodology is given in Popper 1959):

(a) Formulate a hypothesis (H).
(b) Deduce an empirical consequence (C) from H.
(c) Test C directly.
(d) If C is acceptable under the scrutiny of sense experience, return to step (b) and obtain another C for the further testing of H.
(e) If, on the other hand, C is rejected, H should be rejected as a consequence of modus tollens.

Regardless their differences in the specific steps included in the process of scientific investigation, the inductive empiricists and deductive empiricists are common in believing that there should be a general process of scientific investigation.

Chinese science teacher educators argued that 'science accumulates and progresses in its own self-correcting process. Most of other subjects do not have such a feature...The typical example is art. If anyone wants to revise de Vinci's Mona Lisa, who can produce a one better than the original?' (Wan et al. 2012, p. 14). Stating the development of scientific knowledge is an accumulative or a progressive process is a typical stance originating in empiricism philosophy of science. The empiricists commonly conceive the development of science as a process that the replaced theories are reduced (and thus absorbed) into the replacing theory. For example, they take Newton's theory as being reduced to Einstein's theory. On the basis of such kind of understanding, the development of scientific knowledge is naturally considered as cumulative and continuous. In contrast with the understanding of the development of scientific knowledge as a process of replacement and absorption, it was thought by Kuhn (1970) that *scientific revolutions* also exist in the development of scientific knowledge, during which a process of replacement and displacement happens, rather than replacement and absorption.

Some interview data indicated that the popularity of Chinese science teacher educators' inclusion of empiricist NOS elements in their instruction may also be partly caused by Marxism. As explicitly stated by two educators during the interview:

> I believed dialectical materialism since I was young...I think it's consistent with the realist and empiricist philosophy of science, so I choose to focus on these classical NOS elements in my teaching. (Wan et al. 2012, p. 18)

Practice is the crucial concept in Marxism, which is considered as the starting point, the basis, the criterion and even the purpose of all knowledge (Mao 1986).[5] As stated by Marx (1976), 'the question whether objective truth can be attributed to human thinking is not a question of theory but is a *practical* question … the dispute over the reality or non-reality of thinking that is isolated from *practice* is a purely scholastic question' (p. 3) (authors' emphasis). Here, *practice*, in plain words, means the activity of applying mind into reality. For example, when some hypotheses are generated during the process of doing scientific investigation, we need to predict what phenomena are to be observed if the hypotheses are correct. On the basis of such predictions, appropriate scientific experiments are designed. The results of the scientific experiments are the evidence of supporting or refuting such hypotheses. The action of designing and doing experiments is a process of applying ideas into reality, so it is a kind of *practice*.

During the process of teaching, we use a certain kind of educational theory to guide the design of our teaching activities. If we find that students learn better, such a result is the evidence to support this educational theory. This kind of teaching activity is also *practice*. In fact, during the process of applying our mind into reality, we can, at the same time, get the responses from the reality through our sense experience, which can be not only empirical evidence to support or refute our mind, but the resources of generating new ideas. It is not difficult to note that empirical evidence, which is emphasised by empiricist epistemology, is implicitly integrated into the concept of *practice* in Marxism. Thus it is believed that Marxist philosophy, to a large extent, is consistent with empiricist epistemology (Creaven 2001). Of course, except the logics in scientific investigation, no addition clue in the wording can be found for the other three NOS elements to link their origin of Marxism. They may be influenced by Marxism in a more indirect manner.

Unlike the West where most science teacher educators focus on the contemporary views of NOS, most Chinese science teacher educators participated in Wan's (2010) study tend to put greater emphasis on realist views of science, though many of them are also knowledgeable about the contemporary views of NOS.

66.5 Preparing Science Teachers to Teach NOS in Hong Kong

In stark contrast to mainland China, Hong Kong science teacher educators have received continuous funding and support from the Education Bureau (EDB) of the Hong Kong Government in conducting a series of teacher professional development (TPD) and research projects to prepare teachers to develop students' understanding of NOS. As said in the previous section, science teacher educators and teachers in

[5] Mao Zedong first elaborates the concept of practice in his classic article 'On Practice: on the relation between knowledge and practice, between knowing and doing' written in July 1937, which is collected in various versions of Selected Works of Mao Zedong. The specific article cited here is this one included in Mao (1986).

mainland China put the highest priority in practising inquiry-based teaching and learning in science classrooms. Most teacher seminars and national teaching competitions are mainly centred at promoting teachers' pedagogical skills in carrying out inquiry-based teaching and learning. Most resources are allocated to equip the school laboratories with modern apparatus and equipment for conducting inquiry activities. It may be a reason why there have been no reports about the regular practice of teaching NOS in school classrooms in mainland China.

For Hong Kong, learning science through laboratory-based activities has been common at school level (albeit the activities are often carried out through 'cookbook' approach). School-based assessment on practical work or investigative studies have also been implemented in the 1980s starting from chemistry, followed by biology in the 1990s and physics in the 2000s. The emphasis on hands-on practical activities in school science was largely a result of the influence of the British educational system in which the Nuffield Curriculums had been influential since the 1970s until 2012 when the Hong Kong Advanced Level Examination[6] was replaced by the Hong Kong Diploma of Secondary Education Examination. Thus the school laboratories are mostly well equipped already. For the recent reform in science education which aims to promote students' scientific literacy, NOS turns out to be an area that most science teachers are unfamiliar with and hence EDB sees an urgent need to provide funding in support of TPD projects along the direction of the new curricular goals.

In this section, we review the decade of efforts by the group of science educators of the University of Hong Kong (HKU) in promoting teachers' understanding of NOS and pedagogical skills in implementing NOS teaching in their classrooms. Our sharing will situate in the 8-year experience in learning to teach NOS of a physics teacher, Henry. Henry first started learning about some general NOS ideas since 2003 when he studied for his Postgraduate Diploma in Education (PGDE) course, major in physics at HKU. Through the reflection on Henry's learning journey based on the detailed records of his learning and our TPD programs, we highlight some critical stages of his learning journey. We hope our in-depth reflection will provide insights on the design of effective TPD for similar future initiatives.

66.5.1 A Decade of Effort in Preparing Teachers for the Reformed Curricular Goals

In response to the curriculum reform in Hong Kong, science teacher educators have initiated a series of research and TPD projects since early 2000 to support pre-service and in-service science teachers to teach NOS. We are cognisant of the

[6] The examination was similar to the British General Certificate of Education Advanced Level Examination except that it was much more competitive due to the very low university admission rate in Hong Kong.

numerous problems and challenges identified in the West. Lederman's (1992) critical review came to a disappointing conclusion that both students and science teachers have inadequate understanding of NOS (Lederman 1992). There is however empirical evidence that can inform ways to improve NOS understandings. Explicit and reflective approaches and strategies using historical account of scientific development and/or through scientific inquiry in teaching NOS can support learner development of sophisticated NOS ideas (Abd-El-Khalick and Lederman 2000; Khishfe and Abd-El-Khalick 2002). Yet, teachers with good understanding in NOS still face many constraints including concerns for student abilities and motivation (Abd-El-Khalick et al. 1998; Brickhouse and Bodner 1992), lack of pedagogical skills in teaching NOS (Schwartz and Lederman 2002) and lack of teaching resources (Bianchini et al. 2003) particularly those in local contexts and language (Tsai 2001). Effective NOS teaching also depends on teachers' belief in the importance of teaching NOS (Lederman 1999; Tobin and McRobbie 1997) and their conception of appropriate learning goals and teaching role (Bartholomew et al. 2004).

66.5.1.1 Inadequate NOS Understanding: Missing the Target NOS

In support of the implementation of the revised junior secondary science curriculum, Tao, Yung, Wong, Or and Wong (2000) wrote a new series of textbooks which included four science stories on penicillin, smallpox, Newton's Law of Universal Gravitation and the treatment of stomach ulcers. Although these stories were designed such that teachers could highlight the NOS aspects through an explicit approach (Abd-El-Khalick and Lederman 2000), it was found that students' learning of NOS based on these stories was disappointing (Tao 2003). In fact many teachers, including Henry, considered the stories good for arousing student interest without attending to the primary purpose of teaching relevant NOS idea, as shared by a junior science teacher who came to realise his oversight after he attended the NOS sessions in our PGDE courses in early 2000s:

> I found the story on stomach ulcers very interesting....Marshall tested his hypothesis by trialing out himself....Students all enjoyed the story… I only realise now that there are deeper meanings behind the story and other important learning outcomes to be achieved through it and other stories.

We learnt from this experience that availability of teaching resources would not by itself result in teachers making use of the materials to teach NOS unless teachers had the ability to understand and appreciate the intended learning outcomes of the instructional materials. It is likely that they would overlook the targeted learning objectives (McComas 2008) and cling onto the parts which were more appealing to them (dramatic stories which promote students' interest). We also reckoned that there were some inadequacies of these relatively 'old' stories. Teachers and students expressed that though these stories aroused their interests, they happened quite a while ago. Those who did not have the historical and cultural backgrounds of the scientific discoveries and inventions would fail to develop an in-depth understanding of, and hence appreciate, the thought processes of the scientists related to what they encountered at their time.

66.5.1.2 Effective Learning of NOS Contextualised in the SARS Story

In summer 2003, when the crisis due to the severe acute respiratory syndrome (SARS) in Hong Kong was coming to an end, we saw a golden opportunity to turn the crisis into a set of meaningful instructional resources which might help address the issues raised above. The SARS incident was a unique experience that everyone in Hong Kong had lived through and the memories of which would stay for years. At the beginning of the outbreak, the causative agent was not known, the pattern of spread was not identified, the number of infected cases was soaring, yet an effective treatment regimen was uncertain. It attracted the attention of the whole world as scientists worked indefatigably to understand the biology of the disease, develop new diagnostic tests and design new treatments. Extensive media coverage kept people up to date on the latest development of scientific knowledge generated from the scientific inquiry about the disease. As anticipated we identified many interesting aspects of NOS based on our interviews with key scientists who played an active role in the SARS research, analysis of media reports, documentaries and other literature published during and after the SARS epidemic.

The SARS incident illustrated vividly some NOS features advocated in the school science curriculum. They included the tentative nature of scientific knowledge, theory-laden observation and interpretation, multiplicity of approaches adopted in scientific inquiry, the interrelationship between science and technology and the nexus of science, politics, social and cultural practices. The incident also provided some insights into a number of NOS features less emphasised in the school curriculum. These features included the need to combine and coordinate expertise in a number of scientific fields, the intense competition between research groups (suspended during the SARS crisis), the significance of affective issues relating to intellectual honesty, the courage to challenge authority and the pressure of funding issues on the conduct of research. The details on how we made use of the news reports and documentaries on SARS, together with episodes from the scientists' interviews to explicitly teach the prominent features of NOS, can be found in Wong et al. (2009). Since January 2005, we have been using the SARS story in promoting understanding of NOS of hundreds of pre-service and in-service science teachers. The learning outcomes have been encouraging (Wong et al. 2008).

66.5.1.3 Integrating NOS Teaching into Teaching of Science

While the SARS story has been effective in promoting NOS understanding, there was still a lack of NOS instructional materials closely related to the school science curriculums. We also planned to produce instructional materials grounded in more local contexts and in both English and Chinese language to suit the need of local schools using either English or Chinese as the media of instruction. Thus, in September 2005, we embarked on a 2-year project, Learning and Teaching about Science Project (LaTaS), which aimed to produce local NOS curriculum resources[7] and prepare teachers for teaching NOS. We took the advice of Hodson (2006) that:

> Curriculum materials need to have a "street credibility" that can only be gained when they are developed *by* teachers *for* teachers. (p. 305)

Thus we deliberately involved teachers at the beginning stages of the design process of teaching materials. More than 50 senior secondary science teachers worked together with the university team members to develop 12 sets of teaching resources which integrate NOS knowledge with subject knowledge for the new senior secondary biology, chemistry and physics curricula (CDC 2007). Efforts were made to include as many local examples as possible on top of some classic stories of science. The topics included development of teaching materials on topics including *Disneyland in Hong Kong* to be integrated in the teaching of ecology, an abridged version of the *SARS Story* to be integrated in the teaching of infection diseases, *Nature of Light* to be integrated in the teaching of wave properties of light, *Ban of Eating Shark Fin* to be integrated with the teaching of reaction rate, etc. By specifically choosing relevant socioscientific issues or well-known classic stories of science, teachers would be more ready and willing to practise NOS teaching by making use of the materials developed by the project team in their own science classrooms. Teachers' classroom practice helped us refine the teaching materials.

Henry was among one of the teachers who helped to try out the set of instructional materials on LASIK. Like other teachers, Henry found the in-depth analysis of the lesson video of their own practice most valuable in addressing some areas for improvement. Henry noted that he had missed a number of golden opportunities in receiving student good answers on questions related the interaction of science-technology-society (STS). Unlike the usual discussion on physics questions in which teachers would guide his students to come to 'the solution', in the discussion of questions related to STS, there could be many reasonable responses. Although he was not satisfied with his implementation of the NOS activities, he was pleased to have identified the areas of improvement and was confident that he could do a much better job in his next NOS teaching. An important encouraging factor for Henry to continue NOS teaching was his realisation of students' ability and interest in learning NOS as evident in the good responses by the students (though missed by him during the trial lessons).

Henry further expressed that he found his students liked the hands-on activity that simulated the LASIK surgery[8] (Wong 2004) very much, but he found it awkward (so was his students) to have another whole lesson following the LASIK surgery to cover its related NOS and STS activities. Henry said, 'it wasn't the style of my physics teaching and my students' style of learning physics'.

---

[7] Curriculum materials developed in this project are accessible at http://learningscience.edu.hku.hk/

[8] Notes to technician/teacher on the preparation of the 'cornea' and the set-up can be found in the LASIK instructional materials at http://learningscience.edu.hku.hk/

Henry's reflection told us that it was likely that the ambition in having as many NOS activities trialled and refined within the limited project period might have given teachers an impression that an NOS lesson has to be heavily loaded with NOS activities. Henry's sharing also reminded us that we should not take it for granted that teachers could naturally integrate the NOS materials with the science knowledge covered in the textbook (written before the new curriculum) in a meaningful manner. Good integration of NOS activities with the subject knowledge also requires careful planning and practice by the teacher to match with his own teaching style and his students' learning style. Just like teaching of a science topic using a textbook, each teacher could have his own style in integrating the NOS materials developed in the LaTaS project into his science classroom.

66.5.2 Modifying and Enriching Available NOS Resources

Henry, as many other participating teachers, continued to practise NOS teaching in his lessons even after the LaTaS project ended in late 2007. There are various reasons for teachers to continue their endeavour: (1) prepare students for possible questions assessing on the NOS understandings in public examination, (2) students show interest in learning NOS, (3) aware of students' ability in learning NOS and (4) positive experience in support of the 'science learning argument' by Driver and her colleagues (1996).

In Henry's further attempts in teaching NOS in 2008, he decided to address the areas of improvement identified in his trial of the LASIK materials. Firstly, instead of cramming in too many NOS ideas in one or two lessons, he infused NOS at appropriate places in a series of three double lessons each of 80 minutes during the teaching of *Nature of Light* and *Electromagnetic Wave*. Secondly, instead of bringing up each NOS idea in a fragmented way, he organised the NOS ideas in an interrelated manner. Thirdly, inspired by the experience of further learning about NOS through the SARS story, Henry included some NOS ideas which are not commonly introduced in textbooks, e.g. competition among scientists and peer review, establishment of scientific knowledge as an outcome of consensus reached by scientists and model-building as one of the typical activities of scientists in their endeavour in the pursuit of understanding of the nature. Lastly, instead of just focusing on the widely cited seven NOS aspects advocated by Lederman (1992), he went beyond the list and introduced some apparently contrasting NOS features by situating them in relevant historical contexts.

In the first double lesson, he planned to review the four wave properties, introduce Young double-slit experiment and explain the interference pattern. He tailored some NOS activities developed in the *Nature of Light* materials in the LaTaS project. He made use of the controversial arguments between Newton and Huygens about the nature of light to highlight some 'relatively negative' features about the subjectivity in science. For example, he elaborated that (i) scientists could be very 'stubborn' in holding on their beliefs due to their beloved models in explaining the natural phenomena (theory-laden observation/inference/explanation), (ii) there were two camps of scientists (one supported particle model of light and the other supported the wave model of light) and so more than one scientific models might be found coexisting in explaining a specific observation or natural phenomenon and (iii) Newton was the 'winner' as a result of submission to authority even among scientists (when both models could explain reflection and refraction of light).

In the second double lesson, he planned to review interference of light and introduce single-slit and its associated concept about diffraction. He attempted to balance the 'relatively negative' NOS features with the 'relatively positive' NOS features by highlighting the empirical evidence of Young in support of wave theory of light. He enriched the existing materials by adding the story of Poisson's challenge to Fresnel's wave theory of light. Poisson, who was a believer of particle theory of light, challenged Fresnel's arguments towards wave theory due to the lack of experimental evidence of diffraction effect of light. However, when hard experimental evidence was produced by Fresnel with the help by Arago, Poisson had to convert his belief about the nature of light based on the convincing experimental evidence.

In a subsequent double lesson, Henry then made use of the story of the discovery of DNA as one of the applications of X-ray to highlight the competitive nature of scientists and building of models through scientists' imagination and creativity. Towards the end of the series of lessons on Light Wave, he even linked all the NOS aspects covered in the topic in the form a mind map to strengthen students' appreciation of the interconnectedness of NOS features. Figure 66.2 gives a reproduction of his summary on the blackboard during the last lesson of the series on *Nature of Light* and *Electromagnetic Wave*.[9]

66.5.3 Learning from Henry's Professional Growth in Teaching NOS

Through active and persistent reflection upon each attempt of NOS teaching, Henry, as other teachers having similar commitment, has come a long way from the learning of NOS to successful design and implementation of the lessons that teaching physics as well as NOS ideas in an interconnected manner. His exemplary teaching has received recognition by different stakeholders, including science teachers who are preparing themselves to teach NOS. EDB and University colleagues have also invited him for sharing the lesson series designed by himself. To him, the greatest reward of the long journey of professional development is probably the learning of his students. It was not difficult to tell from Henry's smile when his shared students' responses received in later lessons showing their good understanding of some

[9] Video episodes of the three series of lessons can be accessed at http://web.edu.hku.hk/knowledge/projects/science/qef_2010/d1/1b2_training_workshop.html

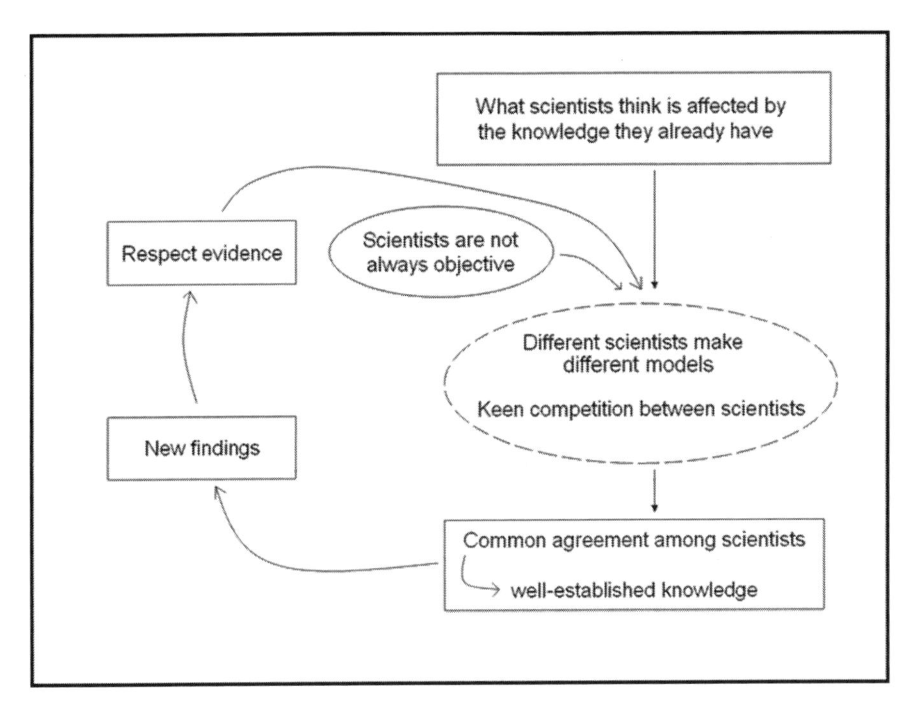

Fig. 66.2 A summary of the interconnectedness of NOS features given at the end of the lesson series

difficult NOS ideas like theory-laden observation. In his word, '*[NOS can] promote student interest in learning the content knowledge*' and '*students do learn NOS*'.

A year after the completion of the LaTaS project, EDB further supported us to conduct another 2-year project from 2008 to 2010 with a focus on the enhancement of teachers' pedagogical content knowledge (PCK) of teaching NOS. We have noted from Henry's 'revolutionary jump' in his learning curve when he attempted to plan his own lessons. Thus in the PCK project, at the outset of the project, we encouraged teachers not to simply modify and adapt the available teaching resources, but to proactively design their own instructional materials. This is indeed a goal that we wish teachers could ultimately achieve. We further explained our intention to the teachers by a Chinese proverb, (授人以魚,三餐之需;授人以漁,終生之用。). In English, it says 'Give a man a fish and you feed him for a day. Teach a man to fish and you feed him for a lifetime'.

66.6 Conclusion

In this chapter, we have compared the NOS ideas as portrayed in the curriculum standards documents, NOS ideas as portrayed in the textbooks and the teacher training for preparing teachers towards their teaching of NOS. We have also shared an exemplary lesson series conducted by a Hong Kong teacher in his classroom. Although there are variations of a few NOS aspects as portrayed in the official curriculum standards documents and school science textbooks as well as those taught in teacher training institutes, most prominent differences occur at the school classroom implementation. There are currently no reports on the implementation of NOS teaching in school science classrooms in mainland China regardless the similar degree of emphasis placed on NOS education in the official curriculum. On the other hand, the NOS teaching has taken root in Hong Kong science classrooms. The key difference mainly originates from the support and resources input from the government in realising the intended goals. We do see a related emphasis on learning science through inquiry-based approach which has taken off in science classrooms in many cities as evident in the annual national competition in mainland China since early 2000s. Schools with teachers gaining the awards will also receive funding as encouragement. In Hong Kong, the support from the government is provided to all schools through supporting institutes with experts in the area of focus and the most relevant and appropriate research and teacher professional development plans.

References

Abd-El-Khalick, F., & Lederman, N. G. (2000). Improving science teachers' conceptions of the nature of science: A critical review of the literature. *International Journal of Science Education*, 22, 665–701

Abd-El-Khalick, F., Bell, R. L., & Lederman, N. G. (1998). The nature of science and instructional practice: Making the unnatural natural. *Science Education*, *82*, 417–437.

American Association for the Advancement of Science (AAAS). (1990). *Science for all Americans: Project 2061*. New York: Oxford University Press.

Bartholomew, H., Osborne, J., & Ratcliffe, M. (2004). Teaching students "ideas-about-science": Five dimensions of effective practice. *Science Education*, 88, 655–682.

Bianchini, J. A., Johnston, C. C., Oram, S. Y., & Cavazos, L. M. (2003). Learning to teach science in contemporary and equitable ways: The successes and struggles of first-year science teachers. *Science Education*, 87, 419–443.

Brickhouse, N. W., & Bodner, G. M. (1992). The beginning science teacher: Classroom narratives of convictions and constraints. *Journal of Research in Science Teaching*, 29, 471–485.

Centre for Research and Development on Physics Curriculum Resources, Curriculum Resources Research Institute, People's Education Press (PEP). (2004). *Physics*. Beijing: People's Education Press.

Chen, Q., & Pang, L. J. (2005). On the nature of science and science education. *Peiking University Education Review (in Chinese)*, 3(2), 70–74.

Cheng, K. L & Wong, S. L. (in print), Nature of Science as portrayed in the Physics official curricula and textbooks in Hong Kong and the mainland of the People's Republic of China. In Catherine Bruguière, Pierre Clément & Andrée Tiberghien (Eds.), ESERA 2011 *Selected Contributions. Topics and trends in current science education.*

Clough, M. P. (2006). Learners' responses to the demands of conceptual change: Considerations for effective nature of science instruction. *Science & Education*, 15(5), 463–494.

Council of Ministers of Education, Canada. (1997). *Common Framework of Science Learning Outcomes*. Toronto, Canada: CMEC.

Creaven, S. (2001). *Marxism and Realism: a Materialistic Application of Realism in the Social Sciences*. New York, USA: Routledge.
Curriculum Development Council (CDC). (1998). *Science syllabus for secondary 1–3*. Hong Kong: Curriculum Development Council.
Curriculum Development Council (CDC). (2002). *Physics/Chemistry/Biology curriculum guide (Secondary 4–5)*. Hong Kong: CDC.
Curriculum Development Council and Hong Kong Examinations and Assessment Authority ("CDC-HKEAA"). (2007). *Physics curriculum guide and assessment guide (Secondary 4–6)*. Hong Kong: Curriculum Development Council and Hong Kong Examinations and Assessment Authority. Retrieved May 20, 2012 from http://www.edb.gov.hk/FileManager/EN/Content_5999/phy_final_e.pdf
Curtis, M. (1970). *Marxism*. New York: Atherton Press.
Department for Education (DfE). (1995). *Science in the National Curriculum*. London: HMSO.
Driver, R., Leach, J., Miller, A., & Scott, P. (1996). *Young people's images of science*. Bristol, PA: Open University Press.
Editorial Group for Physics Textbooks, Guangdong Basic Education Curriculum Resources Research and Development Centre (GEP). (2004). *Physics*. Guangzhou: Guangdong Education Press.
Engels, F. (1976). Dialectics of Nature. In K. Marx & F. Engels, *Collected Works* (Vol. 25, pp. 315–590). London, UK: Lawrence & Wishert.
Farr, J. (1991). Science: Realism, criticism, history. In T. Carver (Ed.), *The Cambridge Companion to Marx* (pp. 106–123). Cambridge University Press.
Hodson, D. (2006). Why we should prioritize learning about science. *Canadian Journal of Math, Science & Technology Education*, 6(3), 293–311.
Irzik, G. & Nola, R. (2011). A family resemblance approach to the nature of science for science education. *Science & Education*, 20, 591–607.
Khishfe, R., & Abd-El-Khalick, F. (2002). The influence of explicit/reflective versus implicit inquiry-oriented instruction on sixth graders' views of nature of science. *Journal of Research in Science Teaching*, 39, 551–578.
Kuhn, T. S. (1970). *The structure of scientific revolutions*. Chicago: The University of Chicago Press.
Lederman, N.G. (1992). Students' and teachers' conceptions of the nature of science: A review of the research. *Journal of Research in Science Teaching*, 29, 331–359.
Lederman, N.G. (1999). Teachers' understanding of the nature of science and classroom practice: Factors that facilitate or impede the relationship. *Journal of Research in Science Teaching*, 36, 916–929.
Lederman, N.G. (2007). Nature of science: Past, present, and future. In S.K. Abell & N.G. Lederman (Eds.), *Handbook of research on science education* (pp. 831–879). Mahwah, NJ: Lawrence Erlbaum Associates.
Lenin, V. I. (1977). *Lenin Collected Works* (Vol. 14). London, HK : Lawrence & Wishart.
Liang, Y.P. (2007) Developmental evaluation on science teachers' teaching behavior of Nature of Science. *Education Science*, 23(3), 48–53.
Liu, E. S. (2004). *Teaching Methods for School Biology*. Beijing, China: Higher Education Press.
Mao, Z. D. (1986). *Selected Works of Mao Zedong*. Beijing, China: People's Publishing House.
Marx, K. (1976). Theses On Feuerbach. In K. Marx & F. Engels, *Collected Works* (Vol. 5, pp. 3–5). London, UK: Lawrence & Wishert.
McComas, W. F. (1998). The principal elements of the nature of science: Dispelling the myths. In W. F. McComas (Ed.), *The Nature of Science in Science Education: Rationales and Strategies* (pp. 53–70). Hingham: Kluwer Academic.
McComas, W. F. (2008). Seeking historical examples to illustrate key aspects of the nature of science. *Science & Education*, 17, 249–263.
McComas, W. F., & Olson, J. K. (1998). The nature of science in international science education standards documents. In W. F. McComas (Ed.), *The Nature of Science in Science Education: Rationales and Strategies* (pp. 41–52). Hingham: Kluwer Academic.

Ministry of Education (MOE). (2001a). *Chemistry curriculum standards (7–9 years) of full-time compulsory education*. Beijing: Beijing Normal University Press.

Ministry of Education (MOE). (2001b). *Physics curriculum standards (7–9 years) of full-time compulsory education*. Beijing: Beijing Normal University Press.

Ministry of Education (MOE). *(2004). Ordinary Senior Secondary Physics Curriculum Standards (Trial)*. Beijing: People's Education Press.

Murray, P. (1990). *Marx's Theory of Scientific Knowledge*. Atlantic Highlands, NJ: Humanities Press.

National Research Council. (1996). *National Science Education Standards*. Washington, DC: National Academy Press.

Osborne, J., Collins, S., Ratcliffe, M., Millar, R., & Duschl, R. (2003). What "ideas-about-science" should be taught in school science? A Delphi study of the expert community. *Journal of Research in Science Teaching*, 40(7), 692–720.

Popper, K. (1959). *The logic of Scientific Discovery*. London: Hutchison.

Schwartz, R. S., & Lederman, N. G. (2002). "It's the nature of the beast": The influence of knowledge and intentions on learning and teaching nature of science. *Journal of Research in Science Teaching*, 39, 205–236.

Stalin, J. (1985). *Dialectical and Historical Materialism*. London, UK: Taylor & Francis.

Tao, P. K. (2003). Eliciting and developing junior secondary students' understanding of the nature of science through a peer collaboration instruction in science stories. *International Journal of Science Education*, 25(2), 147–171.

Tao, P. K., Yung, H. W., Wong, C. K., Or, C. K., & Wong, A. (2000). *Living science*. Hong Kong: Oxford University Press.

Tobin, K., & McRobbie, C. J. (1997). Beliefs about the nature of science and the enacted science curriculum. *Science & Education*, 6, 355–371.

Tong, S. S. W., Kwong, P. K., Wong, Y. L., & Lee, L. C. (2009). *New Senior Secondary Physics in Life*. Hong Kong: Longman Hong Kong.

Tsai, C. C. (2001) A science teacher's reflections and knowledge growth about STS instruction after actual implementation, *Science Education*, 86, 23–41.

Wan, Z.H. (2010). Chinese science educators' conceptions of teaching nature of science to prospective science teachers. Unpublished doctoral thesis of the University of Hong Kong.

Wan, Z.H., Wong, S.L. & Yung, B.H.W. (2011). Common interest, common visions? Chinese science teacher educators' views about the values of teaching Nature of Science to prospective science teachers. *Science Education*, 95(6), 1101–1123.

Wan, Z.H., Wong, S.L., & Zhan, Y. (2012). When Nature of Science meets Marxism: Aspects of Nature of Science taught by Chinese science teacher educators to prospective science teachers. *Science & Education*. DOI: 10.1007/s11191-012-9504-2.

Wei, B., & Thomas, G. (2005). Explanations for the transition of the junior secondary school chemistry curriculum in the People's Republic of China during the period from 1978 to 2001. *Science Education*, 89, 451–469.

Wong, S. L. (2004). Modelling LASIK in a school laboratory – away from 'cookbook' experiments. *School Science Review*, 85(312), 57–61.

Wong, S. L., & Pang, W. C. (2009). *New Senior Secondary Physics at Work*. Hong Kong: Oxford University Press (China).

Wong, S. L., & Hodson, D. (2009). From the horse's mouth: What scientists say about scientific investigation and scientific knowledge. *Science Education*, 93(1), 109–130.

Wong, S. L., & Hodson, D. (2010). More from the horse's mouth: What scientists say about science as a social practice. *International Journal of Science Education*, 32(11), 1431–1463.

Wong, S. L., Hodson, D., Kwan, J., & Yung, B. H. W. (2008) Turning crisis into opportunity: Enhancing student teachers' understanding of the nature of science and scientific inquiry through a case study of the scientific research in Severe Acute Respiratory Syndrome. *International Journal of Science Education*, 30, 1417–1439.

Wong, S. L., Kwan, J., Hodson, D., & Yung, B. H. W. (2009). Turning crisis into opportunity: Nature of science and scientific inquiry as illustrated in the scientific research on Severe Acute Respiratory Syndrome. *Science & Education*, 18, 95–118.

Xiang, H. Z. (2002). On the education of the essentials of science. *Science and Technology Review*, 11, 35–37.

Yu, Z. Q. (2002). *Science Curriculum*. Beijing, China: Educational Sciences Press.

Yuan, W. X. (2009). A review of the research into NOS teaching. *Comparative Education Review*, 288, 7–12.

Zhang, H. X. (2004). *Science curriculum and instruction in elementary school*. Beijing, China: Higher Education Press.

Siu Ling Wong is an Associate Professor and Assistant Dean (Programme) in the Faculty of Education at the University of Hong Kong (HKU). She received a B.Sc. in Physics and PGDE Major in Physics from HKU and received her DPhil in Physics from the University of Oxford. She had worked as a Croucher Foundation Postdoctoral Fellow in the Clarendon Laboratory and a Lecturer at Exeter College, Oxford. She is a National Research Coordinator of the Trends in International Mathematics and Science Study (TIMSS) for Hong Kong. She is a founding and executive member of the East-Asian Association for Science Education. She is also the leading author of the most popular physics textbook in Hong Kong. Her current research interests include nature of science, socioscientific issues and teacher professional development.

Zhi Hong Wan is an Assistant Professor in the Department of Curriculum and Instruction at the Hong Kong Institute of Education. He received a B.Sc. of physics education from the Jingdezhen Institute of Education, a M.Phil. in educational psychology from Soochow University, and a D.Phil. in science education from the University of Hong Kong. Before starting his research in science education, he had taught middle school physics in Mainland China for 5 years. Zhi Hong's research focuses on how nature of science instruction is conducted and perceived in different social-cultural backgrounds and assessment for learning. He has authored more than 20 articles.

Ka Lok Cheng is a lecturer and Project Manager in the Faculty of Education at the University of Hong Kong. He previously worked in the Education Bureau of the Hong Kong SAR Government and facilitated the provision of school-based professional support services to teachers. He served as a school teacher of Chemistry for 6 years formerly. He earned his B.Sc. and M. Phil. in Chemistry and is currently a doctoral candidate in Science Education at the University of Hong Kong. He teaches sociology of education courses for student teachers at BEd and PGDE levels. His main research interest focuses on the history of science and nature of science ideas in the official curriculums and textbooks.

Chapter 67
Trends in HPS/NOS Research in Korean Science Education

Jinwoong Song and Yong Jae Joung

67.1 Introduction

The HPS/NOS in the science education plays roles in stimulating students' motivation, in illustrating humanistic aspects of science (Jung 1994; Matthews 1994), and in helping the understanding of scientific methods (Lederman 2007; McComas and Olson 1998; Matthews 1992). In the recent days, the National Science Curriculum of Korea also highlights the promotion of students' scientific literacy which requires students' understanding of HPS and NOS (MEST 2007b; MEST 2011).

Historically, HPS and NOS have largely been of a minor attention in Korean science education. Following its liberation from Japanese occupation following the World War II, Korea's modern school education considers science as one of the core school subjects. After the Korean War (1950–1953), along the continuous national policies for industrial rebuilding and economic development, science education has continued to be the subject of national attention and public's concern. Because of this relationship with national development, science education in Korea used to focus more on practical usefulness and manpower-building-oriented approaches. In other words, school science education has often been the sources of scientific and technical majors in earlier years and of more advanced and creative scientific and engineering specialists in recent years. As a result, in Korea, HPS and NOS have largely been of a minor attention.

Science education has been a subject of systemic research activities since the establishment of Korean Association for Science Education (KASE) (formerly

J. Song (✉)
Department of Physics Education, Seoul National University,
Seoul, South Korea
e-mail: jwsong@snu.ac.kr

Y.J. Joung
Department of Science Education, Gongju National University
of Education, Gongju, South Korea

M.R. Matthews (ed.), *International Handbook of Research in History, Philosophy and Science Teaching*, DOI 10.1007/978-94-007-7654-8_67,

known as Korean Association for Research in Science Education: KARSE) in 1976. For the last 35 years, KASE has played the central role in researching, developing, and implementing issues of science education, mainly through its official journal, *Journal of Korean Association for Science Education* (JKASE) (formerly known as *Journal of Korean Association for Research in Science Education*: JKARSE) and biannual conferences. Beginning in 1978, JKASE was published every other year. In 1984 JKASE began publishing twice a year. Currently, eight issues (six issues in Korean and two issues in English) are published every year. As the most comprehensive and highly recognized journal, JKASE covers a whole range of areas of science education, i.e., informal as well as formal education and from preschool to tertiary levels.

Besides KASE, there are also several academic societies related to science education which target more specific groups of audience: for example, the Korean Society for Elementary Science Education (KSESE) established in 1970 and its official journal, *Journal of Korean Elementary Science Education*; the Korean Society of Biological Education (KSBE) established in 1968 and its official journal, *The Korean journal of Biology Education*; the Korean Earth Science Society (KESS) established in 1969 and its official journal, *Journal of Korean Earth Science Society*; the Korean Science Education Society for the Gifted (KSESG) established in 2008 and its *Journal of Science Education for the Gifted*; and the Korean Society for School Science (KOSSS) established in 2006 and its journal *School Science Journal*.

Similar to many countries and perhaps more strongly, science curriculum and science education research in Korea have been influenced by those in the United States. The aims of the National Science Curriculum and research trend of science education were largely similar to a decade's earlier versions of US ones. Nevertheless, the practice of school science in Korea has been inevitably dependent on social and educational conditions of the time in Korea. For example, exam-oriented, teacher-centered, and government-driven (science) education are the most prevalent conditions currently influencing Korean science education. These conditions, directly or indirectly, influence the practice and research activities of Korean science education, including HPS/NOS-related aspects. The Korean social environment that emphasizes students' hard work and high scores for entering good schools/universities makes difficult for school science education to be free from knowledge-oriented teaching and learning. And this environment would discourage for students to have ample experience to think and discuss the nature of and the historical content of science.

In this chapter, the status quo of HPS/NOS-related aspects of Korean science education are outlined, beginning with a brief summary of the history of National Science Curriculum which is the basic foundation of school science education. Next, the change of research trend of science education will be analyzed using the HPS/NOS-related papers published in JKASE for the last three decades. Finally, to find out how HPS/NOS has been incorporated in the practice of science education, HPS/NOS-related aspects in science textbooks and science teacher education will be explored.

67.2 A Brief History of National Science Curriculum of Korea

Korea has a centralized education system firmly based on its National Curriculum (Kim 2001). The National Curriculum is provided in the form of official document by the Ministry of Education, Science and Technology of Korea (MEST), and is mandatory to all schools from kindergarten to high schools. The National Curriculum of Korea has been revised regularly by the government since the first revision in 1954 (Park 2011). On average, the National Curriculum has been revised every 6–7 years (Kim 2001). At present, 2007 National Science Curriculum (MEST 2007a) is being applied to primary and middle school science education, and 2009 National Science Curriculum (MEST 2011), which is recently revised, is being applied to high school science education and will be implemented to primary and middle school after 2013.

The past curriculum reforms can be historically divided into three periods: the periods of Teaching the Syllabus (1945–1954), Course of Study (1954–1963), and Formal Curriculum (1963–present) (Kwon 2001). The period of Teaching the Syllabus was from 1945, the year of Korea's liberation from Japan, to 1954. During this period, the Korean government was the temporary government under the control by the US army. In this period, the curriculum (i.e., syllabus) was just a list of teaching items (e.g., "the existence of air," "space occupation by air," "the weight of air," "thermal expansion of air," and "compression and the use of air") without having objectives or any suggestions on teaching methods and evaluation. The syllabus of science contained not only science content but also practical arts content. Despite the inevitable influence from Japanese curriculum during the colonial rule, the school teaching during this period in large focused on the basic content and abilities of the subjects and also on overcoming the Japanese vestiges in people' spirit and everyday life.

The period of Course of Study (1954–1963) was officially named the first Korean curriculum, although there have been some debates over whether or not the Course of Study constitutes a formal curriculum. The objectives of science in this period were categorized into scientific knowledge, scientific inquiry, and attitudes toward science. The titles of content were expressed as questions (e.g., "How do the weather and seasons influence our everyday living?"). The Course of Study, however, did not include any suggestions on teaching and evaluation.

The period of Formal Curriculum can be further divided into three periods of school science curriculum: experience-centered curriculum (1963–1973), discipline-centered curriculum (1973–1981), and humanistic curriculum (1981–present). The experience-centered curriculum, that is the 2nd National Curriculum, was the first well-formalized curriculum which included objectives, content, and suggestions for teaching methods. The subjects on science were "Nature" for elementary school, "Science" for middle school, and 7 subjects (Physics I, II; Chemistry I, II; Biology I, II; and Earth Science) for high school. This curriculum was based on American progressive philosophy and emphasized everyday experience and science content.

The discipline-centered curriculum (1973–1981) was also influenced by the US science education innovation movement. The philosophy, objectives, content, and methods of teaching of the existing curriculum, i.e., the experience-centered curriculum, were changed completely. The new National Science Curriculum profoundly accepted the philosophy of discipline-centered curriculum proposed by J. S. Bruner and the Woods Hole Conference. This curriculum reorganized the content to reflect the basic concepts of science, changed the methods of instruction from rote memory to inquiry, and emphasized the teaching of science through discovery and problem-solving, students' voluntary involvement, and students' inquiry process skills. The science content of this curriculum included the five basic concepts: matter, energy, interaction, change, and life.

The humanistic curricula (1981–present) includes the 4th (1981–1987), 5th (1987–1992), 6th (1992–1997), 7th (1997–2007), 2007 Revised Curriculum, and 2009 Revised Curriculum.[1] The humanistic curricula were influenced mainly from the movements of "Science for All" and Science Technology and Society (STS). The discipline-centered curriculum had been criticized on the ground that the content was isolated from real-life situations and problems that students confront in everyday life and was too abstract and difficult for most students to understand (Kim 2001). The attention to "Science for All" and STS increased with such critiques. However, in the 4th curriculum, science content did not overcome the discipline-centered philosophy although new slogans, "Science for All" and STS, were introduced into Korean science education. The STS spirit did not explicitly appear in the objectives until the 5th curriculum. Since the 5th curriculum, "Science for All" and STS spirits were continuously strengthened until the present curriculum. One of the important changes in this period was the creation of a new combined or integrated high school science subject: "Common Science" in the 6th curriculum and "Science" in the 7th curriculum for G10, both emphasizing everyday science and decision-making. Another important change was the emphasis on scientific literacy including recognition of the relationship between STS and rational decision-making in the context of everyday life. In the 2007 and 2009 Revised Science Curricula, the terminology "scientific literacy" was expressed explicitly as the objectives of science education (MEST 2007a, 2011).

The 2007 Revised Science Curriculum, the latest curriculum applied to schools with textbooks, aims to help students understand the basic concepts of science through inquiry with interest and curiosity toward natural phenomena and objects. Students are expected to be able to develop the scientific literacy necessary for solving the problems of daily life creatively and scientifically (MEST 2009). In this curriculum, the content of "Science" includes the domains of motion and energy, materials, life, and earth and space, linking basic concepts with inquiry processes across grades and domains. In addition, these include Free Inquiry, which would provide students with the opportunities to select their own topics based on their interests, to enhance their interest in science, and to develop creativity.

[1] After the 7th Curriculum, the revision was to be made upon its demand, thus the new Curriculum was named with the revision year instead of calling 8th or 9th.

In addition, in this Curriculum, science learning is centered around various inquiry-based activities including observing, experimenting, investigating, and discussing. Learning emphasizes independent as well as group activities for nurturing scientific attitudes and communication skills, including criticism, openness, integrity, objectivity, and cooperation. Learning also stresses the comprehensive understanding of basic concepts, rather than acquisition of fragmented knowledge, and the ability to scientifically solve problems in daily life. The core concepts of Science are taught with a close relation to learners' experiences, and students are provided with opportunities to apply science-related knowledge and inquiry skills for problem-solving in society and their daily life. By learning about science, students are to be able to recognize the relationships between science, technology, and society as well as the values of science. The particular goals of the 2007 Revised Science Curriculum are as follows:

> The Science Curriculum aims to help students understand the basic concepts of science through inquiry into natural phenomena and objects with interest and curiosity and the development of scientific thinking and creative problem solving abilities. As a result, students will be able to develop the scientific literacy necessary for solving creatively and scientifically the problems of daily life. The objectives of the Science Curriculum are to educate students so they will be able:
>
> (a) To understand the basic concepts of science and apply them to solve problems in daily life.
> (b) To develop the ability to inquire about the nature scientifically and to use this ability for solving problems in daily life.
> (c) To enhance curiosity and interest toward natural phenomena and science learning, and develop an attitude to scientifically solve problems in daily life.
> (d) To recognize the relationship between science, technology, and society. (MEST 2009, pp. 12–13)

In 2011, the 2009 Revised Science Curriculum (MEST 2011) was introduced, in which the main objective was to raise students' scientific literacy for creative and rational problem-solving. In addition, 2009 Revised Science Curriculum emphasizes the fusions among science disciplines (i.e., physics, chemistry, biology, and earth science) as well as different fields (i.e., science, technology, engineering, arts, and mathematics – often called STEAM).

Since the middle of 1980s, the description related to HPS/NOS in Science Curricula has been gradually increased. The 5th Science Curriculum, announced in, began to express the attention to NOS by declaring as its objective "... through the experience with the nature, students are expected to have interest in science and scientific literacy." The 6th Science Curriculum announced in 1992 included STS education in its objectives by saying "... students are expected to know that science influences technological developments and has close links with our everyday life." And its section for Teaching and Learning Methods stated "... to appreciate the relationship between science and life by using things available around our daily life, and to make use of what they learned during science classes in their daily life." This kind of STS-related descriptions has appeared ever since the 6th Science Curriculum. The 6th Science Curriculum also began to include HPS/NOS-related descriptions, such as "... to stimulate students' interest in and curiosity toward science through the appropriate introduction of content related to science, scientists, and current issues…." Furthermore, 2007 Science Curriculum, the curriculum focused on "scientific literacy" and creativity, presented more clearly HPS/NOS-related descriptions, such as "… to prepare the list of science books for the effective teaching of writing and discussion on the basis of materials on cutting-edge science, scientists, the history of science" (MEST 2007a, p. 25), "… to guide students to read materials on science and scientific issues, and, through the writing and discussion based on the materials, to improve their skills of scientific thinking, creative thinking, and communication" (MEST 2007a, p. 25), and "… by introducing content on cutting-edge science, scientists, current issues, to stimulate students' interest in and curiosity toward science" (MEST 2007a, p. 25).

In 2009, the National Science Curriculum (MEST 2011) emphasized scientific literacy, creative and rational problem-solving, and integrated approach, similar descriptions on HPS/NOS. For instance, the objectives included elements of scientific attitudes (such as critical ability, openness, honesty, objectivity, and cooperation) and communication skills. In particular, for G7-9 level, there is a part called "What is science?" through which students learn about science and its relevance to everyday life and develop their interest and curiosity in science. In addition, another part, "Science and Human Civilization," is included to understand scientific contributions to human civilization based on historical facts and to develop a viewpoint to see science in connections with technology, engineering, arts, and mathematics (MEST 2011, p. 66). Furthermore, its Guidance for Teaching and Learning suggests "to teach with appropriate examples the content of NOS-related issues, such as the tentativeness of science, multiplicity of science methods, feature of scientific models, the difference between observation and inference" (MEST 2011, p. 68) and recommends to introduce stories of scientists, history of science, and current scientific issues.

As illustrated so far, in general HPS/NOS-related content in Korean Science Curricula have been largely included in connection with STS approach and became recently to include a wider range of content.

67.3 HPS/NOS in Research Papers in Korea: Based on the Analysis of Papers Published in JKASE

The *Journal of the Korean Association for Science Education* (JKASE, http://www.koreascience.org/) began in 1978 and, ever since, has been the most representative science education journal in Korea, publishing the largest number of research papers on science education in Korea across the whole range of science education. In this section, the results of analysis of the trends of HPS/NOS-related papers published in JKASE for the last three decades will be reported and discussed. JKASE publishes eight times annually. Currently six of these issues are written in Korean and two issues are written in English. The papers analyzed here were from issues in both languages.

67.3.1 Types of HPS/NOS-Related Papers and Classification Method

The HPS/NOS-related papers that appeared in JKASE were classified according to its four dimensions, that is, research theme, research method, target subject, aims, and results. Referring to existing literatures which have collections of HPS/NOS-related papers (such as, *The History & Philosophy of Science in Science Teaching* (Herget 1989), *More History & Philosophy of Science in Science Teaching* (Herget 1990), *The Nature of Science in Science Education: Rationales and Strategies* (McComas 1998)), a rough category framework was developed for the classification. Based on this rough category framework, each paper was classified and then checked to see if the paper properly fit that particular category. In doing so, the framework was repeatedly revised by further dividing or combining together categories. In the process of classifying the papers, a paper could be identified with more than one category, which would provide a more comprehensive understanding of the distribution and features of the papers. As a result, eight categories in the dimension of research theme, six categories in the dimension of research method, and ten categories in the dimension of target subject for the analysis of HPS/NOS-related papers were established (Table 67.1).

When a paper was classified according to this category framework, the decision was made based on a wide consideration of not only its title or keywords but also its abstract, theme, aim, and results. Through this comprehensive process, it was found that HPS/NOS-related papers could be largely grouped into two groups: (a) researches which investigate or discuss "about" themes or theories related to HPS/NOS and (b) researches which applied ideas "based on" the results of HPS/NOS-related studies. All studies belonged to the former group, (a), were identified as HPS/NOS-related studies, whereas some belonged to the latter group were not identified as so. For example, if there is a study which investigated an effective instruction method for students' conceptual change, it is ultimately related to HPS/NOS because instruction for conceptual change is too somehow related to philosophical arguments of science such as the nature of concept, constructivism, and knowledge development to some degree. If this kind of papers was included, in principle there would be no study which is not related to HPS/NOS. Thus, to avoid this situation, when making a judgment on papers of (b) kind, the paper was checked to see if it met the following criteria:

- Does this paper have explicit connects with HPS/NOS-related theories or concepts?
- Does this paper investigate how people think or perceive the nature of, not just simple interest in, science or scientific method or scientific concepts?
- Does this paper discuss explicitly target subject's concept or the nature of their behaviors in terms of HPS/NOS, not just investigate their concepts or behavior?
- Does this paper make the participants or subjects to think about the nature of theories or concepts of science education and about theories or concepts related to HPS/NOS?

Table 67.1 Categories in each dimension

Dimension	Symbol	Categories
Theme	T1	Research on the trends of theories and ideas related to HPS/NOS
	T2	Research on history or historical episodes related to HPS/NOS
	T3	Research on concepts/terminologies and their nature related to HPS/NOS
	T4	Research on historical figures' life, works, and theory related to HPS/NOS
	T5	Research on social issues related to HPS/NOS
	T6	Research on curriculum and educational policies related to HPS/NOS
	T7	Research on participants' views and attitudes related to HPS/NOS
	T8	Research on instructional strategies, program, and textbooks related to HPS/NOS
Method	M1	Investigation of historical materials
	M2	Theoretical review
	M3	Quantitatively analysis on quantitative data
	M4	Quantitatively analysis on qualitative data (drawings, narrative answer, etc.)
	M5	Qualitative research
	M6	Experimental research
Participants/subject/objects	S1	Literature (published paper, historical materials, textbooks, etc.)
	S2	Kindergarten/primary students
	S3	Secondary students
	S4	University/Graduated students
	S5	In-service teachers
	S6	Experts (professor of science education, scientist, etc.)
	S7	Publics
	S8	Gifted students
	S9	Underachieve/disabled students
	S10	Facilities

– In case of studies which deal with instruction methods or programs, does this paper pay attention to their nature and change, rather than to obtaining scientific knowledge or to improving scientific skills?

67.3.2 Results of the Analysis

67.3.2.1 General Trend

Among a total of 1,362 papers published in JKASE between 1978 and 2010, the number of HPS/NOS-related papers was 246 which is 18.1 % of the total papers. Figure 67.1 shows the distribution of the HPS/NOS-related papers across the period.

As can be seen from Fig. 67.1, the number of HPS/NOS-related papers increased gradually until mid-2000s when the number started to decline slightly. However,

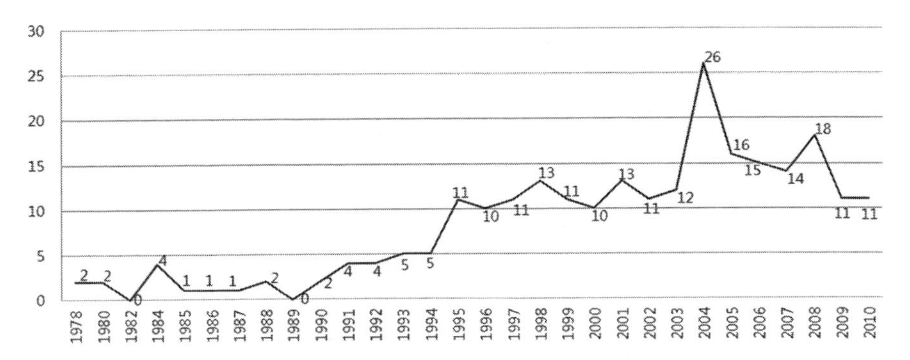

Fig. 67.1 Frequency of the papers related to HPS/NOS published in the JKASE

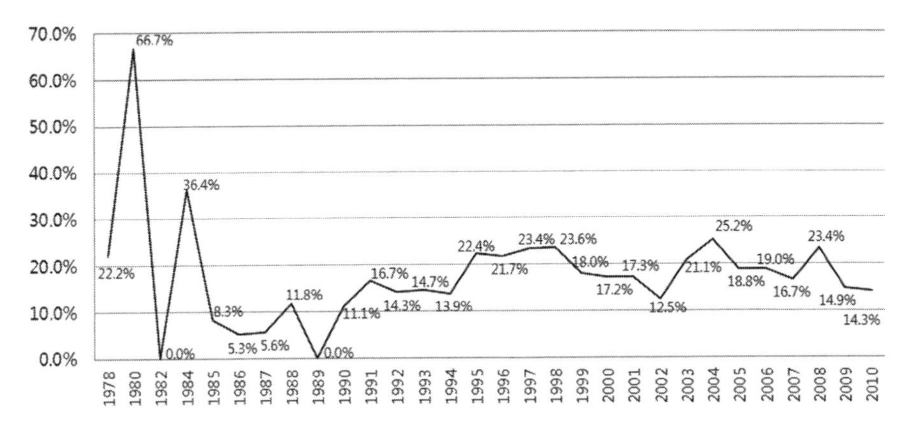

Fig. 67.2 Proportion of the papers related to HPS/NOS to the whole papers in the JKASE

since the number of total papers published by the journal itself has been increased (that is, 12 in 1978–1980, 24 in 1981–1985, 89 in 1986–1990, 171 in 1991–1995, 267 in 1996–2000, 408 in 2001–2005, 391 in 2006–2010), it is not appropriate to just compare the number of HPS/NOS papers over the periods. Figure 67.2 shows the percentage of HPS/NOS-related papers among the total papers published in JKASE in each year.

Based on Figs. 67.1 and 67.2, the proportion of HPS/NOS papers increased until the end of the 1990s, with the exception of the late 1970s and early 1980s when the total number of paper itself was very small. During this period of time when the proportion of HPS/NOS papers were increasing, Korea's 5th (1987–1992) and 6th (1992–1997) National Science Curricula were in practice. These National Science Curricula were based on the critical reflection of the previous National Science Curriculum in which too much emphasis was given to the discipline-centered approach and introduced the ideas of "Science for All" and STS in which not only scientific knowledge and process skills but also scientific attitudes and everyday scientific inquiry were emphasized at the level of the National Science Curricula (MEST, 2007a). It seems that these emphases of the National Science Curricula encouraged studies on the reconsideration of inquiry or concepts (e.g., Cho 1988; Cho 1990), on the needs and implementation methods of STS (e.g., Kwon 1991; Ha 1991; Cho 1991), and on the discussion and surveys of science-related attitude (e.g., Kwon and Park 1990; Hur 1993). In addition, following the first paper on constructivism in JKASE (Cho 1984), active research activities on the nature of science, scientific knowledge, scientific concepts, and scientific method (e.g., Cho 1988; Cho 1992) and on survey studies related to these aspects (e.g., Song and Kwon 1992; Kwon and Pak 1995) were carried out throughout the period. During the 1990s, there had been many empirical survey studies on NOS (e.g., Kwon and Park 1990; Song 1993), which made use of various newly developed instruments (such as, TOSRA, VOSTS, DAST, Nott & Wellington). The abovementioned trend during this time contributed to the growth of HPS/NOS-related papers during the 1990s.

The total number of authors of the 246 HPS/NOS-related papers in JKASE was 580, thus the average number of authors per paper was 2.4. Breaking down the period into three segments, the average author number was 1.5 during 1978–1990, 2.1 during 1981–1990, and 2.6 during 2001–2010. It is clear that the degree of co-studying for papers has been increased over time. Nevertheless, among the 580 authors of HPS/NOS-related papers, there were only 13 foreign authors, suggesting that the international collaborative research work has not been active.

Among the HPS/NOS-related papers, there were only 36 papers, i.e., 14.6 %, which were related to the history of science. Nevertheless, if we compare the numbers of history-related papers over the three decades, there was a noticeable increase: 1 paper during 1978–1990, 10 papers during 1991–2000, and 25 papers during 2001–2010.

67.3.2.2 Analysis Results of the Dimension of "Research Theme"

Table 67.2 shows the results of the analysis of the whole 246 HPS/NOS-related papers in terms of the dimension of research theme, whereas Fig. 67.3 shows the comparison of the results during the three decades.

Among the eight categories of the dimension of "research theme," the most popular one was "T7 Research on participants' views and attitudes related to HPS/NOS" with 43.9 % (i.e., a total of 108 papers). For example, like "Middle school science teachers' philosophical perspectives of science (Soh et al. 1998)" in which the authors found that the teachers predominantly held inductivistic views regardless of their major, gender, and career. Studies on students' and teachers' perceptions on the nature of science, philosophical aspects of science, STS, scientific attitudes, and science concepts occupied almost half of the HPS/NOS-related papers. Together with studies on "T8 Research on instructional strategies, program, and textbooks related to HPS/NOS" (22.8 %), this result indicates that studies on educational application and implementation of HPS/NOS were very popular. The second most popular category (27.2 %) was "T3 Research on concepts/terminologies and their nature related to HPS/NOS," suggesting that Korean science educators' attention has also been given to the

Table 67.2 Dimension of "research theme"

Symbol	Category	N	(%)	Example
T1	Research on the trends of theories and ideas related to HPS/NOS	12	4.9	Science education: constructivist perspectives (Cho and Choi 2002)
T2	Research on history or historical episodes related to HPS/NOS	14	5.7	Historical review in Foundation of London Loyal Society and France Loyal Academy (Kim 1978)
T3	Research on concepts/terminologies and their nature related to HPS/NOS	67	27.2	The role of deductive reasoning in scientific activities (Park 1998)
T4	Research on historical figures' life, works, and theory related to HPS/NOS	13	5.3	John Tyndall (1820–1894), who brought physics and the public together (Song and cho 2003)
T5	Research on social issues related to HPS/NOS	4	1.6	Brief discussion on the scientific creationism critiques (Yang 1987)
T6	Research on curriculum and educational policies related to HPS/NOS	25	10.2	An investigation into "Science-Technology-Society" curricula (Cho 1991)
T7	Research on participants' views and attitudes related to HPS/NOS	108	43.9	Middle school science teachers' philosophical perspectives of science (Soh et al. 1998)
T8	Research on instructional strategies, program, and textbooks related to HPS/NOS	56	22.8	The influence of small group discussion using the history of science upon students' understanding about the nature of science (Kang et al. 2004)
Total		299		

analysis and examination of existing theories or concepts related to HPS/NOS. For example, Park (1998) attempted to clarify the logical structure of the scientific explanation, prediction, and the process of hypothesis testing using syllogism and various concrete examples in his paper, "The role of deductive reasoning in scientific activities."

On the contrary, there were only small numbers of studies on social issues (1.6 %), on persons (5.3 %), and history or episodes (5.7 %) related to HPS/NOS. It would be quite natural that there were so many studies on the educational application or implementation of HPS/NOS in science education. However, too much emphasis on educational application could overlook the critical analysis of existing theories or perspectives and the suggestion of new theories or perspectives.

According to the comparison among the three decades (shown in Fig. 67.3), the proportions of T3 and T6 categories decreased, whereas the proportion of T8 category increased. In other words, studies on terminology/concepts/features of HPS/NOS became less popular, while those on instructional methods/programs/

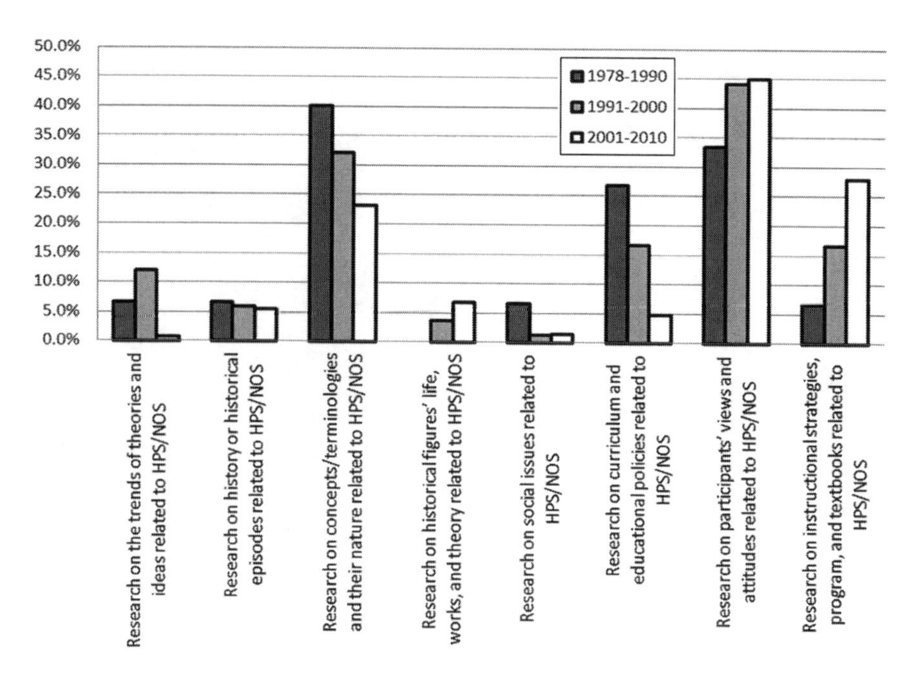

Fig. 67.3 Proportion of each "theme" to the whole HPS/NOS papers in three decades

textbooks became more popular. It seems that this trend reflects the history of Korean science education where in earlier years there were needs to carry out theoretical discussion and educational arguments for introducing HPS/NOS ideas, whereas after the 1990s there were needs to pursue more practical and concrete instructional methods and programs based on earlier more theoretical studies and on the introduction of the National Curricula emphasizing STS approaches and scientific literacy. This overall trend is also reflected in the very low proportion of "T1 Research on the trends of theories and ideas related to HPS/NOS" (only 0.7 %), like "Science education: constructivist perspectives (Cho and Choi 2002)" in which the authors described the characteristics of constructivism through reviewing the literatures from a few schools of constructivism, during the period of 2001–2010. Meanwhile, as discussed earlier, it is presumed that the continuous increase of "T7 Research on participants' views and attitudes related to HPS/NOS" was facilitated by the wide use of various newly developed instruments (such as, TOSRA, VOSTS, DAST, Nott & Wellington).

67.3.2.3 Analysis Results of the Dimension of "Research Methods"

Table 67.3 shows the result of the analysis of HPS/NOS papers in terms of research method, whereas Fig. 67.4 shows the comparison of the results during the three decades.

Table 67.3 Dimension of "research methods"

Symbol	Category	N	%	Examples
M1	Investigation of historical materials	20	8.1	The activity of an interpreter on science education during the enlightenment period in Korea: Focus on Hyun Chae (Park 2009)
M2	Theoretical review	92	37.4	A philosophical study on the generating process of declarative scientific knowledge: Focused on inductive, abductive, and deductive process (Kwon et al. 2003)
M3	Quantitative analysis on quantitative data	98	39.8	Teachers' and students' understanding of the nature of science (Han and Chung 1997)
M4	Quantitative analysis on qualitative data (drawings, narrative answer, etc.)	53	21.5	Teachers' images of scientists and their respected scientists (Song 1993)
M5	Qualitative research	18	7.3	An intensive interview study on the process of scientists' science knowledge generation (Yang et al. 2006)
M6	Experimental research	28	11.4	The effects of decision-making-centered STS (Science-Technology-Society) classes on the students' attitudes toward science and perceptions about STS (Hong 2001)
Total		309		

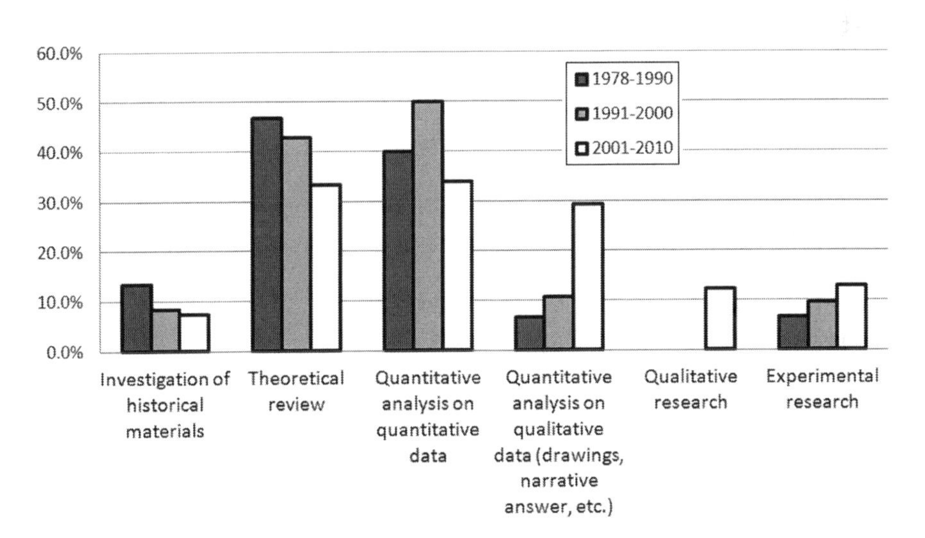

Fig. 67.4 Proportion of each "methods" to the whole HPS/NOS papers in three decades

Through the analysis of HPS/NOS-related papers in terms of research method, the most popular categories were "M3 Quantitative analysis on quantitative data" (39.8 %) and "M2 Theoretical review" (37.4 %). "M4 Quantitative analysis on qualitative data (drawings, narrative answer, etc.)" (21.5 %), exemplified by "Teachers' images of scientists and their respected scientists (Song 1993)" in which the author explored through Draw-A-Scientist Test and analyzed the drawings with frequencies, was also relatively popular. On the contrary, "M1 Investigation of historical materials" (8.1 %) such as "The activity of an interpreter on science education during the enlightenment period in Korea: Focus on Hyun Chac (Park 2009)" and "M5 Qualitative research" (7.3 %) like "An intensive interview study on the process of scientists' science knowledge generation (Yang et al. 2006a)" was not popular. This may illustrate that HPS/NOS-related studies in Korea during the last three decades focused more on reviews or analysis of existing relevant studies or with interpretation or investigation through the analyses of data (esp. quantitative). However, examining the trend along the timeline, a different picture of recent research methods becoming more diverse can be seen.

As shown in Fig. 67.4, recently, the two dominant methods (i.e., M2 and M3) became less popular, whereas M4 and M5 which included qualitative data and analysis became more popular especially during the 2000s. This trend of widening research methodology seems to reflect that from the 1990s Korean science education researchers began to pay attention to the weaknesses of quantitative research methods (e.g., for the investigation people's understanding of the nature of science by Aikenhead and colleagues (1989)) and to advantages of more open and qualitative data analysis (e.g., Merriam 1988).

67.3.2.4 Analysis Results of the Dimension of "Research Subjects"

The results of the analysis of HPS/NOS-related papers in terms of the dimension of research subjects are shown in Table 67.4, whereas the comparison of the results during the three decades is shown in Table 67.5.

In terms of research subjects, the most popular categories were "S1 Literature (published paper, historical materials, textbooks, etc.)" (43.8 %) and "S3 Secondary students" (41.5 %). Conversely, the least popular categories were "S7 The general public" (1.2 %), "S9 Underachieving/Disabled students" (0.8 %), and "S10 Facilities" (0.8 %).

As shown in Fig. 67.5, as the subjects of the studies, teachers became less popular in the 2000s, whereas secondary students became much more popular from the 1990s. It is reasoned that this change was brought about by the increased research interest in students' thinking and ideas influenced by constructivist approach from the end of the 1980s in Korea.

In case of "S8 Gifted students" like "The effects of explicit instructions on nature of science for the science-gifted (Park and Hong 2010)" in which the authors invited 20 science-gifted students as participants, there had been almost no studies until the 1990s, but then suddenly there was a sharp increase in the 2000s. This was mainly because that the Gifted Education Act was passed in 2000 and as a result,

Table 67.4 Dimension of "subjects/participants/objects"

Symbol	Category	N	%	Examples
S1	Literature (published paper, historical materials, textbooks, etc.)	108	43.8	The images of science education illustrated in the books written by modern philosophers of science (Song et al. 1997)
S2	Kindergarten/primary students	22	8.9	Perceptions about science and scientific activity of students in kindergarten and primary school (Kim and Cho 2002)
S3	Secondary students	102	41.5	Perception survey on characteristics of scientific literacy for global Science-Technology-Society for secondary school students (Ryu and Choi 2010)
S4	University/graduated students	29	11.8	Preservice science teachers' understanding of the nature of science (Nam et al. 2007)
S5	In-service teachers	29	11.8	A study on Korean science teachers' points of view on nature of science (Cho and Ju 1996)
S6	Experts (professor in science education, scientist, etc.)	8	3.3	Aims of laboratory activities in school science: A Delphi study of expert community (Yang et al. 2006)
S7	The general public	3	1.2	Science-related attitudes of Korean housewives (Kim et al. 2004)
S8	Gifted students	12	4.9	The effects of explicit instructions on nature of science for the science-gifted (Park and Hong 2010)
S9	Underachieving/disabled students	2	0.8	A investigation of the attitudes toward science and scientific attitude for the underachievers (Yi and Kim 1984)
S10	Facilities	2	0.8	Developing active role of science museum in educating on ethical issues on science and technology: Four case studies (Choi 2004)
Total		317		

government administrative and financial supports were put in place, such as establishing Gifted Education Institutions in universities and local education offices. This might show that studies related to HPS/NOS can also be influenced by national policies and governmental supports.

In sum, the analysis of HPS/NOS-related papers published in JKASE (*Journal of Korean Association for Science Education*) from 1978 to 2010 shows that HPS/NOS-related studies have been continuously carried out largely on HPS/NOS-related people's ideas/views/attitudes, on instructional methods or programs, and on terms/concepts/features of HPS/NOS. In general, although there has been some concentration on specific areas in terms of research method, research subjects, and purpose and results, recently research with diverse purposes and methods became more and more popular.

Table 67.5 Analysis of the type and organization of the historical information

Subdimensions	Textbooks G3	G4	G5	G6	G7	G8	G9	G10	Total
1.1. Scientists									
1.1.1. Scientists' life									
1.1.1.1. Biographic data	–	–	2	7	6	15.5	8	43.0	81.5
1.1.1.2. Personal characteristics	–	–	–	–	–	–	–	0.5	0.5
1.1.1.3. Episodes/anecdotes	1	–	1	5	0.5	1.5	1	0.5	10.5
1.1.2. Scientists' characteristics									
1.1.2.1. Famous/genius	2	1	2	7	1	–	2	0.5	15.5
1.1.2.2. Ordinary	–	–	–	–	–	–	–	–	–
1.2. Evolution of science									
1.2.1. Type of evolution									
1.2.1.1. Mention to a science discovery	1	1	–	2	6	8	5	32	55
1.2.1.2. Description of a science discovery	3	3	1	6	2	1.5	2	3.5	22
1.2.1.3. Mention to discreet periods					0.5	1	0.5	4.5	6.5
1.2.1.4. Linear and straightforward	4	–	4	1	1.5	11	1.5	12	35
1.2.1.5. Real evolution	–	–	3	1	–	–	–	0.5	4.5
1.2.2. Responsible people									
1.2.2.1. Individual scientists	2	1	5	10	8.5	16.5	8	39.5	90.5
1.2.2.2. Group of scientists	1	1	2	–	1	2.5	1	10.5	19
1.2.2.3. Scientific community	2	–	3	–	1.5	5	0.5	2.5	14.5

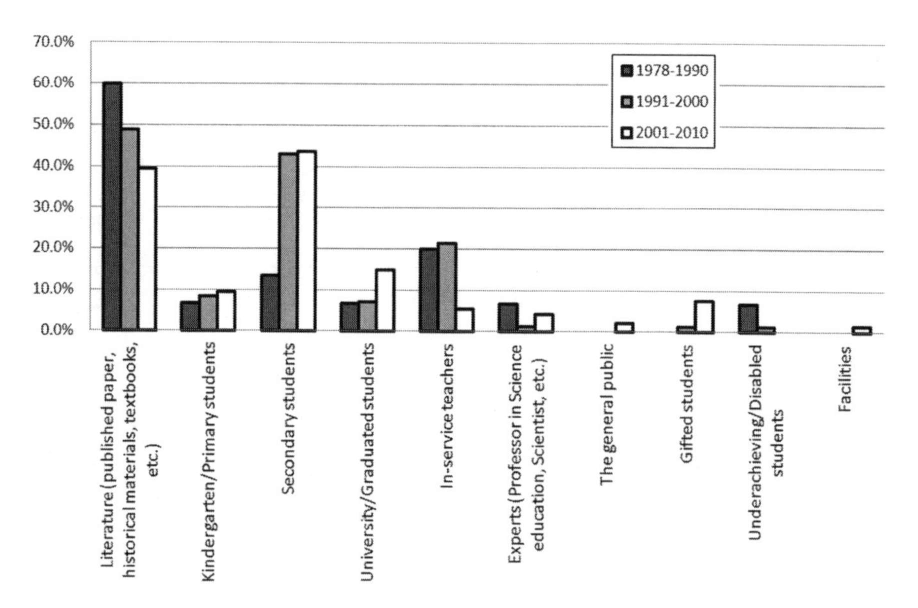

Fig. 67.5 Proportion of each "subjects/participants/objects" to the whole HPS/NOS papers in three decades

67.4 HPS/NOS in Textbooks in Korea

The science textbook is the most basic instructional material used in schools, particularly countries like Korea where school education is considerably centralized. Thus, the analysis of the content related to HPS/NOS in science textbooks would provide a good indication showing how much of HPS/NOS-related content are taught in schools. Existing studies on HPS/NOS content in textbooks (e.g., Cho 2008; Lee and Shin 2010; Leite 2002; Niaz 2000) also provide some meaningful references to this analysis. In this section, the analysis is done with the most recent science textbooks in Korea, which were published between 2009 and 2011 according to 2007 National Science Curriculum. It is expected that the results of the analysis illustrate the current practice of HPS/NOS-related instruction in Korean school science education.

67.4.1 The Method of Analysis

67.4.1.1 Textbooks Analyzed

This analysis was done with textbooks for G3 to G10 Science subject in Korea. The textbooks were written based on 2007 National Science Curriculum, except for G10 of which National Science Curriculum was revised once again in 2009. For G3 to G6 which are parts of elementary school years, there is only one kind of Science textbooks because there is a system of national textbooks for elementary schools in Korea (see textbooks of G3-1–G6-2 in the References). For G7 to G10 for which different kinds of Science textbooks are available through the government approval, the most two popular Science textbooks for each grade were chosen and analyzed (see textbooks G7-A–G10-B in the References).

67.4.1.2 Analysis Framework and Method

The content related to the history of science was analyzed based on the framework of Leite (2002). However, since in Korea there are no official student workbooks provided for G7 to G10 students and no separate lists of content according to learners' levels in science textbooks, it was not easy to use directly the analysis framework of Leite (2002). Thus, the frameworks of Choi, Yeo, and Woo (2005) and of Lee and Shin (2010) were consulted, which modified Leite's framework for better adjustment to Korean situation, and added some new items which were turned out to be necessary through the actual analysis. Tables 67.5, 67.6, 67.7, and 67.8 in the Sect. 67.4.2 show the framework for analyzing historical content of Korean science textbooks used in this study and the results.

There were four dimensions for analyzing historical content of the science textbooks: "Type and organization of the historical information" (see Table 67.5 in the

Table 67.6 Analysis of the content of the history of science

Subdimensions	Textbooks								
	G3	G4	G5	G6	G7	G8	G9	G10	Total
2.1. Contexts to which historical information is related									
2.1.1. Scientific	2	3	8	7	3.5	12	3.5	13	52
2.1.2. Technological	7	3	4	4	1.5	4	–	6	29.5
2.1.3. Social	1	–	2	1	0.5	1.5	0.5	2	8.5
2.1.4. Political	–	–	–	–	–	0.5	–	–	0.5
2.1.5. Religious	–	–	–	–	–	–	–	–	–
2.2. Domain of science content									
2.2.1. Physics	4	1	2	5	1	1.5	3	13.5	31
2.2.2. Chemistry	2	1	2	3	3.5	7	2	5.5	26
2.2.3. Biology	1	1	–	–	1	1	1	17.5	22.5
2.2.4. Earth science	1	1	4	2	7	12.5	3	16	46.5
2.3. Domain related to the purpose of science education									
2.3.1. Cognitive domain	2	3	6	7	9.5	17.5	8	34	87
2.3.2. Affective domain	–	–	2	2	–	–	–	–	4
2.3.3. Process skill domain	–	–	3	1	–	2	0.5	–	6.5
2.3.4. STS	8	3	4	4	2.5	2.5	0.5	17.5	42

Sect. 67.4.2), "Content of history of science" (see Table 67.6 in the Sect. 67.4.2), "Source of history of science" (see Table 67.7 in the Sect. 67.4.2), and "Role of the historical content in science teaching and learning" (see Table 67.8 in the Sect. 67.4.2). The dimension of "Type and organization of the historical information" is to see what kinds of and how the historical information is treated in science textbooks and is consisted of two subdimensions: "Scientists" and "Evolution of science." "Scientists" subdimension is divided further into two, "scientists' life" and "scientists' characteristics," each of which is further divided into two or three items. The dimension of "Content of history of science" is to see the features of historical content in science textbooks and is consisted of three subdimensions – "Context to which historical information is related," "Domain of science content," and "Domain related to the purpose of science education" – each of which is divided further into several sub-subdimensions. The dimension of "Source of history of science" is to see the features of the sources of historical information and materials and is consisted of three subdimensions – "Materials used to present the historical information," "Culture/Nations related to history of science," and "Times related to history of science" – each of which is divided further into several sub-subdimensions. Lastly, the dimension of "Role of the historical content in science teaching and learning" is to see the actual roles of the historical content shown in textbooks in terms of science teaching and learning and is consisted of two subdimensions – "Fundamental" and "Complementary – each of which is divided further into several sub-subdimensions.

On the other hand, the content related to the philosophy of science or NOS was analyzed based on the framework of Choi, Choi, and Jin (2010). This framework is based on that of Leite (2002) but with a revision after considering VOSTS items in

Table 67.7 Analysis of the source of the history of science

Subdimensions	Textbooks								
	G3	G4	G5	G6	G7	G8	G9	G10	Total
3.1. Materials used to present the historical information									
3.1.1. Scientists' pictures	1	–	5	8	3	2.5	2	11.5	33
3.1.2. Pictures from machines, laboratory equipment, etc.	3	1	2	2	1	−1	–	1	11
3.1.3. Original documents/texts	–	–	–	–	–	1.5	–	0.5	2
3.1.4. Historical experiments	–	–	3	3	0.5	–	1.5	0.5	7.5
3.1.5. Secondary sources	–	–	–	–	–	1	–	–	1
3.1.6. Texts by the textbook author(s)	5	2	8	9	2	5	0.5	3	34.5
3.1.7. Pictures of science cultural heritage	1	1	2	2	1	0.5	0.5	0.5	8.5
3.1.8. Pictures of historical event	–	1	1	–	1.5	4.5	0.5	2.5	11
3.2. Culture/nations related to history of science									
3.2.1. The Eastern culture/nations									
3.2.1.1. Korea	1	1	1	3	3	2.5	0.5	2	14
3.2.1.2. China	–	–	–	–	–	–	–	0.5	0.5
3.2.1.3. Japan	–	–	–	–	–	–	–	–	–
3.2.1.4. General Orient	–	–	–	–	–	0.5	–	–	0.5
3.2.1.5. Other Eastern culture/nations		–	–	–	0.5	1	0.5	–	2
3.2.1.6. Not mention to particular Culture/nation		–	–	–	–	–	–	–	–
3.2.2. The Western culture/nations									
3.2.2.1. United Kingdom	1	1	–	–	3.5	3.5	1.5	9	19.5
3.2.2.2. France	–	–	1	–	0.5	2	1.5	3	8
3.2.2.3. Germany	–	–	–	–	3.5	1.5	1.5	7	13.5
3.2.2.4. Italy/Greece	–	–	2	2	–	4	2.5	2	12.5
3.2.2.5. USA/Canada	1	2	–	1	0.5	1	1.5	9	16
3.2.2.6. General West	–	–	–	–	–	–	–	0.5	0.5
3.2.1.7. Other Western culture/nations	1	–	1	–	2	2.5	1.5	9	17
3.2.1.8. Not mention to particular Culture/nation	1	–	3	4	1	11	3	17.5	40.5
3.2.3. Others									
3.2.3.1. Other culture/nations	–	–	–	–	0.5	2	0.5	0.5	3.5
3.2.3.2. Not mention to particular Culture/nation	4	–	–	–	–	–	–	4.5	8.5
3.3. Times related to history of science									
3.3.1. Before around 14C	2	1	–	–	1.5	6.5	2	2.5	15.5
3.3.2. Around 14C–16C	2	–	–	2	1.5	4	1	2.5	13
3.3.3. Modern times (17–19C)	3	2	1	6	3.5	10.5	4.5	22.5	53
3.3.4. Contemporary times (20C-)	6	2	1	3	6	6.5	2.5	29	56
3.3.5. Not mention to particular times	–	–	–	–	1	4	0	4	9

Aikenhead, Ryan, and Fleming (1989); national standards of several countries analyzed by McComas and Olson (1998); and VNOS items in Lederman, Abd-El-Khalick, Randy, and Renée (2002). Items of "Scientists' life" and "Scientists' attribute"

Table 67.8 Analysis of the role of historical content in science teaching and learning

	Textbooks								
Subdimensions	G3	G4	G5	G6	G7	G8	G9	G10	Total
4.1. Fundamental	–	–	5	3	4.5	17.5	3.5	37	70.5
4.2. Complementary	8	4	3	7	8	6.5	5.5	15.5	57.5

Table 67.9 Analysis of the responsibility and role of scientists, the development of scientific knowledge, and scientific methods

	Textbooks								
Dimensions and subdimensions	G3	G4	G5	G6	G7	G8	G9	G10	Total
1. Responsibility and role of scientists									
1.1. Personal level	2	1	4	5	2	6.5	1	8	29.5
1.2. Social level	6	4	6	5	0.5	–	–	0.5	22
2. Development of scientific knowledge									
2.1. Model of development									
2.1.1. Cumulative model	1	–	2	3	2	4	0.5	3	15.5
2.1.2. Evolutionary model	–	–	1	1	0.5	2	–	3.5	8
2.1.3. Revolutionary model	–	–	–	–	–	0.5	–	2.5	3
2.1.4. Gradual model	–	–	1	–	1	2	0.5	3.5	8
2.2. Introduction scope of "development of science"									
2.2.1. Introduction of scientific developments only	1	–	–	–	1	4	–	2	8
2.2.2. Introduction of scientific developments with the background and consequences of them	–	–	4	4	2	4.5	1	10.5	26
3. Scientific methods									
3.1. Inductive method	7	3	3	2	2	6	2	5.5	30.5
3.2. Deductive method (including the method of testing hypothesis)	1	3	6	7	2.5	4.5	2	7.5	34
3.3. Abductive method (included in generating hypothesis)	1	3	5	5	1	1.5	0.5	4	21
3.4. Social consultation method	–	–	1	–	–	–	–	0.5	1.5

in the framework of Choi, Choi, and Jin (2010), which earlier appeared in the framework for the content related to the historical content of science, were excluded. In addition, through the analysis of textbooks, it was found that there was a need to introduce a few new classifications and to make a slight revision of the framework. Tables 67.9 and 67.10 in the Sect. 67.4.3 show the framework used in this study to analyze textbook content related to the philosophy of science or NOS and the results.

The framework of analyzing philosophical/NOS content in science textbooks is consisted of 6 dimensions: "Responsibility and role of scientists" (see Table 67.9 in the Sect. 67.4.3), "Development of scientific knowledge" (see Table 67.9 in the Sect. 67.4.3), "Scientific methods" (see Table 67.9 in the Sect. 67.4.3), "Science-Technology-Society relation" (see Table 67.10 in the Sect. 67.4.3), "Domain related

Table 67.10 Analysis of the Science-Technology-Society relation, domain related to the purpose of science education, and role of the NOS content in science teaching and learning

	Textbooks								
Dimensions and subdimensions	G3	G4	G5	G6	G7	G8	G9	G10	Total
4. Science-Technology-Society relation									
4.1. Science-Technology relation									
4.1.1. Positive relation	–	2	3	2	1.5	1	–	1	10.5
4.1.2. Negative relation	–	–	–	–	–	–	–	–	–
4.2. Science-Society relation									
4.2.1. Positive relation	2	1	3	–	–	–	0.5	0.5	7
4.2.2. Negative relation	–	–	–	–	–	–	–	–	–
4.3. Science-Technology-Society relation									
4.3.1. Positive relation	6	4	–	4	0.5	–	0.5	0.5	15.5
4.3.2. Negative relation	–	1	–	–	0.5	–	–	–	1.5
5. Domain related to the purposc of science education									
5.1. Cognitive domain	2	2	5	5	3.5	8.5	0.5	12.5	39
5.2. Affective domain	1	–	2	1	1	–	–	0.5	5.5
5.3. Process skill domain	7	5	9	6	1.5	1	1	–	30.5
5.4. STS	6	6	5	6	0.5	–	0.5	0.5	24.5
6. Role of the NOS content in science teaching and learning									
6.1. Fundamental	–	1	6	7	1.5	6.5	–	10.5	18.5
6.2. Complementary	12	11	6	7	3.5	3	2	3	47.5

to the purpose of science education" (see Table 67.10 in the Sect. 67.4.3), and "Role of the historical content in science teaching and learning" (see Table 67.10 in the Sect. 67.4.3). The dimension of "Responsibility and role of scientists" is to check the viewpoints of the textbooks describing scientists' roles and is consisted of two subdimensions: "Personal level" and "Social level." The dimension of "Development of scientific knowledge" is to check on what basis of the model textbooks describe the development of scientific knowledge and is consisted of two subdimensions: "Model of development" and "Introduction scope of 'development of science.'" The dimension of "Scientific methods" is to check the viewpoints of textbooks toward scientific methods and is consisted of four subdimensions: "Inductive method," "Deductive method (including the method of testing hypothesis)," "Abductive method (included in generating hypothesis)," and "Social consultation method." The dimension of "Science-Technology-Society relation" is to check whether textbooks describe the positive or negative relations among science, technology, and society and is consisted of three subdimensions: "Science-Technology relations," "Science-Society relation," and "Science-Technology-Society relation." The dimensions of "Domain related to the purpose of science education" and of "Role of the historical content in science teaching and learning" are consisted of the same subdimensions as earlier in the framework used for the analysis of historical content of textbooks.

For the analysis, the frequencies of items or sub-subdimensions were checked. If a part of a textbook dealt with one discovery or event or theme (either of the history of science or of the philosophy of science or NOS), it was counted as one occasion, regardless of the number of pages. In other words, two separate historical contents can be identified from a single page or more than one page can be regarded as dealing with a single historical content. On the contrary, in case of having two or more sub or sub-subdimensions or items in a single content related either to the history of science or to the philosophy of science/NOS, the frequency was counted as many as needed. The frequency was counted as a year total. That is, in the case of science textbooks for G3–G6 where there is only one kind of government-approved textbook, the frequency was calculated as a sum for two semesters. In the case of science textbooks for G7–G10 where the most popular two textbooks were analyzed, the frequency was calculated as an average of the sums of the two textbooks for two semesters.

67.4.2 Results of the Analysis of the Historical Content of Science Textbooks

Table 67.5 shows the results of the analysis of the type and organization of the historical content of Korean science textbooks for G3–G10.

As shown in Table 67.5, in the case of "scientists' life," the historical content of scientists' life was mostly on biographical data (such as their names, years of birth, or death) which appeared with 81.5 cases, while the content on the episodes/anecdotes or personal characteristics of scientists was relatively very few appearing with 10.5 cases and 0.5 cases, respectively. This result is in accordance with the results of previous studies on Portuguese textbooks by Leite (2002) and on Korean textbooks based on the previous National Science Curricula by Choi and colleagues (2005) and by Lee and Shin (2010). This result shows that the descriptions of scientists tend to provide fragmented information, rather than meaningful narrative elements, of the scientists. Nevertheless, the appearance (i.e., 10.5 cases) of the episodes/anecdotes is in fact much higher than the result of the study by Lee and Shin (2010) which showed only 0.7 cases with the textbooks (for G3–G7) based on the 7th National Science Curriculum (MOE 1997). In fact, the elementary science textbooks (for G3–G6) based on the 7th National Science Curriculum had no appearance of episodes/anecdotes. The big increase of the episodes/anecdotes in the elementary science textbooks based on 2007 National Science Curriculum was made by introducing new sections, called "Science Stories" and "Inquiry by Scientists," to the elementary science. These sections encouraged to have ample and diverse descriptions on scientists.

In the case of "scientists' characteristics," scientists were described in 15.5 cases as being of different characteristics – such as with full of curiosity and endless efforts to achieve exactness or with talented ideas – from ordinary people, while there was no description of scientists as ordinary people. For example, the process of which A. L. Wegener claimed the theory of continental drift and was searching for the grounds of his claim in a very rigorous way (G6-1, pp. 188–191) and the story of which Dr. Jangchun Woo (a famous pioneering Korean scientist) saved the people from the shortage of food through his talented ideas and enormous passion (e.g., G10-A, p. 294) are some examples of scientists as persons with exceptional abilities and attitudes. This kind of the description of scientists as exceptional abilities should be reconsidered, especially if we wish to encourage people's science-friendly attitudes and students' careen guidance toward science and engineering.

For "type of evolution" in "Evolution of science," the most common items described were "mention to a science discovery" (55 cases), "linear and straightforward" (35 cases), and "description of a science discovery" (22 cases), while "real evolution" and "mention to discreet periods" were only 4.5 cases and 6.5 cases, respectively. However, compared with that (i.e., 1.3 cases shown in Lee and Shin (2010)) of the science textbooks based on the 7th National Science Curriculum, "real evolution" became more popular, and this was also due to the introduction of "Science Stories" and "Inquiry by Scientists" sections. For example, the science textbook for the 2nd semester of G6 describes the whole process of the discovery of oxygen across its four pages, which includes A. L. Lavoisier's criticism to and experiment against J. Priestley's claim, discovery of the nature of air as the mixture of various gases, and the naming process of oxygen (G6-2, pp. 162–165).

In the case of "responsible people," the description of "individual scientists" (90.5 cases) was much more popular than those of "group of scientists" (19 cases) and of "scientific community" (14.5 cases), implying that historical content in the textbooks was largely focusing on the achievements of individual scientists.

Table 67.6 shows the analysis results of the content of the history of science. Popular contexts to which historical information is related were "scientific context" (52 cases), "technological context" (29.5 cases), and "social context" (8.5 cases), while "political context" and "religious context" were rarely or no described. The only "political context" was the case of the development of the Western calendar in which the role of the rulers is described as an important factor (G8-A, p. 351). Thus, in terms of the contexts, it can be argued that the textbooks still have rather narrow descriptions of the history of science, exclusively focusing on its scientific and technological contexts.

The analysis in terms of "Domain of science content" shows that the historical content was from "earth science" (46.5 cases), physics (31 cases), chemistry (26 cases), and biology (22.5 cases). Despite a higher ratio from "earth science," the distribution across the domains of science seems not to be overly biased. One thing to pay attention is that the vast majority of "biology" domain was from G10, 17.5 out of 22.5.

In the case of "Domain related to the purpose of science education," "cognitive domain" and "STS" were found to be most common, 87 cases and 42 cases, respectively, while "affective domain" and "process skill domain" were less common, 4 cases and 6.5 cases, respectively. This result with current science textbooks is rather similar to that of the result with previous science textbooks based on the 7th National Science Curriculum (Lee and Shin 2010) (that is, cognitive domain 96.7 cases, STS

16.9 cases, process skill domain 5.3 cases, and affective domain 1 case), except for the fact that “STS” became more popular. This result shows that although historical content can be effective in stimulating students’ motivation, in illustrating humanistic aspects of science (e.g., Jung 1994; Matthews 1994) and in helping the understanding of scientific methods (e.g., Matthews 1992), the treatment of historical content in science textbooks is still biased toward its “cognitive” and “STS” aspects.

Table 67.7 shows the results of the analysis of “Source of History of Science” in terms of its materials form, cultural/national background, and historical period. The most popular forms of historical information were “texts by the textbook author(s)” (34.5 cases) and “scientists’ pictures” (33 cases), such as M. Faraday, G. Galilei, and R. J. E. Clausius. On the contrary, original documents/texts were little used, as is seen in only 2 cases (e.g., the chemistry textbook and a part of element table by A. L. Lavoisier – G8-A, p.54). Despite the difficulty of the translation into Korean, for the credibility of historical information, it would be better to have more original documents/texts in science textbooks. On the other hand, in the case of photos and pictures of scientists, while there was no case in the elementary science textbooks (for G3–G6) based on the 7th Science Curriculum (Lee and Shin 2010), there were 14 cases in those based on 2007 National Science Curriculum, such as C. Alessandro Volta (G5-1, p. 86), L. Pasteur (G5-1, p. 166), and M. Faraday (G6-1, p. 182).

The “culture/nation” which had been the background of the history of science turned out to be overwhelmingly Western (in total 127.5 cases) compared with those from Eastern cultures (in total 17 cases). Furthermore, among those from Eastern cultures, 14 cases were Korean while the rest (3 cases) were from the rest of the Eastern region. This distribution, heavily biased toward the Western, would reflect the situation that most of school science content is in fact rooted into the Western science. It seems that this result is somehow related to the heavy emphasis on “cognitive domain” over “affective domain” in “Domain related to the purpose of science education” (see Table 67.7). That is to say, since the purpose of introducing of the history of science was mainly to help students’ understanding of scientific concepts, it might have been natural to focus on the Western history which would be more directly linked with scientific concepts, rather than the Eastern history which would be more useful in terms of affective domains.

The analysis of “Times related to the history of science” shows that those of “contemporary times (20C-)” (56 cases) (e.g., stories of Apollo 8 & 11 (G8-A, p. 320), the development of solar cells (G10-A, p. 424)) and “modern times (17C–19C)” (53 cases) (e.g., Boyle’s discovery of indicator (G6-1, pp. 62–63), and Lavoisier’s naming of oxygen (G6-2, pp. 162–156)) were the most popular sources of the history of science. The high appearance of “modern times” is easily understood when it is noticed that the majority of school science content are on the historical developments of that particular period of the history. The appearance of “contemporary times” is in fact higher than that (38.4 cases (Lee and Shin 2010)) of the 7th National Science Curriculum, and this seems to be the result of 2007 National Science Curriculum’s emphasis on scientific literacy (MEST 2007a, 2009).

Table 67.8 shows the result of the analysis of the role of historical content. The role of textbooks historical content in science teaching and learning was found to be slightly more frequently used as "Fundamental" than as "Complementary," although the trend was reversed at the elementary school level especially for G3–G4.

As indicated in Tables 67.5, 67.6, 67.7, and 67.8, for the most part, science textbooks for G10 appear to have much more historical content than those for other grades. This is because science textbooks for G10 were developed on the basis of the most recent curriculum, 2009 National Science Curriculum, which emphasizes the improvement of students' scientific literacy through various materials for the teaching of meanings, values, and roles of science instead of fragmented knowledge of each discipline of science.

67.4.3 Results of the Analysis of the Philosophy of Science/NOS-Related Content

The analysis of the philosophy of science/NOS content of Korean science textbooks for G3–G10 resulted in a variety of dimensions and subdimensions. The results of the analysis for "Responsibility and role of scientists," "Development of scientific knowledge," and "Scientific methods" are shown in Table 67.9.

For the "Responsibility and role of scientists," the "personal level" (29.5 cases) was found to be more popular than the "social level" (22 cases). However, at the elementary level, the trend was reversed. In elementary science textbooks, 21 cases of the "social level" were presented while 12 cases of the "personal level." For example, there was a great deal of content on the social roles and responsibility of scientists, such as, the story of Jane Goodall's study of chimpanzees. Jane Goodall's study was described in details in conjunction with the activities of environmental and animal protection groups (G3-2, pp. 66–67). Another example was a description of the beneficial contributions of scientists to the development of artificial internal organs based on their knowledge on human body (G5-2, p. 54). Once again, this increase illustrates the effect of introducing the sections for "Science Stories" and "Inquiry by Scientists" in 2007 National Science Curriculum.

In the case of "Mode of development" of the "Development of scientific knowledge," the "cumulative model" (15.5. cases) was found to be most common, followed by "evolutionary model" (8 cases), "gradual model" (8 cases), and "revolutionary model" (3 cases). This bias toward the "cumulative model" of scientific knowledge development can be an obstacle in introducing the diverse and complicate nature of scientific knowledge. Nevertheless, considering the result of the previous study by Choi and colleagues (2010) with science textbooks based on the previous 7th National Science Curriculum showing that "cumulative model" was 89.1 % and that there was no description on the model of scientific knowledge development in elementary science textbooks, it is a remarkable improvement. In addition, the current science textbooks include many more cases of "introduction of scientific developments with the background and consequences of them" (26 cases) than those of "introduction of scientific developments only" (8 cases).

In the case of "Scientific methods," "Deductive method (including the method of testing hypothesis)" (34 cases) and "Inductive method" (30.5 cases) were found to be much more popular than "Social consultation method" (1.5 cases). The extremely low description of "Social consultation method" illustrates well how current science textbooks overwhelmingly adapt traditional models of scientific development. Nevertheless, considering the result of Choi and colleagues (2010) in which, with no classification of "Abductive method (included in generating hypothesis)," "Inductive method" (84.5 %) was the vast majority compared with "Deductive method (including the method of testing hypothesis)" (15.5 %) and "Social consultation method" (no case), it can be said that the current science textbooks adapt more diverse views of scientific development. This tendency can also be found in the fact that there were 21 cases of the descriptions on "Abductive method (included in generating hypothesis)." In the case of "Abductive method (included in generating hypothesis)," the cases in which the process of abductive inference (Peirce 1878) based on similarity were mentioned or in which students were guided to have tentative explanations after observation activities were counted.

Meanwhile, in contrast to the situation that there had been no description on scientific method in the previous elementary science textbooks based on the 7th National Science Curriculum (Choi et al. 2010), the current elementary science textbooks contain 47 cases of descriptions on scientific method. It is presumed to be caused by the 2007 National Science Curriculum's introduction of "Free Inquiry" in which students are expected to carry out their own long-term investigations and thus to have experience of choosing inquiry topics, selecting inquiry methods, transforming inquiry data, and reporting inquiry results (MEST 2007a). Together with "Science Stories" and "Inquiry by Scientists," the introduction of "Free Inquiry" encourages the description of scientific methods.

Table 67.10 shows the result of the analysis of "Science-Technology-Society relation," "Domain related to the purpose of science education," and "Role of the NOS content in science teaching and learning."

For "Science-Technology-Society relation," there were 10.5 cases of "Science-Technology relation," 7 cases of "Science-Society relation," and 17 cases of "Science-Technology-Society relation." Among them, while the cases of the first two were only positive ones, 1.5 cases out of the 17 cases of the third were negative ones, such as "… the development of industry and the growth of population demand more fresh water. However, due to the environmental destruction and water pollution, the amount of usable water decreases and thus we are making a great deal of efforts to secure the water resource. We need water conservation" (G4-1, p. 125).

For "Domain related to the purpose of science education" which was further classified according to the major purpose areas stated in the 2007 National Science Curriculum, 39 cases of "cognitive domain," 30.5 cases of "process skill domain," 24.5 case of "STS," and 5.5 cases of "affective domain" were identified. Compared with that with historical content (6.5 cases), the number of cases of "process skill domain" with philosophical content/NOS here, i.e., 30.5 cases, is much higher. This is thought to be caused by the introduction of "Free Inquiry" activities in the 2007 National Science Curriculum (MEST 2007a).

For "Role of the NOS content in science teaching and learning," there were 47.5 cases of "Complementary" which is much higher than that of "Fundamental." This tendency was more apparent with lower grades, implying that philosophical content/NOS descriptions are treated of more importance with higher grade students.

In sum, in Korean science textbooks for G3 to G10, the historical and philosophical/NOS content is still represented with a bias in areas like the descriptions on scientists, the evolution of science, the sources of historical materials, the developmental model of scientific knowledge, scientific methods, and the roles in science teaching learning. Nevertheless, as the new 2007 (or 2009) National Science Curriculum introduced new sections like "Science Stories," "Inquiry by Scientists," and "Free Inquiry," new science textbooks emphasized scientific literacy and inquiry activities and thus included more frequent and diverse descriptions of the history of science and the philosophy of science/NOS-related aspects.

67.5 HPS/NOS in Teacher Education Programs in Korea

The practice of education is heavily depending on teachers' actions. Thus the professional development of teachers must be one of the most important elements of improving the quality of education. The teachers' professionalism includes the speciality in teaching methods (Shulman 1987) and the specific abilities required in school practice (Castetter 1986). The professional development of teachers is considered as the starting point of a systemic innovation of education (Desimone et al. 2002; Seo et. al. 2010; Smith and O'day 1991), and the quality of science education depends on the professionalism of teachers who teach science. The professional development of teachers needs to be considered at the two different levels, preservice and in-service levels. In this respect, HPS/NOS-related practice of teaching science can also be examined with preservice as well as in-service teacher programs. Thus, this section will analyze the programs related to HPS/NOS for preservice and for in-service teacher education programs and investigate the efforts to improve the HPS/NOS-related teachers' professionalism.

67.5.1 The Method of Analysis

67.5.1.1 Programs Analyzed

The analysis of the ratio of HPS/NOS-related preservice programs was done with the curricula of national universities. In Korea there are ten National Universities which provide preservice teacher education programs for secondary science and ten National Universities of Education which provide preservice teacher education programs for elementary science.[2] These universities are scattered evenly across the nation and thus provide a good representation of the nationwide situation.

The analysis of the ratio of HPS/NOS-related in-service programs was made with the programs provided by eight major Offices of Education (of seven metropolitan cities and one province) among sixteen regional Offices of Education. Among the eight, Offices of Education in Seoul and in Gyeonggi Province cover nearly a half of the total population of the nation, while the remaining six Metropolitan Offices of Education are scattered across the nation.[3] Therefore, the in-service programs of the eight major Offices of Education provide a good representation of the whole nation.

67.5.1.2 Analysis Method

The analysis of preservice programs was carried out with the curriculum information shown on the universities' home pages in 2011. The titles and credits of the courses which appeared to be HPS/NOS-related were identified, and the ratios of the credits of the courses out of the total credits for a successful completion of the degree (i.e., B.Sc.) were calculated. Since the information used in the analysis was obtained from university home pages, the actual practice in the universities might be somewhat different from the data obtained in this study.

Similarly, the analysis of in-service programs was carried out based on the information obtained from the home pages of the Offices of Education. Generally, there are two kinds of in-service programs: one conducted directly by the Office of Education or its official in-service training institution and the other conducted usually by schools or teacher associations or private sectors. While the former is listed in the science education plan of the Office of Education, the latter is often classified as the specialized institutions' in-service programs. For each Office of Education, the whole in-service programs were first checked, and among them science programs and HPS/NOS-related programs were identified based on their titles. During this process, in order to focus on science or science education programs, programs for teachers' general qualifications, gifted education, environment, invention, and visiting from domestic or foreign universities were excluded.

Offices of Education provide the information of their in-service programs with different levels of classification, some with broader classifications while others with narrower classifications. Thus it would be impossible to directly compare the numbers of programs provided by different Offices of Education. As a result, the analysis of HPS/NOS-related in-service programs in this session is sensible only in terms of ratio compared with the whole programs of a particular Office of Education and any comparison in terms of its number between Offices of Education would be meaningless.

---

[2]The names and URLs of the National Universities and of the National Universities of Education are listed in the Appendix.

[3]The names and URLs of the eight Offices of Education are listed in the Appendix.

Table 67.11 Titles and credits of HPS/NOS-related courses in the curricula of National Universities in Korea

University	Credits for graduation	Titles of HPS/NOS-related courses (credit)	%[a]
A	130	Historical Development of Physics Concepts (3)	5.4
		History of Science for Teachers (2)	
		Philosophy of Science for Teachers (2)	
B	140	History and Philosophy of Physics (3)	4.3
		Philosophy of Science (3)	
C	150	History of Earth Science and Inquiry Method (3)	2.0
D	150	History of Physics (3)	2.0
E	140	History of Science (3)	2.1
F	140	History of the Earth (2)	1.4
G	141	History and Philosophy of Science (3)	5.0
		Education of Earth History (3)	
		History of Chemistry (1)	
H	150	History of Chemistry and Chemistry Education (2)	7.3
		History of Earth Science and Earth Science Education (3)	
		History of Earth and Practice (3)	
		Philosophy of Science and Science Education (3)	
I	140	History of Physics and Philosophy of Science (3)	4.3
		History of Biology and Biology Education (3)	
J	140	History of Physics (3)	2.1
Total	1421		3.5

[a]The ratio of the credits of HPS/NOS-related courses to the total required credits for graduation

67.5.2 Results of the Analysis of HPS/NOS-Related Preservice Teacher Education Programs in Korea

Table 67.11 shows the ratios of the credits of HPS/NOS-related courses in the ten National Universities which provide preservice teacher education for secondary science.

As seen in Table 67.11, all ten National Universities which provide preservice teacher education for secondary science offer some sorts of HPS/NOS-related courses, such as "History of Science for Teachers," "Philosophy of Science for Teachers," and "History of Physics and Philosophy of Science." However, about a half of the universities offer only one course of two or three credits. Only three of them offer three or more HPS/NOS-related courses. The average ratio of the credits of HPS/NOS-related courses to the total required credits for graduation was only 3.5 %. Although as seen earlier HPS/NOS-related papers and content in textbooks increased over time, teaching of HPS/NOS in preservice teacher education seems still not to be popular. This tendency is even more apparent in National Universities of Education, as shown in Table 67.12.

As shown in Table 67.12, only four National Universities of Education offer HPS/NOS-related courses, and among the four three offer only one course of two credits.

Table 67.12 Titles and credits of HPS/NOS-related courses in the curricula of National Universities of Education in Korea

University	Credits for graduation	Titles of HPS/NOS-related courses (credit)	%[a]
A	140	–	–
B	144	–	–
C	152	Science in Life and in History (2) Understanding Earth Science with History (2)	2.6
D	140	–	–
E	145	–	–
F	145	History of Science (2)	1.4
G	145	History of Science (2)	1.4
H	145	–	–
I	147	History of Science (2)	1.4
J	145	–	–
K	150	–	–
Total	1448		1.9

[a]The ratio of the credits of HPS/NOS-related courses to the total required credits for graduation

In sum, the average ratio of credits of HPS/NOS-related courses to the total required credits for graduation was only 1.9 %. Furthermore these courses are only optional, not mandatory. For example, in "G" National University of Education, there are 12 departments and only about a half of the students of science education department usually take the course of History of Science, implying that only about 5 % of the students of G National University of Education take the course of History of Science. Although this analysis was only based on the information shown at the universities' home pages, which might be quite different from their actual practice, and some HPS/NOS-related content can be taught in more general courses like "Theories of Science Education" or "Methods of Teaching and Learning of Science," the fact that only less than a half of the universities offer HPS/NOS-related courses clearly shows that the teaching of HPS/NOS in Korean National Universities of Education is very limited.

Tables 67.13 and 67.14 also show that the provision of HPS/NOS-related in-service programs is also extremely limited.

As shown in Table 67.13, out of a total of 130 programs appeared in the in-service training plans for science education of the eight Offices of Education, there were 72 science-related programs, and among them "History of Science for Teachers," provided for 32 teachers schedules for 15 h, was the only program related to HPS/NOS. Similarly, out of a total of 2083 programs in special institutions' in-service training programs provided by the Offices of Education, there were 81 science-related programs, and among them "Teaching Practice of Science Ethics Lessons," provided for 25 teachers scheduled for 30 h, was the only program related to HPS/NOS. Even considering the possibility that there would be some more programs which would include HPS/NOS-related content or activities, the HPS/NOS-related content or activities in Korean preservice as well as in-service teacher education programs are very much limited.

Table 67.13 HPS/NOS-related programs appeared in the in-service training plans for Science Education of Offices of Education

Office of Education	Number of the total in-service programs[a]	Number of science-related in-service programs	Number of HPS/NOS-related in-service programs (title, target group, participants, hours)	Year
A	42	18	1 (History of Science for Teachers, Secondary Science Teachers, 32 persons, 15 h)	2011
B	17	4	–	2011
C	16	12	–	2010
D	9	9	–	2011
E	14	8	–	2011
F	15	9	–	2011
G	4	3	–	2011
H	13	9	–	2008
Total	130	72	1	

[a]Since each Office of Education provides the information of its in-service programs with a different level of classification, it would be meaningless to compare the numbers of the programs provided by different Offices of Education

Table 67.14 HPS/NOS-related programs in special institutions' in-service training programs provided by offices of education

Office of Education	Number of the total in-service programs[a]	Number of science-related in-service programs	Number of HPS/NOS-related in-service programs (title, target group, participants, hours)	Year
A	1144	54	–	2011
B	160	7	1 (Teaching Practice of Science Ethics Lessons, K-12 Teachers, 25 persons, 30 h)	2011
C	111	1	–	2010
D	60	2	–	2011
E	264	5	–	2011
F	186	9	–	2010
G	55	1	–	2011
H	103	2	–	2011
Total	2083	81	1	

[a]Since each Office of Education does or does not provide the information of its in-service programs done by special institutions, it would be meaningless to compare the numbers of the programs provided by different Offices of Education

67.6 Conclusion

In this chapter, the status quo of HPS/NOS-related aspects of Korean science education was outlined. To do this, the historical changes of Korean National Science Curricula were summarized. Then the change of research trends of science education by analyzing the HPS/NOS-related papers published in JKASE for the last three decades was examined. After that the HPS/NOS-related aspects in science textbooks and in the programs of (preservice and in-service) science teacher education were analyzed.

Since the middle of 1980s, science in Korean National Curriculum began to have more of concrete and expressive descriptions related to HPS/NOS. The 5th National Science Curriculum announced in 1987 included "scientific literacy" as one of its major objectives, while the 6th National Science Curriculum announced in 1992 emphasized STS-related content and instruction methods. 2007 National Science Curriculum included once again "scientific literacy" as its one of major objectives and suggested to use materials related to scientists, the history of science, and to science-related or socio-scientific issues.

The analysis of the papers published in *Journal of Korean Association for Science Education* (the most representative academic journal of the field) for the last three decades shows that there has been a continuous substantial body of HPS/NOS-related studies on the areas, such as HPS/NOS-related recognition and ideas, instruction methods or programs, and the nature and features of terms or concepts. Although in general science education research in Korea has been more inclined to some specific areas (such as, quantitative studies, review studies, studies on students' or teachers' HPS/NOS-related perceptions or attitudes), recently studies with more diverse research purposes or methods or new perspectives or new theories became more popular.

The HPS/NOS-related content in Korean science textbooks for G3–G10 appeared to be biased in areas like the descriptions of scientists, the evolution of science, material of historical content, scientific method, and the developmental model of scientific knowledge and in terms of the purposes or roles in science education. However, the newly introduced sections (such as "Science Stories," "Inquiry by Scientists," "Free Inquiry") in the 2007 National Science Curriculum encouraged more and diverse HPS/NOS-related content in the textbooks.

The analyses of the HPS/NOS-related courses or programs in preservice and in in-service teacher education programs showed that there is a need to have more programs related to HPS/NOS both in preservice as well as in-service programs.

Based on the findings of the analyses reported in this paper, the following implications to Korean science education can be made.

Firstly, there is a need to have a continuous emphasis and concrete guidelines for HPS/NOS in the National Science Curriculum. As we have seen in the analyses of research papers in *JKASE* and of science textbooks, the National Science Curricula have played important roles in encouraging relevant studies and in including related content in textbooks. Especially, in a nation like Korea where a centralized education system is firmly placed, the inclusion and concrete descriptions of HPS/NOS aspects in the National Science Curriculum are vitally important in bringing real changes in research activity, in textbook content, and in teacher education programs. Although this cannot be the sufficient condition, it would be a necessary condition or an effective way for the actual change.

Secondly, there is a need to apply a wide range of methods and tools of research. As seen earlier, HPS/NOS research activities in Korea have been influenced by the available research methods and tools of the time. For example, the increase of studies on students' or teachers' perceptions or attitudes around the 1990s was possible by applying newly developed tools like TOSRA, VOSTS, DAST, and Nott and Wellington's. On the other hand, the increase of qualitative studies in Korea was affected by the worldwide trend of qualitative studies. Recently, Lederman (2007) began to combine the NOS survey tool with the qualitative approach, thus is expected to produce more comprehensive research results. In addition, since research results are influenced by their sociocultural backgrounds, the development of research tools, especially revised or newly developed in connection with Korean sociocultural context, would provide more meaningful and comprehensive results for HPS/NOS-related studies.

Thirdly, there is a need to develop and to put into practice more HPS/NOS-related programs for preservice and in-service science teachers. HPS/NOS-related programs for preservice and in-service science teachers in Korea have been insufficiently provided, especially for elementary science teaching. According to the study of Lee and Shin (2011) which carried out a survey with specialists of science education and of the history of science, the most common responses to "why is the history of science not used actively in schools?" were "no sufficient appropriate materials for teachers in schools," "teachers' ignorance of how to use the history of science," "no proper preservice education programs for the usage of the history of science," and "no proper in-service training programs for the usage of the history of science." In order to have more active teaching and learning of HPS/NOS in school education, not only its emphasis in the National Science Curriculum and in science textbooks, the development and implementation of HPS/NOS-related preservice and in-service programs are needed.

Fourthly, there is a need to have comparative studies across nations. Although in this chapter the findings of National Science Curricula, science textbooks, and teacher education programs of Korea were loosely compared with related international trends, the discussion was very limited due to the lack of the authors' understanding and information of the world trend and cases of other nations. Further studies addressing this comparative analysis are in need. In particular, regions in East Asia where much of the educational system and school practice is shared would be the easiest and should be the first area to be targeted in this kind of comparative studies.

Finally, there is a need to develop the collaboration between the communities of science education and science studies. HPS/NOS is the very area where the two (or more) different academic communities need to work closely. However, it is in fact true that the two different communities have rarely shared their expertise and experience. Perhaps like in many other countries, in Korea, many of science educators frequently refer to the theories and concepts of the history, philosophy, and communication of science but with little relevant professional training, while the historians and philosophers of science have strong interest in science education but with no relevant experience. Together with today's great demands for informal science education, PUS (public understanding of science), and science culture, the active communication and collective efforts between the two communities are called for more than ever before. In this respect, the activity of IHPST (International History, Philosophy, and Science Teaching group) and its regional activities (like 2012 IHPST Asian Regional Conference) are with high expectations.

In this chapter, the authors focused on describing the recent and current situations of HPS/NOS-related aspects in Korean science education. Thus, due to some practical reasons, it was inevitable not to discuss the issues in depth and to cover the whole range of related issues. Despite all these limits, it is hoped that this chapter will help the readers to better understand some aspects of Korean science education, in particular those related to HPS/NOS.

Acknowledgement The authors of this chapter wish to thank Jinsun Park, a doctoral student of Seoul National University, for her kind help in analyzing the data reported here.

Appendix

Textbooks Selected for the Analysis in the Study

G3-1: Ministry of Education, Science and Technology of Korea (2010). *Science 3–1*. Kumsung, Seoul.
G3-2: Ministry of Education, Science and Technology of Korea (2010). *Science 3–2*. Kumsung, Seoul.
G4-1: Ministry of Education, Science and Technology of Korea (2010). *Science 4–1*. Kumsung, Seoul.
G4-2: Ministry of Education, Science and Technology of Korea (2010). *Science 4–2*. Kumsung, Seoul.
G5-1: Ministry of Education, Science and Technology of Korea (2011). *Science 5–1*. Kumsung, Seoul.
G5-2: Ministry of Education, Science and Technology of Korea (2011). *Science 5–2*. Kumsung, Seoul.
G6-1: Ministry of Education, Science and Technology of Korea (2011). *Science 6–1*. Kumsung, Seoul.
G6-2: Ministry of Education, Science and Technology of Korea (2011). *Science 6–2*. Kumsung, Seoul.
G7-A: Kim et al. (2010). *Middle School Science 1*. Doosandonga, Seoul.
G7-B: Lee et al. (2010). *Middle School Science 1*. Kumsung, Seoul.
G8-A: Kim et al. (2011). *Middle School Science 2*. Doosandonga, Seoul.
G8-B: Lee et al. (2011). *Middle School Science 2*. Kumsung, Seoul.
G9-A: Kim et al. (2012). *Middle School Science 3*. Doosandonga, Seoul.
G9-B: Lee et al. (2012). *Middle School Science 2*. Kumsung, Seoul.
G10-A: Jeon et al. (2011). *High School Science*, Mirae N, Seoul.
G10-B: An et al. (2011). *High School Science*, Kumsung, Seoul.

National University (Secondary Education) Selected for the Analysis in the Study

Chonbuk National University (http://www.jbnu.ac.kr/)
Chonnam National University (http://www.jnu.ac.kr/)
Chungbuk National University (http://www.chungbuk.ac.kr/)
Gyeongsang National University (http://www.gnu.ac.kr/)
Jeju National University (http://www.jejunu.ac.kr/)
Kangwon National University (http://www.kangwon.ac.kr/)
Kongju National University (http://www.kongju.ac.kr/)
Korean National University of Education (http://www.knue.ac.kr/)
Pusan National University (http://www.pusan.ac.kr/)
Seoul National University (http://www.snu.ac.kr/)

National University of Education (Primary Education) Selected for the Analysis in the Study

Busan National University of Education (http://www.bnue.ac.kr/)
Cheongju National University of Education (https://www.cje.ac.kr/)
Chinju National University of Education (http://www.cue.ac.kr/)
Chuncheon National University of Education (http://www.cnue.ac.kr/)
Daegu National University of Education (http://www.dnue.ac.kr/)
Gongju National University of Education (http://www.gjue.ac.kr/)
Gwangju National University of Education (http://www.gnue.ac.kr/)
Gyeongin National University of Education (http://www.ginue.ac.kr/)
Jeonju National University of Education (http://www.jnue.kr/)
Seoul National University of Education (http://www.snue.ac.kr/)

Office of Education Selected for the Analysis in the Study

Busan Metropolitan City Office of Education (http://www.pen.go.kr/)
Daegu Metropolitan City Office of Education (http://www.dge.go.kr/)
Daejeon Metropolitan City Office of Education (http://www.dje.go.kr/)
Gwangju Metropolitan City Office of Education (http://www.gen.go.kr/)
Gyeonggi Provincial Office of Education (http://goe.go.kr/)
Incheon Metropolitan City Office of Education (http://www.ice.go.kr/)
Seoul Metropolitan Office of Education (http://www.sen.go.kr/)
Ulsan Metropolitan City Office of Education (http://www.use.go.kr/)

References

Aikenhead, G., Ryan, A., & Fleming, R. (1989). *Views on science-technology-society*. Saskatoon, Canada: Department of Curriculum Studies, University of Saskatchewan.

Castetter, W. B. (1986). *The personal function in educational administration*. Macmillan Publishing Co., NY.

Cho, H. –H. (1984). A study of philosophical basis of preconceptions and relationship between misconceptions and science education. *Journal of the Korean Association for Science Education, 4*(1), 34–43.

Cho, H. –H. (1988). Analysis of theoretical background for current research on science curriculum and teaching/learning and implications for future science education. *Journal of the Korean Association for Science Education, 8*(2), 33–41.

Cho, H. –H. (1992). An analysis of the nature of scientific inquiry and a study on the instructional method for promoting inquiry competence. *Journal of the Korean Association for Science Education, 12*(2), 61–73.

Cho, J. (2008). Analysis of Inquiry Unit of Science 10 in terms of nature of science. *Journal of the Korean Association for Science Education, 28*(6), 685–695.

Cho, J. –I. (1990). The meanings of teaching science as inquiry and change of conditions for inquiry science education. *Journal of the Korean Association for Science Education, 10*(1), 65–75.

Cho, J. –I. (1991). An investigation into "Science-Technology-Society" curricula. *Journal of the Korean Association for Science Education, 11*(2), 87–101.

Cho, J. –I., & Ju, D. K. (1996). A study on Korean science teachers' points of view on nature of science. *Journal of the Korean Association for Science Education, 16*(2), 200–209.

Cho, H. –H., & Choi, K. (2002). Science education: constructivist perspectives. *Journal of the Korean Association for Science Education, 22*(4), 820–836.

Choi, G. (2004). Developing active role of science museum in educating on ethical issues on science and technology: Four case studies. *Journal of the Korean Association for Science Education, 24*(1), 109–120.

Choi, C. I., Yeo, S. –I., & Woo, K. W. (2005). Analysis of the contents of science history introduced into Chemistry II textbooks based on the 7th curriculum. *Journal of the Korean Association for Science Education, 25*(7), 820–827.

Choi, Y. –H., Choi, G., & Jin, H. –K. (2010). An Analysis of the Nature of Science Incorporated in the History of Science in School Science Textbooks Based on the 7th Curriculum. *New Physics: Sae Mulli(The Korean Physical Society), 60*(3), 273–282.

Desimone, L. M., Porter, A. G., Garet, M. S., & Birman, B. F. (2002). Effects of professional development on teacher's instruction: Results from a three-year longitudinal study, *Educational Evaluation and Policy Analysis, 24*(2), 81–112.

Ha, M. –K. (1991). The attempts to introduce Science-Technology-Society (S-T-S) education to Korean science education. *Journal of the Korean Association for Science Education, 11*(2), 79–85.

Han, J. –S., & Chung, Y. L. (1997). Teachers' and students' understanding of the nature of science. *Journal of the Korean Association for Science Education, 17*(2), 119–125.

Herget, D. E. (Ed.) (1989). *The History & Philosophy of Science in Science Teaching*. Science Education and Department of Philosophy, Florida State University, Tallahassee, Florida.

Herget, D. E. (Ed.) (1990). *More History & Philosophy of Science in Science Teaching*. Science Education and Department of Philosophy, Florida State University, Tallahassee, Florida.

Hong, J. –L. (2001). The effects of decision-making centered-STS(Science-Technology-Society) classes on the students' attitudes towards science and perceptions about STS. *Journal of the Korean Association for Science Education, 21*(2), 422–432.

Hur, M. (1993). Survey on the attitudes toward science and science courses of primary and secondary students. *Journal of the Korean Association for Science Education, 13*(3), 334–340.

Jung, W. (1994). Toward preparing students for change: A critical discussion of the contribution of the history of physics in physics teaching. *Science & Education, 3*, 99–130.

Kang, S., Kim, Y., & Noh, T. (2004). The influence of small group discussion using the history of science upon students' understanding about the nature of science. *Journal of the Korean Association for Science Education, 24*(5), 996–1007.
Kim, D. P. (1978). Historical review in Foundation of London Loyal Society and France Loyal Academy. *Journal of the Korean Association for Science Education, 1*(1), 29–34.
Kim, H. –B., M. J. –S., Park, J. –Y., Heo, N., & Song, J. (2004). Science-related attitudes of Korean housewives. *Journal of the Korean Association for Science Education, 24*(1), 183–192.
Kim, J. –H. (2001). The characteristics of the 7th national science curriculum of the Republic of Korea. *Journal of the Korean Association for Science Education, 21*(5), 1012–1026.
Kim, J., & Cho, B. (2002). Perceptions about science and scientific activity of students in kindergarten and primary school. *Journal of the Korean Association for Science Education, 22*(3), 617–631.
Kwon, J. S. (1991). Problems of discipline centered science education and a method of the utilization of everyday materials in science education. *Journal of the Korean Association for Science Education, 11*(1), 117–126.
Kwon, J. S. (2001). The problems of Korean school science curriculum and suggestions for a new direction. *Journal of the Korean Association for Science Education, 21*(5), 968–978.
Kwon, J. S., & Park, D. –Y. (1990). A survey on attitudes toward science among the elementary school students. *Journal of the Korean Association for Science Education, 10*(2), 39–47.
Kwon, S., & Pak, S. (1995). Elementary preservice teachers' conceptions about and its changes in the nature of science and constructivist' view of learning. *Journal of the Korean Association for Science Education, 15*(1), 104–115.
Kwon, Y. –J., Jeong, J. –S., Park, Y. –B., & Kang, M. –J. (2003). A philosophical study on the generating process of declarative scientific knowledge: Focused on inductive, abductive, and deductive process. *Journal of the Korean Association for Science Education, 23*(3), 215–228.
Lee, B. W., & Shin, D. H. (2010). Analysis of the historical content of 'Science' textbooks on the 7th Curriculum in Korea. *New Physics: Sae Mulli(The Korean Physical Society), 60*(5), 488–496.
Lee, B. W., & Shin, D. H. (2011). Professionals' opinion of science education using history of science. *Journal of the Korean Association for Science Education, 31*(5), 815–826.
Lederman, N. G. (2007). Nature of science: Past, present, and future. In S. K. Abell and N. G. Lederman (eds), *Handbook of Research on Science Education*, Routledge, NY, pp. 831–879.
Lederman, N. G., Abd-El-Khalick, F., Randy, L. B., & Renée, S. S. (2002). Views on nature of science questionnaire: Toward valid and meaningful assessment of learners' conceptions of nature of science. *Journal of Research in Science Teaching, 39*(6), 497–521.
Leite, L. (2002). History of science in science education: Development and validation of a checklist for analyzing the historical content of science textbooks. *Science & Education, 11*(4), 333–359.
Matthews, M. R. (1992). History, philosophy, and science teaching: The present approach. *Science & Education, 1*, 11–47.
Matthews, M. R. (1994). *Science teaching: The role of history and philosophy of science.* Routledge, NY.
McComas, W. F. (Ed.) (1998). *The Nature of Science in Science Education: Rationales and Strategies*. Kluwer Academic Publishers, Dordrecht.
McComas, W. F., & Olson, J. K. (1998). The nature of science in international science education standards documents. In W. F. McComas (ed), *The Nature of Science in Science Education*, Kluwer Academic Publishers, Dordrecht, pp. 41–52.
Merriam, S. B. (1988). *Case study research in education: A qualitative approach.* Jossey-Bass Publishers, San Francisco.
Ministry of Education of Korea (MOE) (1997). *The Science Curriculum.*
Ministry of Education, Science and Technology of Korea (MEST) (2007a). *The 2007 Revised Science Curriculum.*
Ministry of Education, Science and Technology of Korea (MEST) (2007b). *The guide book of the 2007 Revised Science Curriculum.*
Ministry of Education, Science and Technology of Korea (MEST) (2009). *The development of English version of the 2007 Revised Science Curriculum.*

Ministry of Education, Science and Technology of Korea (MEST) (2011). *The 2009 Revised Science Curriculum.*

Nam, J., Mayer, V. J., Choi, J., & Lim, J. (2007). Pre-service science teachers' understanding of the nature of science. *Journal of the Korean Association for Science Education, 27*(3), 253–262.

Niaz, M. (2000). A rational reconstruction of the kinetic molecular theory of gases based on history and philosophy of science and its implications for chemistry textbooks. *Instructional Science, 28*(2), 23–50.

Park, E. –I., & Hong, H. –G. (2010). The effects of explicit instructions on nature of science for the science-gifted. *Journal of the Korean Association for Science Education, 30*(2), 249–260.

Park, J. (1998). The role of deductive reasoning in scientific activities. *Journal of the Korean Association for Science Education, 18*(1), 1–17.

Park, J. (2009). The activity of an interpreter on science education during the enlightenment period in Korea: Focus on Hyun Chae. *Journal of the Korean Association for Science Education, 29*(6), 741–750.

Park, S. K. (2011). The national curriculum of Korea: Current practice and future prospects for school curriculum policies. International Symposium: Current Practice and Future Prospects for School Curriculum Policies, April 8th 2011, Seoul.

Peirce, C. S. (1878). Deduction, induction, and hypothesis. In C. Hartshorne and P. Weiss (Eds.) *Collected Papers of Charles Sanders Peirce, vol. 2 (1931–1958)* (pp. 372–388). Cambridge, MA: Harvard University Press.

Ryu, H. –S., & Chol, K. (2010). Perception survey on characteristics of scientific literacy for global science-technology-society for secondary school students. *Journal of the Korean Association for Science Education, 30*(6), 850–869.

Shulman, L. S. (1987). Knowledge and teaching: Foundations of the new reform. *Educational Review, 57*, 1–22.

Smith, M., & O'Day, J. (1991). Systemic school reform. In S. Fuhrman & B. Malen (Eds.), *The politics of curriculum and testing: The 1990 yearbook of the politics of education association.* Falmer Press, Bristol, pp. 233–267.

Seo, H. –A., Kim, Y., Kim, J., Paik, S. –H., Son, J., Song, J., Lee, K. –Y., Jeong, Y., Cho, W., Cha, J., Han, I., & Heo, N. (2010). *Lifelong professional development system for in-service teachers of science and mathematics.* Korean Foundation for the Advancement of Science & Creativity.

Soh, W. –J., Kim, B. –k, & Woo, J. –O. (1998). Middle school science teachers' philosophical perspectives of science. *Journal of the Korean Association for Science Education, 18*(2), 221–231.

Song, J. (1993). Teachers' images of scientists and their respected scientists. *Journal of the Korean Association for Science Education, 13*(1), 48–55.

Song, J., & Kwon, S. –G. (1992). The change of postgraduate students' conceptions towards the nature of science through the course related to philosophical of science. *Journal of the Korean Association for Science Education, 12*(1), 1–9.

Song, J., & cho, S. –K. (2003). John Tyndall(1820–1894), who brought physics and the public together. *Journal of the Korean Association for Science Education, 23*(4), 419–429.

Song, J., Chung, B. –H., Kwon, S. –G., & Park, J. –W. (1997). The images of science education illustrated in the books written by modern philosophers of science. *Journal of the Korean Association for Science Education, 17*(2), 209–224.

Yang, I. H., Jeong, J. S., Kwon, Y. J., Jeong, J. W., Hur, M., & Oh, C. H. (2006a). An intensive interview study on the process of scientists' science knowledge generation. *Journal of the Korean Association for Science Education, 26*(1), 88–98.

Yang, I. H., Cho, H. J., Jeong, J. W., Hur, M., & Kim, Y. S. (2006b). Aims of laboratory activities in school science: A Delphi study of expert community. *Journal of the Korean Association for Science Education, 26*(2), 177–190.

Yang, S. –H. (1987). Brief discussion on the scientific creationism critiques. *Journal of the Korean Association for Science Education, 7*(2), 89–95.

Yi, B. –H., & Kim, Y. –M. (1984). A investigation of the attitudes toward science and scientific attitude for the underachievers. *Journal of the Korean Association for Science Educationn 4*(1), 26–33.

Jinwoong Song is a Professor at the Department of Physics Education, Seoul National University, Korea. He took his BA in physics and MEd in science education at Seoul National University and received Ph.D. in science education at King's College London (1990). In addition to his previous position at Daegu University, Korea, he had opportunities to work as a visiting academic to King's College London, Institute of Education, and the University of Oxford. He served as the editor in chief of *Journal of the Korean Association for Science Education*, as the president of EASE (*East-Asian Association for Science Education*) (2009–2011), and more recently as the chair of the organizing committee of the First IHPST Regional Conference in Asia held on 18th–20th October, 2012, at Seoul National University, Korea. His research interests cover a wide range of topics in science education, including the contextual dimension of physics learning and teaching, informal science education, socio-scientific issues, science culture, the history of science education, and linking HPS with science education.

Yong Jae Joung is an Assistant Professor of Gongju National University of Education, Korea. He received his Ph.D. in science education from the Seoul National University and was a postdoctoral research fellow, 2006–2007, at Monash University. His research interests include the nature of science and scientific methods in science education, conceptual learning of science with philosophical views, and the relationship between individuals and community in science learning. Some of his recent published papers are Y. J. Joung (2008) Cases and features of abductive inference conducted by a young child to explain natural phenomena in everyday life, *Journal of Korean Association for Research in Science Education*; Y. J. Joung (2009) Children's Typically-Perceived-Situations of floating and sinking, *International Journal of Science Education*; Y. J. Joung, & R. Gunstone (2010) Children's Typically-Perceived-Situations of force and no force in the context of Australia and Korea, *International Journal of Science Education*; and H. –G. Yoon, Y. J. Joung, & M. Kim (2012) The challenges of science inquiry teaching for preservice teachers in elementary classrooms: difficulties on and under the scene, *Research in Science Education* (Online First article).

Chapter 68
History and Philosophy of Science in Japanese Education: A Historical Overview

Yuko Murakami and Manabu Sumida

68.1 Precursors

68.1.1 Science and Technology in the Edo Period

68.1.1.1 Under the Isolation Policy

Japan developed its own science and technology during the Edo period (1600–1867). At the time the government countered colonization efforts by Christian ministries by adopting a policy of national isolation (1633–1854) and by banning Christianity (1587–1858). Diplomatic ties were, however, continued with Korea, the Netherlands, and China. Such ties were nonetheless still regulated by the government. Some translators did manage, though, to import scientific knowledge from the Western world. For example, the Dutch version of Johan Adam Kulmus's *Anatomische Tabellen* was translated into Japanese (1774) with illustrations by Odano Naotake (1750–1780).

Scholars are still examining the philosophical foundations of science in the Edo period. Some principal figures in scientific thought in the period include Arai Hakuseki and Miura Baien. Tsuji (1973) points out the roles of neo-Confucianism combined with rationalism in shaping the nature and practical emphasis of Edo science where articulating the laws of nature was connected to social justice. Nakayama (1977), by contrast, argues that the influence of Chinese thought on the understanding of

Y. Murakami
Graduate School of Science, Tohoku University, Sendai, Japan

M. Sumida (✉)
Faculty of Education, Ehime University, Matsuyama, Japan
e-mail: msumida@ed.ehime-u.ac.jp

M.R. Matthews (ed.), *International Handbook of Research in History, Philosophy and Science Teaching*, DOI 10.1007/978-94-007-7654-8_68,

scientific laws was political: if observation did not fit a theory, it was considered an omen requiring governmental response. There is thus no consensus on the influence of Confucianism upon Japanese science.

Research in astronomy, initially imported from China, was conducted to develop the locally adjusted calendar system for Japan. Shibukawa Harumi became the first official astronomer in 1685 responsible for a Japanese calendar. The astronomy office dealt with the importation of all scientific knowledge, including maps and translations from the Dutch. This office became the Bansho Torishirabesho, one of the origins of the University of Tokyo.

Wasan, or Japanese mathematics, was developed independently from Western mathematics. Seki Takakazu (1642–1708) discovered differentiation and developed an equivalent of Bernoulli numbers, an approximation of pi, and trigometric functions.[1] Wasan was widely accepted in Japanese society before the Meiji restoration as a form of entertainment. People from various social classes donated wasan problems to temples and shrines.

Interest in wasan reflected the notable literacy rate of the Edo period even in the general public. It is said that 50 % of males and 20 % of females were literate nationwide in late Edo period. Generally speaking, the public was eager to learn to read and write and to perform basic calculations with an abacus. Literacy of this kind was facilitated by thousands of small private schools open to the general public.

Almost everyone in the Samurai class was literate. The government operated various kinds of schools for elites. Shoheizaka Gakumonsho in Edo (current Tokyo) was the main school, which aimed to teach Confucianism. It later developed into the Kaiseisho, which in turn merged with the University of Tokyo. There were also schools for the study of China, medicine, and foreign languages. Each local government also had a school. Altogether there were 270 governmental schools in the late Edo period.

68.1.1.2 Tension Between Nationalists on Science and Technology

Discussion on Japanese isolation ended when war vessels of Russia, the British Empire, France, and the United States approached Japan (1787–1854) to negotiate unilateral treaties, forcing Japan to open its borders. Nationalists argued for "Western technology with the Japanese mind (和魂洋才)," while advocates for opening the country coined the slogan "Secede from Asia and join Europe (脱亜入欧)." Both parties recognized that Western technology and scientific knowledge were the keys to avoiding colonization and maintaining national independence. The tension between nationalists who favored isolationism and those who insisted on opening the country continued for a long time.

---

[1] Research on the meaning of wasan is ongoing. The concept, to date, has been largely misunderstood. For example, some argue that wasan lacks the notion of functions (Ueno 2006).

68.1.1.3 Technology Transfer from Europe Around the Meiji Restoration (1862–1880)

Western technology was imported to some regions in the end of Edo period. Satsuma (current Kagoshima area) introduced British cotton and sugar industries, for example. Technology transfer was extended after the Meiji restoration (1868). The Japanese government then took steps to import Western technology and scientific knowledge.

The foundation for a strong emphasis on engineering was laid by a younger generation that studied abroad before the restoration, such as the "Choshu five," five young samurais who left Japan illegally in 1862 and smuggled themselves to England in 1863. Kido Takayoshi, who was one of the main actors of the Meiji restoration and later became an important figure of the new government, supported their travel. Jardine Matheson & Co. arranged their stay in London. They first learned English and then began studying in the University College of London. Among the five students, Ito Hirobumi and Inoue Kaoru went back to Japan earlier than the other three when Choshu's fight against Europe was reported on a newspaper in 1864. They became important figures in the new government: Ito Hirobumi became the first minister of industry in 1876; and Inoue Kaoru became the second minister of industry. The other three continued their study in England. Endo Kinsuke followed them due to health problem in 1866. Inoue Masaru learned railroad systems in London. Yamao Yozo, interested in shipbuilding, moved from UCL to Glasgow. He worked at Napier shipbuilding in the daytime and studied at Anderson College at night. He met Henry Dyer there, who later came to Japan where he became the first chairperson of the new school of industry. Yamao and Inoue were called back to Japan in 1868.

It was not only Choshu who sent students abroad. Satsuma sent 19 students to England in 1865 with full financial support from the Satsuma feudal government. The Edo government also sent 17 students to study abroad with Enomoto Takeaki as its leader; he later became an admiral. Most of the students were naval cadets, but Nishi Amane and Tsuda Mamichi received training in law and political science in the Netherlands. They left Edo in 1862 and arrived at Rotterdam in 1863. Nishi and Tsuda went to Leyden and attended lectures of Simon Vissering and Cornelius Willem Opzoomer. They went back to Japan via Paris in 1865, where they met other intellectuals and the future political leaders of Japan. One of them was Mori Arinori, who later became the first minister of education in the Meiji government.[2]

After their arrival, Nishi and Tsuda were assigned to Kaiseisho, the main school of the Edo government. Kaiseisho merged with the University of Tokyo after the Meiji restoration. Nishi and Tsuda became government officers and translated many academic books as well as developing their own original philosophy. It was a critical time for Japan to integrate Western ideas to the native Japanese framework of thought. The words "Kagaku" and "Tetsugaku" were coined by Nishi to mean "science" and "philosophy."

[2] See Piovesana (1963).

"Kagaku" is most widely used as a general term for "science." The word "Kagaku" literally means "section-study," which reflected the Western propensity to develop specialized scientific fields in the late-nineteenth century. The literal meaning may even have precipitated sectionalism in the development of science and technology in Japan.

The word "science" is currently translated into four different words, "gakujutu," "kagaku," "rika," and "rigaku." "Gakujutu" is almost similar to academic activities or the Latin word *scientia*. "Kagaku" can be paralleled with "shakai" (society) or "gijutu" (technology) as general terms, where "rika" and "rigaku" sound inappropriate. "Rika" is currently used for the name of subject in elementary and secondary education, while "rigaku" is used for the academic area of natural science in higher education in the same level as "kogaku" (engineering) or "hogaku" (legal science). A difference is that "rigaku" often includes mathematics, while "rika" does not. "Rika" as an academic subject covers physics, chemistry, geology, and biology.

The word "rika," however, appeared first as the name of an academic field in the same level of "bunka" (humanities) and "ika" (medical), the same as the current use of "rigaku." Thus, "Rika Daigaku" meant the college of science in the early modern period of higher education in Japan. The first Education Law (教育令) (1879) specified physics, physiology, and natural history as optional subjects, i.e., the word "rika" appeared only to indicate one of specialized areas in higher education.

The word "rika" was then introduced into primary and secondary education in an 1886 ordinance of ministry of education (Ministry of Education 1986) following an 1885 amendment to educational law. The ordinance was the first official curricular guideline, and it specified "Rika" as an educational subject in upper elementary school. The contents were not explicitly stated, but the emphasis was placed on scientific phenomena in everyday life.

Such multiple meanings for the Western term "science" are indicative of the conceptual struggles at stake in adapting ideas and concepts from the Western world in Japan.

68.2 Science and Technology in the Meiji Period

68.2.1 Institutionalization of Education in Science and Technology

After the Meiji Restoration (1868), the new government was eager to introduce Western science and technology in order to maintain national independence in an age of imperialism. This decision was for the most part successful, but it neglected NOS.

The ministry of education was established in 1870. In 1871 it announced a plan for 4-year lower elementary schools, 4-year higher elementary schools, middle schools, and universities. Despite the ambitiousness of their plan, only elementary schools were inaugurated, with the number of schools growing to 26,000 within a few years. Although the ministry of education set the curriculum, most elementary schools taught only basic writing, reading, and calculation as they had in the Edo period because most teachers were unable to instruct in much more. In 1882 the ministry of education changed the initial educational system by dividing elementary schools into three levels: 3 years for the first stage, 3 years for the second stage, and 2 years for the last stage. Science was taught from the second stage onward. The first 4-year education became compulsory in 1887. The school attendance rate was about 50 % at that time and grew to more than 99 % by 1917. The majority of the population did not attend school beyond what was compulsory, however. Those who learned science, therefore, were of a relatively small number due to the position of the subject in the curriculum.

The number of middle schools grew only slowly. Curriculum guidelines for middle schools were published by the ministry of education in 1882. There were vocational-oriented middle schools and academic-oriented middle schools.

Governmental examination of school textbooks began in 1887 in order to determine compliance with national standards. This textbook examination system continues to the present.

68.2.2 The School of Engineering and the Imperial University of Tokyo

The establishment of a modern university was a main goal of late-nineteenth century educational reforms. Only one university, though, was launched by 1877 following the 1870 law governing university education. Its predecessor was the school of European culture the Edo government had established in 1855.

The basic idea behind the higher education was nationalistic. The first item of the Imperial University Law explicitly stated that Imperial Universities were for teaching and research in the arts and sciences to meet the country's needs (帝国大学令第一条　帝国大学ハ国家ノ須要ニ応スル学術技芸ヲ教授シ及其蘊奥ヲ攷究スルヲ以テ目的トス). It envisioned the Imperial Universities as contributing to national power and prestige through practical research and basic science. Consequently, certain dimensions of NOS were not emphasized, and Western notions of science and technology were still in the process of being integrated into the Japanese conceptual framework for science education.

The University of Tokyo began offering courses with four schools in 1877: the schools of law, literature, science, and medicine. The school of science had opened in 1876 with ten departments: mathematics, physics, chemistry (basic chemistry and applied chemistry), biology (zoology and plant biology), astronomy, engineering (mechanical engineering, civil engineering), and mining. Emphasis was placed on applied science and engineering at first.

With a ministry of industry opening in 1870, a school of engineering opened in 1873 before the University of Tokyo. The school of engineering began to offer classes in 1875. Yamao Yozo advocated national educational institutions of engineering.

An elementary school of industry was also proposed but unrealized. The ministry of industry planned to hire a chairperson and six lecturers from industrialized Scotland; in the end nine lecturers were hired. Henry Dyer, who learned mechanical engineering in Anderson College in Glasgow, came to Tokyo to work as the chairperson of the new school of industry in 1875 when he was 25.

Dyer crafted the school's mission: an institution with a comprehensive coverage of engineering. Thus, the school had seven departments from the beginning: civil engineering, mechanical engineering, architecture, electronics, chemistry, metal engineering, and mining. Each department offered both theoretical and practical aspects of its field. The school had a 4-year course of study, each with a modest tuition fee of ten Japanese Yen a year. Excellent students were sent to Europe after graduation.

The four departments of engineering and applied science at the University of Tokyo were separated from the school of science in 1885 to be merged with the school of engineering of the Imperial University of Tokyo and the school of engineering earlier established by the ministry of industry. Those lecturers from Scotland moved to the Imperial University of Tokyo. The school of science of the Imperial University of Tokyo had seven departments: mathematics, astronomy, physics, chemistry, zoology, plant biology, and geology. It also had a research institute of marine science and a school of agriculture. The schools of engineering and agriculture were intended to promote industry, while the school of science concentrated on basic science.

The Imperial University then invited more lecturers from European countries, while young Japanese elites were sent to European countries to study industrial systems and scientific and technical knowledge. Some of them taught in universities after coming back to Japan, but most went to the government to establish social systems. Initially students of engineering were sent to England, but later more students went to the United States. Medical students along with students in the humanities, social sciences, and physics were mainly sent to Germany.

More Imperial Universities were established after Tokyo: Kyoto, Tohoku, Kyushu, and Hokkaido. Furthermore, educational institutes for engineers were established. Tokyo Institute of Technology was established in 1881 as the Tokyo Vocational School. The name was changed to Tokyo Technical School (1890) and later Tokyo Higher Technical School (1901). In 1929, Tokyo Higher Technical School was elevated to a degree-conferring University as Tokyo Kogyo Daigaku (Tokyo Institute of Technology). The introduction of history and philosophy of science and technology as subjects of instruction had been discussed since 1930s at Tokyo Institute of Technology, whose institutional model was Massachusetts Institute of Technology.

Private schools attracted students who wished to learn modern science as well as foreign languages. Tokyo professional school, which is the precursor of Waseda University, was established in 1882 with four schools: political science and management, law, science, and English. The school of science disbanded after 3 years, but a new school of science and engineers was established in 1908.

Fujiwara University of Engineering was the first private university dedicated to science and engineering. It was established in 1939 by Ginjiro Fujiwara, the first president of Oji paper manufacturing company. He intended to donate the university to his alma mater, Keio University. It eventually merged with Keio University as a school of science and engineering in 1944.

Due to the increasing number of institutions of higher education, the number of engineers continued to grow. There were 1,500 graduates from colleges and their equivalents in 1900; 5,000 in 1910; and 55,000 in 1934. These technical graduates found employment in the government as well as in industry. The transfer for technology government to industry began around WWI.

Still, NOS and HPS remained the exception rather than the rule in Japanese higher education. Tohoku University had a position of philosophy of science in the school of science. It also planned to have a permanent position for Japanese mathematics (wasan), but budgetary limitations did not allow for it. Only a wasan archive was introduced in the university library. Tokyo Institute of Technology had HPS positions; the tradition continues to the present as a part of the department of management engineering. Those positions were considered as part of a liberal arts education for scientists and engineers, but they were always secondary to specialized fields of science and engineering.

68.2.3 Nature of Science in Japanese Literature of the Meiji Period

Although nature of science was not widely taught in Japanese education in the Meiji and Taisho periods, there were popular novels and essays which dealt with nature of science.[3] They influenced Japanese ideas of science.

Terada Torahiko (1878–1935) was a geophysicist, essay writer, and haiku poet. He taught physics in the Imperial University of Tokyo and had studied in Berlin. At the same time, he was a core member of the literary community with Natsume Soseki. He published a number of popular essays whose main topic was nature of science. Terada criticized lectures at all educational levels that crammed knowledge of science into them. "When teaching science," he argued,

> teachers must be most careful to nurture children's minds for research. That goal is not reached if they are merely given pieces of knowledge. Today even university students who are science majors are not motivated to do research by themselves. Their knowledge tends to be superficial. So it seems that they remember from their education in elementary schools to be content only with remembering knowledge (Terada 1918, p. xx).

[3] The importance of popular culture in public's interest in science in Japan continues to the present. The atomic bomb case before and after WWII was discussed in detail in Ito (2010), for example.

Terada was close to Ishihara Jun (1881–1947), a physicist. Ishihara studied physics under Einstein at Zurich Institute of Technology in 1914 and taught in Tohoku University, where he resigned due to love affair in 1921. He then became a professional author of NOS topics to gain popularity from a wide range of audiences. When Einstein came to Japan in 1922, Ishihara served as an interpreter. He published many popular articles in which he emphasized the analysis of observations and experimentation in terms of quantification and measurement as an effective means of dealing with abstractions and new phenomena (Ishihara 1936).

68.2.4 Ashio Copper Mine Accident (1885–current)

The Ashio copper mine accident turned into a social movement that changed opinions about the value of Western science. The mine was discovered in 1550, and its operation began 1610. Its production decreased after the early seventeenth century, but the Edo government kept it open. After 1877 when its ownership moved to Furukawa Ichibei, its production grew drastically with modern technology. The amount of sulfur dioxide produced during the refinement process jumped up, and the surrounding environment was heavily polluted by 1890. Polluted water contaminated crops in widespread areas.

The pollution accident was the first civil movement against the government's role in producing pollution in Japan. Tanaka Shozo, a politician of the affected area, led residents to demand compensation from the government. They also wanted the government to stop mining in Ashio. The government kept the mine open, however, due to the first Sino-Japanese War (1894–1895), the Russo-Japanese War (1904–1905), and WWI (1914–1918).

68.2.5 Impacts of WWI to NOS in Japan

Technology was transferred from the governmental sector to industry during WWI in Japan. Scientist and engineers from universities were assigned to the industrial sector as production shifted for wartime needs. As the war escalated the government planned domestic centers for research and development. In 1917 RIKEN was established. It was the first comprehensive research institute of basic and applied science in Japan. While practical goals were emphasized due to the government's policy of industrialization, RIKEN also devoted considerable resources to basic science. It supported Japanese physicists in international research in quantum physics and particle physics especially between 1925 and 1950.

The physicist Nishina Yoshio (1890–1951) built the first cyclotron in Japan (the second in the world) at RIKEN in 1937 for nuclear physics and its application to radioisotopes in medicine and biology. Research in nuclear physics was connected to potential military applications of atomic energy: Japan's atomic bomb program in

WWII was located at RIKEN. The development of uranium mines in Japan began in 1938, but mining ended after the WWII. Nuclear development resumed for non-military purposes in 1954 as a result of the "Atoms for Peace" program. The first postwar uranium mine in Japan, the Ningyo Toge mine, opened in 1955. It produced uranium until 1987, but the amount was not enough for practical uses.

68.2.6 Kyoto School of Philosophy

The Kyoto school of philosophy,[4] represented by Nishida Kitaro (西田幾多郎) and Tanabe Hajime (田辺元), argued for the close relationship of science, technology, and philosophy. Under the influence of neo-Kantian philosophy,[5] this school developed the philosophical position that science and philosophy have connections to an epistemology based on Zen Buddhism and Husserl's phenomenology. Nishida's main work, *An Inquiry into the Good* (1911), includes an acknowledgement of the restricted nature of science as a form of knowing: "A scientist's way of explanation is slanted toward just one aspect of knowledge, whereas for a complete explanation of reality we must satisfy intellectual demands as well as the demands of feeling and the will" (Nishida 1911, p. 50).

The school of science at Tohoku University was one of the few that regularly offer courses in NOS/HPS, a practice it continued until the 1990s. Philosophers took the lead in the early period. The first lecturer of NOS was Tanabe Hajime. He began his career in the school of science at Tohoku University after graduating from the University of Tokyo. He taught NOS there and published *Saikin no Sizenkagaku* (*Recent National Science*) (1915) and *Kagaku Gairon* (*Outline of Science*) (1918). He also translated Poincare's *La Valeur de la Science* into Japanese (1916).

He moved to Kyoto and succeeded Nishida's position.[6] There he published *Suri tetsugaku kenkyu* (*A Study of Philosophy of Mathematics*). His held two views of the philosophy of science (Sawada 1997), but both were different from present notions. The first was a collection of scientific approaches to philosophy that included aesthetics and religious studies; the second was the idea that the philosophy of science was part of a philosophy whose subject was science. In the latter sense he wrote: "Science today . . . has reached the stage where scientific theory has gone beyond the position where its subjects are the entities and existents dealt with in science. Science is now in the position to realize things beyond philosophy. Science has become philosophy; philosophy is a subject to be described in a scientific theory.

[4] The development of HPS in Japan in its current sense is almost absent in the Kyoto School, as the school was accused of contribution to the Japanese navy after WWII.

[5] Manuscripts of Nishida and Tanabe can be examined online at the Kyoto school archive. http://www.kyoto-gakuha.info/

[6] Tanabe's position in Tohoku was taken by the philosophers Takahashi Satomi and Miyake Goichi, both of whom studied under Edmund Husserl.

In fact the new quantum theory of physics represents such a state of knowledge" (Tanabe 1963, p. 285).

Tanabe's illustrations of quantum mechanics as well as Ishihara Jun's commentaries on the theory of relativity, where their intended audiences were the general public, inspired young Yukawa Hideki, who later specialized in particle physics.

68.2.7 *Science Education and NOS During WWII*

Every aspect of Japanese society turned to the wartime efforts during WWII. Education was no exception. The purpose of mathematics and science in elementary schools was stated in a nationalist way: "to train the ability to think precisely and process normal events and phenomena and to apply this ability to everyday practice and to nurture rational and creative mind in order to prepare for contributions to the prosperity of the nation." (通常ノ事物現象ヲ正確ニ考察シ処理スル能ヲ得シメ之ヲ生活上ノ実践ニ導キ合理創造ノ精神ヲ涵養シ国運ノ発展ニ貢献スル素地ヲ培フコト[7])

The possibility of an exemption from military service motivated students to study the natural sciences, engineering, and medicine even in the later stages of the WWII when university students of social science and humanities were still being ordered to the front lines.

Paradoxically, though, the word "scientific" became a code word for left-wing movements and so was banned under the wartime militarism. Discussions concerning the social meaning of technology went underground during the war and stayed there for some time after the war.

At the same time, Sogensha, an Osaka-based publisher, published a series of classics of science. It includes a Japanese edited version of the Unity of Science (*Einheitswissenschaft*) work, but it was neglected.[8]

68.3 Higher Education in Japan After the WWII

68.3.1 *Reform of Education After the WWII*

Before and during WWII, the notion of science had been distorted to emphasize its nationalistic aspects. Every activity turned to wartime efforts.[9] Japan's surrender in 1945 led to a restructuring of the entirety of Japanese society and largely along

[7] Ministry of Education (1941).

[8] Sawada (1997), p. 3.

[9] Okazaki and Okuno-Fujiwara (1999) points out that the social systems of Japan during the WWII essentially continued after the war under different names.

American lines. The Japanese educational system, formerly shaped by the strong influence of the German educational system, was no exception. Reform was realized in the fundamental education act (1947), the school education law (1947), and the standards for establishment of universities (1956).

The reform was in fact an introduction of American educational system with some lead by CIE-GHQ. The Standards for the Establishment of Universities was announced 1949. The standard general education in universities as occurred in the United States—which required every student to take some courses in humanities, social sciences, natural science, language study, and physical education – was introduced to Japan. The main body of the instructors of general education, however, came from former high schools, while university professors had more power with governmental policy to enhance science and technology.

This idea of a standard general education included basic courses in the natural sciences, including for students of the humanities and social sciences. Specialists taught these natural science courses. In this context, NOS-like courses were welcomed by students in humanities and social science as courses requiring less background knowledge in the sciences.

Students studying science and technology jumped in national universities during the 1950s: the number of entering students in science and engineering increased from 142,546 in 1952 to 202,334 in 1957. Popularization of higher education in Japan occurred in the late 1960s, when baby-boomers approached college-entrance age. The number of first-year students in universities and colleges jumped up to 598,872 in 1975 from 273,098 in 1963.

While science and technology grew in national universities as a result of a concentration of investment by the ministry of education, private universities were established to meet the growing demands of higher education especially for majors of business and the social sciences. National and public universities focussed on engineering, natural science, and medical departments. Yet, each type of university – public and private – had comparable tuition and fees due to subsidies.

Many private universities did not include science and technology in their upper divisions, although they still had to offer natural science courses for students of the humanities and social sciences. They often hired historians of science for such courses on a budget smaller than that for researchers in science and engineering who required expensive laboratories and experimental equipment.

Research universities with schools of science and engineering did not put much emphasis on NOS. Scientists taught natural science courses for students of the humanities and social sciences in general education. They were not necessarily courses on NOS but basic introductory courses. Professors of the natural sciences in the upper division focussed on research more than education. The training of students in natural science and engineering emphasized research skills. This emphasis has not changed since the end of WWII.

University standards were revised in the Deregulation of University Act (taikoka) of 1992. The relaxation of regulations did not specify general education requirements, the student/faculty ratio, or the requirements for facilities. Most universities disbanded their general education divisions as a result of deregulation. Former general education teachers were moved to other departments or to a new faculty or stayed in general education. The number of universities electing retaining general education, however, was small. Hence, the number of teaching positions of general education has decreased because retired faculty members have not been replaced; the positions have moved to upper division. Faculty members appointed in the previous system (appointed in or before 1949) retired by 1992 due to age restrictions. Students are now not required to take any natural science courses, and NOS/HPS courses have been either taught by part-time lecturers or discontinued.

68.3.2 Yukawa on Science in Late-Twentieth Century in Japan

Science essays were popular even after WWII when the formal education system in Japan was under a total reform. Yukawa Hideki, a physicist and the first Nobel laureate of Japan (1949), committed himself to public awareness of science. His writings on NOS issues were influential.

For example, Yukawa (1945) asked:

> What should be done to promote science in Japan? There are two typical answers. One: practical research in short-term goals has been overemphasized in Japan; basic science should be stressed from now; for that, scientific mind must be nurtured; research and education of history of science, for example, will be effective. The other typical answer: science education has focused on theory without connection to the real world; knowledge of the facts are essential; students should act by themselves; they should first get accustomed to operate various machines although they may not fully understand principles. Those answers seem to be totally opposite. The direction of science education will be significantly affected which answer we think right.
>
> Then, what have we lacked? I would say, in short, we lacked "thought". This sort of consideration would arise various oppositions. Theoretical research and imagination will just have no relationship to sound development of science and no advantage on science. There actually are various thought in Japan. Native philosophy has grown in Japan. (Yukawa 1945, p. 10)

Yukawa was vocal in arguing against the military uses of atomic energy. He signed the Russell-Einstein manifesto in 1955 and attended the Pugwash conference in 1957. His thought on science and society resulted as the Yukawa-Tomonaga manifesto in 1979.

68.3.3 NOS/HPS Societies in Japan

Yukawa also set the stage for interdisciplinary discussion on NOS. He was one of the founders of the Japan Association for Philosophy of Science in 1954. The association publishes the Japanese journal *Kagaku Kisoron Kenkyu* (1954-) and the English journal *Annals of the Japan Association for Philosophy of Science* (1956-). The title of this association does not have the Japanese word for philosophy "tetsugaku";

its literal translation is "Association for the Foundational Studies for Science." The reason of the name sometimes is said that some scientists did not like the word. Philosophers of science were unhappy. Uchii Soshichi explained:

> Thus, the Japan Association for Philosophy of Science started in 1954; but that was not the end of the matter. Many philosophers were frustrated. There were two groups, one with the name "American Philosophy Group," the other with the name "Logic of Science Group;" and they decided to meet annually with the title "The Meeting for Philosophy of Science" beginning in 1957. And these meetings led, eventually, to founding another association, with the literal title of "Philosophy of Science" (in Japanese) in 1967; that is the Philosophy of Science Society Japan (PSSJ). Thus we now have two associations for philosophy of science, which is quite unusual in the world. And this reflects the relationship between philosophers and scientists in Japan. (Uchii 2002)

The Japanese Philosophy of Science Society partially overlaps with the Japan Association of Philosophy of Science in terms of memberships especially among philosophers. The latter covers analytic philosophy and philosophy in the English-speaking world in general.

Such a history led to the current situation of Japan, where HPS-related domestic academic societies currently are mixed. Most were established in the late-twentieth century and remain small to middle sized like other academic societies of humanities and social sciences in Japan. The History of Science Society Japan (1,000 members) was established in 1941 and publishes *Kagakushi Kenkyu* 「科学史研究」 (Japanese) and its English journal *Japanese Studies in the History of Science* (1962), which was followed by *Historia Scientiarum* (1980-). It has offered seminars to the public since 1975.

Other national-level societies include the Japanese Society for Science and Technology Studies and the Japanese Society for the History of Chemistry. There are many other regional societies and societies of each area, such as industrial history.

68.3.4 Department of History and Philosophy of Science in the University of Tokyo

HPS was introduced to Japan as an interdisciplinary subject in the course of educational reform after the WWII by Professor Tamamushi Bun-Ichi, who visited the United States. At Harvard University in 1950, he found the strong influence of Alfred North Whitehead and George Sarton. The delegates including Tamamushi agreed that such an interdisciplinary field would be of strong social needs in restoration of the country.

Tamamushi launched the Department of History and Philosophy of Science at the University of Tokyo in 1951, as a part of trial to introduce general education in the upper undergraduate program. Another founding member was a biochemist-historian of science, Kimura Yuichi. Later the botanist-historian of science Kimura Yojiro and philosopher Omori Shozo joined. Omori was an assistant of Morton White at Princeton IAS in 1950s. Its graduate program was established in 1970.

Saegusa Hiroto (Yokohama City University) and Yajima Suketoshi (Tokyo University of Science) supported the inauguration of the department.

Nakayama Shigeru studied astronomy in the University of Tokyo and moved as a Fulbright scholar to the Department of History of Science at Harvard from 1955 to 1959. After obtaining his Ph.D. from Harvard, he went back to Japan to teach history of science at the University of Tokyo.

The HPS department in the University of Tokyo was the only institutionalized HPS department in Japan until 1993. The liberal arts section of Tokyo Institute of Technology also had philosophers and historians of science. Yoshida Natsuhiko, who translated Ayer's *Language, Truth, and Logic* into Japanese in 1955, was the main figure there to attract many students of a wide range of backgrounds from physics to management engineering.

There were other groups of philosophers of science in Japan, however.[10] For example, Ichii Saburo (1922–1989) was a philosopher of history with a physics background. He studied in University of Manchester and then studied under Karl Popper. He later adopted the British Marxist approach to history. After returning to Japan in 1954, he translated Bertrand Russell's work into Japanese. Another example is Sawada Nobushige (1916–2006), who taught philosophy of science in Keio University. He also studied at Harvard as a visiting scholar. He wrote a popular introduction to philosophy and did original research in epistemology and logic.

68.4 NOS/HPS in School Science (Rika) in Japan After the WWII

After World War II a national education standard was issued in 1947. Textbooks were authorized by Ministry of Education in accordance with the prescribed course of study. Teachers had to teach the recommended content. The course of study for elementary schools and junior high schools and the course of study for high schools have been revised eight and nine times, respectively.

68.4.1 NOS/HPS in the Early Postwar Period of School Science Education

After the war the 1947 national education standard, science was referred to as Rika. This nomenclature has not changed for the past half century. In the 1947 course of study, elementary schools and junior high schools were handled as a single unit. The initial stated purpose of science (Rika) was "to equip students with the following three qualities related to problems of the students' environments, so as to ensure that

[10] Sawada (1997).

all people can live a rational life and can enjoy better lifestyles": (1) the ability to look, think about, and deal with things scientifically, (2) knowledge related to the principles and applications of science, and (3) an attitude oriented toward finding and promoting the truth and creating new things. The following 13 sub-purposes were also identified (Ministry of Education 1947):

1. An attitude of being familiar with nature and an interest in scientific works
2. The ability to observe objects and phenomena in the natural world
3. The ability to think in a logical way
4. The ability to use machines and instruments
5. An interest in cultivating living things with care
6. Health-maintaining habits
7. Perseverance, willingness to help others, the habit of pursuing scientific work or research on one's own will
8. A desire to follow the truth and seek out the unknown
9. The ability to read easy science books
10. Knowledge of major scientific principles and their applications, allowing one to better understand the property of surrounding things and the relationships between them
11. Knowledge of the harmony, beauty, and bounty of nature
12. Respect for the work of scientists
13. Preparations for advanced science learning and necessary occupational preparations

Fujii (2005a) makes the following argument regarding the initial postwar purpose of science (Rika) education. The Japanese people had endured poor food, clothing, and housing conditions during the difficult years during and following World War II. Because it was a major social goal to somehow overcome these hardships and improve their living conditions, science education took on the characteristics of "Science (Rika) of everyday living." Nonetheless, it is interesting that the world "kagaku" is frequently used in the list of the purposes of Rika and that goals like cultivating "respect for the work of scientists" were included among its purposes.

The preliminary course of study for high schools, published in 1948, included separate proposals for physics, chemistry, biology, and earth science but contained no descriptions related to the science as a whole or anything related to the nature of science (Ministry of Education 1948).

In the 1951 revision, the junior high school and curriculum was coordinated with the high school curriculum, emphasizing the integrated nature of junior and senior high school education. Thus, the following were included among the purposes of science (Rika): science as a method and the use of scientific methods to solve problems, scientific attitudes and habits, the role of scientists and science in promoting human welfare and the development of contemporary civilization, and cultivating respect for specialists. These phrases applied to physics, chemistry, and biology (Ministry of Education 1951).

In the revised course of study for elementary school science prepared in 1952, science courses were expected to convey basic ideas on the nature of science, including the reality of natural phenomena and the objectivity and universality of science. Simple examples, such as the life cycle of the tadpole, were recommended as exercises suited to convey these notions. Beyond recommending the content of courses, revisions to the curriculum also specified the need for elementary children to understand the argumentative qualities of scientific knowledge, which the reforms viewed as linked to the students' engagement in simple investigations, in making predictions, in conducting experiments, and in comparing results. In this way students came to understand, the reformers believed, not only the persuasive qualities of scientific knowledge but also how new knowledge was created and competing results reconciled (Ministry of Education 1952).

68.4.2 NOS/HPS in School Science During the High-Growth Period

Japan became an independent nation in 1952 and in 1957 became a new member of the international community by joining the United Nations. This move was accompanied by developments in science and technology, including the enactment of the Vocational Education Promotion Act in 1951 and the Science Education Promotion Act in 1953. Revisions made to the course of study for elementary schools and junior high schools in 1958 and for high schools in 1955 (Ministry of Education 1955) represented an important turning point that shifted the focus of science education in Japan toward the natural sciences.

The stated purpose of elementary school science (1958) included the term "natural scientific" rather than just "scientific," and in high schools (Ministry of Education 1955), two of the four science subjects (Physics, Chemistry, Biology, and Earth Science) were made mandatory (further revisions in 1960 made four of the following courses mandatory: Physics A and B, Chemistry A and B, Biology, Earth Science). With these revisions, which can be viewed as a shift toward natural science fundamentalism, the content related to NOS was eliminated from the purposes and content of the course of study. What remained were descriptions of the "methods of natural science." It was in this revision that the term "scientific inquiry (*tankyu*)" first appeared within the stated purposes of Science (Rika) in the course of study for junior high schools and high schools.

In the revisions made to the course of study for elementary schools in 1968 (Ministry of Education 1968), for junior high schools in 1969 (Ministry of Education 1969), and for high schools in 1970 (Ministry of Education 1970), which occurred in the middle of the high-growth period, systematic learning that conformed to the systems of natural science was steadily promoted, and the emphasis was placed on understanding scientific methodology and the process of scientific inquiry. Given that the four subjects listed above were mandatory in high school science programs, a "Basic Science (Kiso Rika)" introductory course was established to ease the burden of taking the more specialized courses later on. One of the stated goals of this course was "to teach the methods of science" and "to make students aware of the contributions that natural science has made to improving human welfare" (Ministry of Education 1970).

68.4.3 NOS/HPS in School Science Education During the Stable Growth Period

As both the positive and negative aspects of the high-growth period began to be revealed, Japanese society found itself facing a variety of social problems that directly affected education. In a period of material affluence and prosperity, the issue of the pressures placed on students and student overwork – both the result of strict national standards – surfaced. At the same, Japan's low birth rate emerged as a central social issue. The keyword in the revised course of study around this time consequently was "yutori" (relaxed/pressure-free).

In the 1977 revisions to the course of study for elementary schools and junior high schools, the purpose of science (Rika) was condensed into a highly simplified form and the total class hours devoted to science reduced. For example, the purpose of elementary school science was "To cultivate the skills and attitudes needed for exploring nature through observation and experimentation, to facilitate understanding of natural events and phenomena, and to cultivate a deep sense of appreciation for nature" (Ministry of Education 1977a). In junior high school, students in all grade levels had been required to take 4 h a week in science classes up to this point, but the requirement was changed to 3 h per week for first- and second-year students. It is important to note that at this time, the word "science (*kagaku*)" was eliminated from the stated purposes of school science (Rika) in the course of study for elementary schools and junior high schools (Ministry of Education 1977a, b).

In the course of study for high schools published in 1978, science courses were completely reorganized and a standard number of credits were established (Ministry of Education 1978). These were as follows: Science (Rika) I (4 units), Science (Rika) II (2 units), Physics (4 units), Chemistry (4 units), Biology (4 units), and Earth Science (4 units). Science I – which was established to cover the content students had to learn prior to junior high school and to prepare them for further advanced learning – became mandatory, but no NOS content was included in the subject.

On the other hand, the stated purpose of Science (Rika) II, which was newly established by this revision, was "to identify issues related to events and phenomena that can be seen in the natural world and related to historical examples of science; and through scientific inquiry, to teach the methods of science and to cultivate problem-solving skills." In this revision the history of science entered into the science curriculum through the examination of historical examples of important scientific discoveries that demonstrated how principles and theories were established. Because Physics, Chemistry, Biology, and Earth Science, as well as Science (Rika) II had been made elective courses, the number of students electing Science (Rika) II was not so high compared to the number of students electing other courses.

In revisions made to the course of study for elementary schools (Ministry of Education 1989a) and junior high schools (Ministry of Education 1989b) in 1989, when the Showa era gave way to the Heisei era and educational reform for the twenty-first century was just beginning, the stated purpose of science (Rika) included terms like "the cultivation of scientific ways of looking and thinking" and "science (*kagaku*)," but statements related to NOS were still absent. In junior high school science (Ministry of Education 1989b), an elective science course, offered once a week, was established, and efforts were made to allow students with a particular interest in or passion for science to engage selectively in scientific inquiry.

Following the trend among elementary and junior high schools, major changes were made to the 1989 course of study for high schools with regard to science (Rika). The basic policy was to "establish course diversity." This change occurred in an environment where there was little freedom of selection in science courses, and courses were not able to accommodate sufficiently differences in students' skills, aptitudes, and preferred plan of study (Fujii 2005b). Thus, 13 elective courses were established: General Science (Rika) (4 units), Physics IA, Chemistry IA, Biology IA, and Earth Science IA (2 units each), Physics IB, Chemistry IB, Biology IB, and Earth Science IB (4 units each), Physics II, Chemistry II, Biology II, and Earth Science II (2 units each) (Ministry of Education 1989c).

Content related to NOS/HPS was still covered in the General Science class, but it only appeared as the "Study of cases of experiments in scientific history," which was one of three items in a list of "Research Topics." Considering the scope and level of the content, it was noted that "regarding important discoveries in the history of science, students should learn about the process by which principals and theories are established through the repetition of experiments and review of the scientific literature." Miyashita (2006) proposed using an earth science textbook with emphasis of history on geology and NOS for high school students and first-year college students. His idea was that the history of plate tectonics – a twentieth-century discovery – gave students a good grasp of how geological ideas had developed and, by implication, how science operated. The textbook also covered topics of national importance, including astronomy during the Edo period and the development of seismology as a native Japanese science.

68.4.4 NOS/HPS in School Science Education in the New Century

The key phrase found in the revised course of study of 1998 (elementary and junior high school) and 1999 (high school), at the dawn of the new century, was "a zest for life (*ikiru-chikara*)." Schools were asked to cultivate in students a rich sense of humanity and the ability to learn and think on their own. While descriptions of the purpose of science at the elementary school and junior high school levels included the usual wording concerning "carrying out observations and experiments," it was emphasized that students would do so "with their own prospectuses" (Ministry of Education 1998a) or "with a sense of purpose" (Ministry of Education 1998b). While this wording emphasized active problem-solving by students, other sections also reflected a departure from the simple empiricism-oriented teaching methods through which scientifically appropriate knowledge was cultivated through verification and falsification (Kadoya 1998).

Subjects were again reorganized in the 1999 revisions to the course of study for high schools (Ministry of Education 1999). With regard to NOS/HPS, a Basic Science (Rika Kiso) class (2 units) was newly established, with the stated purpose mentioning "the relationship between science and human activity" and "the scientific inquiry and investigation of nature and the process of scientific development." In terms of content, the guidelines included references to the "beginning of science" and "scientific inquiry into nature and the development of science." They addressed the inquiry into the origin of matter, which ultimately led to the development of science, cell discovery and theories of evolution, the process of establishing ways of thinking about energy, and the Copernican theory and plate tectonics. Students were required to select two of the following courses: Basic Science (Rika Kiso), General Science (Rika Sogo) A and B (2 units each), Physics I, Chemistry, Biology I, and Earth Science I (3 units each). To enable students to gain a broader range of basic science skills, they were also allowed to include one or more of the following courses in their curriculum: Basic Science, General Science A, or General Science B. Basic Science was no longer included as an entrance examination subject.

At the same time the Science and Technology Basic Law was enacted in 1995. It called for Japan to be "a nation based on the creativity of science and technology." The justification for this legislation cited three expectations of science and technology (S&T) for the twenty-first century. Science and technology were expected to lead to creative, cutting-edge developments and the creation of new technologies; contribute to solutions to the various problems that humanity will face in the future, including environmental problems, food, and energy problems, and AIDS; and create new cultures related to human life, society, and nature. The Science and Technology Basic Law also contained references to science education. Chapter 5, Article 19 stated that "the nation should implement necessary policy measures to promote the learning of S&T in school and social education, to raise awareness of S&T and to disseminate knowledge of S&T, so that all Japanese people, including the young, have every opportunity to deepen their understanding of and interest in S&T."

The Fourth Science and Technology Basic Plan,[11] approved by the Cabinet in 2011 based on this law, called for Japan "to consistently and systematically nurture talented children to lead the next generation" and "to enhance the interest of children in science and mathematics starting in elementary and secondary education so as to increase the population of children interested in such subjects, and to identify talented children and develop their abilities." The Super Science High School Project undertook various efforts to cultivate student interest in science and technology based on plans created by each school, such as developing classes based on an independent curriculum, forming partnerships with universities and research institutions, and conducting research on issues that take advantage of local conditions. Launched in 2002, the budget for this project grew to more than three times its original size, and it is still growing.

[11] Also see Sect. 68.5 of this chapter.

The 2008 version of the course of study for elementary schools and junior high schools continued to highlight the key phrase "a zest for life (*ikiru-chikara*)" but incorporated major reforms. Specifically, the number of classroom hours devoted to elementary school science was increased from 350 to 405 (MEXT 2008a). The number of classroom hours devoted to junior high school science was increased from 290 to 385 (MEXT 2008b). These changes were expected not only to increase the quantity of science content but to improve qualitatively the process of scientific inquiry through the introduction of observational experiments and report writing by students. Emphasis was once again placed on the connection between science and everyday life. In the 2009 version of the course of study for high schools (MEXT 2009), "Science (Kagaku) and Our Daily Life" was established as the subject equivalent to the former "Basic Science (Rika Kiso)," and course names that included the word "General" were eliminated. The curriculum now consisted of Basic Physics, Basic Chemistry, Basic Biology, and Basic Earth Science (2 units each), and Physics, Chemistry, Biology, and Earth Science (4 units each) (MEXT 2009).

The purpose of the Science (Kagaku) and Our Daily Life subject was "to understand scientific view and methods and to inspire interests in science via experiences of observation and experiments of phenomena in everyday life in a fashion fit for the needs of vocational high school students." History and philosophy of science were thereby eliminated from the official school curriculum of Japan, and science had only weak connections to other subjects. The course content was divided into three major segments: the development of science and technology, science in everyday life, and science and everyday living in the future. Students could either (1) take two of the following subjects, as long as one of them is Science and Our Daily Life: Science and Our Daily Life, Basic Physics, Basic Chemistry, Basic Biology, and Basic Earth Science or (2) take three of the following classes: Basic Physics, Basic Chemistry, Basic Biology, and Basic Earth Science.

University entrance examinations have excluded NOS/HPS-related content in favor of focusing on specialized subjects. The National Center for University Entrance Examination again announced in April 2011 that its examinations would not include "Science (Kagaku) and Daily Life" because that topic now appeared too general for high school instruction. As a result, high schools evaluated by the number of students accepted to prestigious universities would probably elect to eliminate NOS/HPS in order to spend more time on subjects considered to fit to entrance examinations.

This revision moved Japanese science education away from what was intended in the immediate postwar period, which at the elementary school level addressed the basic idea of science and at the high school level had taken up the systematic nature of natural science. Given that contemporary society is grounded in science and technology and the world has become a place where scientific knowledge is becoming increasingly globalized and technologies developed through continuous innovation, it is important for all people, regardless of gender or age, to have a wide range of knowledge of science and technology, as well as flexible ways of thinking and making decisions. This development may be called neo-scientism in the school science of the twenty-first century.

68.5 The Science and Technology Basic Plans and Their Impacts on NOS

The Science and Technology Basic Law (1995) led the Japanese government to issue four Science and Technology Basic Plans (1996–2000, 2001–2005, 2006–2010, and 2011–2015), which aimed to enhance the roles of science and technology in Japanese society. Each included NOS components, with foci on public awareness of science and enforcement of school science.

68.5.1 The First Science and Technology Basic Plan

The first Science and Technology Basic Plan, issued in 1999, lamented the public's low esteem of science and technology. So the first plan was designed to "gain the public's deep and broad understanding for the promotion of science and technology with full respect towards harmony with humans, society, and nature" by implementing government measures to improve public understanding of science and technology, a task for which they expected the cooperation of specialists in the production of "easy-to-understand information on science and technology" (pp. 14–15).

The first basic plan targeted the improvement of science and technology education in school education by focusing on teaching methods and new facilities, such as computer-aided learning facilities, as well as by emphasizing the practical aspects of science and technology (p. 42). A National Museum of Emerging Science and Innovation (*Kagaku Miraikan*) was established in 2001 in the framework of this basic plan for science and technology as the center for promoting science and technology. NOS/HPS content did not appear in the first basic plan, though.

Independently of the basic plan, the Japan Accreditation Board for Engineering Education was established in 1999 for accrediting engineering programs complying with international standards of engineering certification. As the international standard for engineering education includes an ethics requirement, universities which intended to offer JABEE-accredited programs were forced to introduce engineering ethics courses. The impact on NOS/HPS was limited, however. Most "ethics" courses were dedicated to practical aspects of engineering, such as compliance with regulations or general workplace ethics as a part of the social accountability and responsibility of engineers. Only a few programs invited NOS/HPS researchers to include other dimensions of science and technology in the curriculum.

The Japanese Society for Science and Technology Studies was established in 2001. Kobayashi Tadashi, the first president of the society, remarked in the prospectus of the society that twenty-first century technoscience entailed sociopolitical and philosophical challenges. He pointed out that the

> uncontrolled production of artificial goods is fast overwhelming the natural world, aggravating an already precarious environmental crisis; developments in biotechnology and information technology threaten the survival of traditional lifestyles and value systems. Human societies and individuals need to rethink their relationship with technoscience. (Kobayashi 2006)

68.5.2 The Second Science and Technology Basic Plan

The Second Basic Plan (2001) cited the importance of communication between science and society, but its main focus was mainly on communication between researchers in science and engineering and those in the social sciences and humanities. The public's understanding, assessment, and acceptance of science were mentioned, but the major actors charged with evaluating science and technology were natural scientists, technological experts, and experts in the social sciences and humanities. This second plan cast the reform of science education in the context of international competition, a long-standing motivation for training scientific specialists.

The role of the public in the advancement of science and technology was specified in the fifth chapter of this plan as embracing a responsibility to understand the role of science and technology in daily life, a recognition that science and technology are synergistic activities, and an obligation to develop a level of scientific understanding enabling one to make rational and independent judgments. Interestingly the plan believed that the development of this kind of scientific understanding could occur outside the educational area in institutions like museums.

Yet in spite of the attention paid to public understanding of science, the plan overwhelmingly emphasized the responsibilities of scientists and engineers to act in accordance with ethical standards. Hence bioethics, research accountability, and risk management among scientists and engineers were stressed to a greater degree than the public's responsibility to make rational assessments about science and technology.

68.5.3 The Third Science and Technology Basic Plan

The attitude toward NOS changed drastically in the third basic plan for science and technology (2004). An entire chapter was dedicated to science communication, especially through the proactive participation of the public in scientific and technological issues and the active engagement of scientists and engineers in communicating their work through outreach activities.

The emphasis on science communication in the third basic plan was a reaction to the discussion of issues related to bovine spongiform encephalopathy (BSE) in Japan. Although the Japanese government had discussions on BSE in the United Kingdom since the 1980s, it failed to respond adequately to the threat of the disease. Not until March 2001 did the government act, and by the end of the summer, a BSE-infected cow was found in Japan. An independent investigation in 2002 found the government lax in disseminating information on BSE and preparing the public for the possibility that the disease might be found in Japan. Ill-prepared and confused statements by the government generated public distrust as had also occurred in the UK. As a result the government tried to enact good practices of science communication.

Implementation of the third basic plan for science and technology began in 2005, which Kobayashi (2006) called the first year of science communication in Japan. The following steps were taken to improve the situation:

1. Three universities (Hokkaido University, the University of Tokyo, and Waseda University) introduced science communicator training programs with governmental support.
2. Osaka University launched the Communication Design Center, an interdisciplinary-oriented institute with a strong focus on research, development, and personnel training in science communication.
3. The National Museum of Emerging Science and Innovation began offering a training course for science communicators. It also hired science communicators with a fixed term.
4. The National Science Museum conducted a research project (2005–2008) on training and evaluation of science communicators.
5. Science cafes became common in Japan.

Nevertheless, it turned out that the activities and efforts of science communication had a limited effect in the Japanese society as the whole, with catastrophic results.

68.5.4 The East Japan Earthquake and the Fourth Science and Technology Basic Plan

One of the strongest earthquakes in human history hit eastern Japan at 14:46 on March 11, 2011. The epicenter was 130 km from the Miyagi prefecture coast. The earthquake's magnitude was 9.0, the fourth strongest of recorded earthquakes. It was the first gigantic earthquake where the earthquake waves were almost captured by recorders and subsequent events were broadcast to the world. Within 30 min at the earliest after the earthquake, a tsunami up to 40 m high hit Japan's Pacific coastal areas from Hokkaido to Chiba. About 20,000 people were killed in the tsunami and the earthquake; 118,621 buildings were razed; and 802,814 buildings suffered partial damage. Thousands of people lost their homes and livelihood as a result of the earthquake and tsunami.

The tsunami-affected area included the site of the Fukushima 1 nuclear power plant. Due to loss of electricity, the power plant lost control and four of its six nuclear generators exploded. Radioactive particles dispersed in the air and the Pacific Ocean. Over 113,000 residents in the Fukushima Prefecture evacuated due to high levels of radiation (Cabinet Office 2011). Broadcast in real time over television and the Internet, the earthquake, tsunami, and nuclear accident shocked the world.

The consequences of the Fukushima accident reverberated in the years that followed, and assessments of the damage to society, politics, and the economy were difficult. From an NOS/HPS perspective, the main issue concerned public trust, especially trust in experts. Initially professional opinions on the dangers from Fukushima varied. Some claimed, for instance, that the radioactive level would not immediately affect human body, while others insisted that everybody should evacuate eastern Japan immediately. Due to these contradictions and others like them, the public came to distrust the government, scientists and engineers, and even mass media (and some say, social media too). The public classified some researchers and other professionals as *goyo gakusha* (governmental junk scientists), while it identified others as "sound" professionals. But the public's criteria for placing someone in one category or another were not clear, indicating a profound inability to assess independently "official" statements about the nature and extent of the crisis. Scientific communication had malfunctioned despite the science communication programs that had been inaugurated in the decade before the accident. Improving public understanding of science and the expert's ability to communicate science were high priorities in the years following Fukushima.

A complicating factor was the unusual nature of the accident – the probability of a nuclear disaster following an earthquake was thought to be very low – making the crisis one of determining an "unknown unknown." The environmental and health effects of low levels of radiation were not known, yet decisions had to be made in order to take action on the relocation of the population after the quake. For instance, before beginning the reconstruction of the area, wrecked homes and ships had to be examined for the level of radiation contamination, but it was not clear at what level the radiation was harmful. Reconstruction itself was difficult because debris from the tsunami was left on the roadside more than a year after the earthquake, and roads themselves were often so damaged that they became barriers to the heavy construction vehicles needed to clean up the sites. Finally, the sheer amount of debris exceeded the capacity of plants in the area. Thus, the restoration process needed informed expert *and* public decision-making at the national level.

Furthermore irrational nuclear fears surfaced. Some residents outside the Tohoku area reacted against Tohoku debris, believing that anything in Tohoku was contaminated by radioactive particles. While professional measuring of radiation levels was extensive, the public simply did not believe the data any longer due to the distrust of professionals formed in consequence of miscommunication after the earthquake, the tsunami, and the nuclear reactor accident at Fukushima.

The Fukushima disaster had an added dimension: how to manage the consequences of the tsunami and how to handle reconstruction so that the impact of a future tsunami could be minimized. Thousands of residents lost their homes and had to live in temporary housing. Residents wanted to return to their homes, but whether or not their homes were safe was difficult to determine. Hilly areas are naturally relatively tsunami-proof and it would have been appropriate to move the displaced population to them, but these areas were already built up with little room for expansion. Municipal governments often owned other available plots of land, but they were designated as parks or historical landmarks and therefore could not be occupied by survivors. An added complication was the loss of gainful employment in the area after the tsunami.

Discussions on those issues might have been a little easier if the Japanese public were well educated on NOS and if the public as well as the experts acknowledged that science cannot solve all problems by itself. Tadashi Kobayashi strongly argued for the importance of NOS education for the public in a cabinet meeting in April 2011 in the hope of changing the situation.

The fourth Basic Plan of Science and Technology (2011) was crafted with the problems caused by the earthquake in mind. Greater stress was placed on the role of science communication in social decision-making as well as in the public understanding and awareness of science than had been the case in previous basic plans. The fourth plan explicitly stated that the evaluation of scientists and their work now had to include consideration of the scientist's outreach to the public, including the communication of research results to society.

68.6 Conclusion

The promotion of science and technology in Japan after the Meiji restoration has been motivated by external pressures and global competitions. Engineering was favored over science in education in order to increase innovation through practical results.

NOS in school science (Rika) was the most clearly articulated in the course of study for elementary science revised in 1952 in the early postwar period. The word "science (kagaku)" disappeared from the course of study for elementary science revised in 1977. Shindo (1995) noted that "science," "scientific method," and "scientific inquiry" in Japanese science education should be reconsidered from the point of view of modern science studies. NOS has hardly appeared in any stage of education since it was not included on university entrance examinations, and, with few exceptions, it has been largely excluded from higher education programs in science.

In Japan university courses for science teacher training, even for science majors, usually do not have NOS/HPS-related content. So NOS/HPS is not required to be a science teacher. Toda (1992) investigated the views of science and science teaching methods held by science teachers in training and concluded that teachers in training were never fully conscious of their own specific views regarding science. They tended to hold traditional views of science as common sense, and their choice of science teaching methods was based on "familiarity" and "orthodoxy" rather than informed and up-to-date views of science.

International developments may compel Japanese science teachers to take a different view of their subject and their profession. In 1999 a declaration on science and the use of scientific knowledge was issued under the aegis of the United Nations Educational, Scientific and Cultural Organization (UNESCO), and the International Council for Science (ICSU). The declaration, which addresses scientific knowledge and science education, stated that "Science curricula should include science ethics, as well as training in the history and philosophy of science and its cultural impact" (The World Conference on Science 1999).

In Japan HPS grew in response to nationalist movements and educational reform after WWII. HPS courses since then have been offered mainly for non-science students, with the number of universities offering HPS courses decreasing after successive reforms in higher education. Crisis events such as BSE resulted in an emphasis on improving science communication rather than enhancing instruction in HPS. The earthquake of March 2011 and the subsequent nuclear accidents have drawn attention to NOS/HPS in Japan, but the integration of NOS/HPS courses into the Japanese educational system has been uneven, with most educational sectors ignoring NOS/HPS.

Thus, Japanese society began to realize the importance of NOS in the wake of the East Japan Earthquake and the nuclear accident that followed. There is still a long way to go to change Japanese attitudes in the direction of a deeper understanding of nature and of ways of studying and managing it. NOS and HPS can be of assistance in changing attitudes, but there is much to be done. Concerned parties need to be persistent.

References

Cabinet Office (2011). Hinansho seikatsusha no suii (in Japanese. 避難所生活者の推移, the number of evacuating people). http://www.cao.go.jp/shien/1-hisaisha/pdf/5-hikaku.pdf. Accessed 21 October 2012.

Fujii, H. (2005a). Development of the course of study for junior high school science, In. N. Tomoyuki (Ed.), *Rika kyouiku gairon* (in Japanese. 理科教育概論, *Introduction to science education*) (pp. 95–101). Okayama: Daigakukyouiku-Shyppan.

Fujii, H. (2005b). Development of the course of study for high school science, In. N. Tomoyuki (Ed.), *Rika kyouiku gairon* (in Japanese. 理科教育概論, *Introduction to science education*) (pp. 102–109) . Okayama: Daigakukyouiku-Shyppann.

Ishihara, J. (1936). Shizenkagaku no Hoho to Shiteno Jikken Kansatsu ni Tsuite. (In Japanese. 自然科学の方法としての実験・観察について, On experiments and observation as methods of natural science). First appeared in *Riso* in 1936. Reprinted in Tsuji, T. (1980) Sekai no Naka no Kagaku Seishin: Nihon no Kagaku Seishin 5 (In Japanese. 世界のなかの科学精神:日本の科学精神5. Ideas of science in the world: Ideas of science in Japan vol. 5). Kosakusha.

Ito, K. (2010). Robots, A-bombs, and War: Cultural meanings of science and technology in Japan around World War II. In R. A. Jacobs (Ed.), *Filling the hole in the nuclear future: art and popular culture respond to the bomb* (pp. 63–97) . Plymouth, UK: Lexington Books.

Kadoya, S. (1998). *Rika-gakusyu-shido no kakushin* (in Japanese. 理科学習指導の革新, Innovation in science teaching). Tokyo: Toyo-kan Syuppannsya.

Kobayashi, T. (2006). Konsensasu kaigi no genjo to kongo. (In Japanese. コンセンサス会議の現状と今後, The consensus meeting: now and future) Lecture at Sokendai. http://sas.soken.ac.jp/ja/wp-content/uploads/lecture/gouikeisei/kobayashi.pdf. Accessed 25 May 2012.

Ministry of Education (1941) Ordinance国民学校令施行規則http://www.mext.go.jp/b_menu/hakusho/html/others/detail/1318024.htm Accessed 6 January 2013.

Ministry of Education (1947). *The course of study*. Tokyo: Tokyo Shoseki.

Ministry of Education (1948). *The course of study for high school*. Tokyo: Dainihon Tosyo.

Ministry of Education (1951). *The course of study*. Tokyo: Meiji Tosyo

Ministry of Education (1952). *The course of study for elementary school*. Tokyo: Dainihon Tosyo

Ministry of Education (1955). *The course of study for high school*. Tokyo: Dainihon Tosyo.

Ministry of Education (1958). *The course of study for junior high school*. Tokyo: Teikoku-chiho-gyosei Gakkai.

Ministry of Education (1968). *The course of study for elementary school*. Tokyo: Meiji Tosyo.
Ministry of Education (1969). *The course of study for junior high school*. Tokyo: Ministry of Finance Printing.
Ministry of Education (1970). *The course of study for high school*. Tokyo: Ministry of Finance Printing.
Ministry of Education (1977a). *The course of study for elementary school*. Tokyo: Ministry of Finance Printing.
Ministry of Education (1977b). *The course of study for junior high school*. Tokyo: Ministry of Finance Printing.
Ministry of Education (1978). *The course of study for high school*. Tokyo: Ministry of Finance Printing.
Ministry of Education (1986). Ordinance 小学校ノ学科及其程度(抄) (明治十九年五月二十五日文部省令第八号) http://www.mext.go.jp/b_menu/hakusho/html/others/detail/1318012.htm. Accessed 24 May 2012.
Ministry of Education (1989a). *The course of study for elementary school*. Tokyo: Ministry of Finance Printing.
Ministry of Education (1989b). *The course of study for junior high school*. Tokyo: Ministry of Finance Printing.
Ministry of Education (1989c). *The course of study for high school*. Tokyo: Ministry of Finance Printing.
Ministry of Education (1998a). *The course of study for elementary school*. Tokyo: Ministry of Finance Printing.
Ministry of Education (1998b). *The course of study for junior high school*. Tokyo: Ministry of Finance Printing.
Ministry of Education (1999). *The course of study for high school*. Tokyo: Ministry of Finance Printing.
MEXT (Ministry of Education, Culture, Sports, Science & Technology) (2008a). *The course of study for elementary school*. Tokyo: Tokyo Shoseki.
MEXT (Ministry of Education, Culture, Sports, Science & Technology) (2008b). *The course of study for junior high school*. Tokyo: Higashiyama Shobo.
MEXT (Ministry of Education, Culture, Sports, Science & Technology) (2009). *The course of study for high school*. Tokyo: Higashiyama Shobo.
Miyashita, A. (2006). *Zeminaru chikyu kagaku nyumon* (in Japanゼミナール地球科学入門:よくわかるプレートテクトニクス, Seminar of earth sciences: Introduction to Plate Tectonics). Nihon Hyoronsha.
Nakayama, S. (1977). *Nihon jin no kagaku kan*. (in Japanese. 日本人の科学観, Japanese views on science). Sogen Shinsho.
Nishida (1911) Zen no Kenkyu (｢善の研究｣) Kodokan 弘道館. English translation: Kitaro Nishida, An Inquiry into the Good, Yale University Press, 1990.
Okazaki-Okuno-Fujiwara eds. (1999), The Japanese Economic System and its Historical Origins. Clarendon Press, Japan Business and Economics Series.
Piovesana, G. K. (1963). *Recent Japanese philosophical thought*: 1862–1962. Enderle Bookstore.
Sawada, N. (1997). Nihon ni okeru Kagaku Tetsugaku no Rekishi. (in Japanese. The History of the Philosophy of Science in Japan, 日本における科学哲学の歴史) *Kagaku Tetsugaku* (Philosophy of Science) 30, 1–13. doi: dx.doi.org/10.4216/jpssj.30.1 Accessed 28 May 2012.
Shindo, K. (1995). The advocacy and principal of science education, In T. Terakawa (Ed.), *Rika kyouiku sono dainamikusu* (in Japanese. 理科教育そのダイナミクス, Science education and its dynamics) (pp. 191–222). Okayama: Daigakukyouiku-Shyppann.
Soshichi UCHII (2002) IS PHILOSOPHY OF SCIENCE ALIVE IN THE EAST? A Report from Japan. ,40th Anniversary Lecture Series, Center for Philosophy of Science, University of Pittsburgh, March 14, 2002. Script available at: http://philsci-archive.pitt.edu/585/1/cntr-lec.html

Tanabe (1963) 田辺元全集第2巻 (Complete works of Tanabe Hajime, vol 2) Chikuma Shobo (筑摩書房). 1963.

Terada (1918) Kenkyuteki Taido no Yosei (研究的態度の養成, Nurturing mindset to research). Rika Kyoiku (「理科教育」, Science Education) 1918. Reprinted in寺田寅彦全集 第五巻 (Complete Works, Terada Torahiko vol.5) Iwanami Shoten 岩波書店 1997

The Education Law (教育令) Dazyoukanhukoku (太政官布告) Act No. 40 (1879). http://www.mext.go.jp/b_menu/hakusho/html/others/detail/1317966.htm. Accessed 24 May 2012.

The World Conference on Science (1999). Declaration on science and the use of scientific knowledge. http://www.unesco.org/science/wcs/eng/declaration_e.htm. Accessed 10 October 2012.

Toda, K. (1992). Pre-service science teachers' view of science and its influence on their selection of science teaching methods. *Journal of Japan Society for Science Teaching*, 33(1), 59–70. (in Japanese)

Tsuji, T. (1973). *Nihon no kagaku shiso: Sono jiritsu he no mosaku* (in Japanese. 日本の科学思想—その自立への模索, Scientific thoughts in Japan: Grope for independence). Chuko Shinsho.

Ueno, K. (2006). From Wasan to Yozan. In Leung, F. K. S., Graf, Klaus-D., & Lopez-Real, F. J. (Eds.), *Mathematics education in different cultural traditions: A comparative study of East Asia and the West : the 13th ICMI study* (pp. 72–75). Springer.

Uchii, S (2002). Is philosophy of science alive in the east? A report from Japan. Center for Philosophy of Science, University of Pittsburgh, 40th Anniversary Lecture Series, March 14, 2002. http://www.bun.kyoto-u.ac.jp/~suchii/cntr-lec.html. Accessed 28 May 2012.

Yukawa, H. (1945). Kagaku nihon no saiken. (in Japanese. Restoration of science in Japan, 「科学日本の再建」) October 1945. Reprinted in Yukawa, *Zenshu* (*Collected Works*, 全集) 4.

Yuko Murakami is an associate professor, Graduate School of Science, Tohoku University. Ph.D (Philosophy, Indiana University), M.S. (HPS, University of Tokyo), and B.A. (HPS, University of Tokyo). Her research areas cover logic, analytic philosophy, philosophy of science, science, technology and society, and higher education. She is a member of the boards of the Japan Association for Philosophy of Science and the Japan Association of Contemporary and Applied Philosophy and of the editorial board of Philosophy of Science Society Japan.

Manabu Sumida is an associate professor of science education in the Faculty of Education, Ehime University, Japan. He holds a BS in chemistry from Kyusyu University and a Ph.D in science education from Hiroshima University. His research interests are focused on relating the cognitive and philosophical approach to science education and the cultural effects on science education. He is the author of "The public understanding of pendulum motion: From 5 to 88 years old" (In M. R. Matthews et al. (eds.) *The pendulum*, pp. 465–484, 2005) and "The Japanese and Western views of nature: Beyond cultural incommensurability" (In T. Papatheodorou (ed.) *Debates on early childhood policies and practices*, pp. 123–135, 2012). He is also the co-translator of "Science for All Americans" into Japanese. He was a committee member of the TIMSS 2003 study, TIMSS 1999 Video Study, and OECD PISA 2006. He received the Young Scholars' Award from the Japan Society for Science Education in 1996 and from the Society of Japan Science Teaching in 1999 and the Best Paper Presentation Award in 2007 and 2008 from the Japan Society for Science Education. He is currently Director of the Japan Society for Science Education, Executive Board of East-Asian Association for Science Education, and Country Delegate of the Asia-Pacific Federation of the World Council for Gifted and Talented Children.

Chapter 69
The History and Philosophy of Science and Their Relationship to the Teaching of Sciences in Mexico

Ana Barahona, José Antonio Chamizo, Andoni Garritz, and Josip Slisko

69.1 Introduction

Why is science so important in today's societies? Science (along with technology) is one of the salient endeavours of the contemporary world and, more than any other human activity, distinguishes the current period from previous centuries. According to Stehr, it is a widely shared assumption among contemporary social scientists that the immense impact of science and technology on society has become one of its defining characteristics (Stehr 1994).

Nowadays, we are experiencing the fourth, postindustrial, technoscientific revolution, where science and technology play an increasingly important role in most spheres of life and where our dependence on knowledge-based occupations is considerably growing (Böhme 1988). Contemporary society may be described as a knowledge society, based on the penetration of all its spheres by scientific and technological knowledge (Stehr 1994).[1]

[1] Some authors consider the first of the technoscientific revolutions to be the agricultural revolution; the second, the industrial revolution, (these two revolutions emerged from applying new sources of energy to mass production of goods and the transfer of information theory to industrial processes); the third the informatics and robotics revolution; and the fourth the postindustrial revolution. These revolutions were manifestations of the ever-increasing capacity of human beings to

A. Barahona (✉)
Departamento de Biología Evolutiva, Facultad de Ciencias,
Universidad Nacional Autónoma de México, UNAM, México
e-mail: ana.barahona@ciencias.unam.mx

J.A. Chamizo • A. Garritz
Seminario de Investigación Educativa en Química, Facultad de Química,
Universidad Nacional Autónoma de México, UNAM, México

J. Slisko
Facultad de Ciencias Físico Matemáticas, Benemérita Universidad
Autónoma de Puebla, México

M.R. Matthews (ed.), *International Handbook of Research in History, Philosophy and Science Teaching*, DOI 10.1007/978-94-007-7654-8_69,

Advances in science and technology deeply influence natural and social processes. Science and technology as an instrument of mediation between nature and society have transformed people's lifestyles and their relationship with the cultural and natural environment.

Now we know more about the way the world and the universe work; in matters of health many diseases have been eradicated, and many therapies have been found for others. On the technological side, modern agriculture and industry have been developed to cover the needs of more and more inhabitants of the planet, as well as increasing the possibilities to access information in real time with worldwide coverage. Modern societies cannot function without the products of science and technology; they are now so commonplace that they have become largely invisible.

Changes due to science and technology have generated transformations in the way knowledge is organized and have transformed societies into knowledge societies, where information is manifold, decentralized and available to more and more people around the world. This is why it has become essential to modify the current education system regarding science and technology in countries like Mexico. In order to do so, the inclusion of reflections produced by historical and philosophical studies has been a cornerstone over the last three decades.

The first part of this chapter is about the relationship between the history and philosophy of science and the teaching of science. It will allow us to emphasize the value that recent studies on the history and philosophy of science have had in science education in Mexico. On one hand, in it we stress the importance of the history and philosophy of a discipline in the teaching of science, and on the other, we insist in the role of science in modern societies and encourage science teaching within a historical and philosophical perspective. In the second part we will review the latest Mexican educational reforms in 1993 and 2006 and acknowledge the advances regarding the teaching of biology, physics and chemistry in basic education (elementary and junior high school) as well as the inclusion of the history and philosophy of science.[2]

69.2 The History and Philosophy of Science and Their Relationship to the Teaching of Science

Science, like other human activities, is a complex and social one (Longino 1990). We can say that science is a way of knowing about and explaining the world around us. It differs from other forms of knowledge in its particular ways of observing, thinking, experimenting and testing, which constitute the fundamental aspects of its nature. From a scientific perspective, things and events in the universe

control and manipulate their environment and resulted in important social and political changes (Hirschhorn 1986; Stehr 1994).

[2] A more broad approach had been boarded by two of the authors (Chamizo and Garritz 2008).

present consistent patterns which can be understood by means of systematic study. Scientists attempt to make sense of the observation of phenomena by formulating explanations based on scientific principles accepted by the community that are compatible with these phenomena.

Science can be understood as a process of knowledge production that not only has instruments which expand the senses and allow careful observations and interventions in phenomena but also establishes the theories which make sense of them (see, e.g. Golinski 1998; Hacking 1983).

Science has a history of elucidating many processes; the way human beings have observed and explained nature has changed through history. At the beginning of the nineteenth century, for example, the existence of genes was unknown, though it was known what happened when one crossed certain plant varieties. Nowadays we have the sequence of the human genome. Change in knowledge is evident and inevitable. Scientists reject the idea that one goal of science is to reach the absolute truth and agree that there is some uncertainty that is part of its nature and modification of knowledge is one of its norms; however, it can be said that most of scientific knowledge is long-lasting. What we know now can be modified or rejected by future observations or theoretical proposals. Therefore, stability and change are integral parts of the nature of science.[3]

Science is not only a collection of data. Concepts, scientific theories and methodologies, along with goals, values, aptitudes and abilities (which are handed down from generation to generation), are an integral part of science. When one teaches or learns science, one does not only teach or learn 'scientific knowledge' but also goals and values (objectivity, honesty, collaboration, conservation of nature), abilities (to observe, manipulate, calculate, measure, estimate) and aptitudes (curiosity, openness to new ideas, confrontation of different positions before problems, informed scepticism, communication). Scientific education can and must contribute towards enhancing people's knowledge as well as to develop scientific values and/or social values in general positive aptitudes and abilities that help improve quality of life. In this sense, schools have an unavoidable social duty, as they are in charge of distributing scientific knowledge to the population.[4]

What is the importance of teaching science? Human beings have everyday principles, which allow them to interact with the world. However, science enables us to have a better quality interaction. In modern societies, active participation and a

[3] Thanks to recent studies on the history and philosophy of science, it can be said that the different ways in which humanity has explained phenomena, i.e. the different patterns of scientific explanation, have been modified over time (see, e.g. Martínez 1993).

[4] Values have been basic elements of the twentieth-century educational perspective in Mexico, for they have social, political and pedagogical content that expresses the standards of comprehensive human education. For this reason, values have been considered an asset whose conveyance and quality must be promoted. Their presence in the social milieu has been linked to the development of the Mexican educational system since the end of the nineteenth century (Latapí 2003). Nevertheless, as Wuest Silva and collaborators (1997) mention, the study of the role played by the values associated with science and pedagogy did not begin until the 1980s.

sense of critique[5] are essential before the magnitude of the problems we face. For example, in nuclear power, climate change, the loss of biodiversity, atmospheric pollution, serious diseases such as AIDS or cancer, to name a few, scientific knowledge has become valuable in itself, and these issues have caught our attention regarding the relationship between science and society (Shortland and Warwick 1989). The teaching of science and the acquisition of scientific knowledge have value because knowing science allows us to have explanations about natural or social phenomena and develop the capacity to solve problems with efficiency (Matthews 1994/2014).

Over the last three decades, the importance of the history of science in scientific education has been gaining recognition. Below are a few of the most important reasons. The study of the history of science:

- Helps us understand the nature of science as a complex cultural enterprise that can be presented as part of a wider cultural heritage (Jenkins 1989) and therefore helps place professional education appropriately within a broader cultural context. It is not about forming scientists at an early age (which may be a positive effect), but to form informed citizens with the capacity to decide, observe and manipulate their surroundings.
- Gives us a better understanding of the methods and concepts associated with goals and values which are characteristic of different times and that remain stable for long periods.
- Can enable future scientists to improve their response to the challenges posed by the rapid globalization of science and technology (Wilson and Barsky 1998); according to Gooday and collaborators (2008), the history of science has particularly important forms of knowledge and understanding concerning science that cannot be obtained so effectively by any other means, like the ability to read and interpret primary sources and formulate and defend a cogent argument (see also Solomon 1989).
- Allows us to understand how scientific goals and values go beyond disciplinary boundaries and contribute to the reorganization of disciplines and the development of technological advancement, important to the understanding of modern science. For example, the Human Genome Project would have been impossible without the participation of the most important technological firms in charge of making the sequencers, the big philanthropic foundations in charge of financing, the universities and higher learning centres where scientific knowledge is produced and disseminated, etc.
- Allows us to find suppositions which are shared by students and whose critique and abandonment are associated with important scientific advancements. The teaching of the history of science will allow students to locate these presuppositions (or previous ideas) and be in a position to abandon them rationally. For example, a serious

[5] The role of critical discourse in science is not a peripheral feature, but rather it is at the core of its practice, and without it, it would be impossible to construct reliable knowledge; for authors like Osborne (2010), scientific education must include critical discourse in the teaching of science to foster the ability to reason and argue scientifically.

problem in students at a higher learning level is their lack of post-Lamarckian evolutionary thought. Many explanations of evolutionary processes in these students are those that correspond to Lamarckism which explained, in the nineteenth century, that species were modified due to the needs imposed by the environment: the necks of giraffes were very long because these animals had to continuously stretch them in order to reach the foliage of trees, wisdom teeth do not come out because we do not use them, etc. This kind of Lamarckian thought, where the need creates the organ, is an idea no longer shared by scientists after the theory of evolution by natural selection that Charles Darwin proposed in 1859 in *On the Origin of Species* (see, e.g. Ayala 1977, 1994, 1994b; Ruse 1979, 1996).
- Enables the idea that students put forward their explanations and are in a position to modify them to acquire modern scientific knowledge. In this way, the study of the history of science will help them understand that some of the explanations they provide, though inaccurate, can provoke a conceptual change.
- Constitutes a strong source of suggestions about how the contents and concepts of a course must be organized according to their complexity and can be used to define the pertinent didactic sequences in the development of a topic.
- Finally, allows us to locate scientific and technological developments within the general outlook of the history of humanity, which is useful for understanding the link between a scientific approach and social problems (UNESCO 1999).

One of the authors of this chapter has promoted an initiative to include the history of science in the basic education curriculum of Latin America schools (Chamizo 1994, 2007).

69.3 The 1993 and 2006 Reforms and the Transformation of Science Teaching in Mexico

Mexico has constructed a significant and high-quality scientific and technological system over the last 20 years. However, this system is insufficient before the new challenges imposed by novel problems and international competition. For these reasons, our scientific and technological system must be consolidated and expanded in a very particular way: through the teaching of science and technology in the early stages of individual development. We must emphasize that the development of science and technology in Mexico has public institutions at its foundation. Any project that comes from the State will have as a starting point the cultural, scientific, professional and historical capital generated in said institutions.

Until a few decades ago, basic-level student education regarding science was concentrated on presenting a rigid structure of subjects which tended to promote the idea that science is a great deal of information that, when processed, offers scientifically correct answers about the phenomena in our surroundings. Thanks to the development in historical and philosophical studies of science, it is now thought that the disciplines that make up science were historically formed through posing problems, not the other way around.

69.3.1 The 1993 Reform

During the presidency of Carlos Salinas de Gortari (1988–1994), the Educational Modernization Programme (Programa para la Modernización Educativa) was proposed and enacted in 1993 (known as the 1993 Reform). It contained a diagnosis of the country's situation and proposed a deep structural change. This model implied radical structural changes and the innovation of practices to modify educational content, the ongoing training of teachers, the organization of different educational levels and the integration of basic education in one cycle that would include preschool and basic education (elementary and junior high).[6] All this in order to elevate the quality of education, to reduce backwardness and decentralize the education system.[7,8]

Methodological, conceptual and epistemological aspects were included in the 1993 Reform of the science curriculum and the study programmes for elementary and junior high schools, which meant an advance regarding the conception of modern science in national curricula. The new natural sciences programmes were based on a formative perspective according to the goal of helping students 'to acquire knowledge, capacities, attitudes and values that can be expressed by the development of a responsible relationship with the environment… and to educate children not as scientists in a disciplinary and formal way; instead, students are encouraged to observe, question, and formulate simple explanations about what happens in their surroundings' (Barraza 2001).

Thanks to this reform, there was progress regarding the teaching of science in basic education,[9] for not only elementary and junior high school curricula were modified but also new textbooks[10,11] and new materials were prepared for the

---

[6] Elementary or basic education includes compulsory preschool, primary and junior high education. Preschool lasts for 2 years (4–5 years old), primary education lasts for 6 years (6–11 years old) and junior high education lasts for 3 years (12–15 years old).

[7] On March 4, 1993, the Article 3 of the Constitution was amended, assigning a mandatory character to junior high school. This fact provoked one of the most important changes in the 70-year life of junior high school since its foundation. This reform was incorporated into the General Education Act (Ley General de Educación), enacted on July 12, 1993. In this way the government, through the Ministry of Public Education (Secretaría de Educación Pública, SEP), together with the states, committed to the decentralization of education, to 100 % coverage and to raising its quality levels.

[8] The SEP was founded in 1921 by the Mexican government. Since then, this ministry has designed the content of the national curricula for all subjects for basic education.

[9] The teaching of science in elementary school includes biology, physics and chemistry.

[10] In 1959 the SEP launched a new program, the Free-Text Program (Gilbert 1997), which established the National Commission for the Free Textbooks (Comisión Nacional de Libros de Texto Gratuitos, Conaliteg), and the production of the national textbooks for all basic education subjects, which are based on the national curricula. These textbooks, official and distributed for free, are still being handed out to every basic-level student, teacher and school (private and public, urban and rural) in the country, giving access to all basic-level students to education. These textbooks provide specific guidelines for each grade and are considered excellent sources of information.

[11] It is worth mentioning that some science educators were engaged in the production of the elementary textbooks around 1996 and added a good deal of history and philosophy of science to them.

teachers, with a focus that attempted to centre the teaching of science according to the modern ideas of the history and philosophy of science mentioned above.

Bonilla and colleagues (1997a, b) and Chamizo (2005) have documented this reform. The natural sciences' programme for primary school included five major topics: living beings; human body and health; environment and environmental protection; raw material, energy and change; and science, technology and society (STS). The STS dimension of teaching science corresponds to a large need of innovation in science education. As early as 1971 Jim Gallagher proposed a new goal for school science: 'For future citizens in a democratic society, understanding the interrelationships of science, technology and society may be as important as understanding the concepts and process of science' (Gallagher 1971, p. 337).

As is it outlined by Aikenhead (2003) in his synopsis on the origins and dispersion of this new approach of teaching science, the name STS was coined by John Ziman (1980) in a book titled *Teaching and Learning about Science and Society*. In spite of its title, the book consistently referred to STS in its articulation of the rationale, directions and challenges for STS in school science. It is important to mention that Aikenhead mentioned the following sentence about the relationship of history and philosophy of science with the STS scheme: 'A more comprehensive treatment of STS includes the internal social context (the epistemology, sociology and history of science itself) as well as the external social context of science' (2003, p. 63). It must be emphasized that the recent inclusion of STS in Mexican education means recognition of the importance of history and philosophy of science (Garritz 1994).

Peter Fensham (1985), in his famous paper *Science for All*, contributed directly to the evolution of STS by forging links between science education and technology education, embedded in social contexts relevant for all students. Fensham (1995) has mentioned in the Mexican *Chemistry Education Journal* that in 1984 the Science Council of Canada reported on a 4-year study of school science in that country. The title was 'Science for Every Citizen'. A year later the Royal Society in London published a manifesto, 'Science for Everybody', as part of a larger report on the public understanding of science. In 1988, Australia's Curriculum Development Centre put out a national discussion document entitled 'Science for All', and in 1989 the American Association for the Advancement of Science summarized phase 1 of its Project 2061 under the title *Science for All Americans*. Finally, before the Mexican reform, UNESCO and ICASE had launched 'Project 2000+: Scientific and Technological Literacy for All' (ICASE 1993).

In the Mexican reform of junior high school, the diverse methodologies of each one of the sciences (biology, physics and chemistry) were acknowledged, and the curriculum changed from 'natural sciences' to 'biology', 'physics' and 'chemistry'.

69.3.2 Biology: The Teaching of Evolution

For natural sciences in elementary education, it was established that biology (its first three topics: living beings, human body and health, environment and environmental protection), from the third to the sixth grade, should be taught from an evolutionary perspective. Evolution itself became a subject in the sixth grade.[12]

> Diverse themes with an evolutionary focus were introduced in the beginning of the third grade. For example, there is a discussion of plants' capacity to nourish themselves and how this relates to the oxygen that we breathe today, which comes from photosynthesis of plants that existed thousands of years ago [...] Throughout the development of themes regarding the study of plants and animals, there are multiple references to the importance of adaptations that are a result of the evolution of the species [...] In the fourth grade, the study of evolution is reinforced when, among many examples, students learn about the role of human beings in changing ecosystems. In the fifth grade the subject of "cells, one-cell and multi-celled organisms" is introduced. Fifth-graders also learn about the first grand division between one-celled organisms with a nucleus and one-celled organisms without a nucleus or bacteria. (Barahona and Bonilla 2009, p. 16)

The sixth-grade programme extensively included evolution: the origins of the earth, the transformation of ecosystems (throughout time and due to continental drift), fossils, the extinction of species, geological eras, Darwin and his book *Voyage of the Beagle*, the concepts of natural selection and adaptation, among others. This resulted in a fundamental transformation of the curriculum and textbooks, as previous materials had discussed knowledge about the origin of species in a purely descriptive manner. This change constituted a great challenge for the design and elaboration of the new third-to-sixth-grade Mexican textbooks (Barahona and Bonilla 2009).

As Shortland and Warwick (1989) have shown, historical case studies draw attention to the failures and disappointments that often follow long years of work or to the communal effort that goes into the production of new scientific knowledge. This viewpoint was particularly crucial for the teaching of evolution in elementary and junior high schools. The inclusion in the six-grade programme of Darwin's Voyage of the Beagle is an example of how historical case studies can show not only on evolution and Darwinism but also on the teaching of the nature of science, the scientific method and the role of evidence in science.

In sum, the 1993 curriculum and textbooks were an important leap forward and indeed a great advance over other educational systems that still question the value of including the Darwinian theory in elementary school.[13]

---

[12] This was already a requirement in the 1970s but only as a junior high school subject among many. For example, the discussion was limited to the study of fossils as evidence of life in the past, with illustrations that showed the gradual evolution of horses as well as the differences between contemporary humans and their ancestors; the references to Darwin were minimal (Barahona and Bonilla 2009).

[13] According to the 1993 Reform, the federal authorities launched a new curriculum including these new perspectives in 1997 for teacher's colleges; 4 years later, in 2001, the first group of elementary school teachers graduated with this training. However, there has been no evaluation as to whether the training truly is enabling them to teach natural sciences with an evolutionary focus or, even more importantly, if the students manage to develop an evolutionary mindset.

69.3.3 Chemistry and Its Social Benefits

In junior high school there were three courses in which chemistry was involved: in the first year 'introduction to physics and chemistry', in the second 'chemistry I' and in the third 'chemistry II'. A thorough revision of the curriculum changes and the teachers' training effort needed for this reform is detailed in Chamizo, Sánchez and Hernández (2006). The major theme in chemistry I is the identification of the particulate nature of matter until its concretion in Bohr's atomic theory. The third course is centred on energy and environmental topics (Chamizo 1992).

The most important change in the chemistry curriculum of the 1993 Reform surely was the focus on the STS dimension. The main purpose of the two last chemistry courses is quoted as being one where 'pupils preserve the main elements of basic culture, to enrich their vision of Mexico and the world and assess social benefits that represent the contribution of this science, as well as the risk of its inappropriate utilization' (SEP 1993, p. 95).

The six units in which the courses chemistry I and II were divided had the following names:

Unit 1. You and chemistry
Unit 2. Matter: its manifestations. Mixtures: its separation. Compounds and chemical elements
Unit 3. The discontinuous nature of matter
Unit 4. Water, dissolutions and chemical reactions
Unit 5. Burning fuel. Oxidations
Unit 6. Electrochemistry

The necessity to include environmental education topics is emphasized often. The following can be mentioned as examples: acid rain, ozone and low atmosphere contamination; management of industrial residues; sulphur and nitrogen oxides produced by internal combustion machines; chlorofluoroalkanes; and the ozone hole in the stratosphere. And the STS focus insists in integrating the same critical stance on everyday chemical products such as acids like vinegar, lemon juice, gastric juice; bases like antacids or drain cleaner; colloids like gelatin, mousse, mayonnaise or egg white; hydrocarbons like gasoline, candle, gas cooker, asphalt; and gases solubility like soda and fish tanks.

The introduction of historical facts and biographies of scientists is welcome, because 'science is not a mystery, but a product of human activity...It is not about fulfilling an encyclopaedic commitment, but about giving science a vitality focus' (Chamizo and Garritz 1993, pp. 136–7).[14] A relevant point of this reform is that an ambitious updated programme accompanied it for teachers, which included readings from various issues of history and philosophy of chemistry and physics (Chamizo et al. 2006). An interesting impact of the 1993 Reform in chemistry was documented by applying a 'chemistrymeter' to a set of students just finishing its secondary studies (Tirado et al. 2001).

[14] A couple of more references on the philosophical bases of this reform can be found in Chamizo (1994, 2001).

69.3.4 School Physics and Philosophy of Science

Like chemistry, physics had its curricular presence in three junior high school years: introduction to physics and chemistry, physics I and physics II, taught in the first, second and third grades, respectively. Some aspects of modern philosophy of science are clearly presented among general aims of the subject, such as:

1. The students should think about the nature of scientific knowledge and how it is generated, developed and applied (SEP 1993, p. 77).
2. Formulations of an alleged scientific method, unique and invariable and formed of successive phases should be avoided in teaching. That version of the method is hardly adequate for the students and does not correspond to the real steps which scientists follow in carrying out their work. It is more valuable that students have a vision according to which scientific knowledge production from systematic and rigorous procedures and from intellectual flexibility derive in a capacity to plan adequate questions and search for unconventional explanations (SEP 1993, p. 78).
3. Physics should be presented as a product of human activity and not as an accidental result of work of a few exceptional persons. To this aim, it is convenient to propose examples of scientific developments motivated by challenges and problems which appear in social life and to stress concrete cases in which scientific advances are results of the accumulative work of many people, although they may have worked independently and in different places (SEP 1993, p. 78).

The inclusion of physics' history is rightly suggested as a way to exemplify the nature of science: 'It is convenient to study and discuss biographies of important persons in physics history, not as an encyclopedic recount, but stressing the forms of reasoning, inquiry, experimentation and error correction which leaded to some relevant discoveries and inventions' (SEP 1993, p. 78).

Although the importance of philosophy and history of science is clearly stressed among the general aims of the physics' curriculum, it is not explicitly materialized and specified at the content level. Only three obligatory topics have this historical and philosophical perspective:

Physical view of the world
Analysis of the Galileo Galilei's experiments and their relevance in scientific work
The ideas of Copernicus, Galileo, Kepler, Newton and Einstein

Without clear curricular indications about philosophical and historical themes, further developments of the intended curriculum were left to the textbook authors. Common models of curricular processes in science education fall into three levels (Robitaille et al. 1993; Valverde et al. 2002):

1. Intended curriculum (aims and goals)
2. Potentially implemented curriculum (textbooks and other organized resource materials) and factually implemented curriculum (teachers' classroom strategies, practice and activities)
3. Attained curriculum (students' knowledge, ideas, constructs and schemes)

At the level of the potentially implemented curriculum and in the absence of further guidelines, historical themes can have very distinct and arbitrary presentations. This was the case in 15 authorized physics' textbooks, written according to the 1993 curriculum reform, where three famous experiments by Galileo had diverse presentations. Regarding the Pisa Tower experiment, five authors did not mention it, five authors described it in a relatively acceptable way and five authors treated it completely wrong. Namely, these last authors present it as an experiment in which times and positions of a body in free fall were measured exactly. Obviously, the authors ignored that such measurements were technologically impossible in Galileo's time. Precisely due to this impossibility, Galileo designed and carried out his groundbreaking inclined plane experiment!

Eleven authors did not mention the thought experiment, one author treated it properly by using Galileo's account of it and three authors presented it as a real experiment.

The inclined plane experiment also had diverse presentations in physics' textbooks. Five authors omitted to mention it; only two authors gave it a satisfactory treatment, while eight authors presented that historically important experiment either wrongly or incompletely.

As all authorized textbooks passed an expert evaluation by the Mexican Ministry of Public Education, the authors are not the only ones to blame. It means that real content and meaning of historical episodes should be disseminated among textbook authors and reviewers (maybe via workshops organized by educational authorities), especially when such episodes form part of the intended national curriculum's objective in order to give students a reliable information about how science works. Furthermore, for an adequate curricular impact in Mexican classrooms, a pedagogical analysis and implementation strategies of such historical episodes should be included in professional programmes for in-service and prospective teachers.

69.3.5 The 2006 Reform

During the presidency of Vicente Fox Quesada (2000–2006), the Junior High School Reform (Reforma de la Escuela Secundaria, RES) was undertaken by the federal government in the National Programme of Education (Programa Nacional de Educación) 2001–2006. It established that the 'Mexican State must offer democratic, national, intercultural, secular and mandatory education that favours the development of the individual and his community, as well as a sense of belonging to a multicultural and multilingual nation, and the awareness of international solidarity of the educated' (SEP 2006).

In 2000, Mexico dedicated 100 dollars to each one of its elementary students. This amount that can be compared with the 600 dollars spent in the USA, the 130 USD used for Argentineans and the 220 USD spent by Chileans (Chamizo et al. 2006).

The designing group of this reform spent a lot of sessions deciding the order in which the three natural sciences should be presented. The decision was centred in a

Table 69.1 2006 Reform. Secondary science contents

Sciences I (emphasis in biology)
Unit I. Biodiversity: result of evolution
Unit II. Nutrition as the base for health and life
Unit III: Respiration and its relation with the environment and health
Unit IV. Reproduction and the continuity of life
Unit V. Health, environment and quality of life
Sciences II (emphasis in physics)
Unit I. The description of movement and force
Unit II. Laws of motion
Unit III: A model to describe the structure of matter
Unit IV. Internal structure of matter manifestations
Unit V. Knowledge, science and technology
Sciences III (emphasis in chemistry)
Unit I. The characteristics of materials
Unit II. Properties of materials and their chemical classification
Unit III: Materials transformation: chemical reaction
Unit IV. Formation of new materials
Unit V. Chemistry and technology

Project 2061 document (AAAS 2001) that recommended biology first, physics second and chemistry third. The natural sciences programmes were called sciences I, II and III, in accordance with the three grades in junior high school. In the first grade the students take sciences I (biology), in the second sciences II (physics) and in the third sciences III (chemistry). The scientific contents of the 3 years of education are represented by the titles of its units in Table 69.1.

At the end of each unit or at the end of the course, projects are developed by each student or groups of students as a good way to develop competencies because 'it favors integration and application of knowledge, skills and attitudes, giving the study a social and personal meaning' (SEP 2006).

Often, the projects select aspects related with the everyday life of students and their interests. Projects must favour attitudes as curiosity, creativity, innovation, informed scepticism and tolerance towards different ways of seeing the world. Each project requires the consideration of historical aspects as well as experimental work, and at the end students have to share their results. This objective was based on Stone and Tripp (1981), SATIS (1986) and Chamizo and Garritz (1993).

Some studies made a diagnosis of the scientific curriculum in basic education in Mexico prior to the 2006 Reform. For example, among the problems detected in the teaching of science, Flores and Barahona (2003) found a split between elementary and junior high schools; problems associated with the conception, development and decoupling of science and technology; the inadequate incorporation of the history of science in some subjects; little exploration of values and, finally, that science had not been inserted into the frame of culture.

It is necessary to emphasize that the teaching of science and technology was not marginalized, but played an important role in the focus of the curricula and new textbooks. This reform, despite requiring improvement in the future regarding the teaching of science and technology, promises to be a necessary step for the consolidation of a national science and technology programme and establishes graduation profiles related to science. Some of the most sensible decisions are mentioned below:

> The history of science employs a line of argument and reasoning to analyse situations, identify problems, formulate questions, pass judgment and propose diverse solutions. The teaching of science selects, analyses, evaluates and shares information from diverse sources and takes advantage of technological resources within reach to deepen and widen the learning of science in a permanent manner. It employs knowledge acquired with the purpose of interpreting and explaining social, economic, cultural and natural processes, as well as to make decisions and act, individually or collectively, to promote health and care for the environment as ways to improve the quality of life. (SEP 2006)

Also, the RES mentions the need to take advantage of information and communication technologies in general education, and particularly in scientific education, for this is a powerful tool in the socialization of knowledge and holds important pedagogical and didactic possibilities. The RES starts with a broader vision of technological education, understood as a social, cultural and historical process, which allows students to develop knowledge to solve problematic situations in an organized, responsible and informed manner, as well as to meet needs of a diverse nature. Technological education must contribute to the training of students as competent and critical users of the new technologies, in order to face the challenges of today's society.

It was established in this reform that scientific training is a goal for boosting cognitive development, strengthening individual and social values in teenagers as well as learning to reflect, exercising curiosity and using informed critique and scepticism, which will allow them to decide and, when necessary, act. A fundamental epistemological focus of the teaching of science relates to the understanding of science and technology as historical and socially constituted activities performed by men and women from different cultures.

The way in which different cultures in Mexico explain and construct knowledge about nature constitutes a practice that arrives nowadays through knowhow, folk knowledge and techniques in which different logics for building knowledge are mixed. From there, it is important to know, recognize and value such perspectives (SEP 2006). The history of science, according to this point of view, gained particular importance in the modification of the study programmes.

The RES expects that when students finish junior high school:

1. They have broadened their conception of science, of its processes and interactions with other areas of knowledge, as well as its social and environmental impact, and value in a critical manner their contributions for the betterment of the quality of life of people and the development of society.

2. They have advanced in the understanding of explanations and arguments of science about nature and use them to better understand the natural phenomena of their surroundings, as well as to place themselves within the scientific and technological development context of their time. This implies that students build, enrich or modify their first explanations and concepts, as well as develop abilities and aptitudes that provide them with elements to configure an interdisciplinary and integrated vision of scientific knowledge.
3. They can identify the characteristics and analyse the processes that separate living beings, relating them to their personal, family and social experience, to know more about themselves, their potential, their place among living beings and their responsibility in the way they interact with their surroundings, so they can participate in promoting health and the sustainable conservation of the environment.
4. They progressively develop knowledge that favours the understanding of concepts, processes, principles and the explanatory logic of science and its application to diverse common phenomena. They should go deeper into basic scientific ideas and concepts and establish relationships among them so they can build coherent explanations based on logical reasoning, symbolic language and graphic representations.
5. They have boosted their capacity to handle information, communication and social coexistence. This implies learning to value diverse ways of thinking, discern between founded arguments and false ideas and make responsible and informed decisions, at the same time as strengthening self-confidence and respect for themselves and for others (SEP 2006).

69.3.6 Biology: The Essence of Evolution

In the case of biology, evolution and genetics appear as central pillars in its teaching. For this reason the teaching of biology in junior high school starts with integrative theories such as evolution by natural selection, referring to biology as a scientific discipline from a historical perspective. Many references to Darwin's construction of the theory are taught in order to focus the attention of students on the historical and epistemological aspects of this discipline. Following the elementary school curriculum, the teaching of evolution is reinforced in junior high school. Regarding genetics, Mendel's laws are taught using his famous experiments with peas to show the manifold aspects of the experimental method in biology.

According to the RES in junior high school, as in elementary school, the scientific learning method must be encouraged, not as the scrupulous monitoring of a series of steps to be followed mechanically (observation, hypothesis, experimentation), but as a flexible and applicable method for the construction of knowledge over a whole course, not only in biology but also in other subjects such as physics, chemistry and geography.

In the 1993 Reform, the changes to the content of the educational programmes represented progress considering the epistemological and pedagogical references, but social aspects remained much diluted. For this reason the intercultural perspective was included in the RES, based on the idea that the diversity of forms in which human beings build knowledge about nature is of a cultural, social and historical order. In our country, cultural diversity has been the source of multiple ideas, explanations and interpretations, which have enriched, complemented and sometimes strained the development of scientific and technological knowledge. It is very important to recognize the diversity of ways to interpret the world and how, in some cases, these have aided scientific developments (like herbalism), or native technological development, which is beneficial to communities' relationship with the environment (SEP 2006).

69.3.7 School Physics and Philosophy of Science

In general terms, the 2006 curriculum framework is much better articulated than the 1993 version to move school physics activities closer to authentic science practices (Chinn and Malhortam 2002). Namely, it is planned that students gain basic scientific culture, in resonance with actual constructivist views on school science learning, through various (and even ambitious) learning tasks:

(a) Select and relate, in a causal and functional way, adequate variables to explain phenomena.
(b) Establish relationships between fundamental concepts which make it possible to construct coherent interpretative schemes in which logical reasoning, symbolic language and graphical representations are involved.
(c) Pose questions, elaborate hypothesis and inferences and construct explanations of some ordinary physical phenomena.
(d) Carry out experiments, get information from diverse sources, use different means to make measurements, analyse data and look for alternative solutions.
(e) Communicate, listen to and discuss ideas, arguments, inferences and conclusions related to physical concepts and their applications in scientific, technological and social contexts (SEP 2006, pp. 65–66).

Explicit curricular spaces and times for such activities are dedicated to develop projects which students are supposed to carry out at the end of each of five blocks.

As in the 1993 curriculum, philosophical aspects of science are not among explicit general aims. Nonetheless, the historical development of physics, the nature of scientific knowledge construction, the integration of science and relationships between science, technology and society are supposed to be taken into account, together with different students' comprehension levels, conceptual problems and previous ideas, as criterions for selection, organization and continuity of the course content (SEP 2006, p. 66)

Such intention is clearly visible in the general structure of the physics course parts, summarized in the Table 69.2.

At the content level, historical and philosophical aspects of physics have a more visible presence than in the 1993 curriculum. Besides the Galileo's

Table 69.2 Relationships between physics domains, representational means and thematic blocks

Physics domain	Elements for representations of physical phenomena	Thematic blocks
Study of motion	Descriptive schemes	Block I. Description of the changes in nature
Analysis of forces and changes	Relationships and sense of mechanism	Block II. The forces. Explanations of the changes
Particulate model	Images and abstract models	Block III. Interactions of matter. A model for description of unseen
Atomic constitution	Images and abstract models	Block IV. Manifestations of internal structure of matter
Universe, interaction of physics, technology and society	Integrated interpretations and relationships with environment	Block V. Knowledge, society and technology

contribution to science (SEP 2006, p. 76), students are supposed to know not only about motion laws but also about the role Newton had in the development of scientific thinking (SEP 2006, p. 86). In addition, the historical development of kinetic model and the atomic model of matter (SEP 2006, p. 102) are mandatory contents.

However, the main difference regarding the 1993 curriculum is the central place given to scientific models. Students are supposed to learn about the general role of models in the construction and verification of scientific knowledge. This intention is explicitly stated in the subthemes (what is the use of models, the models and the ideas they represent, the role of models in science) and curricular goals (the role of models in explanations of physical phenomena, as well as their advantages and limitations).

Nevertheless, the features of scientific models presented might be misleading for the expected learning results. It is said that students should 'recognize that a model is an imaginary and arbitrary representation of objects and processes which include rules of its function and is not the reality itself' (SEP 2006, p. 94). Strictly speaking, although theoretical models in physics are abstract representations and not copies of reality, they are not arbitrary because their predictions must be in concordance with observations.

Regarding textbook presentations of Galileo's work, the situation is similar as it was with the 1993 curriculum. Majority of authors treat the inclined experiment either inadequately or wrongly (Miguel Garzón and Slisko 2010).

69.3.8 Chemistry Presented as Projects and Models

Following the proposal expressed in the course of physics regarding the use of models, an important emphasis on the features of models in scientific explanation is made in chemistry (Gilbert and Boulter 1998, Chamizo 1988). Teachers generally ignore the issue, as exposed in educational research whose products have been books (Chamizo and García 2010) and articles on training experiences (Justi et al. 2011) and the reconceptualization of the subject (Chamizo 2011).

Nevertheless, this reform took into serious account Jensen's proposition of three Chemical Revolutions (1998) and explicitly mentioned them even in the programme (SEP 2006). The types of projects mentioned in the curriculum are scientific, technological and of citizenship. At the end of the units, the following projects must have been developed:

Unit I. The characteristics of materials
Projects related with separation methods to purify substances from mixtures. Or the work developed in a salt installation and its impact to the environment. Discussion, evidence research, information and communications technology (ICT) use, measurement, information analysis, interpretation of results and argumentation are to be fostered.

Unit II. Properties of materials and their chemical classification
The suggested projects point to the identification of elements of the human body, its health and environmental implications.

Unit III: Materials' transformation: chemical reactions
Projects related with soap production, energy release and absorption by human body are suggested.

Unit IV. Formation of new materials
In the framework of sustainability, the projects suggested have to do with avoiding corrosion or with fuel efficiency.

Unit V. Chemistry and technology
These technologic projects are developed to integrate the four previous units. The following topics are suggested: synthesis of an elastic material, Mexican contributions to the chemistry of fertilizers and pesticides, cosmetic products, Mesoamerican construction materials, chemistry and art and the importance and impact of petroleum products.

69.3.9 *The Recent Years*

During the Presidency of Felipe Calderón Hinojosa (2006–2012), the Mexican authorities launched a new reform in 2009 that has not yet concluded. It began in 2009 with curriculum changes to the first and sixth grades, in 2010 the changes affected the second and fifth grades, and finally in 2011 they included all the grades of elementary school. In July 2011, the SEP announced a new junior high school reform that intended to link all basic education levels (preschool, elementary and junior high school) and the production of new textbooks accordingly. Much of the progress made in previous reforms regarding the introduction of history and philosophy of science in science education was lost, particularly in the production of the most recent science textbooks.[15]

This new reform, called the Integral Reform for Basic Education (Reforma Integral de la Educación Básica, RIEB), intended to give continuity to the curricula and study programmes of all basic education. The organization of the subjects remained the same, although big changes were introduced in the natural sciences curricula (SEP 2011).

The national standards for science are the acquisition of scientific literacy, the use of scientific and technological literacy, development of skills associated with science and attitudes towards science. It does not mention the use of the history and philosophy of science in the teaching of sciences, and in the case of sciences I (biology), many topics about evolution are missing, but most importantly biology is not taught from an evolutionary perspective. The references to Darwin are very few, the Voyage of the Beagle is not mentioned, and fossils are seen as evidence of living beings in the past (not as relatives of present organisms). Although the topic of biodiversity is seen as the result of evolution, little is said about the processes that make up biological diversity and the evolutionary history of organisms. The teaching of biology in this reform is descriptive in comparison with the two previous ones.

69.4 Conclusion

We have tried to illustrate how science and technology are essential and at the same time constitutive parts of the modern society known as the knowledge society. Their importance demands, on one hand, reflection on the impact and scope of knowledge and, on the other, the modification of the educational agenda to make scientific and technological knowledge available to everyone. This strategy goes beyond the introduction of natural sciences as mandatory subjects; it implies a different focus on the selection, organization and sequencing of contents and the way to work with them.

It is decisive to collaborate in the change of the public perception of science and technology. In knowledge societies it is necessary that citizens have a positive attitude towards science and technology. This means that they have scientific and technological literacy that allows them from an early age to understand the potentials, benefits and risks of technoscientific products. This way, citizens, as well as local, municipal or federal officials, can make informed decisions before the problems technological changes produce in society take effect.

In this sense, the educational reforms in Mexico, the 1993 Reform and the RES, manifest the significance that the history and philosophy of science have had in the conception of teaching of science. Particularly, evolution and STS in Mexican education constituted an important advancement. It is worth saying that the

[15] These two reforms (2009 and 2011) are so recent that it is impossible for us to make an evaluation that provides a comparison with regard to the reforms referred to in this document.

introduction of the history and philosophy of science into the formal science curriculum in Mexico took the country some steps forward and some backward. For instance, the 1993 natural sciences programme for elementary education was more progressive regarding the history and philosophy of science than the 2006 programme for junior high education, and contrary to these advances, the 2009–2011 Reform lacked the teaching of science from a historical perspective and the evolutionary one regarding biology. This is to say that in the latest reform 2009–2011, the history and philosophy of science in relation with the teaching of biology is absent. We must wait for results to modify glitches and consolidate progress.

Acknowledgements This paper was supported by the projects Ciencia Básica 2012 SEP-CONACyT 178031 'La enseñanza de la evolución en el contexto de la historia y la filosofía de la ciencia en México'; SEP-CONACyT 49281 'La enseñanza de los modelos y el modelaje en la enseñanza de las ciencias naturales'; and DGAPA/UNAM, IN403513 'El tema de la evolución en los libro de texto de secundaria en México desde la historia y la filosofía de la ciencia, 1974–2012'. The authors also want to thank M. A. Alicia Villela González for her research assistance and the comments and suggestions of the four anonymous reviewers on an earlier version of this manuscript.

References

AAAS (2001). Atlas of Science Literacy, Project 2061, Washington D. C., USA: American Association for the Advancement of Science.

Aikenhead G. (2003). STS education. A rose by any other name, in Roger Cross (ed.), *A vision for science education. Responding to the work of Peter Fensham*. New York, USA: RoutledgeFalmer.

Ayala, F. J. (1977). Nothing in Biology Makes Sense Except in the Light of Evolution. Theodosius Dobzhansky 1900–1975. *Journal of Heredity,* 68, 3–10.

Ayala, F. J. (1994). *La Teoría de la Evolución* [The Theory of Evolution] Madrid: Ediciones Temas de Hoy.

Ayala, F. J. (1994b). On the Scientific Method, Its Practice and Pitfalls. *History and Philosophy of the Life Sciences,* 16, 205–240.

Barahona, A. & Bonilla, E. (2009). Teaching Evolution: Challenges for Mexican Primary Schools. *ReVista. Harvard Review of Latin America*, 3(3), 16–17.

Barraza, L. (2001). Environmental Education in Mexican Schools: the Primary Level. *The Journal of Environmental Education*, 32(3), 31–36.

Böhme, G. (1988). Copying with science. *Graduate Faculty Philosophy Journal*, 12, 1–47.

Bonilla E., Sánchez A., Rojano T. & Chamizo J.A. (1997a). Mexique. Démographie scolaire et réforme de l´enseignement, *Revue internationales d´education*, 14, 10–15.

Bonilla E., Rojano T., Sánchez A. & Chamizo J.A. (1997b) Curriculum scientifique et innovation, *Revue Internationale, d´education*, 14, 53–61.

Chamizo J. A. (1988). Proyectos de investigación como una alternativa a la enseñanza de la química en el bachillerato [Research Projects as an Alternative to High School Chemistry Teaching], *Contactos*, 3, 26–29.

Chamizo J. A. (1992). La química en secundaria, o por qué la enseñanza moderna de la química no es la enseñanza de la química moderna [Chemistry in Junior High School, or why Modern Teaching of Chemistry is not Modern Chemistry Teaching], *Información científica y tecnológica*, 14, 49–51.

Chamizo J. A & Garritz A. (1993). La enseñanza de la química en secundaria [Junior High School Teaching of Chemistry], *Educación Química*, 4(3), 134–139.

Chamizo J. A. (1994). Hacia una revolución en la educación científica [Towards a revolution in Science Education], *Ciencia*, 45, 67–79.

Chamizo J. A. (2001). El currículo oculto en la enseñanza de la química [The Hidden Curriculum in Chemistry Teaching], *Educación Química*, 12(4), 194–198.

Chamizo J. A. (2005). The teaching of Natural Sciences in Mexico: New Programs and textbooks for elementary School, *Science Education International*, 16, 271–279.

Chamizo, J. A, Sánchez, A. & Hernández, M. E. (2006). La enseñanza de la química en secundaria. El caso de México [Chemistry teaching in secondary school. The case of Mexico], en Quintanilla M. y Adúriz-Bravo A. (eds) *Enseñar ciencias en el nuevo milenio. Retos y propuestas* [Teaching sciences in the new millenium. Challenges and proposals], Santiago: Universidad Católica de Chile.

Chamizo J. A. (2007). La historia de la ciencia: un tema pendiente en la educación latinoamericana [History of science: a matter still pending in Latinoamerica's Education], in Quintanilla M. (ed). *Historia de la ciencia. Aportes para la formación del profesorado.* Santiago de Chile, Chile: Arrayan editores.

Chamizo J. A. & Garritz A. (2008). Reseña sobre la enseñanza escolar de la ciencia (1990–2006). El caso de México. Editorial [Review on Scholar Science Teaching (1990–2006). The Mexican case. Editorial], *Educación Química*, 19(3), 174–179.

Chamizo, J. A. & García, A. (Eds.). (2010). Modelos y modelaje en la enseñanza de las ciencias naturales. [Models and Modelling in Natural Sciences Teaching]México: Universidad Nacional Autónoma de México. Available in the URL http://www.modelosymodelajecientifico.com/

Chamizo, J. A. (2011 on line). A new definition of models and modelling in chemistry' teaching. *Science & Education*, published Online 31th October

Chinn, C. A. & Malhorta, B. A. (2002), Epistemologically Authentic Inquiry in Schools: A Theoretical Framework for Evaluating Tasks. *Science Education* 86, 175 – 218,

Fensham, P. J. (1985). Science for all: A reflective essay. *Journal of Curriculum Studies*, 17(4), 415–435.

Fensham, P. J. (1995). Science for all: Theory into Practice. *Educacion Quimica*, 6(1), 50–54.

Flores, F. & Barahona, A. (2003). El Currículo de Educación Básica: Contenidos y Prácticas Pedagógicas [Elementary and Junior High School Education Curriculum: Contents and Pedagogical Practices]. In G. Waldegg, Barahona, A. Macedo, B. & Sánchez, A. (Eds.), *Retos y Perspectivas de las Ciencias Naturales en la Escuela Secundaria* [Challenges and perspectives of Natural Sciences in Secondary School], (pp. 13–35). México: SEP/OREALC/UNESCO.

Gallagher, J. (1971). A broader base for science education. *Science Education*, 55, 329–338.

Garritz A. (1994) Ciencia–Tecnología–Sociedad. A diez años de iniciada la corriente [Science-Technology-Society. Ten years after this dimension began], *Educación Química*, **5**(4), 217–223.

Gilbert, D. (1997). Rewritting History: Salinas, Zedillo and the 1992 Textbook Controversy. *Mexican Studies/Estudios Mexicanos*, 13(2), 271–297.

Gilbert, J. & Boulter, C. (1998). Learning science through models and modeling. In B. Fraser & K. Tobin (Eds.), *The international handbook of science education*. Dordrecht: Kluwer.

Golinski, J. (1998). *Making Natural Knowledge. Constructivism and the History of Science.* Cambridge: Cambridge University Press.

Gooday, G., Lynch, J. M., Wilson, K. & Barsky, C. K. (2008). Does Science Education Need the History of Science? *Isis*, 99, 322–330.

Hacking, I. (1983). *Representing and Intervening. Introductory Topics in the Philosophy of Natural Science*. Cambridge: Cambridge University Press.

Hirschhorn, L. (1986). *Beyond Mechanization: work and technology in a postindustrial age.* Cambridge: MIT Press.

ICASE (1993) *Project 2000+: Scientific and Technological Literacy for All,* Paris, International Council of Associations for Science Education.
Jenkins, E. (1989). Why the History of Science? In M. Shortland & Warwick, A. (Eds.), *Teaching the History of Science* (pp. 19–29). Oxford: Basil Blackwell Ltd.
Jensen, W. B. (1998). Logic, History, and the Chemistry Textbook.I. Does Chemistry Have a Logical Structure? *Journal of Chemical Education*, 75(6), 679–687; II. Can We Unmuddle the Chemistry Textbook? 75(7), 817–828; III. One Chemical Revolution or Three? 75(8), 961–969.
Justi R., Chamizo J.A., Figueiredo C. & García A. (2011) Experiencia de formación de profesores de ciencias latinoamericanos en modelos y modelaje [Experience of Latino-American Science Teachers' Training in Models and Modelling], *Enseñanza de las Ciencias*, 29, 413–426.
Latapí, P. (2003). *El debate sobre los valores en la escuela Mexicana* [Discussion over values in the Mexican school]. México: Fondo de Cultura Económica.
Longino, H. (1990). *Science as Social Knowledge*. Princeton: Princeton University Press.
Martínez, S. (1993). Método, Evolución y Progreso en la Ciencia, 1ª. Parte [Method, Evolution and Progress in Science, 1st. Part]. *Crítica*, 25(73), 37–39.
Matthews, M. R. (1994/2014). *Science Teaching. The Role of History and Philosophy of Science*. New York: Routledge.
Miguel Garzón, I, & Slisko, J (2010). Uso de la historia en la enseñanza de la física en los libros de texto de Ciencias 2 para segundo de secundaria [Use of History in Physics Teaching through Science Textbooks for Junior High School Second Degree], *Latin American Journal of Physics Education*, 4, Supplement 1, 987 – 993.
Osborne, J. (2010). Arguing to Learn in Science: the Role of Collaborative, Critical Discourse. *Science*, 328, 463–466.
Robitaille, D. F., Schmidt, W. H., Raizen, S. A., McKnight, C. C., Britton, E. D. & Nicol, C. (1993). *Curriculum frameworks for mathematics and science* (Vol. TIMSS Monograph No.1). Vancouver: Pacific Educational Press.
Ruse, M. (1979). *The Darwinian Revolution. Science in Red and Tooth and Claw*. Chicago: The University of Chicago Press.
Ruse, M. (1996). *Monad to Man. The Concept of Progress in Evolutionary Biology.* Cambridge, Mass: Harvard University press.
SATIS (Science and Technology in Society, 1986). Herts, UK: Association for Science Education.
Secretaría de Educación Pública, SEP. (1993). Educación Básica. Secundaria. Plan y Programas de Estudio 1993. México, D. F.: Secretaría de Educación Pública.
Secretaría de Educación Pública, SEP. (2006). *Educación Básica. Secundaria. Plan de Estudios 2006* [Junior High School Curriculum 2006]. México: Dirección General de Desarrollo Curricular, Subsecretaría de Educación Básica de la Secretaría de Educación Pública. http://www.reformasecundaria.sep.gob.mx/doc/programas/2006/planestudios2006.pdf. Accessed June 2008.
Secretaría de Educación Pública, SEP. (2011). *Plan de Estudios 2011. Educación Básica*. [Basic Education School Curriculum 2011], México: Secretaría de Educación Pública.
Shortland, M. & Warwick, A. (1989). Introduction. In M. Shortland & Warwick, A. (Eds.), *Teaching the History of Science* (pp. 42–53). Oxford: Basil Blackwell Ltd.
Solomon, J. (1989). Teaching the History of Science: Is Nothing Sacred? In M. Sorthland & Warwick, A. (Eds.), *Teaching the History of Science* (pp. 1–16). Oxford: Basil Blackwell Ltd.
Stehr, N. (1994). *Knowledge Societies*. London: Sage.
Stone R.H. & Tripp D.W.H. (1981). Projects in Chemistry, London: Routledge & Kegan Paul.
Tirado F., Chamizo J. A., Rodríguez F. & Pérez A. (2001). La enseñanza de la química. Conocimientos, actitudes y perfiles [Chemistry teaching. Knowledge, attitudes and profiles], *Ciencia y Desarrollo*, 159, julio-agosto 2001, 59–71.
Valverde, G. A., Bianchi, L. J., Wolfe, R. G., Schmidt, W. H. & Houang, R. T. (2002). *According to the Book. Using TIMSS to investigate the translation of policy into practice through the world of textbooks*. Dordrecht: Kluwer Academic Publishers.

UNESCO (1999). *Declaration on science and the use of scientific knowledge*. http://www.unesco.org/science/wcs/eng/declaracion_e.htm. Accessed 18 March 2011.
Wilson, K. G. & Barsky, C. (1998). Applied Research and Development: Support for Continuing Improvement in Education. *Daedalus*, 127, 233–258.
Wuest Silva, T., Jiménez Silva, M. P., et al. (1997). *Formación, representación, ética y valores* [Training, representation, ethics and values]. México: Coordinación de Humanidades, Centro de Estudios sobre la Universidad, UNAM.
Ziman, J. (1980). *Teaching and Learning about Science and Society*, Cambridge, UK: Cambridge University Press.

Ana Barahona is full-time Professor at School of Sciences of the National Autonomous University of Mexico, UNAM. She followed undergraduate studies on biology and graduate studies on history and philosophy of biology at the UNAM. She made postdoctoral studies at the University of California, Irvine. In Mexico, as one of the pioneers in the historical and philosophical studies of science since 1980, she founded the area of Social Studies of Science and Technology at the School of Sciences of the UNAM. Some of her interests focus on the history and philosophy of evolution and genetics (especially in Mexico), and in the relationship between epistemology and science teaching, and science education. She has published more than 80 specialized articles and book chapters, several research books and textbooks on biology and history and philosophy of science for elementary and middle school (among which the Mexican free textbook of natural sciences edited by the Education Ministry have been published to date more than one hundred million copies) and also for college education. She has been President (2009–2011) of the International Society for the History, Philosophy and Social Studies of Biology and Associated Editor of the journal *History and Philosophy of the Life Sciences*; currently she is a member of the editorial board of *Biological Theory* and a member of the editorial committee of *Almagest, History of Scientific Ideas* and *Science & Education*. She has been technical advisor of the National Institute for the Assessment of Education in Mexico for the period 2002–2010. Currently she is a Council Member of the Division of History of Science and Technology of the International Union of History and Philosophy of Science and has been recently acknowledged as a Corresponding Member of the International Academy for the History of Science.

José Antonio Chamizo is full-time Professor at the School of Chemistry of the National Autonomous University of Mexico, UNAM. Since 1977 he has taught over 100 courses from high school to Ph.D. and published over one hundred refereed articles in chemistry, education, history, philosophy and popularization of science. His BS and MS degrees were in chemistry from the National University of Mexico and his Ph.D. from the School of Molecular Sciences at the University of Sussex. He is author or coauthor of over 30 chapters in books and of more than 50 textbooks and popularization of science books among which the Mexican free textbook of natural sciences, coordinated by him and edited by the Education Ministry which have been published to date more than one hundred million copies.

Andoni Garritz is full-time Professor at the School of Chemistry of the National Autonomous University of Mexico, UNAM, where he teaches and researches 'Didactics of Chemistry', 'Structure of Matter' and 'Science and Society'. He studied chemical engineering at UNAM and got his Ph.D. with a stay in the Quantum Chemistry Group at Uppsala University, Sweden, where he received the Hylleraas' Award. He has been giving lectures for 40 years at different levels, from high school to postgraduate. He is a consultant of UNESCO, coordinating seven chemistry experimental education projects in Latin America. His most relevant books are *Chemistry in Mexico: Yesterday, Today and Tomorrow* (UNAM, 1991); *From Tequixquitl to DNA* (Fondo de Cultura Económica, 1989); *Atomic Structure: A Chemical Approach*; and *You and Chemistry* and *University Chemistry* (Pearson Education 1986, 2001, 2005). He is the founder Director of the Journal *Educación Química* that now has 23 years of circulation, and it is indexed by Scopus.

Josip Slisko (B.Sc. in physics, M.Sc. in philosophy of physics, Ph.D. in philosophical sciences) teaches courses on physics and mathematics education at Facultad de Ciencias Físico Matemáticas of the Benemérita Universidad Autónoma de Puebla. His research interests include student explanatory and predictive models of physical phenomena, student strategies for solving untraditional physics and mathematics problems, presentation of knowledge in textbooks and design of learning sequences which promote cognitive development of students. Josip Slisko is author or coauthor of 80 journal articles and 12 physics textbooks. Since 1994 he joins the Sistema Nacional de Investigadores (National System of Researchers), a programme of Mexican Government through which scientific production of one is evaluated and a stipend is paid to recognized scientists. He is a member of American Association of Physics Teachers and serves in editorial boards of various educational journals (*Latin American Journal of Physics Education*, *European Journal of Physics Education*, *Metodicki ogledi* and *Eureka*). Since 1993, every last week in May, he is the president of the committee which organizes an international workshop called new trends in physics teaching.

Chapter 70
History and Philosophy of Science in Science Education, in Brazil

Roberto de Andrade Martins, Cibelle Celestino Silva, and Maria Elice Brzezinski Prestes

70.1 Introduction

This paper addresses the context of emergence, development, and current status of the use of history and philosophy of science in science education in Brazil. Its main scope is the application of this approach to teaching physics, chemistry, and biology at the secondary school level.

The first Brazilian researches and projects along this line appeared in the decade of 1970, although before that time it is possible to find scattered claims of the relevance of history and/or philosophy of science in science teaching. From the decade of 1980 onwards, the importance of this approach became widely accepted, and the subject became a common theme of educational dissertations and theses, appearing with a considerable frequency in papers presented at conferences on science education and published in educational journals. Since 1998, the use of history and

R.A. Martins (✉)
State University of Paraíba (UEPB), Campina Grande, Brazil

Group of History, Theory and Science Teaching (GHTC),
University of São Paulo (USP), São Paulo, Brazil
e-mail: roberto.andrade.martins@gmail.com; http://www.ghtc.usp.br

C.C. Silva
Institute of Physics of São Carlos (IFSC), University of São Paulo (USP), São Carlos, Brazil

Group of History, Theory and Science Teaching (GHTC),
University of São Paulo (USP), São Paulo, Brazil
e-mail: cibelle@ifsc.usp.br; http://www.ghtc.usp.br

M.E.B. Prestes
Institute of Biosciences, University of São Paulo, São Paulo, Brazil

Group of History, Theory and Science Teaching (GHTC),
University of São Paulo (USP), São Paulo, Brazil
e-mail: eprestes@ib.usp.br; http://www.ghtc.usp.br

M.R. Matthews (ed.), *International Handbook of Research in History, Philosophy and Science Teaching*, DOI 10.1007/978-94-007-7654-8_70,

philosophy of science was included among the government recommendations for secondary school science teaching in Brazil. Nowadays, this is an important line of research in graduate programs on science and mathematics education. However, the actual use of this approach in secondary education is still a desideratum.

Before entering in the main subject, this paper will present a short overview of the development of philosophy of science, history of science, and science education in Brazil, especially from 1960 onwards.

70.2 Philosophy of Science

The main development of philosophy of science in Brazil, in the twentieth century, began after the creation of the University of São Paulo (USP) in 1934. From its very inception, this university established a practice of bringing to Brazil foreign researchers to help starting new disciplines and research lines. In the case of philosophy, the main foreign professors were French: Jean Maugüe, from 1935 to 1943; Giles Gaston Granger, from 1947 to 1953; Martial Guéroult, from 1948 to 1950; Claude Lefort, from 1955 to 1959; and Gérard Lebrun, from 1960 to 1966 and from 1973 to 1980 (Hopos 2000). There were other strong influences, too. For instance, in 1942, Willard Van Orman Quine spent a few months in Brazil. However, the main influence was French, and Guéroult's approach to history of philosophy dominated the University of São Paulo for decades (Lefebvre 1990).

During his short stay at the University of São Paulo, Quine learned Portuguese and wrote in this language his book *O Sentido da Nova Lógica* (*The Meaning of the New Logic*), published in Brazil in 1944 (Quine 1944; Stein 2004, p. 376).[1] However, Granger was the main reference for philosophy of science at USP, for a long time. His first book was published in Brazil, in Portuguese, in 1955: *Lógica e Filosofia das Ciências* (*Logic and Philosophy of Science*) (Granger 1955).

Notwithstanding those precedents, philosophy of science would only begin to bloom in Brazil in the 1970s (Salmerón 1991). At the University of São Paulo, the main Brazilian professors who began to develop this line of research were Oswaldo Porchat Pereira da Silva and João Paulo Monteiro. Monteiro, a specialist in Hume (Monteiro 1967), was a strong influence in the development of philosophy of science at USP, attracting new philosophers to this field and creating in 1979 the journal *Ciência e Filosofia* (*Science and Philosophy*).

Oswaldo Porchat, who had studied with Victor Goldschmidt in France, wrote a Ph.D. thesis on Aristotle, but obtained a broad acquaintance with philosophy of

[1] Quine stayed in 1942 at the *Escola Livre de Sociologia e Política* (*Free School of Sociology and Politics*), which at the time was an "autonomous complementary institution" of the University of São Paulo. At that time, one of the few Brazilian scholars who could interact with him on equal grounds was the philosopher of logic Vicente Ferreira da Silva, who published an important book on logic in Brazil, *Elementos de Lógica Matemática* (*Elements of Mathematical Logic*) (Silva 1940), and later became a Heideggerian.

science (Silva 1967). Although his first connection was with French history of philosophy, he also had a postdoctoral stage at the University of California, Berkeley (1969–1970). In 1975 Porchat left the University of São Paulo to found the Philosophy Department at the State University of Campinas (Unicamp) and the Center for Logic and Philosophy of Science (CLE) at the same university. This center was a main influence in the development of philosophy of science, in Brazil, organizing meetings and publishing two journals: *Manuscrito* (*Manuscript*), since 1977, and *Cadernos de História e Filosofia da Ciência* (*History and Philosophy of Science Notepads*) since 1980. The Philosophy Department at Unicamp started the first Brazilian graduate program on logic and philosophy of science.

Leônidas Hegenberg, after graduating in mathematics, physics, and philosophy in Brazil, spent two years working with Alfred Tarski at the University of California (1960–1962). Returning to Brazil, he completed his Ph.D. in philosophy at the University of São Paulo, in 1968 (Hegenberg 1968). During most of his professional life, he taught mathematics, logic, and philosophy of science at the Technological Institute of Aeronautics (ITA). For that reason, he never supervised any M.Sc. or Ph.D. student. However, he published many papers and books and was highly influential in Brazil. Besides that, he kept up to date with the development of philosophy of science abroad and translated about 60 books to Portuguese, including works by Karl Popper, Paul Feyerabend, Max Weber, Wesley Salmon, Mario Bunge, Derek J. de Solla Price, Charles S. Peirce, and others.

In Rio de Janeiro, a strong tradition in philosophy of science began with the arrival of the Brazilian researchers Raul Ferreira Landim Filho and Oswaldo Chateaubriand Filho, in the 1970s. Chateaubriand obtained his Ph.D. in 1971, at the University of California, Berkeley, on ontology and semantics (Chateaubriand Filho 1971). After teaching at Cornell University from 1972 to 1977, he returned to Brazil, where he finally settled at the Catholic University of Rio de Janeiro (PUC-RJ). Landim, who obtained his Ph.D. at Louvain (Landim Filho 1974), also arrived to Rio at about the same time and was influential in the development of philosophy of science at the Federal University of Rio de Janeiro (UFRJ). Another important philosopher of science, Alberto Oscar Cupani, born in Argentina, obtained his Ph.D. at Córdoba in 1974, moving to Brazil in 1977 (Cupani 1974). He first worked at the Federal University of Santa Maria (UFSM) and then settled at the Federal University of Santa Catarina (UFSC).

Around 1980 the area of philosophy of science was well established in Brazil and had attained an international standard of research. There are currently 12 graduate programs in philosophy of science, several journals, and regular meetings over the country.

70.3 History of Science

History of science, in Brazil, developed later than philosophy of science. Up to 1970 there were few universities with regular courses on history of science (in general) or history of specific scientific disciplines. Those who taught history of science had no specific training in this discipline; they were commonly senior scientists who had a broad cultural background, such as Mario Schemberg and Francisco Magalhães Gomes (physics), Antonio Brito da Cunha (biology), Leopoldo Nachbin (mathematics), and Simão Mathias (chemistry).[2] Up to the 1970s, works on history of science written by Brazilian authors were, in general, descriptive and laudatory accounts of Brazilian researchers and institutions. One of the best productions of this period was a book organized by Fernando de Azevedo, *As ciências no Brasil* (*Sciences in Brazil*), published in 1956.[3]

Research on the history of Brazilian science gradually improved. In 1971 the graduate program of the History Department of the University of São Paulo (USP) established the first research line on history of science. The main focus was the study of Brazilian science, although the group has also produced research and supervised dissertations and theses on the conceptual history of international science. In 1979–1981 they published the three-volume work *História das Ciências no Brasil* (*History of Sciences in Brazil*), organized by Mário Guimarães Ferri and Shozo Motoyama (Ferri and Motoyama 1979–1981).

One stimulus to the study of history of Brazilian science was the expectation that it could help to develop scientific policies. In Rio de Janeiro, the sociologist Simon Schwartzman, who had obtained his Ph.D. in political science at the University of California, Berkeley (Schwartzman 1973), developed an ambitious project to study the Brazilian scientific community. He conducted a large series of interviews with leading Brazilian scientists and in 1979 published the analysis of this work in an influential book: *Formação da Comunidade Científica no Brasil* (*The Development of the Scientific Community in Brazil*) (Schwartzman 2007).

In 1982 the *Sociedad Latinoamericana de Historia de las Ciencias y la Tecnología* (SLHCT) (*Latin American Society of History of Science and Technology*) was founded in México, with the participation of Brazilian historians of science. In the next year, the group of the University of São Paulo created the *Sociedade Brasileira de História da Ciência* (SBHC) (*Brazilian Society of History of Science*) (Motoyama 1988; Bassalo 1992). This association soon began to organize biennial meetings (beginning in 1986) and in 1985 started the publication of the first Brazilian journal devoted to the history of science.

The largest research institution in history of science, in Brazil, is devoted to the study of Brazilian medicine and related subjects: *Casa de Oswaldo Cruz*, in Rio de Janeiro (founded in 1986). Another important institution is the *Museu de Astronomia e Ciências Afins*, MAST (*Museum of Astronomy and Related Sciences*), founded in the same year.

Historical researches on Brazilian science had no impact in science teaching, because the curriculum of the scientific disciplines does not include any topic related to the development of national science.

The Brazilian researches on the history of international science followed another line of development that is more difficult to track down. Diverging from the

[2] See, for instance, Mathias (1975), Gomes (1978), Nachbin (1996), and Schemberg (1984).

[3] See, for instance, Beltran (1984), Goldfarb (1994), and Vergara (2004).

situation that occurred in philosophy of science, the main stimulus was not the contact with foreign researchers. However, there have been some influential foreign scholars, such as Michel Paty, who spent several periods at the University of São Paulo and also supervised Brazilian students in France.

Around 1980, several scattered scholars who taught history of mathematics, physics, chemistry, and biology began to devote a larger effort to the study of history of science. At this time, there were no graduate courses in Brazil where one could obtain the adequate training for research in the history of international science. Although they had no specific training in this field, they began to produce better research by employing primary sources and to stimulate younger scientists to dedicate themselves to this field. Among other researchers of this generation, we can cite Ubiratan D'Ambrosio and Guilherme de La Penha (mathematics), Aécio Pereira Chagas and Carlos Alberto Lombardi Filgueiras (chemistry), and Roberto de Andrade Martins and Penha Maria Cardozo Dias (physics).[4] Two philosophers who began to devote themselves to the history of science in this period should also be mentioned: Pablo Mariconda and Carlos Arthur Ribeiro do Nascimento (Mariconda 2003; Nascimento 1995).

In 1985, a series of annual meetings on history of science was started at the State University of Campinas, and in 1990 a first attempt was made to establish a graduate program on history of science at the same university. However, internal problems suspended this development. Although dissertations and theses on history of science had been produced at several Brazilian institutions, since the decade of 1980, the first specific graduate course on history of science was created at the Catholic University of São Paulo (PUC-SP) in 1997, the second at the Federal University of Bahia (UFBA) in 2000 (with strong emphasis in science teaching), and the third one at the Federal University of Rio de Janeiro (UFRJ) in 2002. There is also a graduate program on the history of medicine, at Casa de Oswaldo Cruz, Rio de Janeiro, founded in 2001.

From the late 1980s onwards, the number of researchers in conceptual history of science who had received specific training gradually increased in Brazil. Among them, we may cite Olival Freire Jr. and Antonio Augusto Passos Videira (physics); Anna Carolina Regner, Lilian Al-Chueyr Pereira Martins, Gustavo Caponi, and Nelio Bizzo (biology); Ana Maria Alfonso-Goldfarb (chemistry); and Sérgio Nobre (mathematics).[5] Most of them received part of their training abroad. In the decade of 1990, research in history of conceptual science attained an international level, in Brazil.[6]

The strong development of the history of mathematics led to the creation of the *Sociedade Brasileira de História da Matemática* (SBHMat) (*Brazilian Society for the History of Mathematics*) in 1999. This society has its own journal and regular

[4] See, for instance, Chagas (2001), D'Ambrosio (1996, 2008), Dias (1994, 1999), Filgueiras (1994, 2002), La Penha (1982), La Penha et al. (1986), and Martins (1996, 1997).

[5] See, for instance, Alfonso-Goldfarb (1999), Bizzo (1991, 2004, 2009), Caponi (2010, 2011), Freire Jr. (1995, 2002), Martins (2005, 2007), Nobre (2001), Regner (1995, 2003), and Videira (1994).

[6] For more recent developments, see Krause and Videira (2011).

biennial meetings. In 2000 the *Associação de Filosofia e História da Ciência do Cone Sul* (AFHIC) (*South Cone Association for Philosophy and History of Science*) was created, bringing together scholars from Brazil, Argentina, Chile, and Uruguay. This society gave a new impetus to interchanges between historians and philosophers and stimulated the formation of thematic groups. The idea of creating AFHIC was initially discussed at the *First Meeting of Philosophy and History of Science of the South Cone*, which took place in 1998 at the Federal University of Rio Grande do Sul (UFRGS), Brazil, under the aegis of the Research Group on Philosophy and History of Science (GIFHC), from that university.

The development of research in history and philosophy of biology started with the first *Meeting of Philosophy and History of Biology* that occurred in 2003, in São Paulo, followed by a series of annual conferences. During the fourth meeting, in 2006, the *Associação Brasileira de Filosofia e História da Biologia* (ABFHiB) (*Brazilian Association for Philosophy and History of Biology*) was founded and in the same year began the publication of the journal *Filosofia e História da Biologia*. It is possible to notice a conspicuous interest on the use of history and philosophy of science in biology teaching in the meetings and publications of this society.

Although history of physics is a very strong research area in Brazil, no specific society for its study has been created, neither for the history of chemistry.

The largest part of the researches in history and philosophy of science developed in Brazil had no impact in science education. High-level researches in those fields, written in specialized jargon, published in professional journals or in foreign languages, are seldom read by Brazilian science educators. Besides that, philosophers and historians of science hardly ever write textbooks or popular works in this country.

70.4 Science Education in Brazil

This section will present an overview of the development of science education in Brazil, in the second half of the twentieth century. Although science education can be understood as including all levels from elementary school to graduate studies, the focus here will be the Brazilian equivalent to high school or secondary education, that is, the 3 last years of basic education, preceding higher education. We may confine our analysis to this level, since most researches on the use of history and philosophy of science in science teaching, in Brazil, deal with secondary education.

Before describing the historic qualitative changes in science education in Brazil, it is relevant to remark that in this country only a very small part of the population was able to attain secondary education in the first half of the twentieth century and that this proportion has been increasing up to the present. Around 1995, the Brazilian gross secondary school enrolment ratio reached about 50 %, being much worse than that of other South American countries, such as Argentina (76 %), Chile (73 %), and Uruguay (81 %) (Rigotto and Souza 2005). This unpleasant situation was one of the reasons for the educational reform announced by the Brazilian government in 1996, which proposed several policies for improving the enrolment ratio of students between 15 and 17 years old at secondary schools. The huge quantitative increase of the secondary school enrolment was achieved with a significant deterioration of material conditions and teaching and learning quality. Improving the overall quality of education is the current challenge for educational authorities in the country. Nevertheless, let us go back and review the development of science teaching in that country.

Until the Second World War, the main educational influence in Brazil was European (especially French). Textbooks were translated, laboratory equipment used in demonstrations was imported, and educational methods were copied. Secondary education was not compulsory and there were few public schools offering this level. In general, only people who intended pursuing higher education would enroll in high school. Access to the universities requires both the completion of secondary education and approval in competitive entrance examinations.

Shortly after the end of the war, several changes occurred. The American influence expanded very fast; there was a stronger concern with the scientific development of the country; and there arose the first attempts to develop national teaching projects. The prevalent view was that scientific development was a necessity for industrial and economical development of the country; and the government of President Getúlio Vargas was deeply concerned with those issues.

Two important institutions were created in 1951: the *National Research Council* (CNPq), to stimulate and support scientific researches, and the *Campaign for the Improvement of Personnel for Higher Education* (CAPES), belonging to the *Ministry of Education and Culture* (MEC), with the aim of improving the level of university professors by the creation of graduate courses and international exchange.[7]

In the early 1950s, under the leadership of Isaías Raw, the recently founded *Brazilian Institute of Education, Science and Culture* (IBECC) produced the first laboratory kits developed in this country. This was a nice innovation in education in Brazil, because it introduced low-cost equipment that could be used by students (not for class demonstrations, as the former imported equipments) and had a strong positive influence in science teaching. In the late 1960s, the industrial dimensions of the production of laboratory equipment led to the creation of the *Fundação Brasileira para o Desenvolvimento de Ensino de Ciências* (FUNBEC) (*Brazilian Foundation for the Development of Science Teaching*) to cope with the large-scale production of teaching materials (Villani et al. 2009; Nardi 2005).

In 1965, six centers of science for teaching training and development of educational materials were created in Brazil: in Pernambuco (CESINE), Rio Grande do Sul (CECIRS), Minas Gerais (CECIMIG), Rio de Janeiro (CECIGUA), São Paulo (CESISP), and Bahia (CECIBA). The leaders of those centers were trained at IBECC, in 1966 (Nardi 2005). Most of these initiatives, however, were not maintained to the present day.

[7] The names of these institutions were later changed to *National Council for Scientific and Technological Development* and *Coordination of Improvement of Personnel of Higher Education*, respectively, although their acronyms were maintained.

Nowadays many critics point out that the empiricist view behind those projects was naïve and inadequate – and that is a correct appraisal (Villani et al. 2009; Nardi 2005). However, positive features cannot be denied. Brazilian science educators had been deeply influenced by John Dewey's work, especially his book *How We Think* (1910). This book was first translated into Portuguese in 1933 and republished in 1953 and 1959. Following Dewey's ideas, science educators were striving to provide a more active involvement of students with science, attempting to develop their reasoning capacity and critical attitude (Freire Jr. 2002).

Notice that, parallel to any innovatory trends, there was a conservative undercurrent in Brazilian education. Essentially, in the 1960s as now, science and mathematics teaching in secondary schools was grounded upon books written with a very simple aim: to train the students to obtain a good performance at the universities' entrance examinations.[8]

In 1961, an educational reform increased the weight of scientific disciplines in both elementary and secondary education. In 1964 a military *coup d'état* and the beginning of a long dictatorship in Brazil (1964–1985) led to deep changes in the educational system, but the previous impetus to improve science education was maintained. The educational influence of the United States increased, and an agreement established in 1965 between the Brazilian *Ministry of Education and Culture* (MEC) and the *United States Agency for International Development* (USAID) led to introduction of many North American educational materials in Brazil (Nardi 2005).

It is well known that in the late 1950s, due to the Cold War between the United States and the Soviet Union, five outstanding American educational projects were developed to improve the teaching of mathematics, physics, chemistry, and biology at high school level: *Physical Science Study Committee* (PSSC), *Biological Science Curriculum Study* (BSCS), *Chemical Bond Approach* (CBA), *Chemical Education Material Study* (CHEMS), and *School Mathematics Study Group* (SMSG). Those projects were introduced in Brazil in the decade of 1960 and they had a strong impact. The textbooks were translated and the experimental kits were reproduced, with small adaptations, by *Instituto Brasileiro de Educação, Ciência e Cultura* (IBECC) (*Brazilian Institute for Education, Science and Culture*). In the United States, about 200,000 students used the PSSC and CHEMS materials, 600,000 used the BSCS texts, and 1,350,000 students used SMSG books. In Brazil, about 400,000 copies of PSSC volumes were published and a similar number of copies of BSCS (Barra and Lorenz 1986, *apud* Nardi 2005). Other foreign products, such as the Nuffield Foundation project, were also translated and used in Brazil (Krasilchik 1992; Villani et al. 2012).

[8] Universities are free to create any kind of entrance examination. The most traditional one is called "vestibular" and assesses the student's knowledge on the subjects studied in the secondary school. Except the last decade exams conducted by University of Campinas (UNICAMP), "vestibular" in general has a strong inertia regarding the style and content. In recent years, the performance at *Exame Nacional do Ensino Médio* (ENEM) (*High School National Exam*), designed to assess scientific contents and other competencies as reading and comprehension, has also been used as entrance examinations by over 300 institutions. This is a nonmandatory exam, attended by 5.8 million students in 2012.

In the American projects, the use of the "scientific method" was emphasized, with a strong empiricist bias. Those approaches were typical in science education during the 1960s and were still influential in the 1970s, in Brazil.

The reception of the American projects among secondary school teachers was not altogether positive. They had difficulties in dealing with the new methods and contents (Villani et al. 2009). In January 1970 the *Sociedade Brasileira de Física* (SBF) (*Brazilian Physics Society*) sponsored the *First National Symposium on Physics Teaching*. The PSSC project was much criticized, and the participants of the event reached the conclusion that it was necessary to develop Brazilian projects to elaborate new textbooks and laboratory materials. The first initiatives were already on the move, at the University of São Paulo (USP) under the leadership of Ernest Hamburger and at the Federal University of Rio Grande do Sul (UFRGS), by initiative of Marco Antonio Moreira (Oliveira and Dias 1970). Educators such as Pierre Henri Lucie, who was one of the main supporters of the introduction of PSSC in Brazil, were already looking for alternatives. In 1969 Lucie started the publication of an original line of physics textbooks for secondary schools, with a deeper conceptual discussion, using cartoons: *Física com Martins e eu* (*Physics with Martins and I*) (Lucie 1969).

70.5 Brazilian Research and Projects in Science Education

There were new educational reforms in Brazil in 1968 (higher education) and 1971 (elementary school and secondary education). In 1972 the *Ministry of Education and Culture* (MEC) started the *Project of Expansion and Improvement of Education* (PREMEN) to promote an enhancement of education with the development of teaching materials for science and mathematics in the country and adapted to the national context, also providing adequate teaching training for the use of those materials. In the early 1970s, stimulated by government guidelines and financial resources, the formation of research groups and the development of science teaching projects started up in Brazil (Barra and Lorenz 1986).

At that time, three physics teaching projects were under way: *Physics Teaching Project* (PEF) at the University of São Paulo (coordinated by Ernst W. Hamburger and Giorgio Moscati), *Self-Instructive Physics* (FAI) by the *Group of Studies in Physics Teaching Technology* (Fuad Daher Saad, Kazuo Watanabe, Paulo Yamamura, and others), and *Brazilian Project for Physics Teaching* (PBEF) at FUNBEC (organized by Rodolpho Caniato, Antonio Teixeira Jr., José Goldenberg). The three projects were student centered, without expositive classes. The first project had a stronger experimental emphasis which generated a difficulty for its application at secondary schools that could not acquire laboratory equipment. In the two other projects, experiments were secondary activities to illustrate knowledge that had already been learned using self-instructive techniques. The FAI project included passages describing the history of physics, written by Shozo Motoyama. Its books sold 490,000 copies, between 1973 and 1976 (Flores et al. 2009).

In the early 1970s, educational projects on chemistry and biology were also developed in Brazil: *The National Project for the Teaching of Chemistry* (1972), developed by the *Northeast Coordination of Science Teaching* (CECINE), and the Project of Biology Applied to Secondary School (1976), an initiative of the *São Paulo State Center for Science Teaching* (CESISP) (Barra and Lorenz 1986, *apud* Nardi 2005). The most relevant educational initiative in mathematics was developed under the guidance of Ubiratan D'Ambrosio at the State University of Campinas (Unicamp): *New Materials for the Teaching of Mathematics*, for the fundamental school level.

In 1972 the *Brazilian Foundation for Science Teaching* (FUNBEC) and the publisher *Abril Cultural* launched a new project called *Os Cientistas* (*The Scientists*): a series of 50 experimental kits for the study of chemistry, biology, and physics, which were sold in newsstands. Each edition contained an experimental kit, instructions for performing the experiments, and the biography of a famous scientist related to the experiment (Newton, Pasteur, Lavoisier, etc.). This initiative was planned by two professors of University of São Paulo, Isaías Raw and Myriam Krasilchik, and was coordinated by the latter. The project was highly successful and sold three million units (a mean of 60,000 copies of each edition). It was later translated into Spanish, English, and Turkish and sold in other countries (Krasilchik 1990). Although the project included material related to the history of science (the biographies), this was circumstantial: the authors of the biographies and of the experimental kits had no interaction,[9] and the emphasis of the project was an empiricist approach to science.

The development of the area of science and mathematics education along the last decades in Brazil can be noticed in the beginning of regular conferences, founding of societies, creation of journals devoted to school teachers and researchers, and establishment of graduate courses.

The first regular series of congresses devoted to physics education, called *Simpósio Nacional de Ensino de Física* (SNEF) (*National Symposium on Physics Teaching*), started in 1970, one decade before the creation of general congresses devoted to science education in general or to the teaching of the other scientific disciplines such as *Encontro Nacional de Ensino de Química*, ENEQ (*National Meeting on Chemistry Teaching*) (1982); *Encontro Perspectivas do Ensino de Biologia*, EPEB (*Meeting on Perspectives of Biology Teaching*) (1984), discontinued; *Encontro de Pesquisadores de Ensino de Física*, EPEF (*Meeting of Researchers of Physics Teaching*) (1986); *Encontro Nacional de Educação Matemática*, ENEM (*National Meeting on Mathematics Teaching*) (1987); *Encontro Nacional de Pesquisa em Ensino de Ciências*, ENPEC (*National Meeting on Research in Science Teaching*) (1997); *Colóquio de História e Tecnologia no Ensino de Matemática*, HTEM (*Conference of Technology and History of Mathematics Teaching*) (2002); and *Encontro Nacional de Ensino de Biologia*, ENEBIO (*National Meeting of*

[9] Roberto de Andrade Martins, one of the authors of this paper, can safely state that the two sides of the project were widely independent, because of his personal involvement with the project: he was the author of some of the biographies of *Os Cientistas*.

Biology Teaching) (2005). Most of these are biennial meetings. There are also many regional and local relevant conferences. Besides that, for a long time other general scientific and educational conferences have also included sessions on science and mathematics teaching.

Since 1976, when *Boletim Gepem* devoted to mathematics education was created, many other Brazilian journals on general and specific areas of science education appeared along the years, for instance, *Revista de Ensino de Física, Caderno Catarinense de Ensino de Física, Química Nova na Escola, Investigações em Ensino de Ciências, Ciência & Educação, Revista Brasileira de Pesquisa em Educação em Ciências, Revista de Ensino de Biologia,* and *Alexandria: Revista de Educação em Ciência e Tecnologia,* among many others.

Several scientific societies began to sponsor educational activities, and later on, specific associations were created. The *Sociedade Brasileira de Física* (SBF) (*Brazilian Physical Society*) established a Commission for Physics Teaching in 1970. In 1988 the *Sociedade Brasileira de Educação Matemática* (SBEM) (*Brazilian Society of Mathematical Education*) was founded. In 1997 the *Associação Brasileira de Pesquisa em Educação em Ciências* (ABRAPEC) (*Brazilian Association for Research in Science Education*) and the *Associação Brasileira de Ensino de Biologia* (SBEnBio) (*Brazilian Association of Biology Teaching*) were founded. There are no specific societies related to the teaching of chemistry and physics. In both cases, there are teaching divisions belonging to the corresponding national scientific societies.

Although there were educational initiatives in the several disciplines, physics took the leadership in the establishment of a definite enterprise for the improvement of science teaching. In 1967 the graduate program in physics of the Federal University of Rio Grande do Sul (UFRGS) established the area of physics teaching. The first specific graduate course in science teaching was created in Brazil in 1973, at the University of São Paulo (USP), as a joint initiative of the Physics Institute and the Faculty of Education; about 20 years later, the areas of chemistry and biology were also introduced. There was also an attempt to establish the area of physics teaching in the graduate program created at the Catholic University of Rio de Janeiro (PUC-RJ) in 1967, but it did not succeed. In 1975, the first graduate course in mathematical education started at the Catholic University of São Paulo (PUC-SP) and the second one in 1984, at the Rio Claro campus of the São Paulo State University (UNESP). The latter was the first one to offer a Doctor degree, in 1993. Other graduate courses in science education started at the Federal Rural University of Pernambuco (1995), Federal University of Rio de Janeiro (1995), and at the Bauru campus of UNESP (1997) (Moreira 2004). From 1972 to 1995, 572 theses and dissertations had been finished on science and mathematics education, including those that have been produced at other graduate programs. There was a very fast increase of graduate programs on science and mathematics education from 2000 to 2010, reaching a total of 78 programs in the area at the end of 2010 (CAPES 2010).

In 1999 the first Brazilian graduate program devoted to history and philosophy of science and science education was founded in Bahia State: the *Graduate Program in Teaching, Philosophy and History of Sciences* (EFHC). It is an interinstitutional program between Federal University of Bahia and State University of Feira de Santana, with master and doctorate courses (Freire Jr. and Tenório 2001).

70.6 History and Philosophy of Science in Science Education

According to Susana de Souza Barros, it is possible to find several uses of history and philosophy of science in different periods, in Brazil. In the decade of 1970, history of physics was regarded as an important component of the teaching training. The transposition of this historical knowledge to physics teaching at the secondary school level was not discussed, however. During the 1980s, the study of concept formation and conceptual change led to the combined use of psychological, epistemological, historical, and sociological approaches. There were researches using classroom experimentation, and the new line of attack was regarded as useful for the training of physics teachers. Nevertheless, it was not directly applied to introduce educational changes at the secondary school level. During the next decade (1990), there was a strong emphasis on the relation of science teaching and the education for citizenship, using the science, technology, and society approach (STS). The idea that science educators should teach not only science but also about science (i.e., the inclusion of a metascientific level) launched a new use for history and philosophy of science in physics teaching (Martins 1990). Towards the end of the last century, the Ministry of Education established new educational guidelines recommending the use of history and philosophy of science at the secondary school level (Barros 2002).

70.6.1 The Beginning: Physics

It was already remarked that physics was the first area where science education research began, in Brazil. Let us see how the use of history and philosophy of science in physics teaching started and was understood in this discipline, in the early period. This occurred at the University of São Paulo (USP), and therefore our focus, here, will be that institution.[10]

History of physics was taught from the very beginning of the establishment of its undergraduate physics course, at USP. One of the most influential early professors who taught this discipline was Mário Schenberg (1914–1990), a Brazilian physicist with strong political interests. He was a member of the Brazilian Communist Party before this political organization was banned from the Brazilian politics, and he was

[10] The other early group, at the Federal University of Rio Grande do Sul (UFRGS), did not develop this line of research at first, and its leader complained in 1970 that there was no one available at that institution to teach history of physics (Oliveira and Dias 1970, p. 106).

twice elected deputy of the State of São Paulo (1946 and 1962). He was arrested twice for his political involvement. He was highly influential, and his interest for Marxism and history of science was shared by many other physicists.

Several books on history of science with a Marxist outlook were well known in Brazil, around 1970 (Azevedo and Costa Neto 2010). Friedrich Engels' *Dialectics of Nature* had been translated into Portuguese in 1946, with John Burdon Sanderson Haldane's introduction, and it was republished in 1962 and 1964. It was a very popular book among physicists, at USP, in the 1960s and 1970s. John Desmond Bernal's works were also very influential. Bernal's *Science in History* (1954) was translated in México, in 1959, and in Portugal, in 1965. Both translations, as well as the original English version, were familiar to many Brazilian physicists. Boris Hessen's "The Socio-economic Roots of Newton's Principia" was also well known and highly praised. The Marxist play writer Bertolt Brecht's version of Galileo's life was very popular in Brazil, and parallels were drawn between his struggle with the Catholic Church and the scientists' resistance to the Brazilian military government of that period. The play was enacted in São Paulo, in 1968, with the priests using olive dresses – olive being the color used by soldiers, in Brazil. At that time, the study of the relations between science, history, politics, society, etc. was regarded as a means to denounce the alleged neutrality of science, leading the students to have a more critical view of the scientific endeavor, and also critical of the Brazilian political situation of the time. Students were regarded as citizens that should be educated to deal, among other things, with the political and economic forces surrounding them.

This was a very influential trend, in the early development of the use of history (and sociology) of science in science education in Brazil (Villani et al. 2010). Even now, long after Marxism had become outmoded, its inspiration is still present as an undercurrent in the *science, technology, and society* (STS) approach, in Brazil, although the recent generation is not aware of its early history.

Besides this politically motivated interest in the history of science, there was another view about its educational value. In 1970 the *Harvard Project Physics* was being introduced in Brazil, and its strong use of history of science was described by Giorgio Moscati, who emphasized the motivational aspect of the historical and humanistic approach (Oliveira and Dias 1970). There was a widespread belief that mere contact with history of science could enhance the motivation of students and also to improve their learning of scientific concepts.

This was also, seemingly, the opinion of Pierre Lucie, who taught physics at the Catholic University of Rio de Janeiro. As mentioned above, Lucie had been a strong supporter of the PSSC project in Brazil. However, in 1970 he was devoting himself to other educational projects. In that year he delivered a course on the history of mechanics during the first *National Symposium on Physics Teaching* that occurred in São Paulo. He published a book on this subject in 1978, called *Gênese do Método Científico* (*Genesis of Scientific Method*). This work was not an adaptation of history of physics for science teaching; it was just a plain work on history of science, written by an outstanding educator (Lucie 1978). Several papers published in the early volumes of *Revista de Ensino de Física* (*Journal of Physics

Teaching), such as those authored by José Maria Filardo Bassalo, were also devoted to the presentation of historical information, without any special application to physics teaching.

Although there was a widespread interest in the history of science at the USP group from its very beginning, its effective influence only began to produce noticeable results in the late 1970s. Let us notice some instances of works produced by the group. In 1978 Amélia Império Hamburger produced a study of physics textbooks, including "the concept of physics and science" among the several features that should be analyzed. In the same year, she wrote a historic and philosophical analysis of mechanics and electricity to help circumventing conceptual learning difficulties. In 1979, João Zanetic wrote a work on the role of history of physics in education. Ernest Hamburger and Joaquim Nestor de Morais presented a historical analysis of the concept of electrostatic potential, comparing it with its textbook presentation. Amélia Hamburger proposed a project using historical examples for teaching physical concepts. In 1980 Alberto Villani started a research on the history of the theory of special relativity. In 1981 Amélia Hamburger began the development of a series of didactic booklets, and one of them contained texts on "science, technology, and society." Zanetic and José D. T. Vasconcellos proposed the introduction of the Popper-Kuhn debate in physics teaching (Gama and Hamburger 1987).

From 1979 onwards, the USP group invited several foreign visitors who delivered courses on history and philosophy of physics: Marcelo Cini (in 1979 and 1980), William Shea, in 1979, and Michel Paty, in 1982, (Gama and Hamburger 1987; Robilotta et al. 1981).

In 1981 Alberto Villani published the first paper, in Brazil, that referred to the so-called spontaneous concepts, mentioning the recently published works (1979) of Laurence Viennot and John William Warren on this subject. Viennot was invited to Brazil and delivered a course on this subject at USP, in 1981. In the following years, this became a very strong line of research of the USP group, involving several researchers such as Villani, Jesuina Lopes de Almeida Pacca, Anna Maria Pessoa de Carvalho, and Yassuko Hosoune, together with graduate students. Three M.Sc. dissertations on this subject were finished between 1982 and 1985 (Gama and Hamburger 1987).

Independently of the USP group, Arden Zylbersztajn, who was a professor at the Federal University of Rio Grande do Norte (UFRN) and had strong interest in history and philosophy of science, started in 1979 his doctoral studies in the University of Surrey, under the guidance of John Gilbert, and began the study of physical "spontaneous concepts," publishing his first paper on this subject (with Michael Watts) in 1981 (Gilbert and Zylbersztajn 1985; Watts and Zylbersztajn 1981). After his return to Brazil, in 1984, Zylbersztajn continued to develop this research line at UFRN and, after 1987, at the Federal University of Santa Catarina (UFSC).

The comparison between the students' concepts and the historical evolution of science became one of the main uses of history of science in physics education, in Brazil, for a significant period. This trend was soon linked to the work of Piaget and Garcia (1982) on the parallels between psychogenesis and history of science.[11]

---

[11] There were some critics of this kind of parallelism, for instance, Franco and Colinvaux 1992.

The study of the students' previous concepts and of the strategies to produce conceptual change became more and more sophisticated during the decades of 1980 and 1990, with the development of analogies between science education and the ideas of Feyerabend, Laudan, Bachelard, and other philosophers of science.

In the 1990s, educational experiments developed by Anna Maria Pessoa de Carvalho and Ruth Schmitz de Castro introduced historical texts in secondary school classrooms, to explore the similarity between the student's concepts and the ideas presented in those texts (Castro and Carvalho 1995). Many students were stimulated by noticing the similarity between their own concepts and those of important scientists. The discussion of historical texts also helped in producing a conceptual change in the students. The texts and the description of their use were later incorporated in teachers training courses, showing a specific useful application of history of science in science teaching.

There were other different trends. Towards the end of the 1970s, another group of the University of São Paulo, including Luis Carlos Menezes, João Zanetic, and the graduate students Demétrio Delizoikov Neto and José André P. Angotti, endeavored to apply the educational ideas of Paulo Freire (Freire 1970) to science teaching, linking some of his ideas to Thomas Kuhn's views. Delizoikov, Angotti, and other educators put to practice this proposal during a stay in Guinea-Bissau, one of the countries where Freire had worked during his exile from Brazil, after the 1964 military *coup d'état* (Delizoikov Neto et al. 1980). The dissertations of Delizoikov and Angotti, supervised by Menezes, were completed in 1982 (Delizoikov Neto 1982; Angotti 1982). Although Kuhn's ideas had been a starting motivation, the stronger emphasis of those works is neither historical nor philosophical.

Menezes also supervised Alexandre José Gonçalves de Medeiros, who finished in 1984 his M.Sc. dissertation on the sociocultural and economic influences that acted upon the development of physics up to the end of the seventeenth century. As pointed out above, this line of social history of science had been strongly influenced by Marxist authors.

Although several researchers of USP involved with history and philosophy of science participated in projects that produced educational materials for secondary schools, those projects did not include the use of history and philosophy of science.

Since that time, the uses of history, philosophy, and sociology of science in teaching are a thematic area in graduate programs and conferences on physics and science teaching. There are several research groups devoted to this topic exploring different approaches and methodologies. Along the last years, several books and papers were published on this topic, attesting that history and philosophy of physics is embedded within physics education in Brazil since its very beginning. One strong trend is that several authors hold that the study of historical episodes of science can help students to form a more accurate view of the nature of science and to learn about scientific concepts.[12]

[12] Assis (2008), Batista (2004), Braga et al. (2012), Carvalho and Vannucchi (2000), Forato et al. (2012), Greca and Freire Jr. (2003), Martins and Silva (2001), Pagliarin and Silva (2007), Rosa and Martins (2009), Moura (2012), Silva (2006), Silveira et al. (2010), Silveira and Peduzzi (2006), and Teixeira et al. (2012), Silva and Moura (2012).

70.6.2 Chemistry

The development of a research line in chemistry teaching, in Brazil, had its beginning at the University of São Paulo (USP). When this university was created, in 1934, a German chemist called Heinrich Rheinboldt (1881–1955) was invited to begin the chemistry department. Rheinboldt had a strong interest in the education of teachers and also in history of chemistry, having published in 1917 a study about Johann Baptist van Helmont. He was very influential in stimulating the study of history of chemistry and chemical education (Schneltzler 2002). Under his inspiration, Simão Mathias devoted himself to the history of chemistry, especially after retiring from the Chemistry Institute, in 1972, when he became a professor at the History Department and was responsible for the discipline of history of chemistry.

In the decade of 1960, Ernesto Giesbrecht, of USP, became involved with the translation and adaptation of the North American projects of chemistry (CBA, CHEMS). In the next decade, a group began to form at the Chemistry Institute of USP, under the leadership of Luiz Roberto de Moraes Pitombo and Maria Eunice Ribeiro Marcondes, devoted to the formation of chemistry teachers. José Atílio Vanin also began the development of activities of popularization of chemistry, with the help of students. Those activities finally led to the creation of the Group of Research in Chemical Education (GEPEQ), which is very active. In the decades of 1980–1990, the approach of the group was contributing to chemistry teaching at the secondary school level, with an emphasis in experimentation, relations between chemistry and everyday life, and the use of cognitivist proposals (Ausubel, Piaget).

Although there were some early activities related to chemistry education, such as those described above, the expansion of the area is strongly linked to the creation of the *Sociedade Brasileira de Química* (SBQ) (*Brazilian Chemical Society*) in 1978. During the first meeting of this society, there was a session devoted to the discussion of chemistry teaching, and interest in this line increased in the following years. The journal of this society *Química Nova* (*New Chemistry*) started the publication of papers on chemistry teaching in 1980.

At the same time, at the south edge of the country (State of Rio Grande do Sul), a series of regional annual meeting started in 1980: the *Encontro de Debates sobre Ensino de Química* (EDEQ) (*Meeting of Debates on Chemistry Teaching*), organized by Attico Chassot. In 1988 the *Brazilian Chemical Society* founded its Teaching Division and in 1982 began the series of biennial conferences called *Encontro Nacional de Ensino de Química* (ENEQ) (*National Meetings on Chemical Teaching*). In 1995 the *Brazilian Chemical Society* founded a new journal: *Química Nova na Escola* (*New Chemistry at School*), the main target of this publication being secondary school chemistry teachers.

One of the strongest lines of research in the 1990s was the study of previous concepts of students and the way of dealing with those ideas. The earlier idea that those spontaneous concepts should be transformed or replaced by the standard scientific concepts was given up by Eduardo Fleury Mortimer, who developed a new approach of conceptual profiles, inspired by Gaston Bachelard (Mortimer 1995, 2000).

The new attitude allows the students to keep their previous concepts, being aware of the difference between the scientific and popular cultures.

During the decade of 1990 the analysis of the epistemological beliefs of chemistry teachers led to the conclusion that they adopted a naïve empiricism and transmitted this attitude to their students. It became evident that the training of chemistry teachers should include not only the knowledge of chemistry but also its historical and epistemological features, as well as the social, economic and political context of the development of this science (STS approach).

Contributions from history and philosophy of science are not very common in research on chemistry teaching. Among several different approaches that can be found, we can point out an emphasis in the analysis of epistemological views presented in textbooks and by students and teachers and proposals of strategies to provide a more adequate view on the nature of science using historical studies of chemistry. The science, technology, and society approach is also deemed important, and historical examples (such as the development of dyes) are suggested to introduce this issue. Some of the works also claim the improvement of learning of chemical concepts using a historical approach.

Many works that cannot be classified as "research in chemistry education" should also be mentioned. From the 1990s onwards, several Brazilian chemists have published popular books on history of chemistry (Vanin 1994; Chassot 1994) and many papers on specific subjects. More recently, Juergen Heinrich Maar is producing a three-volume work on the history of chemistry (two parts have been published: Maar 2008, 2011). The production of papers on specific episodes of the history of chemistry has also provided the Brazilian teachers with some nice works that can be put into use in their educational practice.[13]

70.6.3 *Biology*

As noticed above, the BSCS project was introduced in Brazil in the 1960s. Besides many innovations, the project included a historical study of biological concepts. However, this did not lead to any stimulus for the development of studies of history of biology applied to science education in Brazil, at that time.

Brazilian educational projects related to biology teaching were produced from the decade of 1960 onwards. They attempted to produce textbooks with new content, together with laboratory materials. Until the next decade, the main concern was the production of teaching materials and the teaching training, and there was no concern with educational research in biology. Around 1990 the area begins to produce researches on spontaneous concepts of students and teachers and on the use of history and philosophy of science.

[13] Among the articles devoted to the use of history of chemistry on education see, among others, Bagatin et al. (2005), Baldinato and Porto (2008), Chassot (2001), Farias (2001), Flôr (2009), Oki (2000), Paixão and Cachapuz (2003), Porto (2004), Tolentino and Rocha Filho (2000), and Vidal et al. (2007).

In the cases of physics and chemistry, the respective scientific institutes of the University of São Paulo (USP) played a relevant role in the development of the area of science education. The national physical and chemical societies also gave strong support to this area. In the case of biology, the situation was widely different. Research in biology education was strongly developed at the Faculty of Education of USP, especially under the leadership of Myriam Krasilchik – not at the Institute of Biology.[14] Besides that, since a *Brazilian Society of Biology* never existed, there was no association that could support the area. Indeed, there are several biological societies in Brazil, related to genetics, zoology, etc. but none that could assume the improvement of biology teaching as its concern.

In 1984 the Faculty of Education of USP began a series of conferences, called *Encontro Perspectivas do Ensino de Biologia* (*Meeting on Perspectives of Biology Teaching*). Although they did not have a national character, those events attracted researchers from other institutions, starting a process of organization of the area. The creation of the *Sociedade Brasileira para o Ensino de Biologia* (SBEnBio) (*Brazilian Society for the Teaching of Biology*) in 1997 led to a decentralization of the events, with regional conferences on biology teaching promoted at other states. The first *National Meeting for the Teaching of Biology*, organized by SBEnBio, occurred only in 2005. The journal of this society, *Revista de Ensino de Biologia* (*Journal of Biology Teaching*), started in 2007. In 2008, the *Brazilian Association for Philosophy and History of Science* (ABFHiB) created a Commission for Biology Teaching and produced a series of case studies for application in secondary schools.[15]

Up to 1996 the number of theses and dissertations on biology teaching was very small. In the last years of the twentieth century, there was a strong increase, parallel to the creation of the *Brazilian Society for the Teaching of Biology*, but not as an effect of this society (Teixeira et al. 2009). Around 1990 the area begins to produce researches on spontaneous concepts of students and teachers and on the use of history and philosophy of science.[16] The first works on the STS approach in biology education appeared a few years later.

Works using this approach usually stress the importance of introducing history of science in biology teaching to present science as a human construct, subject to mistakes, influenced by external factors, producing provisory knowledge. They denounce the inductive view of science, the idea that biology was produced by a few bright minds, and the view that science is the attainment of absolute truth. Those researches also emphasize the need to introduce history and philosophy of science in the teaching training. The STS approach is also recommended, adding the environment dimension. Besides discussion and recommendations, there are several

[14] It is worth mentioning that the situation has changed along the last 5 years. Due to the Teacher Education Program of USP, the Institute of Biosciences hired professors on biology teaching and history of biology, and new similar positions for science teaching were created in other science institutes (Universidade de São Paulo 2004).

[15] Andrade and Caldeira (2009), Batisteti et al. (2009), Bizzo and El-Hani (2009), Carmo et al. (2009a), Brandão and Ferreira (2009), Martins (2009a), (2009b), and Prestes et al. (2009).

[16] Among others: Bastos (1998), Cicillini (1992), Martins (1998), Slongo (1996).

works that include the production, application, and analysis of teaching activities using history and philosophy of biology.[17]

70.7 National Educational Guidelines

The relevance of history and philosophy of science in science teaching was officially recognized, in Brazil, at the end of the twentieth century. In 1996, the Brazilian Ministry of Education (MEC) began an educational reform. The first official step was the promulgation of the *Leis de Diretrizes e Bases* (*Law of Brazilian Education Guidelines and Bases*), followed by a *Resolution* of the National Education Council that established the *National Curricular Guidelines for Secondary Education*, in 1998 (Brasil 1996, 1998). This *resolution* describes, in its tenth article, some of the abilities and competencies that the students should acquire in their study of mathematics and natural sciences, including:

> (a) To understand the sciences as human constructs, recognizing that they develop by accumulation, continuity or paradigm rupture, correlating the scientific development to the transformation of society; [...] (i) To understand the relation between the development of the natural sciences and the technological development and to associate the different technologies to the problems that they intended to solve; (j) To understand the impact of the technologies associated to the natural sciences in the student's personal life, in the production processes, in the development of knowledge and in social life (Brasil 1998, p. 4–5).

The first of those items is directly related to the nature of science issues, associated to history and philosophy of science, and the other ones, to history of science and technology and science-technology-society issues. That document did not suggest any other roles for history and philosophy of science.

A group of educators, invited by the Ministry of Education, produced in 1997–1998 a document explaining how the general guidelines should be applied by teachers: *Parâmetros Curriculares Nacionais para o Ensino Médio* (PCNEM) (*National Curriculum Parameters for Secondary Education*) (Brasil 1997). This was followed by a more detailed complement, published in 2002: *PCN + Ensino Médio: Orientações Educacionais Complementares aos Parâmetros Curriculares Nacionais* (*Educational Complementary Guidelines to the National Curriculum Parameters*) (Brasil 2002). The sections of those two official documents concerning mathematics, physics, chemistry, and biology point out, at several places, the relevance of history and philosophy of science to science education.

The elaboration of the two documents on natural sciences and mathematics was coordinated by Luís Carlos de Menezes. For each discipline, the group provided

---

[17] Among others, see Almeida and Falcão (2005), Batista and Araman (2009), Baptista and El-Hani (2009), Bastos and Krasilchik (2004), Caldeira and Araújo (2010), Carmo et al. (2009b), Carneiro and Gastal (2005), El-Hani and Sepulveda (2010), El-Hani et al. (2004), Goulart (2005), Justina (2001), Leite (2004), Meglhioratti (2004), Pereira and Amador (2007), Rosa and Silva (2010), Santos (2006), Santos et al. (2012), Scheid et al. (2005), and Slongo and Delizoikov (2003).

specific instances of the use of history and philosophy of science, especially related to the issues of the nature of science and science, technology, and society. The previous official documents understood the contextualization of science education in the sense of the cognitive approaches to education. The group interpreted the contextualization in a much broader sense: "In general terms, contextualization in science education includes competencies related to the insertion of science and its technologies in a historic, social and cultural process and the recognition and discussion of practical and ethical features of science in the contemporaneous world [...]" (Brasil 2002, p. 31). The *Educational Complementary Guidelines* (PCN+) provide a large number of specific instances that can be used by teachers in addressing those issues.

Besides those features related to history and philosophy of science, there were many other new proposals that cannot be described here. If the guidelines could be put into practice, they would greatly improve science teaching, in Brazil. Unfortunately, more than 10 years after the educational reform and the publication of the above-described documents, one cannot recognize any definite transformation in science education. Secondary school science teachers could hardly understand all the changes that have been recommended. One can attribute this failure to a lack of effective public policies to improve the school system as a whole. One central aspect that is rarely addressed is the fact that teachers had no adequate training for coping with the new proposals that could help them to attain the new aims.

70.8 Conclusion

Nowadays, the relevance of history and philosophy of science in science education is widely recognized in Brazil. Although the effective classroom practice has not yet incorporated its use, the official educational guidelines are stimulating its development. A large number of books and papers, theses and dissertations, have been produced on this theme. This approach is an important trend in graduate programs and in educational conferences.

Two specific meetings on this subject were held in Brazil, in 2010: the *8th International Conference for the History of Science in Science Education* (ICHSSE) and the *1st Latin American Conference of the International History, Philosophy and Science Teaching Group* (IHPST-LA), with the participation of about 150 researchers (Silva and Prestes 2012). Some specific books on the subject, providing information and suggestion of classroom activities, were published in recent years. There is also an increasing international and regional collaboration.

The Brazilian contributions in this area are not well known worldwide, because most works are published in Portuguese. Although it is relevant to participate in global conferences and projects and to publish in other languages that can reach a wider public abroad, it is very important to produce works in Portuguese for local use. The vast majority of secondary school teachers cannot understand books and papers in English, and even those who can do it prefer reading works published in our national language.

There are several types of publications in this area. The academic ones (those presented at conferences or published in scholarly journals and books) describe the several approaches and defend (or criticize) their use; they review researches published abroad and in Brazil; they analyze textbooks; they present surveys of the concepts of students and teachers concerning the nature of science and other subjects related to history and philosophy of science; they provide information about history and philosophy of science that can be used in science teaching; they describe specific developments of syllabi, texts, and other educational materials applying history and philosophy of science in education; they report classroom experiments using history and philosophy of science; and they present proposals of new initiatives in the field, such as teaching learning sequences based in application of history and philosophy of science in science education (Peduzzi et al. 2012). Nowadays, the focus of history and philosophy of science in science education is the discussion of the nature of science and science, technology, society and environment issues. The uses of history of science to improve the learning of scientific ideas, to increase the motivation of students, its intrinsic cultural value, and other uses that were proposed in the 1970s and 1980s, are nowadays seldom mentioned.

The vast majority of authors of the abovementioned academic works are either university professors or graduate students. Most of this production will never reach in-service science teachers, but might be used for preservice teaching training at the university, or in specific courses for in-service teachers.

On the other hand there are publications targeted at in-service teachers (and also students), such as the journals that have already been described (*Revista do Professor de Matemática*, *Química Nova na Escola, Física na Escola*) and books. Those journals have a wide penetration, but their function is mainly informative. It is doubtful that they have contributed to effective changes in science teaching, because they only contain short papers. There are many books on history and philosophy of science published in Portuguese. However, teachers interested in this subject usually read popular, out-of-date books that reproduce the old views on the nature of science.

There is a shortage of educational materials using history and philosophy of science in the secondary school classroom. As described above, as a rule science textbooks include mistaken information about history and philosophy of science. There have been attempts to produce supplementary texts on history of science for use by students, but their effective utilization has been very limited.

Much remains to be done in consolidating the effective use of the history and philosophy of science in Brazilian science education. Nevertheless, it is possible to say that there is a solid ground, in research and graduate courses; there is clear official interest in the use of history and philosophy of science in education; there is a growing awareness and interest in this subject by teachers; there is a pressure by the Ministry of Education upon publishing houses in order to improve the quality of history views on nature of science in textbooks; in a nutshell, there are nice conditions to take off and fly. Of course, government support is essential and is sometimes unavailable; but many initiatives that depend only on researchers can start a snowball effect and produce important results.

It is necessary to consider the complexity of the education system in order to create a successful implementation of history and philosophy of science at schools. There is a gap between research and practice to connect curricular innovation with teaching practice; it is not restricted only to history and philosophy of science (see, among others, Pekarek et al. 1996; Pena and Ribeiro Filho 2008; Höttecke and Silva 2011). Thus, working collaboratively with in-service teachers is essential. In order to foster the use of history and philosophy of science in science education, it is important to take the teachers' perspective into account; otherwise curricular innovations will be hard to implement.

The introduction of disciplines in teachers training courses that combine scientific content and historical and philosophical issues with didactical aspects is highly desirable. It can allow future teachers to develop some of the needed skills to implement this approach in practice and to develop an awareness of the real worth of the use of history and philosophy of science in their teaching (Höttecke and Silva 2011).

It is also important to establish stronger cooperation between scholars in this field: one swallow does not make a summer. Collaboration may occur at several different levels. The creation of a national society for history and philosophy of science in science education, together with a specific journal and periodic meetings, might enhance the visibility of the area and provide a better forum for debate of researches and for the planning of national strategies. This does not mean that researchers working in this area should keep apart from other societies, of course. The creation of national databases for dissertation, theses, electronic versions of books, conference proceedings, and papers would be a nice instrument both for the improvement of research and for the dissemination of works. A teamwork including many groups and institutions could propose and lead ambitious projects – both research projects and educational ones. In a lower scale, a researcher producing new educational materials should ask the help of other researchers, from different institutions, to test and comment on his/her work; a researcher doing a survey of concepts on the nature of science at one institution should ask the cooperation of colleagues from other institutions to do a similar survey at other places, to compare their results. Researchers should think great: could my current work improve if I asked other people to collaborate?

Another front to be developed is a combination of research and application. We mean *real* application, such as posting on the Internet educational materials, together with suggestions for their use and additional materials (such as a video uploaded to *YouTube*). Of course, quality should always be a concern – both in research and application. A careful transposition can produce high-quality popularization and educational materials. The combination of those last two attitudes – trying to cooperate and to be useful in a broader sense – could greatly improve the situation in the area.

References

Alfonso-Goldfarb, A. M. (1999). *Livro do Tesouro de Alexandre: um estudo de hermética árabe na oficina de história da ciência*. Petrópolis: Vozes.

Almeida, A. V., & Falcão, J. T. R. (2005). A estrutura histórico-conceitual dos programas de pesquisa de Darwin e Lamarck e sua transposição para o ambiente escolar. *Ciência & Educação*, 11(1),17–32.

Andrade, M. A. B. S., & Caldeira, A. M. A. (2009). O modelo de DNA e a biologia molecular: inserção histórica para o ensino de biologia. *Filosofia e História da Biologia*, 4, 101–137.

Angotti, J. A. (1982). *Solução alternativa para a formação de professores de ciências: um projeto educacional desenvolvido na Guiné-Bissau*. São Paulo. Dissertação (Mestrado em Ensino de Ciências) – Universidade de São Paulo.

Assis, A. K. T. (2008). *Arquimedes, o centro de gravidade e a lei da alavanca*. Montreal: Apeiron.

Azevedo, D. N., & Costa Neto, P. L. (2010). A história da publicação das obras de Marx e Engels no Brasil de 1930 a 1964. *Monografias, Universidade Tuiuti do Paraná*, 95–143.

Bagatin, O., Simplicio, F. I., Santin, S. M. O., & Santin Filho, O. (2005). Rotação da luz polarizada por moléculas quirais: uma abordagem histórica com proposta de trabalho em sala de aula. *Química Nova na Escola*, 21, 34–38.

Baldinato, J. O., & Porto, P. A. (2008). Michael Faraday e *A História Química de uma vela*: estudo de caso sobre a didática da ciência. *Química Nova na Escola*, 30, 16–23.

Baptista, G. C. S., & El-Hani, C. N. (2009). The contribution of ethnobiology to the construction of a dialogue between ways of knowing: a case study in a Brazilian public high school. *Science & Education*, 18, 503–520.

Barros, S. S. (2002). Reflexões sobre 30 anos da pesquisa em ensino de física. *Atas VIII EPEF. Proceedings VIII Meeting on Research in Physics Teaching*. CD ROM. São Paulo: Sociedade Brasileira de Física.

Bassalo, J. M. F. (1992). A importância do estudo da história da ciência. *Revista da SBHC*, 8, 57–66.

Bastos, F. (1998). *História da ciência e ensino de biologia*. São Paulo. Tese (Doutorado em Educação) – Faculdade de Educação, Universidade de São Paulo.

Bastos, F., & Krasilchik, M. (2004). Pesquisas sobre a febre amarela (1881–1903): uma reflexão visando contribuir para o ensino de ciências. *Ciência & Educação*, 10(3), 417–442.

Batista, I. L. (2004) O ensino de teorias físicas mediante uma estrutura histórico-filosófica. *Ciência & Educação*, 10(3), 461–476.

Batista, I. L., & Araman, E. M. O. (2009). Uma abordagem histórico-pedagógica para o ensino de ciências nas séries iniciais do ensino fundamental. *Revista Electrónica de Enseñanza de las Ciencias*, 8(2), 466–489.

Batisteti, C. B., Araújo, E. S. N., & Caluzi, J. J. (2009). As estruturas celulares: o estudo histórico do núcleo e sua contribuição para o ensino de biologia. *Filosofia e História da Biologia*, 4, 17–42.

Beltran, E. (1984). La historia de la ciencia en América Latina. *Quipu*, 1(1), 7–23.

Bizzo, N. (2004). Educational systems: case studies and educational indices: Central America. In Natalia Pavlovna Tarasova (Ed.). *Encyclopedia of Life Support Systems*. Oxford: EOLSS Publishers.

Bizzo, N. (1991). *Ensino de evolução e história do darwinismo*. São Paulo. Tese (Doutorado em Educação) – Faculdade de Educação, Universidade de São Paulo.

Bizzo, N., & El-Hani, C. (2009). O arranjo curricular do ensino de evolução e as relações entre os trabalhos de Charles Darwin e Gregor Mendel. *Filosofia e História da Biologia*, 4, 235–257.

Braga, M., Guerra, A. & Reis, J. C. (2012). The role of historical-philosophical controversies in teaching sciences: The debate between Biot and Ampère. *Science & Education*, 21(6), 921–934.

Brandão, G. O., & Ferreira, L. B. M. (2009). O ensino de genética no nível médio: a importância da contextualização histórica dos experimentos de Mendel para o raciocínio sobre os mecanismos da hereditariedade. *Filosofia e História da Biologia*, 4, 43–63.

Brasil. Conselho Nacional de Educação. (1998). Resolução CEB No 3, de 26 de junho de 1998. Institui as Diretrizes Curriculares Nacionais para o Ensino Médio.

Brasil. Ministério da Educação. (1996). Lei No 9.394, de 20 de dezembro de 1996. *Diário Oficial da União*, Brasília, v. 134, n. 248, p. 27.833–27.841, 23 dez. 1996, Seção I. Estabelece as diretrizes e bases da educação nacional.

Brasil. Ministério da Educação. Secretaria de Educação Fundamental. (1997). *Parâmetros Curriculares Nacionais para o Ensino Médio: introdução*. Brasília: MEC/SEF.

Brasil. Ministério da Educação. (2002). *PCN+ Ensino Médio: orientações educacionais complementares aos parâmetros curriculares nacionais. Ciências da natureza, Matemática e suas tecnologias*. Brasília: MEC.

Caldeira, A. M. A., & Araújo, E. S. N. N. (2010). *Introdução à didática da biologia*. São Paulo: Escrituras.

CAPES (2010). *Relatório de avaliação 2007–2009*. http://trienal.capes.gov.br/wp-content/uploads/2011/01/ENSINO-DE-CM-RELAT%C3%93RIO-DE-AVALIA%C3%87%C3%83O-FINAL-jan11.pdf Acessed 21 November 2012.

Caponi, G. (2010). *Buffon: breve introducción al pensamiento de Buffon*. México, DF: Universidad Autónoma Metropolitana.

Caponi, G. (2011). *La segunda agenda darwiniana: contribución preliminar a la historia del programa adaptacionista*. Mexico, DF: Centro de Estudios Filosóficos, Políticos y Sociales Vicente Lombardo Toledano.

Carmo, R. S., Nunes-Neto, N. F., & El-Hani, C. N. (2009). Gaia theory in Brazilian high school biology textbooks. *Science & Education*, 18, 469–501.

Carmo, V. A., Bizzo, N. & Martins, L. A.-C. P. (2009). Alfred Russel Wallace e o princípio de seleção natural. *Filosofia e História da Biologia*, 4, 209–233.

Carneiro, M. H. S., & Gastal, M. L. (2005). História e filosofia das ciências no ensino de biologia. *Ciência & Educação*, 11(1), 33–39.

Carvalho, A. M. P., & Vannucchi, A. I. (2000). History, philosophy and science teaching: some answers to "how?". *Science & Education*, 9(5), 427–448.

Castro, R., & Carvalho, A. M. P. (1995). The historic approach in teaching: Analysis of an experience. *Science & Education*, 4(1), 65–68.

Chagas, A. P. (2001). 100 anos de Nobel: Jacobus Henricus van't Hoff. *Química Nova na Escola*, 14, 25–27.

Chassot, A. (1994). *A ciência através dos tempos*. São Paulo: Moderna.

Chassot, A. (2001). Outro marco zero para uma história da ciência latino-americana. *Química Nova na Escola*, 13, 34–37.

Chateaubriand Filho, O. (1971). *Ontic commitment, ontological reduction and ontology*. Berkeley. Tese (Doutorado em Filosofia) – University of California.

Cicillini, G. A. (1992). A história da ciência e o ensino de biologia. *Ensino em Re-Vista*, 1(1), 7–17.

Cupani, A. O. (1974). *Exámen de la razón dialéctica*. Córdoba. Tese (Doutorado em Filosofia) – Universidad Nacional de Córdoba.

D'Ambrosio, U. (1996). *Educação matematica: da teoria á prática*. Campinas: Papirus.

D'Ambrosio, U. (2008). *Uma história concisa da matemática no Brasil*. Petrópolis: Vozes.

Delizoikov Neto, D. (1982). *Concepção problematizadora para o ensino de ciências na educação formal: relato e análise de uma prática educacional na Guiné-Bissau*. São Paulo. Dissertação (Mestrado em Ensino de Ciências) – Universidade de São Paulo.

Delizoikov Neto, D., Zanetic, J., Angotti, J. A. P., Menezes, L. C., & Takeya, M. (1980). Uma experiência em ensino de ciências na Guiné-Bissau. *Revista Brasileira de Ensino de Física*, 2, 4, 57–72.

Dias, P. M. C. (1994). A path from Watt's engine to the principle of heat transfer. In: D. Prawitz & D. Westerstahl (Ed.). *Logic and Philosophy of Science in Uppsala*. (pp. 425–438). Dordrecht: Kluwer Academic Publishers.

Dias, P. M. C. (1999). Euler's 'harmony' between the principles of '"rest" and "least action". *Archive for History of Exact Science*, 54, 67–88.

El-Hani, C., Tavares, E. J. M., & Rocha, P. L. B. (2004). Concepções epistemológicas de estudantes de biologia e sua transformação por uma proposta explícita de ensino sobre história e filosofia das ciências. *Investigações em Ensino de Ciências*, 9(3), 265–313.

El-Hani, C. N., & Sepulveda, C. (2010). The relationship between science and religion in the education of protestant biology pre-service teachers in a Brazilian University. *Cultural Studies of Science Education*, 5, 103–125.

Farias, R. F. (2001). As mulheres e o prêmio Nobel de Química. *Química Nova na Escola*, 14, 28–30.

Ferri, M. G., & Motoyama, C. (Eds.) (1979–1981). *História das ciências no Brasil*. São Paulo: E.P.U./Edusp, 3 vols.
Filgueiras, C. A. L. (1994). Orígenes de la química Luso-Brasileña: realizaciones, vicisitudes y equívocos. In P. A. Pastrana (Ed.). *La química en Europa y América (siglos XVIII y XIX)* (pp. 201–210). México: Universidad Autónoma Metropolitana.
Filgueiras, C. A. L. (2002). *Lavoisier e o estabelecimento da Química Moderna*. São Paulo: Odysseus.
Flôr, C. C. (2009). A história da síntese de elementos transurânicos e extensão da tabela periódica numa perspectiva Fleckiana. *Química Nova na Escola*, 31(4), 246–250.
Flores, D. T., Silva, L. G., & Will, R. (2009). *O Projeto FAI – Física Auto-Instrutivo*. http://www.fsc.ufsc.br/~arden/fai.doc. Accessed 25 November 2011.
Forato, T. C. de M., Martins, R. de A., & Pietrocola, M. (2012). History and nature of science in high school: Building up parameters to guide educational materials and strategies. *Science & Education*, 21(5), 657–682.
Franco, C., & Colinvauz-de-Dominguez, D. (1992). Genetic epistemology, history of science and science education. *Science & Education*, 1, 255–271.
Freire Jr., O. (1995). *A emergência da totalidade: David Bohm e a controvérsia dos quanta. São Paulo*. Tese (Doutorado em História Social) – Faculdade de Filosofia Ciências e Letras, Universidade de São Paulo.
Freire Jr., O., & Tenório, R. M. (2001). A graduate programme in history, philosophy and science teaching in Brazil. *Science & Education*, 10(6), 601–608.
Freire Jr., O. (2002). A relevância da filosofia e da história das ciências para a formação dos professores de ciências [The relevance of philosophy and history of science to science teachers training]. Pp. 13–30, in: Silva Filho, Waldomiro José da (org.). Epistemologia e ensino de ciências. Salvador: Arcadia.
Freire, P. (1970). *Pedagogia do oprimido*. Rio de Janeiro: Paz e Terra.
Gama, H. U., & Hamburger, E. W. (1987). *O "Grupo de Ensino" do IFUSP. Histórico e atividades*. São Paulo: Instituto de Física da Universidade de São Paulo. http://repositorio.if.usp.br/xmlui/bitstream/handle/1396/358/pd623.pdf?sequence=1. Accessed 20 November 2012.
Gilbert, J. K., & Zylbersztajn, A. (1985). A conceptual framework for science education: the case study of force and movement. *European Journal of Science Education*, 7(3), 107–120.
Goldfarb, J. L. (1994). Mário Schenberg e a história da ciência. *Revista da SBHC*, 12, 65–72.
Gomes, F. A. M. (1978). *História do desenvolvimento da indústria siderúrgica no Brasil*. 6 vols. Belo Horizonte, CETEC.
Goulart, S. M. (2005). *História da ciência: elo da dimensão transdisciplinar no processo de formação de professores de ciências*. In J. C. Libaneo & A. Santos (Eds.). Campinas: Alínea.
Granger, G. G. (1955). *Lógica e filosofia das ciências*. São Paulo: Melhoramentos.
Greca, I. M., & Freire Jr., O. (2003). Does an emphasis on the concept of quantum states enhance students' understanding of quantum mechanics? *Science & Education*, 12(5–6), 541–557.
Hegenberg, L. (1968). *Mudança de linguagens formalizadas*. São Paulo. Tese (Doutorado em Filosofia) – Universidade de São Paulo.
Höttecke, D., & Silva, C. C. (2011). Why implementing history and philosophy in school science education is a challenge. An analysis of obstacles. *Science & Education*, 20, 293–316.
Hopos. (2000). Report on HOPOS-related resources in Brazil. *Newsletter of the History of Philosophy of Science (HOPOS) Working Group*, 6(1), 7–16, 2000.
Justina, L. D. (2001). *Ensino de genética e história de conceitos relativos a hereditariedade*. Florianópolis. Dissertação (Mestrado) – CED, Universidade Federal de Santa Catarina.
Krasilchik, M. (1990). The scientists: an experiment in science teaching. *International Journal of Science Education*, 12(3), 382–287.
Krasilchik, M. (1992). Caminhos do ensino de ciências no Brasil. *Em Aberto*, 11(55), 3–8.
Krause, D., & Videira, A. (Eds.). (2011). *Brazilian studies in philosophy and history of science: an account of recent works*. Dordrecht, Springer.
La Penha, G. M. S. M. (1982). *Sobre as histórias da história da mecânica do século XVII*. [Rio de Janeiro]: LCC/CNPq/IM-UFRJ.

La Penha, G. M. S. M., Bruni, S. A., Papavero, N., & Rodrigues, A. M.E. G. O. (1986). *O Museu Paraense Emilio Goeldi*. [São Paulo]: Banco Safra.

Landim Filho, R. F. (1974). *Completude et decidabilité de quelques systèmes de logique du temps*. Louvain. Tese (Doutorado em Filosofia) – Université Catholic de Louvain.

Lefebvre, J.-P. (1990). Les professeurs françaix des missions universitaires au Brésil (1934–1944). *Cahiers du Brésil Contemporain*, 12.

Leite, R. C. M. (2004). *A produção coletiva do conhecimento científico: um exemplo no ensino de genética*. Florianópolis. Tese (Doutorado) – CED, Universidade Federal de Santa Catarina.

Lucie, P. (1978). *Gênese do método científico*. Rio de Janeiro: Campus.

Lucie, P. (1969). *Física com Martins e eu*. Rio de Janeiro: Raval Artes Gráficas.

Maar, J. H. (2008). *História da Química*. Parte 1: dos Primórdios a Lavoisier. Rio de Janeiro: Conceito Editorial.

Maar, J. H. (2011). *História da Química*. Parte 2: de Lavoisier à Tabela Periódica. Florianópolis: Papa-Livro.

Mariconda, P. R. (2003). Lógica, experiência e autoridade na carta de 15 de setembro de 1640 de Galileu a Licete. *Scientiae Studia*, 1(1), 63–73.

Mathias, S. (1975). *Cem anos de Química no Brasil*. São Paulo, [s.n.].

Martins, L. A.-C. P. (1998). A história da ciência e o ensino da biologia. *Ciência & Ensino*, 5, 18–21.

Martins, L. A.-C. P. (2005). História da ciência: objetos, métodos e problemas. *Ciência & Educação*, 11(2), 305–317.

Martins, L. A.-C. P. (2007). *A teoria da progressão dos animais, de Lamarck*. Rio de Janeiro: Booklink/FAPESP.

Martins, L. A.-C. P. (2009). Pasteur e a geração espontânea: uma história equivocada. *Filosofia e História da Biologia*, 4, 65–100.

Martins, R. A. (1990). Sobre o papel da história da ciência no ensino. *Boletim da Sociedade Brasileira de História da Ciência*, 9, 3–5.

Martins, R. A. (1996). Sources for the study of science, medicine and technology in Portugal and Brazil. *Nuncius – Annali di Storia della Scienza*, 11(2), 655–667.

Martins, R. A. (1997). Becquerel and the choice of uranium compounds. *Archive for History of Exact Sciences*, 51(1), 67–81.

Martins, R. A. (2009). Os estudos de Joseph Priestley sobre os diversos tipos de "ares" e os seres vivos. *Filosofia e História da Biologia*, 4, 167–208.

Martins, R. A., & Silva, C. C. (2001). Newton and colour: the complex interplay of theory and experiment. *Science & Education*, 10(3), 287–305.

Meglhioratti, F. A. (2004). *História da construção do conceito de evolução biológica: possibilidades de uma percepção dinâmica da Ciência pelos professores de Biologia*. Bauru. Dissertação (Mestrado em Educação para a Ciência) – Universidade Estadual Paulista.

Monteiro, J. P. G. (1967). *Ensaios Políticos de David Hume*. São Paulo. Dissertação (Mestrado em Filosofia) – Universidade de São Paulo.

Moreira, M. A. (2004). Pós-graduação e pesquisa em ensino de ciências no Brasil. *Atas IV ENPEC. Proceedings IV National Meeting of Research in Science Education*. CD ROM. Bauru: ABRAPEC.

Mortimer, E. F. (1995). Conceptual change or conceptual profile change. *Science & Education*, 4(3), 267–285.

Mortimer, E. F. (2000). *Linguagem e formação de conceitos no ensino de ciências*. Belo Horizonte: Editora UFMG.

Motoyama, S. (1988). História da ciência no Brasil: apontamentos para uma análise crítica. *Quipu*, 5(2), 167–189.

Moura, B. A. (2012) *Formação crítico-transformadora de professores de física: uma proposta a partir da história da ciência*. Doctorate Thesis. São Paulo: Instituto de Física – Universidade de São Paulo.

Nachbin, L. (1996). *Ciência e sociedade*. Curitiba, UFPr.

Nardi, R. (2005). *A área de ensino de ciências no Brasil: Fatores que determinaram sua constituição e suas características, segundo pesquisadores brasileiros*. Postdoctoral thesis, Bauru: Department of Education, Unesp.
Nascimento, C. A. R. (1995). *De Tomás de Aquino a Galileu*. Campinas: IFCH-UNICAMP.
Nobre, S. (2001). *Elementos historiográficos da matemática presentes em Enciclopédias Universais*. Rio Claro: Unesp.
Oki, M. C. M. (2000). A eletricidade e a química. *Química Nova na Escola*, 12, 34–37.
Oliveira, A. G., & Dias, C. A. (1970). *Boletim da Sociedade Brasileira de Física*, 4, 1–335.
Pagliarini, C., & Silva, C. C. (2007). History and nature of science in Brazilian physics textbooks: some findings and perspectives. *Proceedings of Ninth International History, Philosophy & Science Teaching Conference*. http://www.ucalgary.ca/ihpst07/proceedings/IHPST07%20papers/2122%20Silva.pdf . Accessed 20 November 2012.
Paixão, F., & Cachapuz, A. (2003). Mudança na prática de ensino de química pela formação de professores em história e filosofia das ciências. *Química Nova na Escola*, 18, 31–36.
Peduzzi, L. O. Q., Martins, A. F. P., & Ferreira, J. M. H. (Orgs.). (2012). *Temas de História e Filosofia da Ciência no Ensino*. Natal: EDUPRN.
Pekarek, R., Krockover, G., & Shepardson, D. (1996). The research/practice gap in science education. *Journal of Research in Science Teaching*, 33, 111–113.
Pena, F. L. A., & Ribeiro Filho, A. (2008). Relação entre pesquisa em ensino de física e a prática docente: Dificuldades assinaladas pela literatura nacional da área. *Cadernos Brasileiros de Ensino de Física*, 25, 424–438.
Piaget, J., & Garcia, R. (1982). *Psicogénesis e Historia de la Ciencia*. Mexico: Siglo XXI.
Pereira, A. I., & Amador, F. (2007). A história da ciência em manuais escolares de ciências da natureza. *Revista Electrónica de Enseñanza de las Ciencias*, 6(1), 191–216.
Porto, P. A. (2004). Um debate seiscentista: a transmutação de ferro em cobre. *Química Nova na Escola*, 19, 24–26.
Prestes, M. E. B., Oliveira, P. P., & Jensen, G. M. (2009). As origens da classificação de plantas de Carl von Linné no ensino de biologia. *Filosofia e História da Biologia*, 4, 101–137.
Quine, W. V. O. (1944). *O sentido da nova lógica*. São Paulo: Livraria Martins.
Regner, A. C. K. P. (1995). A natureza teleológica do princípio darwinano de seleção natural: a articulação do metafísico e do epistemológico na Origem das Espécies. Porto Alegre: Tese (Doutorado em Educação) – Universidade Federal do Rio Grande do Sul.
Regner, A. C. K. P. (2003). Uma nova racionalidade para a ciência? In: B. S. Santos (Org.). *Conhecimento prudente para uma vida decente* (pp. 273–303). Porto: Afrontamento.
Rigotto, M. E., & Souza, N. J. (2005). Evolução da educação no Brasil, 1970–2003. *Análise*, 16(2), 339–358.
Robilotta, C. C., Robilotta, M. R., Kawamura, R., & Kishinami, R. I. (1981). Entrevista com o Prof. Marcello Cini. *Revista de Ensino de Física*, 3(1), 67–75.
Rosa, K., & Martins, M. C. (2009). Approaches and methodologies for a course on history and epistemology of physics: analyzing the experience of a Brazilian university. *Science & Education*, 18(1), 149–155.
Rosa, R. G., & Silva, M. R. A (2010). História da ciência nos livros didáticos de biologia do ensino médio: uma análise do conteúdo sobre o episódio da transformação bacteriana. *Alexandria, Revista de Educação em Ciências e Tecnologia*, 3(2), 59–78.
Salmerón, F. (1991). Nota sobre la recepción del análisis filosófico en America Latina. *Isegoria*, 3, 119–137.
Santos, C. H. V. (2006). História e filosofia da ciência nos livros didáticos de biologia do ensino médio: análise do conteúdo sobre a origem da vida. Londrina. Dissertação (Mestrado) – Universidade Estadual do Paraná.
Santos, V. C., Joaquim, L. M., & El-Hani, C. N. (2012). Hybrid deterministic views about genes in Biology textbooks: a key problem in genetics teaching. *Science & Education*, 21, 543–578.
Scheid, N. M. J., Ferrari, N., & Delizoikov, D. (2005). A construção coletiva do conhecimento científico sobre a estrutura do DNA. *Ciência & Educação*, 11(2), 223–233.

Schenberg, M. 1984. *Pensando a Física*. São Paulo: Brasiliense.
Schneltzler, R. P. (2002). A pesquisa em ensino de química no Brasil: conquistas e perspectivas. *Química Nova*, 25(1), 14–24.
Schwartzman, S. (2007). Entrevista Simon Schwartzman: o crítico da ciência. *Pesquisa Fapesp*, 12–17.
Schwartzman, S. (1973). *Regional cleavages and political patrimonialism in Brazil*. Berkeley. Tese (Doutorado em Ciência Política) – University of California Berkeley.
Silva, C. C. (ed.) (2006). *Estudos de história e filosofia das ciências: subsídios para aplicação no ensino*. São Paulo: Editora Livraria da Física.
Silva, C. C., & Moura, B. A. (2012). Science and society: the case of acceptance of Newtonian optics in the eighteenth century. *Science & Education*, 21(9), 1317–1335.
Silva, C. C., & Prestes, M. E. B. (Eds.). (2012). Special issue: First IHPST Latin American Regional Conference: select contributions. *Science & Education*, 21(5), 603–766.
Silva, O. P. P. (1967). *A teoria aristotélica da ciência*. São Paulo. Tese (Doutorado em Filosofia) – Universidade de São Paulo.
Silva, V. F. (1940). *Elementos de lógica matemática*. São Paulo: Cruzeiro do Sul.
Silveira, A. F., Ataíde, A. R. P., Silva, A. P. B., & Freire, M. L. F. (2010). Natureza da ciência numa sequência didática: Aristóteles, Galileu e o movimento relativo. *Experiências em Ensino de Ciências*, 5(1), 57–66.
Silveira, F. L., & Peduzzi, L. O. Q. (2006). Três episódios de descoberta científica: da caricatura empirista a uma outra história. *Caderno Brasileiro de Ensino de Física*, 23, 2006.
Slongo, I. I. P. (1996). *História da ciência e ensino: contribuições para a formação do professor de Biologia*. Florianópolis. Dissertação (Mestrado) – Universidade Federal de Santa Catarina.
Stein, S. I. A. (2004) Willard van Orman Quine: a exaltação da 'nova lógica'. *Scientiae Studia*, 2(3), 373–379.
Slongo, I. I, & Delizoikov Neto, D. (2003). Reprodução humana: abordagem histórica na formação dos professores de Biologia. *Contrapontos*, 3(3), 435–447.
Teixeira, E. S., Greca, I. M., & Freire Jr. (2012). The history and philosophy of science in physics teaching: a research synthesis of didactic interventions. *Science & Education*, 21(6), 771–796.
Teixeira, P. M. M., Silva, M. G. B., & Anjos, M. S. (2009). 35 anos de pesquisa em ensino de biologia no Brasil: um estudo baseado em dissertações e teses (1972–2006). *Anais do VII Enpec*, Florianópolis.
Tolentino, M., & Rocha-Filho, R. C. (2000). O bicentenário da invenção da pilha elétrica. *Química Nova na Escola*, 11, 35–39.
Vanin, J. A. (1994). *Alquimistas e químicos: o passado, o presente e o futuro*. São Paulo: Moderna.
Vergara, M. R. (2004). Ciência e modernidade no Brasil: a constituição de duas vertentes historiográficas da ciência no século XX. *Revista da SBHC*, 2(1), 22–31.
Vidal, P. H. O., Cheloni, F. O., & Porto, P. A. (2007). O Lavoisier que não está presente nos livros didáticos. *Química Nova na Escola*, 26, 29–32.
Videira, A. A. P. (1994). Ciência, técnica e filosofia da ciência. *Cadernos de História e Filosofia da Ciência*, 4(1), 93–108.
Villani, A., Dias, V. S., & Valadares, J. M. (2012). The development of science education research in Brazil and contributions from the history and philosophy of science. *International Journal of Science Education*, 32(7), 907–937.
Villani, A., Pacca, J. L. A., & Freitas, D. (2009). Science teacher education in Brazil, 1950–2000. *Science and Education*, 18(1), 125–148.
Universidade de São Paulo (2004). Pró-Reitoria de Graduação. Programa de Formação de Professores. São Paulo, Universidade de São Paulo.
Watts, D.M., & Zylbersztajn, A. (1981). A survey of some ideas about force. *Physics Education*, 15, 360–365.

Roberto Andrade de Martins is currently a Visiting Professor at the Physics Department, State University of Paraíba, Brazil. His first degree was in physics (University of São Paulo) and his Ph.D. in logic and philosophy of science (State University of Campinas). He did postdoctoral researches in Oxford and Cambridge. His main research areas are history and philosophy of science (especially of physics and biology) and their use in science education. He was President of the Brazilian Society for History of Science (SBHC) and of the South Cone Association for Philosophy and History of Science (AFHIC). His publications include the following books: *Commentariolus: Pequeno Comentário de Nicolau Copérnico Sobre Suas Próprias Hipóteses Acerca dos Movimentos Celestes* (1990; 2003), *O Universo: Teorias Sobre Sua Origem e Evolução* (1994; 2012), *Contágio: História da Prevenção das Coenças Transmissíveis* (1997), *Os "Raios N" de René Blondlot: Uma Anomalia na História da Física* (2007), *Teoria da Relatividade Especial* (2008; 2012), and *Becquerel e a Descoberta da Radioatividade: Uma Análise Crítica* (2012).

Cibelle Celestino Silva is a Professor at the Institute of Physics of São Carlos, University of São Paulo, Brazil. She earned her M.Sc. and Ph.D. degrees on history of physics at the State University of Campinas and was a Dibner Fellow at Dibner Library in Washington DC, USA. Her main research interests are history of optics and electromagnetic theory. She is also concerned with physics teacher education, chiefly on bridging history and philosophy of science to classrooms and on nonformal science education. She is currently the head of the Group of History, Theory, and Science Teaching (GHTC), University of São Paulo. She is currently a member of the governing bodies of the South Cone Association for Philosophy and History of Science (AFHIC) and of the International History, Philosophy and Science Teaching Group (IHPST). One of the books she published is *Estudos de História e Filosofia das Ciências: Subsídios para Aplicação no Ensino* (2006).

Maria Elice Brzezinski Prestes is a Professor at the Department of Genetics and Evolutionary Biology, Bioscience Institute, University of São Paulo, Brazil. She achieved her first degree in Biological Sciences at Pontifícia Universidade Católica do Paraná (1982), M.Sc. in Environmental Sciences at University of São Paulo (1997), and Ph.D. in Education at the same university (2003). She did postdoctoral researches in Paris and Montreal. She has a wide experience in history of science, doing research on the following subjects: history of biology with emphasis in the eighteenth century and the uses of history of science in the teaching of biology. Her publications include the following books: *A Teoria Celular: De Hooke a Schwann* (1997) and *A Investigação da Natureza no Brasil Colônia* (2000) e *Florestas Brasileiras* (2003). She is currently the President of the Brazilian Association of Philosophy and History of Biology (ABFHiB) and editor of the journal *Filosofia e História da Biologia* (*Philosophy and History of Biology*).

Chapter 71
Science Teaching and Research in Argentina: The Contribution of History and Philosophy of Science

Irene Arriassecq and Alcira Rivarosa

71.1 Introduction

The analysis of the possible contributions to physics education made by the history and the philosophy of science constitutes a formidable task. Before we go on, a short note about some abbreviations that will be used throughout the text. From now on we will abbreviate the history of science as HS and the philosophy of science as PS; whenever we refer to both areas together, we will write HPS. In order to understand how complex this task is, we should first identify the multiple theoretical aspects that converge here.

On the one hand, physics as a scientific discipline has among its main objectives the aim of explaining, understanding and predicting natural phenomena. These objectives are achieved by intervening in those phenomena using specific methodologies and also specific language to communicate findings.

On the other hand, HPS, in spite of not strictly being considered a meta-science, is a hybrid field constituted by contributions from different meta-sciences, or second-order criteriology, which have other scientific disciplines as their objects of study (Losee 1972; Klimovsky 1994). HPS is a theoretical reflection of scientific knowledge and activity from an internalist and logical-linguistic perspective which focuses on the study of the processes, conditions and results of innovation, justification, systematization, application, evaluation and communication in science (Adúriz-Bravo et al. 2006).

I. Arriassecq (✉)
CONICET – Núcleo de Investigación Educación en Ciencias con Tecnologías (ECienTec) – Departamento de Formación Docente – Facultad de Ciencias Exactas, Universidad Nacional del Centro de la Provincia de Buenos Aires, Buenos Aires, Argentina
e-mail: irenearr@exa.unicen.edu.ar

A. Rivarosa
Profesor Adjunto Epistemología y Didáctica de las Ciencias (PIIAC). Departamento de Ciencias Naturales – Facultad de Ciencias Exactas-Físico-Químicas y Naturales, Universidad Nacional de Río Cuarto, Río Cuarto, Argentina

M.R. Matthews (ed.), *International Handbook of Research in History, Philosophy and Science Teaching*, DOI 10.1007/978-94-007-7654-8_71,

More recently, a line of work under the name 'nature of science' has developed. It is made up of a group of meta-scientific laws which are valuable to natural science teaching, and it is precisely the 'nature of science' which constitutes its object of study.

The term 'meta-science' refers to all the disciplines which have science as their object of study: epistemology, history of science and sociology of science, among others. These disciplines provide different frameworks for the study of science, the aim of which is to answer questions such as 'what are scientific knowledge and scientific activity like?', 'how does science change through time?', 'who have been the most relevant scientists in history and in what way are they relevant?', 'which are the values the scientific community adheres to?' and 'how does science relate to the other disciplines (humanities, technology, art) and other ways of interpreting the world (religion, myths, even popular beliefs)?' (Aduriz-Bravo 2005).

Multiple contexts coincide in the area of science education: a discipline to be taught, theoretical frameworks, processes for teaching and learning, teaching proposals, teaching contexts, conceptions of such a discipline and its own nature on the part of both the teachers and the students, the teachers' background, etc. Research into these aspects of HPS throughout its evolution has not always considered the possibility that the analysis of the topics of this subject could contribute to physics education. However, in recent years, this has started to be taken into account for the design of syllabuses, in light of the contributions made by researchers like Salomon (1988) and Lakin and Wellington (1994), who consider that the teacher's view of science, whether explicit or not, affects what and how he/she teaches. In connection with this opinion, many other authors such as Lantz and Kass (1987), and Duschl (1997), state that a science teacher's background should involve not only a science but also aspects related to the nature of science, such as knowledge about its purposes, methods and its relationship with technology and society.

In a time when scientific literacy has become one of the goals of science education in many countries, it is of paramount importance to gain a deeper understanding of the history and nature of science in order to achieve such an objective. It is expected that a scientifically literate individual should be able to distinguish between scientific and non-scientific knowledge, science and pseudoscience, to know the limits and scope of science as well as what science can or cannot explain and to identify the scientific methodologies. As an individual and as a member of a particular society, he should also be able to tell what is relevant to the scope of science, taking the positive as well as the negative aspects into account.

However, the concept of the nature of science (NOS), as Acevedo-Diaz and colleagues (2007) point out, is complex and dialectical and therefore difficult to define with accuracy and by general consent. Specialists often discuss descriptions and representations of the NOS which are as dynamic as scientific knowledge itself, and so it is impossible to support the idea of only NOS capable of representing either this knowledge or all the scientific disciplines. Therefore, any representation of the NOS will be partial and will compete with other incomplete representations.

With regard to science didactics, even though there is consensus about the importance of the NOS in science education (Bell et al. 2001), the means of achieving their own objectives are not clear.

There exist at present several international groups that study the applications of the NOS to science education at different levels. Among these groups one of the most important is the International History, Philosophy and Science Teaching Group (IHPST). This group has been holding conferences since 1989 and has encouraged international magazines with great prestige within the scientific community and among science education teachers – such as the *International Journal of Science Education*, among others – to dedicate special editions to the NOS and its relation to science education. Another major landmark was the creation of *Science & Education: Contributions from History, Philosophy and Sociology of Science and Mathematics*, which since 1992 has promoted the inclusion of history and philosophy of science and mathematics courses in science and mathematics teacher education programmes. Moreover, it promotes the discussion of the philosophy and the purpose of science and mathematics education and their place in, and contribution to, the intellectual and ethical development of individuals and cultures. It is associated with the International History, Philosophy and Science Teaching Group, and Michael Matthews is its editor.

The First IHPST Latin American Conference was held in 2010. This conference focused on the presentation and discussion of papers about the use of history and philosophy in science education, in accordance with the guidelines drawn up by the IHPST group for international conferences.

This chapter analyses the historical evolution of the HPS in connection with science teaching and learning, the different lines of research that have developed over time and some examples of teaching proposals – aimed at students, teachers and trainee science teachers – which were designed by taking research results into account. A critical analysis of the present situation regarding the incorporation of the HPS in science education is carried out, and some ideas are suggested in order to make progress in this direction. More precisely, the last two sections evaluate the situation in Argentina with regard to the incorporation of HPS contributions into physics and biology education. Some of the aspects analysed are the underlying epistemological features in educational laws and the natural science curriculum design at secondary level in Argentina, the incorporation of HPS into research and the dissemination of HPS contributions to physics teaching among in-service teachers.

71.2 Design, Implementation and Assessment of Teaching-Learning Sequences That Incorporate Contributions from the HPS

Teixeira and colleagues (2009), in a thorough and methodologically rigorous work, investigate teaching experiences of applying HPS in physics classrooms, with the aim of obtaining critical and reliable information on this subject.

The vast majority of the studies selected for analysis support the idea of similarity between students' spontaneous understanding of scientific concepts and the historical development of these concepts. The aim was to obtain a conceptual change, despite the large amount of criticism found in the literature about this type of approach. In spite of the presence of a variety of teaching strategies based on HPS, comparatively few of them provided the pedagogical references to justify the use of these strategies, and few were concerned with assessing the students' prior knowledge of HPS.

The studies analysed by the authors presented various ways of utilizing HPS in the teaching of physics: in relation to teaching objectives (learning concepts, nature of science (NOS), attitudes, argumentation and metacognition), in relation to teaching strategies (integrated with the subject of physics, integrated with another teaching strategy and not integrated) and in relation to didactic materials (historical narratives, biographies, replicas of historical experiments, historically contextualized problems and stories of scientists' lives).

The results revealed the occurrence of positive effects in the didactic use of HPS in the learning of physics concepts, despite there being no consensus about this, and they also indicated a lack of agreement about the occurrence of conceptual change. Greater research efforts are therefore needed to investigate these aspects, especially when the aforementioned limitations in research procedures are taken into account. In the same way, no consensus was found as to how HPS promotes improvements in the students' attitudes to science, which also leads the authors to conclude that this subject needs further investigation.

This type of approach appears to promote a more mature vision in respect of the students' understanding of NOS, which should be taken into consideration when planning curricula and/or physics teaching strategies. Favourable results were also found when looking at the effects of the didactic employment of HPS on the areas of argumentation and metacognition, despite the dearth of studies in the analysis dealing with these areas. Potentially important areas are being explored which warrant a higher position on the HPS-based physics teaching research agenda.

A recent thesis (Arriassecq 2008) dealt with the problem of meaningful teaching of the special relativity theory (SRT) at secondary school level in Argentina. Several studies have been carried out, focusing on the epistemological difficulties presented by the content of the SRT itself, the teachers' difficulty in approaching the task of dealing with such a theory at that level and the textbooks that both teachers and students would have as a teaching resource. The results of such studies showed that there is a wide gap between the proposals presented in the documents from the ministry, as well as some research reports, and their actual practice in class. In order to narrow this gap, a teaching proposal was developed in which the SRT is approached within a historic and epistemological context.

Part of the design of this teaching proposal – designed within a framework that comprises epistemological, psychological and didactic aspects – consisted of producing written material (in textbook format) to be used by teachers and students, taking into account the deficiencies identified through this approach.

Several studies conducted before the design of the teaching proposal and the supplementary teaching material for its implementation adopted an approach that assigns the use of the history of science and epistemology for the design of concrete class proposals, a role as important as that of working within a psychological and didactic framework. This contextualized approach places great conceptual emphasis on the topics discussed, so that the historical-epistemological discussions can be meaningful to students.

The use of the history of science and epistemology is considered to allow, among other aspects, the determination of the epistemological obstacles that serve as a guide for the selection of the relevant content to be taught as well as to favour the discussion of the production of scientific knowledge, the role of the social and cultural context at the time in which that knowledge is developed and its impact within and outside the scope of science. The aim should be to eradicate stereotypes about science that keep students away from this discipline.

With regard to the epistemological aspect, the thesis draws on elements of Bachelard's epistemology (1991) to produce an epistemological analysis of the content of the SRT (Arriassecq and Greca 2010). This analysis defined the central concepts that students should learn in a meaningful way. These are space, time and notions related to reference systems, observer, simultaneity and measurement, which are essential for a relativistic understanding of space-time.

The results of the assessment of the implementation of the teaching proposal show that the acquisition of key concepts of the SRT seems to be much better than those obtained when the SRT is approached in a 'traditional' way, in which the traditional textbook is the main teaching resource used by the teacher (Arriassecq 2008; Arriassecq and Greca 2010).

As for the text produced as part of the design of the teaching proposal, it treats in greater depth a topic that, despite its importance, has not been sufficiently investigated within the area of physics education in Argentina. In addition, it has been produced within an innovative theoretical framework that comprises epistemological, psychological and didactic elements.

71.3 Researches on HPST in Latin America

In Latin America, Brazil was the first country to consolidate the science education area and then to incorporate the study of the contributions made by HPS as a line of research within science education itself. Researchers in the science education area in Argentina have been establishing links with researchers in Brazil since this area started developing in this country. This has been done on the one hand through the guidance given to Argentinean researchers on their theses by prestigious Brazilian researchers such as Marco Antonio Moreira, from UFRGS. On the other hand, it has been accomplished by Argentinean researchers spending time at Brazilian institutes and universities as well as by disseminating and publishing Argentinean research projects in pioneering magazines issued in Brazil.

As Villani and colleagues (2010) point out in a thorough review, the development of science education research in Brazil was very similar to that which took place in many other Western countries, at least until the early 1990s. Research in science education first appeared systematically 40 years ago, as a consequence of an overall renovation of the field of science education. This evolution was also related to the political events taking place in the country.

After this period, Brazilian researchers became less dependent on foreign sources, and original lines of research were introduced, differentiating the development of this period from the one preceding it. Finally, the role of the history and philosophy of science appeared as an important intermediary during the stabilization process, not only unifying a great number of the research projects carried out during that period, but also serving as a means of participating in the political and ideological arena.

During the first phase, which included the founding of the area, the history and philosophy of science played only a limited role in the process of institutionalizing science education research. On the one hand, articles and books showed thinking based on the simplistic idea that the mere knowledge of the history of science would in itself stimulate students' motivation and facilitate their learning of scientific concepts. Other books and articles seemed unrelated to teaching and showed very little progress in exploring the history and philosophy of science more efficiently in the classroom. However, the community of researchers involved in educational projects considered that contributions from the history of science were very important, including its practical results, such as the production of teaching material to complement projects for physics teaching.

During the 1970s, some of the researchers went on to study further and publish a history of science that included the connections and commitments between scientific development and economic and political power. This task was considered a way to combat the military dictatorship then in power in Brazil and thus implicitly to denounce the pact between universities and the government.

The contribution of the history and philosophy of science to the consolidation of the area was stronger during the 1980s. First, journals founded during this period disseminated their ideas to others who were also interested in the history and philosophy of science. The *Journal of Physics Teaching* was launched in 1979, but until 1993 it had no specific section dedicated exclusively to the history and philosophy of science. Nonetheless, each edition contained individual articles dedicated to the theme. In contrast, right from its beginnings in 1984, the *Santa Catarina Journal of Physics Teaching*[1] included a section entitled 'The History and Philosophy of Physics'.

Another important contribution came from abroad, where the students could develop studies about the history and philosophy of science.

During the 1990s, a more theoretical contribution of the philosophy of science also fostered the development of variations in the Conceptual Change Model, giving special emphasis to other philosophers besides Kuhn and Lakatos. Specifically, some researchers developed analogies between science learning in school and the development of science as understood by Feyerabend, Laudan, Popper and Bachelard. Villani (1992) explored the flexibility of Laudan's approaches to the progress of science

[1] *Revista Catarinense de Ensino de Física.*

(Laudan 1984) for understanding the changes that take place in students' ideas in school. Based on Bachelard's idea of 'epistemological profile' (Bachelard 1978), Mortimer (1995) proposed a new version of the same concept, which became known as conceptual profile. He posited that cognitive evolution intrinsically joins old ideas with new ones and considered that teaching should promote a change in students' profiles by broadening their spectra of useful ideas.

During the last phase of the institutionalization process of science education research, two trends were developed in relation to the history and philosophy of science: towards reforming the cultural scientific knowledge on which high school education was based and towards training teachers to develop corresponding lines of research.

The first trend was a reform in scientific knowledge, focused on information concerning the genesis of conflicts and the evolution of scientific theories as well as on successes and failures. The work was carried out at schools themselves. But time should also be devoted to reflection on the presuppositions, images and basic intuitions of scientific advances. The training of science teachers should foster the acquisition of this knowledge for two purposes: first, from the cultural standpoint, to enrich the quality of the content to be taught and, second, to adhere to the methodological requirements that were considered the most appropriate.

In the second trend, researchers in the history and philosophy of science comprised a specific group, with their own graduate and postgraduate courses, journals and congresses.

The researchers' work became more technical (Martins 2000; Pietrocola 1992), their methodology became more precise, and case studies were included, such as information about Becquerel (Martins 1997) and the alchemist Sendivogius (Porto 2001). At the same time, efforts were made to keep in contact with the area of science teaching through attention to studies such as Newton's theory of colours, which is full of technicalities but useful for teaching purposes (Martins and Silva 2001).

In the case of Argentina, as Orlando and colleagues (2008) point out, the physics teachers' community received great encouragement in the 1980s. More precisely in 1983 in Cordoba, the physics education meeting (REF),[2] which attracts more than a thousand teachers of this subject matter, was held for the first time in several years. At that meeting the Association of Physics Teachers of Argentina (APFA)[3] was created. This association would be in charge, among other things, of organizing periodical meetings and events. The success of such events was reflected not only in the number of people present and the variety of topics that were discussed but also in the increasing number of research projects presented. Throughout the years the increasing number of research papers that were presented in that meeting showed the need for the creation of another meeting focused on research on the area. This new meeting was called the 'Symposium on Physics Education Research' (SIEF), and the first one took place in Tucumán in 1992.

---

[2] Reunión de Educación en Física.

[3] Asociación de Profesores de Física de la Argentina.

APFA is responsible for summoning researchers to an alternate REF and a SIEF every 2 years. The researchers' community has been growing in number and its tasks have become more specific. At the beginning the number of researchers with a doctoral degree was small, and this was done either abroad or under the guidance of foreign supervisors. Nowadays it is possible to form human resources and take postgraduate courses in our country.

The growing number of projects presented at the various symposiums throughout the years would reflect the researchers' increasing interest in working on topics related to physics education in the different education areas.

The increase in the number of articles that follow those criteria for research work since the first SIEF up to the present shows growth in the construction of knowledge in the field of physics education as well as an incremental change in specific training for researchers in science education.

It should be noted that the research projects presented at the symposiums deal with heterogeneous topics. Most of them refer to problems connected with teaching, learning, curricular aspects and context problems or about the teacher training or with educational transfer. A small percentage of them deal with topics related to theoretical frameworks, epistemological aspects or methodological development. However, there has been an evolution in research related to the use of HPS as a theoretical framework. This will be addressed further.

There are still some questions to be answered: Have the curricula – especially secondary school curricula – included HPS? Do the curricula of colleges of education provide for training in HPS? What happens with practicing teachers who as undergraduates did not have the chance to become familiar with the contributions made by the HPS to science education? Do class textbooks, both the teachers' and the students', incorporate such contributions? If they do, how do they do so? Are there any teaching proposals based on research results?

The next section evaluates HPS contributions to science education in Argentina. Some of the aspects analysed are the underlying epistemological features in education laws and natural science curriculum design at secondary level in Argentina, the incorporation of HPS in research and the dissemination of HPS contributions to physics teaching among in-service teachers.

71.4 Incorporation of Contributions from the HPS into Physics Education in Argentina

This section evaluates the incorporation of HPS into physics education in Argentina. The analysis focuses on education laws and on curriculum designs – particularly at secondary level – on the most relevant aspects of the curriculum designs for physics teacher training and on research and articles published in *Physics Teaching Magazine*, which is a publication issued by the Association of Physics Teachers of Argentina (APFA).

The selection of the materials analysed was based on the fact that the APFA magazine is the main channel of communication between researchers and teachers.

It should be noted that most of the researchers who attend national events and publish in the aforementioned magazine also attend international events and publish in prestigious magazines in other countries. However, only a small number of teachers have access to them.

71.4.1 Epistemological Features Underlying Educational Laws and Curricular Designs for Natural Science Teaching at Secondary Level in Argentina

This section analyses several national education laws that have been, and are still, enforced in Argentina, the regulations on some of these laws which provided a frame of reference at certain times in Argentinean history and in curriculum designs – for natural sciences at secondary level. Curriculum designs were formulated by the Ministry of Education and the Buenos Aires Provincial Ministry of Education in order to define possible epistemological conceptions in the history of Argentinean education, based on the stances underlying, sometimes implicitly, the aforementioned ministry documents (Framework for the Oriented Bachelor in Natural Sciences 2011).[4] This province was chosen because it accounts for 33 % of all the educational institutions of the country and accounts for 38 % of students.

Society has changed throughout the history of the country, from an economic, as well as a political and a cultural, point of view. Education policies are therefore expected to evolve and create favourable conditions for teachers, students and institutions so that objectives are achieved, adjusting to changes in society.

Particular consideration is given to whether there has been an evolution in the epistemological criteria that promote certain models for natural science teaching at secondary level and the development of notions, not only about science but also regarding scientific activity in different types of secondary schools. Furthermore, the aim is to identify certain characteristics present in the evolution of educational laws and in curriculum designs that may indicate an evolution from the so-called standard epistemologies to nonstandard ones.

The descriptive analysis is based on the identification of the characteristics of standard and nonstandard epistemologies in the following documents:

- The four national education laws: Law 1420 dating from 1884, Law 4874 from 1905, Law 24195 from 1993 and Law 26206 from 2006. Each was duly ratified and promulgated at a different time in Argentinean history, which means that each was conceived within different social, economic, political and cultural contexts. Only in the last two laws are there explicit references to epistemological debates.

[4] Marcos de Referencia. Educación Secundaria Orientada. Bachiller en Ciencias Naturales. Aprobado por Res. CFE N° 142/11. Consejo federal de Educación.

- Two magazines: *Annals of Education of Buenos Aires Province* and *of Santiago del Estero Province*. Since, in the past, curriculum designs were not developed, State guidelines on teaching practice are only found in this kind of magazine on education.
- National and Buenos Aires Province curriculum designs for natural sciences. This province was chosen because it includes 33 % of the educational institutions in the country and 38 % of the students.

The analysis focuses especially on education laws and the corresponding curriculum designs at the secondary level.

Law 24195, dating from 1993, sets the guidelines for the national educational policy under the so-called Education Federal Law (EFL) and determines the structure of the educational system as consisting of early education; basic general education; the 'polymodal' level, with final (preuniversity) oriented cycles, among which is the natural science-oriented one; and higher and postgraduate education. The objectives of each of these levels are also defined in the different articles of the official curricular document. The regulatory style of the previous laws was abandoned in order to extend compulsory education to 10 years.

With regard to teachers, their rights and duties are established in Art. 46 and 47 of the EFL. The pedagogical and curricular guidelines mention academic freedom and freedom of education.

This law is accompanied by a national curriculum design, and each province develops its own with the different polymodal orientations.

In the document entitled 'Curricular Support Document N°1' (Petrucci 1994), about the natural sciences area – in particular, physics curriculum design in Buenos Aires Province – the author states that the document '… is a useful theoretical support when planning, implementing and assessing teaching practice …'. Teaching is considered a professional practice in which theoretical foundations underlie decisions, and therefore there are neither recipes nor methods to implement the teaching-learning process.

The view of science expressed here contradicts the traditional one which considers that science uses a unique method consisting of a 'recipe': several steps to obtain a product. Science is regarded as '… an open process, the stages of which are determined by the issues under study, the aims of the study, the historical context and the interests of the community'. As regards the way in which research work should be presented, the document states that '… according to present epistemological conceptions, scientific knowledge is built through a process of development of theories and models that aim to give meaning to a reference experimental field'.

In addition, the document recommends some authors who, based on the present theoretical framework for science teaching, stress the dynamic and provisional character of scientific knowledge, highlighting therefore its dependence on the historical context (Kuhn 1962).

The latest law, the National Education Law, sanctioned in 2006, modifies the organization and selection of the curriculum contents of the national education system. Four levels of education were established, early, primary, secondary and upper secondary/higher, whereas the period of compulsory education was increased to 13 years.

The national curricular design corresponding to this law refers to that regarding the natural sciences orientation within upper-secondary level. In the introduction, teachers are asked to present science '... as a social construction that is part of culture, with its own history, communities, agreements and contradictions ...', to treat models and scientific theories as attempts to answer real problems and to take into account the role specific teaching methodologies play in understanding teaching-learning processes.

One of the general objectives is to form scientifically literate citizens who, during their school years, develop both scientific knowledge and a view on scientific activity.

In present designs, HPS takes on a different status, as the core thematic contents include science history and philosophy in order to approach the subject within its own curricular area at secondary level.

To sum up, the epistemological characteristics underlying educational laws and documents regarding natural science teaching at secondary level in Argentina have evolved since the first laws were sanctioned. They have developed from a teaching approach with positivist characteristics to models underlaid by nonstandard epistemological conception.

This evolution of the view on science promotes at the same time a change in the science teaching and learning approach, offering new roles for teachers and students, from a model in which the student is 'recipient' of knowledge 'imparted' by the teacher to one in which the student 'builds' 'academic knowledge' through the teacher's mediation.

71.4.2 The Incorporation of the HPS in Teacher Training

The general guidelines that physics teachers follow to learn topics of epistemology are derived from the regulations laid down by the organs responsible for education policies at the different levels of education in Argentina. In her thesis, Islas (2010) compiles all the documents that give advice on the incorporation of the HPS into the syllabuses of physics teacher training colleges.

The document '2008-Science education year' extracts the main points of the Report of the Argentinean National Commission for Improvement in Natural Science and Mathematics Education. The report stresses

> the need to overcome both the simplistic views on science and scientific work as well as those views of scientific work as something extremely difficult which lead to school failure. [...] At the same time the program aims at arousing interest in those disciplines that follow from understanding what producing science and producing mathematics mean, their usefulness and importance for citizenship; at demystifying the process of knowledge development for students and teachers at different education levels, encouraging them to value it as an activity for social construction; at promoting future scientific vocations.

In the recommendations section of the document, those objectives are translated into some suggestions that are transferable to teaching practice. It is recommended that the different aspects of scientific knowledge should be taken into account, some of which are 'its empiricism, the need to build models, compulsory debate and discussion of the results and their interpretations'.

There are other points in the document regarding classroom work, of which it is worth mentioning the following one:

> Generally, in order for the students to build solid knowledge there should be experimentation, high frequency of questions, socratic dialogue, and rigorous, logically sound and simple reasoning. All these are characteristics of "proper thinking" in the science class. But they are also distinctive characteristics of scientists' thought when doing research.

Some lines below, in order to differentiate the contexts where knowledge is developed (the class and the researchers' community), it is stated that 'the most significant difference between both activities is that, whereas the scientific community generates new knowledge on the borderline between the known and the unknown, students in class build concepts that, despite being new to them, have already been validated by science'. Finally, it is worth pointing out that the Report of the Argentinean National Commission for Improvement in Natural Science and Mathematics Education states that one of the obstacles that have been detected in the diagnostic is the 'stereotyped picture of science and scientists, also shared by teachers'.

Most of the professors who are members of universities or state institutes have some curricular time at their disposal to study topics related to the HPS. Moreover, all courses include time for seminars in which it would be possible to approach topics connected with the construction of scientific knowledge.

Islas (*op. cit.*) has also done a review of the syllabuses, taking into account, apart from the list of contents, other elements such as the bibliography, the objectives of the subject and the requirements for passing the subject.

A characteristic that all programmes under analysis have is that they include nonstandard epistemologies. Even more, the section entitled 'tendencies among contemporary epistemologies', which appears in almost all syllabuses, is considered an indicator of the presence of innovating explanations. Authors like Kitcher, Giere, van Fraassen, Habermas and Gadamer appear in the references together with the more common ones like Lakatos, Feyerabend, Laudan, Toulmin and, to a lesser extent, Bachelard.

71.4.3 The Communication Among In-Service Teachers with Regard to HPS Contributions to Physics Teaching

The magazine *Physics Teaching,*[5] first published in 1985, is an undisputed reference on this issue at every level and is also a vehicle for communication among the members of the Association of Physics Teachers of Argentina.

---

[5] *Revista de Enseñanza de la Física.*

The magazine has been published twice a year since 1992, and there are some special issues, with research articles, proposals for the classroom, reflections, notes and information on different aspects related to physics and physics teaching. It makes possible the communication of research works in the area of science education, providing insight into theoretical foundations, analysing the state of the art and making headway in understanding significant issues. It provides elements to enhance teaching practice, allowing the publication of teaching proposals, discussion of particular activities and analysis of experiences. It promotes events of interest, and it offers information and news that help members of APFA, as well as any readers, to keep up to date.

Three publishing teams have been in charge of the magazine during its 27 years of existence. The first one was in charge from 1985 to 1992, the second from 1993 to 2002 and the present one from 2003. The same spirit has been present in all of them, and it is possible to find articles that match the objectives mentioned above. However, the organization of the magazine has gone through different changes.

The first issues did not have permanent sections. During the second publishing period, the following sections were introduced: research and development, teaching issues, physics topics, history, science philosophy, information and news. Not all these sections were present in every issue and were replaced by others such as laboratory work and extracurricular activities.

The magazine now has permanent sections in all the issues. Each section is briefly discussed below so as to analyse those that deal with HPS contributions.

The 'Research' section includes articles on education research, related to teaching and learning of physics and other experimental sciences in general. It covers the incorporation of empirical research that answers paradigms and diverse approaches, as long as this is carried out with scientific rigour and systematicity, the development of theoretical frameworks, advances in methodology or revisions that deal with the state of the art on topics regarding research in science education. The 'Proposals' section includes articles offering teaching alternatives, innovations and curriculum formulation, among other topics, in connection with the teaching of physics and other experimental sciences at different levels. It may include contributions made in order to integrate physics with other sciences, specific classroom proposals and analysis of the curriculum and of resources. In the 'Essays' section special thematic articles are published (articles on physics topics, on the science-technology-society-environment relationship, philosophical and epistemological reflections, historical accounts, description of projects or programmes, etc). The 'Miscellaneous' section deals with discussions of problems, challenges and paradoxes, comments on books and/or software, teaching materials on the web, projects, classroom resources, etc. Finally, the 'News' section includes information on events and news of interest, workshops, development and information about postgraduate level courses, thesis abstracts and book reviews.

In the second period of the magazine, two types of articles coexist in the 'History' and 'Science Philosophy' sections. On the one hand, there are those that have been written by teachers and physics researchers who have taught or done research in this subject at different levels of education. Even though they lack a formal education in history and epistemology, they have been in a way pioneers in the incorporation of such debates in teaching. The first issue in 1997 featured an article in this respect written by Dr. Alberto P. Maiztegui – for many years president of the National Academy of Sciences and without any doubt remembered for his contributions to science teaching: his physics books for students and teachers, the promotion of science fairs in Argentina and the steady support to these topics that he provided from IMAF (nowadays FaMAF) and later the National Academy of Sciences. The article was entitled 'Archimedes and Hieron's crown' and was based on the well-known anecdote that describes Archimedes running naked around the streets of Syracuse exclaiming '¡Eureka! … ¡Eureka!' because of the joy he felt when finding a solution to the scientific problem he was working on. The article develops a series of detailed calculations, in connection with the problem in question, which are not frequently found in secondary textbooks. It should be noted that the article cites a source where the anecdote can be read, contrary to other texts that try to incorporate 'historical cartoons', without any mention of sources, only because students may find them attractive.

In the same issue, Guillermo Boido published the article 'The reconstruction of experiments in the history of science: Galileo under discussion' which states that the reconstruction of experiments carried out by scientists in the past has contributed to a better understanding of certain historical episodes and that this method has gained popularity especially when applied to Galileo's work, even though the results have received controversial interpretations. The possibilities and limitations of experimental history are explored, and there is a brief analysis of the state of the discussion – at the end of the 1990s – on the character of Galilean science in light of the reconstruction of some of his experiments. Prof. Boido does research into epistemology and science history and has been associate editor of the prestigious Argentinean scientific magazine *Science Today.*[6] The professional profile of this author is clearly different from that of the author of the article mentioned above. This article focuses on science history research problems. At the same time, the editors point out that the article

> … is valuable material for our readers, particularly now that science history demands attention on the part of science education and a better place in the curricula in general, both at secondary and higher education levels.

Finally, in the present stage of the magazine, there has been a significant change with regard to the kinds of publications that consider the incorporation of HPS contributions to science and particularly to physics education. They share in common the fact that they are research articles written by researchers who have been specifically trained in science teaching. This means that they are not historians or philosophers making contributions from their own disciplines or physicists exploring history and philosophy topics. They are physics teachers and physics teaching researchers who, based on a solid background as regards the nature of science,

---

[6] *Revista Ciencia Hoy.*

identify problems in physics teaching and learning and deal with them from a historically and epistemologically contextualized approach.

The following are examples of articles of such a kind (all of which were published in the 'Research' section, except one that appeared in the 'Thesis' section):

'Teachers' conceptions of the role of scientific models in physics classes' (Islas and Pesa 2004)
'The Scientific Revolution in the Argentinean Education System' (Cornejo 2005)
'Analysis of relevant aspects to deal with the Special Relativity Theory in the last years of secondary school from a historically and epistemologically contextualized approach. First part 1 and 2' (Arriassecq and Greca 2005)
'Laboratory work design within an epistemological and cognitive framework: Physics Teacher Training College case' (Andrés Zuñeda 2006-doctoral thesis-)
'Mechanics teaching at secondary school: historical evolution of the texts (1840–2000)' (Cornejo and Nuñez 2006)
'The view of a group of science teachers on science and schooling' (de la Fuente et al. 2006)
'Teaching the philosophical components and the different views on the world of science: some considerations' (Matthews 2009)
'Epistemology for physics teaching training courses: transpositive operations and the creation of a 'metascientific school activity'' (Adúriz-Bravo 2011)

71.4.4 *The Incorporation of Contributions Made by the HPS into Argentinean Textbooks*

There is research that studies how to incorporate the HPS into the physics textbooks used by both teachers and students. Arriassecq and Stipcich (2000) follow this line of scientific enquiry. Their work offers a critical analysis of the incorporation of the HPS into physics secondary textbooks that were written after the educational reform that took place in Argentina in 1993. One of the main conclusions these authors point out is that the HPS is not incorporated into school textbooks within a theoretical framework but it is rather reduced to mere anecdotal vignettes or, at best, to the inclusion of a chapter on epistemological aspects which do not bear any relationship with the rest of the text structure and approach. They also draw attention to the need to incorporate concrete proposals to deal with the contents of the HPS in class.

In later studies (Arriassecq and Greca 2004, 2007), it is argued that the results of different investigations point out that:

- The textbook appears to be the main resource that teachers use for preparing their classes, especially at secondary level. The same textbooks are recommended to students.
- The way in which topics are approached may seriously condition the results achieved by students when learning them.

In the same studies, which refer specifically to the treatment textbooks give to the special relativity theory (SRT), it is argued that the teachers who face the task of approaching the SRT for the first time will generally resort to the textbooks as a guide for their classes. Considering that in many cases the teacher has not had the opportunity to reflect on which concepts are the most relevant to understand the theory, he will probably follow the plan offered by the textbook or several textbooks he has selected to prepare his class, without adapting the material to fit his own criteria.

On the other hand, the results presented in this article coincide with those obtained on the same topic in other countries. This demonstrates that in order for secondary teachers to approach topics, such as the SRT in this case, from an epistemological and contextualized perspective, the available teaching materials are inefficient.

Based on that fact, it seems necessary to produce teaching material to be used by teachers and students, which provides for students' meaningful learning by introducing contents appropriately from a conceptual and motivational point of view.

This material could offer a serious discussion based on the contributions made by research into physics education, about the contextual aspects that are relevant to several physics theories.

Kragh's contextualized, or 'anti-Whig', approach analyses historical events in light of the beliefs, theories and methods that belong to the time when the ideas were conceived. This view offers a more realistic idea of history that does not fail to take into account obstacles and mistakes in scientific work. This view of science that textbooks may therefore reflect would be more realistic since it would give the same importance to successes as to failures.

71.5 Incorporation of HPS into Biology Education in Argentina

This section evaluates the incorporation of HPS into biology education in Argentina. The analysis focuses on the most relevant aspects of the curriculum for biology teacher training and research and the dissemination articles published in *Biologics Teaching Magazine*, which is a publication that has been issued by the *Argentinean Biologics Teachers Association* (ADBiA) for the last 15 years.

HPS has been incorporated into biological sciences and their teaching, both in the curriculum and in teacher training colleges, on the basis of complementary lines of development. On the one hand, there has been an advance and an epistemological turn in the development of knowledge in biology in the twentieth century (the molecular revolution, genetics and biotechnology, ecosystemic studies regarding human production, economy and consumption). On the other hand, the global/local crisis in scientific education has given birth to a second line that questions the science curricula, and a third line refers to science teaching in search of identity as a field of education research.

71.5.1 Evolution and Changes in Biology

Biology studies life and its organization in unifying principles of different levels of complexity: biosphere, ecosystem, population, individual, organism, organs, tissues, cells, macromolecules and biochemical level (biodiversity, taxonomy, Mendelian and population genetics, embryology, biology of the organism, molecular biology).

It is a mainly historical and evolutionary science, which develops explanatory models based on different research methods – comparative, systemic, hypothetical-deductive, genetic and historical ones – within structural, functional and behaviourist approaches (Ruiz and Ayala 1998; Barberá and Sendra 2011).

The conceptual and methodological development of biology differs from the research paradigm for physics, since life processes – self-regulation, unstable equilibrium and invariant and irreversible evolution – are informed by diachronic and synchronic perspectives, articulating internal and external interactions in an open system. On the other hand, its explanatory models are connected with various social and human practices (Giordan 1997; García 2006).

Advances resulting from technological research and applications to improve the quality of life, such as the digitalization of the genetic code and its molecular delimitation, led to numerous developments in biochemistry, medicine, technology and science. These, in turn, brought new and complex ethical problems with an economic and social impact (transplants, medicines, biochemical weapons, food production and biotechnology) (Testart and Godin 2002; Geymonat 2002).

With that in mind, and as a result of the multiple ethical conflicts, studies have been promoting new research and educational directions that combine new approaches (environmental, STSE, humanistic ones) with the bioecological, social, economic and political dimensions (Gudynas 2002; Sacks 1996). As a result, different areas of knowledge have produced proposals of an interdisciplinary nature linking science, cultural practices and the natural and social environment, as is the case for health and environmental education.

Biological knowledge in textbooks and curricula was not greatly updated until 15 years ago when the popular communication of the issues offered an opportunity to revise knowledge in interaction with health, economy, agricultural production, food and medicine industries, etc. (Datri 2006; Memorias ADBia 1993–2000).

71.5.2 Scientific Education Crisis

The international education movement following this line (Fourez 1997; Jenkins and Pell 2006; Hodson 2003), as well as the dissemination of magazines dealing with science education during the last 40 years, is evidence of the ideological turn which took place from the 1970s, affecting the objectives of scientific training. Some of them that centred on the development of theories and concepts pertaining only to the discipline were gradually modified by the incorporation of new objectives and strategies – the scientist's doings; the question of method and disciplines; and the incorporation of history, the sociocultural context of science and the ideological, economic and ethical assumptions (Latour and Woolgar 1995; Matthews 1991, 2009).

The science teacher associations, the AAAS and the NSTA (National Science Teachers Association, USA), have been contributing with their harsh criticism since the 1970s: they recommended in their documents of the years 1979 and 1986 that students should be given the chance to explore the history and the philosophy of science. In other words, students should reach basic understanding of how sciences and technologies were developed in the context of humanity. The long-standing recommendations and projects for the incorporation of HPS into science education constitute a tradition that dates back to the mid-nineteenth century (Paul Tannery, Pierre Duhem, Ernst Mach), the analysis of the historical cases of Harvard's project (Conant, James) or the British Nuffield project (Duschl 1995; Datri 2006; Martínez and Olivé 1997).

Therefore, the so-called science for all takes up the 1960s and 1980s challenge (Secondary Curriculum Review 1983 Great Britain; Learning in Science in New Zealand) to try to cover educational deficiencies that are the consequences of a scientific and cultural heritage from the mid-century: the break with the certainties of scientific progress and the manipulation of its intellectual product (Fensham 2002; Ramontet 1997). Such contributions lead to educational proposals based on three lines: 'Science in society', 'Science in a social context' and 'Science and Technology in Society', developing a range of topics that give rise to teaching booklets on scientific work: the role of government and industry in science, the commercialization of scientific breakthroughs, the involvement of researchers in food production, the fight against disease, nuclear weapons, technology in everyday life, etc.

These events interact with an educational crisis in biology teaching at school level which establishes the need for science for all, from the point of view of its role in society rather than science itself, thus adding weight to the arguments that claim that the curricula present a very limited vision of science, without any historical and cultural contextualization or any reference to its social impact. This approach aims to motivate the student not only to study scientific disciplines but also to view themselves as citizens and to be able to take a stand on the value and the use of scientific knowledge.

It is in this spirit that the *Argentinean Biologics Teachers Association* (ADBIa 1993) began active participation and played a crucial role in establishing and defining the epistemological scope of the contents set by the education reform in Argentina (1995, 1997) and in highlighting the need to include contextual, historical and ideological approaches in the biology curricula. After the return of democracy (1983), teachers' opinions on biology teaching became visible during the discussions within the framework of the curricular reforms, pointing out as weaknesses the dissociation between what is taught and real everyday problems, the encyclopedic approach, the atomization, the lack of the history of ideas and cross-cutting topics (ADBia 2002). In this respect, secondary teachers are the ones who have made constant demands for the updating of the subject in line with the results of research in the field of biology (De Longhi and Ferreyra 2002).

In the 1990s, the Science, Technology, Society and Environment (STSE) approach pervaded biology educational programmes in different ways: (a) CTS[7] was incorporated into a year or course; (b) from *scientific problems to concepts* was taught from the CTS approach; and (c) the scientific content in CTS proposals had 'a subordinate role' (Marco Stiefel 2005). It should be noted that the STSE approach was expressed in the biology curricula through the teaching of environmental conflicts, facilitating a break with the traditional scientific content and introducing new strategies for understanding and dealing with the environmental conflict, the study of the social reality and a conscientious citizenship.

Concurrently with the contributions from the new philosophy of science, psychogenetic studies on biological notions (Giordan and De Vecchi 1987; Piaget and Garcìa 1982), in 1990 the ADBiA Journal presented an interesting alternative to explore the historical perspectives of the ideas and the socio-cognitive processes in students, in this way providing the educational debate with new epistemological criteria for establishing the curricular science contents. In this context, it therefore became essential to reintroduce the history of science and its conceptualization, especially the two characterizations regarding the nature of science, expressed as (a) science as a process of justification of knowledge (what we know) and (b) science as a process of discovery of knowledge (how we know).

In this regard, the first characterization has dominated the contemporary teaching of biological sciences, providing incomplete knowledge of its conceptual and axiological field. There is at present a need to broaden the science curriculum and to design and implement teaching proposals that deal with the other aspect, that is to say, 'how' the model for the transmission of hereditary traits or its evolutionary process became known (Wolovelsky 2008). This involves going beyond a simple historical account as the central theme of our classes to deal with philosophical views which serve as tools for analysis and meta-reflection, allowing a better understanding of the aspects of scientific practice and showing the different lines of argument in the context of technological and sociocultural breakdown (e.g. the spontaneous generation theory of the twelfth century, the fixism theory in the sixteenth century or the synthetic theory of evolution in the twentieth century). There are many beliefs and pseudosciences present in our culture, especially regarding topics like the origin, continuity and evolution of life, to which the historical and philosophical contextualization may significantly contribute by favouring less radicalized and less dogmatic positions (Schuster 1999; Palma & Wolovelsky 2001).

In this regard, we now know that the meta-scientific approach allows the establishment of relationships between the knowledge taught and its historical and evolutionary context, problematizing it within a cultural moment, with the strategies, ideas and problem-solving approaches available at that time. This meta-scientific component in the definition of science teaching practices on the

[7] Ciencia, Técnica y Sociedad.

one hand informs strategies for didactic transposition and, on the other, broadens the scope of traditional science teaching and learning models (Adúriz-Bravo et al. 2002b; Aduriz-Bravo 2005; Quintanilla et al. 2005).

Following from this, objectives have been set for the teaching of the biological sciences which relate to those skills students should develop:

- To learn the concepts within the context of the models and theories that created them. That is to say, bring the intention behind the phenomena closer to the models put forward by the scientific community. Such interpretation calls for the development of cognitive and scientific reasoning skills, usually referred to as 'doing sciences'.
- To promote conceptual change, practical and argumentative reasoning and understanding of the problematic, historical and cultural condition of scientific practice.
- To develop critical and projective thinking, which enables students to give opinions and take decisions. All of the aforementioned should provide for a non-neutral image of science under constant review, with technological applications and immersion in a sociocultural context.
- Promote, in turn, scientific literacy that provides basic culture and enables students to take decisions, analyse information, raise questions and detect deception.

71.5.3 The Identity of Teaching as a Research Area

On the one hand, the studies developed from the teaching of science provide an area of epistemological enquiry and revision of scientific knowledge, developing processes of transposition of communication and teaching hypotheses.

In this regard, the history and evolution of the major theories of the last 150 years confer identity to biology concerning not only its content and explanatory models but also the impact such knowledge causes, the attitudes it fosters and its relationship with different ways of life and culture (Memorias de las V Jornadas Nacionales de Enseñanza de la Biología). Instead of an academic practice, the teaching of science becomes therefore a learning process oriented to daily life, community, work and production activities.

There emerges the view that the student should acquire an idea of science connected with social issues, especially specific present ones. Among these, we can mention those related to the main pillars of development and sustainability, the management of natural resources, the oil and petrochemical industry, the mining industry and technologies, agriculture and agroindustries, metal mechanics, the food industry, the environment, health and biotechnology (Meira 2006).

Present societies are gradually becoming more and more dependent on technological knowledge, restating the relationship *production-information-education*. A new kind of citizen literacy is then suggested since the communication and appropriate significance of topics related to health, nutrition and pollution demand individuals who can be critical of problems and their solutions and able to improve the quality of life and their environment. In this direction, research into teaching has also proved that learning centred only around the incorporation of conceptual content fosters a distorted and limited view of scientific activity and its real production practices.

A wide range of research works agree on the fact that students gain a better conceptual understanding when they try to understand the origin and nature of knowledge, the argumentative conflicts and the sociology of research, as well as the ethical and attitudinal dilemmas among individuals and institutions (Jiménez and Sanmarti 1997; Lemke 2006; Rivarosa et al. 2011).

Moreover, the systematization of research into biology teaching at the national level (De Longhi et al. 2005; Berzal 2000; Astudillo et al. 2008) highlights the need for innovation in teaching practice based on issues related to (a) the history of biological notions and their epistemological development and the analysis of students' notions and conceptual obstacles and the problematic issues in connection with biology and culture, (b) the teacher's background and thinking and the transposition of communication and (c) the teaching models and the curricular materials.

It should be noted that the main concern in the last decade (2000–2012) has been mainly to promote curricular and training opportunities (postgraduate courses, master's degrees, seminars) for teachers to develop a real understanding of scientific activity and its relationship with genuine issues, to be used as a relevant criterion to rethink the teaching of biology. For this purpose, time should be devoted to education research into major biological notions, the analysis of documentary sources, biographical texts and historical events, the levels of curricular complexity as well as tracing back instances of scientific performance, all of which will help to question and change conceptions regarding biological theories and their teaching.

Furthermore, the bringing together of the meta-scientific knowledge (HPS) and the teacher's subject and teaching knowledge provides complementary options for problematizing themes. These allow for the combination of conceptual history and experimental design, theory and argumentation, metacognition and educational transposition and the relationship among science, culture and society.

In this regard, knowledge of the historical evolution of the issues that make up scientific culture provides a greater opportunity to 'devise' new strategies and educational objectives, the purpose and the reason for educating in science, that is to say, what is the importance of scientific education in our present society? Who is science for? What ideas and values underlie research practice? Does the available scientific and technological knowledge foster a way of thinking and acting for social change and the improvement of the quality of life and the environment?

71.6 Conclusion

The last 20 years has seen a significant rapprochement between the area of science education and HPS. Advocates of the incorporation of aspects of the HPS into science teaching, even though they are aware of the existence of differing opinions, stress the importance of a contextualized approach to this topic. That is to say, teaching science in a way that would allow students to carry out a critical analysis of the fact that the social, historical, philosophical, ethical and technological context is closely linked to the development, validation and application of scientific knowledge.

However, it is worth pointing out that, as Matthews (2000) states,

> It is unrealistic to expect students or preservice teachers to become competent historians, sociologists, or philosophers of science. We should have limited aims in introducing questions about the nature of science in the classroom: a more complex understanding of science, not a total or even a very complex, understanding.

Nevertheless, it is essential to consider in greater depth the reformulation of curricular projects at the different education levels – including science teacher colleges of education – taking into account the evaluation of the results obtained from them. At the same time, it should, on the one hand, be necessary to consider the possibility of training practicing teachers who as undergraduates never had the opportunity to approach the issue from this perspective. On the other hand, a greater production of written teaching material for students as well as for teachers should be beneficial. In recent years, there have been a growing number of research studies of science education that present teaching units for different science topics within a historical and epistemological context and that take account key aspects of the NOS. However, only a few of them transcend and reach teachers, pre-service teachers and students.

If one examines the aspects of NOS that have been incorporated into curricular designs during the last twenty years in Argentina, one notices a significant development. At the same time, research work in the subject has increased, and teachers are kept informed about the results of such research through national magazines, among others. However, it is necessary to promote other training instances for those in-service teachers who have no access to these publications or conferences. This is essential if it is our aim to form scientifically literate citizens.

References

Acevedo-Díaz, J., Vázquez-Alonso, A., Manassero, M. & Acevedo-Romero, P. (2007). Consensos sobre la naturaleza de la ciencia: fundamentos de una investigación empírica. *Revista Eureka* 4(1), 42–66.

Aduriz-Bravo, A. (2005). *Una introducción a la naturaleza de la ciencia*, Fondo de Cultura Económica.

Adúriz-Bravo, A. (2011). Epistemología para el profesorado de física: Operaciones transpositivas y creación de una "actividad metacientífica escolar". *Revista de Enseñanza de la Física*, 24(1), 7–20.

Adúriz-Bravo, A., Couló, A., Kriner, A., Meinardi, E., Revel Chion, A. & Valli, R. (2002). Three aspects when teaching the Philosophy of Science to science teachers, http://www1.phys.uu.nl/esera2003/programme/listofauthors2.htm. Accessed 25 June 2007.

Adúriz-Bravo, A., Perafán, G. & Badillo, E. (2002b). Una propuesta para estructurar la enseñanza de la filosofía de la ciencia para el profesorado de ciencias en formación, *Enseñanza de las Ciencias* 20(3), 465–476.

Adúriz-Bravo, A., Salazar, I., Mena, N. & Badillo, E. (2006). La epistemología en la formación del profesorado de ciencias naturales: aportaciones del positivismo lógico, *Revista Electrónica de Investigación en Educación en Ciencias* 1, 6–23.

Arriassecq, I. & Stipcich, S. (2000). La visión de Ciencia en los textos de Física de nivel básico, *Memorias V Simposio de Investigadores en Educación en Física*, Santa Fe, 18 al 20 de octubre de 2000 (CD).

Arriassecq, I. & Greca, I. (2004). Enseñanza de la Teoría de la relatividad Especial en el ciclo polimodal: dificultades manifestadas por los docentes y textos de uso habitual, *Revista Electrónica de Enseñanza de las Ciencias* (http://www.saum.uvigo.es/reec) 3(2), Artículo 7. ISSN: 1579 - 1513.

Arriassecq, I. & Greca, I. (2005). Análisis de aspectos relevantes para el abordaje de la Teoría de la Relatividad Especial en los últimos años de la enseñanza media desde una perspectiva contextualizada histórica y epistemológicamente. *Revista de Enseñanza de la Física*, 18(1), 17–24.

Arriassecq, I. & Greca, I. (2007). Approaches to Special Relativity Theory in School and University Textbooks in Argentina, *Science & Education* 16(1), 65–86.

Arriassecq, I. (2008). *La Enseñanza y el Aprendizaje de la Teoría Especial de la Relatividad en el nivel medio/polimodal*, tesis de doctorado, en prensa (Universidad de Burgos, España).

Arriassecq, I. & Greca, I. (2010). A Teaching–Learning Sequence for the Special Relativity Theory at High School Level Historically and Epistemologically Contextualized, *Science & Education*, now online.

Astudillo, C., Rivarosa, A. & Ortiz, F. (2008). El discurso en la formación de docentes de Ciencias. Un modelo de intervención, *Revista Iberoamericana de Educación* N° 45/4, 1–13. http://www.rieoei.org/deloslectores/2107OrtizV2.pdf Accessed 30 November 2010.

Bachelard, G. (1978). *A filosofia do não*, Abril Cultural (Col. Os Pensadores), São Paulo.

Bachelard, G. (1991). *La formación del espíritu científico*, Siglo XXI.

Barberá, O & Sendra, C. (2011). La biología y el mundo del siglo XXI, en Pedro Cañal (coord), *Biología y Geología. Complementos de formación disciplinar*, vol. I, Edit. Grao-Barcelona, pp. 77–94.

Bell, R. L., Abd-El-Khalick, F., Lederman, N.G., McComas, W.F. & Matthews, M.R. (2001). The Nature of Science and Science Education: A Bibliography, *Science & Education* 10(1/2), 187–204.

Berzal, M. (2000). La investigación en didáctica de la Biología: temas problemas y tendencias, mesa debate II Congreso Colección Revista de Educaciòn en Biología (ADBiA), Personería Jurídica 201/A 96. ISSN 0329-5192-Indexada Latindex. Argentina.

Cornejo, J. (2005). La Revolución Científica en el Sistema Educativo Argentino. *Revista de Enseñanza de la Física*,18(1), 55–67.

Cornejo & Nuñez (2006). La enseñanza de la mecánica en la escuela media: la evolución histórica de los textos (1840–2000). *Revista de Enseñanza de la Física.*, 19(2), 35–46.

de la Fuente, A., Gutiérrez, E., Perrotta, M. , Dima, G., Botta, I., Capuano,V., & Follari, B. (2006). Visión de ciencia y de lo escolar en un grupo de docentes de ciencias. *Revista de Enseñanza de la Física*, 19(2), 47–56.

Datri, E. (2006). Una interpelación desde el enfoque CTS a la privatización del conocimiento. Política, Ideología y Tecnociencia, *Colección de Cuadernillos para pensar la enseñanza universitaria,* Año 1, N° 7, Universidad Nacional de Río Cuarto, Río Cuarto.

De Longhi, A. & Ferreyra, A. (2002). La formación de docentes de Ciencia en Argentina. Problemáticas asociadas a su transformación, *Journal of Science Education*, 2(3), 95–98.

De Longhi, A. (Coord.), Ferreyra, A., Paz, A., Bermudez, G., Solis, M., Vaudagna, E. & Cortez, M. (Integrantes) (2005). *Estrategias Didácticas Innovadoras para la Enseñanza de las Ciencias Naturales en la Escuela,* Ed. Universitas, Córdoba.

Duschl, R. (1995). Más allá del conocimiento: los desafíos epistemológicos y sociales de la enseñanza mediante el cambio conceptual, *Enseñanza de las Ciencias*, 13(1), 3–14.
Duschl, R. (1997). *Renovar la enseñanza de las ciencias*, Narcea.
Fensham, P.J. (2002). Time to change drivers for scientific literacy, *Canadian Journal of Science, Mathematics and Technology Education* 2(1), 9–24.
Fourez, G. (1997). Scientific and Technological Literacy, *Social Studies of Science* 27, 903–936.
Framework for the Oriented Bachelor in Natural Sciences (2011). http://www.me.gov.ar/consejo/resoluciones/res11/142-11_cs_naturales.pdf.
García, R. (2006). *Sistemas Complejos, abordajes interdisciplinares y fundamentos epistemológicos*, Ed. Gedisa, México.
Geymonat, L. (2002). (trad.) *Límites actuales de la filosofía de la ciencia*, Edit. Gedisa, Barcelona.
Giordán, A. (1997). Las ciencias y las técnicas en la cultura de los años 2000, *Kikirikí*, 44–45, 33–34.
Giordan, A. & De Vecchi, G. (1987). *Les origines du savoir*, Delachaux, Paris.
Gudynas, E. (2002). *Ecología, Economía y Ética del Desarrollo Sustentable*, Ediciones Marina Vilte, Buenos Aires.
Hodson, D. (2003). Towards a philosophically more valid science curriculum, *Science Education* 72(1), 19–40.
Islas, S. & Pesa, M. (2004). Estudio comparativo sobre concepciones de modelo científico detectadas en Física. *Ciencia, Docencia y Tecnología*, 15(29), 117–144
Islas, S.M. (2010). *La formación epistemológica de los profesores de Física respecto de los debates producidos al interior de la comunidad de investigadores en la disciplina*, tesis de doctorado, Universidad Nacional de Córdoba.
Jenkins, E.W. & Pell, R.G. (2006). *The Relevance of Science Education Project (ROSE) in England: A summary of findings*, University of Leeds, Leeds (online).
JIMÉNEZ ALEIXANDRE, M. P., y SANMARTI, N. (1997): "¿Qué ciencia enseñar: objetivos y contenidos en la educación secundaria?". In CARMEN, L. del (Ed.), *La enseñanza y el aprendizaje de las ciencias de la naturaleza en la educación secundaria* (pp. 17–23). Barcelona: ICE-Horsori.
Klimovsky, G. (1994). *Las desventuras del conocimiento científico. Una introducción a la epistemología*, AZ Editores, Buenos Aires.
Kuhn, T. (1962). *The Structure of Scientific Revolutions*. University of Chicago Press.
Lakin, S. & Wellington, J. (1994). Who will teach the "nature of science"?: teachers' views of science and their implications for science education, *International Journal of Science Education* 16, 175–190.
Lantz, O. & Kass, H. (1987). Chemistry Teachers' Functional Paradigms, Science Education 71, 117–134.
Latour, B. & Woolgar, S. (1995). *La vida en el laboratorio: la construcción de los hechos científicos*, Alianza Editorial, Madrid.
Laudan, L. (1984). *Science and values*, University of California Press, Berkeley, CA.
Lemke, J.L. (2006). Investigar para el futuro de la educación científica: nuevas formas de aprender, nuevas formas de vivir, *Revista Enseñanza de las Ciencias* 24(1), 5–12.
Losee, J. (1972). *Introducción histórica a la filosofía de la ciencia*, Alianza Editorial, Madrid.
Marco Stiefel, B. (2005). La naturaleza de la ciencia, una asignatura pendiente en los enfoques CTS: retos y perspectivas, en Membiela, P. & Y. Padilla (Eds.), *Retos y perspectivas de la enseñanza de las ciencias desde el enfoque Ciencia-Tecnología-Sociedad en los inicios del Siglo XXI*, Educación Editora, Vigo, pp. 35–39
Martínez, S. & Olivé, L. (1997). *Epistemología evolucionista*, Paidós, México.
Martins, R.A. (1997). Becquerel and the choice of uranium compounds, *Archive for History of Exact Sciences* 51(1), 67–81.
Martins, R.A. (2000). Que tipo de história da ciência esperamos ter nas próximas décadas? [What kind of history of science do we expect having in the next decades?], *Episteme: Filosofia e História das Ciências em Revista* 10(1), 39–56.

Martins, R.A. & Silva, C.C. (2001). Newton and colour: The complex interplay of theory and experiment, in Bevilacqua, F., Giannetto, E. & Matthews, M.R. (Eds.), *Science & Education*, Kluwer Academic, Dordrecht, The Netherlands, pp. 273–291.

Matthews, M.R. (2000). *Time for Science Education: How Teaching the History and Philosophy of Pendulum Motion can Contribute to Science Literacy*, Kluwer Academic, Springer.

Matthews, M.R. (2009). Science, worldviews and education: an introduction. In Matthews, M. (Ed.), *Science, Worldviews and Education: Reprinted from the Journal Science and Education*, Springer Science/Business media, Sydney, pp. 1–25.

Matthews, M.R. (1991). Un lugar para la historia y la filosofía en la enseñanza de las Ciencias, *Comunicación, Lenguaje y Educación* 11(12), 141–145.

Meira, P. (2006). Crisis ambiental y globalización: una lectura para educadores ambientales en un mundo insostenible, *Trayectorias* 8(20/21), 110–123.

Memorias de las V Jornadas Nacionales de Enseñanza de la Biología. ADBIa (Nacional/ Internacional. Años1993; 1994; 1996; 1999; 2002; 2005; 2008; 2010).

Mortimer, E.F. (1995). Conceptual change or conceptual profile change, *Science & Education* 4(3), 267–285.

Orlando, S., Gangoso, Z., Lecumberry, G. & Ortiz, F. (2008). Educación en Física, ¿Qué?, y ¿Dónde investigamos?: Una mirada a la producción nacional. *Memorias del SIEF 9.*

Palma, E. & Wolovelsky, H. (2001). *Imágenes de la racionalidad científica*, Eudeba, Buenos Aires.

Petrucci, D. (1994). Documento de Apoyo Curricular N° 1, Espacio Curricular Física.

Piaget, J. & Garcìa, R. (1982). Psicogénesis e Historia de las Ciencias. Edit. Siglo XXI-Mexico.

Pietrocola, M.P.O. (1992). Élie Mascart et l'optique de corps en mouvement [Élie Mascart and the optics of moving bodies], Ph. D. Thesis, Université Paris VII, France.

Porto, P.A. (2001). Michael Sendivogius on nitre and the preparation of the philosopher's stone, *Ambix*, 48(1), 1–16.

Quintanilla, M., Izquierdo, M. & Aduriz–Bravo, A. (2005). Avances en la construcción de marcos teóricos para incorporar la historia de la ciencia en la formación inicial del profesorado de Ciencias Naturales, *Revista Enseñanza de las Ciencias*, Número extra. VII Congreso Internacional sobre Investigación en la Didáctica de las Ciencias, Granada, España, 1–4.

Ramontet, I. (1997). *Un mundo sin rumbo. Crisis de fin de siglo.* Madrid: Editorial Debate.

Rivarosa, A., De Longhi, A. & Astudillo, C. (2011). Dilemas sobre el cambio de teorías: la secuenciación didáctica de una noción de alfabetización científica. *Revista Electrónica de Enseñanza de las Ciencias* 10(2), 368–393.

Ruiz, R. & Ayala, F. (1998). *El método de las Ciencias: epistemología y darwinismo,* Fondo de Cultura Económica, México.

Sacks, O. (1996). Escotoma: una historia de olvido y desprecio científico en *Historias de la Ciencia y del Olvido,* Editorial Siruela S.A., Madrid.

Salomon, P. (1988). *Psychology for Teachers. An Alternative Approach*, Hutchinson, London.

Schuster, F. G. (1999). Los laberintos de la contextualización en ciencia, en Althabe, G. & Schuster, F.G. (comps.), *Antropología del presente,* Edicial, Buenos Aires.

Teixeira, E., Greca, I. & Freire Jr, O. (2009). The History and Philosophy of Science in Physics Teaching: A Research Synthesis of Didactic Interventions, *Science & Education*, now online.

Testart, J. & Godin, Ch. (2002). *El racismo del gen: Biologia, medicina y bioetica bajo la fèrula liberal*, Fondo de Cultura Economica. Bs. As.

Villani, A. (1992). Conceptual change in science and science education, *Science Education* 76(2), 223–238.

Villani, A., Silva Dias, V. & Valadares, J. (2010). The Development of Science Education Research in Brazil and Contributions from the History and Philosophy of Science, *International Journal of Science Education* 32(7), 907–937.

Wolovelsky, E. (2008). *El siglo ausente: manifiesto sobre la enseñanza de la ciencia,* Editorial Zorzal, Buenos Aires.

Irene Arriassecq is professor in physics and mathematics, and member of the Teacher Training unit at Universidad Nacional del Centro de la Provincia de Buenos Aires (UNICEN), Argentina. She received her B.Sc. in Physics, M.Sc. in epistemology and methodology of science at Universidad Nacional de Mar del Plata and Ph.D. in science education from Universidad de Burgos, Spain. Her present research interests include the use of the history of science and epistemology in physics education and the design of teaching-learning sequences for physics education with focus on teaching and learning of special relativity theory in high school level. She has published in the area of physics teaching in journals such as *Science & Education, Revista de Enseñanza de las Ciencias, Caderno Catarinense, Ciência & Educação, Enseñanza de la Física, Revista Electrónica de Enseñanza de las Ciencias and Revista Investigações em Ensino de Ciências*, among others.

Alcira Rivarosa is senior professor of didactic and epistemology and history of science in the Department of Natural Sciences in the Faculty of Exact, Physical, Chemical and Natural Sciences at Universidad Nacional de Rio Cuarto (UNRC), Argentina. She received her B.Sc. in biology, M.Sc. in epistemology and methodology of science and Ph.D. in science education from Universidad Autónoma de Madrid, Spain. Her present research interests are related to teaching and learning biology. She has published six books, ten chapters as co-author and more than 35 papers in the area of science education in national and international refereed journals.

Part XIV
Biographical Studies

Chapter 72
Ernst Mach: A Genetic Introduction to His Educational Theory and Pedagogy

Hayo Siemsen

> *An effect size of d=1.0 should be regarded as large, blatantly obvious, and grossly perceptible difference.* Jacob Cohen (1988), "inventor" of the meta-analysis
>
> *[...] one gets large numbers when one weighs an elephant with gram weights.* An eminent statistician in defence of the dissertation of Wertheimer's assistant Luchins after one year of discussion about a possible error in it

72.1 Introduction and How to Read the Article

This article will attempt to provide a genetic introduction to the ideas of Ernst Mach, especially concerning education. It does not include a biographical overview on Mach as this can be found in many other sources (see end of the article). The first part of the article will give an introduction on how to read it (and why it is important to know). The next part will then provide an overview on the teaching phenomena associated with Machian teaching. These phenomena tend to be far away from most teachers experience with teaching. Even supporters of Mach often do not believe that such phenomena are possible. Therefore, the following part is concerned with the empirical evidence showing why the difference from an exponential teaching model to a standard linear teaching phenomena is so large ("[...] one gets large numbers when one weighs an elephant with gram weights", see Luchins 1993, p. 162). After that, the main idea of Mach is elaborated, i.e. why he changed from the antique understanding of genesis and adapted to a post-Darwinian concept of genesis. The main implications of this adaptation are then elaborated: sensualism,

H. Siemsen (✉)
Ernst Mach Institute for Erkenntnistheorie, Masaryk University Brno, CZ, Wadgassen, Germany
e-mail: hayo_siemsen@yahoo.com

M.R. Matthews (ed.), *International Handbook of Research in History, Philosophy and Science Teaching*, DOI 10.1007/978-94-007-7654-8_72,

gestalt[1] (economy of thought) and *erkenntnis* theory.[2] Finally, Mach had several successors, whose teaching one can use as empirical examples of Mach's educational meta-method.

Understanding Mach's ideas is very challenging. This can be seen from the many, often contradictory attempts of eminent scientists to do so. Why is it so difficult? Mach integrates facts, phenomena and observations from many different sciences as well as pre-scientific ideas. His goal is to provide *general* concepts, applicable for and consistent with the knowledge from all sciences and all experience. They are thought economical, i.e. one concept can be applied to many or – if possible – all fields of knowledge.

In order to achieve this, Mach takes a fundamentally different world view. As he emphasises, this is not a necessary view. Other views are possible. Like one can see the movement of the solar system and the stars from earth in a geocentric, Ptolemaic view, or one can view it (as a fictional thought experiment) from the sun, a Copernican heliocentric view. For Mach the difference is mostly an economic one. Each view serves as a yardstick, a "currency" by which concepts are "exchanged", i.e. related to each other. For some measures, one view will be the more economical, e.g. practical; for other purposes, another view might be more central. The sun – because of its high gravity – influences the other planets much more than they influence each other, except for those close to each other, like the moons to their respective planets. Therefore, the movements of all planets are easier to understand from a heliocentric view than from a geocentric one. But for the observation of the moon relative to the earth (e.g. for sending a rocket to the moon), a geocentric point of view is often the easier, more pragmatic point of view. Flying over the Atlantic, one rarely bothers to imagine or even calculate the distance via the sun (*Helios*). Likewise, Newtonian physics is mostly sufficient for such a flight, though maybe not anymore for moon or space travel.

Points of view thereby serve as "currencies of exchange" between concepts. They provide a single theoretical point, through which one can compare the relations of many or all observations, just as one can value all kinds of goods via one currency. Without a currency, i.e. in a barter economy, one has to remember how many oxen have the value of a horse and how many sheep have the value of two horses as well as how many sheep one can trade for an ox. If one now imagines that there are not only oxen, sheep and horses but also different qualities of cloth of different width to trade with the service of bringing the oxen to my farm passed the raiders on the road, etc., the value of a standard currency for the economy of memory and calculation becomes obvious.

Such a currency of course is most valuable, if one can use it very broadly. The Euro and US dollar have their respective "usage" value not only because of some gold reserves but also because they are used in many places by many people. Applying this

[1] For Goethe a *gestalt* used to be a "whole". Mach changed this into a genetic understanding of the concept, where a gestalt is like a species. It can adapt and transform; it is a process and a product (plasticity). Why it is thought economical will be explained later.

[2] *Erkenntnistheorie* in German means something in-between theory of knowledge, cognition and epistemology, though it is neither of these.

analogy to making concepts comparable over many disciplines, Mach's world view is optimised to represent the most economical view.[3] Mach's world view consistently comprises most known facts from all different areas of human experience. Experiences from different domains can thereby not only be compared but also be used complementary; reciprocally, they can "inform" each other. The relation between concepts is always possible via the currency. The basic concepts forming the world view are explicit as far as possible, including pre-scientific, intuitive concepts. If one does not do this, like in many speculative theories, a single mistake or inconsistency in the beginning can metastasise manifold and – as in the case of the Euro in 2012 – come back to haunt the founders much later.

A singular view furthermore generalises many concepts so that few concepts can stand for many facts. One idea, once thought, just needs to be reproduced from memory and applied everywhere. This application is not just analogous, but is based on the most general facts and experiences. Like the Euro, it requires a larger investment in the beginning, because many existing currencies need to be replaced and – also like the Euro – it as well requires "political" (philosophical) adaptations of many pet ideas held dear.[4]

In what way is the psychophysical[5] world view which Mach proposes more consistent than other views? The view comprises the physical view (about the world) and the psychical view (about ourselves and how we individually experience the world) into a singular meta-perspective. This singular general perspective is called monistic.[6] In order to be able to say something about the world, one needs to have a physical theory. This theory is first of all a private theory. It might – depending if and how much one studied physics – be naive or more sophisticated. Sophistication here only denotes how much the theory will cover regarding potential experiences. Will it still work when applied to a double-slit experiment? Will it still apply to the workings of CERN? Few people in the world will have developed their physical theory to the latter. If the physics of tomorrow is different from today's, such a view might also then become quickly outdated.

[3] This is why Mach acknowledges that there are several world views possible, but others are optimised for other criteria. It would of course be scientific to always make these criteria explicit so that one can compare the motives.

[4] This often causes – sometimes very sudden and fierce – emotional rejection of the ideas. One feels like the previously sturdy floor on which one learnt to walk is suddenly pulled away underneath one's feet. Siemsen et al. (2012) call it the "rug of horror" symptom, describing the emotional trouble of some students facing up to some of their lifelong inconsistencies in learning. The mathematician Weyl (1928) put it in a different metaphor when the sturdy mathematician's "tower of knowledge" is "turned to mist". Kurki-Suonio recalls that the in-service teachers he taught always showed one year (in a two-year course) of "resistance" before a gestalt switch in their world view (for the technical details how the course was implemented, see Lavonen et al. 2004). The older one gets, the more one tends to become emotionally attached to often used habits of thought.

[5] Mach's generalised interpretation of psychophysics as a world view is unusual, also in psychophysics.

[6] This is not necessarily contradictory to what William James called a "pluralistic world view", because Mach's view deliberately comprises different perspectives. Mach makes them comparable, without claiming absoluteness, just economy.

On the other hand, as humans we also require a psychical view. Otherwise we cannot conceptualise other humans (and ourselves as different from them). Such a "theory" in its rudiments is formed in very early childhood (when the child starts to communicate and interact with its fellow human beings).[7] Later, this "naive" theory might become more sophisticated through systematic psychological observation and experimentation, i.e. by a "scientific" psychical theory.[8]

What Mach now proposes is to combine the physical and the psychical view[9] into a singular world view (general or meta-perspective). Although such a singular view is again very close to the initial perspective of a baby, we could not have acquired it without having previously gone through many human experiences. This view is an abstraction derived from these experiences.

One needs to shift one's view away from a mainly dualistic view in which one sees the "world" and the "self" as separate entities, towards a monistic view. In a Machian monism, one assumes the relations between the "world" and the "self" as the basic elements. One can only know that there is such a relation. What one learned to know as the "world" and the "self" are then only intuitive constructs of such relations from childhood.[10] In order to know, what the relation is, one has to apply the physical view and the psychical view, like standing on and walking with the left as well as the right leg.[11]

[7] Thereby, Mach's psychical theory is first of all a theory of the individual, not of the "self". The "self" in his view is just one outcome, a construct of individual (sensual) experiences.

[8] The process of science for Mach consists in a higher economising by systematically using and testing the experience of many people and making them comparable though a more standardised framework of observation.

[9] For this, one should take more sophisticated and consistent versions of these views in order to consistently cover many facts and human experiences.

[10] "Every human discovers within himself, when waking up to his complete consciousness, already a completed image of the world, to which accomplishment he did not at all willingly contribute to and which he accepts on the contrary as a gift from nature and of the civilization and as something immediately intelligible. This image was built up under the pressure of the practical life; extremely valuable, in this regard, it is inerasable and never ceases to act upon us, no matter which are the philosophical views that we will later adopt". "Here everyone has to start" (Mach 1905, p. 5).

[11] "For most natural scientists and many philosophers, who do not admit it, the thought that all psychical could be deducible to the material is very congenial in private. Even if this materialism has a catch, it is not the worst possibility; it stands at least with one foot on secure ground. But if all psychical should be understandable physically, why not the other way round? […] Is the other [psychical] foot standing in the air? I would prefer […] to stand on both feet. […] There is no necessity to become dualist thereby for the one, who considers both feet as equal and both floor spaces under the soles to belong not to two different worlds" (Mach 1920, p. 434; 1883/1933/1976; 1883/1893/1915). In his book *Knowledge and Error* (1905), Mach uses such a dual approach, switching back and forth between a physical and a psychical perspective. This (and other unusual features) seem to have been unfortunately too much for instance for the translators (in English as well as in French) to cope with. Especially the translations of his books have led to many unfortunate misunderstandings of Mach's ideas (except for the excellent translation of *The Mechanics* by McCormack and Pierce, though that has suffered as Jourdain complained from not being updated to the many further developments Mach included in the later German editions).

Mach calls this relation "sensory element". He states that he could interchangeably call it "physical element", though then people would easily mistake it for physicalism. The relation is very broadly defined, including the sun lighting the perceived object as well as the memory (and prior experience), which makes the object recognisable. These psychophysical relations thus have a genesis, which needs to be researched. What one perceives as simple relation might actually under scrutiny turn out to be a cluster of relations. The sensual elements are therefore recursively defined. Although the relation is "immediately given" as a gestalt, this perceived "immediately given" represents only one side of the relation. The "given" is not simple, but requires many perspectives of research of which none can ever supposed to be final. Even the habitual separation of sense elements into senses might already be too much reductionist or too little. One cannot know a priori, one can know only genetically. Sensual elements are gestalts, i.e. they adapt and transform like species.

Other monisms of course exist, namely, physical monism (materialism or physicalism), which presumes that all psychical phenomena are epiphenomena, and the "panpsychism", which supposes that everything is initially psychical. There are also more or less arbitrary combinations of the two, delimiting the rule of one relative to the other (James noted that some monisms are actually dualisms under disguise). Finally, there are the "parallelisms", which suppose that the physical and the psychical always go together "in parallel", like a tandem. But as one might notice, Mach's feet will rarely work "in tandem", i.e. hopping together, but rather complementing each other while walking (Mach used the metaphor of a "tunnelling" between the two perspectives, i.e. a process, which brings them continuously closer together). Mach's psychophysical relation is only the starting point of genetic (and other) enquiry.

The "problem" of this shift in world view is one of learning in general. As one learns a different way of learning, one has to discard previous ways of learning or at least to adapt them to the new framework. Learning methods tend to be well associated (neurally connected), because of the high frequency of learning in general. Changing these associations is more difficult than regular learning. It is not only learning something new but resembles changing early childhood experiences, like sometimes done in psychotherapy. There are hints that such meta-learning involves not only changes in neuronal connections but also actual formation of new neurons on a scale not regularly observable in learning processes (see, e.g. Buonomano & Merzenich 1998).

The view presented here is not a standard (formal) account of Mach's ideas. It is a genetic introduction to Mach's ideas. New concepts will be introduced genetically, i.e. starting with a more standard meaning and then successively being transformed into Mach's usage. As several of the main concepts, such as *erkenntnis* theory, genesis and world view, are reciprocal, i.e. have to be developed interdependently, they need to be introduced before each concept is fully developed in its new meaning. One thereby has to read the text several times. The first time, the footnotes can be left out. In the second reading, they provide additional facts and information for further transforming some important main concepts. In a third reading, one can therefore focus mainly on the footnotes and skip through the main text. If possible, one should also leave a night in-between these readings. On the next day, one will be able to identify the less well-understood parts more clearly.

72.2 The Phenomenon of Genetic-Adaptive Learning

William James called the lecture he heard from Mach "one of the most artistic lectures I ever heard" (Thiele 1978, p. 169).[12] What constitutes this artisanship of teaching? James (Thiele 1978, pp. 169/173) later tried to copy this method as well as Mach's world view in teaching his students: "I am now trying to build up before my students a sort of elementary description of the construction of the world as built up of 'pure experiences' related to each other in various ways, which are also definite experiences in their turn. "There is no logical difficulty in such a description to my mind, but the *genetic* questions concerning it are hard to answer." I wish you could hear how frequently your name gets mentioned, and your books referred to". James thereby encounters some effects of genetic teaching, which – when compared to other cases of Machian teaching – seem to be typical, including the difficulty of implementing the genetic questions (as Mach had noted – see later quotation – this can be as challenging as finding the idea in the first place). What are the typical effects of genetic-adaptive learning?

Intensive (transformative) experience: There are accounts of people who have been influenced by Mach and genetic-adaptive learning by just two hours (Siemsen 1981), an oral exam (Heinrich Gomperz), even of 10 min (in the case of Karel Lepka[13]) or hearing one lecture from Mach (Fritz Mauthner, one of the founders of linguistic critique, see Siemsen 2010a). This short time was sometimes enough to change the lives of the people involved (for instance, when Karl Hayo Siemsen met Joachim Thiele[14] once for two hours, but now considers himself following in his footsteps). Also people who have been influenced by Mach, such as Schumpeter, are known to produce this effect.[15]

---

[12] Mach's way of learning will in the following be called "genetic adaptive". Such learning should be viewed like a Darwinian species: a process and a product. The growth processes are exponential rather than linear; they adapt and transform. Therefore, some empirical results from such learning might look extreme and unbelievable to anybody who has never observed them. Nevertheless, over the last 150 years, there have been many such empirical observations in many contexts, mostly influenced directly or indirectly by Mach.

[13] Lepka is professor for mathematics education at Masaryk University, Brno, CZ. He described this occurrence in detail in several interviews 2011/2012. Several other students of Černohorský have described similar, though a bit less extreme effects.

[14] Thiele first systematically published Mach's letters. From this work he gained important insights on Mach missed by many other Machian scholars.

[15] This effect and several of the following effects can certainly be observed in other circumstances, which have no Machian background. Nevertheless, some of the following effects are unique. Others are unique in their combination, frequency of occurrence or reproducibility. In all known

Inspiring eminent new ideas in multiple areas: In principle, this seems to be a corollary of the transformative experience described before (though in many cases, it has not been properly described). Few people have influenced so many eminent other people than Mach. Maybe the last time this has happened in human history were the many schools of thought founded after Socrates (and maybe the many ideas Galileo inspired; though in that case it is difficult to trace this effect mainly to Galileo). Mach has directly or indirectly influenced not only a handful but also many Nobel Laureates, even much after his time and not just from one area, but all areas in which such prizes are awarded.[16] Einstein (1916, p. 102) provides a metaphor for this type of intuitive influence. Regarding his own generation of physicists, "I think that even those who think of themselves as enemies of Mach, don't remember how much of Mach's approach they have – so to speak – imbibed with their mother's milk".

Aspects of this "mother's milk effect" can be found in several areas, methodology, epistemology, praxis, etc. For instance, according to Einstein (1916, p. 102/103), it "trained [the physicists] to analyze the long-time prevalent concepts and to show, of which circumstances their eligibility and usefulness depends upon, how they have grown in detail out of the conditions of experience. Thereby, their excessive authority is broken".

The mother's milk did not start with Mach's famous *Mechanics*. There was a genetically even earlier mother's milk for many eminent scientists in German-speaking countries. The effect was transmitted through Mach's schoolbooks, which they grew up with (and certainly could not remember consciously) between 1886 and 1919.[17] They already learnt the concepts the Machian way. Even though some, like Planck, later criticised aspects of Machian ideas, they still intuitively retained and used nearly all of his *erkenntnis* theory (see Siemsen 2010c).

This did not only happen in physics but also in chemistry (where, for instance, Wilhelm Ostwald was close friend and admirer of Mach) and biology. Especially in medicine, Mach's teaching of "physics for medical students" in his *Compendium* (Mach 1863) had long-term effects, as many medical students were travelling all over Europe during their study times. As other eminent scholars – even such as the likes of Wundt – did not have a consistent *erkenntnis* theory of their own (as James had noted), the students coming out of their labs were mainly Machians, at least in their *erkenntnis* theory (such as Kuelpe or Titchener, see Boring 1950/1957).

Machian teaching phenomena, several of the effects have been observed, while other effects might have been overlooked as the effects have never been systematised.

[16] For example, Einstein, Pauli, Bohr, Planck, von Laue, Raman, Heisenberg, Rabi, Bridgman, Ostwald, Arrhenius, de Broglie, Landau, Ramsey, Polanyi, Wilczek, Eigen, Lorenz, Musil, Bergson, Hayek, Samuelson, Simon and Coase. No systematic inquiry has yet been made on many others. Especially, it would be interesting to know how many winners of the "physiology or medicine" prize have been in Mach's "physics for medical students" lecture. Probably there are several. Mach himself was suggested for the prize but probably was too much of a generalist for a specific prize. There is no Nobel prize for *erkenntnis* theory.

[17] See Siemsen (2012a) or Hohenester (1988).

Also many eminent economists, such as Schumpeter, Hayek, Georgescu-Roegen, Samuelson and Polanyi, took a deep zip of Machian mother's milk. Even many artists were not free from this influence. Some, like Schnitzler or Hofmannsthal, had even been attending Mach's last lectures in Vienna, while the literature Nobel Laureate Musil wrote his dissertation on Mach. There were also influences on linguists (Mautner), philosophers (Wittgenstein, Feyerabend, Popper, Vaihinger, Radulescu-Motru), sociologists (Zilsel, Cohen, Neurath), mathematicians (Poincaré, Nevanlinna, Haret, Hadamard, Hahn, Menger, Weyl, Brouwer, Mandelbrot, Marcus and 65 recent mathematicians and maths teachers, see Ahlfors et al. 1962), historians (Sarton, Koselleck), anthropologists (Boas, Lowie, Malinowski and through their students also Jerome Bruner), biologists (Loeb), logicians (Pierce), philosophers of science (Frank), etc. (see Holton 1992; Siemsen 2010a, d).

Why has Mach's central role in these general aspects often not been properly described or even recognised? Einstein gives a hint when he tells his friend Besso in 1948 (Speziali 1972, Doc. 153), "Now, as far as Mach's influence on my development is concerned, it was certainly great. [...] How far [Mach's writings] influenced my own work is, to be honest, not clear to me. In so far as I can be aware, the immediate influence of D. Hume on me was greater. [...] However, as I said, I am not in a position to analyze what is anchored in unconscious thought".

Teaching is mostly intuitive: As Einstein had so aptly described with his "mother's milk" metaphor, much of Mach's teaching "becomes anchored" on the intuitive (nonconscious) level. This is a corollary of the intensive sensualism involved. The sensual relations belong to some of the oldest biological setup and are therefore mainly connected with consciously not easily accessible parts of thinking. To make them accessible to reflection is a task of *erkenntnis* psychology, which is unfortunately the least understood and reproduced element of Machian teaching. Kurt Lewin aptly called the result "practical theory".

When Mach's *erkenntnis* theory is acquired early, the transformative effects are not experienced as special. One might only later be surprised that other people's thinking is not the same. What is "normal" and therefore not special about teaching in Finland may be unique in the world. When asked what is special in their teaching, Finnish teachers are often not able to answer this question, or the theory provided might have little correlation with the actual phenomenon. Scientists (influenced by Mach) might intuitively know what to do and what works. When one follows their (later constructed) theory, one unfortunately cannot reproduce their results. What works for them works because of effects not covered by their theory (cases of this are, for instance, Piaget or Wagenschein, see Siemsen 2010b, 2012b).

Teaching becomes mostly independent of age and "stages": One of the theoretical (metaphysical) elements Piaget popularised in science education is the model of the "stages" of development taken from the intelligence scale of Alfred Binet. The stages are teleological, i.e. depend on a specific definition of culture and intelligence. They are not genetic. Mach instead describes the genetic process of popularisation:

> Once a part of science belongs to the literature, a second task remains, which is to popularize it, if possible. This second task also has its importance, but it is a difficult one. It has its importance, because – regardless of the distribution of knowledge that increases its value – it

is not unimportant either for the further development of science itself how much knowledge has been disseminated into the public. The difficulty is to know the soil very well in which one wants to plant the knowledge.

It is a prevalent but wrong opinion that children are not able to form sharp concepts and come to the right conclusions. The child is often more sensible than the teacher. The child is very well able to comprehend, if one does not offer too much new at a time, but properly connects the new to the old. The adult is a child when facing the completely new. Even the scholar is a child when confronted with a foreign subject. The child is a child everywhere, as everything is new to him. The art of popularization lies in avoiding too much of the new at one time.[18] (Mach 1866, p. 2–3)

Can this be observed empirically? For example, the physics Nobel Prize winner Wolfgang Pauli was godchild of Mach (born Pascheles, the family, especially the father, were close friends of Mach). He received *The Mechanics* (Mach 1883) as a gift from Mach when he was 8 years old. Another example is Peter Drucker, an eminent management "guru", who describes in his autobiography, how his 1 year of school in fourth grade in the Schwarzwald School[19] changed his life (Drucker 1979, pp. 62). He later adapted and used the methods he learnt in this year in school to management (see Eschenbach 2010).[20] They became some of the most influential ideas in modern management.

"Bad" students suddenly become "good" or "very good": The "shift" of students improving in grades is not only by one to three grades (on a 5–1 or 0–6 scale), but can happen in "jumps". In many instances, for example, in Finland, but also for Siemsen et al. (2012) or Gabriel Szász,[21] initial "laggards" suddenly become excellent students. They thrive with the new method.

Alfred Binet, the "inventor" of the intelligence test and the concept of age-related "stages", which was later perfected by his student Piaget, late in his life worked specifically on the "laggards", i.e. the lowest 5–10 % on his intelligence scale. After a discussion with Mach, whom he had invited to write an article for his "*L'Anneé*

[18] Instead of age groups and stages, the pre-knowledge ("not too much new at a time") becomes a more important factor. Some pre-knowledge might be enabling, some obscuring for the new *erkenntnis* process.

[19] One of the schools inspired by Mach.

[20] Another example is a girl of 3 1/2. She was supposed to learn about opera. After intensively using first very simple versions of the "Magic Flute" by Mozart (abridged puppet and children's versions), connecting these to previous experiences and repeating scenes as requested by the child, it became possible to watch a regular version of the Magic Flute. Now the question was if one could directly present her a "difficult" opera. One of the most challenging operas is certainly Wagner's "Ring des Nibelungen" (at least concerning the length and the complexity of the story, though also the music is challenging compared to Mozart). She loved it and though she was allowed to watch only smaller parts of the total 16 h at once, she always wanted to continue and remembered the scene where she had to stop before. Also she developed favourite scenes to be repeated and started to recognise gestalts from the opera in daily life. What was new for her was the medium of opera. Once she knew the basis of it very well, expanding (exponentially) from this basis was not a problem, neither of learning, nor of age.

[21] In courses on astrophysics at the Masaryk University in Brno, Czech Republic, 2008 and 2010 (see http://astro.physics.muni.cz/iwssp2008/ and http://astro.physics.muni.cz/iwssp2010/, accessed 25/08/2012).

Psychologique" (see Siemsen 2010b), Binet developed the concept of "mental orthopaedics" (*orthopedie mentale*) specifically for this problem. Mental orthopaedics is an application of psychophysics to learning. Unfortunately, Binet died shortly after implementing this idea for a larger number of students, but the rector of the school with whom Binet worked published some empirical results posthumously (Vaney 1911). The results show not only a very fast relative but also an absolute "catching up" of most students compared with the regular students (the school got the students only after a minimum of 3 years of "lagging behind" in a regular school, but many of them were on the same level after 2 years of mental orthopaedics). Additionally, although the singular "curves" seem to be linear, taken together they show a clear exponential component in learning.[22]

The phenomenon of the thriving students also shows that empirically, the so-called Matthew effect in education,[23] i.e. that good students tend to get always better and laggards continuously lose out on them, is actually not a fact, but an artefact of a linear learning paradigm.

Learning happens without outside pressure: The students themselves perceive this way of learning mostly positive (in spite of the previously described "rug of horror" effects), because there is no external pressure necessary for learning. The pressure and motivation becomes intrinsic, i.e. self-organised (see Siemsen et al. 2012). As a result, also motivation is not a problem anymore. All students want to learn, and the problem is mostly to limit the time for learning so that they have enough time for intuitively digesting what they have learnt while doing other recreational activity. Some adaptation of the thoughts to each other needs to take place overnight during sleep.

Time perception changes in learning: One indirect consequence of genetic-adaptive learning is the relativity of time perception in learning. Because the learning process

---

[22] The details will be elaborated in a separate article following more detailed research. Karl Hayo Siemsen stated in a personal message that Binet basically used the same method as he used (Siemsen 1981), but 60 years earlier. Just that Binet used school instead of university students, but with very similar empirical results (Vaney 1911). This finding would imply that Mach, James and Binet, the eminent researchers in the German-speaking countries, the USA and France at the time, were all using in principle the same method as ideal teaching method. It seems unfortunate that the method got lost nevertheless, probably because James and Binet developed it only at the end of their lives, with the chaos of the world wars ensuing shortly afterwards. Thereby, the method was never even properly identified as a specific method with specific phenomena.

[23] The "Matthew effect" was first coined by Robert K. Merton in order to describe, for instance, that eminent scientists tend to get more credited for the same work than unknown ones. "For unto every one that hath shall be given, and he shall have abundance: but from him that hath not shall be taken even that which he hath" (Matthew 25:29, King James Version). This idea was then applied to education by Keith Stanovich regarding reading literacy. But the actual story is probably more intricate. A little bit before, Matthew also entails a "counter-Matthew effect": "But many that are first shall be last; and the last shall be first" (Matthew 19:30, similar in 20:16, King James Version). Maybe one should rename the effect in education instead into "Hit-The-Luke" (translation of "*Hau-den-Lukas*", the German name for a "Ring-The-Bell" at funfairs) for all the students who are never taught to become very good students.

is so intensive, the thoughts related to the learning process become dominant versus other background thought processes, such as the one keeping time.[24]

Learning takes place in genetic "loops": Genetically, many ideas and concepts cannot be introduced at once, but need repeated scaffolding and auxiliary concepts in order to develop. In such "groping phases" (Kurki-Suonio 2011), sometimes little seems to happen on the surface, while "suddenly" gestalt switches of whole ranges of concepts appear. This effect has also been described by Poincaré (1908) for how he found new mathematical functions for which he became famous. Contrary to common belief, such intuitive thought processes can be observed, measured and consistently reproduced (see Siemsen 1981, 2010a; Siemsen et al. 2012).

Genetic "loops" have been used by Mach already in his schoolbooks, but they become imperative for meta-learning as Mach describes: "The value of such general methods lies in the economization of thinking about the single case, in the easy templating of it. This cannot be understood, unless initially a whole host of single cases has been sufficiently dealt with through the consideration of very different details. More general methods in science are a result of much detail work and have to be this also in teaching. Without this, method is an un-comprehended gift. In the method lies the insight that one can think a thought one-and-for-all and does not have to think it again in each case. [...] 'What you have inherited from your forefathers, gain it in order to own it'" (Mach 1876, p. 13).

Long-term and transfer effects: From the observations by Kurki-Suonio, Siemsen and Černohorský, but also from the Schwarzwald School (see, for instance, Siemsen et al. 2012), a student only has to pass through a teacher with genetic learning one time during his life and can afterwards apply the genetic learning as meta-learning on all other learning throughout his or her life. Siemsen et al. (2012), for instance, have shown strong transfer effects, where the *erkenntnis* theory taught in a project study reduced the percentage of students not passing a mathematics exam from 90 % to 40 % without any teaching of the contents of "higher mathematics 1" or the mathematics teaching being a core topic in the project study. The empirical results of the experiment of Siemsen from 1981 suggest that with some genetic mathematics teaching (2–4 h), this with 40 % still high dropout rate can be brought down to nil. Similarly, the long-term effect of genetic-adaptive teaching seems to show in terms of personality development. This is documented for many cases of the Schwarzwald School as well as students of Černohorský and Siemsen. Surprisingly many of them become professors, entrepreneurs, eminent writers, politicians (ministers and prime ministers), etc. When interviewed, they tend to agree that this teaching has had a very important effect on their career.

Exponential learning: If one compares the empirical effects of methods based on linear models of learning, one typically arrives at meta-analysis effect sizes of d between 0.0 and 0.6, sometimes up to 0.8 (see Hattie 2009).[25] Jacob Cohen (1988),

---

[24] This effect is, for instance, also described by Csikszentmihalyi (1990) with his concept of "flow".

[25] Hattie (2009, 2012) comprises more than 60,000 empirical analyses on student achievement from more than 900 meta-analyses onto one scale of comparison, which is the effect size (after-before or

the "inventor" of the meta-analysis, argued that an effect size of d=1.0 should be regarded as "large, blatantly obvious, and grossly perceptible difference" (quoted by Hattie 2009, p. 8). This is as perceivable as more than 20 cm difference in the height of persons. Finland as a field experiment (OECD Pisa 2006) has a d=1.0. The problem can be that many teachers have never observed effects even close to this and therefore may regard such effects as "unbelievable",[26] contrary to Cohen's view.[27]

72.3 Empirical Observations When Changing to Machian Teaching

What happens if one changes to a Machian teaching? As Mach (1886/1893/1986) described, "I know of nothing more terrible than the poor creatures who have learned too much. Instead of that sound powerful judgement which would probably have grown up if they had learned nothing, their thoughts creep timidly and hypnotically after words, principles and formulae, constantly by the same paths. What they have acquired is a spider's web of thoughts too weak to furnish sure supports, but complicated enough to confuse".

Mach sees teaching very empirically. Today, most teachers experience that they have to teach too much in too little time. But this is an empirical illusion created by the standard (linear) teaching model. Martin Wagenschein (1970) suggested a simple empirical test against this misconception. How much knowledge is really taught, i.e. how much "remains" in long-term memory? Everything that is forgotten is obviously superfluous to teach in the first place. One can just repeat the test one does at the end of the course and repeat the same test 2 weeks after without telling the students beforehand. Repeat the same test after a year. How much of the apparent knowledge remains? On average it is just about 5–10 % (even if it would be

experimental-control group divided by standard deviation). Meta-meta-analyses also provide an interesting opportunity for conceptual analysis and finding inconsistencies in the empirical meaning of concepts. If the results of different meta-analyses for one concept vary largely, this might be because the description of the concept or the concept itself is inconsistent or confused, leading the studies about it into different interpretations or into white noise.

[26] For an example, see Luchins (1993), the assistant of Wertheimer in the USA, recalling the problems with his dissertation about rigidity, which seemingly included exponential effects. This is an important reason to make statistical analysis of Mach's teaching method, because it grossly violates the gut feelings (intuitions) of many experienced educators. People with less experience have less trouble accepting it. The problem also effects student's evaluations of this method, because also their intuitive model of learning is linear, which leads to conceptual inconsistencies when confronted with exponential learning effects (see Siemsen et al. 2012).

[27] This "unbelievability" also happens in the method of meta-analysis. Typically the "outliers", i.e. the most extreme effects, are sorted out. Unfortunately thereby, if any Machian teaching effect was ever initially included in such a study, it will probably not have been further researched as best practice, but statistically eliminated.

50 %, still half the teaching is a waste of time).[28] So from an empirical standpoint, one should teach only the knowledge which is later present in memory and use the rest of the time of the course otherwise. The 95 % which is forgotten is a waste of time anyway, of the teacher and of the students. This is how one can teach more within less time.

This "watering down" effect of knowledge taught to (less) knowledge used is actually taking place severalfold throughout the education process. According to inquiries (by Mueller-Fohrbrodt et al. 1978), teachers tend to run into a "shock" of praxis after their theoretical education at university. They adapt mostly by throwing away what they learnt at university (see Mueller-Fohrbrodt et al. 1978). Teacher education has little effect on student performance (d=0.11, see Hattie 2009, p. 110). Professional development might still have a high effect on teacher knowledge (d=0.90 probably measured for short-term and not long-term memory and effects), but this is less implemented (d=0.60) and has even less influence on student learning (d=0.37, probably also short term, see Hattie 2009, p. 120). The experience of Kurki-Suonio (2011) in his in-service teacher training courses are that the Machian world view is more difficult to teach to teachers in the first place (high initial resistance/rigidity), but then the new gestalt is much more stable. It also leads to much higher student learning (d=1 long term for all of Finnish students with only 1/3 of Finnish science teachers trained in this way, see OECD 2007 and Siemsen 2011).

Mach describes the role of the teacher relative to the subject matter: "If now the teacher, who can present to himself the whole subject matter, subsumes a theorem adapting it to his *own* conceptualizations and to his *own* satisfaction of needs, then therein lies, if it happens unconsciously, certainly an error in the person whose satisfaction of needs is central in this case. If it happens on purpose, it is a didactical insincerity. For the student, premature completeness and logical finesse is useless and without any value, often even detrimental" (Mach 1890, p. 2/3). Why are 95 % of teaching used for teaching short-term memory? For the satisfaction and exculpation of the teacher, so that he or she can claim to have "taught everything"?

Empirically speaking, most of lecturing at schools and universities is wasted. Instead, one could free time for effective learning in the way described above and reduce overall time spent at school and university. In logistics, this radical approach of quality improvement is now standard. Quality management is not mainly about certification and formalising processes.[29] It is about reducing (or rather preventing) faulty products and processes.[30] In education, one could say that many products pass

[28] One could call this test "Wagenschein's razor".

[29] "[...] control charts are justified for only a small minority of the quality characteristics" (Juran 1951, p. 308).

[30] Joseph M. Juran (1951, p. 247), one of the early "gurus" of quality management, wrote about "the principle of prevention" that "It needs no argument to conclude that it is better to prevent defects from happening than to sort them out after they have happened. Shop supervisors and personnel are fully aware of this principle. The failure to apply the principle of prevention lies *not* in any disbelief in the principle. Rather the problem is one of *how to achieve* prevention. Any lack of achievement is in turn not the result of deliberate (or even unwitting) resistance by individuals; it is rather the result of the limitations of modern industrial organization". Juran (1951, p. 248) then

the final quality test, but still more than 90 % fail in their first application.[31] This is a huge waste.

Japanese quality improvement calls instead for the identification and reduction of "*muda*", the Japanese word for waste (see Ohno 1978/1988, p. 18).[32] Taiichi Ohno, the Toyota engineer who made the Toyota Production System (TPS or Lean Management) popular, described the success of Toyota through the reduction of three types of waste: *muri* (overburden, beyond power, by force), *mura* (inconsistency, irregularity) and *muda* (waste).[33] Applied to education it would mean that one needs to cut all learning, which (a) has to be forced on the learner, which (b) is too difficult to learn in a singular effort or which (c) entangles and confuses the students rather than enlightening them. One further has to cut all inconsistencies in

---

quotes Howell B. May (1921): "While inspection of the product is of much assistance in improving the processes, nevertheless its greatest usefulness is found in the prevention of loss from defective goods. [...] The success of inspection in decreasing the quantity of defective goods is based upon the fact that to maintain the necessary standards of excellence with minimum loss, the quality of the product must be known at all times". Inspection nevertheless should not be mistaken for exams and exams not with facts. "In a production plant operation, data are highly regarded - but I consider facts to be even more important" (Ohno 1978/1988, p. 18).

For achieving the compliance of students in the process, being open about not understanding something, the barrier between "work" and "fun" has to be broken down (Juran 1951, p. 269): "To the extent that a management can minimize compulsion, give impersonal supervision, allow participation in establishing objectives, provide creativeness, provide a social atmosphere, give meaning to what is being done, and provide incentives over and above money [...] it secures the enthusiasm as well as the compliance of the operator". It thus does not help to give bad grades for not understanding or to use grades for motivation.

[31] T. H. Huxley (1864/1870) warned in his inaugural speech as Rector of the University of Aberdeen that "examination, like fire is a good servant, but a bad master". A constant effort of passing exams has a deteriorating methodological effect on students: "They work to pass, not to know; and outraged science takes her revenge. They do pass, and they don't know". The worst is that students also do not care that they forget more than 90 % shortly after the exam. They should be the first to complain if they do not learn as it is a loss for their lives. Seemingly they do not think that what they learn is somehow helpful for their lives, for them it is just the certificate. Note: This is a part quotation only. Because of the complexity of Huxley's initial text, it is more economic to quote this way. Any reader may feel free to look up the actual quotation.

[32] The concept of "waste" was seemingly adapted from Henry Ford, who wrote a whole chapter on "Learning from Waste" in his *Today and Tomorrow*. "My theory of waste goes back of the thing itself into the labour of producing it. We want to get full value out of labour so that we may be able to pay it full value" (quoted in Ohno 1978/1988, p. 97).

[33] After WWII, the Japanese economy had only about 10 % of the productivity of the USA. The then president Toyoda Kiichiro wanted to catch up within three years in order to survive. "I still remember my surprise at hearing that it took nine Japanese to do the job of one American. [...] But could an American really exert ten times more physical effort? Surely, Japanese people were wasting something. If we could eliminate the waste, productivity should rise by a factor of ten. This idea marked the start of the present Toyota production system" (Ohno 1978/1988, p. 3). The Toyota Production System evolved by repeating *why* five times. "By asking *why* five times and answering it each time, we can get to the real cause of the problem, which is often hidden behind more obvious symptoms" (Ohno 1978/1988, p. 17). The idea is to separate facts from artefacts and to find solutions, which let the problems disappear. In principle, this is relatively close to Mach's idea of enlarging or transforming the view in such a way that the problem disappears.

what is being learnt as well as in the process of learning of the learner (mental orthopaedics). Finally, one should focus on the learning, which has long-term value for life and which remains in the memory of the learner (providing long-term value). The main focus has to be on a sound and consistent foundation of any learning process.[34] Currently instead, quality improvement in education seems mainly (de facto) aimed at enshrining the status quo by formalising it,[35] rather than bringing about such fundamental improvements or even to have such long-term and transformative goals.

Frank Oppenheimer, the brother of Robert Oppenheimer and founder of the Exploratorium – probably the most empirically successful[36] science museum in the world – already in 1981 described the role of the schools in being unintentionally detrimental to learning:

> A large part of the neglect [of learning opportunities such as museums in the popular mind] I think had to do with the extraordinary preoccupation of both school and college teaching faculty with the notion of certification. Educational opportunities, which did not provide any way of certifying the students or of evaluating their performance were relegated to a different domain and were not considered part of the overall process of public education. This preoccupation with certification has been, in fact, a very deadly one. It has produced generations of teachers who teach what they are supposed to teach rather than what they know and want to teach. […] In any event, since museums do not certify and the watching of television shows is not graded or going to libraries cannot be supervised, all of these adjunctive resources for public education have been neglected. [Schools] have taken on, and jealously guarded, the total job of education while having at the same time had to cope with an ever more complicated society and an increasingly mobile population. They have unfortunately failed abysmally. (Oppenheimer 1981, p. 1)

[34]According to Mach, all initial mistakes or unnecessary metaphysics are repeated and thereby metastasised throughout the whole process. "A problem early in the process always results in a defective product later in the process" (Ohno 1978/1988, p. 4).

[35]As Hattie (2009, p. 109/110) concludes from the meta-analyses in this area, much of teacher education is based on the gut feeling, (quoting Walsh) ""that there is presently very little empirical evidence to support the methods used to prepare the nation's teachers." [In my experience] every time the 'core' knowledge decided on by a group has been different. […] it seems surprising that the education of new teachers seems so data-free; maybe this is where the future teachers learn how to ignore evidence, emphasize craft, and look for positive evidence that they are making a difference (somewhere, somehow, with someone!). Spending three to four years in training seems to lead to teachers who are reproducers, teachers who teach like the teacher they liked most when they were at school, and teachers who too often see little value in other than practice-based learning on the job". The challenge is to "unlearn" the perspective of teaching as a student. This requires an *erkenntnis* theoretical reflection of the teacher (who now thinks to know) to imagine the perspective of the learner new to the topic.

[36]I personally know people who have been influenced by the Exploratorium for their lifetime, for instance, becoming rocket scientists, famous writers, etc. Other typical effects and principles of Machian teaching can be found in the observations of K.C. Cole in her biography of Frank Oppenheimer "something incredibly wonderful happens" (Cole 2009). Cole wrote (email 22/08/2012) "many of my other books do [cite Mach], and that thinking came first from Frank [Oppenheimer], but I don't remember how…". The observation is typical for the mother's milk effect described by Einstein regarding Mach. Before intensive research and revisiting old notebooks, also Kurki-Suonio was not aware of where he had his ideas from.

Certification[37] in education has a tendency to promote the production of large groups of below-average performing students. The method by itself does not improve understanding. The low-achieving students are sieved out, while the other part is not taught much new. Certification optimises the maximum passing of a minimum standard. If one sees certification instead as a means of providing a minimum bar *everybody* must cross,[38] finding out who still might require closer attention, at least the method is stripped off its teleology. It would mean that all students must have a minimum understanding of the topic taught. Still such a method is not genetic, though the genetic method is probably the only one providing this effect until now. Nevertheless, standard tests can be used to test genetic teaching versus standard teaching. But such tests will miss out on the most important dimensions of genetic-adaptive learning (the tests are linear, not exponential). They are a poor proxy, though students taught with genetic-adaptive teaching do not tend to have trouble passing, on the contrary.[39] Only if all students pass such tests with the highest possible grade, the tests do not make sense anymore (see Siemsen 1981). The number of students not passing shows the failure of the teacher to teach and not necessarily the failure of the students to learn. The model that the grades in a class must follow a normal distribution is a human convention, not an empirical law. It is not consistent with the teaching goal to teach everybody well.

Mach always saw exams not as a means of testing the knowledge of the student, but to teach the student something general, something the student would be able to use for a lifetime (a method, which in several cases has been described to work by Mach's former students).

For Mach, learning is a continuous activity of all living nature, not restricted to classrooms. Phenomena do not tend to adhere to scientific disciplines, but still require to be observed. In such a view, classroom activity is just a refinement of daily activity into a specific direction. There is no need for artificial borders unnecessarily separating the classroom from learning activities taking place elsewhere. It is the learning, which is the central phenomenon, not the place.

---

[37] "One has to go rather slowly on fixing standards, for it is considerably easier to fix a wrong standard than a right one. There is the standardizing which marks inertia and the standardizing which marks progress. Therein lies the danger in loosely talking about standardization. [...] no body of men could possibly have the knowledge to set up standards, for that knowledge must come from the inside of each manufacturing unit and not at all from the outside. In the second place, presuming that they did have the knowledge, then these standards, although perhaps effecting a transient economy, would in the end bar progress, because manufacturers would be satisfied to make the standards instead of making to the public, and human ingenuity would be dulled instead of sharpened" (Ford quoted in Ohno 1978/1988, p. 99). Currently, no standardisation system in education is made for improving education by a factor of ten.

[38] The goal of quality management as taught, for instance, by Juran in Japan, which led to the Toyota Production System, is to optimise the process so that it has zero defects. As education is about humans and their lives, the goal of zero dropouts and even zero low achievers should be at least as important in education as in the automotive industry.

[39] In Finland, teachers from Kurki-Suonio's courses already have trouble to measure their performance, because even the phenomenology of the PISA-type tests is insufficient to measure their way of teaching.

The freed time and resources can then instead be spent on helping the students to improve their learning and improving teaching so that more than 5 % remains. In such a way, teaching can be easily improved by several hundred percent with no new resources necessary. The OECD estimates that Finnish students gain ¾ of a year on 4 years of school against German students.

72.4 The Historical-Genetic Background of Mach's Ideas on Science Teaching

What is the central change in Mach's world view in comparison to the previous (antique) world view? The main difference is the transformation in the concept of genesis brought about by many new empirical observations, which Darwin has so consistently put together in his *Origin*.

In the previous 2,500-year-old Aristotelian genetic view (based on Empedocles, see Freemen 1947/1971), animal organs (legs, arms, etc.) were initially "puzzled" together, with more or less fitting results from which only the best (the actual species, some anthropomorphic beings like centaurs and especially humans) were retained. These were the "ideal forms" (species, ideas) subsequently serving as "mould" for further copies (like a signet ring is copied into molten wax). This antique concept of genesis had quite some, but limited applicability. It would fit to some of what we would now call evolutionary phenomena, but not to others. It would explain some prenatal defects, such as Siamese twins, or cloven hands as well as anthropomorphic figures of gods with animal heads, maybe deriving from dreams or observations of masked performances. It would not fit to genetic aspects of evolution, such as fossils or species closer or less related (especially to humans). But in ancient times, such phenomena were not easily observable.

Plato developed a philosophy of knowledge in analogy to this Empedocletian hypothesis. Like the Babylonians saw the stars as the eternally unchanging and ideal gods, the Platonian universe god (in *Timaios*) was ideal and eternal. It was the signet ring, from which the earthly (wax) moulds were formed (with some errors in the process). Similarly, the godly ideas were ideal and eternal. As all souls were born with some divine essence, they could sometimes remember these ideas and thus retain them. Plato wanted to include the idea of *metempsychosis* (reincarnation of eternal souls), already much used by the Eastern religions and philosophies, into his philosophy as well.

For knowledge it meant that in humans potentially ideal (godly) thinking was degraded by earthly experience. Many of the ideas transported in myths thus become archetypes of other ideas (see, for instance, Eliade 1975/1981). Knowledge is seen as fixed forms, which do not change over time. They can only be rejected as false and replaced. Thereby, the differentiation between "true and false" becomes central. The ideas that "errors" are an integral part of knowledge and that knowledge and errors can be "good" or "less good" or "bad" (i.e. completely senseless), like gestalt psychologists suggest (see Koehler 1925/1957 or Lipmann and Bogen 1923), are not part of Plato's system.[40] As a consequence, the logical components were given priority within the overall world view as logic was seen as god's ideal way of thinking and detecting erroneous ideas. This view of course tends to be absolutistic rather than democratic, like the society in Plato's Republic (see also Siemsen 2010c). One tends to claim to know the truth, while other ideas are declared aberrations. Plato's ideal ideas cannot be developed further. They are the goal (*telos*) for any learning. For teaching, the Platonian goal is in principle already achieved and just needs to be reproduced.

Darwin (1859) in his *Origin of the Species* replaced this antique "organ puzzle" with the idea of an adaptive and transformative non-teleological process over time. Species constantly evolve in an interactive process with the environment. Organs are not puzzled together, but are formed by a long process of small adaptations, which can open new opportunities and thus "transform" into new applications. There is no godly signet ring, but each "mould' itself creates new "moulds" (though the idea of errors in the reproduction process remains as mutations). No eternal soul needs to be transported from our "inner fish" (see Shubin 2008) to our "human form".

What does Darwin's view on species mean for the area of knowledge and ideas? Mach (1883/1888) in his essay *Transformation and Adaptation in Scientific Thought* generalises

> Knowledge is an expression of organic nature. The law of evolution, which is that of transformation and adaptation, applies to thoughts just as well as to individuals or any living organisms. A conflict between our customary train of thought and new events produces what is called the problem. By a subsequent adaptation of our thought to the enlarged field of observation, the problem disappears and through this extension of our sphere of experience, the growth of thought is possible. Thus the happiest ideas do not fall from heaven, they rather spring from notions already existing. (Mach 1883/1888)

As a result, Mach's method is the genetic method of teaching, i.e. teaching as close as possible to the way how we as humans have biologically and culturally evolved. Biology and culture for Mach are reciprocal processes, which have a *plastic* result.[41]

Plato's hypothetical analogy of biological genesis and human knowledge thus becomes a singular genetic process through a consistent (and thought economical)

---

[40] "The self is no pot into which the blue and the ball just have to drop into so that a judgment [a blue ball] shall result. The self is *more* than a simple unity, and certainly no Herbertian *simplicity*. The same spatial elements, which close to a ball have to be blue and the blue has to be recognized as different from the places, as separable, in order to come to a judgment. [...] If we will see the self not as a monad isolated from the world, but a part of this world and in the midst of its flux, from which it came from and to which it is willing to diffuse again, so we will not anymore be inclined to see the world as something *unknowable*. We are then *close* enough to ourselves and *related* enough to the other parts of the world in order to hope for real *Erkenntnis*" (Mach 1905, p. 462).

[41] He is thus neither a nativist nor a blanc-slate empiricist, but close to modern notions of neuronal plasticity (for instance, by Buonomano and Merzenich 1998).

application of the idea of evolution. There is only biological and cultural evolution as two properties of the same genetic process. In this respect, Plato's conception of ideas is retained.

But then, another aspect of Plato's synthesis must be rejected as inconsistent with the idea of evolution. In the new evolutionary perspective, "forms" are not ideal anymore in an absolute, godly sense, just like species are not ideal, but temporary adaptations. Species as well as ideas are processes and products at the same time. In a Newtonian sense, they "interact over time" with the environment, or they are "reciprocal" as James suggested. Forms are only temporary *gestalts*, resulting from the way humans adapted biologically and culturally (historical genesis). These gestalts become adapted and transformed. They are never final.

From an *erkenntnis* theoretical view, and especially an *erkenntnis* psychological view, this fundamental conceptual change reciprocally requires many other conceptual changes which are based on this question. The empirical meanings of several foundational concepts, such as "knowledge" change, i.e. need to be adapted accordingly. Thus, the *erkenntnis* psychology of education fundamentally changes from the older Platonian/Aristotelian[42] view.

At the time of Mach, the Platonian/Aristotelian view had already been slowly eroded by central facts from various disciplines, which were obviously inconsistent with it. Many scientists (Erasmus Darwin, Spinoza, Herbart, Spencer, etc.) had already tried new syntheses from this, groping for new empirical meanings of concepts and new models regarding humans and nature. Charles Darwin's *Origin* was just a culmination of such efforts. Mach was afterwards the first[43] to develop a fundamentally new world view, including the (partial) synthesis from Charles Darwin and the facts from other sciences (physics, psychology, etc.). But he was unable to finalise (sufficiently stabilise) this view in all aspects during his lifetime.

Although Mach had a strong influence on science education (especially on the development of ideas of eminent scientists), he did not write a complete theory of science education. Thus, his intuition in science education remains more influential than his theory of it. His ideas have nevertheless been very fruitful (see Siemsen 2010a, b, c, d, 2011, 2012a, b). Therefore, an evaluation of the questions leading Mach to his ideas might have a strong effect on the future of science education. In science education, especially regarding the history and philosophy of science teaching, the time might now seem more recipient for his ideas than at Mach's own times.

[42] Aristotle as a student of Plato shared many of Plato's ideas; though as son of a physician, he was certainly more empirical in the physiological details of the theory.

[43] Mach published the first articles and books in this direction already four years (1863) after the publication of the *Origin*. This was much before Haeckel and Darwin himself wrote on the evolution of humans and human knowledge.

72.5 Mach's Central Ideas: Sensualism, Gestalt and *Erkenntnis* Theory

What are Mach's central ideas for science education? From his synthesis between the Platonian/Aristotelian world view and Darwin's world view, there are in principle three reciprocal areas of application: a consequent basis of all knowledge in its *sensual* origins, the thought-economical reduction of sensual elements to conceptual *gestalts* (*Vorstellungen*) and the *erkenntnis* theory, which provides the overall consistency of concepts as well as a meta- perspective and method for optimising one's own learning.

As Mach takes the psychophysical relations as basis for his world view, the sensual and physical perspectives provide the starting points for exploring this relation and therefore the empirical basis of all knowledge. Thus, in line with David Hume and contrary to the aversion of Plato and Aristotle towards sensuality, Mach postulates that the only possible (and empirical) grounds for concepts are the senses and sensual experiences. Reason and reflection are not something apart from this, some higher faculty of thought, but merely an extension and further development of sensualistic origins.[44]

When Aristotle postulates that "*no bodily activity* has any connexion with the *activity of reason*" (Aristotle 1930, p. 2068) and Plato urges to wait for the "opposing currents" of the senses to abate,[45] for Mach these seemingly "opposing [sensual] currents" (from an adult perspective) are the genetic origins of reason. Instead of Platonic waiting, one can encourage the thought-economical process right from its origins. Just because a suckling cannot speak words (i.e. "call names") does not imply that it cannot think or communicate. On the contrary, the suckling's speaking and reasoning are based on bodily activities. For Mach, there is only the "adaptation

[44] "Certainly one can judge internally, *without* linguistic expression or *before* it. [...] One can easily observe this for clever dogs or children, who cannot yet speak" (Mach 1905, p. 112/113). "The basis of all knowing is therefore the *intuition*, which can be related to something sensually perceived, just vividly imagined or potentially imaginable, conceptualizable. The logical knowledge is just a special case of the formerly described, which is only concerned with the finding of consistency or inconsistency and which cannot be brought about without perception or imagination related to former findings. If we come to this new *finding* by pure physical or psychical chance or by planned extension of the experience by thought experiments [...], it is always *this* finding, from which all knowledge [Erkenntnis] grows" (Mach 1905, p. 315).

[45] "... and by reason of all these affections [the "opposing currents" of nutrition and senses], the soul, when encased in a mortal body, now, as in the beginning, is first without intelligence; but when the flood of nutriment abates, and the courses of the soul, calming down, go their own way and become steadier as time goes on, then the several circles return to their natural form [mirroring the ideal thoughts of the Cosmos-god seen in the astronomical observations of circular movement], and their revolutions are corrected, and they call the same and the other by the right names, and make the possessor of them to become a rational being. And if these combine in him with any true nurture [Pythagorean dietary obligations] or education, he attains the fullness and health of the perfect man, and escapes the worst disease of all; but if he neglects education, he walks lame to the end of his life, and returns imperfect and good for nothing to the world below. This however, is a later stage [...]" (Plato 1871/1892, Vol. III, p. 463).

of the thoughts to the facts and the thoughts to each other". But the adaptation of the thoughts to the facts must take priority as otherwise the thoughts have no relation to the world and are thus arbitrary. Furthermore, learning – including human learning – has evolved (according to Darwin) over millions of years from the sensual experience and not from some (relatively recent) metaphysical a priori system. Sensualistic learning must therefore be much more effective than rationalistic learning.

For Mach, reasoning is just a (cultural) result of the economy of thought. As one cannot memorise all singular sensual experiences, one needs to economise the thoughts on experience.[46] To describe the result of this process, Mach uses the concept of "gestalt".

72.6 Gestalt Psychology

Mach is known as the intellectual father of the gestalt[47] concept, which first Christian von Ehrenfels (1890) identified as a specific Machian concept. The gestalt psychologists (Wertheimer, Koehler, Koffka, Kaila, Lewin, etc.) developed Mach's gestalt concept into a full psychology (see, for instance, Ash 1995).

In his *Analysis*, Mach (1914, footnote p. 90) describes the origin of his idea of gestalt, which led to his last shift in world view.[48] "Some forty years ago […], in a society of physicists and physiologists, I proposed for discussion the question, why geometrically similar figures were also optically similar. I remember quite well the attitude taken with regard to this question, which was accounted not only superfluous, but even ludicrous. Nevertheless, I am now as strongly convinced as I was then that this question involves the whole problem of visual gestalts. That a problem cannot be solved which is not recognized as such is clear. In this non-recognition, however, is manifested, in my opinion, that one-sided mathematico-physical direction of thought […]".[49]

These ideas of Mach had intuitive influences on physics and mathematics but also on the development of psychology, for instance, on the concept of gestalt and

[46] Actually, nature economises memory in the sense of Hering (1870/1969). Therefore, gestalts are a plastic psychophysical process between physiological and cultural genesis, not limited to thoughts alone.

[47] The gestalt concept as a holistic concept was already used by Goethe or Herbart, but it did not have the genetic perspective. Only after Darwin, Mach could transform the concept into a "process and product". Therefore, seeing gestalt only as "holism", which one often finds in the literature, is an outdated, pre-Darwinian and in the sense of gestalt psychology inadequate understanding.

[48] There were at least two more shifts in his world view (see Mach 1914, p. 30). Mach had several such gestalt shifts in world view, but this shall not be of concern in this article. One nevertheless needs to be careful when reading Mach, which view he held at the time of writing. Some of the views are not consistent to each other.

[49] This quotation also bears reference to the prior-discussed question of methodological specialisation in science and the (unintended) intuitive training of one's thoughts.

gestalt psychology.[50] The article *On Gestalt Qualities* from 1890 by von Ehrenfels was regarded by Wertheimer and Koehler as the foundational article of gestalt psychology (see Wertheimer 1924, Koehler 1920/1938, and Ash 1995). In the beginning of this article, von Ehrenfels states that his ideas initiated from an intuition he had after reading Mach's *Analysis*:

> My starting point arose from several remarks and hints from E. Mach's "Contributions to the Analysis of Sensations" (Jena 1886) […]. Mach sets up the, for some people certainly paradoxically sounding, hypothesis that we can immediately "sense" [*empfinden*] spatial patterns [*Gestalten*] and even "sound-patterns" or melodies. And indeed at least the second of these theses not only seemingly, but also from its contents should be undisputedly absurd,[51] if it would not be immediately intelligible, that "sensation" here is used in a different than the usual sense. […] But if Mach, by using the term "spatial and tonal gestalts", also wanted to stress its simplicity, it becomes clear that he […] viewed these "gestalts" not as mere combination of elements, but as something new (relative to the elements on which they base) and up to a certain degree independent. […] I hope to be able to show in the following that Mach [in his reflections] has shown us the way of resolving the problem mentioned. (von Ehrenfels 1890, pp. 249–251)

The problem von Ehrenfels mentions is what William James called *The Knowing of Things Together* (1895).[52] If one takes a "regular" concept of sensation, it is a question of descriptive psychology. For Mach, it becomes a question of genetic psychology. What is "simple" might therefore appear to be simple in its current sensational gestalt but is actually a result of a "complex", i.e. reciprocal genetic process between sensation, physiology, memory and background. Gestalts are not linear (additive); they are adaptive and transformational.[53]

Abstraction therefore is not necessarily a "higher development", but a specialisation of thoughts. This specialisation might be thought economical in some contexts or hindering in others. What is "higher" and "lower" in a cultural sense cannot be judged anthropomorphically from the point of view of a specific culture.[54] Such categories often make no sense in anthropology, as the whole conceptual frames (world views) are different. All concepts might be based on completely

---

[50] His other influences, for instance, on the development of genetic psychology (Siemsen 2010b) and Boring (1950), shall not be of concern here.

[51] The German word used here is *widersinnig*, which literally translates as "counter sensual".

[52] Hadamard (1945, p. 65) quotes a description from Rodin: "Till the end of his task, it is necessary for [the sculptor] to maintain energetically, in the full light of his consciousness, his global idea, so as to reconduct unceasingly to it and closely connect with it the smallest details of his work. And this cannot be done without a severe strain of thought". The gestalt (statue) integrates the many details (ideas) into a *global idea.*

[53] Many gestalts are not best recognised in reality, but in caricatures, i.e. when certain properties are overemphasised, while others are neglected. Greek statues in the "classic" times are thus not just idealised bodies. They are idealised beyond the point a body could look like. Egyptian obelisks are not built straight, but slightly curved. Equally, "natural laws" cannot be observed in reality. They approximately describe many facts under idealised circumstances which can never be "achieved" for each single fact.

[54] Similar ideas have been developed by Franz Boas and have been very influential in US anthropological thinking until today (see Stocking 1968). On the closeness of the development of these ideas to Mach and their later influence on the initial ideas of Jerome Bruner, see Siemsen (2010b).

different sets of empirical meanings.[55] Also logic and our current Western concept of number came out of practical requirements and needs. They can thus neither be considered universal nor culturally independent. Because of the dominance of Western culture and the expansion of modern science, they have just been developed by more people over longer time than any other similar conceptual systems from other cultures.

The concept of gestalt also resolves the unproductive questions regarding the initial primacy of "nature or nurture". From a gestalt perspective it makes no sense to try to draw a triangle in one dimension and ask which of its sides has a larger share in its area. If one takes the concept of memory as a general function of evolution (see Hering 1870/1969), the problem disappears as a pseudoquestion.

72.7 Mach's *Erkenntnis* Theory in Science Education

The Machian world view is psychophysical, i.e. its basic "conceptual currency" of comparison (elements) requires switching between a physical and a psychical perspective. From the psychical perspective, the basic elements can be seen as sensual (defined recursively as gestalts). In a historical genesis process, clusters of sensual elements from different domains are synthesised into new gestalts, such as space, time and measurement. The syntheses in turn change the focus of attention to new observations, while neglecting others. Depending on the syntheses made, foundational empirical meanings of pre-scientific concepts are laid, which can help or hinder the acquisition of scientific concepts.

In order to keep all these processes consistent, to integrate them, one needs an *erkenntnis* theory. The *erkenntnis* theory provides the conceptual toolkit, the generality, but even more important the meta-method of consistently combining different methods of learning. It provides the observational meta-perspective on oneself as a learner so that one can optimise one's own learning, detect systematic errors, etc.

For implementing this in science education, Mach suggests, for instance, the use of history as (one) method, "[…] It should be shown in all analogous cases, how the concepts have originated historically, which observations have urged towards them. The most naïve historical exposition is always the best. The discoverer of a truth in natural science and the able student both stand before a new theorem without presuppositions. For both the same way is therefore most natural" (Mach 1876, p. 6). "The teacher using historical material […] will not run into the danger of expecting from the student to understand in one attempt, what could develop in the most

[55] The difference in observed facts can be so intuitively fundamental that even with long training, one cannot observe them in principle. For instance, Boas observed that he consistently heard phoneme in Inuit language one time as one sound, another time as a different sound. His interpretation was that psychophysically, the factual sound gestalt was in between, but impossible for him to hear in the way the natives would hear it.

important heads but slowly and gradually" (Mach 1876, p. 6/7).[56] "The historical exposition will lead to the comparison of different chapters and thereby separate the fundamental from the accidental and conventional" (Mach 1876, p. 8). "The historical [genetic] exposition will apart from clarity have the advantage that the teaching stops to be dogmatic. Science appears as something evolving, not finished, still to be shapable in the future. Mental educability instead of simple education should be the result of such a teaching" (Mach 1876, p. 9). The use of history for Mach thus mainly serves an *erkenntnis* theoretical function.

Mach elaborates the idea of how to teach in a way to optimise the learning for science education: "[…] The formal education is a seed, which by itself will develop fruitfully. This process is not to be underestimated relative to the positive skills, when the student immediately transfers to the practical life, because then he can easily acquire what is missing by himself. On the other hand, an unnecessarily stuffed head cannot help to lead him in times of need out of the critical situation. Formal education is on the other hand the main goal if the teaching is continued in higher education. Then all elements have to be transformed anyway and have to be taken as material and foundation of the new. It is therefore practical to limit the subject matter as far as possible, but work on it in a *many-sided way*. Only by dealing differently with the same issue, one understands the basis of the method and achieves ability in its usage" (Mach 1876, p. 1/2).

The genetic method is thereby different from the axiomatic method, which starts from basic axioms; it is different from the Socratic method, which starts from (linguistic) definitions (already presupposing a frame or a "background", see Rubin 1921), but as described before, it is also not simply historical, though it makes intensive use of history. "One can certainly provide the student with a broader knowledge by starting from ready-made definitions and concepts, placing ready-made doctrines in front of him and proving them. This method is not even so bad in mathematics, because there the step from the experience [*blosse Anschauung*] to the definition and to the theorem might be very short and can therefore be complemented by each able student. The physical knowledge which is acquired in this way nevertheless always appears as externally imposed. Especially for the student who reflects, sudden gaps of clarity will appear which he will not be able to resolve, if he does not know, how one has arrived at the concepts put at the top" (Mach 1876, p. 2/3).

One way of implementing this is by letting the students guess the "success" of an experiment before it is conducted. "Not only thereby the attention is increased, but youths will from the errors which occur in this also draw the lesson that natural laws cannot be philosophized-forth [*lassen sich nicht herausphilosophieren*]. A body does not, as most will guess, in double the time also fall double the way. The pendulum of fourfold length does not also show the fourfold duration of oscillation. One here does not have to construct *a priory*, but to observe" (Mach 1876, p. 20/21).

---

[56] Quoted with permission of the Philosophical Archive of the University of Konstanz.

72.8 Mach's Successors

The successors of Mach, who had Machian effects in their teaching, are William James, Alfred Binet, Eino Kaila, Kaarle Kurki-Suonio, Karl Hayo Siemsen, Martin Černohorský, Rudolf Laemmel, Eugenie Schwarzwald, Frank Oppenheimer, Alexander Israel Wittenberg, Max Wertheimer, Abraham S. Luchins, Adolf Hohenester and John Bradley.

There are several people who used at least part of Mach's teaching ideas in their teaching philosophy, but probably not enough to produce the exponential effects (though this has not been studied in all detail in every case): John Dewey, Otto Blueh, Jerome Bruner, Benchara Branford, Catharina Stern, Eric M. Rogers, Peter Drucker, Paul A. Samuelson, Wilhelm Ostwald, Henry Edward Armstrong, Edgar W. Jenkins, Martin Wagenschein, Georg Kerschensteiner, Spiru Haret, Efraim Fischbein, Solomon Marcus, Joachim Thiele, Robert Wichard Pohl, Walter Jung, Fritz Siemsen, Hugo Kükelhaus, Hans Freudenthal, George Sarton, Alfred N. Whitehead, James Conant, Gerald Holton, Frank Wilczek, K. V. Laurikainen, Edouard Claparède, Théodore Flournoy and Jean Piaget. Many more educators have been inspired by Machian ideas; though after several generations, these "mother's milk" effects become increasingly difficult to trace and to make explicit.

72.9 Sources on Mach's Educational Method

Binet, A. (1911/1975). *Modern Ideas about Children.* Suzanne Heisler, Albi.

Blueh, O. (1967). Ernst Mach as Teacher and Thinker. *Physics Today*, 20: 32–42.

Bradley, J. (1975). Where Does Theory Begin? *Education in Chemistry* March 1975: 8–11.

Drucker, P.F. (1979). Miss Elsa and Miss Sophy. In: *Adventures of a Bystander*, Harper & Row, New York: pp. 62–82.

Hoffmann, D. & Laitko H (1991). *Ernst Mach. Studien und Dokumente zu Leben und Werk.* Deutscher Verlag der Wissenschaften, Berlin, p. 369–370.

Hohenester, A. (1988). Ernst Mach als Didaktiker, Lehrbuch- und Lehrplanverfasser. In: Haller, R. & Stadler, F. (Eds.), *Ernst Mach Werk und Wirkung.* Vienna: Hölder-Pichler-Tempski.

Koller, S. (1997). *Kommentare zu den physikdidaktischen Schriften Ernst Machs.* Diploma Thesis at the Karl-Franzens-Universität Graz, supervised by Adolf Hohenester. It contains a collection of Mach's articles for the teacher's journal *Zeitschrift für der physikalischen und chemischen Unterricht.*

Kurki-Suonio, K. (2011). Principles Supporting the Perceptional Teaching of Physics: A "Practical Teaching Philosophy". *Science & Education*, 20: 211–243.

Laemmel R (1910). *Die Reformation der Nationalen Erziehung.* Mit einem Vorwort von Ernst Mach in Wien. Zürich: Speidel.

Mach, E. & Odstrčil, J. (1887). *Grundriss der Naturlehre für die unteren Classen der Mittelschulen. Ausgabe für Gymnasien.* Prag: Tempsky.

Mach, E. (1893/1986). *Popular Scientific Lectures*. La Salle: Open Court.

Mach, E. (1915). *Kultur und Mechanik*. Stuttgart: Spemann.

Matthews, M. R. (1990). Ernst Mach and Contemporary Science Education Reforms, *International Journal of Science Education*, 12/3: 317–325.

Rogers, E. M. (1960/1977). *Physics for the Inquiring Mind. The Methods, Nature and Philosophy of Physical Science*. Princeton: Princeton University Press.

Streibel, R. (ed.) (1996). *Eugenie Schwarzwald und ihr Kreis*. Wien: Picus.

Swoboda, W. (1973). *The Thought and Work of the Young Ernst Mach and the Antecedents to Positivism in Central Europe*. dissertation, University of Pittsburgh.

Wittenberg, A. I. (1968). *The Prime Imperatives: Priorities in Education*. Toronto: Clarke, Irwin & Company.

Kaarle Kurki-Suonio wrote a series of schoolbooks titled *Galileo* and a book on his method, which are currently available only in Finnish.

References

Ahlfors, L.V. et al. (1962). On the Mathematics Curriculum of the High School. American *Mathematical Monthly*, 69: 425–426. The Memorandum was signed by 65 mathematicians, among them Wittenberg, Polya, Kline, Weyl, etc.

Aristotle (about 350 B.C./1930). *The Works of Aristotle. Organon and Other Collected Works edition*. Ross, W. D. (Ed.). Oxford: Oxford University Press.

Ash, M. G. (1995). *Gestalt Psychology in German Culture 1890–1967: Holism and the Quest for Objectivity*. Cambridge: Cambridge University Press.

Boring, E. G. (1950/1957). *A History of Experimental Psychology*. Prentice Hall: Englewood Cliffs.

Buonomano, D. V. & Merzenich, M. M. (1998). Cortical Plasticity: From Synapses to Maps. *Annual Review of Neuroscience* 21: 149–186.

Csikszentmihalyi, M. (1990). *Flow: The Psychology of Optimal Experience*. New York: Harper.

Cohen, J. (1988). *Statistical power analysis for the behavioral sciences* (2nd ed.) Hillsdale, NJ: Erlbaum Associates.

Cole, K. C. (2009). Something incredibly wonderful happens. Frank Oppenheimer and the World He Made Up. Boston: Harcourt.

Drucker, P. F. (1979). *Adventures of a Bystander*. New York: Harper & Row.

Ehrenfels, C. von (1890). Über Gestaltqualitäten. *Vierteljahresschrift für wissenschaftliche Philosophie*.

Einstein, A. (1916). Ernst Mach. *Physikalische Zeitschrift*, 17/7, 1st of April: 101–104.

Eliade, M. (1975/1981). *History of Religious Ideas: Vol. 1*. Chicago: University of Chicago Press.

Eschenbach, S. (2010). From inspired teaching to effective knowledge work and back again: A report on Peter Drucker's schoolmistress and what she can teach us about the management and education of knowledge workers. *Management Decision*. 48/4: 475–484.

Freemen, K. (1947/1971). *Ancilla to the Pre-Socratic Philosophers*. Translated after the fifth edition of Diels, *Fragmente der Vorsokratiker*. Blackwell: Oxford.

Hadamard, J. (1945/1954). *The psychology of invention in the mathematical field*. New York: Dover.

Hattie, J. (2009). *Visible Learning. A Synthesis of Over 800 Meta-Analyses Relating to Achievement*. London: Routledge.

Hattie, J. (2012). *Visible Learning for Teachers. Maximizing Impact on Learning*. London: Routledge.

Hering, E. (1870/1884/1969). *Über die spezifischen Energien des Nervensystems*. New print, Amsterdam: Bonseth.

Hohenester, A. (1988). Ernst Mach als Didaktiker, Lehrbuch- und Lehrplanverfasser. In: Haller, R. & Stadler, F. (Eds.), *Ernst Mach Werk und Wirkung*. Vienna: Hölder-Pichler-Tempski.

Holton, G. (1992). Ernst Mach and the Fortunes of Positivism in America. *Isis, 83/1, March,* 27–60.

Huxley, T. H. (1864/1870). Criticisms on "The Origin of Species". In: *Lay Sermons, Addresses, and Reviews*. New York: Appleton.

James, W. (1895/1967/1977). The Knowing of Things Together. In: McDermott, J. J. (Ed.), *The Writings of William James – A Comprehensive Edition* (152–168). University of Chicago Press: Chicago.

Juran, J. M. (Ed.) (1951). Quality-Control Handbook. New York: McGraw-Hill.

Koehler, W. (1925/1957). *The Mentality of Apes*. London: Penguin.

Koehler, W. (1920/1938). 'Physical Gestalten'. In: Ellis, W. D. (Ed.) A Source Book of Gestalt Psychology (17–54), London: Kegan, Trench, Trubner & Co.

Kurki-Suonio, K. (2011). Principles Supporting the Perceptional Teaching of Physics: A "Practical Teaching Philosophy". *Science & Education*, 20: 211–243.

Lavonen J, Jauhiainen J, Koponen IT, Kurki-Suonio K (2004). Effect of a long-term in-service training program on teacher's beliefs about the role of experiments in physics education. *International Journal of Science Education*, 26/3: 309–328.

Lipmann, O. & Bogen, H. (1923). Naive Physik. Theoretische und experimentelle Untersuchungen über die Fähigkeit zu intelligentem Handeln. Leipzig: Barth.

Luchins, A. S. (1993). On Being Wertheimer's Student. A contribution to the eighth international *Scientific Convention of the Society for Gestalt Theory and its Applications* (GTA) at the University of Cologne, Friday, March 26, 1993.

Mach, E. (1863). *Compendium der Physik für Mediciner*. Wien: Braumüller.

Mach, E. (1866). *Einleitung in die Helmholtz'sche Musiktheorie – Populär für Musiker dargestellt.* Graz: Leuschner & Lubensky.

Mach, E. (1876). *Entwurf einer Lehrinstruction für den physikalischen Unterricht an Mittelschulen.* Document from the Philosophical Archive at the University of Konstanz.

Mach, E. (1883/1933/1976). *Die Mechanik – historisch-kritisch dargestellt.* (reprint 9th ed), Darmstadt: Wissenschaftliche Buchgesellschaft.

Mach, E. (1883/1893/1915). *The Science of Mechanics – A Critical and Historical Account of Its Development. Supplement to the Third English Edition by Philip E. B. Jourdain.* Chicago: Open Court.

Mach, E. (1883/1888). Transformation and Adaptation in Scientific Thought. *The Open Court*: Jul 12 1888, 2/46, APS Online IIA.

Mach, E. (1886/1919). *Die Analyse der Empfindungen und das Verhältnis vom Physischen zum Psychischen*. Jena: Fischer.

Mach, E. (1890). Über das psychologische und logische Moment im Naturwissenschaftlichen Unterricht. *Zeitschrift für den physikalischen und chemischen Unterricht*, 4/1, October 1890: 1–5.

Mach, E. (1905/1926/2002). *Erkenntnis und Irrtum: Skizzen zur Psychologie der Forschung*. 5th Edition, Leipzig, Berlin: reprint by rePRINT.

Mach, E. (1905/1976). *Knowledge and Error*. Dordrecht: Reidel.

Mach, E. (1914). *The Analysis of Sensations and the Relation of the Physical to the Psychical.* Chicago: Open Court.

Mach, E. (1920). letters to Gabriele Rabel. In Rabel, G.: Mach und die „Realität der Außenwelt". *Physikalische Zeitschrift, XXI*, 433–437.

May, H. B. (1921) The Part Inspection Plays in Good Management. *Factory* 26/2: 175–179, 26/3: 335–339.

Mueller-Fohrbrodt, G., Cloetta, B. & Dann, H.-D. (1978). *Der Praxisschock bei jungen Lehrern.* Stuttgart: Klett.

OECD (2007). *PISA 2006 – Science Competencies for Tomorrow's World, Volume I and II*, http://www.oecd.org. Cited 12 Jun 2008.[57]

Ohno, T. (1978/1988) Toyota Production System. Beyond Large-Scale Production. Portland: Productivity Press.

Oppenheimer, F. (1981). Museums, Teaching and Learning. Paper prepared for the AAAS meeting, Toronto 1981, downloadable at http://www.exploratorium.edu/about/our_story/history/frank/articles/museums/index.php, accessed 27/08/2012.

Plato (360-380B.C./1871/1892). *The Dialogues of Plato*. Translated and commented by Jowett, B. Third Edition. Oxford: Oxford University Press.

Poincaré, H. (1908). L'invention mathématique. *L'Année Psychologique, XV,* 445–459.

Rubin, E. (1921). *Visuell wahrgenommene Figuren. Studien in psychologischer Analyse*. Copenhagen: Gyldendalske Boghandel.

Shubin, N. (2008). *Your Inner Fish. A Journey into the 3.5-Billion-Year History of the Human Body*. New York: Vintage Books.

Siemsen, H. (2010a). Intuition in the Scientific Process and the Intuitive "Error"of Science. In: Columbus, A. M. (Ed.) *Advances in Psychology Research*. Vol. 72. Nova Science, Hauppauge, 1–62.

Siemsen, H. (2010b). Alfred Binet and Ernst Mach. Similarities, Differences and Influences. University of Nancy, *Revue Recherches & Éducations*, 3/2010: 352–403.

Siemsen, H. (2010c). The Mach-Planck debate revisited: Democratization of science or elite knowledge? *Public Understanding of Science,* 19/3: 293–310.

Siemsen, H. (2010d). Erkenntnis-theory and Science Education – An intuitive contribution to the Festschrift for the 85th birthday of Solomon Marcus. In: Spandonide, H., Paun, G. (Eds.) *Meetings with Solomon Marcus to his 85th Birthday*. Bucuresti: Editura Spandugino.

Siemsen, H. (2011). Ernst Mach and the Epistemological Ideas Specific for Finnish Science Education. *Science & Education*, 20: 245–291.

Siemsen, H. (2012a). Ernst Mach, George Sarton and the Empiry of Teaching Science Part I. *Science & Education*, 21/4: 447–484.

Siemsen, H. (2012b). Ernst Mach, George Sarton and the Empiry of Teaching Science Part II. *Science & Education*, online first 25/04/2012.

Siemsen, K. H. (1981). *Genetisch-adaptativ aufgebauter rechnergestützter Kleingruppenunterricht. Begründungen für einen genetischen Unterricht*. Frankfurt: Lang.

Siemsen, K. H., Schumacher, W., Wiebe, J. & Siemsen, H. (2012). Dokumentation des Projektstudiums nach 10 Jahren. Working Paper.

Speziali, P. (Ed.) (1972). *Albert Einstein – Michele Besso: Correspondance*. Paris: Hermann.

Stocking, G. W. Jr. (1968; 1982). *Race, Culture and Evolution: Essays in the History of Anthropology*. Chicago: University of Chicago Press.

Thiele, J. (1978). *Wissenschaftliche Kommunikation: Die Korrespondenz Ernst Machs*. Kastellaun: Henn.

Vaney, V. (1911). Les classes pour arriérés. Recruitement, organisation, exercises d'orthopédie mentale. *Bulletin de la Société libre pour l'Etude psychologique de l'Enfant*, 73: 293–300.

Wagenschein, M. (1970). 'Was bleibt?' In: Fluegge, J., *Zur Pathologie des Unterrichts*, Bad Heilbrunn: Klinkhardt, 74–91.

Wertheimer, M. (1924/1938). Gestalt Theory. In: Ellis, W. D. (Ed.) *A Source Book of Gestalt Psychology*. London: Kegan, Trench, Trubner, 1–11.

Weyl, H. (1928). *Philosophie der Mathematik und Naturwissenschaft*. Munich: Leibnitz.

[57]The other PISA studies are equally interesting, but will not be cited, as the general issues discussed here can be taken from any single one of them

Hayo Siemsen is working on his habilitation on Mach's influences and his ideas on the philosophy of education and the psychology of research. He is also member of the Ernst Mach Institute for Philosophy of Science and member of the newly founded international Institute for Processes of Learning. His Ph.D. theses were focused on innovation systems, especially on the question if one can learn to be innovative. In various publications he covered a wide range of areas, including education, psychology, economics, management, history, methodology and philosophy/*erkenntnis* psychology of mathematics. The publications are following Mach's general ideas and broad influences rather than scientific categorisations. Recently he published a series of articles in *Science & Education* on Machian education, especially in Finland in comparison with other, similar science education.

Chapter 73
Frederick W. Westaway and Science Education: An Endless Quest

William H. Brock and Edgar W. Jenkins

73.1 Introduction

> I fear that during my professional career, I advocated the claims of science teaching much too strongly, and I am now quite sure that the time often devoted … to laboratory practice, and to the purely mathematical side of science, more especially chemistry and physics, was far too great. (Westaway 1942a, p.v)[1]

So wrote F. W. Westaway, teacher, headmaster, His Majesty's Inspector (HMI) and eloquent advocate of science education, in the preface of his last book, published in 1942, with the intriguing title *Science in the Dock: Guilty or Not Guilty?* Who was Westaway, what influence did he have upon school science teaching and what had prompted him to raise and address this question?

Frederick William Westaway was born on 29 July 1864 at Cheltenham, Gloucestershire, the first of seven children of William and Caroline Westaway, three of whom died in infancy. It seems clear that the family circumstances were extremely modest. His father was a travelling blacksmith, and his mother (to judge from the mark she made on Frederick's birth certificate) was unable to write. Westaway later recalled receiving his first chemistry lesson at the age of 10 in 1874. It was given by the Gloucester Public Analyst[2]:

[1] The preface was dated September 1941. He had expressed similar doubts following WW1 in a new preface to the 2nd ed. of Westaway (1919, p. xii).

[2] Probably John Horsley, a founder member of the Society of Public Analysts in July 1874. He specialised in the analysis of milk and dairy products. Westaway joined the Chemical Society in June 1892.

W.H. Brock (✉)
Department of History, University of Leicester, Leicester, UK
e-mail: william.brock@btinternet.com

E.W. Jenkins
Centre for Studies in Science and Mathematics Education, University of Leeds, Leeds, UK
e-mail: e.w.jenkins@education.leeds.ac.uk

M.R. Matthews (ed.), *International Handbook of Research in History, Philosophy and Science Teaching*, DOI 10.1007/978-94-007-7654-8_73,

> There was no laboratory available, but there was a well-fitted lecture room, and in later lessons a few gases were prepared. But the first lesson, which extended over an hour, was frankly a lecture on the atomic theory. No experiments whatever were performed, but the formulae and equations which covered the blackboard impressed at least one small boy. (Westaway 1937a, p. 490; Westaway 1929, p. 18)

By 1881 the family had moved to Ruardean in Gloucestershire's Forest of Dean, where Frederick's father was landlord of a public house.[3] By then Frederick was a pupil teacher at the village school from where he enrolled at St John's Training College in Battersea and began his formal teaching career in London in 1886 (Westaway 1929, p. 19). Concerning this experience, he provides a personal anecdote of some historical interest. He was allocated two hours for chemistry and two hours for mechanics.[4] He had no laboratory or demonstration bench and only a balance that he had made himself. The Bunsen burner had to be fed from the gas pendant above the pupils' heads. Using Ira Remsen's revolutionary American textbook, he proceeded to teach (Remsen 1886):

> In the middle of a lesson on equivalents, two visitors whose names I did not catch were shown in, and they sat down and listened. When I had finished they came up and showed what I thought to be a surprising appreciation of what I had been doing, and eventually one of them said: "Do you happen to know Roscoe's book on chemistry?" "Yes," I replied, "and a thoroughly unsatisfactory book it is. The writer makes unjustifiable assumptions about chemical theory before he had established necessary facts. It is the kind of thing that no teacher ought to do." At this stage the second visitor interposed and said, "I think, perhaps, you are asking for trouble. Let me introduce you to Professor [afterwards Sir Henry] Roscoe." However, in spite of the criticized book, I learnt more about the teaching of chemistry in the next quarter of an hour than I might have learnt in the next ten years. In particular, I learnt a much needed lesson – that there is more than one avenue of approach to the teaching of science, and that it is sheer folly to assume that science must be taught according to one pedagogue's prescription. (Westaway 1929, pp. 19–20)[5]

He lodged in Lambeth, joined the rifle volunteer movement, passed his London Matriculation examination in 1887 and graduated BA from the University of London in 1890. It was a career path followed by a significant number of the more able pupil teachers in the last two decades of the nineteenth century, and many of them found employment in the growing number of post-elementary schools, known as Higher Grade Schools, established by the School Boards in larger towns and cities (Vlaeminke 2000). From hints in his later books, it appears that Westaway continued his self-improvement by attending evening classes and the summer schools for science teachers that were run by Frankland (chemistry), Guthrie (physics) and Huxley (biology) at South Kensington under the auspices of the Department of Science and Art (DSA).

---

[3] Later, between 1897 and 1901, William Westaway was landlord of the George Inn, Market Street, Gloucester.

[4] This may imply that the bulk of his timetable was spent in teaching more general subjects such as English and Latin.

[5] For Roscoe's textbook, ironically published in the same Macmillan's Primer Series as Remsen's book, see Roscoe (1866; new ed. 1886). The other visitor was the chemist and educationist John Hall Gladstone, as revealed in Westaway (1936, p. 313).

In May 1892, when teaching at a school at Stockwell in south-west London, Westaway married Mary Jane Collar, the daughter of a pianoforte maker and herself a teacher.[6] Her two brothers, George and Henry, were also teachers, and both eventually became headmasters of London schools. The newly married couple immediately moved to Dalton in Furness[7] where he had been appointed headmaster of the Higher Grade School, the local Board School having been established in 1878. Westaway thereby began his formal connection with South Kensington since the school would have been recognised as an organised science school for the purposes of DSA grants. The chief local industries were iron ore mining and quarrying, and the School Board, like that in other northern towns in England, recognised the need for a better-informed and technically competent workforce. It is alleged that Westaway grew a beard to hide his relative youth. Their only child, Katherine Mary, was born in the School House in 1893, and her father was to exert an important influence upon her upbringing and career. She eventually became a distinguished classical scholar and an outstanding headmistress of Bedford High School from 1924 until her retirement in 1949.

Clearly ambitious, Westaway moved from Dalton in Furness to Bristol in 1894 to become headmaster of the newly built St George's Secondary and Technical School which contained "a thoroughly well-appointed Chemistry Laboratory, a large Science Lecture Room, a Workshop, a Dining Room and accommodation of every kind conducive to the well-being of the scholars".[8] Its finances were entirely contingent upon the school's ability to earn grants by Westaway's pupils gaining high passes in the examinations run by the DSA. A local newspaper reported:

> The [School] Board considered a number of applications for the appointment of the St George Higher Grade School. Mr *F. W. Westaway*, at present holding an appointment at the higher grade schools, Dalton-in-Furness, was appointed unanimously. In his application his University distinction was stated to be (a) B.A. London; Inter B.Sc., final examination, deferred until 1895; (c) Member of Convocation of the University of London. The list of qualifications, particulars of past experience, copies of testimonials, and prospectus of present school were considered by the Board highly satisfactory. (Anon 11 October 1894)[9]

Vlaeminke's study of the school's logbooks reveals that Westaway taught all the science and mathematics himself and triumphantly gained passes for his pupils in the DSA examinations that were the best in the Bristol area. No doubt this was

[6] The ceremony was conducted at St Matthew's Church, West Kensington (Mary Collar's parish), by Edwin Hobson, the principal and chaplain of St Katherine's College, Tottenham, where Mary had trained to be a teacher. The college, which had been set up originally by the Society for the Promotion of Christian Knowledge in 1878, is now part of Middlesex University. Hobson had been vice-principal of St John's College, Battersea, from 1874 to 1877 before Westaway was a student there.

[7] Dalton in Furness lies on the southern edge of the Lake District in Cumbria (until 1974 in Lancashire). It is famed for its castle and for Furness Abbey. The school, which opened in 1877, survives as an infant school.

[8] This was Bristol's first state secondary school created by local entrepreneurs in 1894. Its premises, which are currently used as a Sikh temple, were opened in 1894. Following various changes of name, it moved to new premises in 2005 as St George City Academy.

[9] The school's development is analysed in Vlaeminke (2000).

noticed by the Department of Science and Art for, within a year, Westaway was offered a subinspectorship in the department. On the face of it, accepting this offer was an odd decision, for although the character of the DSA was changing during the 1890s, the task of its inspectors largely remained one of ensuring that its militaristic rules and regulations for the conduct of examinations and payment by results were strictly adhered to (Butterworth 1982, pp. 27–44). The appointment to the inspectorate probably involved a drop in salary, though this would have been compensated by the prospect of a very good pension.[10]

Westaway's movements between 1895 and the passing of the (Secondary) Education Act of 1902 are unclear. He was undoubtedly not content to sit on his laurels but continued his studies at the Royal College of Science at South Kensington in London where he was a prizeman in mathematics and physics. By 1901, however, his post in the civil service had become that of one of Her Majesty's Inspectors of Education with responsibilities for secondary education in the area of Essex.[11] It was here that their daughter Katherine, together with two friends, began her first lessons with a governess (Kitchener 1981; Hunt 2004), and Westaway cultivated a friendship with R. J. Strutt, 3rd Baron Rayleigh, who kept a private laboratory at his home at Terling Place, near Chelmsford. Within a year, however, Westaway took over responsibilities for inspection in the Bedfordshire area. The family moved to Pemberley Crescent in Bedford, close to Bedford School.[12] He remained an HMI until his retirement in 1929, when he moved to the village of Aspley Heath, adjacent to Woburn Sands in Bedfordshire. He was intensely proud of his profession and of the intellectual attainments of the colleagues with whom he mixed. He sometimes indulges in name-dropping: for example, when mentioning the Irish physicist Thomas Preston (1860–1900), he refers to him as "for some time an esteemed colleague of the present writer's" (Westaway 1937a, p. 364).[13] Preston did, indeed, combine his chair of natural philosophy at Trinity College Dublin with a government post as an inspector of science and art for Irish schools, but it seems unlikely that Preston ever came into direct contact with Westaway.

73.2 The Philosophy of Science

During his long life, Westaway authored some sixteen books, many of which ran into several editions, although not all were concerned with science education. His first book, *Scientific Method: Its Philosophy and Practice*, was published in 1912.

[10] On the recruitment of inspectors, see Gosden (1966, p. 25).

[11] The 1870 Education Act abolished the denominational character of the inspectorate and reorganised HMIs territorially to reduce their travel. This system continued after the reorganisation of secondary education following the Education Act of 1902. See Gosden (1966, pp. 27 and 111). The Westaway family moved from Bristol to 87 Camden Villa, Fuller's Road, Woodford, Essex.

[12] According to 1901 and 1911 Census data, the Westaways employed one servant girl.

[13] For Preston, see Weaire and O'Connor (1987).

The date is significant since it coincides with the growing scholarly interest in the history and philosophy of science, evident, for example, in the first publication of the journal *ISIS* by George Sarton in Belgium in 1913. Westaway dedicated his book to the physicist Lord Rayleigh (1842–1919) "to whose work and whose teaching the author is deeply indebted" (Westaway 1912, p. 439).[14] The first edition of *Scientific Method* was in four parts. The first examined philosophical issues and offered a commentary upon the ideas of a range of philosophers including Plato, Aristotle, Francis Bacon, Descartes, Locke and Hume. This was followed by attention to Victorian "methodologists" such as Whewell, Mill and Herschel and a discussion of what might be meant by such terms as induction, deduction, scientific law and hypothesis. The third part of the book turned to the history of science and was devoted to "Famous men of science and their methods". In this section, scientists such as Harvey, Newton, Black, Priestley, Faraday, Wallace, Darwin, Clerk Maxwell, Ostwald and J. J. Thomson were largely allowed to speak for themselves through the form of generous quotations. The book ended with a practical section for science teachers entitled "Scientific method in the classroom" in which Westaway offered examples, drawn from botany, chemistry and physics, of what today would be called teaching by investigation.

Nature thought the book is "a model of clearness" and ideal for both science teachers and the general reader. Its sole fault, if any, was the use of excessive quotations, though the reviewer put this down to the fact that Westaway was exceptionally well read (JAH 1912).[15] The reviewer missed the fact that Westaway was concerned with more than promoting a greater understanding of the history and philosophy of science. As he made clear in his preface, he was anxious to bring humanists and realists (i.e. scientists) closer together and to reconcile the ideals which they represented. The need for such reconciliation was to become particularly urgent during the First World War when something of a battle of the books broke out between the scientific community and those representing the humanities, over the contribution that science could make to liberal education.[16] One outcome was the claim, already promoted in Westaway's book, that the history and philosophy of science offered a means of humanising a narrow, specialised and otherwise dehumanising scientific education.

The second edition was published immediately after the war had concluded in 1919. A new preface blamed Britain's industrial problems on its "continued use of haphazard methods". It would be the undoing of the nation if this were continued:

---

[14] Also Westaway (1919, 2nd ed., p. 426), Westaway (1924, 3rd ed.), Westaway (1931b, 4th ed. "revised and enlarged, present-day methods critically considered") and Westaway (1937a, 5th ed.). Most of these editions are available online. According to the *World Catalogue*, Chinese translations were made in 1935 and 1969.

[15] The reviewer was probably the chemical physicist John Alexander Harker.

[16] There is a large literature on the theme of humanising the science curriculum. See Jenkins (1979, pp. 54–55), Brock (1996, Chap. 19), Mayer (1997), Donnelly (2002, 2004) and Donnelly and Ryder (2011).

> On the one side we have Germany, clear-headed and thorough; America, original and enterprising; Japan, self-denying and observant; France, pain-staking and clever: all four nations believers in *work*. On the other side we have Britain, insular and unsystematic, looking upon work as a nuisance because interfering with pleasures. (Westaway 1919, p. xii)

An appendix, "Retrospect and Reflections 1912–1918", continued this theme, going so far as to assert that Britain had not won the war because of science or education but because of the reawakening of the nation's dormant national qualities. For their part, the Germans had lost because of their servility to authority and inability to think for themselves. Westaway's solution, overtly political, involved the redistribution of wealth and the wholesale application of scientific method. The "Retrospect" surveyed the functions and influence of science and scientific method on national life. This time the *Nature* reviewer, noting how the Thomson Report on Natural Science Teaching had urged science teachers to become acquainted with the history and philosophy of science, recommended the volume enthusiastically as enlightening and helpful. "Clearly presented" with "apt and instructive" examples, "any science teacher, whether at university of school, who reads the book, cannot fail to derive profit and interest from it" (Anon 1920a).[17]

By the time the 4th "revised and enlarged" edition had appeared in 1931 (a third edition in 1924 merely expanded the chapter on the theory of relativity), Westaway was a lot more sanguine about Britain's future. Post-war society, he suggested, had become less rule of thumb, more rational and systematic. He expressed delight in the progress of mass production with its implicit inclusion of specialisation, expertise and machinery in the operations of industry. Even so, there was still need for "the development of the scientific study and impartial examination of all the complex factors, economic, social, political, and racial, involved in controversial problems which are the sources of international friction" (Westaway 1931b, p. xii). Westaway clearly believed that the revised version of his book would help teachers produce a workforce geared to a mass-production society. To that end, he added a fourth section on present-day methods in contemporary science – including nuclear physics and quantum mechanics – and a fifth section on how scientific method could be inculcated in the classroom and lecture theatre.

In the final edition, published well into his retirement in 1937 and which received a Chinese translation, two further sections were added: one giving excerpts from the writings of "distinguished workers of the day" whom he obviously admired[18] the other offering further examples of the application of scientific methods for advanced students (Westaway 1937a). This extraordinary section included the analysis of historical facts using the example of the causes of the decline and fall of the Roman Empire, as well as an open-ended discussion of whether there was a criterion of excellence in aesthetics – a subject that he was to expand in another book.

---

[17] The unsigned review was probably by the editor, Richard Gregory, whose much reprinted book *Discovery* (Gregory 1916) was admired by Westaway.

[18] These included Max Born, Herbert Dingle, Julian Huxley, James Jeans, Hyman Levy, Max Planck and Walther de Sitter.

73.3 Science Teaching

Westaway had included a short section on scientific method applied in the classroom in *Scientific Method* in 1912 and the subsequent editions, but it was not until his retirement that he expanded it as a separate book in 1929. Dedicated to his friend and superior in the inspectorate, Francis B Stead, *Science Teaching: What It Was, What It Is, What It Might Be* was a volume that sought to assert the liberal values of a scientific education, providing always that science was well taught. From this perspective, *Science Teaching* follows the tradition of earlier works by Mach (1893) and Dewey (1900, 1902, 1916),[19] a tradition that was to be sustained in subsequent years by the writings of Schwab (1982), Conant (1947, 1957), Holton (1952) and many others. It also owes something, and not simply as far as its title is concerned, to Edmond Holmes' seminal volume, published in 1911, *What Is and What Might Be* (Holmes 1911),[20] and to Stead's work as secretary (and compiler) of J. J. Thomson's influential wartime report on *Natural Science in Education* (Stead 1918).[21] Drawing upon his experience as an HMI – he speaks of witnessing "1000 lessons a year for over 30 years" (Westaway 1929, p. xii) – Westaway argues in *Science Teaching* for a broadening of school science education to include some biology, astronomy and palaeontology and sets out the case for making science a compulsory part of the school curriculum. But the book is more than this. It is also a primer of practice and a challenge to those whose educational "claims on behalf of science ... are sometimes tinged with a good deal of arrogance and intolerance, and whose advocacy is thus better calculated to make enemies than enlist friends" (Westaway 1942b, back cover).[22] Once again, Westaway is seeking to promote a middle way in education, one in which there is "no natural antagonism between science on the one hand and humanism on the other" and one in which the history and philosophy of science play a key part. Such an ethos was shared with another significant HMI, the historian and positivist, Francis Sydney Marvin (1863–1943).[23]

---

[19] Mach (1893) first appeared in German in 1883. It remained in print throughout Westaway's career. Westaway frequently cited and recommended Mach in his writings, but not Dewey.

[20] Holmes (1850–1936), who became chief inspector of elementary schools in 1905, resigned in 1911 over criticisms of HMIs who had formerly been elementary schoolteachers. See Gordon (1978) and Shute (1998).

[21] Chaired by the physicist J. J. Thomson, the report was actually compiled by the committee's secretary, Francis Bernard Stead (1873–1955), H. M. chief inspector of secondary schools and a close friend of the Westaways. Stead, a Cambridge NST graduate, had worked at Plymouth's Marine Biology Association and Clifton College before joining the inspectorate in 1908. The science report was one of four (science, classics, English, modern languages) eventually prepared by the Board of Education. See *Nature* 175 (1955), pp. 148, 175) and Jenkins (1973, pp. 76–87).

[22] A 4-page publisher's pamphlet of "press appreciations" (c. 1930) in the possession of WHB carries the quotation from *Journal of Education*: "Get the book and read it; it is the best thing yet".

[23] Marvin, August Comte's principal spokesman for positivism in Britain, was an HMI from 1890 to 1924. He played a major role in improving the teaching of history in schools. See Mayer (2004). Curiously, despite his influence on the teaching of history, Marvin has not been included in the *Oxford DNB,* whereas his wife, Edith Mary Marvin (née Deverell) (1872–1958), a fellow HMI,

Westaway's definition of a successful science teacher (ignoring his gender bias) may be demanding and an ideal, but it still suggests what is required in the profession:

> He knows his own special subject through and through, he is widely read in other branches of science, he knows how to teach, he knows how to teach science, he is able to express himself lucidly, he is skilful in manipulation, he is resourceful both at the demonstration table and in the laboratory, he is a logician to his finger-tips, he is something of a philosopher, and he is so far an historian that he can sit down with a crowd of boys and talk to them about personal equations, the lives, and the work of such geniuses as Galileo, Newton, Faraday, and Darwin. More than all of this, he is an enthusiast, full of faith in his own particular work. (Westaway 1929, p. 3)

Little wonder that he admitted that he thought a teacher was not really fully equipped to teach effectively until he was in his thirties.

Science Teaching, with its extensive syllabus suggestions and advice on laboratory accommodation and equipment, as well as its helpful discussion of classroom practice, was well received by the reviewers ("a book of outstanding usefulness", "remarkable, critical and stimulating"), and it became a staple of initial training courses for graduate science teachers.[24] His suggestions that the periodic law and wave motion (not energy) should be the pole stars of the school chemistry and physics syllabuses were undoubtedly influential, as was his emphatic insistence on the introduction of biology into the secondary curriculum. Like the rest of Westaway's books, it is characterised by a directness and lucidity of style and by the author's capacity to engage with a wide range of disciplines and to draw his arguments and examples accordingly. The book was reprinted in its year of publication and again in 1934, 1942 and posthumously in 1947. Even today, much of Westaway's advice to those learning to teach science has the ring of experience. Some of his questions for young science teachers or pupils leaving school remain both challenging and of interest. For example,

> Do you consider that the estimates of stellar distances and electronic magnitudes correspond approximately with actual fact? What part of the available evidence is experimental and what part is inferential? Is the latter evidence convincing?
>
> How would you estimate the number of flaps made in a second by the wings of a flying bluebottle?
>
> Some years ago a science teacher, working single handed, was trying to extinguish the burning woodwork (pitch-pine) of a fume cupboard … when he was called to a boy who had become unconscious through the inhalation of chlorine. How would you have coped with the double emergency?

Further insights into Westaway's views on science teaching are revealed by an appreciation he wrote for *The School Science Review* of Henry Edward Armstrong, following the death of the latter in July 1937. Westaway had come into contact with both Armstrong and Thomas Henry Huxley as early as the 1880s while studying at

has. The same has happened with Westaway and his daughter, Katherine. Both Marvin and Westaway are to be included in a forthcoming update of *ODNB*.

[24] For an appreciative review by the science teacher and historian of science, Eric Holmyard, see Holmyard (1929). Note also the appreciation of the Latymer School science teacher George Fowles (1937, pp. 13, 501, 513, 527).

the Royal College of Science, and he thus constitutes a direct link with some of the leading figures in the late nineteenth-century education.[25] It is clear from Westaway's *Science Teaching* that it was Huxley who exercised the greatest influence upon his ideas, and he was intensely proud to have been personally examined by Huxley in biology. For Westaway, the purpose of a scientific education was the making of Huxley's "cold logic engine", in which "the desire for discriminating evidence" would become a "predominating factor" in thinking.[26] Nonetheless, he found Armstrong's heurism, with its commitment to teaching "scientific method", an approach to science teaching in which "practice simply would not yield to precept" (Westaway 1929, pp. 20–27).[27] The imparting of information was a vital function of teaching, and this was the cardinal fault of heurism in its pure form. Moreover, heurism was "not new" but, in essence, a strategy "used by intelligent teachers all down the ages". For Westaway, therefore, Armstrong's great contribution to science education was not heurism itself but the fact that he compelled others to keep their "defensive weapons keen and bright" and be "ever ready to defend alternative methods". Armstrong, he thought, had given hostage to fortune by calling his system the heuristic method instead of, say, "the search" or "discovery method" (Westaway 1929, p. 20). However, he admired Armstrong as a teacher and had high praise for his account of chemistry in the 13th edition of *Encyclopaedia Britannica* (1926) for the way it cast light on the inner nature of chemistry.

73.4 Mathematics Teaching

It would seem that Westaway's first love as a practising teacher was mathematics. In several places, he lamented the loss of Euclid due to the efforts of the former Association for the Improvement of Geometrical Teaching. Nevertheless, he became a keen member of its successor, the Mathematical Association, and a regular reader of its *Mathematical Gazette* (Price 1994). He published two geometry textbooks for private and state secondary schools and technical colleges which differed only in that the latter contained additional material for middle-form boys up to the age of 14 (Westaway 1928a, b).[28] The texts were addressed directly to both teachers and

---

[25] "Faith in my own old teachers –Thomas Henry Huxley, John Tyndall, John Hall Gladstone, and (the 3rd) Lord Rayleigh – is as strong as ever, all of whom, in season and out of season, insisted on laboratory and field work first and always, on *facts* and ever more facts" (Westaway 1936, pp. ix, 359, and 745). He also recalled the wonderful lecture demonstrations of Charles Vernon Boys in Westaway (1936, pp. 692).

[26] Westaway was quoting from Huxley (1905, pp. 76–110).

[27] See Brock (1973; reprint 2012).

[28] Westaway (1928a) covered the geometry syllabus from age 8 to 13 when the common entrance examination for admission to a public school was taken. Westaway (1928b) must have been published a few months after the other geometry textbook. Both texts mentioned Westaway's admiration for the geometry lessons of his "friends" Frederick William Sanderson (1857–1922), headmaster of Oundle School, and Edward Mann Langley (1851–1933), a teacher at Bedford Modern School and founder-editor of *The Mathematical Gazette*.

pupils, the latter being expected to read a chapter before it was discussed with them. In line with contemporary feelings about geometry teaching, Westaway avoided deductive proofs from first principles. Long chains of reasoning were also avoided, and intuitive reasoning was frequently used, while practical examples from surveying and carpentry demonstrated the practical significance of geometrical reasoning.[29] Typically, Westaway supplied a list of Latin words and the mathematical expressions derived from them in the expectation that teachers and pupils would correlate their mathematical and classical learning.

In 1931, following the tremendous success of his book on *Science Teaching*, Westaway's publishers suggested that a similar book addressed to mathematics teachers was a desideratum. The result was another remarkable handbook offering young inexperienced teachers advice and useful hints on putting mathematics across in the classroom (Westaway 1931a). Although entitled *elementary* mathematics education, the 28 chapters in fact ranged from simple addition and subtraction to teaching the calculus and thus covered the whole of school mathematics from the level of infants to university entrants. As one reviewer remarked (Anon 1931), Westaway's standard for what a pupil should know was very high, citing as evidence the casual remark that a 4th-form boy was apt to forget the factors of $a^4 + 2a^2b^2 + b^4$ [i.e. $(a^2 + b^2)^2$]. (Was this a reflection of a decline in expectations or merely due to the fact that Westaway's treatment smacked of the nineteenth, rather than, the twentieth century, as the same reviewer suspected?) There were also sections on mathematics in astronomy and biology (a field Westaway foresaw as developing in significance), time and the calendar (which he thought was the job of mathematics staff to teach), as well as on non-Euclidean geometry and the history and philosophy of mathematics. Westaway specifically addressed *new* teachers rather than experienced accomplished instructors, thus ensuring, like his treatise on science teaching, the book's heavy use in British teacher training colleges and ready sale among tyros. Its valuable suggestions for organising the math curriculum throughout a school must also have recommended it to senior staff as well. Curiously, and surprisingly, the Mathematical Association ignored the book, though it was reviewed favourably and at length by the Harvard geometer Ralph Beatley who recognised that many of the problems delineated by Westaway were common in America as well (Beatley 1933).

73.5 Language Teaching

Westaway had clearly taught Latin at an early stage of his career and had a deep respect for classical education.[30] Although we think of him primarily as a science educator, he also made an informative contribution to classical teaching in the form of his second book *Quantity and Accent in the Pronunciation of Latin*, which he

[29] This practical aspect was praised by Dobbs (1929).

[30] His love of the classics was inherited by his daughter Kathleen who lectured in classics at Royal Holloway College (1920–1924) before becoming headmistress of her old school. See Westaway K. W. (1917, 1922, 1924).

published in 1914 (Westaway 1914).[31] Westaway was adamant that it was not a textbook and that it was not aimed at schoolteachers. His targets were private students of Latin and those whose knowledge of the rules of pronunciation was rusty. The context was a contemporary debate between an older generation of classicists who wanted to continue using an Anglicised "easy-going" form of pronunciation and the younger generation of classicists whose knowledge was informed by research in philology and phonetics. "Mr Westaway's heart is in the right place", that of a reformer, concluded Edward Adolf Sonnenschein, the professor of Greek and Latin at the University of Birmingham, and "he writes with conviction" (Sonnenschein 1914, pp. 213–214). That the text did well in modernising the teaching of Latin is suggested by the fact that it remained in print in the 1920s and that an enlarged edition appeared in 1930. The text had been read with approval by both John Percival Postgate (1853–1926), professor of Latin at the University of Liverpool and the founder (with Sonnenschein) of the Classical Association in 1903, and Westaway's superior in the Education Department John William Mackail (1859–1945), an eminent Virgil scholar. Westaway became a life member of both the Classical Association and the Modern Languages Association in 1903. Like the British Association for the Advancement of Science, which he also joined in 1903, Westaway must have seen these organisations as a valuable way for an HMI to keep up to date.

In *Scientific Method*, Westaway's concern for logical thinking took him into one of his many other interests, language and clarity of thought and expression, and hence into the use of words to express causality. It seems he regarded English grammar as offering important philosophical insights for the science teacher. In his inspections of schools and technical colleges, not unexpectedly, Westaway came across poorly expressed written reports of experiments. More surprisingly, he came across ungrammatical and illogical prose in reports by scientists in periodicals such as *Nature*, which he evidently read each week. Another periodical he read regularly, but never contributed to, was the *Mathematical Gazette* which had been founded by the Mathematical Association in 1894. Westaway joined the association in 1914. He was impressed and inspired by his chief in the inspectorate, W. C. Fletcher, who published an article in the *Gazette* in March 1924 stressing that mathematicians should write good prose between their symbols (Fletcher 1924).[32] Fletcher's essay, together with the earlier appearance of George Sampson's influential report on the teaching of English in 1921 (Sampson 1921),[33] probably inspired Westaway to publish *The Writing of Clear English* in 1926. While significantly subtitled *A Book*

[31] This was the only one of his books not published by Blackie. Note also Westaway (1933a).

[32] Fletcher (1865–1959) is best known for "Fletcher's trolley", an improvement on Atwood's machine for teaching mechanics. After graduating from Cambridge as 2nd wrangler in 1887, Fletcher taught at Bedford School (1887–1896) before becoming headmaster of the Liverpool Institute (1896–1904). He was appointed Chief Inspector of Secondary Schools in 1904. He retired in 1926 only to teach at the girls' school where his daughter was headmistress. See obituary (Anon 1959).

[33] Sampson (1873–1950) followed a similar path to Westaway – pupil teacher, London Matric, teacher, headmaster, district inspector for London County Council.

for Students of Science and Technology, its clear exposition of the principles of English grammar and advice on sentence and paragraph construction and logical writing style would have made it useful to a generally educated readership that found itself in need of English improvement (Westaway 1926).[34] In retirement, Westaway revised and enlarged the book as *The Teaching of English Grammar Function versus Form* (Westaway 1933b).[35]

In 1932 Westaway persuaded his publishers to launch a series of books offering instruction and advice to teachers under the umbrella title of The Teachers' Library. Although Westaway did not write a book specifically for the series, he recruited the help of other HMIs and acted as the advisory editor for the series. The library was planned "for the guidance of teachers whose daily work is concerned with children of eight and upwards" (Finch & Kimmins 1932, Preface).[36]

73.6 History of Science

As we have seen from both his *Scientific Method* and *Science Teaching*, Westaway believed that teachers should tell stories about great mathematicians and scientists. He also thought sixth-form students should be encouraged and challenged to read, among other original works, Newton's *Opticks*, Faraday's *Researches on Electricity* and Darwin's *Earthworms* (Westaway 1929, p. 383). The interwar years were a particularly productive time for Westaway. In addition to books on the teaching of mathematics, geometry and English (for science specialists), he wrote a large volume of over 1,000 pages entitled *The Endless Quest: Three Thousand Years of Science* (Westaway 1934).[37] This huge, richly illustrated volume is something of a tour de force, and, even today, when specialised degree courses are available in history of science, as a history of science published in 1934, it makes appealing and, in places, challenging reading. Westaway's originality is shown in his deliberate attempt to write a critical appraisal of scientific development and to show its weaknesses as well as its strengths. He acknowledged the influence of his departmental colleague, the positivist historian Sydney Marvin, who was a friend of George Sarton and who was keen to see history of science taught by history teachers. An anonymous reviewer in *The School Science Review* described the book as "a veritable encyclopaedia of science", adding that only "a writer of extreme scientific versatility" could attempt to write it (Anon 1935; Dingle 1935).

Internal inspection shows that Westaway's sources included the British Library and books borrowed from H. K. Lewis, the scientific and medical library

[34] One teacher thought he had "done his work well". See Anon (1927).

[35] For context, see Hudson and Walmsley (2005).

[36] Unfortunately, because in most scholarly libraries books are catalogued by authorship, it is not possible to identify all the books in the series except serendipitously.

[37] 2nd ed. 1935 with minor corrections only; reprint 1936. Chinese translations 1937 and 1966, Czech in 1937. The book of 1,080 pp was dedicated to his daughter.

in Gower Street, the *Encyclopaedia Britannica* and *Nature* as well as his own well-stocked library. The ready availability of Boyle's *Sceptical Chymist* and Harvey's *On the Motion of the Heart* in the Everyman Library enabled him to give detailed accounts of their experiments and reasoning. Westaway's historiography was surprisingly sophisticated for a period when accounts of science were usually triumphalist and Whiggish. When discussing Babylonian mathematics and astronomy, for example, he scorns any reader who condemns the Babylonians for not formulating hypotheses to explain their observations of heavenly events. How would the reader determine the length of a solar year? No reference books allowed, the problem had to be solved by thought alone. By setting the reader such rhetorical questions, the Babylonians' ability to determine that the solar year was $365^1/_4$ days long without using instruments became all that more impressive. The same device is used at the end of the book when Westaway sets the reader 20 thoughtful and probing questions. Two contrasting questions, one based on bookwork and the other more diffuse and philosophical, will suffice to illustrate his purpose:

> What is an explosion? How long does it take a high explosive like T.N.T. to be converted into a gas? Explain why such a gas is so remarkably destructive, why its downward action is so violent, and why it does not expend its force upwards into the atmosphere.
>
> On being established in 1899, the Board of Education adopted the traditional views of the Science and Art Department of the Privy Council, which the Board succeeded, that physics and chemistry were the most suitable subjects of science for teaching in schools, views which still generally survive. To what extent do you consider this to be the cause of (1) the ignorance of, and (2) the lack of interest in, science by the average educated Englishman? If it is the cause, what is the remedy? If not the cause, what *is* the cause? (Westaway 1936, pp. 1037–1040)

73.7 Broadening Science Education

When looking back on his long career as an educator, Westaway was pleased by the way changes in teaching and educational practice had remoulded the minds of the younger generation. He was particularly struck by the way that the creation of sixth forms in secondary schools had led young people to become critical, well informed and opinionated. In one of his last, and perhaps oddest, books, Westaway encouraged intellectual debate among young people by providing raw materials for philosophical discussion and debate. He agreed completely with the views of the political historian Sir Ernest Barker who had written in *The Times*:

> At the end of a life spent in teaching, I am an educational anarchist so far as concerns the growth of true minds. When I find a true mind, I want to let it grow. Conscience used to make a coward of me, and I was once resolved to be a good tutor. Either I have lost my conscience, or it has acquired a finer edge. At any rate, I am now disposed to be very tender to the liberty of young minds. Their liberty includes their freedom from me. I insult them if I tell them first what to read – still more if I tell them what is 'the right view'. They need their own intellectual adventures. If they ask me to go with them, I am proud to be asked: if

> they ask me questions, I will tell them what I think – and I will add that I am far from being sure about it. (Barker 11 August 1937)[38]

Westaway's curiously titled *Obsessions and Convictions of the Human Intellect* (Westaway 1938a) from which this quotation comes contained a variety of informative and unbiased essays on subjects that were likely to interest young people between the ages of 16 and 25. Carefully excluding politics, the subjects ranged from astrology and alchemy, perpetual motion and the fourth dimension to questions about the nature of space and time, miracles, religious persecution down the ages and the concept of Hell ("atrocious" and "immoral"). Much of this curious work, which was aimed at providing a critical synoptic view of modern knowledge for pupils exposed to overspecialisation, was cannibalised (albeit reworked) from his other books. The first impression sold out immediately and was reprinted with a different, and probably more appropriate, title *Man's Search after Truth* (Westaway 1938b).[39] The book, which must have been an essential volume for school libraries, ends typically with a series of questions for the reader, but in this case they were questions that had all been suggested by young people when talking to Westaway.

The idea for *Appreciation of Beauty*, which Westaway published in 1939 soon after Chamberlain had negotiated "peace in our time" with Germany, came from the chapter on aesthetics that he had added to the third edition of *Scientific Method* in 1924. Echoing Robert Bridges' famous poem, dedicated to his wife Mary, on first reading it appears to address science students and scientists who may lack an appreciation of the arts, but a closer reading suggests that he had in mind a more general readership that felt it lacked an appreciation of "high culture" (Westaway 1939).[40] The book is, in fact, a vade mecum of culture offering guidance on how to understand and appreciate painting, sculpture, the history of art, architecture, ornament, arts and crafts, landscape and garden design, literature and music all underpinned by a philosophical discourse on aesthetics. In the final chapter, "What is meant by the beautiful", partly reworked from the third edition of his *Scientific Method* (1924), he concluded that Art "in the highest sense" involved "the reproduction of the phenomenon of nature" (e.g. sights and sounds), an "expression of the thoughts and emotions of the artist" and the embodiment of both these factors in "an external product like a painting, a statue, a cathedral, a piece of ornament, a garden, a poem, or a symphony" (Westaway 1939, p. 195). At the end of the day, beauty, he decided, had nothing to do with accepted canons of beauty or the consensus of experts. Nor could it be a Darwinian evolved sense that provided some kind of survival value. The appreciation of beauty was a form of communication from one human spirit to another, and because this communication was individual and personal, it was incapable of objective examination. Although the appreciation of beauty was undoubtedly a variable function of a person's education, experience, beliefs, traditions and

[38] Sir Ernest Barker (1874–1960), as quoted in Westaway (1938a, p. xi)

[39] Pagination and content were identical to *Obsessions and Convictions* (Westaway 1938a). See the enthusiastic review in *Nature* (Anon 1938).

[40] Robert Bridges' final poem, *the Testament of Beauty*, had appeared in 1929.

customs, ultimately Westaway agreed with Robert Bridges that man's appreciation of nature and man-made art was God given – a conclusion he instantly qualified as an unverifiable hypothesis but one that gave him the greatest degree of personal satisfaction.[41]

The Appreciation of Beauty is fascinating in respect to what it reveals about Westaway's own tastes, as well as implying that he had studied art at the Royal College of Art in the 1880s where he had learned to draw accurately. In contrast to his progressive appreciation of modern science, he stands revealed as a very conservative connoisseur of painting. Good art had to be representational and to tell a story. Although well read on Victorian and Edwardian art criticism, Westaway completely ignored artistic developments since the pre-Raphaelites. Indeed, art since the Impressionists deserved to sink without trace. Not for him Kandinsky's appreciation of the spiritual nature of abstraction. The literary canon ended with Arnold Bennett. He detested jazz bands: "what front-rank musician has any real respect for a Jazz Band?" Live music was better than recorded music or music transmitted by wireless. Such opinions make him sound like an elderly schoolteacher whose fixed opinions had never altered. Despite this conservatism, Westaway's guide would have undoubtedly benefited a reader who wanted to improve their appreciation of high culture. Despite some "old-fogey" opinions and a decidedly deficient coverage of culture after about 1890, any reader would have received a lot of sensible advice of how to view pictures (e.g. where best to stand), how painters and sculptors achieved their optical effects, the need to know the Bible, classical myths and saint's lives to fully appreciate an artists' intentions and a contemporary guide to Britain's best museums and art galleries.

73.8 Westaway's Religious Views

Anyone reading Westaway's *Appreciation of Beauty* in isolation from his other writings might well assume that he held extremely orthodox religious views. They would be mistaken. There was a general tendency among thinkers to take stock of the human race after the cataclysmic First World War. Westaway was but one of many philosophers, theologians and scientists who, in the light of evolution and the emergence of modern physics and cosmology, published their views on the relationship between science and theology (Bowler 2001).[42] Like many other early twentieth-century scientists, theologians and philosophers, he was concerned to reintegrate science and religion by demonstrating that the Victorian forms of materialism no longer appealed to scientists. That being the case, the churches had to modernise and bring their teachings into line with contemporary science. *Science and Theology: Their Common Aims and Methods* appeared in 1920 and was

[41] Westaway acknowledged that in writing the final chapter, he had received "great help and friendly criticism" from the philosopher and statesman, Arthur Balfour (1848–1930).

[42] Bowler did not notice Westaway's contribution to the debate.

reissued in an enlarged edition in 1934. It can be considered as an appendix to his *Scientific Method*.[43] His aim was to present the scientific developments of the past 50 years in layman's language and show their provisional nature. Because science was now a fundamental factor in human life and progress, he argued, any religious system had to find ways to accommodate it. Religious divisions were largely caused by a failure to accommodate new knowledge, and he appealed to Christians especially not to consider only their own doctrines as true and valid.

Reviewers were delighted by the book's directness and lucidity of style in dealing with matter, space and time, the genesis of the earth, the evolution of animal species, the antiquity of man and the emergence of life and consciousness (Sarton 1921; Anon 1920).[44] Westaway's conclusions were blunt: the belief that the same atoms of our bodies would reassemble on the day of judgement to form a human being was a pagan superstition; to express a belief in the resurrection of the body merely emphasised the material aspect of religion and was unnecessary; what really counted was a belief in the survival of personality. It followed that a belief in Christ's literal resurrection was unnecessary since what counted was the survival of Jesus' personality. Westaway accepted that the Bible had not been divinely inspired, but in accepting evolution he made it clear that it was meaningless if not teleological. A long, and erudite, section on controversies and heresies within the early church demonstrated how much theological clutter and primitivism needed to be eradicated from Christian doctrines. The "unbending institutionalism" of the church had to be eradicated. Westaway's convictions were clearly those of the Broad Church, an idealist philosophy and theistic belief:

> Theism has been aptly described as a systematized body of doctrine in which it is shown that an intelligent First Cause is the necessary and inevitable presupposition of experimental science, of reasoned knowledge, of aesthetics, of ethics, as well, of course, as of religion. If we want to explain our conceptions of the Real, the True, the Beautiful, or the Good, in each case alike we are inevitably driven to the conclusion that without an intelligent First Cause as a Beginning or Foundation, the whole of our scheme must dissolve and leave not a rack behind. (Westaway 1932, p. vii)

Once again he thanked Lord Rayleigh, who had died in June 1919, for his help with the manuscript, which was completed 2 months later. In a note added in proof, he observed that Einstein's "hypothesis" of relativity had been apparently verified – or, rather, just one of its consequences had. He remained doubtful, however, because he could not see how gravity acted in a void and he found the theory contradictory regarding the variability of time. However, by the time of the second, enlarged edition in 1932, Westaway had come to terms with relativity, stressing that we must carefully distinguish between *abstraction* which subtracts from observations and *hypotheses* (like relativity) that added to observations. The whole tendency of modern science, he noted in the light of quantum mechanics, was away from dogmatism and towards less certainty.

---

[43] Indeed, chapter 1, "problems of philosophy", is largely a reprint of the final chapter of *Scientific Method* (Westaway 1912).

[44] Sarton (1921, p. 120) thought the book "very good indeed". The *Nature* reviewer (Anon 1920, p. 608) thought "the theology student is left without excuse".

73.9 Westaway's Personal Beliefs

What of the man himself? As a teacher, he was evidently both successful and something of a pioneer, using, for example, three-dimensional molecular models in the earliest days of his classroom career – namely, the 1880s:

> When I first taught organic chemistry… I bought … a gross of small wooden balls into which I screwed midget hooks. I wanted the boys [sic] to visualise molecules as three – dimensional things. (Westaway 1937b, p. 311)[45]

Westaway was not, of course, beguiled by this use of molecular models. Noting that his pupils "loved to play about with and interchange the coloured sub-groups of atoms", he asks:

> Which was the more immoral – to let the boys think that those 'molecules' were truly representative of nature, or to waste the school time in that way?

His answer was that "scientific laws are fundamental, and scientific hypotheses are useful, but scientific shams are an abomination".

It is clear from all of Westaway's writing that he was a notable scholar, with an unusual breadth of knowledge and an abiding commitment to understanding and promoting science as an endeavour dedicated to improving the sum of human happiness and well-being. The use and meaning of words were important to him, although he was no pedant, and in several of his writings, he exposes the ways in which convictions and prejudices hold sway in matters that should be governed by reason. His commentary upon the teaching of general science being promoted by the Science Masters' Association in the late 1930s is characteristically balanced. Although he welcomes the broadening of the science curriculum that the innovation represented, he commented that while "all reformers see pretty clearly the effects of action, … they seldom look far enough ahead to see the ultimate effects of a reaction", a warning that general science should not be reduced to serving up "juicy tit-bits all the year round" (Westaway 1937b).

He was also a man of his time, writing favourably of positive eugenics and sarcastically of antivivisectionists,[46] and although careful to deny gender bias and that he consistently referred to boys (rather than pupils) for convenience, the general impression one has from reading all of his books is that he did not think girls benefited much from courses in science. His professional duties and his writing inevitably meant that he worked to a strict timetable, but it is clear that he doted upon his daughter. He found time to teach her mathematics for two hours each Saturday morning to help her prepare for her secondary school entrance examination. The evidence is that both father and daughter found these sessions stimulating and

---

[45] Such "glyptic" models had been first demonstrated by August Wilhelm Hofmann in the 1860s and could be purchased from instrument makers – to the disgust of the anti-atomist, Sir Benjamin Collins Brodie, professor of chemistry at the University of Oxford. See Brock (1967).

[46] "When the Anglo-Saxon or the Celt engages in a war of reason *v.* sentiment, sentiment almost invariably proves the winner" (Westaway 1936, p. 923).

enjoyable, Katherine describing them in later years as the highlight of her week (Kitchener 1981).[47] They were also undoubtedly successful, as Katherine found the mathematics component of the examination "too easy for words". Nevertheless, Westaway made it plain in his writings that he believed the average mathematical ability of girls was lower than that of boys and that their interest in mathematics was less. He even went so far as to suggest that the majority of girls did not need to study mathematics beyond the age of 13.

He was also a language purist, urging teachers to avoid the word "scientist" while admitting that the more correct term *sciencer* (analogous to astronomer) did not trip easily off the tongue,[48] and urged teachers to insist on the indicative mood when pupils wrote up their laboratory notes. As an inspector, he was highly regarded by his colleagues, while those he inspected found him astute in his judgement, constructive in his comments and unfailingly courteous in his dealings with people. He was unusually, perhaps, a welcome guest in many school staff rooms.

What, other than the deployment of science in war, prompted Westaway to place science "in the dock" and invite his readers to judge its guilt? Of what crime did science stand accused? Any answer to this latter question must, in part, be conjectural, but the "verdict" delivered by Westaway at the end of *Science in the Dock* and the text itself provide important clues. Reviewing, from the perspective of 1942, the position of science in relation to war, civilisation, education and religion in a rather muddled manner, Westaway ultimately leaves the verdict for the reader to decide. However, he suggests that a Scottish jury would bring in a verdict of "not proven" and that almost any jury would append to its verdict "the rider that science seems to be rather indifferent to the results of its work on the happiness of humanity". In addition, the jury would not improbably express the opinion that "not a few of the men who devote their lives to unearthing the secrets of nature really think less of the welfare of mankind than of their own enregistration on the roll of personal fame" (Westaway 1942a, p. 128).

Throughout his life, Westaway had tried to bridge the gap between science and the humanities, and, as noted above, he held firmly to the view that scientific knowledge should be directed towards the greater good of the human race. The collapse of humane values represented by the Second World War and its associated technology of destruction were thus a challenge to the ideas which Westaway had espoused and promised for many years. The publication of *Science in the Dock* in 1942 can be seen both as a personal exploration of the issues surrounding the social relations of science in the strained circumstances of a nation at war and as an attempt to reassert the importance of science as a humane endeavour. It can also be seen as an attempt by Westaway to expose the consequences of the early specialisation that marked English secondary education and to emphasise the need for breadth and balance. From the perspective of the early twenty-first century, however, one is also left to wonder about Westaway's attempt to reconcile his personal beliefs about science

[47] For further information on Katherine Westaway, see Hunt (2004) and Godber and Hutchins (1982).

[48] Here he followed Gregory, *Nature's* editor. *Nature* did not use the word scientist until the 1940s.

and the Christian faith. Again, conjecture is inevitable, but his books on the aims and methods of science and theology suggest strongly that he accepted the notions of a first cause and a directing agency behind the natural phenomena that scientists made it their business to explore (Westaway 1920, Chap. 12).

Frederick William Westaway died from bladder cancer on 25 February 1946 at the age of 81. His personal papers have not survived. Unlike many of his colleagues in the inspectorate, he was never accorded an honour, nor did he have an entry in *Who's Who*, and the teaching press ignored his passing. His one and only obituary notice in the *Bedfordshire Times* simply recorded that his "kind and scholarly face will be missed".[49]

73.10 Conclusion: Westaway's Legacy

Westaway was but one of an army of HMIs whose support, criticisms and writings have contributed to science teaching over the past 150 years in ways that deserve fuller investigation by historians of education. We hope that this essay will encourage others to investigate the role of inspectors in encouraging and promoting science education.[50] In this essay we have highlighted the career of just one HMI whose contribution seems to have been of exceptional importance. Westaway was a somewhat unusual recruit to the inspectorate for most were Oxbridge graduates.[51] However, the range of his reading and expertise put him among the best graduates from Cambridge's mathematical and natural science tripos – the graduates he himself believed were the very best qualified to be excellent science and mathematics teachers. As Eric Holmyard commented, after reading Westaway's writings, a harassed schoolmaster will, after all, appreciate the usually dreaded attentions of an HMI (Holmyard 1929).

[49] At the time of his death, the Westaways and their daughter were living in de Parys Road, Bedford, adjacent to Bedford School. At Probate in Oxford, 26 April 1946, his estate was valued at £4,604 (about £138,000 in today's buying power) and willed solely to his daughter, presumably because of his wife's incapacity. Mary Westaway died on 9 March 1947. All three Westaways are buried in the churchyard of All Saints, Renhold, where the family evidently worshipped.

[50] However, HMI reports of science lessons can be disappointing as source material. HMI visits were infrequent and often focused on different aspects of a school's work, making comparisons difficult. For a guide, see Morton (1997).

[51] For example, Charles Thomas Whitmell (1849–1920), BSc (London), NST (Cambridge), taught chemistry and physics at Tonbridge School before joining the inspectorate in 1879 serving in Cardiff (1879–1887) and at Leeds until retirement in 1910; Frederic W. H. Myers (1843–1901) became an HMI in 1872 after teaching classics at Cambridge; Francis Bernard Stead (1873–1954), a close friend of the Westaways, had worked at Plymouth's Marine Biology Association before joining the inspectorate; and Thomas W. Danby (1840–1924), NST (Cambridge), assisted the chemist George Liveing before becoming chief inspector of schools for south-east England. For some amusing reminiscences of inspection in the second half of the nineteenth century, see Sneyd-Kinnersley, E. M. (1908).

But what, it might be asked, is the relevance of Westaway's writing to school science teaching today, given the profound changes in the social context of both science and religion?[52] Part of the answer lies in the fact that many of the issues that Westaway addressed have arguably become more, rather than less, important and that some of his arguments, lucidly and eloquently developed and expressed, continue to offer a challenge to the contemporary mind. Beyond this, some of Westaway's writing about science and education sets a standard that others might seek to follow, notwithstanding the changes that have taken place in the last half century in our understanding of the history and philosophy of science. Ultimately, however, reading Westaway forces the reader to ask questions about the educational function of science, questions that seem to be urgently in need of answers as school science education is increasingly cast in an instrumental mode, serving economic rather than humane ends.

At a time when serious doubts are again being expressed about whether practical work and experimentation should play an essential (but expensive) role in schools, Westaway's arguments are well worth reading again.[53] We can also take to heart Westaway's expectation (and challenge) that

> When a boy leaves school, he should have been so taught and be so informed that he is able to take an intelligent interest in all scientific, technical, and industrial developments. He should be able to turn up technical reports, and obtain at least an intelligent general grasp of their contents. He should be able to discuss in a council chamber the pros and cons of proposed new applications to industrial processes. In short, the former secondary school boy should be the disseminator of new knowledge and the intelligent adviser of the community. (Westaway 1929, pp. 385–386)

Because of the way that the sociology of science has transformed approaches to the history and philosophy of science since the 1960s, Westaway's *Scientific Method* is now only of historical interest. Similarly, his pioneering *Endless Quest* has been superseded by a more nuanced and critical approach to the history of science – though it has to be said that no historian has been able to provide a better encyclopaedic and illustrated guide for teachers on how scientific achievements were brought about. Its message, taken from Westaway's grasp of philosophy, that despite the "passing of dogmatism" there is no reason to suppose "that what is mathematically describable is ultimately real, and the only reality", is one that all teachers should reflect upon. Finally, despite worldwide cultural differences and national varieties of syllabuses and curricula, his *Science Teaching* and *Craftsmanship in the Teaching of Mathematics* remain inspirational and illuminating reading for both novices and experienced teachers.

---

[52] For Westaway's continuing relevance in an American context, see Keller and Keller (2005) and the electronic blog at http://www.smartscience.net. See also Matthews (1998).

[53] See, for example, Taber (2011). The issue is also highlighted by the REACH legislation on the safety of chemicals.

Acknowledgements We are grateful to Nicola Aldridge for information about the Westaway family and to Dr M. P. Price, librarian of the Mathematical Association Library at the University of Leicester, for comment on Westaway's mathematics.

References

Anon (11 October 1894). *Bristol Mercury.*

Anon (1920a). Review of Westaway (1919). *Nature*, 105, 5

Anon (1920b). Review of Westaway (1932). *Nature,* 105, 607–608

Anon (1927). Review of Westaway (1926). *School Science Review*, June 1927, VIII, 289

Anon (1931). Review of Westaway (1931a). *Nature*, 128, 950

Anon (1935). Review of Westaway (1934). *The School Science Review* (1935), XVI, 135

Anon (1938). Review of Westaway (1938b). *Nature* (1938), 141, 766–767

Anon (1959). Obituary of W. C. Fletcher. *Mathematical Gazette*, 42, 85–87

Barker, Sir Ernest (11 August 1937). *The Times.* Quoted by Westaway (1938a), p. xi

Beatley, R. (1933). Review of Westaway (1931a). *American Mathematical Monthly*. 40, 548–553

Bowler, P. J. (2001). *Reconciling Science and Religion. The Debate in Early Twentieth-century Britain.* Chicago: Chicago University Press

Brock, W. H. (Ed.) (1967). *The Atomic Debates.* Leicester: University of Leicester Press

Brock, W. H. (Ed.) (1973; reprint 2012). *H. E. Armstrong and the Teaching of Science.* Cambridge: Cambridge University Press

Brock, W. H. (1996). *Science for All.* Aldershot: Variorum, Ashgate Publishing

Butterworth, H. (1982). The Science & Art Department examinations: origins and achievements. In MacLeod (1982)

Conant, J. B. (1947). *On Understanding Science*. New Haven, CT: Yale University Press

Conant, J. B. (1957). *Harvard Case Histories in Experimental Science*, 2 vols. Cambridge, MA: Harvard University Press.

Dewey, J. (1900). *The School and Society.* Chicago: University of Chicago Press

Dewey, J. (1902). *The Child and the Curriculum.* Chicago: University of Chicago Press

Dewey, J. (1916). *Democracy and Education.* London: Macmillan

Dingle, H. (1935). Review of Westaway (1934). *Nature*, 135, 938

Dobbs, W. J. (1929). Review of Westaway (1928b). *Mathematical Gazette*, 14, 471

Donnelly, J. F. (2002). The 'humanist' critique of the place of science in the curriculum in the nineteenth century and its continuing legacy. *History of Education*, 31 (6), 535–555.

Donnelly, J. F. (2004). Humanizing science education. *Science Education*, 88 (5), 762–784.

Donnelly, J. F. & Ryder, J. (2011). The pursuit of humanity: curriculum change in English school science. *History of Education*, 40 (3), 291–313.

Finch, R. & Kimmins, C. W. (1932). *The Teaching of English and Handwriting.* Glasgow & Bombay: Blackie

Fletcher, W. C. (1924). English and mathematics. *Mathematical Gazette*, 37, 37–49

Fowles, G. (1937). *Lecture Experiments in Chemistry.* London: Bell & Sons

Godber, J. & Hutchins, I. (Eds.) (1982). *A Century of Challenge: Bedford High School, 1882 to 1982.* Bedford: priv. printed

Gordon, P. (1978). The writings of Edmund Holmes: a reassessment and bibliography. *History of Education* (1978), 12, pp. 5–24

Gosden P H. J. H. (1966). *The Development of Educational Administration in England and Wales.* Oxford: Basil Blackwell

Gregory, R. (1916). *Discovery, or, The Spirit and Service of Science.* London: Macmillan

Holmes, E. (1911). *What Is and What Might Be: A Study of Education in General and Elementary Education in Particular*. London: Constable; 2nd ed. 1928, with many impressions

Holton, G. (1952). *Introduction to Concepts and Theories of Physical Science.* Cambridge, MA: Addison-Wesley
Holmyard, E. J. (1929). Review of Westaway (1929). *Nature* 124, 436–437
Hudson, R. & Walmsley, J. (2005). The English Patient: English grammar and teaching in the twentieth century. *Journal of Linguistics,* 41, 593–622
Hunt, F. (2004). Kathleen Mary Westaway. *OxfordDNB*
Huxley, T. H. (1868), A Liberal Education and Where to Find It. In Huxley, T. H. (1905). *Science & Education Essays* (pp. 76–110). London: Macmillan
J. A. H. (1912). Review of Westaway (1912). *Nature*, 90, 277–278
Jenkins, E. W. (1973). The Thomson Committee and the Board of Education 1916–22, *British Journal of Educational Studies*, XXI, 76–87
Jenkins, E. W. (1979). *From Armstrong to Nuffield.* London: John Murray
Keller H. E. & Keller, E. E. (2005). Making real virtual labs. *The Science Education Review*, 4, 2–11; and the electronic blog at http://www.smartscience.net
Kitchener, D. (1981). *Kate Westaway, 1893–1973: A Memoir.* Bedford: priv. printed
Mach, E. (1893). *The Science of Mechanics. A Critical and Historical Account of its Development.* London: Watts
MacLeod, R. (Ed.) (1982). *Days of Judgement.* Driffield: Nafferton Books
Matthews, M. R. (1998). Opportunities lost: the pendulum in the USA National Science Educations Standards. *Journal of Science Education and Technology*, 7, 203–214
Mayer, A-K. (1997). Moralizing science: the uses of science's past in national education in the 1920s. *BJHS*, 30, 51–70
Mayer, A-K. (2004). Fatal mutilations: educationism and the British background to the 1931 International Congress for the History of Science and Technology. *History of Science*, 40, 445–472
Morton, A. (1997). *Education and the State from 1833.* Kew, Richmond: PRO Publications.
Price, M. H. (1994). *Mathematics for the Multitude? A History of the Mathematical Association.* Leicester: The Mathematical Association
Remsen, I. (1886). *An Introduction to the Study of Chemistry.* London: Macmillan
Roscoe, H. E. (1866; new ed. 1886). London: Macmillan
Sampson, G. (1921). *English for the English. A Chapter on National Education.* Cambridge: Cambridge University Press
Sarton, G. (1921). Review of Westaway (1920). *Isis*, 4, 119–120
Schwab, J. J. (1982). *Science, Curriculum and Liberal Education.* Selected essays edited by Ian Westbury & Neil J. Wilkof. Chicago: University of Chicago Press
Shute, C. (1998). *Edmund Holmes and the Tragedy of Education.* Nottingham: Educational Heretics
Sneyd-Kinnersley, E. M. (1908). *Some Passages in the Life of One of H. M. Inspectors of Schools* (reprint 1913). London: Macmillan
Sonnenschein, E. A. (1914). Review of Westaway (1914). *Classical Review*, 28, 213–214
Stead, F. B. (1918). *Natural Science in Education: Being the Report of the Committee on the Position of Natural Science in the Educational System of Great Britain* (1918). London: HMSO
Taber, K. S. (2011). Goodbye school science experiments? *Education in Chemistry*, 48 (June), 128; available at www.talkchemistry.org
Vlaeminke, M. (2000). *The English Higher Grade Schools: A Lost Opportunity.* London: Woburn Press
Weaire, D. & S. O'Connor (1987). Unfulfilled renown: Thomas Preston (1860–1900) and the anomalous Zeeman effect. *Annals of Science*, 44, 617–644
Westaway, F. W. (1912). *Scientific Method. Its Philosophy and Practice.* Glasgow and Bombay: Blackie. 439pp.
Westaway, F. W. (1914). *Quantity and Accent in the Pronunciation of Latin.* Cambridge: Cambridge University Press. Reprinted 1923; 2nd revised ed., 1930
Westaway, F. W. (1919). *Scientific Method.* 2nd ed. Glasgow & London: Blackie

Westaway, F. W. (1920). *Science and Theology: Their Common Aims and Methods.* Glasgow & London: Blackie. 346pp.
Westaway, F. W. (1924). *Scientific Method.* 3rd ed. Glasgow & Bombay: Blackie
Westaway, F. W. (1926). *The Writing of Clear English.* Glasgow & London: Blackie
Westaway, F. W. (1928a). *Geometry for Preparatory Schools.* Glasgow & London: Blackie. 201pp.
Westaway, F. W. (1928b). *Lower and Middle Form Geometry.* Glasgow & London: Blackie. 260pp.
Westaway, F. W. (1929). *Science Teaching.* Glasgow & London: Blackie. 442pp.
Westaway, F. W. (1931a). *Craftsmanship in the Teaching of Elementary Mathematics.* Glasgow & London: Blackie. Reprinted 1934 and 1937. 665pp.
Westaway, F. W. (1931b). *Scientific Method.* 4th ed., "revised and enlarged, present-day methods critically considered". Glasgow & Bombay: Blackie
Westaway, F. W. (1932). *Science and Theology.* 2nd ed. Glasgow & London: Blackie. Reprint 1934
Westaway, F. W. (1933a), The pronunciation of Latin, *Greece & Rome*, 2, 139–143.
Westaway, F. W. (1933b). *The Teaching of English Grammar. Function versus Form.* Glasgow and London: Blackie
Westaway, F. W. (1934). *The Endless Quest: Three Thousand Years of Science.* Glasgow & London: Blackie. 1080pp. 2nd ed. 1935 with minor corrections only; reprint 1936. Chinese translations 1937 and 1966, Czech in 1937
Westaway, F. W. (1936). *The Endless Quest.* 2nd ed. reprint. London & Glasgow: Blackie
Westaway, F. W. (1937a). *Scientific Method.* 5th ed. Glasgow & London: Blackie
Westaway, F. W. (1937b). The teaching of general science. *School Science Review*, XVIII, 310–311
Westaway, F. W. (1938a). *Obsessions and Convictions of the Human Intellect.* Glasgow and London: Blackie. 528pp.
Westaway, F. W. (1938b). *Man's Search after Truth.* Glasgow and London: Blackie. 528pp.
Westaway, F. W. (1939). *Appreciation of Beauty.* Glasgow and London: Blackie. 223pp.
Westaway, F. W. (1942a). *Science in the Dock: Guilty or Not Guilty?* London: Science Book Club. 133pp. Reprint 1944.
Westaway, F. W. (1942b). *Science Teaching.* Reprint. Glasgow: Blackie
Westaway, K. M. (1917). *The Original Element of Plautus.* Cambridge: Cambridge University Press
Westaway, K. M. (1922). *The Educational Theory of Plutarch.* London: University of London Press
Westaway, K. M. (Ed.) (1924). *Selections from Plautus.* Cambridge: Cambridge University Press

William H. Brock is emeritus professor of history of science at the University of Leicester. He read chemistry at University College London before turning to the history of science which he taught at Leicester between 1960 and his retirement in 1998. His publications include *The Atomic Debates* (1967); *H. E. Armstrong and the Teaching of Science* (1973; paperback 2012); *Justus von Liebig und August Wilhelm Hofmann in ihren Briefen 1841–1973* (1984); *From Protyle to Proton. William Prout and the Nature of Matter 1785–1985* (1985); *The Fontana History of Chemistry* (1992); *Science for All. Studies in the History of Victorian Science and Education* (1996); *Justus von Liebig. Gatekeeper of Science* (1997); *William Crookes and the Commercialization of Science* (2008); and *The Case of the Poisonous Socks. Tales from Chemistry* (2011).

Edgar W. Jenkins is emeritus professor of science education policy at the University of Leeds where he was head of the School of Education (1980–1984 and 1991–1995) and director of the Centre for Studies in Science and Mathematics Education (1997–2000). From 1984 to 1997, he edited the journal *Studies in Science Education*. His principal books include *From Armstrong to Nuffield* (1979); *Technological Revolution?* (1986); *Inarticulate Science?* (1993); *Investigations by Order* (1996); *From Steps to Stages* (1998); *Learning from Others* (2000); and *Science Education, Policy, Professionalism and Change* (2001). He is currently co-editing a volume to mark the fiftieth anniversary of the founding of the Association for Science Education.

Chapter 74
E. J. Holmyard and the Historical Approach to Science Teaching

Edgar W. Jenkins

74.1 Introduction

Eric John Holmyard (1891–1959)[1] was a scholar and schoolmaster and a significant figure in the history of science and in science education during the first half of the twentieth century. Although his research into alchemy is well known to historians of science interested in early chemistry, his historical approach to science teaching has received much less attention from the science education community, the study by Kinsman (1985) being a conspicuous but unpublished exception.

Born in 1891 in Midsomer Norton, Somerset, in the West of England, his father, Isaac Berrow Holmyard, was a schoolteacher in a national school, i.e. an elementary school set up by the National Society for Promoting the Education of the Poor in the Principles of the Established Church. After attending Sexey's School in Bruton, Holmyard went up to Sidney Sussex College, Cambridge, to read history and science. He graduated in both these disciplines with a first in natural science and a second in part two of the history Tripos. After working as a Board of Agriculture Research Fellow at Rothamsted Experimental Station, he taught briefly at Bristol Grammar School and Marlborough College before becoming head of science[2] at Clifton College in Bristol in 1919.

Holmyard entered the teaching profession at a time when schooling in England was undergoing a period of significant change. The Education Act of 1902 had created Local Education Authorities (LEAs) with responsibility for publicly funded

[1] Although Holmyard used both his initials in all his publications, he preferred the name John to Eric (McKie 1960).

[2] The terms of Holmyard's appointment were unusual in that he was relieved of all out-of-school duties and, until 1936, lived in Clevedon and travelled daily to Clifton (Kinsman 1985; Williams 2002). His appointment at Rothamsted reflected his study of biology in Part 1 of the Tripos examination.

E.W. Jenkins (✉)
Centre for Studies in Science and Mathematics Education, University of Leeds, Leeds, UK
e-mail: e.w.jenkins@education.leeds.ac.uk

M.R. Matthews (ed.), *International Handbook of Research in History, Philosophy and Science Teaching*, DOI 10.1007/978-94-007-7654-8_74,

elementary and secondary schools, with a small proportion of pupils from the former selected to enter the latter on the basis of an examination at the age of 11. Secondary, i.e. grammar, schools under the control of the LEAs were quickly established across the country, and many chose to organise themselves on the basis commonly associated with the public (i.e. private) schools, creating houses, appointing prefects and promoting competitive sporting activities. However, unlike the public schools, the curriculum of the LEA grammar schools was governed by Regulations issued annually by the Board of Education which required the teaching of both practical and theoretical science. Despite the experience of pioneering schools like Clifton College, much remained to be learnt about best to organise and teach science in the grant-aided secondary schools, 1,027 of which were established by 1914 (Simon 1974, p. 363). There was therefore much debate about the order and manner in which topics should be presented and taught and about the roles to be accorded to expository teaching, teacher demonstration and laboratory work conducted by pupils (Fowles 1937; Jenkins 1979). For reasons that are discussed below, that debate became particularly significant in the years following the end of the First World War.

This essay reviews Holmyard's contribution to the historical method of science teaching and comments upon why the history of science has featured so prominently in the history of school science education.

74.2 The Historical Method of Teaching

Founded in 1862 under the headmastership of John Percival, Bishop of Hereford, Clifton College was somewhat unusual among the 'public' schools in England in the mid-nineteenth century in that it 'took science seriously' (Brock 1996, p. 373). It was a legacy upon which Holmyard was able to build. Under his leadership, the College acquired an astonishing collection of manuscripts and first editions of scientific works for its new library[3] and established an outstanding reputation for its science teaching. According to the author of the relevant entry in the *Dictionary of National Biography*, 'Under his guidance, Clifton established a reputation for science probably unequalled, and certainly not surpassed, by any other British school' (Williams 2002; see also Williams 2004).

Holmyard was an enthusiastic advocate of a historical approach to the teaching of chemistry which helped pupils to learn how discoveries were made and, more particularly, to familiarise themselves with the ideas in the minds of the researchers

[3] The Library, known as the Stone Library, was built with funds from Old Cliftonians. The first editions purchased for the library included Newton's *Opticks* and Darwin's *On the Origin of Species*. Clifton College also purchased material at Sotheby's 1936 sale of Newtoniana. Opened in 1927, Holmyard was justifiably proud of the new Science Library which he believed had 'no rival in any school in the world' (Holmyard, writing in the *Cliftonian* of June 1927 and quoted in Williams 2002, p. 204).

and to understand the methods by which they had overcome difficulties. His advocacy reflected a deep scholarly interest in the early history of chemistry, notably in alchemy, at a time when the history of science was becoming established as a distinct academic discipline: the journal *ISIS*, for example, was founded by George Sarton in 1912 and Department for the History and Methods of Science Department at University College London in 1921. Holmyard taught himself Arabic and acquired a good working knowledge of Hebrew, skills which enabled him to edit several important Islamic alchemical texts and to shed light on the work of early alchemists, research for which the University of Bristol awarded him a D. Litt. in 1928. He was in demand as a reviewer for several journals, including *The Journal of the Royal Asiatic Society*, reviewing publications in French and German as well as English. He served (1947–1950) as a vice-president of the newly formed British Society for the History of Science and later as an ordinary member of its council (1953–1954) and as chairman of the Society for the Study of Alchemy and Early Chemistry. He was also a corresponding member of L'Académie Internationale de l'Histoire des Sciences.

In a detailed review of approaches that had been recommended or used to teach chemistry, the schoolmaster George Fowles[4] suggested that a historical approach could be biographical, recapitulatory or be based upon the evolution of scientific ideas (Fowles 1937, pp. 511–18). The first of these relied on the biographies or published diaries of famous scientists and was sometimes coupled with the anecdotal in an attempt to relate the research of individual scientists to wider social, economic or political concerns. The work of Haber on nitrogen fixation is an obvious example. The recapitulatory approach was adopted by Perkin and Lean in their *Introduction to the Study of Chemistry*, first published in 1896 (Perkin and Lean 1896).[5] Convinced that 'the order in which problems have presented themselves to successive generations is the order in which they may be most naturally presented to the individual', they claimed that many of the chapters had been worked through by elementary students in the laboratories of Owens College, later the University of Manchester (Perkin and Lean*op.cit.*, pp. vii-viii and xi). However, the recapitulatory approach does not seem to have been widely used, partly because it involved devoting time to teaching ideas and processes that had long been superseded or could be taught much more effectively and economically in other ways. Fowles, writing a generation later, suggested (*op.cit.*, p. 515) that the approach could work well only when controversy and blind alleys did not significantly interrupt the presentation of the historical narrative. His overall judgement (*op.cit.*, p. 514) was that the 'much vaunted' recapitulatory method had 'never been faithfully carried out in practice'.

[4]For Fowles, see Jenkins (2000). Fowles was a close friend for over thirty years of John Bradley who regarded him as his 'friend and mentor' and his book as 'indispensable to the teacher of chemistry' (Bradley 1988, p. 2).

[5]Perkin and Lean's book held 'a position of honour' on the bookshelves of John Bradley who was introduced to it by his teacher at Archbishop Holgate's Grammar School in York, Henry Worth (1870–1949). Bradley was a distinguished teacher and teacher educator who was much influenced by the writings of Mach and who taught at Christ's Hospital where he met H. E. Armstrong (Bradley *op.cit.*).

The evolutionary approach was intended to help students understand the personal, intellectual, professional, economic and social factors that characterised the historical path towards the current understanding of natural phenomena. The value of this approach lay in countering the view that former ideas were simply the absurd outcomes of prejudice or mistaken judgement and in understanding the qualified degree of confidence that should be placed in more contemporary explanations of natural phenomena.

> We like to be thought devotees of truth uninfluenced by prejudice, as open-minded and serene students of nature, free from suppositions and welcoming every fact that comes within our ken… When the errors of our predecessors are forced upon our notice we may lament them or be amused or may seek to excuse them, but that the same lamentations and excuses may some day have to be made for us we can hardly think possible. (Lodge 1925, p. i)

Although he never referred to it as such, it is this evolutionary approach that most aptly describes Holmyard's historical method of teaching chemistry. He argued his case at a vacation course for science teachers held in Oxford in the summer of 1924[6] and in an article published in the same year in *The School Science Review*, the journal of the Science Masters' Association, an organisation of which he was a committee member and, a year later, president (Holmyard 1924a, pp. 227–33).

He began by addressing the widely held belief that school chemistry could be taught in two ways, depending upon whether the students would eventually become chemists or not. For the former group, the emphasis was placed on the grammar and syntax of the discipline, on learning the rudiments of canonical chemistry. For the latter, the 'chemistry of everyday life' was usually accorded priority. For Holmyard, this distinction was 'a grave fallacy' that rested upon 'the fundamental misconception that chemistry is a craft, when essentially it is a philosophy' (*ibid.*, p. 227). He was thus scornful of the enduring educational merits of snippets of chemical knowledge that enabled someone later in life to solve an acrostic in a daily newspaper or to understand that 'the 'will o'-the-wisp' is caused by the combustion of marsh gas…produced by the action of little insects, called germs (*sic*), upon dead plants' (p. 228). He does not, however, suggest that knowledge of this kind is to be regarded as having no value. On the contrary,

> …how often, after having cut myself while shaving, have I thanked Heaven that my chemical education had been carried to such a degree of perfection that I knew a trivalent cation was especially effective in the coagulation of colloids!

He then adds, in a remarkable sentence, that

> Merely to have rubbed on alum in an unintelligent way would have robbed the operation of all its ecstasy, and I should not have felt that piquant sense of superiority over my daily companion in the train, Lucas, the stock-broker, who is so ignorant that he doesn't even know the empirical formula for the starch in his own collars! (p. 229)

This is one of several examples in Holmyard's article of what he called 'levity', the intended irony of which could all too easily be misunderstood. However, all the

[6] Holmyard's contribution to the course was subsequently published as a pamphlet entitled *The Teaching of Science* (Holmyard 1924b).

examples served to introduce his main argument, namely, that the chief value of chemical education stemmed from the precise, logical and formal character of the discipline, not from its personal or economic utility. He then went on to address the question why chemistry has an educational advantage over other disciplines, such as mathematics, which can be characterised in similar terms. His answer was that chemistry appeals both to intellect and to emotion and is not a 'cold, discarnate scheme of mental gymnastics' (p. 231).

Claiming that 'the immediate [favourable] reaction to biographical details is a universally recognized trait of youthful psychology', Holmyard asked who 'could fail to be stirred by accounts of Pasteur's romantic search for *l*-tartaric acid, of Priestley's discovery of oxygen, of Moissan's isolation of fluorine?' He claimed that he had successfully used his historical approach with his own students at Clifton.

> Personally, I have never found any difficulty in getting a boy to 'believe' in the 'truth' of the sulphur-mercury theory of metals, to get him to abandon it for phlogiston theory, or to abandon this in turn for the oxygen theory, with the result that the last theory is regarded by him in a very different way from that in which a boy looks at it who has had it taught to him dogmatically (Holmyard 1924a, p. 232).

In addition to arguing for, and illustrating, Holmyard's historical approach to teaching chemistry, his article in *The School Science Review* also provides some insight into his understanding of the philosophy of chemistry. It is clear that, for Holmyard, this understanding did more than support his historical approach to chemical education: it provided its underpinning rationale. Only by adopting such an approach could students be led to understand that 'Science in general and chemistry in particular are but conceptual schemes which must always bear an unknown relation to the precepts they correlate' (p. 231). Elsewhere, he invokes a biological analogy to express his view of the history of science.

> '…the theory of evolution is applicable to the development of science no less than to the world of birds, beasts and flowers. (Holmyard 1925c, preface)

For Holmyard, scientific knowledge is to be regarded as essentially pragmatic, not connected with questions about the nature of reality, and as free of any special ontological assumptions and untainted by narrow questions of economic benefit. For the anonymous reviewer of his pamphlet, *The Teaching of Science*,

> Mr Holmyard is not concerned with a universe of absolute truth and rigid law, but with a humble and tentative hold on the precarious hypothesis of an external world. (SSR 1925a, p. 266)[7]

Almost twenty years later, Holmyard wrote that

> Reality, if it is to be discussed at all, must for the present be discussed as a philosophical problem, and the scientist as a scientist need not adventure into such regions…a theory is merely a conceptual model of perceptual facts…a tool rather than a creed. (Holmyard 1944, p. 126)

Holmyard's concern to put some distance between the discipline of chemistry and its uses brought him into conflict with his fellow members of the Science Masters' Association (Layton 1984, p. 203).

[7] Unless otherwise indicated, all reviewers were anonymous.

…any scheme that sets out to be utilitarian, in the narrow sense of the word, or merely 'interesting' is blatantly immoral, and rightly deserves the censure it invites. (Holmyard 1924a, p. 230)

He therefore had no time for the courses of 'General Science' which the Association was strongly promoting in the interests of 'science for all'.

I am…with those who cry 'Science for All' but I would add 'and Dabbling Science for Nobody'. I have no sympathy with kindergarten schemes of 'general science'. [They are] fallacious and shallow…. (Holmyard 1924a, p. 229)

If students of science are to be helped towards what Holmyard described as the 'truth', only the historical method of teaching could bring this about.

The historical method is not, I believe, one of several equally good alternative schemes of teaching chemistry in schools: it is the only method which will effectively produce all the results at which it is at once our privilege and duty to aim (*ibid.*).

In addition, it was only the historical method that could enable students to appreciate to the full 'the serene joys of the intellectual life' and steer school science education in a much needed new direction.

…the main result of the teaching of science in our schools has been to accelerate the spread of that contempt for and indifference to ethical, moral and aesthetic values, and to spiritual and religious truth…perhaps the chief characteristic of our civilisation. (Holmyard 1925b, p. 490)

Nothing less than a fully worked out historical approach could bring this about and he cautioned that

Many teachers, in all good faith, imagine that they are adopting this method [of teaching] if they drag in a few biographical details…fascinating as it may be, and valuable as it certainly is, in stimulating the child's interest and attention, it does not by itself constitute the historical method. (*ibid.,* p. 492)

Holmyard's views were supported by J. W. Mellor, FRS, author of several textbooks and of a landmark multivolume *Comprehensive Treatise on Inorganic and Theoretical Chemistry.* Asserting that 'every teacher now recognises that it is a sheer waste of time to introduce many abstract ideas into an elementary science course without a previous survey of the facts from which the generalizations can be derived', Mellor[8] argued that 'in most cases the historical mode of treatment is correct, because the generalizations have usually developed from contemplation of the facts' (Mellor 1932, preface).

---

[8] Joseph William Mellor CBE, FRS (1869–1938) was born in Lindley, a suburb of Huddersfield in the West Riding of Yorkshire, but left for New Zealand at the age of 10. He returned to England in 1899 to take up an appointment as chemist to the Pottery Manufacturers Federation in Newcastle – under Lyme. Mellor soon established himself as a researcher of international standing, working on the structure and properties of ceramic materials. In 1934, he was appointed director of the new laboratories of the British Refractories Research Association, named in his honour. In addition to his monumental multivolume treatise on inorganic chemistry, he was a highly competent mathematician, and his book on mathematics for chemists and physicists became widely used (Mellor 1902).

74.3 Criticism of the Historical Approach

Holmyard will have been well aware that the historical approach to teaching science also had its critics. The pioneering Cambridge chemist, Ida Freund, author of a seminal book, *The Study of Chemical Composition* (Freund 1904), had little time for such an approach. She fully understood the importance of understanding the history of a scientific discipline and had written a prize-winning historical study of the constitution of matter, some of which almost certainly influenced the approach taken in her 1904 book. Nonetheless, to those who wanted students to retrace the paths by which scientific discoveries had been made, she replied that

> Even if such a plan could be consistently adhered to …it is better to take the shorter way to the goal, this being after all a way by which the discovery might have been made. (Freund 1904, quoted in Fowles, *op.cit.*, p. 513)

In 1929, in a book entitled *Science Teaching: What it was – What It Is – What it Might Be*, F. W. Westaway[9] warned that

> if the historical method is to be adopted, the general method of the history teacher must be followed…What is the point of discussing Roger Bacon and his work unless a boy first understands something of the spirit of mediaevalism – any person who attempted to unravel nature's secrets must be an emissary of Satan himself, and punished accordingly. (Westaway 1929, p. 32)

For Westaway, this condition meant that 'teaching in accordance with historical sequence' could not 'be recommended for subjects usually taught up to the fifth form [i.e. about the age of 16] –physics, chemistry and biology' (*idem.*). He was cautious of the slow progress that the historical method allowed, advising that 'it simply does not pay to spend a whole lesson over, say, the phlogiston hypothesis'. Even so, he felt that some aspects of school science, such as astronomy, were well suited to being 'developed historically and to great advantage' (*ibid.* p. 31).

In 1930, the historical method and Holmyard in particular were the subject of criticism by H. H. Cawthorne. A graduate of King's College London, Cawthorne had worked in teacher education at the University College of the South West, Exeter, before taking up a post at Firth Park secondary school in Sheffield. In his *Science in Education* (Cawthorne 1930), he devoted a whole chapter (pp. 66–75) to the historical method which he is careful to distinguish from a more narrowly biographical approach to school science teaching. While acknowledging that the historical method of teaching has merits, including preventing a 'dogmatic treatment of present-day thought', Cawthorne challenged the notion that the standpoints of the student and of the scientific discoverer are roughly the same. He concluded that it was not necessary for students to 'wade through all the mire and clay of controversy' which have, at times, been obstacles to the progress of science. His advice to teachers was that in order to prevent confusing students or using excessive time, it was necessary to 'short-circuit' some parts of the full historical argument. As a further obstacle to

[9] For Westaway, see Jenkins (2001) and Brock and Jenkins in this volume.

teaching in the way advocated by Holmyard, he reminded his readers that the mind of the mid-twentieth-century student was packed with 'many odd scraps of information which the most fertile imagination of the seventeenth century philosopher could never have supplied'. He offered the example of liquid air, a concept familiar to many of his students but which would have nonplussed the early members of The Royal Society (*ibid.,* p. 75). Today, when scientific ideas and explanations are widely available in museums and the print and broadcast media, Cawthorne's point is even more telling.

For another schoolmaster, Fowles, writing in 1937, while the historical method of teaching chemistry was 'attractive in theory' and 'appealed to the philosophically minded…few had drawn up a scheme or work or attempted to put the method in practice' (Fowles *op.cit.*, p. 513). He wondered whether students, helped by their teacher to understand that a theory was no longer tenable, might be puzzled why it was retained in the face of new experimental evidence by 'men of the depth of intellect of Priestley and Cavendish'. He asked who, 'Notwithstanding the simplicity of the experiments …decomposes mercuric oxide with the heat of the sun concentrated by means of a 12-inch lens…when oxygen is much more easily prepared by other means?' (*op.cit.*, p. 514). His judgement on the historical method was that 'one is constrained to believe that the students have done little more than accept the belief of the teacher' (p. 517).

However, there was more to Holmyard's advocacy of the historical approach to science education than the insights it could offer into what today is referred to as 'the nature of science'. It also added a much-needed human dimension to the subject.

> If Science is to retain the honourable place it has won in the educational system of this country…we shall have to recognise that is the greatest of the "humanities", and deliberately abandon the so called "utilitarian" standpoint. (Holmyard 1922, preface)

It is interesting to place Holmyard's rejection of the 'utilitarian standpoint' alongside the view of Harold Hartley, FRS,[10] a distinguished physical chemist and contemporary who saw science

> …not merely as an academic subject- nor only as a basis for applied technological development but as a great cultural adventure that comprised both of these: as an historical sequence, having its nourishing roots in the past and its growing branches thrusting constantly into the future. (Ogston 1972, p. 366)

Holmyard was by no means without support in seeking to present science as a humanity. Turner, in her account of the history of science teaching in England, came to the conclusion that

---

[10] Sir Harold Brewer Hartley, C.H., FRS, was taught chemistry by H. B. Baker at Dulwich College in London who encouraged him to read and buy old chemistry books. Hartley went on to Balliol College, Oxford, to develop a lifelong interest in the history of the discipline. During his long life he held a variety of academic, business and industrial appointments. He secured the Lewis Evans collection of scientific instruments for the University of Oxford and was a frequent contributor of historical articles, especially about nineteenth-century English chemists, to the *Notes and Records of the Royal Society* of which he was the editor for eighteen years. The preface to his *Studies in the History of Chemistry* (Hartley 1971) reveals that the book was commissioned nearly 70 years earlier!

> ...if science teaching is to mean anything more than the acquisition of a few tags of knowledge and a certain skill in manipulation we must accord it a place among the humanities... The human side is perhaps best introduced by a carefully selected historical treatment. (Turner 1927, p. 191)

In the aftermath of the First World War in which chemistry had played such a massive and destructive role, recasting science as a humane study was widely regarded as necessary and it helps to explain why, despite the practical difficulties of implementing Holmyard's ideas, the incorporation of at least some elements of the history of science in school curricula received a broadly sympathetic hearing.

> How necessary Science is in War...we have learnt at a great price. How it contributes to the prosperity of industries and trade, all are ready to admit. How valuable it may be in training the judgement, stirring the imagination and in cultivating a spirit of reverence, few have yet accepted in full faith. (Natural Science in Education 1918, para.4)

Reaction to the role of science in the First World War was not the only influence at work in seeking to 'humanise' school science. During the 'battle of the books' that had broken out in the middle of the First World War, supporters of a classical education had reacted vigorously to the claims of a self-styled 'Neglect of Science Committee' established to promote science education and research (Jenkins 1979). Ramsay MacDonald told the House of Commons that this committee was 'practically telling us to clear the humanities out of our schools' (Hansard 1916, col. 906), and an editorial in *Blackwood's Magazine* in 1916 described the 'ferocious attack on the humanities as evidence of the 'unbalanced men of science who wish to kill off all learning other than their own'. Those who had praised German scientific and technological achievements before the war now found their own evidence being used against them.

The clash between the scientific and classical contributions to education was not confined to the UK, and the battle was not always conducted in terms conducive to effecting a rapprochement. In the USA, for example, one commentator opined that 'Largely without the benefit of the Classics, it is not to be expected that [the ordinary scientist] should know what the Humanists are saying or realize his faults' (Glaser 1924, p. 30). In the UK, the distinguished Wykehamist classical scholar and educator, Sir Richard Livingstone, claimed in his *Defence of a Classical Education* that the fundamental weakness in science as a vehicle of a liberal education was that

> ...[science] hardly tells us anything about man. The man who is our friend, enemy, kinsman, partner, colleague, with whom we live and [have our] business, who governs or is governed by us [never once] comes within our view. (Livingstone 1916, pp. 30–31)

It is clear that Holmyard, who regarded science as an integral part of culture, would have rejected Livingstone's claim that science 'hardly tells us anything about man'. It is equally clear, as noted above, that Holmyard thought science as it had been taught had accelerated a contempt for, and an indifference to, ethical, moral and aesthetic values and to spiritual and religious truth. It is noteworthy therefore that the Foreword to his *Inorganic Chemistry*, first published in 1922, was written, presumably at Holmyard's invitation, by a leading classical scholar, Cyril Norwood. Holmyard would have encountered Norwood when the latter served as headmaster at Bristol Grammar School (1906–1916) and Marlborough College (1917–1925), respectively, before moving on to the headship of Harrow (1926–1934). Norwood has been described as the 'quintessential insider of English education in the first half of the twentieth century' (McCulloch 2006, p. 55; see also McCulloch 1991). He was an influential source of government advice, eventually being knighted for his services to education. In 1929, Norwood published *The English Tradition of Education* which offered a deeply complacent and conservative view of the past, a past that was closely linked to the teaching of the classics in the English public and endowed grammar schools. In his Foreword to Holmyard's book, Norwood is deeply critical of the system of Higher Grade Schools that had developed in England during the last quarter of the nineteenth century (Vlaeminke 2000). Funded for the most part by grants from the Department of Science and Art, these locally controlled post-elementary schools attached much greater importance to science than many public and endowed grammar schools and provided an 'alternative' secondary education in all but name.[11] Norwood claimed that the teaching in these schools had been 'one sided' and excessively formal and that it had taken the Great War to forcibly remind the nation that things were not well. He saw Holmyard's *Inorganic Chemistry* as offering a way forward.

> [The author] knows how to teach with breadth and without exclusiveness. Its pages give information and provoke curiosity: at many points they suggest that there are other realms of knowledge of a quite different sort. (Norwood 1922, p. *v*)

It was an endorsement that one can safely assume met with Holmyard's approval.

When the newly formed Science Masters' Association held its annual meeting in 1920, there was widespread agreement that school science courses needed to be both broadened and 'humanised', although there was much less confidence about how this could be achieved. For some, the way forward lay with a broad course of General Science which 'furnished the mind' and 'gave some knowledge of the world in which we live' (Tilden 1919, p. 12). For others, it was essential to capture the spirit and romance of scientific endeavour by incorporating a biographical element into school science education, 'the history of men and the setting forth of noble objects of action' (Sadler 1909, p. *xi*). For yet others, it was Holmyard's evolutionary approach to the history of scientific ideas, or at least some aspects of it, that seemed to have most to commend it.

Unsurprisingly, each of these possible directions for science education reform presented difficulties. Despite the success that Holmyard claimed he had achieved with his own students, his historical method was the subject of ongoing debate and critical commentary throughout the interwar years. Interestingly, there appears to be no evidence that he sought to reply to his critics by, for example, writing for *The School Science Review* or publishing a more detailed and practical account of

[11] See McCulloch (1984). The Higher Grade Schools are seen by some historians as constituting an 'alternative road' to that represented by the traditional grammar school curriculum. They were swept away by a series of legislative changes between 1899 and 1902, although some re-emerged as secondary, i.e. grammar, schools under the control of newly created Local Education Authorities. See Vlaeminke (2000).

his historical method to which science teachers could refer.[12] Fowles (*op.cit., p.* 517) expressed some admiration for Holmyard's success in 'getting a young class to grasp the doctrine of phlogiston' but questioned 'how little of elementary chemistry' could be taught along the lines advocated by Holmyard and how far it was expedient, and actually possible in the time usually available, to secure topical development along such lines. Although it is impossible to be sure how widely Holmyard's approach to school science was adopted, it seems clear that for most science teachers it presented formidable, even insuperable, difficulties. Few teachers could command the knowledge of the history of chemistry or physics required to take students back and immerse them in the period under consideration, and there is little evidence to suggest that the obstacles identified by Cawthorne, Fowles and others were successfully overcome. In a book written 'for teachers and training college students', John Brown, a school inspector for the London County Council, could do no more than advise that 'with older pupils, the study of a certain amount of *historical development'* of the subject will prove profitable (Brown 1925, pp. 45–6).

74.4 'Ships' Surgeons Are Always Truthful!'

While Holmyard's historical method may have failed to find widespread favour among his science teaching colleagues, his scholarly achievements in the history of science were able to find generous expression in his large number of school science textbooks. There are likely to be few, if any, chemistry teachers of an older generation familiar with the English education system who will not be able to recall his name. Some may even be familiar with the above quotation – of which more below.

Holmyard was a prolific author. Throughout the interwar years in particular, publications concerning the history of science appeared alongside a steady stream of school science textbooks. The British Library Integrated Catalogue lists over a hundred entries for Holmyard (although this number includes several different editions or reprints of the same book) and a large proportion of the entries relates to works published between 1922 and 1939. The 1920s alone produced *Chemistry to the Time of Dalton,* published in 1925, two translations of Avicenna (1925c) and

[12] A comparison with H. E. Armstrong is of some interest. Both men sought to give students an insight into how scientific knowledge was obtained and validated, although Armstrong's emphasis on the practical teaching of 'scientific method' could not be more different from Holmyard's more complex, nuanced historical approach. Unlike Holmyard, Armstrong was a vigorous promoter of his *virus heuristicum Armstrongii* and even compiled a book setting out his ideas (Armstrong 1903), although its lack of coherence made this a less useful advocate of his cause than it might have been. See, for example, Browne (1954/1966) and Brock (1973). For both heurism and the historical method of science teaching, science teachers were undoubtedly the weakest point. As individuals, Holmyard was described as a 'quiet' and 'tolerant' man (McKie 1960, p. 5), as having 'a modest and retiring disposition' (*The Times*, October 15, 1959) and an 'imperturbability of temper' (Singer 1959, p. 17), descriptions that could never be applied to the hot-tempered Armstrong who 'always made one think [but] was fond of saying that very few…were capable of doing so' (Hartley 1971, p. 195). For Holmyard's views on heurism, see Holmyard (1925b, p. 490).

a critical and important translation of the Arabic works of Jâbir Ibn Hayyân (1928a), a year that also saw the publication of *The Great Chemists*. The same decade also witnessed the first publication of over a dozen school science textbooks, some of which were reprinted or revised and remained in use for at least the next thirty years.

Some of Holmyard's textbooks were phenomenally successful. His *Elementary Chemistry*, first published in 1925, was reprinted eleven times by 1933, eventually selling over half a million copies worldwide (Holmyard 1925a). Even today, when chemistry has undergone so many profound changes, it remains a remarkably interesting book. The frontispiece is an extract of a dialogue between master and pupil from *Ye Booke of Allchimye* written in the twelfth century, its first two chapters address the questions of what chemistry is and how it arose and the book ends with biographical notes on some famous chemists. Few, if any, modern elementary school chemistry texts begin by introducing pupils to the complex, multicultural origins of the discipline, provide illustrations of Arabic chemical operations, review the multiple uses and benefits of chemistry and explain how chemistry itself came to be so called. While what follows these opening chapters is a well-ordered presentation of familiar information about the occurrence, preparation, properties and uses of the chemical elements and their compounds, few readers would have failed to find something of interest even when this was incidental to the main text. For one reviewer, the book revealed Holmyard's 'adventurous and at once recognizable style' along with his 'store of humorous delights on chemistry' (*J. Chem. Ind.* 1934, p. 882). His *Chemistry for Beginners* prompted another reviewer to comment that 'Mr Holmyard's books always please us', adding that 'he had the rare gift of writing so as to interest the young' (*J. Chem. Ind.* 1931, p. 146).

No less successful as a publication and equally well received by the book's reviewers was his *A Higher School Certificate Inorganic Chemistry* which appeared in 1939. For one reviewer, the reading and rereading of this book had been 'sheer delight', adding that it was 'not possible to write of it save in terms which appear exaggerated'. For the journal *Nature*, it was simply 'excellent'. Following the replacement in 1951 of the system of School Certificate Examinations by the O- and A-level examinations of the GCE, Holmyard collaborated with W. G. Palmer[13] to produce in 1952 a revised edition, entitled *A Higher School Inorganic Chemistry.* The above reference to the truthfulness of ships' surgeons will be found on page 272 of the first edition and

---

[13] Palmer was a Fellow of St. John's College, Cambridge, when Bradley (see note 5) was an undergraduate and gave him 'a taste for the history of chemistry' (Bradley *op.cit.*, p. 1). Holmyard also collaborated with Frederick Arthur Philbrick to produce *A Textbook of Theoretical and Inorganic Chemistry* (Philbrick and Holmyard 1956). Philbrick died at a comparatively early age. Holmyard may have felt the need for collaboration since he had given up his post at Clifton College in 1940 when the school was evacuated to the relative safety of Bude in Cornwall. Although Philbrick is identified as the lead author, the book 'was mostly Holmyard, Philbrick having been responsible for bringing later editions up to date' (Francis 2004, p. 15).

Holmyard subsequently took up the editorship of the magazine *Endeavour*. Published by ICI, the magazine sought to publicise British scientific and technological achievements. Intended as a war time publication, it proved so successful under his editorship that it continued after the war ended when it was produced in five languages (English, French, German, Italian and Spanish). Holmyard remained its editor until 1954. His obituary was published in *Endeavour*, 73, January 1960.

page 275 of the revised volume. Referring to the salvaging of 30 tons of mercury from a Spanish wreck off Cadiz by HMS Triumph in 1810, the reader is told that the symptoms of mercury poisoning quickly became evident among the crew and livestock of the salvage vessel. Lest any should doubt the word of the ship's surgeon, quoted in the text, that he 'had seen mice come into the ward-room, leap up to some height, and fall dead on the deck', the reader is referred in a footnote to the quotation cited above. In a discussion of the colloidal state on page 147 of the earlier volume, the poet Keats is invoked to describe a gel as 'soother than the creamy curd', a description challenged by the footnoted observation that Shakespeare says 'Out vile jelly!' The likening of the smell of phosphine to that of garlic prompts the comment (p. 373) that this is 'A base libel on a plant recommended by the Father of Chemistry, HERMES, to ODYSSEUS, as an antidote to the poisons of CIRCE'.[14] A reference to 'saltpetre', potassium nitrate, is amplified by a note that the word means 'rock salt' and that this is itself a 'reminder that in bygone days chemists have found it very difficult to distinguish between substances of similar appearance' (p. 216). Elsewhere (p. 9), the reader is told that, in France, nitrogen is called azote (a name that indicates that the gas will not support combustion) and that the element cobalt is supposed to get its name from the German *Kobold*, a mischievous subterranean gnome that haunted the mines from which the ore was extracted (p. 508). All these references also appear in the 1952 publication, although the pagination is slightly different.

Holmyard's output as an author of school textbooks was not restricted to chemistry. *Science: An Introductory Book* was published in 1926, *General Science* appeared the following year and a three-volume series (*Physics/Chemistry/Biology for Beginners*) was published in 1930 (Holmyard 1930a, b, c). Intended for the first two years of secondary schooling, a reviewer of the series in *The School Science Review* welcomed it as 'three very excellent books', adding that it was 'a pleasure to meet school textbooks which are not pervaded by an atmosphere of public examination syllabuses' (SSR 1930, p. 190). In the 1930s, Holmyard also co-authored *Electricity and Magnetism for Beginners* (Badcock and Holmyard 1931); *Heat, Light and Sound for Beginners*; and *Mechanics for Beginners* (Barraclough and Holmyard 1931), published as part of a Modern Science Series. Badcock and Barraclough were two of Holmyard's colleagues at Clifton and he collaborated with a third Clifton colleague to write *Elementary Botany* (Graham and Holmyard 1935). He also edited J. A. Thomson's two-volume *Biology for Everyman* (Thomson and Holmyard 1934) and was one of the editors of an eight-volume *History of Technology*, published between 1954 and 1984 (Singer et al. 1954–84; see also Holmyard 1946).

Why were so many of Holmyard's historicised school chemistry textbooks such successful[15] publishing ventures? To some extent, his success stemmed from a

[14] Holmyard believed strongly in the merits of a classical education, encouraging boys to study classics before taking up chemistry in the upper school, and he made frequent use of classical analogies, e.g. the 'passion of Hydrogen and Oxygen for one another causes as much trouble in the chemical world as that of Paris for Helen did in ancient Troy'.

[15] Although Holmyard's textbooks seem always to have been reviewed very favourably, reviewers were usually able to identify some matters that needed attention. Almost all of these related to a

clarity of style that allowed him to express complex ideas in ways that young pupils could readily access. One reviewer of his *Elementary Chemistry* described it as 'written in English which any boy can understand - the sort...which one seldom finds in a textbook' (SSR 1925b, p. 140). Recalling his schooldays at the Crypt School in Gloucester, Keith Francis, a school teacher who graduated in physics from Cambridge in 1956, described Holmyard as having a 'spicy' literary style that 'grabbed me' (Francis 2004, p. 15).[16] Holmyard's obituarist in *Nature* judged that 'many students' of the journal 'must owe their introduction to chemistry' to his inorganic and organic chemistry textbooks which presented the basic facts of chemistry 'as an experimental science, relating them to general principles in a way which gives them significance and interest' (Partington 1959, p. 1360).

In addition, although the illustrations in Holmyard's books were in black and white rather than colour, they were always carefully chosen, well related to the narrative and often appeared for the first time in texts intended for school use.

> The plates and pictures are delightful and make the book very attractive; after seeing page 73 all our boys will want to get out and collect marsh gas –which is as it should be! (*idem.*)

> The illustrations are admirable and almost all unfamiliar. (SSR 1933, p. 126)

His textbooks also had a readily discernible structure that enabled information to be easily located and, if necessary, retrieved for homework or revision purposes. This became increasingly important as the numbers of candidates attending grant-aided secondary schools and entered for public examinations both increased in the interwar years. The number of pupils attending grant-aided secondary schools increased from 269,887 in 1919 to 470,003 by 1938, although by no means all of these completed their secondary schooling.[17] Aided by grants from the Board of Education, the number of 'advanced', i.e. sixth form, courses in science also increased. By 1938, physics and chemistry each accounted for just under one third of all entries for the Higher School Certificate Examination taken at the end of secondary schooling; the biological sciences (botany, zoology and biology) had yet to establish a secure place in the curriculum of many secondary schools, especially those for boys (Jenkins 1979).

The long publishing history of many of Holmyard's textbooks was also facilitated by the fact that school examination syllabuses in chemistry and physics in England

---

few typographical/proof reading errors or other relatively minor defects. One notable exception was a reviewer of Holmyard's *A Junior Chemistry* (Holmyard 1933). While writing that the book would be 'read with delight by any boy or girl without any need for external stimulus', the reviewer wondered whether 'a boy of 14...may be aggravated by the use of an appeal in the second person singular and of exclamation marks' (SSR 1933, p. 126).

[16] Francis also describes Philbrick and Holmyard's *Theoretical and Inorganic Chemistry* as a 'treasure' and recalls how, as a 13-year-old pupil, it enabled him to know everything that Tarzan [the nickname of his chemistry teacher!] wanted to teach him (Francis 2004, p. 15).

[17] These figures need to be set alongside the much larger number of pupils who attended public elementary schools in England and Wales: 5,933,458 in 1920–1921 and 5,087, 485 in 1937–1938. The division between elementary and secondary education reflected significant differences in social class. It was only after the Education Act of 1944 that all pupils passed from a primary to some form of secondary schooling (grammar, secondary modern or technical).

underwent astonishingly little change between 1918 and the curriculum reform movement of the 1960s. Save for the replacement of Imperial units by the centimetre-gramme-second (cgs) system, examination questions in physics set in the 1920s first appeared in much the same form over a generation later. In chemistry, candidates continued to be examined on their detailed knowledge of the manufacture, preparation and properties of the elements and their compounds with questions following a standard format year after year (Jenkins 1979, p. 293). In Bassey's judgement, Ordinary-level chemistry texts had followed 'one familiar and well-worn path' which, by 1960, had become 'a rut rather than a highway' (Bassey 1960, p. 14).[18] There is little doubt that this judgement can be applied, although to a lesser extent, to school texts in physics and biology and those intended for more senior secondary school students.

Holmyard was also able to call upon his experience both as a teacher and as an examiner for the former Northern Universities Joint Matriculation Board (NUJMB). Indeed, the preface to his *A Higher School Certificate Inorganic Chemistry* informs the reader that 'the allotment of space to individual topics is roughly in proportion to the frequency with which these topics appear in the examination papers', although he is very careful to make clear that more is needed in any textbook that 'aspires to do something more than cram', an aspiration that one reviewer readily acknowledged.

> The author is no Polyphemus and his other eye has watched the cultural aspect of chemistry to prevent the work from degenerating into a mere cram book. (*J. Chem. Ind.* 1940, p. 50)

Holmyard's footnotes and historical commentaries were an integral part of this 'something more', and it would be a serious error to dismiss them as idiosyncrasies or regard them as mere ornaments intended to display the author's undoubted erudition. Footnotes and comments they may be, but they formed part of a coherent and distinctive volume that enlivened, enriched and contextualised school chemistry and its nomenclature, then far from standardised, in ways that have not been matched. No one reading Holmyard's texts can avoid learning something of the many roles that chemistry has played and continues to play in recorded history. Reading them today indicates just how much has been lost from chemical education.

74.5 Historical Approaches: The Wider Context

Holmyard was by no means the first to argue for a historical approach to the teaching of school science nor was he alone among his contemporaries in writing historicized chemistry textbooks (e.g. Partington, Cochrane, Lowry).[19] As long ago as 1855, the

---

[18] Bassey used Holmyard's *Elementary Chemistry* as 'the standard treatment' with which to compare other school chemistry texts in print at the time of his survey.

[19] J. A. Cochrane's *Readable School Physics* (1923) and his *School History of Science* (1925) were highly successful publications, both of which have recently been made available once again. The former was part of a Natural Science Series published by Bell and edited by Holmyard. J. R. Partington, professor of chemistry at the University of London, had studied with Nernst and

president of the British Association for the Advancement of Science, the Duke of Argyll, told his audience in Glasgow that what was wanted in the 'teaching of the young, is not so much the mere results, as the *methods* and, above all, the *history of science'* (BAAS 1856, p. *lxxxiii*). Nor was Holmyard to be the last: over a century and a half later, the history of science featured in a national curriculum introduced for the first time in England (DES 1989). Even more recently, the European Commission has argued that 'a renewed pedagogy' which presents 'the processes and methods of science together with its products' is essential for the 'Future of Europe' (EC 2007, p. 16).

Among Holmyard's contemporaries, Edgar Fahs Smith (1854–1928) in the USA, chemist and author of biographies of several American chemists (Smith 1914), was said to be 'fond of theories in an historical way, but used the luxury of their downfall by experimental observation to illustrate the fallacy of theories'. This quotation is cited in Fowles who attributes to Smith the expression 'Facts remain, theories may change overnight'(Fowles *op.cit.,* pp. 516–7). Among Fahs's successors in the USA, Conant's *Case Histories* (Conant et al. 1957; see also Conant 1947), the *Harvard Project Physics*(1970), the *National Science Education Standards* (NRC 1996), *Benchmarks for Science Literacy* (AAAS 1993), the textbooks of Taylor (1941/1959) and Rogers (1961) and initiatives such as McComas's use of historical examples to illustrate key aspects of the nature of science (McComas 2008) all testify to the enduring appeal of the history of science as a pedagogical tool. Given the profound changes that have taken place in science, philosophy, psychology and society since science was first schooled in the mid-nineteenth century, this widespread and enduring desire to call upon the history of science to illustrate something of the 'nature of science' calls for some comment.

Holmyard's historical approach to teaching science should not be equated with simply incorporating biographical or other elements of the history of science within school science curricula, an accommodation with which many science teachers would have felt much more comfortable and which seems to have met with more success. However, an inadequate knowledge of the history of science and its attendant risk of an unhelpfully Whiggish approach to history were problems common to

---

was the author of advanced chemistry textbooks, of *A Short History of Chemistry* (1937) and of a four-volume history of chemistry Partington (1961–4). His elementary text, *Everyday Chemistry*, is divided into three parts of which the first is entitled 'Historical and Theory' (Partington 1929). A reviewer of Partington's *A College Course of Inorganic Chemistry* (1939) warned his readers that 'Professor Partington has written so many books that have been distinguished by their lucidity and accuracy that a reviewer reading a new book by him begins by being prejudiced in his favour' (*J. Chem. Ind.* (1939) 58 (44), p. 974). A similar comment might well have been made of Holmyard. In 1965, the year of his death, Partington was awarded the George Sarton medal. Thomas Martin Lowry, FRS, worked with H.E. Armstrong from 1896 to 1913 and later became professor of physical chemistry at Cambridge. With J.M. Brønsted, Lowry introduced a new and broader definition of acids and bases. Lowry's generously illustrated *Historical Introduction to Chemistry*, first published in 1915, presents 'a historical account of the more important facts and theories of chemistry, as these disclosed themselves to the original workers' (Lowry 1915, preface). Although perhaps more appropriate for teachers than school students, the book is a further indication of the steady stream of educators who have regarded the history of science as an important component of science education.

both. In addition, as Brush was to suggest later (Brush 1974), introducing students to the history of science might not accord with a desired canonical account of scientific discovery. Even so, whether as a pedagogical approach or as a component of a school science curriculum, the history of science is a striking feature of the history of school science education. Why is this?

Part of the answer may lie in the fact that renewed calls to attend to the history of science have often coincided with a perception that the school science curriculum is in difficulty or facing challenges that prompt a need for reform. As noted above, the desire in the interwar years to 'humanise'[20] school science by teaching the history of science owed much to the feeling that in making possible the unparalleled slaughter of the First World War, science – and chemistry in particular – had lost a sense of moral purpose that it urgently needed to regain. The emphasis on science in Germany came to be seen as having led not only to economic prosperity but also to the moral collapse responsible for the war in which the Allies had been engaged. Restoring that sense of purpose seemed to require reform of the school science curriculum and the history of science pointed a way forward.

> ...the teaching of science must be vivified by a development of its human interest, side by side with its material and mechanical aspects... [and] it must never be divorced from those literary and historical studies that touch most naturally the heart and the hopes of mankind. (Natural Science in Education 1918, para. 3)

The contrast with the aftermath of the Second World War is striking.[21] Despite the advent of nuclear war, science and technology emerged from that war with their prestige greatly enhanced. The consequent demand for reform of school science education was governed by the urgent need to increase greatly the supply of qualified scientific and technological personnel, both for civilian and, at the height of the Cold War, military purposes (Rudolph 2002). The history of science was seen as, at best, only marginally relevant to achieving this goal. Supported by a psychology that favoured 'learning by discovery', the emphasis in what became a global movement for science curriculum reform was captured by Bruner's claim that 'a schoolboy learning physics *is* a physicist' (Bruner 1960, p. 14), a claim repeated by Harlen with respect to primary education in the UK almost forty years later: '*Learning science*

[20] For a historical perspective on the humanist critique of the place of science in the school curriculum, see Donnelly (2002), Donnelly (2004) and Donnelly and Ryder (2011). See also Mayer (2005). For a discussion of science as humanism in Denmark in the 1950s, see Lynning (2007). It is worth noting that there have been frequent calls to 'humanise' other abstract disciplines, e.g. economics (Solterer 1972) and mathematics (Guting 2006). Indeed the word 'humanising' has perhaps acquired something of an Alice in Wonderland quality, writers taking it to mean what they wish it to mean. In 1924, the author of an article in *The North American Review* asked 'By the plain light of noon, what is implied in "humanizing" science?' (Glaser 1924, p. 230). Today, as searching the Web reveals, the science education literature is replete with references to humanising school science education, offering a variety of rationales and strategies for bringing it about. See, for example, Watts and Bentley (1994), Kipnis (1998), and Clough (2009).

[21] However, at least one prominent science educator felt much the same about the role that science was also playing in the Second World War. See Westaway's *Science in the Dock: Guilty of Not Guilty?* (1942).

and *doing science* proceed in the same way' (Harlen 1996, p. 5). In the USA, students following *ChemStudy* courses were promised that they would 'see the nature of science' by engaging in scientific activity' and thereby to 'some extent' become scientists themselves (Pimentel 1963, p.1 and preface). In the UK, despite an assertion[22] by the Science Masters' Association and the Association of Women Science Teachers that science was an 'active humanity' (SMA 1961, p. 5), canonical science was to be presented in a way that enabled students to 'think in the way practising scientists do' (Halliwell 1966, p. 242). Much that followed, such as *Process Science* and *Science in Process,* also sought to introduce students to the nature of science by involving them in suitable practical, laboratory-based exercises designed to encourage the acquisition of those allegedly discrete skills and processes (communicating, interpreting, observing, planning investigations etc.) that in sum enabled them to be good at science (Coles 1989, pp. 4–5).

During the 1960s, it gradually became clear, both in the UK and the USA, that the large-scale efforts to reform school science education had not succeeded in increasing, or in some cases even stemming, the numbers of young people wishing to study the physical sciences, especially physics, beyond compulsory schooling. In England, the Council for Scientific Policy set up a committee, chaired by the eminent chemist Frederick Dainton,[23] to 'examine the flow of candidates in science and technology into higher education' (Council for Scientific Policy 1968). Although the committee's conclusion that there was a 'swing from science' in schools was soon challenged (e.g. McPhemon 1969), it received widespread publicity (e.g. AAAS 1968) and prompted a wide range of research studies of possible causes of the 'swing' away from what have become known as the STEM subjects. Such causes remain the focus of ongoing investigation in many countries (Sneider 2011). In the USA, it was clear as early as 1963 that the PSSC course had attracted 'only about 4 % of the two and a half million senior students in high school, and the total fraction taking *any* physics course was under 20 %, and relatively shrinking' (Holton 2001, p. 2).Once again, the history of science was seen as one way of easing, if not overcoming, the problem. Holton, in a review of the *Harvard Project Physics* course, has recalled how he and others were invited to a meeting by the National Science Foundation at which they were asked

> who would come to the aid of the country? For in those days it was thought without more science-literate students the Russians might get us. (*idem.*)

Holton's answer eventually led him to become the principal investigator of *Harvard Project Physics* and thereby to develop a 'humanistic, historically oriented course for schools' that presented physics, as in his original text (Holton 1952), 'not

---

[22] For the Science Masters' Association, if more emphasis were placed upon the cultural and humanistic sides of science education, this might not only lead more young people to study science but encourage them to favour science teaching as a career (SMA 1957).

[23] Dainton, speaking in 1971, used the memorable phrase 'voluntarily withdrawn from human contact; disassociating himself from personal and societal problems… a man who is "objective" to an objectionable degree', to describe how scientists were widely perceived at the time (Dainton 1971, p. 18).

just as one damned thing after another, but a coherent story of the result of the thoughts and work of living beings' (Holton 2001, p. 2). The outcome was the sort of course that Holmyard would have welcomed, especially the illustrations taken from seminal documents in the history of science that appeared in the course book. Interestingly, the initial reaction of the National Science Foundation to the fact that course wasn't going to be pure physics was one of 'horror'! By the time the final edition of the book was published in 1970, there was significant evidence that it was having an impact on the numbers of students studying physics at high school and college. As many as 300,000 students per year were studying some or all of the course materials and the percentage taking physics, particularly among women students, had increased markedly, 'with some 20 % of all high school students taking *Project Physics*, and in use also in some colleges' (Holton 2001, p. 4).

Although further growth of *Project Physics* in schools (the reference to Harvard in the title was abandoned amid concerns over 'elitism!') fell foul of the phasing out of sections of federal support for science education in the 1970s, especially funds for science teacher education, it acquired a new lease of life when the USA once again sought to reform school science education, this time by prioritising the need for a scientifically literate citizenry, although a number of other factors were involved, e.g. the demand from teachers for a revised and updated version of the project materials and the need to improve the understanding of physics on the part of a new generation of physics teachers. As noted above, the AAAS *Benchmarks for Science Literacy* (AAAS 1993) and the *National Science Education Standards* (NRC 1996) both attached importance to the history of science. The former included an entire chapter devoted to 'Historical perspectives' (chapter 10), while the latter advised standards for the history of science from K4–12. The revised edition of the earlier text, published in 2002 under the title *Understanding Physics* (Cassidy et al. 2002), was described by Holton as the 'completion of a great circle, spanning four decades, from the first draft in 1962…through the slough of the early 1980s and now on to the rising of the new Phoenix' (Holton 2001, p. 6).

That circle offers an interesting example of how a federally funded history of science initiative in the USA, initially prompted by manpower and defence concerns, came to be recast and called in aid of the perceived national need to promote science for all and enhance the general level of scientific literacy. There are some parallels here with the UK, not least in the political anxiety about scientific literacy and the levels of public understanding of science, although the picture is complicated by profound differences in the education systems of the two countries and, more particularly, by the development of a non-selective system of secondary schooling and the subsequent introduction of a statutory national curriculum in England and Wales in 1989. By requiring all students to study science from the age of 5 to 16, it was hoped that this would in due course both promote scientific literacy and develop a larger cohort of students choosing to study science, especially physical science, beyond compulsory schooling. The UK government clearly saw compulsion as a more effective policy initiative than the large-scale curriculum reform of a generation earlier. Despite this, the initial version of the national curriculum included a component (known as an attainment target) that drew upon contemporary views from both the history and the sociology of science. Its inclusion prompted the editor of the *British Journal for the History of Science* to offer the following somewhat slightly anxious comment.

> The overriding statement of intent…is one that might be welcomed by the most radical exponent of the view that scientific knowledge is shaped by social, economic and political context. (BJHS 1990, pp. 1–2)

In the event, the first version of the national curriculum almost immediately proved unworkable and very little of the history of science survived into the revised statutory science curriculum introduced in 1991,or indeed into the several revisions that have taken place since (Donnelly et al. 1996;Donnelly and Jenkins 2001). It has largely been left to individuals and non-governmental sources to take initiatives to promote the history of science in science education.[24]

Advocacy of the history of science in aid of science curriculum reform may also owe much to historians of science themselves. Holmyard and Holton stand as two obvious examples, from different generations and countries, of scholars in the history of science giving attention to the form and content of school science curricula. Their pedagogical interest, however, is part of a much longer-standing and wider interest among their fellow professionals that has been well described by Brock (1989) and Sherratt (1980). There has been an equally long-standing interest on the part of many school science teachers in the history of science and in the contribution that it might make to their teaching. Practising and former science teachers were among the founding membership of the British Society for the History of Science in 1946 and their professional association, The Science Masters' Association and its successor from 1963 onwards, the Association for Science Education, collaborated with the BSHS in a variety of curriculum and policy initiatives (Brock 1989, p. 33 *et seq.*). In addition, history of science books was regularly reviewed in *The School Science Review* and the titles included in the suggested list of *Science Books for a School Library* issued from time to time as a supplement to the association's journal.

The motives of both organisations and their members in promoting collaboration are, as always, historically contingent. Closer involvement with the work of the schools has had the advantage of bringing the history of science to the attention of teachers and students who, in turn, were able to take advantage of historical expertise to their mutual benefit. This no doubt was an important consideration, especially in its early years, for the Department of the History and Methods of Science at University College London which directed its efforts 'chiefly at science teachers and thus at secondary education' (Mayer 1999, p. 233). On some occasions, notably in the interwar years, the rhetoric underpinning this mutual interest related strongly to humanising science teaching and to reasserting the cultural value of science. At other times, the language has been that of bridging the 'two cultures' identified by Snow in his 1959 Rede lectures (Snow 1959, 1964), although the

[24] For example, the *Perspectives on Science* course developed at Rugby School in Warwickshire (www1.edexcel.org.uk/Project/POS-Briefing.doc) and the Public Understanding of Science course developed at the University of York (http://www.nuffieldfoundation.org/science-public-understanding), both accessed 30 January 2012.

emphasis upon science as a 'humanity' has never been lost (Council for Scientific Policy *op. cit.*, para. 181). Most recently, the history of science has been cast as an element of the 'nature of science' which has come to figure so prominently in science curricula across the world, while pedagogy (McComas 1998) and the form and practice of scientific research have become fields of scholarly academic interest to historians of science (Kaiser 2005). That is perhaps another 'great circle' that is in the process of being completed.

74.6 Conclusion

How might Holmyard's work be viewed in the light of the changes outlined above? Among his alchemical studies, his work on Geber[25] was of seminal importance, and like all sound scholarship, his research as a historian of science has provided a platform upon which other scholars have been able to build. J. R. Partington, himself a distinguished historian of chemistry and textbook author, judged that Holmyard's alchemical studies were of 'permanent value'. Noting that alchemists and early chemists believed that metals were composed of mercury and sulphur, Partington drew particular attention to Holmyard's finding that this theory, promoted by Geber, derived from a statement in Aristotle's 'Meteorology' (Partington 1959). Holmyard's 1957 book, *Alchemy,* was published in several languages, reprinted in 1990 and remains an important and readable survey of the subject. His obituarist in *Ambix* offers a useful overview of Holmyard's alchemical studies and comments that he added to 'our knowledge of this subject…its great names, its theories, its experiments and its apparatus'. His conclusion is that Holmyard brought about 'a greater interest in and an improved understanding and appreciation of the work and writings… of …Muslim chemists in general' (McKie 1960, p. 5).

Although Holmyard's school chemistry textbooks were written for syllabuses and examinations that differ greatly from those confronting students in the early years of the twenty-first century, they remain an important model for any author wishing to present chemistry, not simply as a discipline, but as an integral part of a wider cultural history. Reading some of his textbooks helps the reader understand that interest in the behaviour of materials and the desire to understand how they may be changed from one to another is at least as old as recorded history. At the end of a chapter entitled 'How Chemistry Arose', Holmyard invites the reader of his *Elementary Chemistry* to answer questions such as the following:

[25] Geber is the Latin equivalent of Jâbir, and the name was used by the author of an influential series of alchemical texts in the fourteenth and fifteenth centuries. It seems likely that this Latinised version of the name was adopted by a Western writer in order to heighten the authority of his own work: Holmyard's 'likely guess' (Holmyard 1957, p. 134) was that these later mediaeval works were written by a European scholar, possibly in Moorish Spain. Holmyard was able to identify the eighth-century alchemist Jâbir Ibn Hayyân who held a court appointment under the Caliph Harun al-Rashid. A lengthy examination of the 'Geber problem' is available at www.history-science-technology.com.

Mention some of the chemical facts and processes known to the Ancient Egyptians.

Explain how it was that the Muslims were able to establish chemistry on a sound basis.

Who was Paracelsus? What service did he render to chemistry?

Anyone who follows the historical journey that Holmyard sets out so skilfully in his texts will therefore travel with him to China, India and Arabia as well as within Europe, glimpsing other languages and cultures on the way and learning that it is all too humanly possible to take a wrong turning while firmly believing it is the way forward. The journey that he offered was much less familiar in the first few decades of the twentieth century than is the case today when societies have become much more multicultural and scholars more aware of non-Western contributions to the development of science.

Although Holmyard's evolutionary approach found limited favour, his advocacy did much to draw attention to the contribution that the history of science could make to school science education and to encourage teachers to incorporate historical components within their teaching. In doing so, the various attempts to 'humanise' school science teaching presented challenges and problems that have a contemporary international relevance as attempts are made to accommodate the history of science within the movement to promote the teaching of the 'nature of science'. As experience with *Harvard Project Physics* confirmed, addressing and overcoming these challenges and problems requires a substantial investment in science teachers' professional development that is grounded in collaboration between those whose scholarly expertise lies in the history of science and those teaching science in schools. That can succeed only if there is agreement about the contribution that the history of science can legitimately make to the science curriculum and if science teachers enable their students to think historically as well as scientifically. Fortunately, the form and content of that contribution are now matters not only for lively but well-informed debate but also for science curriculum development (e.g. Holton 2003; Matthews 1994/2014, 2009; Kokkotas et al. 2009).

References

Armstrong, H.E. (1903) *The Teaching of Scientific Methods and Other Papers on Education*, London, Macmillan.

American Association for the Advancement of Science. (1968), British Youth Swings Away from Science, *Science,* 15 March 1968, 1214–15.

American Association for the Advancement of Science. (1993), *Benchmarks for Science Literacy: Project 2061,* Oxford University Press, New York.

Badcock, W.C. & Holmyard, E.J. (1931), *Electricity and Magnetism for Beginners*, J.M. Dent, London.

Barraclough, F. & Holmyard, E.J. (1931), *Mechanics for Beginners*, J.M. Dent, London.

Bassey, M. (1960), A Field review of O-level Chemistry Textbooks, *Technical Education and Industrial Training*, 2 (12),11–15.

BJHS. (1990), Editorial, *British Journal for the History of Science,* 23, 1–2.

Blackwood's Magazine. (March 1916), *Editorial*, 407–18.

Bradley, J. (1988), *They Ask for Bread: an essay on the teaching of chemistry to young people*, North Ferriby, North Humberside, Privately published.

British Association for the Advancement of Science. (1856), *Report of the Annual Meeting,* John Murray, London.

Brock, W.H. (1973), *H.E.Armstrong and the Teaching of Science 1880–1930,* Cambridge, University Press, Cambridge.

Brock, W.H. (1989), Past, Present and Future. In M. Shortland and A. Warwick (Eds.), *Teaching the History of Science*, BSHS, Blackwell, Oxford, 30–41.

Brock, W.H. (1996), *Science for All*, Ashgate Publishing, Aldershot.

Brown, J. (1925), *Teaching Science in Schools,* University of London Press, London.

Browne, C.E. (1954), *Henry Edward Armstrong,* Harrison and Sons (privately printed), London. Re-issued in 1966 with a memoir of Browne as *Henry E. Armstrong and Charles E. Browne,* Harrison and Sons for Christ's Hospital, Horsham.

Bruner, J. (1960), *The Process of Education*, Harvard University Press, Cambridge, MA.

Brush, S. G. (1974), Should the history of science be X-rated? *Science*, 183, 1164–72.

Cassidy, D., Holton, G. & Rutherford, J. (2002), *Understanding Physics.* New York, Springer-Verlag-New York.

Cawthorne, H.H. (1930), *Science in Education: Its Aims and Methods*, Oxford University Press, London.

Clough, M.P. (2009), *Humanizing science to improve post-secondary science education,* Paper presented at the 10th IHPST Conference, Indiana, June 24–28th.

Cochrane, J.A. (1923), *Readable School Physics,* Bell and Sons, London.

Cochrane, J.A. (1925), *School History of Science*, Edward Arnold, London.

Coles, M. (1989), *Active Science,* Collins Educational, London.

Conant, J.B. (1947), *On Understanding Science: An Historical Approach*, Yale University Press, New Haven.

Conant, J.B. *et al.* (1957), *Harvard Case Histories in Experimental Science*, Harvard University Press, Cambridge, MA.

Council for Scientific Policy (1968), *Report of an Inquiry to Examine the Flow of Candidates in Science and Technology into Higher Education*, HMSO, London.

Dainton, F.S. (1971), *Science: salvation or damnation?* University of Southampton, Southampton.

DES(1989), *Science in the National Curriculum,* DES/WO, London.

Donnelly, J.F. (2002), The 'humanist' critique of the place of science in the curriculum in the nineteenth century and its continuing legacy, *History of Education*, 3 (6), 535–555.

Donnelly, J.F. (2004), Humanizing Science Education, *Science Education* 88 (5), 762–784.

Donnelly J.F., Buchan, A., Jenkins, E., Laws, P & Welford, G. (1996), *Investigations by Order: policy, curriculum and science teachers' work under the Education Reform Act*, Studies in Education, Nafferton, East Yorkshire.

Donnelly, J.F. & Jenkins, E.W. (2001), *Science Education: policy, professionalism and change,* Paul Chapman, London.

Donnelly, J.F. and Ryder, J. (2011), The pursuit of humanity: curriculum change in English school science, *History of Education*, 40 (3), 219–313.

European Commission (2007), *Science Education NOW: A Renewed Pedagogy for the Future of Europe,* Directorate General for Research, Brussels.

Fowles, G. (1937), *Lecture Experiments in Chemistry,* Bell and Sons, London.

Francis, K. (2004), *The Education of a Waldorf Teacher*, iUniverse, Inc., New York.

Freund, I. (1904), *The Study of Chemical Composition: an account of its method and historical development,* Cambridge University Press, Cambridge.

Glaser, O. (1924), On Humanizing Science, *The North American Review*, 219 (819), 230–239.

Graham, A.P. & Holmyard, E.J. (1935), *Elementary Botany*, J.M. Dent, London.

Guting, R. (2006), Humanizing the teaching of mathematics, *International Journal of Mathematics Education in Science and Technology*, 11 (3), 415–425.

Halliwell, H.F. (1966), Aims and action in the classroom, *Education in Chemistry,* 3 (15), 242–246.

Hansard (18 July 1916), 5 (84), col. 906

Harlen, W. (1996), *The Teaching of Science in Primary Schools*, David Fulton, London.

Hartley, H.B. (1971), *Studies in the History of Chemistry*, Clarendon Press, Oxford.

Harvard Project Physics (1970), *The Project Physics Course*, Holt, Rinehart and Winston, New York.

Holmyard, E.J. (1922), *Inorganic Chemistry: A Textbook for Schools and Colleges*, Edward Arnold, London.

Holmyard. E.J. (1924a), The Historical Method in the Teaching of Chemistry, *School Science Review,* V (20), 227–233.

Holmyard, E.J. (1924b), *The Teaching of Science*, Bell and Sons, London.

Holmyard, E.J. (1925a), *An Elementary Chemistry,* Edward Arnold, London.

Holmyard, E.J. (1925b), Historical Method in School Science Courses, *Journal of Education and School World*, July, 490.

Holmyard, E.J. (1925c), *Chemistry to the Time of Dalton,* Edward Arnold, London.

Holmyard, E.J. (1926), *Science: an introductory book*, J.M. Dent, London.

Holmyard, E.J. (1928a), *The Works of Geber by Richard Russell 1678: a new edition with notes by E.J. Holmyard*, J.M. Dent, London.

Holmyard, E.J. (1928b), *The Great Chemists*, Methuen, London.

Holmyard, E.J. (1930a), *Chemistry for Beginners,* J.M. Dent, London.

Holmyard, E.J. (1930b), *Physics for Beginners*, J.M. Dent. London.

Holmyard, E.J. (1930c), *Biology for Beginners*, J.M. Dent, London.

Holmyard, E.J. (1933), *A Junior Chemistry*, J.M. Dent, London.

Holmyard, E.J. (1939), *A Higher School Certificate Inorganic Chemistry*, J.M. Dent, London.

Holmyard, E.J. (1944), The Meaning of Science, *Endeavour*, 3, 124–5.

Holmyard, E.J. (1946), *Makers of Chemistry*, Clarendon Press, Oxford.

Holmyard, E.J. & Palmer, W.G. (1952), *A Higher School Inorganic Chemistry*, J. M. Dent, London.

Holmyard, E.J. (1957), *Alchemy*, Penguin, Harmondsworth, (reprinted by Dover, New York 1990).

Holton, G. (1952), *Introduction to Concepts and Theories in Physical Science,* Addison-Wesley, Cambridge MA.

Holton, G. (2001), *The Project Physics Course, Then and Now*. Paper delivered to the Annual Meeting of the History of Science Society, November 2001.

Holton, G. (2003), What Historians of Science and Science Educators Can Do for One Another, *Science & Education,* 12 (7), 603–616.

Jenkins, E.W. (1979), *From Armstrong to Nuffield: Studies in twentieth century science education in England and Wales,* John Murray, London.

Jenkins, E.W. (2000), Who were they? George Fowles (1882–1969), *School Science Review*, 82 (299) 87–89.

Jenkins, E.W. (2001),Who were they? F.W. Westaway (1864–1946), *School Science Review*, 83 (302) 91–94.

Journal of Chemistry and Industry (1931), 50 (8), 146.

Journal of Chemistry and Industry (1934), 53 (43), 882.

Journal of Chemistry and Industry (1939), 59 (3), 50–51.

Kaiser, D. (Ed.) (2005), *Pedagogy and the Practice of Science: Historical and Contemporary Perspectives,* MIT Press, Cambridge, MA.

Kinsman, J. (1985), *The Contribution of Eric John Holmyard (1891–1959) to science education in England and Wales*. Unpublished project submitted as part of the fulfilment of the requirements of the degree of M.Sc., Council for National Academic Awards.

Kipnis, N. (1998), A History of Science Approach to the Nature of Science: Learning Science by Rediscovering It. In W.F. McComas (Ed.), *The Nature of Science in Science Education: Rationales and Strategies*, Dordrecht, Kluwer Academic Publishers, 177–196.

Kokkotas, P.V., Malamista, K.S. & Rizaki, A.A. (2009), *Adapting Historical Knowledge Production to the Classroom*, Sense Publishers, Rotterdam.

Layton, D. (1984), *Interpreters of Science: A History of the Association for Science Education,* John Murray/ASE, London.

Livingston, R.W. (1916) *A Defence of Classical Education*, London, Macmillan.

Lodge, O. (1925) Introduction. In R.H. Murray, *Science and Scientists in the Nineteenth Century*, Sheldon, London.

Lowry, T.M.1 (1915), *Historical Introduction to Chemistry,* Macmillan, London.

Lynning, K.H. (2007), Portraying Science as Humanism- A historical Case Study of Cultural Boundary work from the Dawn of the 'Atomic Age', *Science & Education*, 16 (3–5) 479–510.
Matthews, M.R. (1994/2014), *Science Teaching: The Role of History and Philosophy of Science*, Routledge, New York.
Matthews, M.R. (2009), Teaching the Philosophical and Worldview Components of Science, *Science & Education,* 18 (6–7), 697–728.
Mayer, A.K. (1999), I have been fortunate. Brief report on the BSHS Oral History Project: the history of science in Britain 1945–65, *British Journal for the History of Science* 32, 223–235.
Mayer, A.K. (2005), When things don't talk: Knowledge and belief in the inter-war humanism of Charles Singer, 1876–1960, *British Journal for the History of Science*, 38 (3), 325–347.
McComas, W.F. (1998), *The Nature of Science in Science Education: Rationales and Strategies,* Kluwer Academic Publishers, Dordrecht.
McComas, W.F. (2008), Seeking Historical examples to Illustrate Key Aspects of the Nature of Science, *Science and Education* 17 (2–3), 249–263.
McCulloch, G. (1984), Views of the Alternative Road: The Crowther concept. In D. Layton, *The Alternative Road,* Leeds, University of Leeds Press, 57–83.
McCulloch, G. (1991), *Philosophers and Kings: Education for Leadership in Modern England,* Cambridge University Press, Cambridge.
McCulloch, G. (2006), Cyril Norwood and the English Tradition of Education, *Oxford Review of Education*, 32(1), 55–69.
McKie, D. (1960), Obituary of Eric John Holmyard, *Ambix*, VIII (1), 1–5.
McPhemon, A. (1969), 'Swing from science' or retreat from reason? *Higher Education Quarterly*, 22 (3), 54–273.
Mellor, J.W. (1902), *Higher Mathematics for Students of Physics and Chemistry: with special reference to practical work*, London, Longmans, Green.
Mellor, J.W. (1932), *Introduction to Modern Inorganic Chemistry*, Longmans, Green, London.
National Research Council (1996), *National Science Education Standards*, National Academy Press, Washington.
Natural Science in Education (1918), *Report of the Committee on the Position of Natural Science in the Educational System of Great Britain*, HMSO, London.
Norwood, C. (1922), Foreword to Holmyard (1922), *op.cit.*
Norwood, C. (1929), *The English Tradition of Education*, John Murray, London.
Ogston, A.G. (1972), Harold Brewer Hartley 1878–1972, *Biographical Memoirs of Fellows of the Royal Society*, 19, 348–326.
Partington, J.R. (1929), *Everyday Chemistry,* Macmillan, London.
Partington, J.R. (1937), *A Short History of Chemistry*, Macmillan, London.
Partington, J.R. (1939), *A College Course of Inorganic Chemistry*, Macmillan, London.
Partington, J.R. (1959), Eric John Holmyard, *Nature* 184, 1360.
Partington, J.R. (1961–4), *A History of Chemistry*, St. Martin's Press, London.
Perkin, W.H. & Lean, B. (1896), *An Introduction to the Study of Chemistry,* Macmillan, London.
Philbrick, F.A. & Holmyard, E.J. (1956), *A Textbook of Theoretical and Inorganic Chemistry*, J.M.Dent, London.
Pimentel, G.C. (Ed.) (1963). *Chemistry: An Experimental Science*, Freeman, San Francisco.
Rogers, E.M. (1961), *Physics for the Inquiring Mind: The Methods, Nature and Philosophy of Physical Science*, Princeton University Press, Princeton NJ.
Rudolph, J.L. (2002), *Scientists in the Classroom: the cold war reconstruction of American science education,* Palgrave, New York.
Sadler, M. (1909), Introduction to F. Hodson (Ed.) *Broad Lines in Science Teaching,* Christophers, London. Sadler is quoting John Ruskin.
Sherratt, W.J. (1980), *History of Science Education: an investigation into the role and use of historical ideas and material in education with particular reference to science education in the English secondary school since the 19th century,* PhD thesis, University of Leicester.
Simon, B. (1974), *The Politics of Educational Reform, 1920–1940*, Lawrence & Wishart, London.

Singer, C., Holmyard, E.J. & Williams, T.I. (1954–84), *A History of Technology*, 8 volumes, Oxford, Clarendon Press.
Singer, C. (1959), Tribute to E.J. Holmyard, *The Times,* 23rd October,17.
SMA (1957), *Science and Education,* SMA/John Murray, London.
SMA/AWST (1961), *Science and Education; A Policy Statement issued by The Science Masters' Association and The Association of Women Science Teachers,* John Murray, London.
Smith, E.F. (1914), *Chemistry in America: Chapters from the history of science in the United States,* D. Appleton & Co., New York.
Sneider, C. (2011), *Reversing the Swing from Science: Implications from a Century of Research.* Paper presented at the ITEST Convening on Advancing Research on Youth Motivation in STEM, September 9–11, Boston College.
Snow, C.P. (1959), *The Two Cultures and the Scientific Revolution*, Cambridge University Press, Cambridge.
Snow, C.P. (1964), *The Two Cultures, and A Second Look: an expanded version of 'The Two Cultures and the Scientific Revolution*, Cambridge University Press, Cambridge.
Solterer, J. (1972), Towards the humanization of economics, *Revue of Social Economy,* 30 (3), 394–400.
SSR (1925a), *Review of Holmyard's The Teaching of Science*. VI (24), 265–266.
SSR (1925b), *Review of Holmyard's An Elementary Chemistry*, VII (26), 140.
SSR (1930), *Review of Holmyard's[three volumes) Chemistry/Physics/Biology for Beginners,* XII (46), 190.
SSR (1933), *Review of Holmyard's A Junior Chemistry*, XV (57), 126.
Taylor, L.W. (1949/1959), *Physics: The Pioneer Science*, 2 vols, Dover publications, New York.
The Times (1959), *Obituary of E.J. Holmyard.* October 15th, 16.
Thomson, J.A. & Holmyard, E. J. (1934), *Biology for Everyman*, J.M. Dent, London.
Tilden, W. (1919), *School Science Review* 1 (1), 12.
Turner, D.M. (1927), *History of Science Teaching in England*, Chapman & Hall, London.
Vlaeminke, M. (2000), *The English Higher Grade Schools: a lost opportunity*, Woburn Press, London.
Watts, M. & Bentley, D. (1994), Humanizing and feminizing school science: reviving anthropomorphic and animistic thinking in constructivist science education, *International Journal of Science Education*, 16 (1), 83–97.
Westaway, F.W. (1929), *Science Teaching*, Blackie, London.
Westaway, F.W. (1942), *Science in the Dock: Guilty or Not Guilty?* Scientific Book Club, London.
Williams. T.I. (2002), Clifton and science. In N.G.L. Hammond (Ed.), *Centenary Essays on Clifton College*. Bristol, 195–212.
Williams, T.I. (2004), Entry for E.J. Holmyard, *Dictionary of National Biography,* Oxford University Press, Oxford.

Edgar W. Jenkins is Emeritus Professor of Science Education Policy at the University of Leeds where he was Head of the School of Education (1980–1984 and 1991–1995) and Director of the Centre for Studies in Science and Mathematics Education (1997–2000). From 1984 to 1997, he edited the research review journal *Studies in Science Education*. His principal books include *From Armstrong to Nuffield* (1979), *Technological Revolution?* (1986), *Inarticulate Science?* (1993), *Investigations by Order* (1996), *From Steps to Stages* (1998), *Learning from Others* (2000) and *Science Education, Policy, Professionalism and Change* (2001). He has recently coedited a volume published to mark the fiftieth anniversary of the founding of the Association for Science Education in 1963.

Chapter 75
John Dewey and Science Education

James Scott Johnston

75.1 Introduction

John Dewey is perhaps the most well-known philosopher of education of the twentieth century. His output was prodigious and he is distinctive for placing education at the centre of important philosophical discussions. Dewey has in addition a series of attentive accounts on traditionally philosophical topics, such as theory of knowledge, meaning, experience, reality, ethics, political and social theory and aesthetics—all of which count education as important. Dewey's most important educational treatises include *The School and Society* (1899), *How We Think* (1910), *Democracy and Education* (1916) and *Experience and Education* (1938), and the former three were in use as textbooks for hundreds of thousands of teachers in teacher-training programmes in America over several decades in the first half of the twentieth century. Dewey continues to have an influence on teacher training and, important for our purposes here, science education.

In this chapter, I will concentrate on three themes: Dewey's theory of experience and the role of reflection or thought, Dewey's theory of inquiry (scientific method) and Dewey's claims for science education. The three are interrelated and discussing Dewey's claims for science education presupposes discussing his theories of experience, reflection and inquiry. I will follow with a brief discussion of recent developments in science education that have invoked and used Dewey: constructivism and science education, science education and models of inquiry in the curriculum, and the teaching of science. I will finish the chapter with my assessment of the scholarship on Dewey and science education.

J.S. Johnston (✉)
Jointly Appointed Associate Professor of Education and Philosophy,
Memorial University, Newfoundland, Canada
e-mail: scott.johnston@queensu.ca

M.R. Matthews (ed.), *International Handbook of Research in History, Philosophy and Science Teaching*, DOI 10.1007/978-94-007-7654-8_75,

75.2 Brief Biographical Sketch

John Dewey was born in Burlington, Vermont, on April 22, 1859. He grew up in Burlington and attended the University of Vermont. Dewey taught high school for 2 years in Oil City, Pennsylvania, and a further half-year in Charlotte, Vermont, prior to attending graduate study in philosophy at newly minted Johns Hopkins University under advisor, George Sylvester Morris. Having graduated in 1884, Dewey began his university career at the University of Michigan (1884–1888; 1890–1893) and the University of Minnesota (1888–1889), followed by the University of Chicago (1894–1903) (where he ran the famed laboratory school) and, finally, Columbia University in New York (1904–1927) where, in addition to an appointment as professor of philosophy, he was also professor of pedagogy at Teachers College. Dewey married Alice Chipman in 1886, and together, they raised seven children, though two died at very young ages. Dewey would also adopt two children. Dewey travelled and lectured extensively, especially in later life, most famously in Russia, China, Turkey and Mexico. After retirement from active teaching in 1930, his already prodigious output increased; when he died in 1952 in New York City at the age of 92, he had published some 47 books and 1,500 articles.

75.3 Dewey's Philosophical Project

Dewey's philosophical project was to aid in the bringing of the attitudes, criticisms, methods and results of philosophers to bear on practical concerns (Middle Works (MW) 10, 47). As well, it was to make philosophy and philosophers a valued enterprise for the task of democratically associated living (MW 9, 91). Dewey famously claimed that philosophy was the 'criticism of criticisms' and that this ought to be put to work in social and cultural transformation (Later Works (LW) 1, 309). Central to his philosophical project was the 'method of intelligence' (MW 10, 45), variously understood as 'inquiry', 'problem-solving', 'how we think' and 'scientific method'. Though Dewey sometimes distinguished between these (largely on the basis of context), they were coterminous in the project of getting philosophy (and other disciplines) to bear on what Dewey considered 'the problems of men' (MW 10, 48).

75.4 Dewey's Educational Project

Formal education has many important aims and functions, but in terms of science and science education, this one stands out: education was for Dewey the chief means with which to inculcate in the species the problem-solving method of intelligence needed to overcome extant social problems. Having said this, other aims and functions of education are coterminous with the development of the method of intelligence. Social and cultural transmission of past matters of fact and the techniques to solve what Dewey refers to as 'problematic situations' (LW 12, 112) is one example. Growth, which is the natural end of individuals and the human species (and is sometimes said by Dewey to be *the* end of education (e.g. MW 9, 47–49; LW 13, 19)), is another. To utilize the method of intelligence is to draw on past matters of fact and techniques, and to grow is to transform oneself and one's environment in accordance with a set of problem-solving methods, as I will discuss further in the section on inquiry.

75.5 Deweyan Inquiry

Dewey discusses inquiry variously, and in several key texts and articles. What I will do here is provide a summary exegesis of his various claims, beginning with the relation between experience and inquiry, followed by a discussion of Dewey's claims for the stages of thought, the role of specifically scientific thinking in inquiry and the role of scientific inquiry in the social sciences.

75.5.1 The Foundation of Inquiry: Experience

Dewey rejects the empiricist notion that we assemble bits of information through our sense-perceptive faculties and reorder them as ideas in the mind. He also rejects the rationalist notion that ideas are innate or otherwise necessary and pure principles that bear down on our sense-perceptive apparatus to fashion objects of the mind. What Dewey offers is an account of experience in which we exist in a world of what he would call 'brute facts' and 'thats' (MW 3, 164), a world that is first felt rather than cognized (LW 5, 249). What we experience as felt is a whole, not an object. Nor is a felt whole a mere feeling. It is rather that the felt whole is what we experience—as an immediate quality. The human organism exists in a qualitative state, and when she experiences, she does so qualitatively, through feeling. Feeling, in turn, is dependent on generic 'traits of existence', which arise out of every encounter of the human organism with the world (LW 1, 308). These traits are 'qualitative individuality and constant relations, contingency and need, movement and arrest…' (LW 1, 308). Dewey adds to this list 'rhythm and regularity' and 'our constant sense of things as belonging or not belonging, of relevancy…' (LW 10, 198). Not every experience will have these traits as rich and defined as others. Most experiences are 'mundane'; they evince some traits of existence but not of sufficient quality to be notable. Those experiences that do, Dewey considers to be consummatory. A consummatory experience is one in which the experience is felt and taken as a qualitative whole, a unity and a totality.

Inquiry begins and ends in experience. Experience in turn supplies inquiry with the situations or events from which inquiry forms what Dewey calls 'thought' or 'reflection'. In the attempt to reproduce the event or situation that led to the consummatory experience, the human organism will be poised to attempt reflection. Dewey in one place calls this a 'judgment of appreciation' (LW 12, 177–178). Successful reflection will eventually lead to the isolation of the factors involved in the event, together with their control and prediction for further events. However, Dewey cautions we must not confuse the resultant appreciation which is felt, with the operations that bring the material into harmony. To do so is to hypostatize feeling into generalizations (LW 12, 179). Such, historically, is the case with the good, the true, the beautiful and other 'absolutes'.

Dewey discusses the way in which we are induced to reflect upon our experiences. This occurs when we have and undergo an unsatisfying experience when we expect a satisfying or otherwise unproblematic one; an experience that has a paucity or (as Dewey sometimes says) imbalance of the traits of existence and that does not lead to a qualitative whole. In the context of inquiry, Dewey calls the event or situation that concludes in an unsatisfying experience a 'felt difficulty' or 'indeterminate situation' (MW 6, 237; LW 12, 108). We then undertake inquiry to 'determine' the situation. When we are able to order and control the qualities or traits of the experience of an indeterminate situation, we form an experience that is complete, satisfying and qualitatively whole. Indeed, this is the basis of Dewey's most famous definition of inquiry: 'Inquiry is the controlled or directed transformation of an indeterminate situation into one that is so determinate in its constituent distinctions and relations as to convert the elements of the original situation into a unified whole' (LW 12, 108).

75.5.2 The Rise of Science and Scientific Thinking

The self-awareness of the role of reflection and its various techniques and stages obviously did not arise at once: it was the result of thousands of years of investigation (much of it haphazard or misleading) into the forces of nature, the constitution of material objects of use (technologies) and the increasing role of methods in ascertaining instrumental results. For much of this history, the establishment of methods and principles of science fell to philosophy, and Dewey characteristically concentrates his examination of the history of the rise of science and scientific thinking to developments within philosophy. Dewey's thesis is that a previously stable set of affairs (both individually and species wide) is upset, unsettled or otherwise rendered dubious. The tension leading to instability is the impetus for further questioning, investigation and inquiry, and this leads (if successful) to a readjustment, a resettlement. We understand this through recourse to a historical/developmental and evolutionary account of the rise of science—what Dewey calls 'the genetic method' (MW 1, 150; MW 2, 300–301). We have here the pattern Dewey would use in his discussion of inquiry generally—a pattern made famous in his discussion of the stages of thought in the two editions of *How We Think* (1910: 1932) and in his definition of inquiry as the settlement of an unsettled or indeterminate situation in *Logic: the Theory of Inquiry* (1938a).

Dewey's concentrated attention on the rise of scientific thinking through history is best exemplified in the early, extended essay, 'Some Stages in Logical Thought' (1903) together with his book, *Reconstruction in Philosophy* (1920). In his essay, Dewey notes four stages in the development of logical thinking: the magical thinking of the earliest societies, the dawning awareness of the need for method in ancient civilizations, the isolation and abstraction of principles and standards in ancient Greek (Hellenistic) societies and, finally, the rise of a scientific method in the seventeenth century and beyond (MW 1, 172–173). In *Reconstruction in Philosophy*, Dewey alters his stage account of the rise of scientific thinking, noting specifically religious, metaphysical and scientific eras. In the religious era, Dewey claims, 'Savage man recalled yesterday's struggle with an animal not in order to study in a scientific way the qualities of an animal or for the sake of calculating how better to fight to-morrow, but to escape from the tedium of to-day by regaining the thrill of yesterday' (MW 12, 80). In the metaphysical era, 'The growth of positive knowledge and of the critical, inquiring spirit undermined those in their old form…' (MW 12, 89). What was left, according to Dewey, was to 'Develop a method of rational investigation and proof which should place the essential elements of traditional belief upon an unshakeable basis; develop a method of thought and knowledge which while purifying tradition should preserve its moral and social values unimpaired' (MW 12, 89–90). Finally, in the scientific era proper, the recognition of the genetic method—a method at once empirical, experiential, and scientific, historical and developmental—is the best assurance of producing practically valuable investigative results (MW 12, 93).

75.5.3 The Stages of Thought

In *How We Think* Dewey outlined and then expanded upon the stages of thought. Thought for Dewey is synonymous with reflection and inquiry: to think is to pass through the stages or phases of inquiry, from a problematic situation first felt, to the articulation of a problem, to the imagination of various anticipated outcomes, to the actual testing of the results using the various techniques and operations of isolation, ordering and control and prediction of phenomena, and finally, to the objective settlement and felt resolution or satisfaction of the original problematic situation. Thinking or inquiry is in other words, loosely circular. Here, I will spell out the various stages of thought. Dewey first gives a summary estimation of the five stages of the '…complete act of thought' (MW 6, 236). 'Upon examination, each instance [of genuine thought] reveals, more or less clearly, five logically distinct steps: (i) a felt difficulty; (ii) its location and definition; (iii) suggestion of positive solution; (iv) development by reasoning of the bearings of the suggestion; (v) further observation and experiment leading to its acceptance or rejection; that is, the conclusion of belief or disbelief' (MW 6, 236–237).

All genuine thought begins with a problematic situation, which for Dewey manifests as 'a felt difficulty' (MW 6, 237). This is the experiential basis of all thinking: a genuine inquiry can only take place if a genuine problem is found and established. Genuine problems are those that are felt, rather than abstracted or given ready-made to students, and Dewey hammers this point home in all of his educational writings. Felt difficulty may take various shapes. It may exist as 'emotional disturbance, as a more or less vague feeling of the unexpected, of something queer, strange, funny, or disconcerting' (MW 6, 238). Again however, what is important is that 'observations deliberately calculated to bring to light just what is the trouble, or to make clear the specific character of the problem' are undertaken (MW 6, 238).

In the second stage of thought, a clearly articulated problem is set forth. It is imperative that the problem be set forth consciously, because the techniques and operations of further stages of inquiry cannot be properly marshalled unless 'the trouble--the nature of the problem—has been thoroughly explored' (MW 6, 238). Failure to fully articulate the problem prior to engaging in experimentation with ideas and techniques of order and control of variables leads to frustration and dead ends—a sort of 'mission creep' that very often nets results for no particular problem, leading to failure for inquiry. As Dewey puts it, 'The essence of critical thinking is suspended judgment; and the essence of this suspense is inquiry to determine the nature of the problem *before* proceeding to attempts at its solution. This, more than any other thing, transforms mere inference into tested inference, suggested conclusions into proof' (MW 6, 238–239, emphasis mine).

The third stage Dewey calls 'suggestion'. Suggestion is that idea in which 'the perplexity…calls up something not present to the senses....Suggestion is at the very heart of inference; it involves going from what is present to something absent. Hence, it is more or less speculative, adventurous…The suggested perception so far as it is not accepted but only tentatively entertained constitutes an idea' (MW 6, 239). In this stage, what Dewey elsewhere calls a 'dramatic rehearsal', wherein we reconstruct the situation in thought and attempt to control for the various conditions, takes place (LW 10, 81–82). We are literally putting forth scenarios of possible outcomes, through varying the conditions of experimentation in thought. Here, tentative hypotheses are formed, thought through and passed or rejected according to the results of the thought experiment.

In the fourth stage of thought, more extensive inferential operation is carried out on those hypotheses that bear fruit in the earlier stage. In the earlier stage, we produce ideas—anticipated consequences of possible measures of control. 'As an idea is inferred from given fact [isolated from the situation], so reasoning sets out from an idea…' (MW 6, 239). Dewey continues

> …intimate and extensive observation has [its way] upon the original problem. Acceptance of the suggestion in its first form is prevented by looking into it more thoroughly. Conjectures that seem plausible at first sight are often found unfit or even absurd when their full consequences are traced out. Even when reasoning out the bearings of a supposition does not lead to rejection, it develops the idea into a form in which it is more apposite to the problem…The development of an idea through reasoning helps at least to supply the intervening or intermediate terms that link together into a consistent whole apparently discrepant extremes. (MW 6, 240)

Finally, the fifth stage or the conclusion of the 'complete act of thought' is 'some kind of *experimental corroboration*, or verification, of the conjectural idea…If we look and find present all the conditions demanded by the theory, and if we find the characteristic traits called for by rival alternatives to be lacking, the tendency to believe, to accept, is almost irresistible' (MW 6, 240). This is the conclusion to the problematic situation that initially led to the complete act of thought. That is, it is the *existential and qualitative termination* of the original problematic situation. Such terminations are existentially satisfying; they are the terminations of complete experiences—the sort that lead us to further investigation of the traits of existence of which they are constituted.

Dewey's stages of thought have sometimes been taken to be fixed steps that all thinkers must ascend in linear fashion. This is a mistaken interpretation of Dewey's complete act of thought. Dewey's stages are recursive: whenever one of the steps is engaged, the entire procedure is engaged. The recursive nature of the stages allow for the possibility that a complete act of thought occur even if the stages are not engaged in a linear manner. This allows for inquirers to enter the procedure without beginning at the first stage. Likewise, inquirers may exit the procedure prior to closure.

75.5.4 The Logic of Inquiry and Scientific Thinking

As discussed with reference to Dewey's *How We Think*, inquiry begins with a 'felt difficulty' (MW 6, 237); in *Logic: the Theory of Inquiry*, Dewey calls this an 'indeterminate situation' (LW 12, 108). It is an existential *and* felt difficulty that initiates an inquiry. However, a doubt does not carry us far: 'Organic interaction becomes inquiry when existential consequences are anticipated; when environing conditions are examined with reference to their potentialities; and when responsive activities are selected and ordered with reference to actualization of some of the potentialities, rather than others, in a "final existential situation," is inquiry properly speaking, begun' (LW 12, 111). A felt difficulty and indeterminate situation must be followed by a judgment for inquiry proper to be undertaken, a judgment that this situation is problematic (LW 12, 111–112).

Once a problem has been identified and a judgment rendered, the investigator searches out 'the constituents of a given situation which, as constituents, are settled' (LW 12, 112). We begin with these because they are settled (Dewey uses the example of the location of aisles and exists in a hypothetical fire at an assembly hall). The first step in the determination of a problem is to assemble these settled constituents in observation. Dewey says 'A possible relevant solution is then suggested by the determination of actual conditions which are secured by observation' (LW 12, 113). This solution, in turn, becomes an idea: anticipated consequence that is then carried out in practice (LW 12, 113). Observed facts are existential, in that they directly bear on the phenomena to be examined and/or tested; 'ideational subject matters' are of conceptual import. They concern other ideas and, specifically, the way these relate to one another (LW 12, 115). 'Ideas are operational in that they instigate and direct further operations of observation; they are proposals and plans for acting upon existing conditions to bring new facts to light and to organize all the selected facts into a coherent whole' (LW 12, 116). Existential facts and ideational subject matters are to be operationalized—tested out in an existential situation. Once the investigator has established a set of anticipated consequences, she tests these. This involves submitting ideas, as hypotheses, to evaluation on the basis of documented, existential results. Those hypotheses that 'pass the test' are confirmed; those that do not are jettisoned or reconstructed. The determination of a previously indeterminate situation—a situation that is felt, rather than abstracted—is the termination proper of inquiry.

This logic of inquiry applies equally to both common-sense and scientific inquiries. What distinguishes these is not the general method or pattern of inquiry. In both, we experiment; we feel an indeterminate situation; we articulate a problem; we develop anticipated consequences that are then formed into hypotheses; and we test these hypotheses in existential settings. Successful hypotheses are those that satisfy or otherwise determine an indeterminate situation. However, scientific inquiries differ from common-sense inquiries in a number of ways, some of more and some less import; the greatest difference, aside from the contexts in which scientific inquiries operate (often under laboratory or otherwise rigorously controlled environmental conditions), is the level of abstraction common to them (LW 12, 119). Common-sense inquiry and the conclusions it develops very often end in habit formation—a stock of habits is built up that we then use in solving our day-to-day or otherwise mundane problems. These habits are used as well in scientific inquiry; they provide the basis for further psychomotor (in the case of experimental manipulation of phenomena) and conceptual (in the case of objects under investigation) development. However, scientific inquiry must also avoid as much as possible becoming routine, as it then promotes the tendency to complacency, with the result that it overlooks crucial steps in experimentation or changes in phenomena. Even time- and experience-tested habits, including psychomotor skills and habits of thought, must be amenable to reconstruction in scientific inquiry.

It will do to examine some of the differing features of scientific inquiry. Here, I will look at four critical features of scientific inquiry that are self-consciously articulated in Dewey's *Logic: the Theory of Inquiry*: induction and deduction; the nature of propositions; theories, laws, and causality; and the role of mathematics and symbols. In terms of induction and deduction, Dewey tells us:

> Whatever else the scientific method is or is not, it is concerned with ascertaining the conjunctions of characteristic traits which descriptively determine kinds of relation to one another and the interrelations of characters which constitute abstract conceptions of wide applicability…The methods by which generalizations are arrived at have received the name "induction:" the methods by which already existing generalizations are employed have received the name "deduction…." Any account of scientific method must be capable of offering a coherent doctrine of the nature of induction and deduction and of their relations to one another, and the doctrine must accord with what takes place in actual scientific practice. (LW 12, 414)

With induction, we generalize a set of common characteristics, features or attributes. We take the conclusions from those generalizations and we infer what must be the case on the basis of these. Dewey calls the typical understanding of induction 'a psychological process' in which we are 'induced to apprehend universals which have been necessarily involved all the time in sense qualities and objects of empirical perception' (LW 12, 419). In calling the typical understanding of induction a 'psychological process', Dewey means to distinguish his understanding of it from this. Whereas induction is typically understood as the movement of particular to universal, this is heavily qualified by Dewey. 'Induction I take to be a movement from facts to meaning; deduction a development of meanings, an exhibition of implications, while I hold that the connection between fact and meaning is made only by an act in the ordinary physical sense of the word act, that is, by experiment involving movement of the body and change in surrounding conditions' (MW 13: 63). Elsewhere, Dewey says induction is an 'existential determination' (LW 12, 478); he also calls it a determination of meaning(s) (MW 13, 63). 'They make possible the operation of mathematical functions in deductive discourse' (LW 12, 478). In induction, we grasp what is general in the existential situation through examination of the sense qualities of that situation. Dewey calls the generalizations we develop in induction 'generic propositions' (LW 12, 253). These are propositions of kinds or classes. For example, the generalization of the class, 'liquid', consists in all of the like phenomenal attributes or characteristics that are grouped together under this rubric.

In induction, we have 'the complex of experimental operations by which antecedently existing conditions are so modified that data are obtained which indicate and test proposed modes of solution…' (LW 12, 423). Induction draws generalizations from the existing phenomena culled from the existential situation, which are then transformed into hypotheses that are actively brought to bear on the existing phenomena. This generalization, or generic proposition, then takes the form of an 'If-then' statement—what Dewey calls a 'universal conception' (LW 12, 253). In deduction, on the other hand, we make inferences on the basis of the conclusions of hypotheses we generate (LW 12, 422). There is thus 'a functional correspondence' between induction and deduction. 'The propositions which formulate data must, to satisfy the conditions of inquiry, be such as to determine a problem in the form that indicates a possible solution, which the hypothesis in which the latter is formulated must be such as operationally to provide the new data that fill out and order those previously obtained' (LW 12, 423).

As we have noted, if-then conceptions are claims about what will happen to phenomena under certain circumstances. Though they make claims on behalf of phenomena, they are symbolic, rather than referring to phenomena through existential operations. There is yet another class of propositions, a class that operates at a level of abstraction from these; Dewey calls these 'abstract conceptions' (LW 12, 258). Abstract conceptions help to regulate our universal conceptions. As such, these operate *deductively*, rather than inductively. An example will be helpful here. Consider the following if-then claim:

> I hypothesize that all liquids of the kind (an existential proposition) H_2O will evaporate (an abstract concept) over time T when the temperature (an abstract concept) rises above 100 C.

This hypothesis relies on abstract concepts: evaporation and temperature. Abstract concepts do rely on propositions for their content (e.g. we need to know what physical characteristics water consists of at a given temperature); however, the concept is itself an abstraction that relates only to other concepts (such as evaporation). As such, we can assess abstract concepts in one of two ways: the first is how well they hang together with other abstract concepts (such as the case with evaporation and temperature); the second is how well they help to generate working generic propositions (such as the hypothesis above). When we assess abstract concepts in terms of their ability to relate with one another, we are testing deductively.

Some conceptions are simple and concrete. These Dewey calls conventions. It is a convention, for example, not to place your open hand over a flame. Other conventions consist in particular propositions we use for carrying out experiments; for example, 'when attempting to ascertain the degree of evaporation of water in your experiment, be sure to fill the beaker half way for each attempt to maintain consistency across experiments'. However, though they are obviously important, conventions serve scientific inquiry only in the most mundane matters. Far more important for scientific inquiry are 'hypotheses and their meanings', which are 'developed in ordered discourse, observation and assemblage of data…'. Otherwise, 'observation and assemblage of data are carried on at random' (LW 12, 428). Dewey calls these clusters of hypotheses 'theories' (LW 12, 428). Theories rely on both abstract and generic conceptions and are formed when sets of these are banded together into a general claim about a range of phenomena. So, for example, the theory of natural selection encompasses both abstract concepts (various concepts relating to organelles, tissues and organs; various concepts of change across species; various environments), together with universal conceptions (hypotheses that must have been provable for a theory of natural selection to arise) and existential propositions (classes of fauna used in formulating universal conceptions). Laws concern interactions and interrelations that allow a 'comprehensive system of characters'. This in turn allows for 'ordered discourse' to be possible (LW 12, 439). So, for example, Boyle's Law, Charles's Law, Henry's Law and other gas laws are laws precisely because they concern the interactions of the content of the qualitative traits that determine the relations (say, increases or decreases in volume or pressure of gases, gases coming out of solution, etc.) under investigation.

Causation is a particular type of regularity. It is a means of instituting a single, unique, continuous history of events under investigation, rather than a claim for a fixed and final sequence of occurrences (LW 12, 440). For example, consider the law

> All liquids of the kind (an existential proposition) H_2O will evaporate (an abstract conception) over time T at temperature (abstract conception) X, if certain circumstances hold (result of universal conception).

The law denotes the events that take place when qualitative traits are so ordered through propositions and conceptions to render them amenable to a continuous historical series.

Dewey is keen, however, to insist on the absence of fixity or finality with respect to laws or causality. 'The fallacy vitiating the view that scientific laws are formulations of uniform unconditioned sequences of change arises from taking the function of the universal [if-then] proposition as if it were part of the structural content of the existential [classes] propositions' (LW 12, 439). If we mistake the universal conception (the proposition that tells us what we can expect under existential conditions) with the propositions of classes actually used in carrying out the particular experiment, we will mistake regularity under strict scientific conditions for a fixed and final regularity.

Finally, scientific inquiry uses symbols and mathematics to a far greater degree than other forms of inquiry. These are tools and have arisen out of the various attempts at solving existential problems (LW 12, 392). Symbols and mathematics operate at a realm distinct from the existential and the generic; indeed, they are wholly abstract. 'When, however, discourse is conducted exclusively with reference to satisfaction of its own logical conditions, or, as we say, for its own sake, the subject-matter is not only nonexistential in immediate reference but is itself formed on the ground of freedom from existential reference of even the most indirect, delayed, and ulterior kind. It is then mathematical' (LW 12, 393). However, symbols and mathematics are not fixed and final things in themselves, nor are they pure forms in some Platonic universe. They *are* abstract, and they do *not* generally participate in existential contexts; however, they have both their genesis and their purpose(s) in existential inquiries. Not only this, but they must operate to generate workable universal conceptions that can then order existential phenomenon, as Dewey says, 'The necessity of transformation of meanings in discourse in order to determine warranted existential propositions provides, nevertheless, the connecting link with the general pattern of inquiry' (LW 12, 393).

75.5.5 Scientific Inquiry and Social Science

Of central import to scientific inquiry is that it be of use in solving human problems. All inquiry takes place in social contexts, the contexts of human relations (LW 12, 481). What Dewey thinks is wrong with certain scientific research programmes is the failure to reinsert the results of scientific experimentation back into the existential conditions out of which all inquiry arises. Certain fields and programmes have become so esoteric as to have fractured the connection between findings and the existential situations out of which these findings emerged. (Logical positivism is one example of this.) This is to be deplored for Dewey; scientific research has an obligation to help aid in the solution to social problems, nowhere more so than with respect to the social sciences. Classes in particular are often treated as if they were universal conceptions, with the result that they are rendered fixed and final. In these situations, 'At best, inquiry is confined to determining whether or not objects have the traits that bring them under the scope of a given standardized conception—as still happens to a large extent in popular "judgments" in morals and politics' (LW 12, 264). Some of this state of affairs is due to the faulty methods which the social sciences pursue. Dewey admits the social sciences do not as yet possess the sophistication of methods and techniques common to the physical and natural sciences.

Dewey puts the matter this way: 'The question is not whether the subject-matter of human relations is or can ever become a science in the sense in which physics is now a science, but whether it is such as to permit of the development of methods which, as far as they go, satisfy the logical conditions that have to be satisfied in other branches of inquiry' (LW 12, 481). The existing methods of the social sciences have yet to allow for the logical conditions of the particular inquiries therein to be as rigorously developed and articulated as those in the physical and natural sciences.

However, it should not be concluded that the logical conditions themselves are different between the two kinds of inquiry. It is rather that it is much more difficult to convert the indeterminate situation in a social scientific inquiry into a determinate one than it is in a physical scientific inquiry, because of the complexity of social subject matters and the limitations of workable techniques and tools available to social scientists (LW 12, 485). Problems articulated in social scientific inquiry are 'gross' and 'macroscopic' in distinction to those in physical scientific inquiries (LW 12, 414). And though Dewey reminds that it is important to recognize and deal with 'the physical conditions and laws of their interactions', this is obviously not enough to go on for social scientific inquiry (LW 12, 486). New methods and techniques have to be developed to cope with the unique nature of social scientific subject matters. This cannot be a reductive social science, in which the techniques of the physical and natural sciences are transposed onto social scientific subject matters. 'The assumption that social inquiry is scientific if proper techniques of observation and record (preferably statistical) are employed…fails to observe the logical conditions which in physical science give the techniques of observing and measuring their standing and force' (LW 12, 492). Dewey continues, 'Any problem of scientific inquiry that does not grow out of actual…social conditions is factitious; it is arbitrarily set by the inquirer instead of being objectively produced and controlled. All the techniques of observation employed in the advanced sciences may be conformed to, including the use of the best statistical methods to calculate probable errors…and yet the material ascertained be scientifically 'dead', i.e., irrelevant to a genuine issue, so that concern with it is hardly more than a form of intellectual busy work' (LW 12, 492–493).

75.6 Inquiry and Science Education

In addition to Dewey's claims for the importance of science and scientific inquiry in his philosophical and logical works, Dewey had a good deal to say about the role of scientific inquiry in education (see Chap. 42). As well, an increasing number of science educators have turned to Dewey to formulate better pedagogies and curricula for science education. Here, I will discuss Dewey's claims for the role of scientific inquiry in science education; I will then turn to some of the more recent developments in science education conducted along Deweyan lines.

75.6.1 Science and Science Education

Science (and mathematics) education poses a unique and complex predicament for educators: the heightened and abstract nature of scientific findings (including the theories and laws of natural and physical science) are an impediment to learning. What makes scientific education a daunting task is not the lack of innate capabilities of children to master increasingly abstract conceptions, or various propositions; it is the haphazard way in which much science education is taught, together with the differences in background knowledge and techniques of students. A child's genuine inquiry cannot begin with such reified conclusions; rather, it must begin with simple observations and manipulation of the environment present-to-hand. 'What the pupil learns he at least understands. Moreover, by following, in connection with problems selected from the material of ordinary acquaintance, the methods by which scientific men have reached their perfected knowledge, he gains independent power to deal with material within his range, and avoids the mental confusion and intellectual distaste attendant upon studying matter whose meaning is only symbolic' (MW 9, 228). Rather than producing experts in scientific methods, the point of science education is to familiarize students with 'some insight into what scientific method means than that they should copy at long range and second hand the results which scientific men have reached' (MW 9, 229).

Dewey recommends what he calls the 'chronological method' in educating children to scientific inquiry. This is the method '…which begins with the experience of the learner and [that] develops from the proper modes of scientific treatment…' (MW 9, 228). Children are led up from simple observations and conclusions about the workings of their environment, through more sophisticated analyses and syntheses of isolated natural and physical phenomena. In these analyses and syntheses, children are encouraged to develop and apply the various forms of propositions and conceptions involved in actively ordering and controlling specific traits of phenomena under experimental circumstances. So, for example, what begins as a simple experiment to identify the conditions under which water evaporates in a puddle adjacent to the school (conditions such as a sunny day, increased temperature) becomes, as the student ages and is introduced to increasingly sophisticated techniques, an examination of the evaporation of water under strict laboratory conditions, with the use of such techniques as classes (liquids), universal conceptions (experimental hypotheses) and abstract conceptions (temperature, evaporation) and laws for liquids and gases, as well as tools for the examination of tendencies amongst large groups of phenomena, such as mathematics and statistics. Dewey surmises that if children were taught to use the 'chronological method' from the beginning of their formal education and consistently thereafter, much of the confusion that occurs in increasing the level of abstraction in science education would gradually diminish.

In order for students to engage experimentally with phenomena, they must actively engage the world around them. They must experiment with materials in various way and order and control traits of existential phenomena, beginning in a trial-and-error manner. This requires that the material to be tested have some connection to the child's life beyond the classroom. Otherwise, genuine problems—problems that are felt rather than deduced or prescribed—will not materialize. One of the best ways for this to occur is to begin scientific inquiry with available technologies and with existing social problems (MW 9, 232). Of technologies, Dewey says, 'The wonderful transformation of production and distribution known as the industrial revolution is the fruit of experimental science. Railways, steamboats, electric motors, telephone and telegraph, automobiles, aeroplanes and dirigibles are conspicuous evidences of the application of science in life. But none of them would be of much importance without the thousands of less sensational inventions by means of which natural science has been rendered tributary to our modern life' (MW 9, 232). Of social problems, Dewey has in mind the problems common to children at their particular developmental age and stage, as well as the particular problems of circumstance and context—including the barriers children face, such as those of race, class, gender and geography (MW 9, 91).

Properly conducted, inquiry in science education will begin with a 'felt difficulty'; an 'indeterminate situation' that then stimulates genuine interest. This is the crucial stage for inquiry in science education and, indeed, inquiry in all contexts. If genuine interest is not captured, any inquiry that results will be an externally motivated one certain to result in poor focus and haphazard conclusions. This problem is the child's, not the teacher's. It is a waste of time and energy for the teacher to simply pronounce on what problem the student will begin with, if the student hasn't come to see this problem on, and as, his or her own. Only with genuine interest can a problem be properly articulated so that it can satisfy the rigours of inference and testing that follow.

Once an articulated problem is produced, anticipatory consequences of acting on phenomena are put forth. This is an imaginative stage or phase of scientific inquiry, in which the student thinks through various possible courses of action. Upon deciding a course of action, the student will make use of tools of inference (as in formulating hypotheses of the 'If-then' sort, to test out specific anticipated outcomes; deduction, especially in terms of abstract concepts; induction and existential propositions of classes (all of this class or only some or none of this class). These are tools in the experimental phase of the inquiry that are put to work in order to achieve the desired outcome. Depending on the nature of the experiment and the tools and techniques available to the students, more or less attention will be spent on making this phase of the experiment and the tools of inference that belong to its self-conscious. Some of these tools and techniques can be taught and discussed; however, to be of value to the student in her experimentation, they must be developed and worked through *in* the experimentation.

Finally, if the experiment is successful, resolution or closure of the indeterminate situation takes place. Again, this is a felt, as opposed to abstract, resolution. Successful termination of scientific inquiry does not merely end in the 'right' result, or to the satisfaction of the teacher. On the objective side, it ends with the successful settlement of the indeterminate situation; on the subjective side, it ends when the student experiences a qualitative closure of the situation or event. Unless and until this qualitative closure or termination takes place, inquiry is not complete. And if inquiry is not complete, the impetus for further investigation that arises as a result of the satisfaction of having closure of an indeterminate situation is denied. Closure is the 'aha' moment when an experience is at its most satisfying. And this is the moment of genuine growth. To deny or otherwise not conclude with this moment is to blunt the motive force for future scientific inquiries.

However, Dewey recognizes that these changes to the structure and content of science education are not, by themselves, enough; differences in background knowledge are another matter entirely. These cannot be so easily remedied and require changes in existing social and political realities—realities that have resulted in the barriers of race, class and gender and lack of education that keep some disenfranchised (MW 9, 91). Unfortunately, individual schools are hampered in their ability to mitigate circumstances such as poverty, racial and gender bigotry and socioeconomic divides. Inquiry, though inestimably important to students' intellectual development and their capacity to solve problems, needs to be heavily supplemented with cooperative social institutions and programmes if it is going to overcome the paucity of genuine intellectual growth that results from social barriers and divides.

75.6.2 Deweyan Inquiry and Constructivism in Science Education

Dewey has been a central figure in the great mass of scholarship written on behalf of constructivism in science and science education (Von Glaserfeld 1984, 1998; Phillips 1995; Garrison 1995, 1997a; Vanderstraeten 2002; Kruckeberg 2006; Gordon 2009). While some are content to note Dewey's historical influence on later philosophers (such as Richard Rorty) is squarely in line with constructivist beliefs and others are content to point out that Dewey's anti-metaphysical insistence in matters of knowledge is of a piece with their constructivist values, still others have gone further and developed Deweyan accounts of constructivism. Here, I will examine briefly constructivism in science education, noting how and where Dewey is invoked. I will then turn to a very brief exposition of three accounts of Dewey in the name of constructivism and science education: Ernst Von Glasersfeld's 'radical constructivism', Jim Garrison's 'social pragmatic constructivism' and Raf Vanderstraeten's 'transactional constructivism'.

It will do to outline briefly the thinking behind the constructivist movement in education. Constructivism is a label for a variety of various, like-minded models of cognition and knowledge that share a history, as well as a central tenet: the rejection of metaphysical realist and empiricist theories of knowledge, often in favour of developmental and transactional accounts that stress the organism of person's own role in knowledge acquisition (LaRochelle and Bednarz 1998, p. 7). In education there are 'radical constructivisms', 'pragmatic social constructivisms', 'didactic constructivisms' and 'poststructuralist constructivisms', amongst others. Dewey's involvement in constructivism comes largely as a result of his own rejection of metaphysical accounts of knowledge creation and production. Inquiry in constructivist accounts of science education invoking Dewey generally stresses the anti-metaphysical and anti-dualist dimensions of his thinking in their respective projects for science education. However, some thinkers have been compelled to go further and investigate the particular areas where Dewey's account of knowledge creation and production reaches beyond complementarity with existing understandings of constructivism in science education to form a distinctive constructivism in its own right.

Both Garrison and Ernst Von Glasersfeld draw on Dewey in their respective understandings of constructivism, and it will do to discuss their differences. Von Glasersfeld considers himself a 'radical' constructivist, a position Von Glasersfeld says he owes to Kant, Piaget and the notion of 'variability'—the idea that all conceptual schemes are premised on their utility or purpose (Von Glaserfeld 1984, p. 12). Von Glasersfeld appropriates certain of Kant's understandings to elaborate his own articulations of conceptual schemes. For example, Von Glasersfeld appropriates Kant through saying his understanding of reciprocity (as discussed by Kant as the Third Analogy of Experience) is beneficial to understanding how shared conceptual schemes develop. In the Third Analogy, 'All substances, insofar as they can be perceived in space as simultaneous, are in thoroughgoing interaction' (Kant 1998, p. A 211). This has obvious utility to Newton's third law of motion. And it has utility as well to Von Glasersfeld, who quotes Kant approvingly; 'It is manifest that, if one wants to imagine a thinking being, one would have to put oneself in its place and to impute one's own subject to the object one intended to consider…' (Von Glaserfeld 1998, p. 124). This 'reciprocity', analogized further from Kant's Third Analogy, is said by Von Glasersfeld to be akin to how we develop our shared, conceptual schemes. Von Glasersfeld does not intend to claim Kant's understanding and use of reciprocity as completely his own; he is rather content to accept the analogy and turn to empirical factors such as adaptation to understand constructivism. His is a "cognitive constructivism," which also has its locus in the seminal works of Piaget. Yet, this 'cognitive constructivism', as a theory invoking the transcendental substrates of Kant, sharply distinguishes Von Glasersfeld's project from empirically inclined projects such as Garrison's (Grandy 1998, p. 115).

Garrison's account of Dewey's 'pragmatic social constructivism' presents Dewey as a social behaviourist who is inimical to mental representations (Garrison 1995, p. 717). While Garrison clearly endorses a Dewey that evinces a strong role for natural (non-mentalistic) conceptions, his endorsement does not extend to wholly subjective or mentalistic representations, which he sees operating in Von Glasersfeld's account of 'radical constructivism'. Garrison's Dewey stresses language as a social construction, together with 'dialogicality and multiple authorship' (Garrison 1995, p. 727). Unlike Von Glasersfeld's understanding of constructivism, Dewey's account is anti-dualist and anti-subjectivist (Garrison 1997a, p. 305). Indeed, Dewey's theories of mind and meaning were *entirely* behaviourist (Garrison 1995, p. 725; 1997a, p. 308), and mind for Dewey was entirely a social construction. Social behaviour—the behaviour existing between and within social units—transforms organic behaviour as a result of (shared) language into meaningful behaviour (Garrison 1997a, p. 308). Garrison claims that rather than focusing on traditional epistemic tasks, education and science education in particular should be attempting to construct stronger models of dialogicality and promote active listening.

Vanderstraeten's account of Dewey's 'transactional constructivism' begins with an analysis of Dewey's organic account of coordination and perception (Vanderstraeten 2002, p. 236). This leads to an analysis of habit, to thinking and to knowledge formulation and the ways these form an integral whole. Communication (as with Garrison) is key in all of this. Education is a means of communication and education is 'a participatory, co-constructive process' (Vanderstraeten 2002, p. 241). Objects and practices acquire shared meaning because they are part of larger and shared sets of experiences (Vanderstraeten 2002, p. 241). Knowing is an active construction that takes place in the organism-environment transaction. This Vanderstraeten contrasts to Von Glasersfeld's 'radical constructivism', which he seems to see as insufficiently 'deep' to thwart the accusations that it is one more example of an account of correspondence to reality (Vanderstraeten 2002, p. 243). Garrison and Vanderstraeten thus both share an antipathy towards mentalistic representations and a regard for a 'transactional' accounting of constructivism in education.

It is difficult to evaluate the various constructivisms developed in Dewey's name. They clearly concentrate on central themes and concerns of Dewey and to do so is correct. The focus on transaction of the child and environment and the social dimension(s) of experience are vitally important to any further pedagogy of science, as is the stress on communication and dialogue. Beyond this, it is unclear how helpful it is to include Dewey in the pantheon of constructivists, let alone develop a constructivism from his various educational and philosophical writings. While Dewey shares many of the intellectual proclivities of other constructivists (including constructivist science educators), it remains to be seen what tangible benefits to science education come from including Dewey in this pantheon. Much if not most of the work in constructivism is content to present accounts of Dewey that do little to advance our understanding of him beyond what has elsewhere been articulated.

I attribute much of this lack of advance to the superficial nature of the readings of Dewey. In my opinion, seldom do constructivists dig deep enough to reveal the connections between Dewey's theories of knowledge and logic and his theory of aesthetic experience, to say nothing of his social and political writings (Garrison is the exception, here). More problematically, running Dewey together with Kant and Piaget as Von Glasersfeld does serves to mask tremendous underlying differences in their respective epistemologies: if Kant, as a transcendentalist, was a constructivist and Dewey, as a naturalistic empiricist, was a constructivist, it becomes a huge task to explain what foundation constructivism rests upon. Sadly, most constructivists (Garrison is again the exception) do not probe the foundations, and the resultant constructivisms they champion are but castles made of sand (see Chap. 31).

75.6.3 Dewey and the Science Curriculum

Dewey's written output while serving as the head of the University of Chicago Laboratory School (1896–1903) is well known to most educators. It includes *School and Society* (1899) as well as numerous monographs on various aspects of pedagogy and curriculum. Additionally, historical, biographical and philosophical work on the Laboratory School (Camp Mayhew and Camp Edwards 1936; Wirth 1979; Tanner 1997; Johnston 2006) has invoked inquiry directly. Recently, curriculum leaders and scholars have developed novel understandings of science education with a strong role, if not a focus, for inquiry to play. I will discuss some of these further.

Inquiry has become a dominant theme in science education (Rudolph 2005, p. 43). Aside from its logical features, this generally extends to 'hands-on' manipulation of objects, the emphasis on 'real world' activities and the use of associated technologies. As well, projects, where a number of related exercises are undertaken as part of a larger curricular whole, are often stressed. The continuity between a child's past experiences and those the teacher wishes the child to undergo is also maintained (Howes 2008, p. 538). Models of science education for early childhood education that propose drawing on a child's basic impulses to create further situations for the development of intelligent dispositions and skills have also been developed (Howes 2008, p. 538). Beyond this, more specific accounts of Dewey's holism and particularly how science and science education merges with art and aesthetic experience in an organic accounting have been offered. These latter accounts have been developed simultaneously by philosophers of education (Garrison 1997) and curriculum scholars such as those working at the Dewey Ideas Group at Michigan State (Wong and Pugh 2001; Girod and Wong 2002; Pugh and Girod 2007). All stress the importance of inquiry in regards to science education.

Garrison stresses the importance of aesthetic experience and particularly the precognitive background we draw upon when we inquire (1997b, p. 101–102). The logics that we use when we solve educational problems rely on this precognitive background and inform them. Garrison's point is to underscore the situation-specific nature of logic; there is no one logical method right for all situations; the situation itself will very often dictate what logical processes are to operate (Garrison 1997b, p. 98). Attention to the background conditions of experience, as well as the particular experience a child forms, is thus vital for science educators if they are to gauge successfully the student's learning.

Wong, Girod and Pugh also stress the precognitive dimension of experience. This drives instrumental understanding and forms the context out of which instrumental understanding operates (Girod and Wong 2002, p. 200). Like Garrison, these scholars concentrate on the motivation for learning in the first place, rather than the specific logical steps or processes undertaken. This necessitates concentrating on the experiences students have, rather than the outcomes of the particular lesson. A focus on the relationship between the concepts and ideas developed in the lesson and the experience the student has, together with attention to the satisfaction and interest inherent in genuine experiencing, is key (Pugh and Girod 2007, p. 14–16). Aesthetic experience is at once the highest and deepest of the forms of experience had and undergone. Helping to bring a child to have an aesthetic experience involving scientific experimentation is to ensure that genuine learning is taking place. Doing so releases the 'transforming' and 'unifying' elements of aesthetic understanding (Girod and Wong 2002, p. 208). The satisfaction had in having and undergoing the aesthetic experience of a scientific experiment through all of its phases is the basis for the development of real, as opposed to merely rote, knowing. Thus, in crafting science lessons, aesthetic experience must be borne in mind. This is not to say that other forms of experience are valueless; however, facilitating an aesthetic experience is the surest means to the 'teachable moment' that teachers value above all.

75.6.4 Dewey and the Teaching of Science

Problem-solving, discovery and inquiry methods are popular and well represented in theories of teaching science, in part due to Dewey's role in connecting his theory of inquiry to educational situations and events (Glassman 2001). All of these methods insist on taking the needs, desires, attitudes and developmental ages and stages of students as primary in the facilitation of science education, rather than teaching the discipline as a coherent subject matter (Eshach 1997). Concentrating on students' experiences rather than on the dissemination of the subject matter, together with a focus on the active role played by students in investigating natural phenomena, is paramount. This insists the teacher manipulate the environment to help facilitate the students' experiences such that they can actively inquire, in the manner of the stages of inquiry Dewey discusses in *How We Think* and elsewhere.

This facilitation of experiences also insists on conducting science in a manner similar to how scientists conduct it themselves, in their experimental investigations. Science, as Dewey insists, does not occur in isolation: nor should science education. The scientific investigation of natural phenomena is most clearly reproduced when students work in teams or groups, conducting an inquiry from beginning to end (from the first to the final stage of inquiry), rather than learning isolated facts or formulae from a textbook. Models of cooperative learning often cite Dewey as an early exponent (e.g. Brown and Palincsar 1989, p. 397–398). Furthermore, scientific investigation demands corroboration of results and not simply the findings of an isolated experiment (Eshach 1997). Writing up results and (re)testing them are as vital to science as the initial experimentation, and both must be included in a programme of science education. This might include both whole-class discussion and the writing of novel 'texts', particularly in the elementary grades (Howes 2008, p. 545).

Recently, it has been suggested that Dewey's understanding of scientific inquiry and science education be utilized to help integrate existing rival theories such as constructivism and objectivism under the rubric of scientific literacy (Willison and Taylor 2009, p. 32). It is argued that this may help resolve some of the tension that keeps at bay the practical advices nesting in the various theories. Dewey has also been invoked in the ongoing debates about the role of science education in the overall preparation of students: while the standards-based push to increase scientific subject matter and instructional time in public schools is laudable, it is also the case that what gets pushed first is very often abstract concepts, formulae and principles at the expense of context and experimentation (Rudolph 2005, p. 806; Chinn and Malhotra 2002, p. 199). Dewey's understanding of the contexts of scientific inquiry is a valuable corrective in this regard.

75.6.5 Overall Assessment of Dewey's Role in Science Education

Deweyan models of science education have been helpful, particularly in stressing the cooperative nature of learning, the experimental nature of knowledge acquisition and the aesthetic dimensions of children's' experiences. It makes good theoretical sense to concentrate our attention on issues of interest, motivation, satisfaction and other 'traits' of experience that bookend inquiry in science education. Furthermore, invoking Dewey, scholars and practitioners have developed comprehensive models of science education that do not resemble cookbooks or 'how-to' manuals (or lists of objectives and standards). As well, they have contextualized intellectual tools such as concepts, ideas, algorithms and processes. Rather than being set off from the curriculum or subject matter, inquiry is now thoroughly integrated in these models. And rather than being a procedure that is separate from the child or a set of steps or stages the child must plod her way through, inquiry is now seen as a process involving her experience throughout.

However, much of the scholarship on Dewey in science education has been content to draw on Dewey in so superficial a fashion that little of Dewey's actual logical, epistemological or experiential philosophy has been adequately mined. What remains for science educators using Dewey's philosophy of education is to further connect Dewey's theories of experience and art with his theory of logic—especially the accounts of propositions and conceptions he details in *Logic: the Theory of Inquiry*. A suitably comprehensive account of Deweyan inquiry for science education (one that integrates both experience and the tools of logical inquiry) has yet to be fashioned. Beyond this, educators must do a better overall job of convincing sceptics (and there are many) that think Deweyan-inspired methods of science education will reap advantageous results. To give but one example, criticisms regarding the bracketing or ignoring of psychological research that indicates the importance of working memory have been raised against constructivism and inquiry-based teaching:

> Any instructional theory that ignores the limits of working memory when dealing with novel information or ignores the disappearance of those limits when dealing with familiar information is unlikely to be effective. Recommendations advocating minimal guidance during instruction proceed as though working memory does not exist or, if it does exist, that it has no relevant limitations when dealing with novel information, the very information of interest to constructivist teaching procedures. We know that problem solving, which is

> central to one instructional procedure advocating minimal guidance, called inquiry-based instruction, places a huge burden on working memory....The onus should surely be on those who support inquiry-based instruction to explain how such a procedure circumvents the well-known limits of working memory when dealing with novel information. (Kirschner et al. 2006, p. 77)

These concerns are to be taken seriously. As these are empirical concerns, they must be addressed empirically. To address these means not simply providing a re-articulation of Dewey's texts (though this is valuable initially); rather, further articulation of Dewey's understanding, further articulation of the application of Dewey's understanding of science and science education to various teaching practices and further empirical verification of this reconstructed articulation through experimental design must be demonstrated. In other words, the way to address these concerns is through following the stages of inquiry from articulation of the problem to hypothesis testing and the formation of anticipated consequences (the formation of universal conceptions), to rigorous testing (using the 'tools' of induction and deduction, as well as the construction and invocation of abstract concepts) and to evaluation in the empirical setting in which the problem is first felt and articulated. Following the lines of Dewey's theory of inquiry is the only reasonable way to reveal problems with the empirical investigations undertaken in the quest to confirm or disconfirm claims on behalf of inquiry, discovery or problem-solving methods.

75.7 Conclusion: Dewey and Science Education

Dewey has and doubtless will continue to provide fertile ground for explorations into various dimensions of scientific inquiry and science education, including pedagogy and the curriculum. Dewey's account of inquiry offers advantages other accounts very often lack: it is holistic, context-bound and self-correcting; it is rooted in experience and the generic 'traits of existence' that arise out of the transactions of human beings and their environments (including the 'social' environment of other people). Yet, it is rigorous in its logical processes, with a strong and detailed accounting of conceptions, ideas, proposition and other logical functions. Historically, it has been vitally important for various accounts of pedagogy and curriculum.

The state of Deweyan scholarship on science education is another matter; as it stands, Dewey scholarship in science education repeats many of same mantras as Dewey scholarship in other areas of education: the importance of inquiry, discovery and problem-based methods and pedagogies; group, cooperative and team-oriented projects; and an emphasis of experimental learning over and against disciplinary or subject matter learning (at least for the elementary grades). However, it has not progressed much beyond these, despite the invocation of Dewey in constructivism and other fashionable models of teaching and curriculum. In my opinion, what is necessary for further scholarship is to develop a cogent model of Deweyan inquiry for science education that integrates Dewey's accounts of experience and art with his detailed account of the functions of logical inquiry, including the tools of conception, propositions and symbols. Doing so requires us to dig deeper into Dewey's logical, epistemological and experiential theories than is usually done. A few scholars (e.g. Garrison) have begun this scholarship, but much more work on the part of the Deweyan community of scholars remains. A systematic model of science education that is attentive to logical, epistemological, experiential and social-political as well as educational concerns will be of inestimable value for further pedagogical and curricular claims on behalf of science education. Beyond this, it remains to be seen whether any of these accounts of science education invoking the name or scholarship of Dewey are able to penetrate the morass of objectives-driven 'standard' science education.

References

Brown, A. L. & Palincsar, A. S., Guided, cooperative learning and individual knowledge acquisition. *Knowing, Learning and Instruction: Essays in Honor of Robert Glaser*. Edited by L. B. Resnick. Hillsdale, N. J.: Lawrence Erlbaum, 1989.

Camp Mayhew, A. & Camp Edwards, E. *The Dewey School*. New York: Atherton, 1936.

Chinn, C. A., & Malhotra, B. A. (2002). Epistemologically authentic inquiry in schools: A theoretical framework for evaluating inquiry tasks. *Science Education*, 86, 175– 218.

Dewey, J. The School and Society. In *The Middle Works of John Dewey* Vol. 1 1899–1901. Edited by Jo Ann Boydston. Carbondale: Southern Illinois University Press, 1976. 1–112. Cited as MW 1 in text.

Dewey, J. Some Stages of Logical Thought. In *The Middle Works of John Dewey* Vol. 1 1899–1901. Edited by Jo Ann Boydston. Carbondale: Southern Illinois University Press, 1976. 151–174. Cited as MW 1 in text.

Dewey, J. The Postulate of Immediate Empiricism. In *The Middle Works of John Dewey* Vol. 3 1903–1906. Edited by Jo Ann Boydston. Carbondale: Southern Illinois University Press, 1977, 158–167. Cited as MW 3 in text.

Dewey, J. How We Think. In *The Middle Works of John Dewey* Vol. 6 1910–1911. Edited by Jo Ann Boydston. Carbondale: Southern Illinois University Press, 1978. 177–356. Cited as MW 6 in text.

Dewey, J. Democracy and Education. In *The Middle Works of John Dewey* Vol. 9 1916. Edited by Jo Ann Boydston. Carbondale: Southern Illinois University Press, 1979. Cited as MW 9 in text.

Dewey, J. The Need for a Recovery of Philosophy. In *The Middle Works of John Dewey* Vol. 10 1916–1917. Edited by Jo Ann Boydston. Carbondale: Southern Illinois University Press, 1980. 3–48. Cited as MW 10 in text.

Dewey, J. Reconstruction in Philosophy. In *The Middle Works of John Dewey* Vol. 12 1920. Edited by Jo Ann Boydston. Carbondale: Southern Illinois University Press, 1982. 77–202. Cited as MW 12 in text.

Dewey, J. An Analysis of Reflective Thought. In *The Middle Works of John Dewey* Vol. 13 1921–1922. Edited by Jo Ann Boydston. Carbondale: Southern Illinois University Press, 1983. 61–71. Cited as MW 13 in text.

Dewey, J. Experience and Nature. In *The Later Works of John Dewey* Vol. 1 1925. Edited by Jo Ann Boydston. Carbondale: Southern Illinois University Press, 1981. Cited as LW 1 in text.

Dewey, J. Qualitative Thought. In *The Later Works of John Dewey* Vol. 5 1929–1930. Edited by Jo Ann Boydston. Carbondale: Southern Illinois University Press, 1984, 243–262. Cited as LW 5 in text.

Dewey, J. How We Think 2nd Edition. In *The Later Works of John Dewey* Vol. 7 1933. Edited by Jo Ann Boydston. Carbondale: Southern Illinois University Press, 1986. 105–352. Cited as LW 7 in text.

Dewey, J. Art as Experience. In *The Later Works of John Dewey* Vol. 10 1934. Edited by Jo Ann Boydston. Carbondale: Southern Illinois University Press, 1987. Cited as LW 10 in text.

Dewey, J. Logic: the Theory of Inquiry. In *The Later Works of John Dewey* Vol. 12 1938. Edited by Jo Ann Boydston. Carbondale: Southern Illinois University Press, 1986. Cited as LW 12 in text.
Dewey, J. Experience and Education. In *The Later Works of John Dewey* Vol. 13 1938–1939. Edited by Jo Ann Boydston. Carbondale: Southern Illinois University Press, 1986, 1–62. Cited as LW 13 in text.
Eshach, H. (1997). Inquiry-events as a tool for changing science teaching efficacy belief of kindergarten and elementary school teachers. *Journal of Science Education and Technology, 12*, 495–501. Retrieved from JSTOR.
Garrison, J. "Deweyan Pragmatism and the Epistemology of Contemporary Social Constructivism," *American Educational Research Journal*, 32, 4 (1995); 716–740.
Garrison, J. "An Alternative to von Glaserfeld's Subjectivism in Science Education," *Science & Education* 6 (1997a); 301–312.
Garrison, J. *Dewey and Eros: Wisdom and Desire in the Art of Teaching*. New York: Teachers College Press, 1997b.
Girod, M. & Wong, D. "An Aesthetic (Deweyan) Perspective on Science Learning: Case Studies of Three Fourth Graders," *The Elementary School Journal*, 102, 3 (2002); 199–224.
Glassman, M. (2001). Dewey and Vygotsky: Society, experience, and inquiry in educational practice. *Educational Researcher, 30 (4),* 3–14. Retrieved from JSTOR.
Gordon, M. "Toward a Pragmatic Discourse of Constructivism: Reflections on Lessons from Practice," *Educational Studies* 45 (2009); 35–58.
Grandy, R. E. Constructivisms and Objectivity, In *Constructivism in Science Education: A Philosophical Examination*, ed. Michael Matthews (Dordrecht: Kluwer, 1998), 112–21.
Howes, E. "Elementary Experiences and Early Childhood Science Education: A Deweyan Perspective on Learning to Observe," *Teaching and Teacher Education* 24 (2008); 536–549.
Johnston, J. S. *Inquiry and Education: John Dewey and the Quest for Democracy*. Albany, NY: SUNY Press, 2006.
Kant, I. *Critique of Pure Reason*. Translated by Paul Guyer and Allen Wood. Cambridge: Cambridge University Press, 1998.
Kirschner, P., Sweller, J., & Clark, R. "Why Minimal Guidance during Instruction does not Work: An Analysis of the Failure of Constructivist, Discovery, Problem-based, Experiential, and Inquiry-based Teaching," *Educational Psychologist* 41 (2); (2006), 75–86.
Kruckeberg, R. "A Deweyan Perspective on Science Education: Constructivism, Experience, and Why We Learn Science," *Science & Education* 15 (2006); 1–30.
LaRochelle, M., & Bednarz, N. Introduction, in *Constructivism in Education,* ed. M. LaRochelle, N. Bednarz, and J. Garrison (Cambridge: Cambridge University Press, 1998), 3–20.
Phillips, D.C, "The Good, the Bad, and the Ugly: the Many Faces of Constructivism," *Educational Researcher* 24, 7 (1995); 5–12.
Pugh, K., & Girod, M. "Science, Art, and Experience: Constructing a Science Pedagogy from Dewey's Aesthetics," *Journal of Science Teacher Education* 18 (2007); 9–27.
Rudolph, J. "Inquiry, Instrumentalism, and the Public Understanding of Science," *Science Education* 89 (2005); 803–821.
Tanner, L. *Dewey's Laboratory School: Lessons for Today*. New York: Teachers College Press, 1997.
Vanderstraeten, R. "Dewey's Transactional Constructivism." *Journal of Philosophy of Education*, 36, 2 (2002), 233–246.
Von Glaserfeld, E. "An Introduction to Radical Constructivism," in *The Invented Reality*, ed. P. Watzlawick. New York: Norton. (1984); 17–40, 22.
Von Glaserfeld, E. "Why Constructivism Must be Radical," in *Constructivism and Education*, ed. M. LaRochelle, N. Bednarz, and J. Garrison. Cambridge: Cambridge University Press, 1998, 23–28, 25.
Willison, J.W., & Taylor, P.C. Complementary epistemologies of science teaching: Toward an integral perspective. In P.J. Aubasson, et al (Eds.). *Metaphor and analogy in science education*, pp. 25–36. Dordrecht: Springer, 2006.
Wirth, A. *John Dewey as Educator: His Design for Work in Education*. New York: Wiley, 1979.
Wong, D., & Pugh, K. "Learning Science: A Deweyan Perspective," *Journal of Research in Science Teaching* 38, 3 (2001); 317–336.

James Scott Johnston is Jointly Appointed Associate Professor of Education and Philosophy at Memorial University, Newfoundland, Canada. Dr. Johnston has published recently in journals such as the *Transactions of the Charles S. Peirce Society, Educational Theory, Educational Studies, Studies in Philosophy and Education* and the *Journal of Philosophy of Education*. He has single-authored four books (*Inquiry and Education: John Dewey and the Quest for Democracy* (2006), *Regaining Consciousness: Self-Consciousness and Self-Cultivation from 1781–Present* (2008), *Deweyan Inquiry: From Educational Theory to Practice* (2009), *Kant's Philosophy: A Study for Educators* (2013)) and coauthored *Democracy and the Intersection of Religion and Tradition: The Reading of John Dewey's Understanding of Democracy and Education* (with R. Bruno-Jofre, G. Jover Olmeda, and D. Troehler) (2010).

Chapter 76
Joseph J. Schwab: His Work and His Legacy

George E. DeBoer

76.1 Introduction

Most science educators are familiar with Joseph Schwab because of his contributions to the school reform movement in biology in the United States in the 1960s, especially through his connection to the Biological Sciences Curriculum Study (BSCS) (see, e.g., DeBoer 1991). Schwab brought terms like "rhetoric of conclusions" and "narrative of enquiry"[1] to the discussion of school science, and he contributed to the reform of science education as chair of the Teacher Preparation Committee at BSCS and as author of the *Biology Teachers Handbook* (Schwab 1963a). But most of Schwab's work in science education was not focused on the school curriculum, rather on the undergraduate science program while he was on the faculty at the University of Chicago. His ideas about the nature of the science curriculum were shaped as he and his colleagues at Chicago worked out the details of a comprehensive program of general education[2] for the undergraduate college. It was at Chicago that his professional career began and where it ended 36 years later, and it was at Chicago that he thought and wrote about science education, first for undergraduate students and then later as part of the precollege science curriculum reforms of the 1950s and 1960s.

Joseph Schwab was born in Columbus, Mississippi, in 1909; matriculated as an undergraduate at age 15 at the University of Chicago in 1924; and graduated with degrees in physics and biology in 1930. He earned a doctorate in genetics from

[1] Schwab preferred "enquiry" to "inquiry," but in his writing the spelling varies depending on where the work was published. In this chapter, the spelling that actually appeared in a publication will be used, and all my discussions of his work will use the word inquiry.

[2] The terms *general education* and *liberal education* will be used synonymously in this chapter to describe nonspecialized and nonvocational programs of study that offer students a broad base of experience with various modes of thought and knowledge of their culture.

G.E. DeBoer (✉)
American Association for the Advancement of Science, Washington, DC 20005, USA
e-mail: gdeboer@aaas.org

M.R. Matthews (ed.), *International Handbook of Research in History, Philosophy and Science Teaching*, DOI 10.1007/978-94-007-7654-8_76,

Chicago in 1939. In 1937, he spent a year at Columbia University Teachers College, where he was influenced by both John Dewey and Ralph Tyler. Schwab came to Chicago as an instructor in 1938, and he retired as professor of education and the William Rainey Harper professor of natural sciences in 1974. He then joined the Center for the Study of Democratic Institutions, founded by Robert Maynard Hutchins, in Santa Barbara, California, where he continued to think and write about curriculum. He died in Lancaster, Pennsylvania, in 1988.

He began his graduate work at Chicago just as Hutchins was beginning his long tenure, first as president (1929–1945) and then as chancellor of the university (1945–1951). This was also the time that the college was beginning to embark on its decadelong experiment in general education. Hutchins was a vigorous advocate of the Great Books approach to general education and a promoter of liberal education as the best preparation for informed, responsible citizenship (Hutchins 1936). Hutchins believed that undergraduate education should focus on a student's intellectual development through a careful study of classic works of Western civilization, taught through a dialectical Socratic method, rather than on the development of practical skills and professional training, which tended to characterize higher education at that time. His approach was intended to develop citizens with the independence of mind suited for life in a democratic society. The study of a core body of great works would also provide a common educational experience so that citizens could communicate beyond their areas of specialized interest.

Hutchins was joined in 1930 by Mortimer Adler and, with Adler, went on to found the Great Books of the Western World program and the Great Books Foundation in 1947 (http://www.greatbooks.org/about/history). But the faculty rejected Hutchins' plan for a Great Books approach for the undergraduate college, and the program never became the model of undergraduate education at Chicago that it did at St. John's College in Annapolis, Maryland. However, its focus on the intellectual heritage of Western civilization did influence the spirit and forms the general education program took at Chicago, and the approach was used in the university's adult evening extension college, which Schwab chaired when he joined the faculty in 1938.

Although Schwab's primary interest and responsibility was organizing the science curriculum for the general education program at Chicago, the integrative nature of general education also gave him opportunities to think about the role of the social sciences and humanities in general education and the boundaries between those subject areas and the sciences. He had a passion for psychology, social sciences, religion, and humanities, and he addressed issues from these disciplines in his writing on education. In addition to being a member of the science faculty, Schwab was also a respected education theorist. In 1949 he was appointed to the university's education department, where he taught courses in the philosophy of education. And later he did curriculum work at the Melton Research Center of the Jewish Theological Seminary, where he helped develop materials to teach character education to students attending Jewish summer camps.

He had an especially strong interest in psychoanalysis, undergoing analysis himself. In *Eros and Education* (1954) Schwab wrote about the nature of the interactions between faculty and students during classroom discussions from a Freudian perspective. In the late 1950s and early 1960s, his attention shifted to the school science curriculum reform movement through his work with BSCS. In 1969, his attention shifted back to higher education as the student protest movement gained momentum. In response to the protests, he published the *College Curriculum and Student Protest* (Schwab 1969a), in which he focused again on the role of liberal education in society, especially on ideas of "community, of moral choice, and of deliberation and decision-making" (Westbury and Wilkof 1978, p. 30). His final contributions to the field of education were to curriculum development in general. Through a series of papers on the advantages of a practical rather than theoretical approach to curriculum study, he became well known among curriculum theorists for his claim that "the field of curriculum is moribund" (Schwab 1970, p. 1).

76.2 The Undergraduate College at Chicago

Throughout the 1930s, a group of University of Chicago faculty pressed forward on a plan to create a coherent and well-integrated approach to general education for the undergraduate college. In 1937, a four-year program of undergraduate study, completely devoted to general education, was officially approved by the university (Schwab 1950a). When Schwab joined the faculty in 1938, he took a leading role in the development of the science component of the program as chair of the natural sciences sequence in the undergraduate college, and it was his efforts to conceptualize that program to which he devoted most of his professional career.

The early to mid-twentieth century was a time of vigorous debate about the role of undergraduate education at colleges and universities in the United States. At its beginning, higher education in the United States had had a classical character, with a focus on classical literature and languages. But by the mid to late nineteenth century, the model of the German university, with its emphasis on specialization and empirical investigations in the sciences, began to take hold in the United States and elsewhere. By the late nineteenth century, that model began to predominate in universities like Chicago. As Daniel Bell put it:

> The American university, as it emerged in the latter decades of the nineteenth century, brought with it a new religion of research. Even scholarship in the traditional disciplines was conceived, within that purview, as being concerned with detailed and specialized problems. The reaction of the liberal arts college was to strike out against specialism. (Bell 1966, p. 51)

Questions began to be raised in universities across the United States about the appropriate role of the undergraduate experience, coming as it does between the high school and the professional and graduate schools: Should the undergraduate years be spent in preprofessional training for those planning to enter the professional schools, should it focus on early scholarly preparation for those going on to graduate school, or should it simply be preparation for informed citizenry? What, if any, is the importance of having students develop an appreciation for the cultural artifacts that the society thinks a cultivated person should be familiar with, to become aware of basic principles that guide moral behavior, or to gain an understanding of how knowledge is organized and revised? And how important is it for citizens in a democratic society to have a shared intellectual experience that provides them with a common ground for deliberation and debate regardless of their life work or specialization? These were the questions that were being debated.

The programs of general education being developed at Chicago, along with those at places like Columbia and Harvard, became models for colleges throughout the country. (See Bell (1966) for a discussion of the Chicago, Columbia, and Harvard experiments in general education.) Although these programs differed in detail, they all had a commitment to certain general principles and purposes. As Bell said in his study of general education, the function of a general education program in the undergraduate college was:

> …to teach modes of conceptualization, explanation, and verification of knowledge. As between the secondary school, with its emphasis on primary skills and factual data, and the graduate or professional school, whose necessary concern is with specialization and technique, the distinctive function of the college is to deal with the grounds of knowledge: not *what* one knows but *how* one knows. The college can be the unique place where students acquire self-consciousness, historical consciousness, and methodological consciousness. (Bell 1966, p. 8)

General education programs such as this had as their stated purpose the development of enlightened and responsible citizens for life in a free society. They emphasized personal growth and individuality and a universal rather than a provincial or nationalistic world view. Programs usually focused on the humanities and classics, particularly the study of Western civilization, and they avoided connections with utilitarian and vocational aims and with career preparation. But programs were not all the same. They differed in the emphasis they placed on developing moral men and women versus providing students with a broad understanding of multiple ethical perspectives, the importance they placed on learning about the heritage of the society versus studying the contemporary world, and how much they valued the acquisition of a broad base of knowledge across the curriculum compared to providing students opportunities to develop skills as independent thinkers.[3]

For example, the stated purpose of the Chicago program was to develop the intellect. It was not primarily about the knowledge one acquired, but rather the ability to think, to contemplate, and to consider alternatives. To do that well, one had to learn about the complexities of the world in relation to each other. The most important job of the college was to introduce students to positions other than their own and to help them develop the power to form judgments. In the Chicago program, that thinking would take place in the context of cultural elements (works of art, music, literature, and science) that were deemed to be most important by society. The task of curriculum developers was to create curricular content and learning activities that allowed students to investigate these cultural elements thoroughly and in context. The Chicago program also took an analytical approach to knowledge rather than the

[3] See *The Emergence of the American University* by Lawrence Veysey (1981) for an extended discussion of the history of the American University during the time period in question.

historical approach that often characterized general education programs. Especially in the sciences, and largely through Schwab's influence, the goal of the curriculum developers was, in Bell's words, "to find the controlling principles of 'classification' in the definition of subjects or of disciplines within fields" (Bell 1966, p. 33). Referring to the difference between these two approaches in the context of science, Bell said:

> …the question is whether one wants to emphasize historicism, with its doctrine that the understanding of an event can be found only in its unique context, or the analytical approach, which finds meaning in a phenomenon as one of a type-class, and seeks, further, a sense of invariant relationships. …Does one teach science through its history, or by analysis of its models of inquiry? (p. 62)

At Chicago, general education meant learning the special modes of conceptualization that characterized each discipline, not simply reviewing the historical development of a field. Historical texts were read and examined, but the purpose was not just to familiarize students with the knowledge these texts contained but to teach students how different forms of knowledge were created and how the students themselves could be analytical and critical of those intellectual methods and products.

At Chicago, this was to be accomplished by means of an interpretive (hermeneutic) approach in which students extracted meaning from selected written texts, pieces of music, works of art and architecture, and reports of scientific investigations, taking into consideration not only the cultural artifacts themselves but also the purposes and intentions of the creator of those artifacts. Nothing was to be taken as given, but always open to analysis and interpretation. In science, the texts that were subjected to interpretation were original scientific research papers, and the pedagogical approach for teaching them involved having students examine those papers to become familiar with the particular knowledge claims that were made, the investigative methods used, and the broader intellectual and practical contexts in which each investigation was conducted (Schwab 1950a).

The challenge that Schwab and his colleagues faced was how to create an educational experience that would lead to intellectual growth so that students would be open-minded, skeptical, able to think for themselves, and prepared to take on positions of leadership in society. The education that was envisioned emphasized the integration of knowledge from multiple disciplines and a search for and an appreciation of fundamental principles that define human experience, accomplished not through memorization but through discussion and deliberation.

But, even as efforts were under way in places like Chicago to build general education programs, the overall trend in undergraduate education was toward specialization, professional and preprofessional training, and the accumulation of knowledge. As Schwab noted, a "rhetoric of conclusions" dominated undergraduate teaching, where students were presented with knowledge of the disciplines without being required to think critically or make judgments about that knowledge. Many of these issues were addressed in the scholarly writing that Schwab was engaged in while he was the chair of the science program in the undergraduate college at Chicago, and it is to that work that we turn in the next several sections of this chapter.

76.3 The Place of Science in Liberal Education

76.3.1 A Taxonomy of Types of Science

In 1949, Schwab published a paper titled *The Nature of Scientific Knowledge as Related to Liberal Education*. In that paper he argued that all students should be exposed to the breadth and variety of science both in its content and its methods. Science should not be treated monolithically but as a complex and varied study. To accurately represent the complex nature of the physical world and the methods used to study it, liberal education should use pedagogical approaches that reflect that complexity. Diversity exists in the content and methodologies of the separate fields of science, and diversity exists in how philosophers of science view the nature of science, including the nature of causality, the nature of induction, the role of hypothesis testing, and the relationship between mathematical knowledge and the physical world. Schwab argued that because of this diversity of methods that are used and views that are held about the nature of science, no one single set of "epistemic or metaphysical presuppositions" concerning science can cover the variety of ideas that exists (Schwab 1949, p. 248). An accurate portrayal of the nature of science as part of a liberal arts education requires that this diversity of scientific methodology and interpretation be taught as fully as possible.

Schwab proposed a taxonomy of scientific investigation that could serve as an aid to the teaching of science in a general education program, both to support the choice of subject matter and how that subject matter could be examined by students. He identified four types of scientific investigations, which differ from each other in the kind of knowledge that is generated, the kind of data that are collected, and the form of validation that is used in each. The four types, which he believed encompassed most forms of scientific inquiry, were taxonomic science, measurement science, causal science, and relational or analogical science.

Taxonomic science involves the creation of classification schemes for organizing objects and events in the world. These classification schemes exist in virtually all fields of science, including the classification of disease for diagnostic purposes, categorizing living organisms to study their degree of hereditary relatedness, or the classification of types of chemical molecules on the basis of their molecular structures. All of these classification schemes were developed for a purpose, and all of them require difficult decisions at the margins. For the purpose of liberal study, "…a given taxonomic system is understood when it is seen as one of several alternatives" and when "…some of the doubtful areas of the taxonomy are seen and some of the reasons for their doubtful status understood" (Schwab 1949, pp. 255–256).

Measurement science involves measuring and relating changes in two or more objective quantities. Familiar examples include the relationship between the intensity of light and distance from the light source, the frequency of vibration of a plucked string and the length of the string, or the degree of sinking of an object and its density in relation to the density of the medium it is immersed in. For general education purposes, it is important that students understand that assumptions are made when reporting these relationships in a mathematical form, such as the assumption that there is a point source of light (which is an idealization of the real world). Students should also be aware of the possible effects of abstracting only certain variables of interest from a more complex set of related variables that could be studied.

Regarding *causal science*, Schwab argues that much of what is thought of as "causal science" can actually be placed in the other three categories, but even after doing that, there remains a separate type of investigation that deals with systems of mutually interacting and mutually determined parts acting as a whole. He cites physiological and social systems as examples. The defining features of these causal systems involve "interaction, mutual determination, and concerted action" (Schwab 1949, pp. 258). The challenge for students is to grasp the nature of the interacting parts of the system and their relationship to the whole organism or system. Of necessity, because these systems are too complex to be studied as a unity, their parts and pieces must be studied in isolation. For general education purposes, the student:

> ...must be prepared to discover, in the records of such research, answers to the questions of what kinds of "parts" are being treated, what analysis of "functions" are related to the parts and functions of other related researches, and how, if at all, the researcher in question relates his discovered functions and parts to one another to constitute larger units more nearly approaching the unity of the organism as a whole. (Schwab 1949, p. 260)

Finally, Schwab identified *relational science* as a fourth type of scientific inquiry. Regarding this type of inquiry, which relies on models, analogies, and forms of representation, Schwab said:

> By "relational science" I mean those patterns of inquiry which are most fully understood as aiming toward knowledge which attempts to "explain" or "account for" matters previously known by inventing co-related quantities which do not have one-to-one correlates among the phenomena to be accounted for, or by inventing mechanisms not directly accessible to observation but so conceived and applied to the phenomena to be explained that it can be said that certain things behave *as if* these mechanisms existed. (Schwab 1949, p. 260)

These borrowed relationships of relational science, which are applied to the new observations, may come from either physical models or from abstract mathematical and conceptual models.

The educational imperative of these diverse approaches to scientific investigation is that students should have enough familiarity with them to analyze actual research studies in each category and make comparisons between them. Instruction should "...educate, encourage, and exercise the student in applying appropriate canons of comprehension and evaluation to...examples of scientific inquiry" (Schwab 1949, p. 264). This enables students to make judgments about which of a number of possible alternatives is the most appropriate approach to collecting data, drawing conclusions, and linking evidence to conclusions, which in turn will give students a more honest and accurate picture of the physical world and how it is studied. When the nature of science itself is chosen as the subject for students to study, then a variety of historical, philosophical, and methodological interpretations of science should be read, discussed, and analyzed in the same way.

76.3.2 The Tentative Nature of Science

Also key to an understanding of science for liberal education purposes was to appreciate "the ongoing, unclosed character of science" (Schwab 1949, p. 263). Yet, as Schwab observed, colleges still taught "the conclusions of science and definitive solutions to its problems" (p. 263). Teaching the tentativeness of conclusions did not, however, argue for naïve relativism to Schwab. It meant simply that in order to be honest about the nature of science, differences in how the world is viewed by individuals studying the same problem needed to be treated thoroughly. Instruction must teach students "the disciplines of comparison, contrast, choice, and synthesis appropriate to the field in which the diversity takes place" (p. 264).

The pedagogical challenge of such an intellectually sophisticated approach to teaching science was how to get students familiar with and to contemplate the relevance of each these diverse modes of scientific inquiry in the limited time allotted. To Schwab (and his colleagues at Chicago), the answer lay in the analysis of carefully chosen scientific research papers. A scientific research paper is the "bearer of a portion of scientific knowledge in its field," and "…it 'illustrates itself' as an example of scientific investigation" (Schwab 1949, p. 265). All that is required is that the student knows what questions to ask, including what problem is being addressed, the appropriateness of the data, difficulties in obtaining data, how the data were treated, any phenomena that were excluded, and the validity of the conclusions. Each paper "would serve simultaneously to impart subject-matter content and to illustrate aspects of the nature of scientific knowledge at many different levels—from the most specific level at which the paper falls…to the level of science-as-a-whole" (Schwab 1949, p. 251). In the plan developed at Chicago, students would be presented with sets of such papers and with a framework for analyzing them so that they would gain practice in studying those investigations as instances of scientific inquiry, especially how each was similar to and differed from the others.

76.3.3 Science as Constructed Theory

In *Science and Civil Discourse* (Schwab 1956), Schwab elaborates further on the nature of inquiry in science and its importance in liberal education. He says that inquiry is constructive in the sense that conceptions "must be invented…by the investigator" in order to determine what his subject matter and his data will be from the great "complex of things and events" (Schwab 1956, p. 132). According to Schwab, through this process of problem and data selection, the content is inevitably "distorted" and "made incomplete." Therefore, because of this selecting and narrowing of the problem and consequent narrowing of what is observed, a conclusion in science must be thought of as a "taken something, not an objectively given something" (Schwab 1956, p. 132).

This constructive character of scientific knowledge has implications for the liberal arts curriculum. Schwab argued that if a theory is to be taught as a theory about some aspect of the world, it is also important to be clear about which aspects of the subject are not incorporated into that theory:

> We must have something more in the materials of our curriculum than the theories themselves, for the restrictions which define what the theory is about are not readily found in the theory itself. The theory is only the terminal part of an inquiry. We need what comes before the end…to discover what the theory is a theory of…. (Schwab 1956, p. 133)

This means that the student needs to know that scientific problems are constructed out of a much larger array of possibilities, and they should come to appreciate the choices that are made by scientists in the selection of problems, the selection of observations to be made and data to be collected, and how the data are interpreted in terms of existing theory.

76.3.4 Structure of the Disciplines

Although much of Schwab's work involved efforts to integrate scientific knowledge throughout the liberal arts curriculum by showing the interconnections between subject matters across disciplinary boundaries, he also acknowledged the importance of the separate academic disciplines for curriculum development. In fact, Schwab is often associated with the "structure of disciplines" movement, an effort that became popular in the 1960s to describe the structure of knowledge and the relevance of that structure for school curriculum development and content organization. But Schwab's ideas about "structure" were at least as much about disciplinary modes of thought as they were about how content should be organized. Schwab published a number of essays on the topic, including *Structure of the Disciplines: Meanings and Significances* (Schwab 1964). He found support for the idea of disciplinary structure in Aristotle's distinctions between the theoretical, practical, and productive disciplines and in Auguste Comte's hierarchy of scientific disciplines, starting with physics and progressing to chemistry, biology, and finally the social sciences (Schwab 1960). But he also appreciated that these diverse formulations of disciplinary structure provided support for the truism that "if we classify any group of complex things, we are faced with a wide choice of bases of classification" (Schwab 1964, p. 15). In other words, organizational schemes can be helpful for thinking about the curriculum, but they should not be considered to be fixed and absolute.

Schwab distinguished between the substantive structure of the disciplines (their conceptual organization) and their syntactical structure (how knowledge is generated in each field). He argued that because the two structures are necessarily interconnected, students should be taught the conceptual structure of scientific knowledge in the context of the methods of inquiry that produced that knowledge, and they should be taught the methods of inquiry in terms of the conceptual structures:

> In general then, enquiry has its origin in a conceptual structure… It is this conceptual structure through which we are able to formulate a telling question. It is through the telling

> question that we know what data to seek and what experiments to perform to get those data. Once the data are in hand, the same conceptual structure tells us how to interpret them, what to make of them by way of knowledge. Finally, the knowledge itself is formulated in the terms provided by the same conception. (Schwab 1964, p. 12)

But in no way do these structures represent a fixed body of knowledge or a fixed way of organizing that knowledge:

> The dependence of knowledge on a conceptual structure means that any body of knowledge is likely to be of only temporary significance. For the knowledge which develops from the use of a given concept usually discloses new complexities of the subject matter which call forth new concepts. These new concepts in turn give rise to new bodies of enquiry and, therefore, to new and more complete bodies of knowledge stated in new terms. The significance of this ephemeral character of knowledge to education consists in the fact that it exhibits the desirability if not the necessity for so teaching what we teach that students understand that the knowledge we possess is not mere literal factual truth but a kind of knowledge which is true in a more complex sense. (Schwab 1964, pp. 13–14)

And if we do choose to teach just one conceptual structure, Schwab argues that we should at least be honest about what we are doing:

> But if we do, let it be taught in such a way that the student learns what substantive structures gave rise to the chosen body of knowledge, what the strengths and limitations of these structures are, and what some of the alternative structures are which give rise to alternative bodies of knowledge.
>
> If students discover how one body of knowledge succeeds another, if they are aware of the substantive structures that underlie our current knowledge, if they are given a little freedom to speculate on the possible changes in structures which the future may bring, they will not only be prepared to meet future revisions with intelligence but will better understand the knowledge they are currently being taught. (Schwab 1964, pp. 29–30)

Regarding the specific conceptual structures that should be taught in a liberal arts course in science, Schwab admitted that the topics that he was advocating for the Chicago course showed "no notable departure from those which might be found in one or another conventional 'survey' course" (Schwab 1950a, p. 150). For example, the physical science portion of the course included simple Archimedean laws of equilibrium and the lever, phenomena involving chemical and physical change, molecular and atomic theories, and the periodic table. It included concepts of energy, the kinetic molecular theory, the theory of special relativity, and ideas about radiation. In the biological sciences portion of the course, topics included transport and regulation of respiratory gases and the regulation and utilization of food material, the structure of the heart and circulatory system, the levels of organization of organisms, and issues of health and disease. Also included were the developmental history of organisms, Darwinian evolution, Mendelian genetics, embryonic development, and various concepts from the field of psychology.

The reason there were no radical departures from what was traditionally taught in introductory science courses was because the primary focus of the Chicago program was not the content itself but the interconnectedness of knowledge and the nature of scientific inquiry. Much of the content that was taught in traditional survey courses would suffice as long as connections were made between topics and the content was taught in the context of the scientific inquiry that produced it. In the case of physics and chemistry, for example, he said that because concepts of energy are related to various phenomena involving chemical change, "a relation between a problem in physics and one in chemistry is established as illustrative of the unifying function of scientific inquiry" (Schwab 1950a, p. 150). Also, as already noted, original papers would be used to introduce students to both the core ideas of science and to their methods of inquiry, through actual accounts of scientific research. The point is that the science content was seen primarily as a vehicle for teaching about the nature of scientific inquiry rather than as an end in itself. Schwab's interest in the structure of the disciplines had as much or more to do with the modes of thought that characterized science as it did with the products of that inquiry.

76.4 *Eros* and Education

Although Schwab's work at the undergraduate level is most often associated with efforts aimed at intellectual development through the liberal arts, he also appreciated the importance of the affective dimension in education, both as a means to achieve intellectual goals and as a proper educational goal itself. In *Eros and Education* (Schwab 1954), he draws on the concept of *Eros* as the psychic energy of creating and wanting that drives students' desire to learn what is placed in front of them and supplies them with a love of knowledge that makes them want to learn throughout their lifetime. Schwab's notion of *Eros* is akin to Freud's idea of the fundamental life instinct that drives humans to create and be productive (Freud 1975/1920). It also bears similarities to Jung's notion of *Eros* as "psychic relatedness," particularly as Schwab used the idea to describe the interactions between students and teacher during class discussions (Jung 1982, p. 65).

Schwab believed that *Eros* could be nurtured in an educational setting through classroom discussion. To him, discussion was the embodiment of the intellectual skills that define a liberal education. At its best, classroom discussion draws upon an interpersonal relationship between student and teacher that is characterized by liking and respect. The respect of student for teacher comes from the belief on the part of the student that the teacher has something of value to offer that will enable the student to grow toward intellectual maturity. For both student and teacher, the liking and respect comes from shared participation in a problem of genuine interest to the two of them. When done well, classroom discussion stimulates a love of learning that can last a lifetime.

Schwab argued that the truly educative discussion has three functions: the substantive, the exemplary, and the stimulative functions, representing three liberal education aims of knowledge, power, and affection. First, there must be a specified object of knowledge that the discussion is intended to address. Second, the discussion must involve an activity that leads students to an awareness and appreciation of the method of inquiry employed in the generation of the knowledge. Finally, each discussion must serve to motivate students to engage in the activity so that learning can in fact take place.

Discussion satisfies its substantive function when it is focused on a clear knowledge goal. It satisfies its exemplary function when it engages students in an examination of a variety of methods suitable to the questions being addressed, and the students recognize that people can arrive at differing answers to a problem because of differences in how they formulate the problem, differences in the data they collect relevant to the problem, and differences in how they draw conclusions from those data. Discussion satisfies its stimulative function when the *Eros* is activated, as when a teacher inspires students through accounts of personal experience or allows students to share their own insights and opinions. The result of such a balanced approach is the education of students who have both a creative impulse and a desire to engage in a search for knowledge. Schwab noted that the discussions he envisioned share little in common with the all too familiar undergraduate experience in which the discussion is no more than a reorganizing and rearranging of what the students already know with little new knowledge added.

76.5 Character Education

In addition to recognizing the important connection between intellect and emotion in an educational setting, Schwab was also interested in the role of intellect in the development of personal values, ethical behavior, and character. In an early paper titled *Biology and the Problem of Values* (Schwab 1941), he analyzed the relationship between the teaching of biology and the teaching of values in the context of general education. Schwab began by acknowledging that people have a variety of attitudes about criteria for making value judgments. He said there are some who argue that there are *no* useful criteria for judging which of many ethical systems to choose from, others who say that one person's opinion is as good as another's, and still others who choose to follow the ethical position of the majority. Instead, Schwab says, ethical judgments can be made rational and subject to rational test. Value judgments can be made rational to students by having them learn how to think through and analyze ethical problems in the same way that they think through scientific problems. He says:

> …we can take a leaf from the scientist's notebook. A good scientist does not go into the laboratory "cold" to solve a problem. Instead, he reads the available literature by experts in the field—not to believe, of course, nor to reject but to weigh, consider, and verify.
>
> The same can be done in the field of ethics—we can read the experts from Plato and Aristotle, through Bentham and Hobbes, to Dewey; read then, not to swallow what they have to say, nor to reject it—but to see and evaluate the thought and insight and logic…. (Schwab 1941, p. 94)

One way to teach students this connection between the intellectual and the ethical in science classes is to provide them with controversial issues (Schwab suggests soil and water conservation or other bio-economic issues) and to:

> …take them apart for the student to show him that such programs of action involve both data as to means and judgments as to ends, to let him see what ethical principles must be

used to decide the issue, and to give him an opportunity to deduce for himself the appropriate application of these principles to the particular problem. (Schwab 1941, p. 96)

In a later paper (Cohen and Schwab 1965), the idea of an intellectual dimension to value judgments, ethical decision making, and character development was applied in the context of religious education. In that paper, Schwab and his coauthor describe efforts to design curricular activities for character development for students in Jewish summer camps while Schwab was chairman of the academic board of the Melton Research Center of the Jewish Theological Seminary. The authors begin by affirming the connection between character and intellect:

We suspect that one of the chief reasons why educators have been thwarted in devising methods of character education is their failure to consider the possibility that there may be means of advancing the student's character development through his intellect. (Cohen and Schwab 1965, p. 23)

The approach they used was to teach students a familiar set of ethical principles derived from the Bible (e.g., thou shalt not stand idly by while an evil is being committed) and then to ask students to relate these ethical principles to life situations by means of "practical logic" (Cohen and Schwab 1965, p. 24). To Schwab, practical logic involves weighing alternative ethical positions within a logical framework in order to choose the best one. The logical framework provides a structure for deciding which ethical principle is applicable to a particular set of circumstances or for deciding between two or more equally valid but apparently irreconcilable ethical principles.

In one activity, students confront the Biblical dilemma that all of creation is sacred, but yet humans have been given dominion over the earth. They are given a series of situations and asked if they think the action that is described is more consistent with the idea that everything was created for human use and satisfaction or with the idea that everything in nature should be protected by humans because it is sacred and inviolable. The positions that they evaluate range from "every city should have a zoo so that people will have a place to go for picnics" to "we should not…build a dam if this will destroy a beautiful natural vista or displace…wildlife" (Cohen and Schwab 1965, p. 25). These structured activities were meant to provide students with analytical skills that would be useful to them as they applied their logical reasoning to ethical questions. It would give them practice in thinking through real-world cases and experience in using specific analytical structures to identify issues they could then consider when making ethical choices.

Schwab did not believe that there was a single ethical standard that could be used for making value judgments. Rather, ethical inquiry uses the same kind of intellectual approach that empirical inquiry uses. Humans can make ethical judgments using their practical intelligence and in consideration of the consequences of the choices they make.

In 1969, Schwab published *College Curriculum and Student Protest* as a practical example of ethical decision making and the role the college can play in character development. The book was written in response to the student protests of the 1960s and was an attempt to use curricular revision to solve the problems he believed had been created by the existing curriculum. *College Curriculum and Student Protest* takes an analytical approach to solving the problem of student unrest. Who are the protesters? What is it about their education that they are protesting? What could be done differently—both in terms of the content of the curriculum and the way it is taught—to give students greater satisfaction with their college experience or, at least, a more intelligent and informed basis for protest. To each of these questions, he systematically lays out an array of possible answers. He then proposes many of the approaches that he had advocated in his earlier writings. In particular, and especially relevant to the teaching of science, he describes the dissatisfaction that results from "…the neatness and air of inevitability with which we invest our accounts in science textbooks and lectures of the evidences which lead to current theory" (Schwab 1969a, p. 8). As a solution to the alienation that students felt from their college experience, he proposes making better use of the students' own intellectual capabilities by focusing less on the assimilation and use of the products of inquiry and more on how knowledge in each field is acquired. Speaking of how students were being taught, he says:

> Instead of giving experience of the kinds of problems and modes of enquiry characteristic of the field, they provide the student with the experience of assimilating, applying, or otherwise using the fruits of enquiry in the field. Yet these two—assimilation and use as against pursuit of a body of knowledge—are often radically different in the competences they require and the satisfactions they afford. (Schwab 1969a, p. 10)

What students needed, according to Schwab, was experience in the practical art of thoughtful deliberation, opportunities for sharing experiences and ideas, and skill in mutual criticism. Materials should be presented to them not as unqualified assertions but as genuine questions for investigation. And those inquiries should be presented side by side with other inquiries, posing different but similar problems and using different data and arguments so that the student could see the questions, arguments, and conclusions in a broader intellectual context. And students should also have opportunities to engage in the messiness of practical problem solving, not just be presented with problems and the variety of ways of examining and drawing conclusions about those problems. As Schwab put it: "This is essentially the problem of facing the student with 'reality,' that is, of discovering to him the sense and extent to which real cases are not mere instances of general rules or mere members of classes" so that the student can appreciate that "principles are brought to bear on cases only approximately and with great difficulty" (Schwab 1969a, p. 116).

Schwab also argued for having students experience "works in progress," both their own and those of others: "It is one of the most powerful ways—perhaps the only way—to afford experience of the ground of all enquiry: the originating problem, the first idea, the nascent plan, the seminal purpose, from which flow research and scholarship worth the doing" (Schwab 1969a, p. 210). A study of finished products, on the other hand, does not provide a sense of aspects of a problem that are only "half-known" before the project has begun.

Finally, he says, the goal of curricular reform should be to provide students with an intellectual challenge and the opportunity to develop skills in "recovery, enquiry, and criticism appropriate to each discipline." In the sciences, social sciences, history, and philosophy, this means "no 'truth' without the evidence and argument which supports it or from which it grows" (Schwab 1969a, p. 183). This includes the presentation of alternative principles, evidence, and interpretation that give fields of study their competing theories and uncertainties. Instead, the curriculum that protesters were reacting to omitted uncertainty and how decisions are made about what evidence should be counted and which theories should be preferred. "Little wonder," Schwab concludes, "that anxieties, persecution feelings and a wearisome spate of intemperate, stereotyped protest should flood from students' mouths. Still less should we wonder that they so often cite their unexamined impulsions as sufficient ground for choice and, indeed confuse the one with the other" (Schwab 1969a, p. 16).

76.6 Applying Lessons Learned at Chicago to School Science

Beginning in 1959, after more than 20 years of efforts to integrate science into the liberal arts core curriculum at Chicago, Schwab had an opportunity to contribute to the reform of school science through his association with the Biological Sciences Curriculum Study (BSCS). BSCS had been organized by the American Institute of Biological Sciences in 1958 to reform biology teaching in the country. Schwab became chairman of the Teacher Preparation Committee at BSCS and was responsible for developing plans for the preservice and in-service training of teachers who would be teaching new courses that were part of the reform initiative. Under his leadership, the committee produced a *Teacher's Commentary* to accompany each of the three versions (blue, yellow, and green) of the BSCS biology texts (Hurd 1961), and he authored the first *Biology Teacher's Handbook* for BSCS in 1963.

Also in keeping with his interest in precollege science education, Schwab was invited to deliver the Inglis Lecture at Harvard University in 1961. The talk, published as *The Teaching of Science as Enquiry* (Schwab 1962), serves as a summary of his thinking about the nature of science and the teaching of science at the school level. The lecture focused on the nature of scientific investigation, on ways for students to develop an appreciation for science as a process of inquiry, and the intellectual skills involved in inquiry. There were three main themes: First was the importance of offering students a realistic portrayal of the nature of science so that as citizens they would understand that scientific investigations yield theoretical constructions that are tentative and ever-changing. The second focused on the pedagogical approaches that would give students practice in the intellectual skills involved in inquiry so they would be capable of independent critical reasoning throughout their lifetimes. And the third was the idea that science is not just an intellectual activity but also a study of actual events in the world. An educational program, therefore, needs to link the scientific principles and intellectual skills taught in the school curriculum to concrete phenomena in the physical world.

Schwab also argued that schools could play a role in educating the public about the importance of science in society. For citizens to be supportive of science, they must first understand *why* scientific knowledge continues to shift and *why* ideas that were once thought to be true may later be discarded. If the public is expected to support science, they need to understand the revisionist nature of science and appreciate that much of the language of science describes ideas and models, not actual physical reality. To Schwab, the key to having students develop an accurate picture of science was for them to understand that science rests on "conceptual innovation" (Schwab 1962, p. 5) and that scientific understanding changes as new ideas are conceived. This view of science cannot be achieved if students are taught in ways that suggest to them that knowledge is fixed and certain.

In many ways, these ideas about the nature of science are mirrored in Thomas Kuhn's[4] *The Structure of Scientific Revolutions* (Kuhn 1962), published the same year as Schwab's *The Teaching of Science as Enquiry*. Just as Schwab was deeply involved in developing the liberal arts core at Chicago beginning in the 1940s, Kuhn taught a comparable course for undergraduates at Harvard in the 1950s as part of its General Education in Science curriculum. Schwab's thinking did not go quite as far as Kuhn's notions about the incommensurability that results from "paradigm shifts," but a similar idea that significant conceptual shifts occur that make previous thinking obsolete can be seen in Schwab's writing:

> With each change in conceptual system, the older knowledge gained through use of the older principles sinks into limbo. The facts embodied are salvaged, reordered, and reused, but the knowledge which formerly embodied these facts is replaced. There is then, a continuing revision of scientific knowledge as principles of enquiry are used, tested thereby, and supplanted. (Schwab 1962, p. 15)

In Schwab's terms, science enjoys periods of "stable enquiry," during which agreed upon fundamental principles are used to guide research. But occasionally a shift occurs during which the principles that previously guided scientific investigations no longer are relevant. These periods of change are periods of "fluid enquiry." Fluid enquiry is not about filling in the missing pieces of the earlier models and conceptions. Instead, it involves the creation of new conceptions to guide scientific research (Schwab 1962).

Schwab thought that all citizens, not just future scientists, needed to be educated to think in this critical and creative way and that this was a contribution that schools could make to an informed citizenry. Drawing on his experience with undergraduate education at the University of Chicago, he believed this approach to school science would produce leaders who would both understand the nature of scientific inquiry and be able to think reflectively and creatively themselves.

For this approach to be successful, students would have to be active learners, fully engaged intellectually in the study of science. Rather than being told that the textbook and teacher are unquestioned sources of authoritative information, students would be encouraged to challenge teacher and text and to view what was said by them as something to be analyzed and critiqued. The student's attention

---

[4] A comparison between Schwab's and Kuhn's ideas, especially the implications of those ideas for science teaching, can be found in *Kuhn and Schwab on Science Texts and the Goals of Science Education* (Siegel 1978).

should not be on scientific statements as words and assertions to be learned, but on "...what the words and assertions are about: the thoughts and the actions of a scientist which have gone into the making of a piece of scientific research" (Schwab 1962, p. 66). It is the responsibility of the teacher to teach the students how to engage in these intellectual activities—what to look for, the kinds of questions to ask, and when to ask them.

To increase the breadth of their thinking about the various ways that scientific statements can be interpreted, students should also be asked to compare answers from different students and make judgments about those answers based on the evidence that is provided in support of them. In this way, the student learns that "...there is room for alternative interpretations of data; that many questions have no 'right' answer but only most probable answers or more and less defensible answers; that the aim of criticism and defense of alternative answers is not to 'win the argument' but to find the most defensible solution to the problem" (Schwab 1962, p. 70).

As he did when writing about undergraduate education at Chicago, Schwab proposed class discussion as the best way to engage school students in challenging intellectual discourse. And as he did for undergraduate students, Schwab suggested that original scientific papers offered "the most authentic, unretouched specimens of enquiry which we can obtain" (Schwab 1962, p. 74). His primary goal for students at the precollege level as with undergraduate students was the development of broad intellectual competence.

76.7 The Practical in Curriculum Development

Toward the end of his long academic career—which included efforts to create a program of liberal studies at the University of Chicago, his work with the Great Books Program with Hutchins and Adler, his work at the Jewish Theological Seminary, and his contributions to curriculum reform in school science at BSCS—Schwab wrote a series of six essays (four were published) that synthesized his understanding of the essential processes involved in curriculum development (Schwab 1969b, 1970, 1971, 1973, 1983). The first, *The Practical: A Language for Curriculum,* was written for the National Education Association's Center for the Study of Instruction and was published in 1969 (Schwab 1969b, 1970). The last was published in 1983, nine years after Schwab had retired from Chicago. That essay was titled *The Practical 4: Something for Curriculum Professors to Do*.

In *The Practical: A Language for Curriculum,* Schwab begins with an indictment of the present state of the curriculum field:

> The field of curriculum is moribund. It is unable, by its present methods and principles, to continue its work and contribute significantly to the advancement of education. ...The curriculum field has reached this unhappy state by inveterate, unexamined, and mistaken reliance on *theory*. (Schwab 1970, p. 1)

According to Schwab, whether theories are borrowed from disciplines such as philosophy, psychology, or sociology or constructed explicitly as educational theories of curriculum and instruction, they are "ill-fitted and inappropriate to problems of actual teaching and learning" (Schwab 1970, p. 1):

> Theory, by its very character, does not and cannot take account of all the matters which are crucial to questions of what, who, and how to teach; that is, theories cannot be applied, as principles, to the solution of problems concerning what to do with or for real individuals, small groups, or real institutions located in time and space—the subjects and clients of schooling and schools. (Schwab 1970, pp. 1–2)

Simply put, to Schwab education is much too complex an activity to be captured by a unified theory of teaching and learning. Inevitably, all theories create abstractions or idealizations of the particulars of the real world. And, because human behavior—which is what educational theories theorize about—is so complex, educational theories of necessity leave out much of the variation that occurs in the world. Schwab says: "It follows that such theories are not, and will not be, adequate by themselves to tell us what to do with actual human beings or how to do it" (Schwab 1970, pp. 28–29).

In an earlier essay, *On the Corruption of Education by Psychology* (Schwab 1957), Schwab demonstrated how certain theoretical positions from psychology create problems when applied in educational settings. The three theories he focused on were group dynamism, non-directivism, and autonomism. In the case of group dynamism, the group becomes the determiner of knowledge and the central focus of education; in the case of non-directivism, it is the individual who is supreme as a knowledge maker; and in the case of autonomism, the emphasis is on individuals' struggle for autonomy against the hegemony of society. According to Schwab, in each of these three cases the application of the theory goes well beyond what is reasonable or useful, and leads to practical conclusions that are opposite the others. "All three doctrines, beginning as normative or descriptive views of behavior, end by inventing an epistemology which tailors the intellectual aims of the curriculum to fit the terms of their incomplete theories of behavior" (Schwab 1957, p. 44).

He suggests, instead, that education should be seen as a practical enterprise, having many individual components that need to be analyzed separately, not as a unified activity that can be explained by and organized around a single all encompassing theoretical position. These ideas about the practical in curriculum are consistent with his ideas about the use of practical rationality that pervade all of his work.

It's not that Schwab thought that educational theory was useless or irrelevant, but rather that theory needed to be used judiciously to explain individual aspects of the educational enterprise and without overreaching in its attempt to create a grand synthesis.

He proposes three related and overlapping alternative approaches to a purely theoretic approach: what he calls the practical, the quasi-practical, and the eclectic. First is the *practical*. About the practical, he says: "The subject matter of the practical…is always something taken as concrete and particular and treated as indefinitely susceptible to circumstance, and therefore highly liable to unexpected change: this student, in that school…" (Schwab 1970, p. 3). The method of the practical is deliberation, which is a "complex, fluid, transactional discipline"

(Schwab 1970, p. 5). Deliberation involves the use of practical rationality by paying attention to particular events in particular places, recognizing the importance of the particular context in which education takes place, and having an openness of mind about the range of possible explanations for what takes place in each educational setting.

The *quasi-practical* approach shows particular awareness of the diversity that exists in schools and school communities. It is "an extension of practical methods and purposes to subject matters of increasing internal variety" (Schwab 1970, p. 5). It is *quasi*-practical because of its added complexity, which sometimes renders it less effective and, therefore, less practical, than what was desired. A practical solution might be found for a problem in one part of the system, only to find that it was not really a solution at all because of unforeseen and undesirable effects that the solution has on another part of the system. Thus solving a practical problem in the science portion of the curriculum may create problems in another part of the curriculum. Therefore, solving practical problems in complex systems requires coordination of efforts and sharing of information and expertise beyond what is required in simpler systems.

Finally, the *eclectic* approach is an approach that pays attention to a variety of theories or parts of theories that might be used in a practical analysis to inform particular aspects of curricular decision making, while at the same time being aware of the limitations of those theories. To Schwab, it is not that all theory is useless. But because of their enthusiasm to explain human behavior in general terms, educational theorists often inappropriately use theories to explain more than they in fact do explain, and they recommend or prescribe educational practices that are not warranted. It is important to know what a given theory can explain and what it cannot explain. With an understanding of the limits of each theory, it may be possible to use those theories to explain various parts of the educational experience.

76.8 Schwab's Legacy

Schwab can rightly be called a humanist, a constructivist, and a Deweyan progressive, and he lent his considerable support to those streams of thought in his educational writing. Regarding his humanism, according to Eliot Eisner, Schwab, along with educators such as Phillip Jackson (*Life in Classrooms,* 1968), helped to initiate a trend toward the "humanization of educational inquiry" through practical rationality, by his acknowledgement of the idiosyncrasies of educational contexts and his valuing of deliberation as "the exercise of the human's highest intellectual powers" (Eisner 1984, p. 204). He was a constructivist in how he viewed scientific theory as resulting from conceptual innovation, a process by which theoretical structures are constructed and revised in the context of still larger bodies of interconnected observation and theory. Schwab believed that scientists, operating in a milieu of interconnected theory, make choices about what to study, what data to collect, and which theoretical framework to use to make sense of their data. Schwab's writing in this area is still viewed as exemplary. Regarding his progressivism, Schwab showed great admiration for Dewey's work, as he shows in *Dewey: The Creature as Creative* (Schwab 1953) where he praises Dewey's ideas about the human role in generating truth in both philosophy and science. He also took on the role of apologist for Dewey, explaining misunderstood concepts as he did in *The "Impossible" Role of the Teacher in Progressive Education* (Schwab 1959) where he explains and defends Dewey's notion of the dialectic. Schwab himself was a Deweyan progressive in how he valued informed and reflective practice, in his belief in intellectual growth through the continuous reconstruction of experience, and the importance he placed on science as a way of thinking about the world rather than simply as a body of knowledge of the world.

When we look at his legacy at a finer grain size, the success of some of his more specific proposals for science education is somewhat mixed. Schwab devoted a lifetime to thinking and writing about the role of science in a liberal arts setting, first for undergraduate students and then for students at the precollege level. His recommendations were for rigorous intellectual preparation in science so that students could come to know what is known about the world and how the natural world works, but even more important, how we know what we know. His hope was that such an education would give students the capability and desire to learn throughout their lifetime. Among science educators whose interest is the precollege level, he is most well known and appreciated for the application of these ideas to the school curriculum, especially the work he did at BSCS during the 1960s and his very well-received *The Teaching of Science as Enquiry* (Schwab 1962).

On the surface, it is fair to say that his contributions to general education at the college level were short lived. Efforts to create a common experience for undergraduate students at Chicago and to organize the undergraduate college around that common experience eventually gave way to an organization of the curriculum around the disciplines and a requirement that students specialize in one of those disciplines, the very concerns that motivated the general education movement in the first place. The Chicago plan, which was one of the most radical experiments in general education, initially offered a complete program of general education courses in the undergraduate college, but the pressures for specialization led to a reorganization of that program in 1957, and under the reorganized program students were required to major in one of four academic divisions as a requirement of the degree (Bell 1966).

In a 1963 editorial comment, *A Radical Departure for a Program in the Liberal Arts*, Schwab acknowledged the failure to achieve the goals of general education: "It need hardly be said that the most formidable barrier to an effective program of liberal education at the moment is constituted of the concerted pressures toward specialization" and that "the pressure toward specialization has resulted in acute curtailment of the time allotted to a liberal arts program" (Schwab 1963b, no page number). Schwab offered what he saw as a "radical proposal" that the liberal arts could still be communicated to students through a student's area of specialization if they were offered seminars that focused on the development of core ideas in each of those specialties. Not surprisingly, given the courses he had helped to develop in the 1940s, his *radical* proposal included the idea that the study of the development of core ideas in each field of science could be accomplished by way of the students' own examination and comparison of original papers. But, for the most part, that kind of intellectual treatment of the sciences did not find its way into the undergraduate curriculum in any significant way. The products of science, organized by disciplines, or sometimes through interdisciplinary study, continue to be the primary content of the vast majority of undergraduate level science courses today.

Chicago was hardly alone in its inability to maintain a comprehensive program in general education. By 1950, there had been significant erosion of most general education programs, and by the end of the 1950s, those large-scale, comprehensive efforts had for the most part been abandoned.[5] As Daniel Bell pointed out in *The Reforming of General Education* (1966), Harvard's program, whose development was stimulated by the 1945 publication of *General Education in a Free Society* (Harvard Committee 1945) and mandated by the faculty to take effect in 1949, began to come apart almost immediately. Instead of being required to take common courses in each of the humanities, social science, and natural science and mathematics divisions, as initially proposed, students were given lists of courses in each area that could be taken to meet the general education requirement. As Bell observed, the failure was most evident in the sciences:

> The change from the original intention was sharpest in the sciences. In 1949, a faculty committee headed by Jerome Bruner repudiated the idea that the teaching of science could be done through the history of science or by a case-method approach. Instead of a historical emphasis, the Bruner Committee proposed that a student be given a "knowledge of the fundamental principles of a special science," and an "idea of the methods of science as they are known today." The difference between a general education and a departmental course in science would consist, then, only in the selection and coverage of topics, not in approach. (Harvard Committee 1945, p. 48)

Although the Bruner Committee's arguments revealed a fundamental ideological difference in what the nature of the educational experience in the sciences should be, according to Bell (1966) these grand schemes for general education that Schwab was part of were done in as much by practical problems of staffing as they were by intellectual concerns. It was just too difficult to find faculty who were willing to devote their careers largely, if not completely, to teaching undergraduate students the relationships between knowledge in science and the ways that knowledge was generated. In his commitment to do just that, Schwab was unique.

But even though the general education movement that he was a central part of did not last much beyond mid-century, as a thinker in this area, Schwab's ideas had lasting impact. One of the strongest testaments to his work came from Bell who pointed to two books that were most important to him in thinking about the development of the college curriculum for his 1966 work on general education: "One is Ernest Nagel's (1961) *The Structure of Science*, which lays out a 'logic of explanation' dealing with the nature of inquiry. The other is Joseph J. Schwab's *The Teaching of*

[5] A number of undergraduate colleges in the United States continue to require a common liberal arts core, although the grand experiments of the first half of the twentieth century at places like Chicago for the most part no longer exist.

Science as Enquiry, which discusses in a wonderfully lucid way the dependence of science upon conceptual innovation, and applies these ideas to the problems of teaching" (Bell 1966, p. xxiv). There is much wisdom to glean from Schwab's writings on a liberal arts approach to the study of science, especially in the value he places on the development of human intelligence through a study of the complexities of human thought and inquiry.

Concerning his contributions to precollege science education, especially the curriculum reforms of the 1960s, some of Schwab's ideas still resonate with us today, but others have been overtaken by ideas that he argued against. For example, his description of the nature of science and its implications for science teaching that appears in his 1961 Inglis Lecture (Schwab 1962) is one of the best expositions that we have, and it can still serve as a model of what science is and how it should be taught. But other of his ideas have been overridden by an emphasis on standardization and accountability, ideas that have recently taken hold as dominant themes of school science education. Beginning in the 1980s, science educators in the United States began to create, with much more precision than they had ever done before, detailed specifications of what all students should know in science and to hold students accountable for those ideas through standards-based assessments. The first national efforts to describe what all students should know began in 1989, in mathematics, with the publication of *Curriculum and Evaluation Standards for School Mathematics* by the National Council of Teachers of Mathematics (1989) and, in science, with the publication of *Science for All Americans* by the American Association for the Advancement of Science (AAAS 1989).

The primary goal of these publications was to provide more clarity about what the goals of the curriculum in these areas should be, including an appreciation for the methods and processes of inquiry that were used in science, but they also helped move science education in the United States toward a standardization of content, at least at the state level. Federal legislation required that all states develop explicit statements of what students should know and to develop tests to assess that knowledge. Although the national level documents included recommendations for the inclusion of the methods and processes of science along with the subject matter content, most state standards and state assessments focused on the details of the content and not on an examination of scientific inquiry.

At first glance, Schwab's writing seems to offer support for such a focus on subject matter. After all, Schwab is linked to the "structure of the disciplines movement," which typically gives primacy to subject matter and how it is organized. But Schwab's focus was not on prescribing particular conceptual structures for students to learn as much as having them analyze competing knowledge structures and how those competing knowledge structures were created. The implication of his approach for curriculum development is that subject matter should be seen as *useful*, in fact *critical* for curriculum development, but that is not the only thing to be considered. As Fox (1985) put it:

> Schwab argues that it is not the role of curriculum to simplify or to parrot a favored or accepted conception of a discipline, but to reflect on what contribution the various conceptions within a discipline can make to the thinking, the feeling and the behavior of the student.

> Thus, he establishes the basis for his distinction between subject matter and subject-matter-for-education. (Fox 1985)

To Schwab, it is true that the selection of subject matter is critical because it is central to understanding a particular field of study and because of its cultural significance. But, when thinking about subject matter for education, curriculum makers also should take account of the demands of the learner, the teacher, and the school environment (milieu) when deciding what to teach. Schwab argues for a balanced approach to curriculum development and warns against the possible corruption of education by placing too much emphasis on subject matter alone (Schwab 1973).

Therefore, although subject matter is essential to understanding the nature of a discipline, the particular details of that subject matter can and should vary depending on the capabilities of the teacher, the interests of the student, and the constraints of the educational environment. What should be included is subject matter that can act as a vehicle for teaching students the syntactic structure of the disciplines, that is, the ways in which particular knowledge has been generated from a range of possible alternatives. Using this approach, students learn that conceptual structures are dynamic and that there is a knowable logic to the decisions that scientists make about the problems they study, the theories they use to drive their investigations, the data they collect, and how they interpret their findings in terms of the theories that drive those investigations. In such a system, students are challenged to appreciate the complexity of scientific knowledge, the range of existing competing theories, and the variety of methodologies used to generate knowledge in the various science fields. As with his recommendations for undergraduate science education, these ideas about how precollege students should learn science are not generally reflected in the dominant mode of instruction in most schools today, where the focus continues to be largely on the content per se.

Schwab was also concerned about the "objectives" focus that was beginning to drive curriculum development. Objectives were a way of specifying with a high degree of precision what all students should know and be held accountable for in various areas of the curriculum (see, e.g., Mager 1962). The approach was often linked to the psychological theory of behaviorism as applied to education. In 1983, after he had retired from the University of Chicago, Schwab published the fourth in his series of essays on the "practical." In that essay, he identified three limitations of using learning objectives to drive curriculum. First, was that objectives tended to:

> ...anatomize matters which may be of great importance into bits and pieces which, taken separately, are trivial or pointless. Lists of objectives...anatomize, not only a subject-matter, but teachers' thoughts about it, the pattern of instruction used to convey it, the organization of textbooks, and the analysis and construction of tests. (Schwab 1983, p. 240)

Second, he believed that the lists of objectives were of little use if consideration was not also given to the means and materials available for their implementation: "...reflection on curriculum must take account of what teachers are ready to teach or ready to learn to teach [and] what materials are available or can be devised" (Schwab 1983, pp. 240–241). Third, there must be consideration paid to the unintended consequences that might result from pursuing those objectives, "...not merely how well they yield intended purposes but what else ensues" (Schwab 1983, p. 241). In sum, none of this can be accomplished unless "…ends or objectives are tentatively selected and pursued. Hence, curriculum reflection must take place in a back-and-forth manner between ends and means." According to Schwab: "A linear movement from ends to means is absurd" (p. 241). Here, too, curricular development in the United States is more likely to follow a linear approach than the tentative, iterative, back-and-forth approach that Schwab recommended, in which ends and means are continuously reexamined in relation to each other.

Schwab's criticisms about assessment were similar to those he had for curriculum development, and they were equally broad in scope and practical in nature. He questioned, first, whether it is even possible to create an assessment that is both highly valid in that it conforms to the content of the curriculum and is useful at the same time. He said that such a test lacks "usefulness" because it tells the teacher nothing of what else besides the prescribed curriculum the student might be learning, what alternative constructions of the curriculum might be possible, or how other forms of testing might produce different results. He proposed the use and comparison of different types and forms of testing, which could then serve as multiple embodiments of, or reflections on, the ends and outcomes of education. Testing can communicate information about the curriculum and, therefore, should not be "…mere 'valid,' and therefore static, measures of a static curriculum, but as centers of and foci for the discussion and improvement of…the curriculum, including tests" (Schwab 1950b, p. 281, cited in Westbury and Wilkof 1978). He concluded his essay on testing and the curriculum by saying:

> The end of such analysis is, however, simple. It is to bring into the vivid meaningfulness afforded by contrast what it is that each participant curriculum does and does not do for its students. The ultimate aim is the same as before: to initiate thought, experiment, and improvement of the participating curriculums. (Schwab 1950b, p. 286, cited in Westbury and Wilkof 1978)

Once again, for the most part, this dynamic approach to assessment that Schwab recommended, as a tool for evaluating not only the students' knowledge but also the effectiveness of the curriculum and classroom instruction, is not the approach that is currently used.

76.9 Conclusion

Joseph Schwab was an important figure in science education. He tackled difficult subjects, often in a forceful way. Sometimes he was successful and sometimes he was not. He was often critical of mainstream ideas and the status quo. But in everything he proposed, he tried to make us more open to alternative ideas and more practically rational in how we see the world. As Elliot Eisner said of his former teacher: "He tries to make [life] more intelligent" (Eisner 1984, p. 201). When it came to the content of the science curriculum, he was not concerned so much with the particular subject matter that was learned as that people would continue to love and pursue knowledge throughout their lifetimes and that they would have both the intellectual skills needed to analyze the artifacts of our culture and the ability to analyze the claims that experts and fellow citizens make. Whether aimed at the undergraduate college or at precollege education, his writings leave us with a wealth of ideas about how science education could be better and with a good deal to think about.

References

AAAS (1989). Science for all Americans. New York: Oxford University Press.

Bell, D. (1966). *The Reforming of General Education: The Columbia Experience in its National Setting*. New York: Columbia University Press.

Cohen, B. & Schwab, J. J. (1965). Practical logic: Problems of ethical decision. *American Behavioral Scientist*, 8 (8), 23–27.

DeBoer, G.E. (1991). A history of ideas in science education: Implications for practice. New York: Columbia University Teachers College Press.

Eisner, E. (1984). No easy answers: Joseph Schwab's contribution to curriculum. *Curriculum Inquiry*, 14 (2), 201–210.

Fox, S. (1985). Dialogue: The vitality of theory in Schwab's conception of the practical. *Curriculum Inquiry*, 15 (1), 63–89.

Freud, S. (1975). *Beyond the Pleasure Principle* (J. Strachey, ed. and trans.). New York: W.W. Norton. (Originally published in 1920)

Harvard Committee (1945). *General Education in a Free Society*. Cambridge, MA: Harvard University Press.

Hurd, P. (1961). *Biological Education in American Secondary Schools 1890–1960*. Washington, DC: American Institute of Biological Sciences.

Hutchins, R. M. (1936). *The Higher Learning in America*. New Haven: Yale University Press.

Jackson, P. W. (1968). *Life in Classrooms*. New York: Holt, Rinehart, and Winston.

Jung, C. (1982). *Aspects of the Feminine*. Princeton, NJ: Princeton University Press.

Kuhn, T. S. (1962). *The Structure of Scientific Revolutions*. Chicago: University of Chicago Press.

Mager, R. F. (1962). *Preparing Instructional Objectives*. Palo Alto, CA: Fearon Publishers.

Nagel, E. (1961). *The Structure of Science: Problems in the Logic of Scientific Explanation*. New York: Hartcourt, Brace, & World.

National Council of Teachers of Mathematics (1989). *Curriculum and Evaluation Standards for School Mathematics*. Reston, VA: Author.

Schwab, J. J. (1941). The role of biology in general education: I. Biology and the problem of values. *Bios*, 12 (2), 87–97.

Schwab, J. J. (1949). The nature of scientific knowledge as related to liberal education. *Journal of General Education*, 3, 245–266.

Schwab, J. J., (1950a). The natural sciences: The three-year program. In *The Idea and Practice of General Education: An Account of the College of the University of Chicago*, by present and former members of the faculty (pp. 149–186). Chicago: University of Chicago Press.

Schwab, J. J. (1950b). Criteria for the evaluation of achievement tests: From the point of view of the subject-matter specialist, in *Proceedings of the Invitational Conference on Testing Problems*, Princeton, N.J.: Educational Testing Service, pp. 82–94.

Schwab, J. J. (1953). John Dewey: The creature as creative. *The Journal of General Education*, 7, 109–121.

Schwab, J. J., (1954). Eros and education: A discussion of one aspect of discussion. *Journal of General Education*, 8, 54–71.

Schwab, J. J., (1956). Science and Civil Discourse: The uses of diversity. *Journal of General Education*, 9, 132–43.
Schwab, J. J. (1957). On the corruption of education by psychology. *Ethics*, 68 (1), 39–44.
Schwab, J. J. (1959). The "impossible" role of the teacher in progressive education. *School Review*, 67, 139–159.
Schwab, J. J., (1960). What do scientists do? *Behavioral Science*, 5, 1–27.
Schwab, J. J. (1962). *The Teaching of Science as Enquiry*. Cambridge, MA: Harvard University Press.
Schwab, J. J. (1963a). *Biology Teachers' Handbook*. New York: John Wiley and Sons.
Schwab, J. J. (1963b). Editorial comment: A radical departure for a program in the liberal arts. *The Journal of General Education*, 15 (1), (no page numbers).
Schwab, J. J. (1964). Structure of the disciplines: Meanings and significances. In G. W. Ford and L. Pugno (Eds.), *The Structure of Knowledge and the Curriculum* (pp. 6–30). Chicago: Rand McNally.
Schwab, J. J. (1969a). *College Curriculum and Student Protest*. Chicago: University of Chicago Press.
Schwab, J. J. (1969b). The practical: A language for curriculum. *The School Review*, 78 (1), 1–23.
Schwab, J. J. (1970). *The practical: A language for curriculum*. Washington, DC: National Education Association, Center for the Study of Instruction.
Schwab, J. J. (1971). The practical: Arts for eclectic. *School Review*, 79, 493–542.
Schwab, J. J. (1973). The practical: Translation into curriculum. *School Review*, 81, 501–22.
Schwab, J. J. (1983). The practical 4: Something for curriculum professors to do. *Curriculum Inquiry*, 13 (3), 239–265.
Siegel, H. (1978). Kuhn and Schwab on science texts and the goals of science education. *Educational Theory*, 28 (4), 302–309.
Veysey, L. (1981). *The Emergence of the American University*. Chicago: University of Chicago Press.
Westbury, I. & Wilkof, N. J. (1978). Introduction. In I. Westbury & N. J. Wilkof (Eds.), *Science, Curriculum, and Liberal Education. Selected Essays by Joseph J. Schwab* (pp. 1–40). Chicago: The University of Chicago Press.

George E. DeBoer is deputy director for Project 2061, a long-term initiative of the American Association for the Advancement of Science to improve science teaching and learning for all. He has served as program director at the National Science Foundation and is professor of educational studies emeritus at Colgate University in Hamilton, NY. He is author of *A History of Ideas in Science Education: Implications for Practice* (Columbia University Teachers College Press, 1991) and *The Role of Public Policy in K-12 Science Education* (Information Age Publishing, 2011), as well as numerous articles, book chapters, and reviews. His primary scholarly interests include science education policy, clarifying the goals of the science curriculum, researching the history of science education, and investigating ways to effectively assess student understanding in science. He holds a Ph.D. in science education from Northwestern University, an M.A.T. in biochemistry and science education from the University of Iowa, and a B.A. in biology from Hope College. He is a member of the American Association of University Professors, the National Association for Research in Science Teaching, and the National Science Teachers Association and a fellow of the American Association for the Advancement of Science and the American Educational Research Association.

Name Index

M.R. Matthews (ed.), *International Handbook of Research in History, Philosophy and Science Teaching*, DOI 10.1007/978-94-007-7654-8,

C

H

I

K

L

N

P

Q

R

S

T

W

Subject Index

M.R. Matthews (ed.), *International Handbook of Research in History, Philosophy and Science Teaching*, DOI 10.1007/978-94-007-7654-8,

D

E

F

H

J

K